Developments in Applied Phycology

Volume 15

Series Editor

Michael A. Borowitzka, Algae R&D Centre, School of Veterinary and Life Sciences
Murdoch University
Murdoch, Australia

Aims and Scope

Applied Phycology, the practical use of algae, encompasses a diverse range of fields including algal culture and seaweed farming, the use of algae to produce commercial products such as hydrocolloids, carotenoids and pharmaceuticals, algae as biofertilizers and soil conditioners, the application of algae in wastewater treatment, renewable energy production, algae as environmental indicators, environmental bioremediation and the management of algal blooms. The commercial production of seaweeds and microalgae and products derived there from is a large and well established industry and new algal species, products and processes are being continuously developed.

The aim of this book series, Developments in Applied Phycology, is to present state-of-the-art syntheses of research and development in the field. Volumes of the series will consist of reference books, subject-specific monographs, peer reviewed contributions from conferences, comprehensive evaluations of large-scale projects, and other book-length contributions to the science and practice of applied phycology.

Prospective authors and/or editors should consult the Series Editor or Publishing Editor for more details. Series Editor: Michael A. Borowitzka - M.Borowitzka@murdoch.edu.au Publishing Editor: Éva Loerinczi - eva.loerinczi@springer.com

Alan T. Critchley • Anicia Q. Hurtado
Iain C. Neish • Leonel Pereira
Editors

Advances in Temperate Phyconomy: Algal Harvest and Cultivation in Globally Distributed Temperate Waters

Part I

Springer

Editors
Alan T. Critchley
Verschuren Centre
Sydney, NS, Canada

Iain C. Neish
Bagong Dalan, Barangan Yati, Liloan
SeaKITShub
Cebu, Philippines

Anicia Q. Hurtado
Integrated Services for the Development of
Aquaculture and Fisheries (ISDA)
Jaro, Iloilo City, Philippines

Leonel Pereira
Department of Life Sciences
University of Coimbra
Coimbra, Portugal

ISSN 2543-0599 ISSN 2543-0602 (electronic)
Developments in Applied Phycology
ISBN 978-3-032-02954-6 ISBN 978-3-032-02955-3 (eBook)
https://doi.org/10.1007/978-3-032-02955-3

This Springer imprint is published by the registered company Springer Nature Switzerland AG
The registered company address is: Gewerbestrasse 11, 6330 Cham, Switzerland

Introduction to Advances in Temperate Phyconomy

In 2019, the term "*phyconomy*" was coined and published, subject to peer-review, with two long-term colleagues and applied seaweed cultivation practitioners, namely, Drs Anicia Hurtado and Iain Neish (Hurtado et al. 2019). The definition of phyconomy encompasses the sciences and practical arts of successful utilization, management, and cultivation of various types of tropical seaweeds for their economic benefits. There are tremendous socioeconomic benefits derived from these activities in the Blue Economy. The term is conceptually adapted from terrestrial economic models; agronomy also encompassed the volumes and values of commercial seaweeds, or phyconometrics. Given the socioeconomic benefits of eucheumatoid farming, we focused on major issues arising in global production of this important commercial resource.

We like physical books! Together we co-edited a book for Springer entitled *Tropical Seaweed Farming Trends, Problems and Opportunities: Focus on Kappaphycus and Eucheuma of commerce* (Hurtado Critchley and Neish 2017; https://link.springer.com/book/10.1007/978-3-319-63498-2). It is also available in digital format.

Subsequently, these efforts culminated in a multinational workshop on red seaweeds for carrageenan production (*Tropical Phyconomy Coalition Development*), at the University of Makassar, Indonesia, in 2022. The presentations from a number of the world's leading experts in eucheumatoids remain a freely accessible, online resource of >25 hrs (see: www.phyconomy.org) for the benefits of tropical seaweed farmers globally. A co-edited book was published in early 2024 as a proceedings of this meeting: Critchley, Hurtado and Neish, *Tropical Phyconomy Coalition Development: Focus on Eucheumatoid Seaweeds* (https://link.springer.com/book/10.1007/978-3-031-47806-20).

The phyconomic snowball was rolling and we could not stop there. We saw a gap in the collated literature available. We cast our nets much wider to produce two new, co-edited books for 2025 under the umbrella: *Temperate Phyconomy* (i.e., cold-water seaweeds' use, harvest and cultivation) which is booming under the banner of the United Nations' Sustainable Development Goals (SDGs) for the promotion of economic growth in coastal regions of the world. There was a fantastic response to the initial call for chapters. Given the number of submissions, we welcomed our colleague Prof. Leonel Pereira, a noted author of seaweed articles and books, to help co-edit the new *Temperate Phyconomy* books with us.

We are sure readers from many backgrounds of academia and practical, applied use of seaweeds, production, and management will be interested in the widest scope of excellent submissions we are privileged to have submitted to these books.

We hope you enjoy and benefit from reading this book and learning as much as we have from being the co-editors with the distinct pleasure of making this information as globally available as possible.

Best regards, in phyconomy!
Alan, Anicia, Leonel, and Iain

PS:

A *Tropical Phyconomy Coalition Development (TPCD 2)* workshop on tropical seaweed cultivation (the non-eucheumatoids, e.g., *Ulva, Caulerpa, Gracilaria, Sargassum*, etc.) is being planned for the end of 2026.

Sydney, NS, Canada — Alan T. Critchley
Jaro, Iloilo City, Philippines — Anicia Q. Hurtado
Cebu, Philippines — Iain C. Neish
Coimbra, Portugal — Leonel Pereira

Reference

Hurtado AQ, Neish IC, Critchley AT (2019) Phyconomy: the extensive cultivation of seaweeds, their sustainability and economic value, with particular reference to important lessons to be learned and transferred from the practice of eucheumatoid farming. Phycologia 58:472–483. https://doi.org/10.1080/00318884.2019.1625632

Preface: A 60-Year Journey from Cold-Water Wild Seaweed Shortages to Cultivated Tropical Seaweed Abundance

As an adaptive phyconomist and seaweed farmer, I have experienced a 60-year journey on the path of valorization development in tropical red seaweed farming. My phyconomy vocation followed on work of my father, Arthur Neish, who was a pioneer of red seaweed farming when it was almost exclusive to China and Japan.

I started my phyconomic life in 1965 in Nova Scotia, Canada, in the frigid, stormy waters of the North Atlantic Ocean. By 1974, I had experienced the joys of warm water in the Sultanate of Oman and in East Indonesia. By 1977, it was clear to me that carrageenan seaweeds could not be grown at an acceptable cost in cold waters with a frigid winter climate. I have lived and worked in tropical waters ever since.

Commercial farming of tropical red seaweeds was stimulated by "market pull" from carrageenan and agar producers who were short of raw material from traditional cold-water sources (Fig. 1). Ultimately, a major industry developed through collaboration of entrepreneur phyconomists with impoverished coastal people who were hungry for cash crops in areas with limited economic opportunities.

Until 1974, the carrageenan industry was almost totally dependent on wild cold-water seaweeds, mostly *Chondrus crispus* (Irish moss) from Atlantic Canada. Efforts to grow *Chondrus* as a carrageenan sources at an affordable cost have failed in Atlantic Canada, and elsewhere, until the present day.

The carrageenan industry was headed for stagnation unless farmed raw materials could be obtained, so major carrageenan processors commenced diligent farm development efforts starting in the mid-1960s. A driving force was Ocean Resources Development Associates (ORDA), which was based at the Goodwood Hotel Arcade in Singapore and was led by Mr. Hans Porse from 1972 to 1976. ORDA was a collaboration of Marine Colloids Inc. (USA), Copenhagen Pectin Factory Ltd. (Denmark), and Pierrefitte-Auby-CECA (France). Through

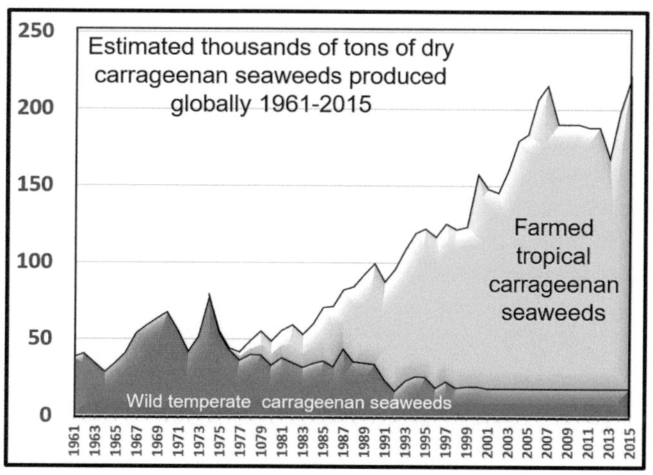

Fig. 1 Author's estimates of thousands of metric tons of global carrageenan raw, dried seaweed production from 1961 to 2015

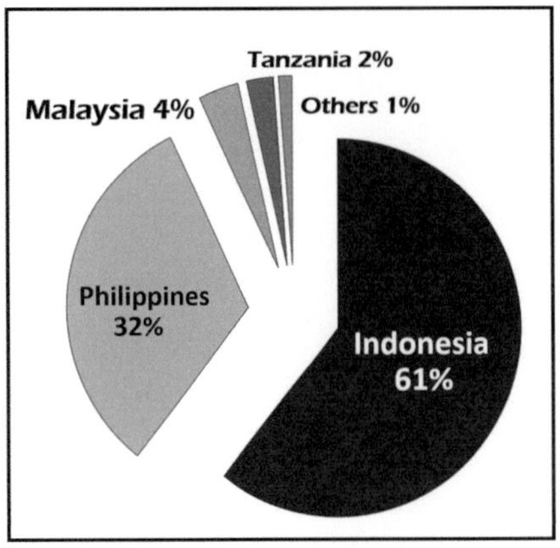

Fig. 2 Author's estimates of approximate percentage contribution of countries to global production of raw, dried farmed eucheumatoid seaweeds

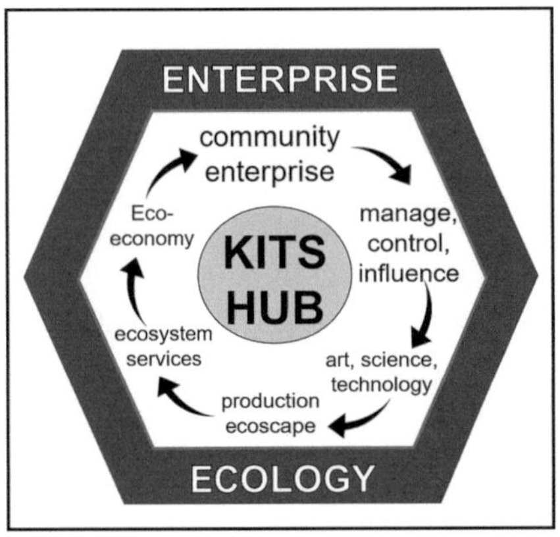

Fig. 3 By applying enterprise ecology, phyconomists are enabling enterprises in ecoscape communities to apply arts, sciences, and technologies that influence, manage, and control biosphere ecosystem services valorized through fair eco-economic market systems

the 1970s, these three companies dominated the carrageenan business with upwards of 90% of global markets.

By the late 1970s, farmed eucheumatoid seaweeds became a significant raw materials source for the carrageenan industry. Commercial farming of eucheumatoid seaweeds commenced in 1974 in Tawi-Tawi province of the Philippines, but Indonesia attained leadership in global tropical eucheumatoid seaweed production by the mid-2000s (Fig. 2).

Tropical red seaweed farm development was funded and driven by entrepreneurial value chain players. Market system providers and governance enforcers got active only after eucheumatoid seaweed farming and value chains were well established. By 2025, phyconomists are moving beyond seaweed monocropping to holistic coastal area management that applies enterprise ecology. At the time of writing, phyconomists were restoring tropical seashores one enterprise at a time with enterprise ecology (Fig. 3).

Cebu, Philippines Iain C. Neish

Acknowledgments

As co-editors of this book, Anicia, Leonel, and I would like to sincerely thank all contributors. A simple fact—without the authors there would be no book! To us, a physical paper-based entity was an important way to record for posterity the tremendous developments and achievements being undertaken in Temperate Phyconomy on a global basis. We are particularly pleased that a number of first and co-authors represent early career scientists, working under the oversight of their supervisors and mentors. Thanks to all. A sincere merci beaucoup to all of the applied phycologists (phyconomists) of *both* micro- and macroalgae!

Some chapters needed extra pairs of eyes, other than the co-eds, and we thank the anonymous reviewers (you know who you are!) for their phyconomic advice.

Similarly, we want to thank the staff at Springer, including Michael Borowitzka and Eva Loerinczi, for seeing the value in our initial proposal and giving us the green light to produce this book. In addition, we are most grateful for the tireless work of Ramya Ravichandran for gentle chivvying and keeping us on track and signing up the authors as we progressed. We also appreciate Ramya Ravichandran and Bibhuti Sharma's inputs into the final stages of the book preparation.

Alan T. Critchley
Anicia Q. Hurtado
Iain C. Neish
Leonel Pereira

Contents

About the Editors

Alan T. Critchley grew up in Birmingham, UK. In 1981, he graduated with a PhD from Portsmouth Polytechnic, now University, in marine ecology. He was based at the university's Hayling Marine Laboratory, Ferry Road, studying the invasion of the brown seaweed *Sargassum muticum*. This was followed by a Royal Society European Postdoctoral Fellowship at the Delta Institute, Yerseke, the Netherlands, also looking at invasives in coastal waters and construction areas there. He moved to South Africa to teach as a Lecturer in Cryptogamic Botany in the Botany Department of the University of KwaZulu- Natal, Pietermaritzburg (7 years), and then as Senior Lecturer/Ass. Prof. at the University of the Witwatersrand, Johannesburg (10 years). He researched sub-tropical seaweeds to the east and cold-water, upwelling-influenced species to the west and developed a much keener interest in the benefits of seaweeds and their applications. His work there, with Klaus Rotmann of Taurus Chemicals, was the start of a career moving towards cultivation and applications of various types of seaweeds.Furthermore, a major influence and start of a lifelong collaboration was with Professor M. Ohno of Kochi University, Usa, Japan, at first to co-edit volumes of *Seaweed Cultivation and Marine Ranching* and *Seaweed Resources of the World*. In 1998, he moved to the University of Namibia, Windhoek, to become a Research Professor at the Multidisciplinary Research and Consultancy Centre; seaweeds and their uses were at the center of a number of novel projects there. In 2001, he made the transition from academia to the commercial world. The call of carrageenan-bearing seaweeds took him to Normandy, France, and a processing facility, then operated by Degussa Texturant Systems, to be in charge of their New Raw Materials Laboratory. It was at that time that his passion for the *Kappaphycus* and *Eucheuma* spp. seaweeds, their biology, production, utilization, and socioeconomic importance was fostered. It was here that he was most fortunate to begin collaborative research with Dr. Anicia Q. Hurtado. In 2005, he moved to Nova Scotia, Canada, to work with a small seaweed processing company. This role involved the management of multi-disciplinary, collaborative projects based on commercial seaweeds (browns and reds). Ensuring sustainable production techniques for seaweed resources and the scientific, peer-reviewed validation of biostimulant and bioactive properties of extracts was a major focus. Carrageenophyte seaweeds continued as a research focus. It was through these activities that Dr. Iain Neish and Alan were introduced. A common goal led Anne, Iain, and Alan to co-edit

Tropical Seaweed Farming (2017) and *Topical Phyconomy Coalition Development (TPCD): Focus on Eucheumatoid Seaweeds* (2024) both for Springer. Current collation of the broad spectrum of science associated with the *Eucheuma* seaweeds, their production, processing, applications, and most importantly socio economic contributions to coastal communities is a keen focus of efforts. He helped the Global Seafood Alliance publish its first standard on seaweed cultivation in 2023, and he has been a member of the science advisors for the Safe Seaweed Coalition, now Global Seaweed Coalition, and also the Global Seaweed Super Star and Seaweed Protect projects. Alan is an Associate Editor of the Journal of Applied Phycology.

Anicia Q. Hurtado was a Senior Scientist at the Aquaculture Department, Southeast Asian Fisheries Development Center (SEAFDEC-AQD), Tigbauan, Iloilo, Philippines, for more than 20 years. She spearheaded the Seaweed Program of AQD during her entire career at the Center. She is at present the Chair of the Integrated Services for the Development of Aquaculture and Fisheries (ISDA Inc.), an organization of past and present scientists of SEAFDEC-AQD. She finished her Doctor of Agriculture (Phycology) at Kyoto University, Kyoto, Japan, in 1988 as a Monbusho scholar.

She was and is still involved in different seaweed engagements such as:

1. GCRF-UKRI GlobalSeaweed* Project, as the *Developing Country Partner (Phil.)* (Oct 2017–Dec 2021) on the *Sustainability of Seaweed Aquaculture in the Developing Countries* at the University of the Philippines Visayas, Miag-ao Iloilo.
2. Evaluator—international research proposal (5)
3. Member—Graduate School Thesis Committee, local and international (8)
4. Member, journal editorial board (6)
5. Referee/reviewer, scientific journals (18)
6. Senior and co-author of books (4)
7. Senior and co-author book chapter (19)
8. Author, monographs/brochures/training manual/flyers/posters (22)
9. Senior and co-author in peer-reviewed journals (82)
10. Senior and co-author of proceedings/magazines/newsletter/terminal reports (20)
11. Consultant/technical adviser to several international agencies
12. Plenary, resource speaker, oral and poster presenter in several national and international seaweed congresses

She is a member of the UK International Peer Review College since February 2018 to the present as a reviewer of proposals submitted for possible funding by UKRI-GCRF. She is one of the Fellow members of the European Algae Biomass Association (EABA) since Oct 2020 to the present, with a VIP status for her significant contributions to the algal sector, a clear recognition of her works in promoting the sector. She is a member of the Executive Committee of the International Society of Applied Phycology. Above all, her unwavering and undaunting passion for seaweeds, directly working with the seaweed farmers, is her priceless gem on her belt.

Iain C. Neish is a Canadian from Nova Scotia who has worked in seaweed aquaculture since 1965. In doing so he has followed the path of his father Arthur Charles Neish. He first experienced tropical conditions in Oman and Indonesia during 1973–74. Since 1977 he has lived and worked full time on small holder phyconomy development, integrated aquaculture and process and product development in Indonesia, the Philippines, Malaysia, and other tropical venues. He is currently based in Olango Island near Cebu City, Philippines, where he focuses on applying ecosystem approaches to restoration of seafolk enterprises, in formerly prolific seaweed production areas that have failed in the wake of natural disasters and tragedy of the commons.

Leonel Pereira has a degree in Biology (Scientific branch), a PhD in Biology (specialty in Cellular Biology), and aggregation in Biosciences, specializing in Biotechnology, from the Faculty of Sciences and Technology of the University of Coimbra, where he is currently Associate Professor, and Coordinator of the Master's in Biodiversity and Plant Biotechnology. In addition to teaching at this university, he is an integrated researcher at CFE—the Center for Functional Ecology: Science for People and Planet. His interests are mainly in the areas of Marine Biodiversity (algae), Marine Biotechnology (bioactive compounds from algae), and Marine Ecology (environmental assessment). Since 2008, he has been the author and editor of the electronic publication MACOI—Portuguese Marine Algae Site (http://www.flordeutopia.pt/macoi/). Since 2018 he has been a member of the editorial board of the scientific journals: *Marine Drugs* and *Applied Sciences*. He is the editor of several special issues of scientific journals (*Journal of Applied Phycology*, *Botanica Marina*, *Marine Drugs*, *Applied Sciences*); he is the editor or author of more than 20 books and author of more than 200 indexed articles and more than 79 book chapters. He has given more than a hundred lectures and oral communications at various national and international scientific events. He is the coordinator of several national and international research projects. In 1998 he received the Francisco de Holanda Award (Honorable Mention) and, more recently, the Mar Rei D. Carlos Award (18th edition), and winner of the CHOICE Award—outstanding academic title award 2016—book title: *Edible Algae of the World*, CRC Press. He is part of the list of the top 2% scientists in the world in 2021, 2022, 2023, and 2024, published by Stanford University and Elsevier, and has an h40 Scopus index. He is currently Coordinator of the Master's Degree in Plant Biodiversity and Biotechnology, and coordinator of ACOI—Coimbra Collection of Algae.

Harvesting for a Lifetime: Celebrating 25 Years of Seaweed Educating

1

Prannie Rhatigan

Abstract

It was a sunny July day in the seaside village of Strandhill, County Sligo, on the west coast of Ireland. Glittering white surf-edged waves danced in off the Atlantic in front of me, while to my back towered iconic Knocknarea, tomb of the legendary warrior queen Maedbh of Connaught.

Keywords

Seaweed history and education · Irish folklore · Seaweed cuisine

1.1 Introduction

Between the two, in the Strandhill Golf Club Facilities, the stuff of legends was also taking place: all of Ireland's leading culinary journalists were gathered to celebrate the launch of a pioneering book which would bring the fantastic vegetable bounty of the sea to the kitchens of chefs and cooks of all abilities worldwide in ways that had been almost forgotten or invented for the purposes of this book (Fig. 1.1). The event was organized as a pre-launch of the *Irish Seaweed Kitchen* due to be published in November of that year (https://irishseaweedkitchen.ie/ accessed February 2024).

One of the reasons my husband Johnny Waters and I, along with a close team of friends and family, decided to get food writers tasting seaweeds in advance of the publication was that for six years I had been unable to find a publisher willing to take on my book on seaweeds (Fig. 1.2a and b). Many letters of rejection containing words such as: 'Futuristic', 'Much too ahead of its time', or 'lovely idea but much too niche' had flowed in. We were worried about their words coming true but knew that if writers could taste the delicious forms of seaweeds they would be hooked (Fig. 1.3).

At the time, I had no idea that the *Irish Seaweed Kitchen* would eventually reach people on all five continents on the planet, or that international TV crews would arrive to film Irish seaweed, writers come to write about it, celebrity cooks come to cook with it. I've lost count of the number of people I've taken on seaweed walks along beaches on the west coast of Ireland (Figs. 1.1, 1.2a and b, 1.3, and 1.4a–c). In the quarter of a century since I began to share my seaweed knowledge and skills, it's gratifying to see how many of those who took their first walks with me are now working professionally with seaweeds themselves and equally heartening to salute the raft of seaweed cookery/lifestyle books that the *Irish Seaweed Kitchen* model has inspired and encouraged worldwide.

So back to that summery day in July 2009.

To launch '*Irish Seaweed Kitchen*,' the team had decided to invite as many Irish journalists as possible for a day by the Sea in Sligo. Incredibly, most of them accepted the invitation. When they arrived in Sligo, we sent them on a tour of the area, before inviting them to dine on seaweed recipes from the new book, in the shadow of Queen Maedbh's cairn. Eithna O' Sullivan, chef extraordinaire of the much-decorated 'Eithna's by the Sea' restaurant in Mullaghmore (sadly now closed), led the kitchen team, and in a nod to Maedbh and her queenly court of yore, we had harp music to accompany the feasting. I remember distinctly telling the harpers and waiting staff that we would sit together afterwards and dine on the fine fare of the day (Fig. 1.5a–f), but that never materialized because there wasn't a scrap of food left! Eithna has a generous hand when it comes to food, and a keen eye when portioning food, but there was still nothing left.

The seaweed chocolate desserts, in particular, had been approached with polite disdain, as very few, if any of our literary guests had tasted that particular mix. I was pretty sure there would be plenty left for my helpers, but the word spread like wildfire that the chocolate fudge nori pudding

P. Rhatigan (✉)
Irish Seaweed Kitchen, Streedagh, Grange, County Sligo, Ireland; https://irishseaweedkitchen.ie/

A. T. Critchley et al. (eds.), *Advances in Temperate Phyconomy: Algal Harvest and Cultivation in Globally Distributed Temperate Waters*, Developments in Applied Phycology 15, https://doi.org/10.1007/978-3-032-02955-3_1

was next level, as was the chocolate fondant, also made with nori (Fig. 1.6a–g). The cinnamon-rolled sea lettuce we served with green tea and nori ice cream also made headlines.

Salads, breads, mains all disappeared, bowls were scraped out, everything vanished.

As for the harpers and helpers, they dined on pizza from one of the very fine Strandhill eateries!

I knew the day was special to my personal journey with seaweeds, and looking back, I think that day was also pivotal as to how seaweeds were perceived in Ireland. After that day, features, articles, recipes, reports on walks, information on baths and lots of detailed information began to appear in the national press regularly. There were interviews on national airways. Shortly after that, feature writers from the international press took an interest in Sligo seaweeds, and in the years following, we welcomed television crews and celebrity chefs to North Sligo. By the time I had published my second book, the *Irish Seaweed Christmas Kitchen*, in 2018 and received a Gourmand Best in the World award for it the following year, seaweed culture had come of age (Figs. 1.7, 1.8, and 1.9).

Fig. 1.1 Sligo

Seaweeds were cool; seaweed culture was becoming a 'thing'.

The first Sligo Food Festival—So Sligo—came into being and ran for a number of years and following that the birth of the Sligo Food Trail. We were lucky enough to have a vibrant organic centre near us in County Leitrim, at the time, and they rolled out some great pioneering courses on seaweed cookery as well as organizing regular seaweed walks. They are still going strong. The current exciting food festival, 'A Taste of Sligo'—again featuring seaweeds—is scheduled to happen again this autumn (2025).

Now, more than 25 years, or so years later, seaweeds have a very definite profile along Irish coastlines in terms of economy, wellness and health.

But if we go back 25 years from that date, how was the Irish seaweed profile some 40 years ago?

The first person to ever hold a licence for seaweed harvesting in Ireland is Frank Melvin, and he first began harvesting in the 1980s in County Sligo. While he was thinking about starting a seaweed business (Fig. 1.10), I was focused on my final clinical years in medical school in Cork, Ireland. He tells a very funny story about having several visits from the gards (police force) in the garda car while getting the licence approved, much to the curiosity of the locals. Garda cars do not appear in a rural area without good reason, and the neighbours sure wanted to know what it was all about! As it was the first licence, there was a lot of paperwork to be submitted to the Irish Department of Marine.

Fig. 1.2 (**a** and **b**) A lifetime harvesting on the shore

Fig. 1.3 A day by the sea in County Sligo

Frank had spotted a niche in the market in the 1980s when seaweed from Asia became available in the local health food store—kombu that he recognized as growing on the shores in County Sligo. He was aware of the growing interest in health and well-being and soon after that he was testing the market with dillisk (*Palmaria palmata*) and carraigin (*Chondrus crispus* or Irish Moss) in the style of a cottage industry. Day by day, it got better and better with journalists taking an interest and influencing an increase in sales. Careful sustainable harvesting has kept the business steadily growing over the last 37 years.

Thinking back over the years we were in agreement with many of the changes that have gradually occurred along the coast with regard to various seaweeds. We both agreed that the huge banks of kelp washed in after storms at all times of the year seem so much smaller now (Fig. 1.11). Ironically, then, these iconic kelp stacks were banked and dried on walls and taken away by lorries to Scotland—to be processed for alginates—which doesn't happen anymore.

Another great stalwart of Irish seaweed is Manus McGonagle (Fig. 1.12) from the neighbouring county of Donegal. His harvesting work has seen him on the shoreline in all seasons, since the 1980s, when he first started work as a part-time seaweed harvester utilizing skills and knowledge learned from his father.

Like all of us, he too has observed small, but steady, changes on the shore over his years working on it, one of which is the almost disappearance of the shelled dulse which grew on small mussels that could be found mid-shore. The dulse started to disappear in the mid-1990s, when it could no longer be found on south-facing rocks, but remained plentiful on north-facing shady areas. In recent years it has all but completely disappeared from these areas too and has now almost become a thing of the past.

As kids, we were discouraged from eating this type of dulse because our father, a dentist, feared that we would destroy the enamel of our teeth if we encountered the small, crunchy mussel shells that the dulse was often attached to.

Manus (Fig. 1.12) remembers that when he started supplying *Palmaria palmata* to local people, they only wanted shell dillisk—the sweet *Palmaria* that grows on the black mussels, known as duilacans. They had no desire to eat the *Palmaria* without the mussels attached.

As the years passed, he gradually got his customers to convert to regular dillisk and weaned them off the shell dillisk. Only the very elderly now still crave the taste of shell dillisk which has become a rarity and has almost gone altogether.

Manus noticed that from the beginning of 2000 onwards some exposed seaweeds were becoming burned or bleached by the sun, at low tide, while growing on the shore during the months of May and June. This sun damage to some exposed seaweeds has been extended to July and August in recent years, although it doesn't happen every year. Dillisk, in particular, is prone to this relatively new phenomenon of sun damage and often cannot be harvested in this condition. Instead of its normal purple/red, it is bleached to a yellow/white (Fig. 1.12).

Manus has other equally interesting comments, mainly to do with the tides, which he observes no longer ebb to the same low levels as they did even 2–3 years ago. This means that access to low-growing seaweeds is limited, and that some of the low water individuals themselves are now more prone to fouling, or other algae growing on them. Manus recalls visiting a French kelp harvester many years ago. It surprised him that the kelp being sent to market had large areas covered with the small white limpet-like shells called *Arthopyrenia sublittoralis*. In 2024, he noticed that these shells are now growing on Irish kelp in greater numbers and need to be cleaned off before the kelp is sent to market.

There are other changes too. Because the tide is not ebbing as far out as it used to, some seaweed species, and in particular the *Fucus* species, are beginning to grow on higher parts of the shoreline. *Fucus vesiculosus* in particular has crept up the shore to areas which were previously bare (Figs. 1.13 and 1.14).

The regular monthly spring tides are not as low as they used to be either. Ten years ago a harvester could harvest the dillisk growing epiphytically on the large kelps almost every month by wading around them. Now this can only be achieved 2–3 times a year, on particularly low tides, for example around the Spring Equinox. We also have noticed it when harvesting or bringing people to the beach for walks, but I attributed it to low pressure and rough seas on the day. High pressure pushes the tide out further.

Is this caused by the ice melting? A shift in the axis of the earth? An increase in temperature?

So far, no one has come up with an explanation. Our tide books still predict a 0.7-m tide but is it really that low?

Another significant change is related to the introduced *Sargassum muticum*. On the northwest coast, it initially occurred in very sheltered locations, but last summer it appeared on more semi-exposed locations and in a much big-

Fig. 1.4 (**a**–**c**) Sligo seaweed experience

ger form. Some rockpools are currently completely choked with the *Sargassum* and native species are blocked out (Fig. 1.15).

During these recent conversations with my neighbouring harvesters, Frank and Manus, which I thoroughly enjoyed, we discussed seaweed matters past and present. A concern we share is the need to maintain our high-quality clean seas, for future generations of harvesters and future generations of seaweeds (Figs. 1.16, 1.17, and 1.18).

Back in the 1990s, when I first started to bring people to the shore to learn about seaweeds, I would check in with them to see what they already knew about seaweeds and what they wanted to get out of the day. They had very little knowledge about seaweeds back then. Perhaps a grandmother who forced them to take an awful drink of carrageen moss when ill with a respiratory infection. Or a grandad who chewed dillisk. There was never a mention of any other seaweed at that time.

Nowadays when I check in with participants, they might have had an *Alaria* salad for dinner the previous evening, or

Fig. 1.5 Few savoury recipes from the Irish Seaweed Kitchen. (**a**) carrot and sea spaghetti salad (**b**) filo pie with sea spaghetti, mushrooms and apples (**c**) oat and seaweed bread (**d**) prawns with land and sea spaghetti (**e**) cannellini bean salad with marinated Alaria (**f**) yeast spelt bread with mixed seaweeds

have made the dillisk and cheese scone recipe from my book, or a green seaweed-containing smoothie. All of them are familiar with seaweeds and want to know more (Fig. 1.19a and b).

Through the years I have encouraged participants to try to keep as minimal a footprint as possible in the form of environmentally friendly laundry and dishwasher products and to recycle.

While walking the shore with seaweed enthusiasts these days, I am often reminded that not so long ago in Ireland there was still some of the legacy of the famine days when people went to the shore to try to find food to survive. In the 1970s, we were among the very few who harvested from the shore: folk memory still associated the practice with extreme poverty, and I remember some folk reacting with exaggerated aversion at the thought of eating any seaweed (Fig. 1.20). We, on the other hand, thanks to my parents, learned to respect seaweeds for what they are.

During my six years of research for the *Irish Seaweed Kitchen*, two critical questions emerged to which there were no clear answers. The first was around the behaviour of carraigin when heated causing concerns over low molecular weight fractions. The issue was first raised by Dr. Joanne Tobacman which also provoked an industry response, and the second concerned the extraordinarily high levels of iodine in kelp which can be ameliorated by simple washing or blanching. Consumers must educate themselves. Years later there is now much more knowledge available to both harvesters and consumers on these issues.

Every year is completely different. When speaking to groups on the shore I say it is like their own garden, if they are gardeners or observers of nature. Some years we have a glut of plums, so great it breaks the branches, and other years a bounty of apples. Some years the hawthorn is spectacular in plumes of white across the countryside, other years it is the turn of the blackthorns with white-tinged pink blossoms. It is the same for the garden of the sea. Wild harvesting of nori is less predictable over the years and very dependent on the timing of storms. During a calm autumn, the sand builds up on the rocks and impedes the growth of nori. A stormy autumn sweeps the sands off the rocks allowing the nori to thrive.

Fig. 1.6 A selection of desserts (**a**) layered chocolate walnut squares with *Alaria* (**b**) chocolate nori molasses meringues (**c**) carrageenan, honey and yogurt panna cotta (**d**) baked lemon cheesecake with nori (**e**) chocolate fondant with nori (**f**) carrot , nori and sugar kelp cake (**g**) green tea and nori ice cream

There is a lot to take in on a shore walk but the main take-home message is that Sustainable Harvesting is critical.

Bringing out the '*Guide to Edible Seaweeds*' in 2013 (https://irishseaweedkitchen.ie/guide-edible-seaweeds/, accessed February 2025) was a game changer for people wanting to get an in-depth look at individual seaweeds with a view to setting up a business. The QR codes printed on each page brought them to the shore to view the seaweeds in situ enabling precise identification, along with the text and photo. It has been widely copied in other countries and even

Fig. 1.7 Gourmand Best in the World award for the *Irish Seaweed Christmas Kitchen* in the fish and seafood category: Awards Ceremony, Macau, China, July 2019

Fig. 1.8 A great night for seaweed!

Fig. 1.9 Seaweed demonstration on the Chefs' Stage, Macau, China

Fig. 1.10 Seaweed distribution on the rocky shore

Fig. 1.11 A smaller kelp bank was notes on the shores in recent years

copied in the world of land herbs - which is lovely—it's all about education.

If seaweed growth patterns, shorelines and tides have changed over the last two decades, the general attitude to seaweeds and their consumption has changed more radically in that time - than it did even in the previous hundred years. People have an appreciation for seaweeds now.

Seaweed consumers these days have vast knowledge of its associated benefits, alleged and otherwise. Research into the various types of seaweeds globally is ongoing and extensive, and the curious can access information with a click.

In an age where people worldwide are obsessed with health, longevity, and in particular food supplements, seaweeds have a vibrant and spectacular role, offering help in areas that are of most concern to a modern population.

And that doesn't even begin to detail the boost to overall well-being and mental health an hour or two by the sea can offer, just checking out the local seaweed flora.

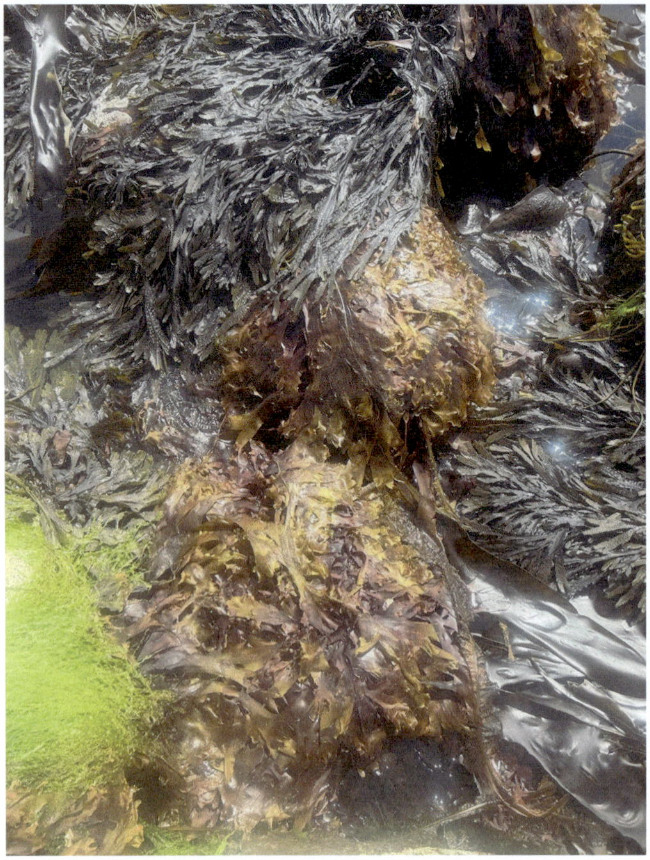

Fig. 1.12 It has been interesting to speculate on the cause of this. (Image credit to Manus McGonagle)

Fig. 1.13 Image credit to Manus McGonagle

The least introspective of people can breathe deeply and just soak in the surroundings, while the more thoughtful can ponder the immense system and structure of the magnificent ocean, and the intricate and complex supports that marine life, and especially seaweeds, offer humans.

Who would have thought that human problems like obesity, inflammation, cancer prevention, digestion, blood pressure, cholesterol and a healthy heart can be positively influenced by specific ingredients found in specific seaweeds (Fig. 1.21a–c)!

Over the last 25 years people have developed an increasing interest in the specific health benefits of seaweeds.

As a medical practitioner, I have to work from an evidence base, but I am always happy to share the emerging research and anecdotal evidence for the basis of increased, responsible consumption of seaweeds. The nutrient-rich profile of seaweeds is undisputed as are the phyto-defensive bioactive components in the form of polysaccharides and polyphenols—unique compounds not found in land plants. I love telling people that 'seaweeds have been described as the most nutritious form of vegetation on the planet'.

I remind people that not all seaweeds are the same and the benefits of one seaweed cannot be attributable to all. There are specific seaweeds for specific applications. Huge variations occur depending on season, geographic location and even within individual thalli. It is very difficult to be specific about the exact nutrient content of seaweeds. However, in general, all seaweeds can be considered to be 'bursting' with minerals, vitamins and trace elements—they are powerhouses of nutrients.

Fig. 1.14 *Himanthalia elongata* at low tide

Fig. 1.15 The invasive brown seaweed *Sargassum muticum*

Fig. 1.16 Sustainable harvesting

Let's take a look at a few of the specifics I share with participants of my courses and shore walks:

Weight loss: Increasingly a current first-world obsession.

The fibre in various seaweeds can promote a feeling of fullness which in turn leads to fewer calories being consumed. Having dillisk handy to snack on—specially when paired with a walnut—and chewed to release the umami mouthfeel is something very special.

Fig. 1.17 *Palmaria palmata* (dillisk, dulse)

Evidence currently points toward facts that members of the genus *Fucus*, in particular, are useful as anti-inflammatory and anti-cellulitic agents and generally for the promotion of weight loss.

Gastrointestinal health: Root cause for many modern illnesses.

Polysaccharides, some of which act as soluble fibre, can have a very positive impact on the gastrointestinal tract, through their interaction with the microbial flora and by helping to regulate and decrease transit time.

Protein source: Great for vegetarians—if it can be utilized.

I am often asked by enthusiastic participants if they could live on seaweeds alone for a few weeks or a month.

Porphyra sensu lato spp. (Fig. 1.22) have one of the highest protein contents, up to 40% of their dry weight. Small amounts along with lots of other land and sea vegetables offer the benefit from the associated anti-oxidant and synergistic effects. The green *Ulva* spp. are also high in protein. While they contain all of the essential amino acids, much more research is needed here regarding digestibility and utilization ....so my answer is always no—not a wise option.

Heart support: Good news for everyone.

Increasingly research is showing that consumption of some seaweeds can be beneficial to heart health.

Claims that seaweeds can lower cholesterol and improve blood flow continue to be investigated, particularly in Japan, and some results are promising.

Brown seaweeds (Fig. 1.23a and b), in particular, are rich in compounds called peptides, which are known to reduce high blood pressure: a major contributor to cardiovascular disease. An Irish review of evidence published in 2011 further supports the hypotensive powers of certain specific seaweeds, while some of the green seaweeds are showing promising anti-lipidemic (lipid-lowering) effects.

Salt substitute: Takes the pressure out of blood pressure.

Too much salt (as sodium chloride) in the diet is also bad news for health. I ask participants in my courses to substitute

Fig. 1.18 (**a** and **b**) *Chondrus crispus* growing and shimmering and being cooked in milk

Fig. 1.19 (**a**) Harvesting seaweeds; (**b**) putting them to good use

Fig. 1.20 Seaweed truffles: a treat on the shore or at the dinner table

table salt for dried, ground seaweeds. I ask them to dry and grind a variety of seaweeds, or buy a ready-made jar of ground seaweeds. It contains potassium salts, so still tastes 'salty' and is much better for you as it has very little sodium chloride—only that what is naturally clinging to the dried seaweeds. An added bonus is that other heart-friendly minerals such as magnesium are also present together with iodine, selenium and calcium, so the whole body benefits. (Of course, anyone with a low tolerance to iodine should be careful.)

Cancer support: One in three Irish people can expect a cancer diagnosis (February 2025).

It is well known that Japanese women have much lower rates of breast cancer than Western women, and experts believe that their seaweed-rich diet may be a factor. Of course, there are many types of cancer, the causes of which are multi-factorial.

Fig. 1.21 (**a–c**) Seaweed pesto of reds, greens and browns. A magic wand in the kitchen!

Fig. 1.22 Nori *Porphyra*

Some brown seaweeds also contain high quantities of naturally occurring lignans, with cancer-fighting phytoestrogens which are of particular interest to researchers at the moment. Lower rates of breast cancer are reported in studies on Japanese women eating a diet high in kelp, and kelps have also been shown to reduce DNA damage induced by several known carcinogens, but more research, and inevitably human clinical trials are required.

I feel that when participants on my seaweed identification and tasting walks hear about current health research and taste the seaweeds, in different forms, it encourages them to continue to experiment and also to be much more aware of ways to protect the ocean, as well as their own well-being.

In conclusion, to emphasize the change in attitude to seaweed harvesting and eating over the last few decades, in my part of Ireland, let me share a little comparison between the way our eldest and youngest daughters were perceived at primary school. Our eldest daughter was teased by her school mates and was called the Seaweed Farmer while in Primary school. (Personally, I loved that name - while she hated it.) While the so-called Seaweed Farmer set off on yet another trip to some far flung land accompanying us, her parents, the school mates were left at school! Perhaps a little bit of the green-eyed monster in action we felt. In recent years, our youngest has been celebrated at school and has written seaweed recipes for the school cookbook. She also does a little trading with her friends who have developed a taste for the *Irish Seaweed Kitchen* blend-trading sachets of dried seaweeds for half a dozen fresh eggs, some Pokeman cards or a nice pot of jam. How times change!

Looking back on all that, and the personal highlights of some educational and media exchanges (Figs. 1.24a–c, 1.25, 1.26, 1.27, and 1.28), I realize that we, as seaweed gatherers, are just a link in a long and basic chain of survival stretching back thousands of years.

Acknowledgement The author thanks Frank Melvin and Manus McGonagle for contributing their time and wisdom to the piece; Johnny Waters for his constant observation of shoreline changes and deep understanding of our coastal environment; Catherine Rhatigan for casting her editorial eye over the finished chapter.

Fig. 1.23 (**a**) *Alaria*; (**b**) Dulaman na binne bhui—*Pelvetia canaliculata*

Fig. 1.24 (**a**) British Broadcasting Corporation (BBC) Northern Ireland film crew for 'Secret Gardens' 2004; (**b**) David Moscow 'From Scratch', filmed 2023; (**c**) Austrian National TV, filmed in Autumn 2024

Fig. 1.25 Launching the Irish Seaweed Association March 2025

Fig. 1.26 Johnny Waters and Prannie Rhatigan going for harvesting

Fig. 1.27 Enjoying the *Irish Seaweed Kitchen* with a cup of tea

Fig. 1.28 (**a** and **b**) Sea views at the start and the end of the day

Prannie Rhatigan is a medical doctor with a lifetime's experience of harvesting, cooking and gardening organically with seaweeds. Born and raised in County Sligo where she still lives, her lifelong interest in the connections between food and health prompted her to publish the *Irish Seaweed Kitchen* in 2009. The book won a Gourmand Award Ireland for Best First Cookbook and has become the ultimate bible for cooking with sea vegetables.

An experienced General Practitioner, she returned to work full time in Public Health Medicine in February 2020 in response to the global COVID-19 pandemic. She spent many years as a motivational tutor with the Irish College of General Practitioners (ICGP) working in the area of patient behaviour change. She also worked on the National Cardiovascular Strategy Programme and represented the ICGP on the National Steering Group for the implementation of 'Smoke Free at Work' in Ireland. She also holds a BA in Psychology.

Prannie's position as Ireland's leading culinary seaweed expert (*Food and Wine)* continues to grow with a swathe of workshops, lectures, talks and sea vegetable cookery demos stretching across the globe from USA and Canada to Sydney, Australia and from the Nordic countries to Spain. She has represented Ireland's finest food abroad on many occasions and has given workshops and lectures on sea vegetables and cooking. She is always delighted to participate in Irish Festivals and is particularly proud and supportive of the strong links being forged between seaweeds and the burgeoning Irish culinary world.

She has been featured on many TV and radio programmes both nationally and internationally, the most recent being the USA-based 'From Scratch' series in 2023, The National Trust podcast series in 2024 and Austrian National TV *Silvia entdeckt Irland* in 2025.

Her enthusiasm for her subject has earned her a well-deserved credit for accessibility and innovation, both of which reflect in her many TV and radio appearances worldwide. She founded Sligo Seaweed Days and The Sligo Seaweed Experience and is a proud member of Sligo Food Trail. In 2012 she was awarded *A Special Acknowledgement for Her Outstanding Contribution to Irish Food* by Euro-Toques. Her second book, *Irish Seaweed Christmas Kitchen*, was the winner of the Best in the World Gourmand Award, Seafood Category 2018. In 2023, she was awarded Local Food Hero for Connaught by The Irish Times and the Irish Restaurant Association.

In addition to spreading the word about seaweeds, Prannie is passionate about sustainability and regeneration, the links between food and genetics, live and fermented foods and the future of food worldwide. She launched her own Irish Seaweed Kitchen Blend of Irish Seaweeds in 2019. She was invited to join The International Seaweed for Health Conference organizing committee.

She lives on the edge of the Sligo coastline with her husband Johnny Waters and their daughters. www.irishseaweedkitchen.ie

Phyconomy Applied to Global Environment Solutions Through Enterprise Ecology Systems

2

Iain C. Neish

Abstract

Human enterprise actions that influence ecosystems are at the front line of humanity's impacts on the global environment, both in temperate and tropical waters. At the time of writing, micro, small, and medium enterprises (MSMEs) comprised 95% of businesses globally and contributed more than half of total employment, according to the WTO, yet many of them languished because of inadequate value chain and market system support. They were commonly family owned and operated and were predominant in seaweed farming worldwide. Enterprise ecology engages MSMEs in ecoscape communities as they apply arts, sciences, and technologies that influence, manage, and control biosphere ecosystem services valorized through fair ecoeconomic market systems. Seaweed farming enterprises are the true practitioners of adaptive phyconomy, hence phycoecology, and thus enterprise ecology. Profitable enterprises in sustainable ecosystems are essential for the success of initiatives associated with global environmental causes, including UN Sustainable Development Goals, blue ocean, climate change actions, biodiversity, social justice in socioeconomies, eco-economics, and a host of others. Enterprise ecology, as described in this chapter, is a practical framework for building robust enterprises in sustainable ecosystems as necessary for such global environmental solutions to succeed. MSMEs are generally deficient in financing, market systems, and technical support and lack the internal Knowledge, Information, Tools, and Solutions (KITS) required to manage ecosystem services in line with most global environmental initiatives. Failed MSME results in failed initiatives, so means must be found to help them succeed.

Keywords

Eco-economics · Ecosystem · Enterprise · MSME · Phyconomy · Phycoecology · Services

I. C. Neish (✉)
Seakitshub Inc., Cebu, Philippines
e-mail: Iain@seakitshub.com

A. T. Critchley et al. (eds.), *Advances in Temperate Phyconomy: Algal Harvest and Cultivation in Globally Distributed Temperate Waters*, Developments in Applied Phycology 15, https://doi.org/10.1007/978-3-032-02955-3_2

Graphic Summary

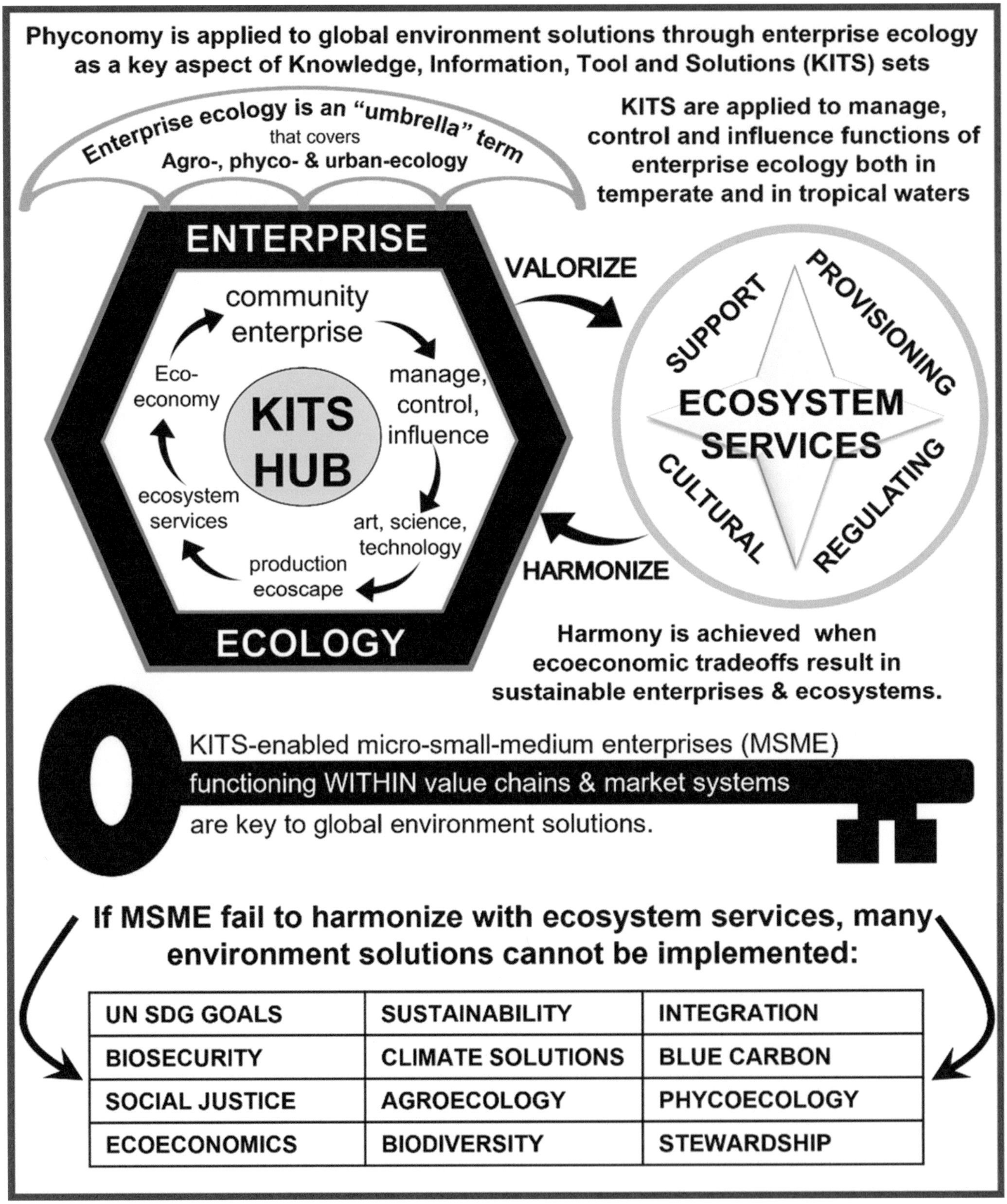

2.1 Introduction: Toward Restoring Seashores One Enterprise at a Time

Human enterprise actions that influence ecosystems are at the front line of humanity's impact on the global environment. Over the past several millennia it has become obvious that the aggregate impacts of human enterprise actions have contributed to system dysfunctions and imbalances not only in Earth's biosphere, but also in the atmosphere, the hydrosphere, and the geosphere (Fig. 2.1). If we accept the premise that enterprise actions can influence global system failures, then we can recognize that they are also instrumental in implementing system solutions. Perhaps we can "save the planet" and restore its seashores, one enterprise at a time.

This book is about phyconomy and is therefore about human enterprises engaged in seaweed farming. Seaweeds have been on planet Earth, possibly for more than a billion years before *Homo sapiens* came to be (Tang et al. 2020), and seaweeds may outlast humanity for billions of years after our extinction, so from a seaweed-centric perspective, it is not the planet that needs saving; it is us. Through phyconomy, we are influencing global seaweed biomass in ways that may help to prolong our existence as a species, and the means for doing that through enterprise ecology are the subject of this chapter.

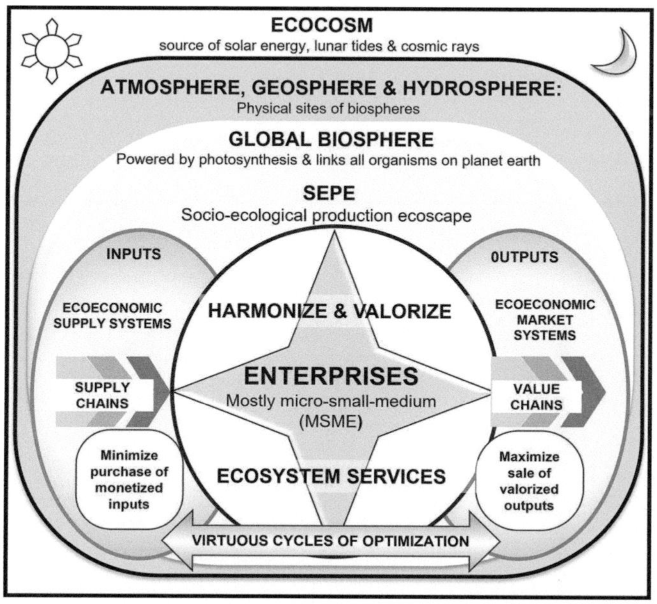

Fig. 2.1 A schematic depiction of how enterprises fit into the world around them. Phyconomy enterprises are in socioecological production ecoscapes (SEPE) where people live and work. Most are micro-, small-, and medium enterprises (MSMEs), and ideally, they optimize transactions according to economic principles (Brown 2001) as they link supply chains to value chains in supply and market systems. On a macro scale, SEPE are nested in the global biosphere within an ecocosm that also includes the atmosphere, hydrosphere, and geosphere. Ultimately, enterprises influence surrounding ecosystems through actions directed toward those spheres

2.2 Can Seaweeds Really Save the Planet?

There is substantial hype through social media that virtually attributes superpowers to seaweeds as heroes coming to the rescue of a planet that must be saved. The dangers of such hype have been the subjects of social media "buzz" and have been discussed comprehensively by Costa-Pierce and Chopin (2021). The essence of their argument is expressed in their header statement as follows:

> We both support sustainable aquaculture practices. However, we should be cautious with the hype we are subjected to daily. We have a collective responsibility to identify, and denounce, the Silver Bullet Solution Salespersons, remembering that in between periods of promising the Moon, there are painful "purgatory" periods of non-constructive regressions

One of the most dubious concepts that has been promoted, and even implemented on modest scales, is the practice of deep-ocean dumping of seaweed biomass as a means for sequestering carbon or other elements. Chopin et al. (2024) addressed this in a many-coauthored "petition paper" with the title *"Deep-ocean seaweed dumping for carbon sequestration: Questionable, risky, and not the best use of valuable biomass."* This title makes the point.

Issues of seaweed hype were also addressed by ETC Group in a 43-page document with the title *"THE SEAWEED DELUSION: Industrial seaweed will not cool the climate or save nature"* (Thomas and ETC Group 2023). A key message of this document was that seaweed projects are beginning to attract major funding from big-money players whom the "good guys" must convince to support the "right things" and must dissuade from backing the delusional "bad things" that ETC vilifies. As it happens, the ETC document eloquently makes much of the case in favor of an enterprise ecology approach to phyconomy, although ETC has a blatant political agenda, while enterprise ecology is politically neutral.

There is a miasma of agenda-motivated rhetoric and hype surrounding issues of the degree to which seaweed can save the planet, and I continue to assert that through phyconomy, we are influencing global seaweed biomass in ways that may help to prolong our existence as a species. No, seaweed cannot save the planet, but phyconomists can move things in the right direction.

2.3 Why Enterprises Are the Key to Global Ecosystem Actions

Per this chapter, an enterprise is defined as *"a unit of economic organization or activity, especially a business organization or a systematic purposeful activity."* (After Merriam-Webster 2024). Thus, an enterprise may be as small as a part-time activity by one individual or as large as the projects of a massive multinational corporation.

In the literature, definitions of enterprise size vary, but generally small and medium enterprises (SMEs) are defined to be those with more than ten and fewer than 200–250 employees. Enterprises with fewer than ten employees may be called micro-enterprises, so there is a designation "MSME" for micro-, small-, and medium enterprises. In this chapter, any enterprise of 1–250 employees was regarded as being an MSME. I believe that few, if any, seaweed farming enterprises in the world have more than 250 employees, and in tropical regions, my experience has been that most have fewer than ten employees, so they could technically be called micro-enterprises.

I posit that most human actions that impact ecosystems are driven by economic motives and are therefore undertaken by enterprises. The interaction of the enterprise with ecology is a key anthropogenic influence on the biosphere, atmosphere, hydrosphere, and geosphere. The focus of this chapter is adaptive phyconomy in open-ocean systems (discussed below), and at the time of writing, I posit that the vast majority of enterprises engaging in seaweed farming around the world were family owned and operated MSMEs.

MSMEs comprise most businesses worldwide and are important contributors to job creation and global economic development. According to the World Trade Organization, MSMEs represent over 90% of the business population, 60–70% of employment, and 55% of GDP in developed economies, and they hold about 20% of patents in biotechnology-related fields in Europe. MSMEs accounted for 95% of businesses globally and contributed more than half of total employment. Around 90% of the world's jobs were generated by the private sector, and some 50% of that was by MSMEs (Bayraktar and Algan 2019; WTO 2016).

"Industrialized seaweed," as vilified in ETC, was not yet conspicuously successful even in temperate regions at the time of writing, and it was virtually non-existent in the tropical regions where most seaweed farming is done. Enterprise ecology, like phycoecology, can therefore be viewed essentially as an MSME activity.

2.4 Phyconomy Defined in Terms Relevant to Enterprise Ecology

Consideration of phyconomy as discussed in Hurtado et al. (2019) led to a notional definition of phyconomy as *"a branch of applied phycology that comprises systems of art, science and technology applied to biosecure production systems that yield crops of algae."* (Neish 2023).

In a context of phycosecurity in multibiomass systems, I proposed that it is more useful to define phyconomy operationally as *"a branch of applied phycology that strives to manage seaweed ecosystem services valorized per ecoeconomic principles in phycosecure, multibiomass, socio-ecological production ecoscapes (SEPE) that are subject to integrated coastal area management (ICAM)."* (Neish 2023), and what I referred to as "operational phyconomy" can more appropriately be described as "phycoecology" and therefore is an aspect of enterprise ecology.

Adaptive and precision phyconomy are compared in some detail in Vadassery and Neish (2023). Adaptive phyconomy is a low-cost, low-control approach to seaweed farming (Fig. 2.2). It is practiced in situations where the biology of algal crops is poorly understood and/or where confounding variables render precision technologies impractical. By contrast, precision phyconomy is a high-cost, high-control approach to algal crop production. It is practiced in situations where the biology of algal crops is well understood, and culture conditions can be comprehensively controlled.

As depicted in Fig. 2.2, there is a continuum of cost/control options that lies between the extremes of adaptive and precision technologies. The salient point for the present chapter is that adaptive phyconomy predominates in open-ocean seaweed farming by MSME undertaking enterprise ecology. It is toward the lowest end of the cost/control range. The only approach to seaweed production that is generally lower in cost and control is managed harvesting and stewardship of seaweed from natural stands, which is still a common seaweed production strategy in temperate production areas, for example, in the case of harvesting *Ascophyllum nodosum* (Phaeophyceae) from natural stands in eastern Canada and Western Europe.

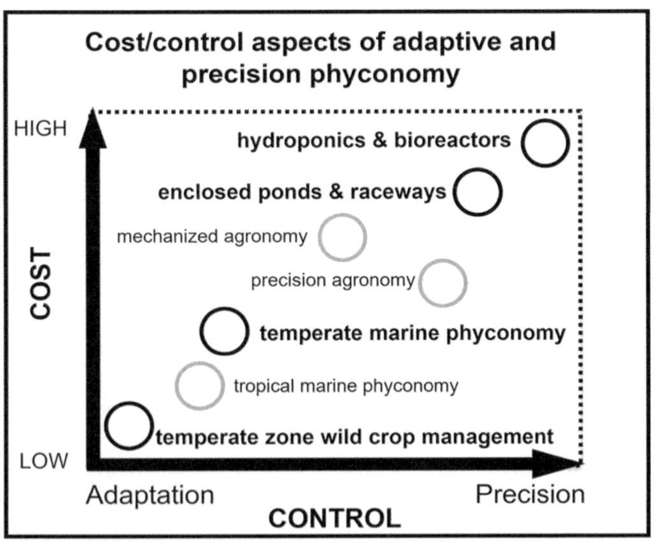

Fig. 2.2 From adaptive phyconomy to precision phyconomy, control comes with costs. In this figure, several types of agronomy and phyconomy are plotted with approximate degree of system control on the abscissa and production cost on the ordinate. Placement of systems on the matrix was based on the author's experience as manager of commercial phyconomy systems

2.5 The Anthropocentric Notion of "Ecosystem Services"

With respect to enterprise ecology, ecosystem services are treated as a bounty of nature bestowed upon humanity with a caveat that we must exercise diligent stewardship of that bounty. Depending on one's religious or philosophical beliefs, this position may or may not be palatable (see Thomas and ETC Group 2023), but many, if not most, human societies view the world as a creation for humanity, and for practical purposes, I regarded it as such when writing this chapter.

According to the Millennium Ecosystem Assessment (MEA), ecosystem services are defined as "*the human gains obtained from ecosystems*" (Carpenter et al. 2006; Cotas et al. 2023). They are divided into four categories designated as supporting, provisioning, regulating, and cultural services. The Common International Classification of Ecosystem Services (CICES) differs from the MEA classification, as it considers supporting services to be organism functions and recognizes only provisioning, regulating, and cultural services per Cotas et al. (2023). The present chapter adopted the MEA classification system and added a fifth category designated as "*ecosystem disservices*" (Fig. 2.3).

Figure 2.3 shows examples of ecosystem services from photoautotrophs in marine ecoscapes. These include seaweeds and all other photosynthesizing marine and aquatic organisms.

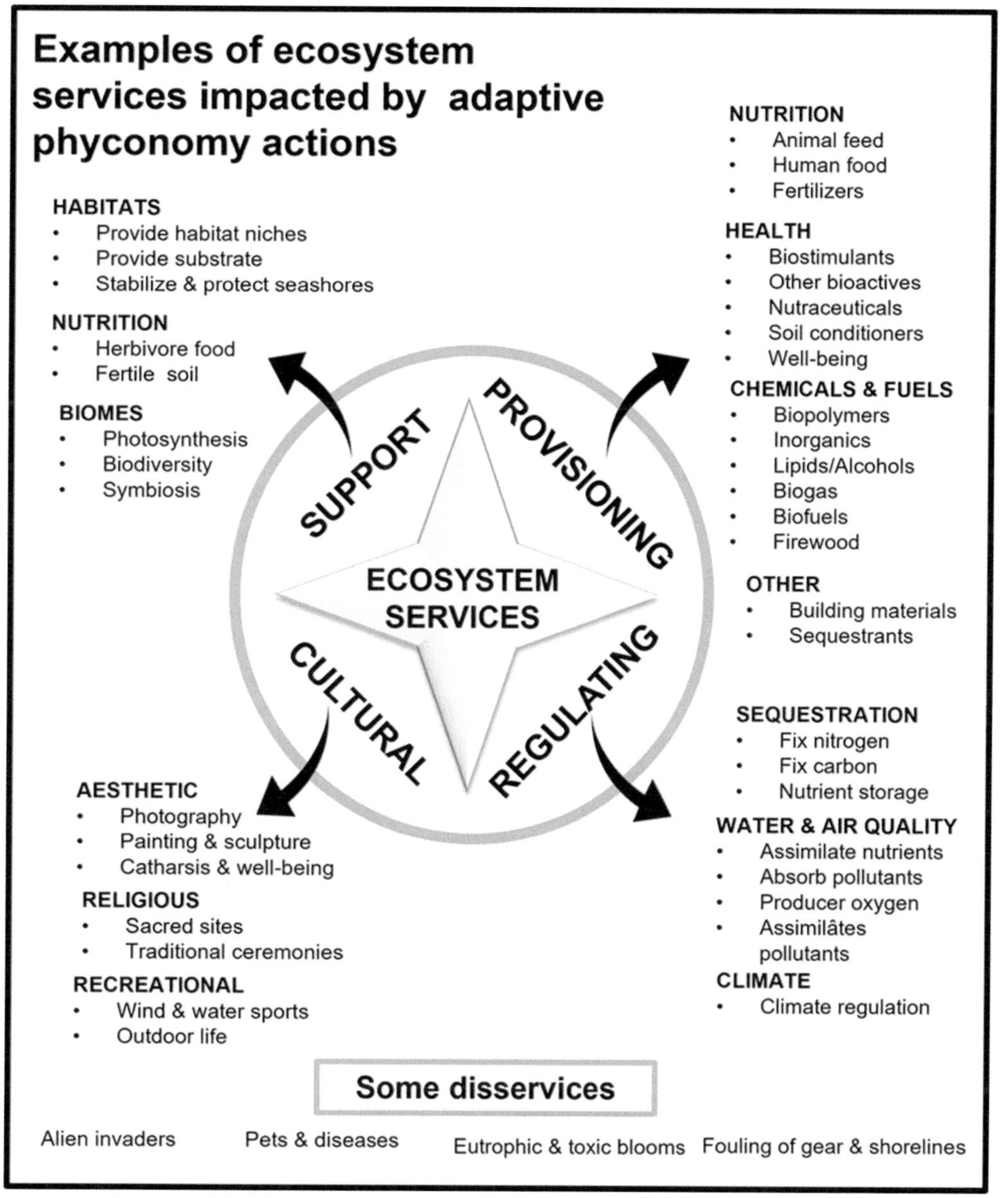

Fig. 2.3 Examples of ecosystem services impacted by adaptive phyconomy actions. As per Fig. 2.1, enterprises harmonize and valorize ecosystem services as they practice enterprise ecology. Harmony results when eco-economic tradeoffs result in sustainable enterprise profitability in sustainable ecosystems

Support services are those required for the development of all other ecosystem services, especially including photosynthesis and other primary productivity processes, so they are valorized through derivative ecological services. Among those, the primary goal of phyconomy systems is to optimize balances among provisioning services that yield financial returns from biomass production. In other words, the prime goal of phyconomy is to produce and sell seaweed crops.

Regulating services comprise ecological processes, such as climate regulation, habitat provision, oxygen production, and geological impacts, whereas cultural services are non-material benefits that people derive from ecosystems, including cognitive development, recreation, and aesthetic experiences. Such services are generally valorized through non-financial costs and benefits per ecoeconomic principles.

It is also necessary to minimize impacts from disservices that photoautotrophs may inflict (Fig. 2.3). For phycomycologists, control of disservices is an inevitable cost of farm operations.

From an enterprise ecology perspective, production of seaweed biomass as a cash crop from seaweed farms is the key provisioning service to be quantified. That must be achieved in a context of systematic balance with services from other socioecological production ecoscape (SEPE) organisms. Methods for doing that are still a work in progress. Many ecosystem regulating and cultural services have intangible value that is difficult to quantify in financial terms.

Overall, the objective of enterprise ecology is to harmonize and valorize ecosystem services. Harmony results when eco-economic tradeoffs result in sustainable enterprise profitability in sustainable ecosystems.

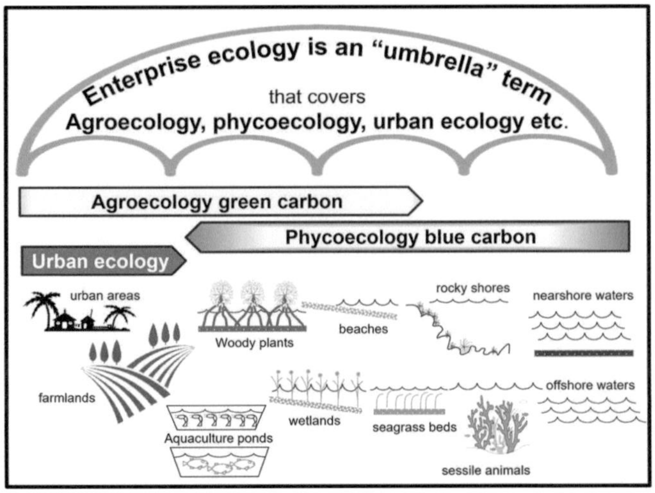

Fig. 2.4 Agroecology and urban ecology (associated with "green carbon") and phycoecology (associated with "blue carbon") are inextricably intertwined and overlapping in coastal ecoscapes, where they function together as enterprise ecology

2.6 Enterprise Ecology Embraces Agro-, Phyco-, and Urban Ecology

The terms "enterprise ecology" and "phycoecology" are neologisms that follow on the well-established frameworks of agroecology (citations below) and urban ecology (Douglas et al. 2011) (Fig. 2.4).

There is copious literature pertaining to agroecology (e.g., CIDSE 2018; ECA et al. 2021; Glessing 2020; HLPE 2019; Sinclair et al. 2019; TWN 2015; Wezel et al. 2020). FAO (2022) has published a document that summarizes ten elements at the core of it (Fig. 2.5a). FAO pointed out that features distinguishing agroecology as a conceptual framework include:

- Agroecology is based on bottom-up (reflexive) and territorial processes, helping to deliver contextualized solutions to local problems.
- Agroecological innovations are based on the co-creation of knowledge, combining science with the traditional, practical, and local knowledge of producers by enhancing their autonomy and adaptive capacity.
- Agroecology empowers producers and communities as key agents of change.
- Agroecology addresses the root causes of problems in an integrated way to provide holistic and long-term solutions.
- Agroecology focuses on social and economic dimensions of food systems, including the rights of women, youth, and indigenous peoples.

The term "phycoecology" is a neologism that derives from the term "phyconomy" as described in Hurtado et al. (2019), Neish (2023), and Vadassery and Neish (2023). Agronomy relates to phyconomy much as agroecology relates to phycoecology in the sense that soil management and land management, by definition, are a core of agronomy (hence agroecology), while phyconomy (hence phycoecology) is concerned with seaweeds that have no roots and grow in flowing seawater rather than in soil. This is in contrast with vascular marine plants such as woody plants (e.g., mangroves) and seagrasses, which arguably can be cultivated in agroecology systems. Anyway, they are all embraced by the term "enterprise ecology."

Agroecology is a discipline that was far advanced relative to phycoecology at the time of writing, and with the copious literature developing around agroecology, it was the basis for concepts that could be generalized as enterprise ecology or particularized as phycoecology. With respect to phyconomy in phycoecology, one can readily see how the FAO and Wezel elements can be adapted to fit. In phycoecology, stewardship over the ocean and seashores is equivalent to land management, and soil management can

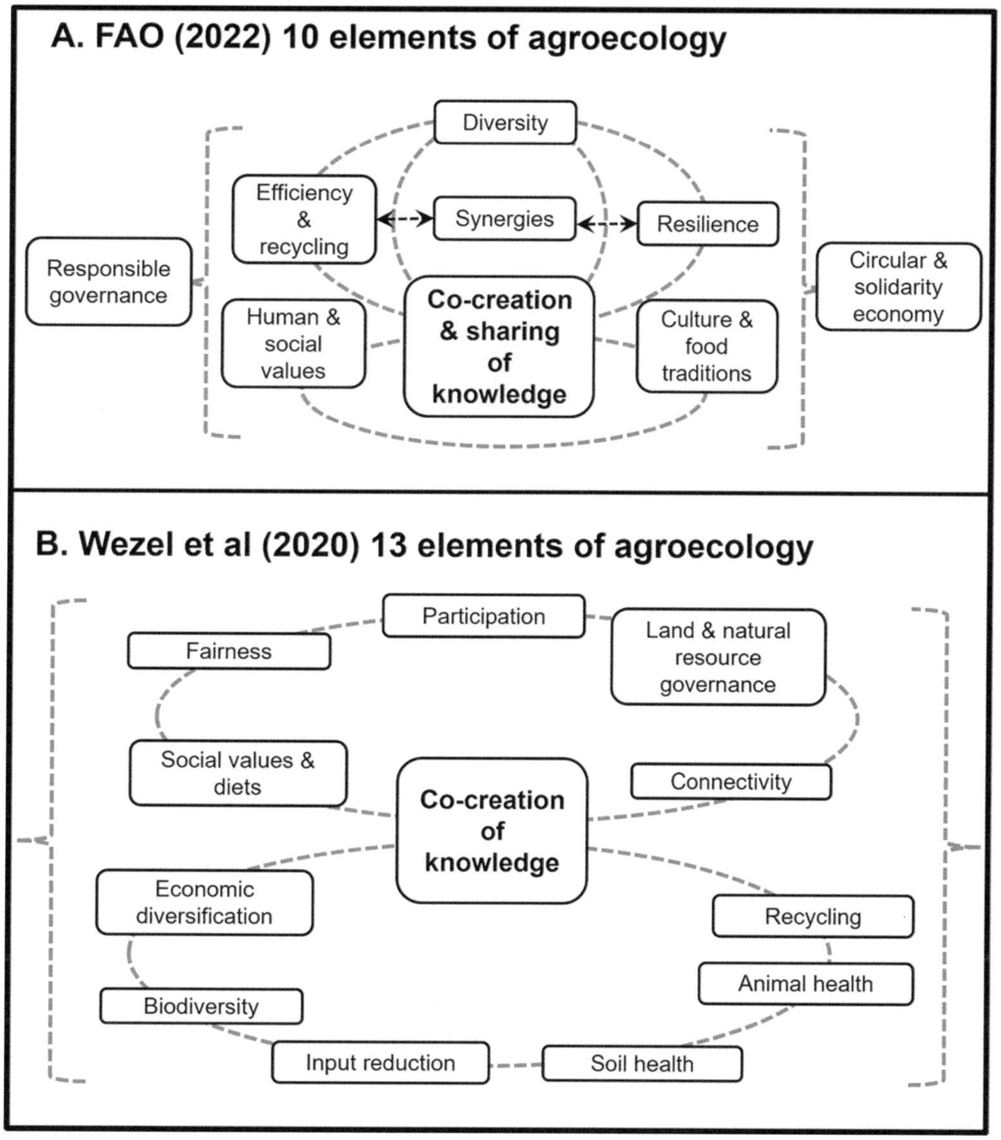

Fig. 2.5 (**a**) The FAO (2022) ten elements of agroecology generally apply to urban ecology and phycoecology and are inextricably intertwined and overlapping in ecoscapes where they function together in KITS sets (Figs. 2.6 and 2.7) of enterprise ecology. (**b**) The Wezel et al. (2020) 13 elements of agroecology are complementary to the FAO elements. Both sets of elements have co-creation of knowledge as a core competency, and that competency is embodied in KITS's sets of enterprise ecology

be contrasted to difficulties associated with the management of flowing seawater in open seas.

2.7 KITS as Co-created Knowledge that Drives Enterprise Ecology Functions

FAO (2022) and Wezel cited "co-creation of knowledge" as a core concept of agroecology. For enterprise ecology, as described in the present chapter, that notion is embodied in sets of Knowledge, Information, Tools, and Solutions (KITS) driving the functions depicted in Fig. 2.6. These functions are meant to sustainably engage enterprises in ecoscape communities to apply arts, sciences, and technologies that influence, manage, and control biosphere ecosystem services valorized through fair ecoeconomic market systems.

Enterprise ecology KITS sets draw from both internal and external sources, and they evolve virtuous cycles of innovation (Fig. 2.7). In the case of adaptive phyconomy, enterprise ecology is an adaptive process that treats implementation of ecosystem services solutions as experimental probes (after Holling 1978, cited in Neish 2023).

2.8 KITS Hub Enterprises Within Enterprise Ecology Fed Value Chains

The version of enterprise ecology described in this chapter was developed as a functional framework that is being applied to real-world seaweed farming in tropical seaweed farms. I assert that it is equally applicable to temperate region seaweed farms.

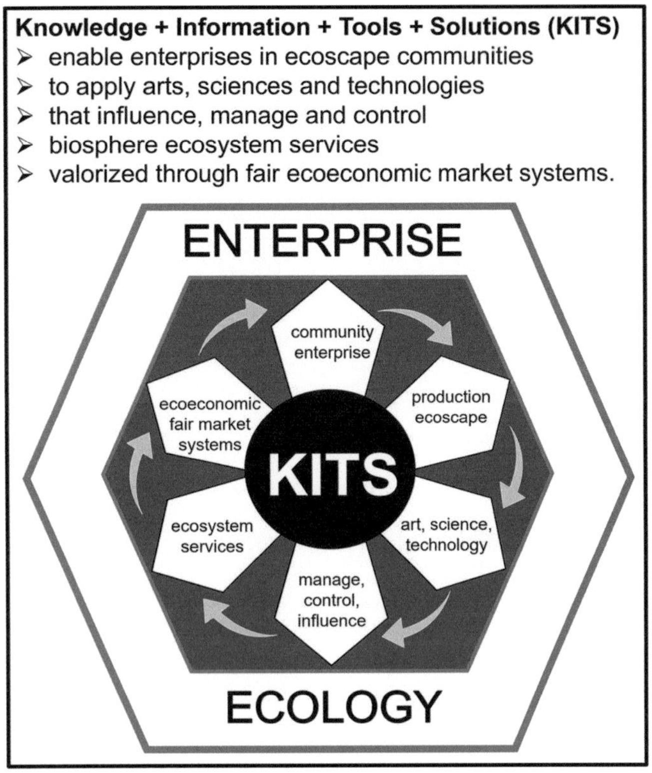

Fig. 2.6 The core of enterprise ecology systems is their sets of Knowledge, Information, tools and Solutions (KITS), which comprise the "co-creation of knowledge" central to the FAO/Wezel et al. elements of agroecology and which are the basis for functions of enterprise ecology shown in this figure

As depicted in Fig. 2.6, enterprise ecology is defined in terms of elements that are relevant to value chains, as shown in Fig. 2.8. From a socioeconomic perspective, the function of a seaweed farming enterprise is to satisfy customers by sourcing from supply chains and farm productivity to provide products through value chains to satisfied customers. This dynamic is depicted in Fig. 2.1, and the position of value chains in market systems is depicted in more detail in Fig. 2.8. With respect to phyconomy, the focus is on diverse ecosystem services provided by seaweeds, especially the provisioning services of crop production.

Value chains function within market systems along lines depicted in Fig. 2.8, which is based on lessons learned from experience with value chain studies (Neish and Suryanarayan 2017) and with guidance from sources including Herr and Muzira (2009), ILO (2016), and UNIDO (2011). As depicted, a KITS hub facilitates farm enterprise access to essential support services and farm interactions with governance enforcement entities. It also facilitates links up and down the value chain. In performing such functions, the KITS hub is a value chain actor that is a strategic ally to the farm enterprise.

During early years of seaweed farming development, in the mid-twentieth century, some KITS hub functions were performed by extension services of government and academia (e.g., in Japan); in some cases, by state enterprises (e.g., in China); and in many cases by processing companies in need of raw material (e.g., for hydrocolloids' producers). Although government and academic extension services are now present in most countries where seaweeds are farmed, industrial enterprise support diminished with privatization in China and with the proliferation of new processors in the hydrocolloid sector elsewhere. New processors tended to not invest in KITS support, and legacy processors ceased supplying KITS support because they could not recoup their investment in it. In my experience, loss of KITS support was a possible contributing factor in value chain failure in many formerly productive farm areas, including in the region where I live and work in the Philippines.

By the time commercial seaweed farming became widespread in tropical Asia during the 1980s through 2000s (Neish et al. 2017), numerous donor-funded non-governmental organizations (NGO) and aid agencies working through governmental organizations (GO) started to channel funding into "interventions" targeting perceived seaweed value chain problems. Many such interventions supplied sets of KITS to farm enterprises, and I was part of many of them in East Africa and Southeast Asia. From my experience, NGO/GO interventions suffered from four constraints that limited their utility to MSME farm enterprises, namely:

1. They were of short duration (usually 5 years or less);
2. Much, if not most, of the funding was consumed by overhead, startup, wind-down, and monitoring/evaluation costs;
3. They were driven by agendas not necessarily aligned with the agendas of enterprise owners and operators; and
4. They failed to leave substantial KITS capacity in value chain enterprises after interventions ceased.

2.9 KITS Provision and Control as a Function of Value Chain Governance

Whoever controls KITS exerts substantial influence over the value chain. Figure 2.9 depicts the KITS provision and control scenarios for five value chain models I have observed over the past four decades (Neish 2013; Neish and Suryanarayan 2017).

Panlibuton et al. (2007) first applied the heuristic models of Gereffi et al. (2005) to seaweed value-chain governance in 2007, and the trajectory of eucheumatoid value chains was discussed in terms of these models in Neish (2013). The models also seem to be relevant to temperate region seaweed farming that I am aware of.

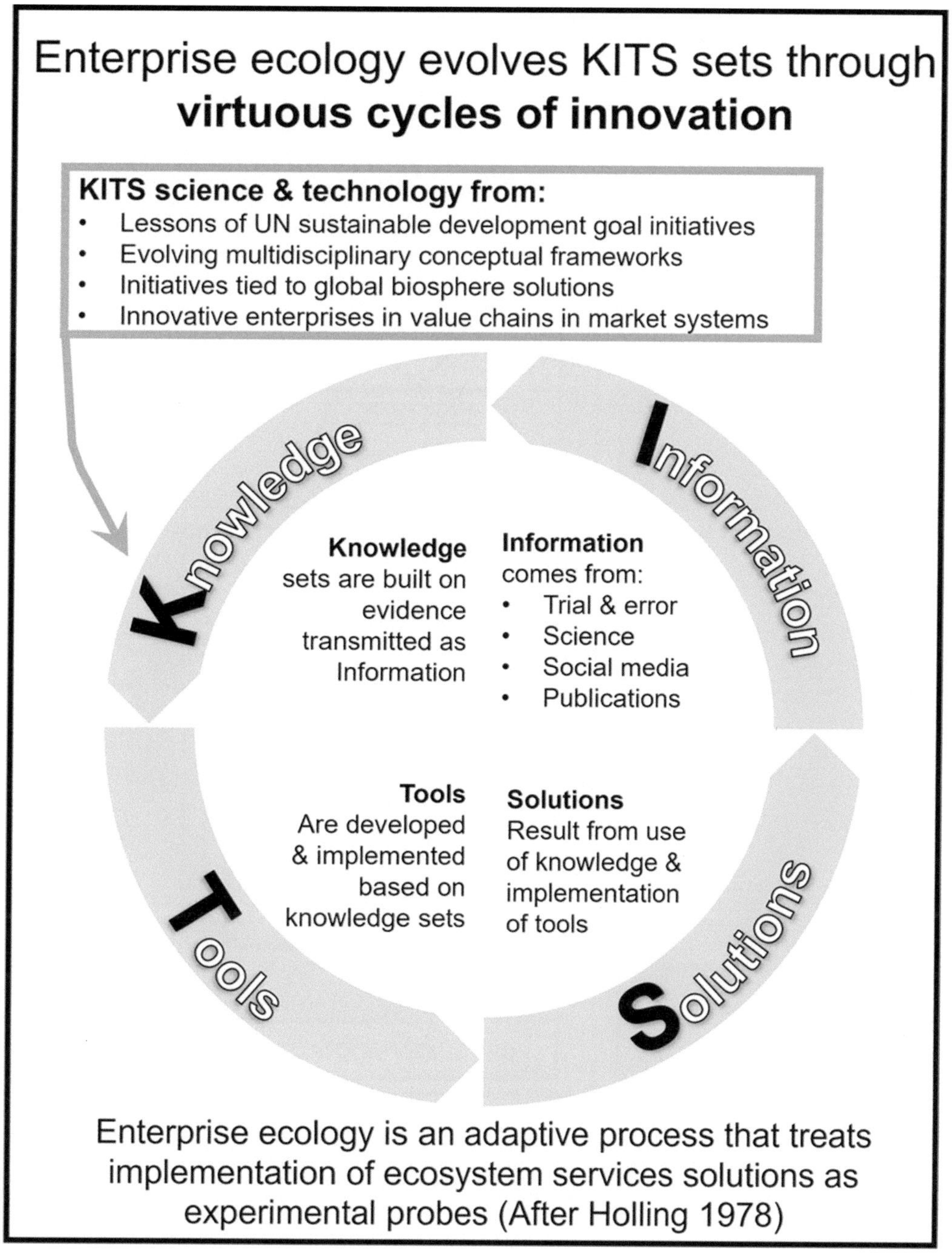

Fig. 2.7 Enterprise ecology evolves KITS sets through virtuous cycles of innovation

Figure 2.9 adds the issue of KITS provision and control to the Gereffi et al. models, and that aspect of value chain governance is placed in the context of trajectories from the cited sources as follows:

1. *Captive or hierarchical governance are systems where KITS tend to be controlled and provided by processors.* They are typical during the early development of farming systems when there tend to be many small sellers and just a few major buyers, so the market is an oligopsony. KITS are developed in-house by processors and/or are obtained from GO, NGO, and academic sources, often with assistance from incentive programs (e.g., the US Seagrant program that subsidized much phyconomy KITS development during the last quarter of the twentieth century). KITS supplies to farmers in captive systems tend to diminish as farming spreads out of their control due to actions of competitors. Hierarchical systems maintain control by enforcing security measures that keep KITS in-house, where the farmers are employees of estate farms.
2. *Modular governance is a system where KITS tend to be controlled and provided by integrated suppliers.* Suppliers insert a value chain link between farm enter-

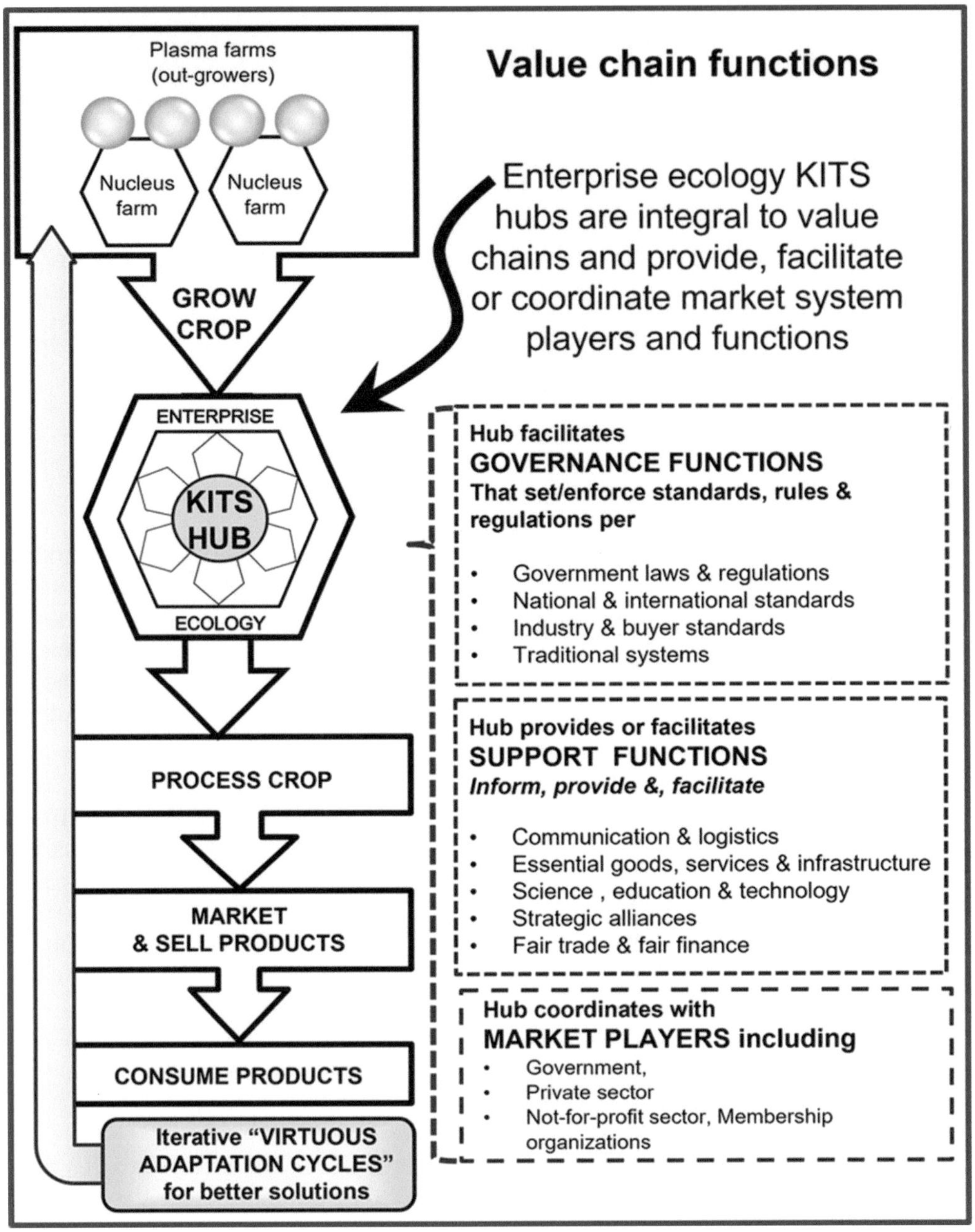

Fig. 2.8 Enterprise ecology KITS hubs are integral to value chains and provide, facilitate, or coordinate farm enterprise interactions with value chain and market system stakeholders and functions

prises and processors. KITS tend to be embedded in farmers' supply agreements. KITS tend to be sourced from processors/buyers and also from GO, NGO, and academic sources, often with assistance from incentive programs. KITS tend to be minimally shared with farmers because possession of KITS is a source of suppliers' competitive advantage.

3. *Market governance is a system where KITS tend to be controlled and provided by market system support providers.* Much, if not most, seaweed is sold in "spot-markets." Although they add little value, agents and officials are able to collect rents and gain a trading advantage through their control of funds and KITS. KITS tend to be minimally shared with farmers because possession of KITS is a source of traders' competitive advantage.
4. *Relational governance is a system where KITS tend to be controlled and provided among allied farming enterprises.* KITS are a core asset of a farm enterprise strategic business alliance. At the time of writing, relational governance systems were rare in seaweed value chains, although some had been achieved by the aggregation of farmers into cooperatives. If the notion of enterprise ecology gains traction, relational governance will be more widespread because relational governance is essential to enterprise ecology systems with co-created KITS.

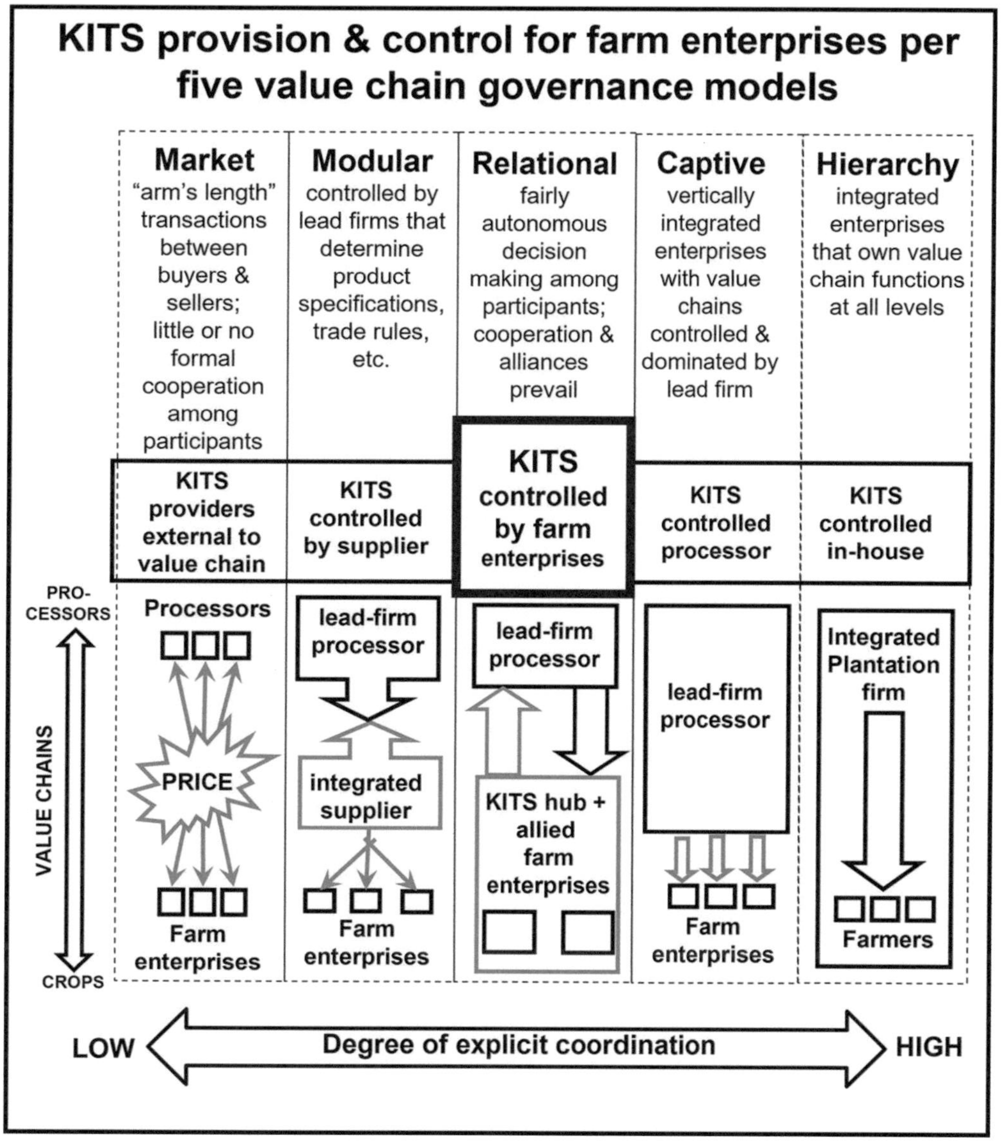

Fig. 2.9 KITS provision and control as a function of value chain governance. (Governance models after Gereffi et al. (2005))

2.10 Enterprise Ecology Essential for Achievement of UN SDG

The United Nations 2030 Agenda for Sustainable Development is a plan of action for people, planet, and prosperity that seeks to strengthen universal peace and freedom. It recognizes that poverty eradication is an indispensable requirement for sustainable development, and the development of profitable small-medium enterprises must surely be essential for this to happen.

The plan includes 17 Sustainable Development Goals (SDGs) and 169 targets that follow on from the Millennium Development Goals and complete what they did not achieve. Figure 2.10 depicts enterprise ecology in the context of SDG. Phyconomy in enterprise ecology does not have a direct bearing on all 17 of the SDGs, but it is relevant to most of them as follows:

Goal 1. End poverty in all its forms everywhere. At the time of writing, especially in tropical regions, seaweed farming was widely undertaken along seashores where poverty was rampant and phyconomy could be a path to prosperity. This may have been less true of temperate regions, although I got my start as a phycomycologist in Nova Scotia, Canada, where my father was developing Irish moss (*Chondrus crispus*) farming specifically as a remedy for the poverty common there in the mid-twentieth century.

Goal 2. End hunger, achieve food security and improved nutrition, and promote sustainable agriculture. Enterprise ecology, including phycoecology, goes beyond promoting agriculture to embrace holistic management of ecosystem services from the totality of ecoscape biomass, whether on land, sea, or in between.

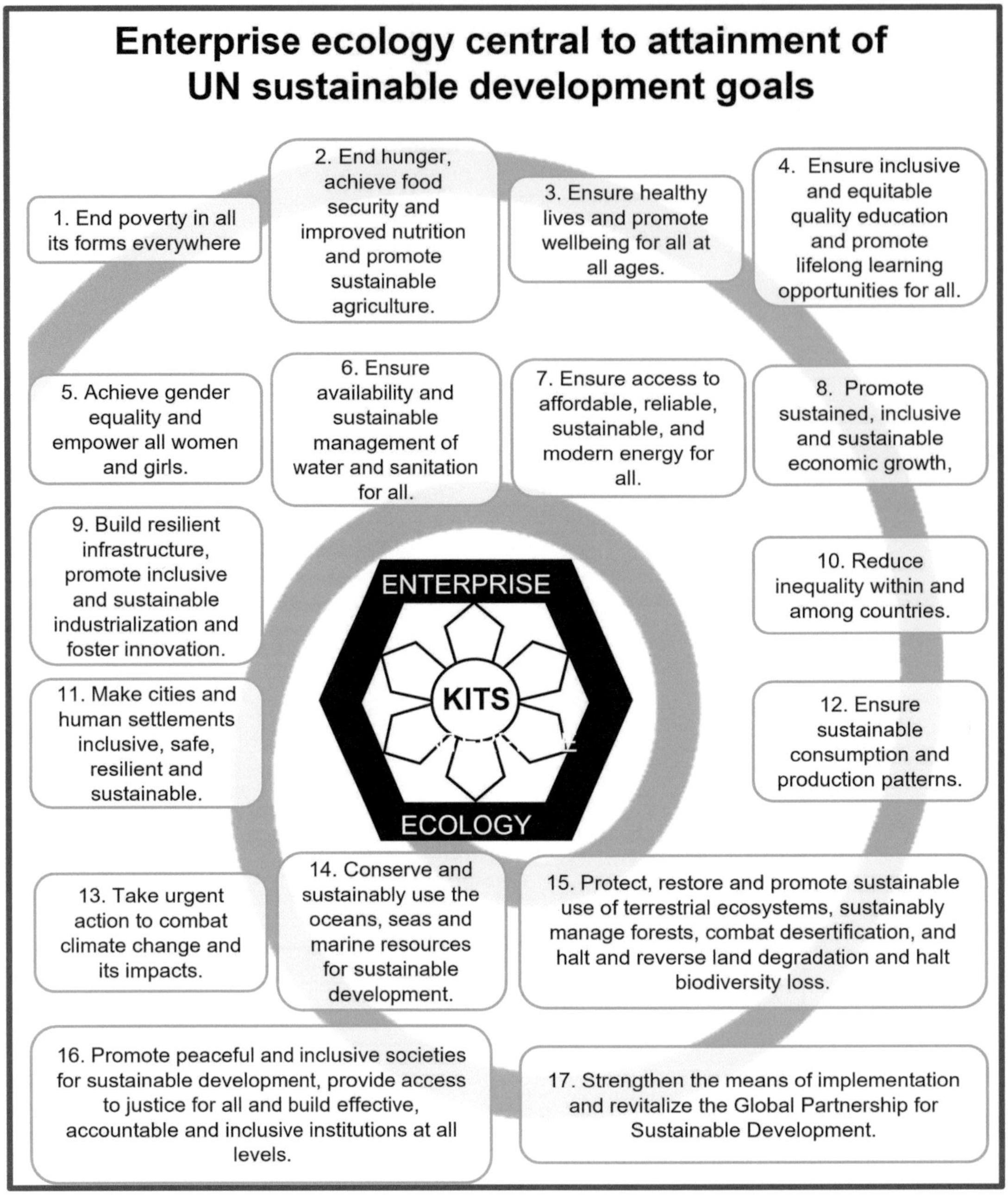

Fig. 2.10 Enterprise ecology depicted in the context of the 17 United Nations Sustainable Development Goals (SDG)

Goal 3. Ensure healthy lives and promote well-being for all at all ages. Phycoecology promotes the well-being of humanity in a context of multibiomass ecosystem services in coastal ecoscapes.

Goal 4. Ensure inclusive and equitable quality education and promote life-long learning opportunities for all. This is a core function of KITS hubs in enterprise ecology.

Goal 5. Achieve gender equality and empower all women and girls. Enterprise ecology in coastal areas is generally a family-centric activity, so gender issues are handled in accordance with the norms, mores, and folkways of coastal communities.

Goal 6. Ensure availability and sustainable management of water and sanitation for all. These are crucial functions that are integral to ecosystem services harmonization in enterprise ecology.

Goal 7. Ensure access to affordable, reliable, sustainable, and modern energy for all. This is also a crucial function that is integral to ecosystem services harmonization in enterprise ecology.

Goal 8. Promote sustained, inclusive, and sustainable economic growth, full and productive employment, and decent work for all. Enterprise ecology strives to reach these goals according to principles of value chain operation in market systems that are operated in conformity with practices of ecological economics.

Goal 9. Build resilient infrastructure, promote inclusive and sustainable industrialization, and foster innovation. These functions fall within the purview of enterprise ecology approaches to value chains in market systems.

Goal 10. Reduce inequality within and among countries. Enterprise ecology is globally applicable in all regions, whether tropical, temperate, or otherwise. Harmonized ecosystem services are consistent with the reduction of inequality.

Goal 11. Make cities and human settlements inclusive, safe, resilient, and sustainable. It is a principle of enterprise ecology in production ecoscapes that human activities are undertaken in harmony with nature and are therefore sustainable.

Goal 12. Ensure sustainable consumption and production patterns. Attainment of this goal is linked with attainment of Goal 11.

Goal 13. Take urgent action to combat climate change and its impacts. Promulgation of enterprise ecology will lead to the development of information technologies bundled with Internet of Things solutions to promulgate systems of environment and biomass monitoring that are crucial to the development and application of climate change solutions.

Goal 14. Conserve and sustainably use the oceans, seas, and marine resources for sustainable development. Coastal regions are the focus of phycoecology, which is directed toward this goal.

Goal 15. Protect, restore, and promote sustainable use of terrestrial ecosystems, sustainably manage forests, combat desertification, halt and reverse land degradation, and halt biodiversity loss. Coastal ecoscapes that are the focus of phycoecology initiatives include both land and sea multibiomass in production ecoscapes.

Goal 16. Promote peaceful and inclusive societies for sustainable development, provide access to justice for all, and build effective, accountable, and inclusive institutions at all levels.

Goal 17. Strengthen the means of implementation and revitalize the Global Partnership for Sustainable Development. MSMEs are the foundation of socioeconomic systems of the world. Enterprise ecology is a development approach that is essential to reaching these goals.

2.11 Conceptual Frameworks Feeding Enterprise Ecology KITS

Enterprise ecology KITS sets must draw from the multitude of conceptual frameworks that are relevant to enterprise profitability and ecosystem well-being (Fig. 2.11). For the most part, these conceptual frameworks are linked to environmental solutions initiatives such as those depicted in Fig. 2.12.

2.12 MSMEs as Essential Drivers of Global Environmental Solutions

Micro-, small-, and medium enterprises (MSME), which are commonly family owned and operated, are key drivers of the "big-picture" global environment solutions that acquire major funding as cures for what ails planet Earth. MSMEs are key, front-line drivers of global environmental solutions (Fig. 2.12).

Failed MSMEs result in failed solution initiatives, yet MSMEs in general and coastal seaweed, aquaculture, and fisheries MSMEs in particular generally receive pathetically inadequate market system support and financing support (Carvajal and Didier 2024). Certainly, in the tropical value chains I know best, KITS at the time of writing were even less available to enterprises than they were when I started in the business in the 1970s. Conclusions reached by Costa-Pierce and Chopin (2021) imply that the situation may be as bad or worse for temperate zone aquaculturists.

2.13 Integration Essential for Ecosystem Services Harmony and Value

Ecosystems are multitrophic (Fig. 2.13), and influences exerted on ecosystem services must be integrated if enterprise ecology is to harmonize and valorize them.

FAO (2021) long advocated an ecosystem approach to aquaculture, and the notion of Integrated Multitrophic Aquaculture originated with Thierry Chopin and Jack Taylor around 2003. Chopin was a tireless promoter of the Integrated Multitrophic Aquaculture (IMTA) brand until his untimely death in mid-2024 (e.g., Chopin 2013). Within a few years of its introduction, the IMTA concept was embraced by many of the world's prominent phyconomists and aquaculturists, and several multi-author papers were published to promote the concept (e.g., Buck et al. 2018; Knowler et al. 2020; Troell et al. 2009).

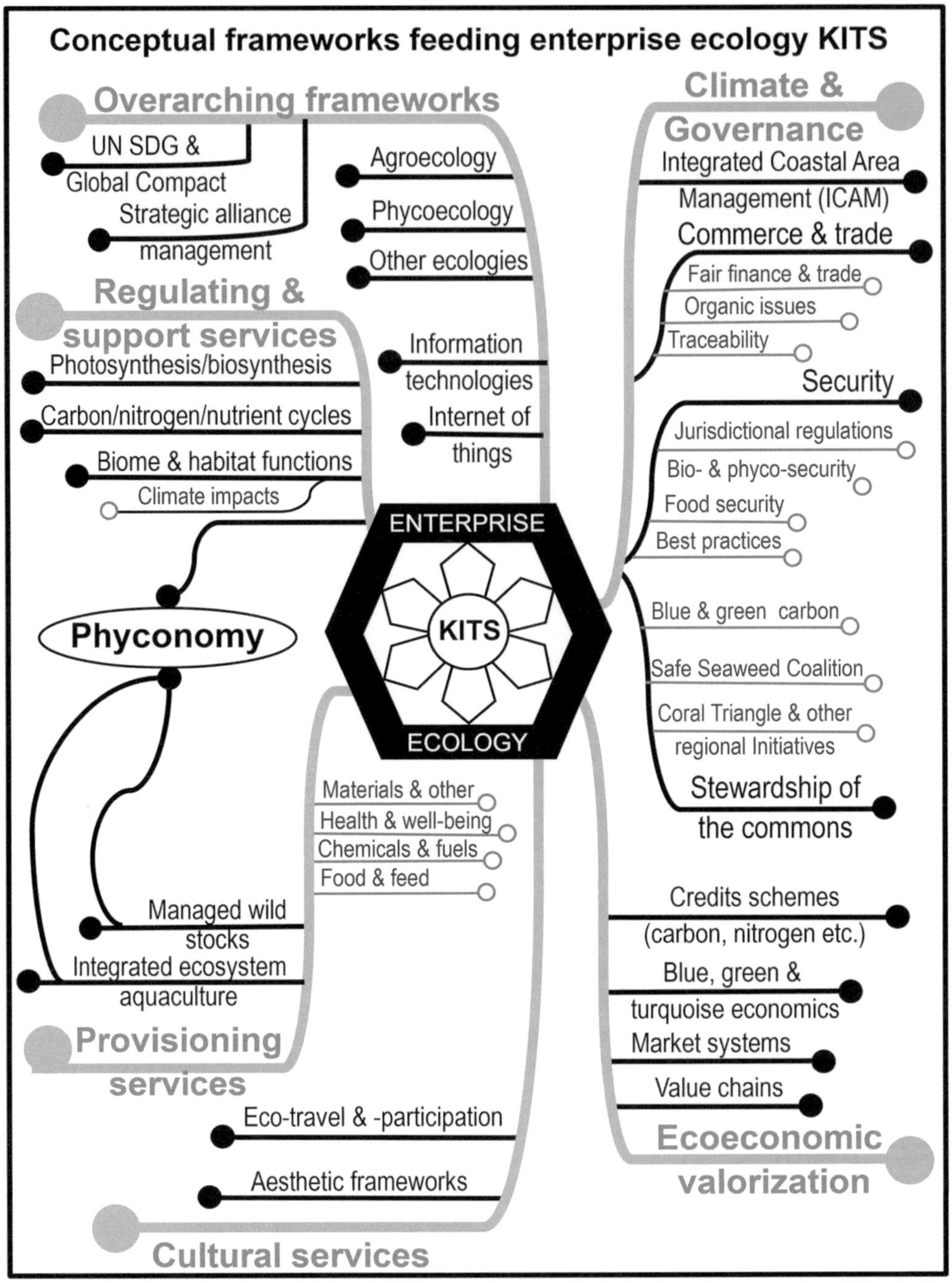

Fig. 2.11 Enterprise ecology KITS sets must draw from the multitude of conceptual frameworks that are relevant to enterprise profitability and ecosystem well-being

Chopin (2013) reviewed several approaches to integrated aquaculture and wrote that:

> Integration should be understood as cultivation in proximity, not considering absolute distances but connectivity in terms of ecosystemic functionalities. It should be made clear that in the minds of those who created the acronym "IMTA," it was never conceived to be viewed with the minimalist perspective of only the cultivation of salmon, kelps and … Saccharina latissima (Alaria), and blue mussels within a few hundred meters: this is only one of the variations … and … the IMTA concept can be extended to very large ecosystems like the Yellow Sea … This also means that IMTA variations include integrated agriculture aquaculture systems (IAAS), integrated sylviculture (mangrove) aquaculture systems (ISiAS), integrated green water aquaculture systems (IGWAS), integrated peri-urban aquaculture systems (IPUAS), integrated fisheries aquaculture systems (IFAS), sustainable systems (SEAS), integrated ecological aquaculture temporal aquaculture systems (ITAS), and integrated sequential aquaculture systems (ISAS, also called partitioned aquaculture systems, or fractionated aquaculture systems. There is no ultimate IMTA system to "feed the world." There is not one world but climatic, environmental, biological, physical, chemical, economic, societal, and political conditions, each of which can lead to different choices of systems for feeding these microworlds.

The IMTA multitrophic approach to influencing ecosystem services extends beyond aquaculture to include all of enterprise ecology. With respect to phyconomy in phycoecology, we see that seaweed fits into various ecological

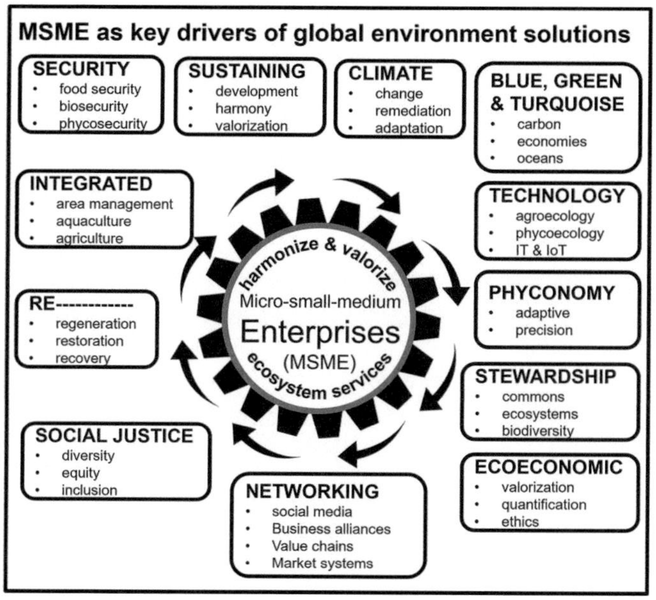

Fig. 2.12 Micro-, small-, and medium enterprises (MSME), which are commonly family owned and operated, are key drivers of the "big-picture" global environment solutions that acquire major funding as cures for what ails planet Earth

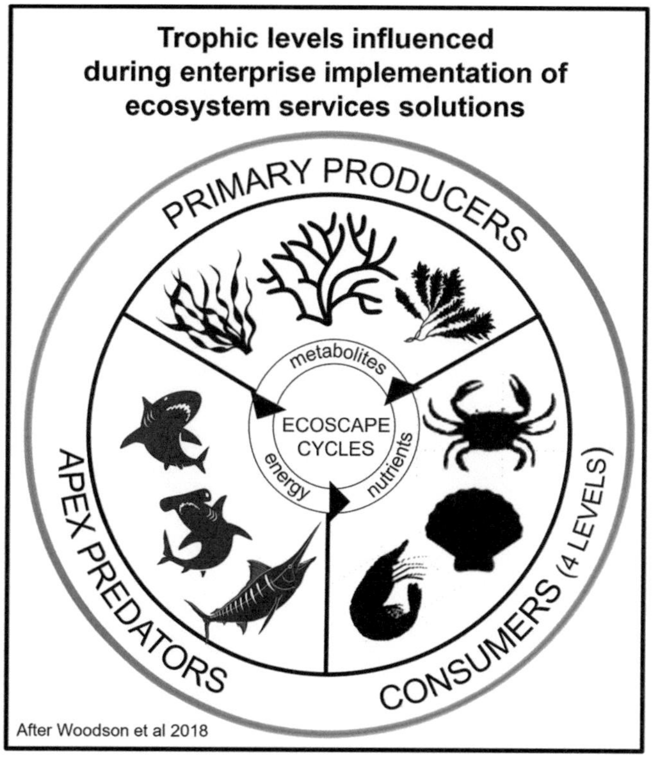

Fig. 2.13 Trophic levels influenced during enterprise implementation of ecosystem services solutions. (After Woodson et al. 2018)

niches. Here are some indications of how seaweed fits into the several enterprise ecology niches shown in Fig. 2.4:

1. *Seaweed farming and land farming*
   (a) Seaweeds share an infrastructure and labor base with land farming.
   (b) Seaweed biostimulants benefit land crops and livestock.
   (c) Micronutrients lost to the sea are returned to the land.
2. *Seaweed farming/stewardship and woody plants*
   (a) Seaweed farms can thrive near woody plants.
   (b) Seaweeds are major primary producers in marine forest ecosystems.
   (c) Marine forests are nurseries for many seaweeds.
3. *Seaweed farming/stewardship and rocky shores*
   (a) Seaweed farms can thrive on the rocky bottom.
   (b) Seaweeds are major primary producers along rocky shores.
   (c) Seaweeds provide refuge/shelter to rocky shore fauna.
4. *Seaweed farming and integrated aquaculture*
   (a) Seaweeds can be major primary producers and companion crops in pond systems.
   (b) Seaweeds provide refuge/shelter to pond fauna.
   (c) Seaweeds share infrastructure and labor bases in pond systems.
5. *Seaweed farming/stewardship, wetlands, and seagrass beds*
   (a) Seaweed farms can have net positive or negative impacts on seagrass beds.
   (b) Wild seaweeds are major primary producers on seagrass beds/wetlands.
   (c) Phyconomy must strive for net-positive impacts.
6. *Seaweed farming/stewardship and sessile animals*
   (a) Seaweed farms can thrive over sessile animal colonies, but phyconomy must strive for net-positive impacts.
   (b) Algal symbionts and synergistic algae must be protected and nurtured.
   (c) Beware of trampling and other destructive farming actions.
7. *Seaweed farming in floating systems*
   (a) Seaweed farms create new habitats as per fish aggregating devices.
   (b) Phyconomy must strive for net-positive impacts.
   (c) Farmers must accept a "fish tax" from grazers that eventually can become human food.

2.14 Discussion

By the time of writing, the volume of literature and multimedia social media products pertinent to enterprise ecology issues had proliferated to the point where it would take the full-time effort of a large team of people to keep current with developments. A substantial flow of hyperbole, misinforma-

Fig. 2.14 Why KITS should be internal to value chains

Why KITS should be internal to value chains

	KITS from external initiatives	KITS internal to value chains
1	Initiatives tend to be of short duration (usually five years or less)	KITS are a permanent core asset of enterprises
2	Much, if not most funding consumed by overhead, startup, wind-down and monitoring/evaluation costs;	Funding is applied entirely to acquisition and deployment of KITS assets
3	Driven by agendas not necessarily aligned with the agendas of enterprise owners and operators; and	Agendas of allied enterprises are fully aligned because their profits are tied to value chain transactions
4	4.Fail to leave substantial KITS capacity in value chain enterprises after interventions cease.	KITS are a permanent core asset of enterprises

tion, and pseudoscience was mixed with old-school evidence-based material, so it was increasingly difficult to sort fact from fiction. It was impossible for most MSME managers to keep current with the KITS that could benefit their enterprises.

Perhaps industrial-scale large seaweed enterprises can have the capacity to keep KITS state of the art, but that will not get KITS applied to the MSMEs that are crucial for the success of global environment initiatives. There must be mechanisms for making KITS integral to MSME enterprise ecology systems (Fig. 2.14), and relational value chain governance is one such mechanism (Fig. 2.9).

ETC asserted that *"Industrial seaweed will not cool the climate or save nature"* (Thomas and ETC 2023). They challenged corporate-controlled techno-industrial systems based on "scientism," discussed the dangers of technological manipulation of life and ecosystems, and argued in favor of building countervailing knowledge and power. They especially upheld peasant and indigenous ways of life and knowledge systems; food sovereignty per ecological food systems; people's control of technology in reflexive (bottom-up) technology development; and just economies and governance. These are values consistent with enterprise ecology, where ecosystem services are not treated with crass monetization but as interactions with nature that demand respectful stewardship by humanity. Enterprise ecology seeks ecoeconomic valorization through harmonized ecosystem services.

I submit that widespread "industrialized seaweed" was an unproven concept at the time of writing and that the feasibility of vast open-ocean seaweed farms was far from proven. Oceanographic factors, weather, nutrient distribution/availability, and valuation of legacy ocean ecosystems were all potential "showstoppers" for industrial seaweed. It was a risky investment with potentially major unanticipated negative consequences.

MSME seaweed farms applying adaptive phyconomy in coastal ecoscapes, on the other hand, have proven feasibility and increasingly well-understood niches among ecosystem services. Existing KITS can readily be applied to the improvement of balanced ecosystem service returns in currently productive ecoscapes; restoration/regeneration of returns in ecoscapes that have been devastated during tragedies of the commons; and sustainable development of ecoscapes with scope for new developments.

The key to successful sustainable development is KITS that are internal to value chains as core competencies of MSME value chain players (see Fig. 2.14).

2.15 Conclusions and Recommendations

During my six decades of involvement with adaptive phyconomy in seaweed value chains, I have witnessed and/or have been a participant in numerous interventions by GO, NGO, and international aid organizations that provided KITS to MSME seaweed farmers with results as listed in Fig. 2.14. Collectively, the funds spent on these interventions were in the tens of millions of USD, yet the results were as follows: After a few years of well-meaning intervention by competent experts, the interventions ended, and farm enterprises were left with few or no useful KITS sets internalized in their enterprises. Unfortunately, such interventions continue to be funded, while KITS development, internal to value chain enterprises, is scarcely funded at all.

At the time of writing, global interest in seaweed developments was at a level unimaginable even a decade ago, and substantial funding was going to industrial seaweed projects, many of which are of dubious value or even likely to cause environmental damage, as discussed in Chopin et al. (2024), Costa-Pierce and Chopin (2021), and Thomas and ETC Group (2023).

Investment in MSME enterprise ecology can give solid, sustainable returns on investment relative to industrial seaweed projects and can provide decent employment to millions of coastal people. Phyconomy in phycoecology can be a major contributor toward realizing those returns.

Acknowledgements I have had the privilege of working with thousands of seaweed farmers, who are the true developers of adaptive phyconomy and enterprise ecology, and I have had the pleasure of working with numerous colleagues from both the private and the public sectors worldwide. I hope that this chapter does justice to their work, and I apologize for any errors or omissions.

As I was writing this chapter, I learned of the shocking and untimely death of Thierry Chopin, whose body of work contributed mightily to the development of integrated ecosystem approaches relevant to phyconomy and to aquaculture enterprise functionality. Integrated Multitrophic Aquaculture (IMTA) was Thierry's passion, and the IMTA concept has been a basis for the development of the enterprise ecology functional framework presented in this chapter.

References

Bayraktar M, Algan N (2019) The importance of SMEs on world economies. International Conference on Eurasian Economies 2019. Available at: https://www.researchgate.net/publication/345358967

Brown LR (2001) Eco-economy: building an economy for the earth. WW Norton & Co. © 2001 Earth Policy Institute

Buck B, Troell MF, Krause G, Angel DL, Grote B, Chopin T (2018) State of the art and challenges for offshore integrated multi-trophic aquaculture (IMTA). Front Mar Sci 5:165. https://doi.org/10.3389/fmars.2018.00165

Carpenter SR, DeFries R, Dietz T, Mooney HA, Polasky S, Reid WV, Scholes RJ (2006) Millennium ecosystem assessment: research needs. Science 314:257–258

Carvajal AF, Didier T (2024) Boosting SME finance for growth; the case for more effective support policies. The World Bank Group, Washington, DC

Chopin T (2013) Aquaculture, Integrated Multi-Trophic (IMTA). http://www.springerreference.com/index/chapterdbid/226358. 18:59 © Springer-Verlag Berlin Heidelberg 2013

Chopin T, Costa-Pierce BA, Troell M, Hurd CI, Costello MJ, Backman S, Buschmann AH, Cuhel R, Duarte CM, Grondahl F, Heasman K, Haroun RJ, Johansen J, Jueterbock A, Lench M, Lindell S, Pavia H, Ricart AM, Sundell KS, Yarish C (2024) Deep-ocean seaweed dumping for carbon sequestration: questionable, risky, and not the best use of valuable biomass. One Earth. https://doi.org/10.1016/j.oneear.2024.01.013

CIDSE (2018) The principles of agroecology towards just, resilient and sustainable food systems. By CIDSE, Brussels, Belgium, www.cidse.org

Costa-Pierce BA, Chopin T (2021) The hype, fantasies and realities of aquaculture development globally and in its new geographies. World Aquacul 52(2):23–35

Cotas J, Gomes L, Pacheco D, Pereira L (2023) Ecosystem services provided by seaweeds. Hydrobiology 2023(2):75–96. https://doi.org/10.3390/hydrobiology2010006

Douglas I, Goode D, Houck M, Wang R (2011) The Routledge handbook of urban ecology. Accessed on: 27 Jan 2020. https://www.routledgehandbooks.com/doi/10.4324/9780203839263.ch3

ECA, ECLAC, ESCAP, ESCWA, ECE (2021) Transforming food systems Regional Policy Brief. http://www.unescwa.org/

FAO (2021) Ecosystem approach to aquaculture management: handbook. Yangon, Myanmar. https://doi.org/10.4060/ca7972en

FAO (2022) The 10 elements of agroecology guiding the transition to sustainable food and agricultural systems. FAO electronic publication per. http://www.fao.org/agroecology

Gereffi G, Humphrey J, Sturgeon T (2005) The governance of global value chains. Rev Int Pol Econ 12:78–104

Glessing SR (2020) Transforming food and agriculture systems with agroecology. Agric Hum Values 37:547–548. https://doi.org/10.1007/s10460-020-10058-0

Herr ML, Muzira T (2009) Value chain development for decent work: a guide for private sector initiatives, governments and development organizations. International Labour Office, Geneva. ISBN 978-92-2-122488-4

Holling CS (1978) Adaptive environmental assessment and management. Wiley, New York

HLPE (2019) Agroecological and other innovative approaches for sustainable agriculture and food systems that enhance food security and nutrition. A report by the High-Level Panel of Experts on Food Security and Nutrition of the Committee on World Food Security, Rome

Hurtado AQ, Neish IC, Critchley AT (2019) Phyconomy: the extensive cultivation of seaweeds, their sustainability and economic value, with particular reference to important lessons to be learned and transferred from the practice of eucheumatoid farming. Phycologia 58(5):472–483

ILO (2016) Value chain development for decent work: how to create employment and improve working conditions in targeted sectors, 2nd edn. International Labour Office, Geneva. 2015 ISBN: 978-92-2-130509-5 (print); 978-92-2-130510-1 (web pdf)

Knowler D, Chopin T, Martinez-Espineira R, Neori A, Nobre A, Noce A, Reid G (2020) The economics of integrated multi-trophic aquaculture: where are we now and where do we need to go? Rev Aquac 12:1579

Merriam-Webster (2024). https://www.merriam-webster.com/dictionary/enterprise#. 3 October 2024

Neish IC (2013) Social and economic dimensions of carrageenan seaweed farming in Indonesia. In: Valderrama D, Cai J, Hishamunda N, Ridler N (eds) Social and economic dimensions of carrageenan seaweed farming, Fisheries and aquaculture technical paper no, vol 580. Rome, FAO. 204 pp

Neish IC (2023) Ten guidelines toward phycosecurity implemented as biosecure ecosystem services management of tropical seaweed farms. Chapter 1. In: Critchley, Hurtado, Neish (eds) Tropical Phyconomy Coalition Development. Springer

Neish IC, Sepulveda M, Hurtado AQ, Critchley AT (2017) Reflections on the commercial development of eucheumatoid seaweed farming. Chapter 1. In: Hurtado AQ, Critchley AT, Neish IC (eds) Tropical seaweed farming trends, problems and opportunities: Focus

on *Kappaphycus* and *Eucheuma* of commerce. Springer. ISBN 978-3-319-63498-2
Neish IC, Suryanarayan S (2017) Development of eucheumatoid seaweed value-chains through carrageenan and beyond. In: Hurtado AQ, Critchley AT, Neish IC (eds) Tropical seaweed farming trends, problems and opportunities: Focus on *Kappaphycus* and *Eucheuma* of commerce. Springer. ISBN 978-3-319-63498-2
Panlibuton H, Porse H, Nadela E (2007) Seaweed/carrageenan value chain assessment. Final Report submitted to IFC/PEP Philippines and GTZ. 47 pp
Sinclair F, Wezel A, Mbow C, Chomba S, Robiglio V, Harrison R (2019) The contribution of agroecological approaches to realizing climate-resilient agriculture. Rotterdam and Washington, DC. Available online at www.gca.org
Tang Q, Pang K, Yuan X (2020) A one-billion-year-old multicellular chlorophyte. Nat Ecol Evol 4:543–549. https://doi.org/10.1038/s41559-020-1122-9
Thomas J, ETC Group (2023) The seaweed delusion: industrial seaweed will not cool the climate or save nature. Published by ETC Group. Available online at: www.etcgroup.org/content/seaweed-delusion
Troell M, Joyce A, Chopin T, Neori A, Buschmann AH, Fang JG (2009) Ecological engineering in aquaculture: potential for integrated multi-trophic aquaculture (IMTA) in marine offshore systems. Aquaculture 297(1):1–9
Vadassery N, Neish IC (2023) Integration of precision technology into adaptive phyconomy systems for extensive tropical red seaweed farming. Chapter 19. In: Critchley, Hurtado, Neish (eds) Tropical Phyconomy Coalition Development. Springer
TWN (2015) Agroecology: concepts, principles and practices. Published by Third World Network, Penang, Malaysia
UNIDO (2011) Industrial value chain diagnostics: an integrated tool. United Nations Industrial Development Organization (UNIDO)
Wezel A, Herren BG, Kerr RB, Barrios E, Gonçalves ALR, Sinclair F (2020) Agroecological principles and elements and their implications for transitioning to sustainable food systems. A review. Agron Sustain Dev 40:40. https://doi.org/10.1007/s13593-020-00646-z
Woodson CB, Schramski JR, Joye SB (2018) A unifying theory for top-heavy ecosystem structure in the ocean. Nat Commun 9:23
World Trade Organization (2016) World trade report 2016: levelling the playing field for SMEs. WTO Publications

Iain C. Neish is a Canadian from Nova Scotia who has worked in seaweed aquaculture since 1965. In doing so, he has followed the path of his father, Arthur Charles Neish. He first experienced tropical conditions in Oman and Indonesia during 1973–74. Since 1977, he has lived and worked full-time on smallholder phyconomy development, integrated aquaculture, and process and product development in Indonesia, the Philippines, Malaysia, and other tropical venues. He is currently based on Olango Island near Cebu City, Philippines, where he focuses on applying ecosystem approaches to restoration of seafolk enterprises in formerly prolific seaweed production areas that have failed in the wake of natural disasters and the tragedy of the commons.

3 A Framework for Understanding the Diversity of the Seaweed Industry

Karlotta Rieve

Abstract

The seaweed sector is remarkably diverse, a complexity often overlooked by newcomers. This article proposes a simple framework, emphasizing that while global issues like warming waters affect many farmed species, unique problems persist across different species and markets. To better understand global seaweed production, we categorize these challenges along two dimensions: temperate versus tropical seaweeds, and emerging versus established farming regions. This approach aims to elevate conversations among stakeholders, enabling the application of targeted solutions tailored to specific contexts within the diverse seaweed industry.

Keywords

Seaweed industry · Cultivation · Mechanization · Temperate seaweeds · Tropical seaweeds

3.1 Introduction

The most commonly cultivated seaweeds primarily represent five generic groups (Fig. 3.1).

The discourse surrounding seaweeds is often oversimplified—with many people speaking of seaweed in general while overlooking the intricate realities of farming selected types of specific seaweeds in different contexts. Although challenges such as warming waters and extreme weather events affect many commercially farmed species, each species and market have unique characteristics. To better understand the current state of global seaweed aquaculture, an approach may be to categorize it into four key dimensions: temperate versus tropical seaweed and emerging versus established farming regions.

Cultivating seaweed in the tropical belt is radically different from temperate zones (Fig. 3.2). Tropical seaweeds, such as *Gracilaria* and eucheumatoids (Rhodophyta), allow for year-round production with growth cycles of only 30–45 days. This quick turnaround enables farmers to initiate, increase, or pause production relatively quickly in response to market trends or external factors. In contrast, temperate seaweeds are typically seasonal crops, primarily farmed during winter. For example, kelp is seeded and harvested only once a year, with relatively short time windows for these activities. While *Pyropia/Neopyropia* (Rhodophyta) can be partially harvested throughout the season, it is highly perishable in its unprocessed form, necessitating processing facilities that can extend operations. Farmers of temperate species must thus meticulously plan to ensure that their harvests align with market demand.

Whether tropical or temperate has many other implications for farming practices. Here are two of the more distinct ones:

In temperate zones, the seasonal nature of production requires a land-based hatchery stage to produce seed material for the following farming cycle. The land-based stage for temperate seaweed ensured investment into seedbanks and hatchery work, with the selection of the best-performing crop and improved seed material being available to farmers, either through cooperative hatchery operations (such as used in Japan) or privately owned hatchery operations (such as used in South Korea; Fig. 3.3). In contrast, tropical seaweeds do not necessarily require this step and have predominantly relied on clonal propagation since farming of eucheumatoids began in the 1970s. The lack of a similar infrastructure for tropical seaweeds has resulted in a decreased pool of genetic diversity, making crops more vulnerable to pests, dis-

K. Rieve (✉)
Hatch Blue, Cork, Ireland
e-mail: karlotta@hatch.blue; https://www.hatch.blue/

A. T. Critchley et al. (eds.), *Advances in Temperate Phyconomy: Algal Harvest and Cultivation in Globally Distributed Temperate Waters*, Developments in Applied Phycology 15, https://doi.org/10.1007/978-3-032-02955-3_3

Fig. 3.1 Five generic groups of cultivated seaweeds

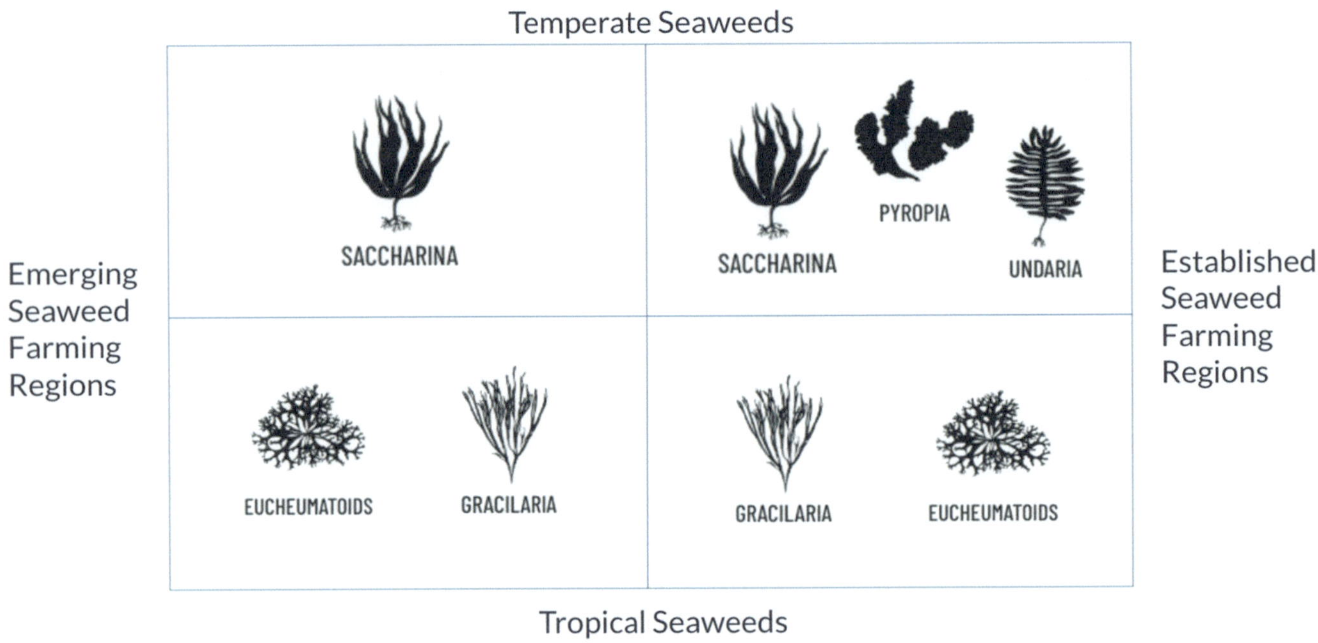

Fig. 3.2 Temperate versus tropical seaweeds

Fig. 3.3 Seaweed farmers attending their long lines of *Kappaphycus striatus* in Semau Island, Indonesia

eases, and climate change effects (Hatch Innovation Services, 2024).

Another significant difference is in post-harvest handling methods. Sun drying is the most common method for stabilizing biomass in tropical regions due to high sunlight exposure year-round. However, this relatively low-cost technique often leads to minimal innovation in processing steps and product development. Furthermore, sun drying can destroy valuable compounds in the various seaweeds. On the other hand, temperate regions face challenges related to having short weather windows for harvesting large amounts of biomass simultaneously. This has driven farmers to explore diverse alternatives to drying—especially given the high energy requirements involved—leading to the exploration of more diverse and innovative product applications.

Lastly, a strong differentiating factor between tropical and temperate seaweed farming lies in their socioeconomic contexts. Tropical seaweed farming often occurs in developing countries within the global South and is dominated by smallholder farmers who operate small farms in less regulated environments. For many coastal communities in these regions, seaweed farming is one of the few sources of livelihood, increasing their dependency on this activity while also exposing them to exploitation risks.

In contrast, temperate regions—typically found in developed countries within the global North—feature more company-type farming businesses due to more regulated access to farming. Farmers usually have other economic opportunities, so seaweed becomes either a "side gig", as we witness in many North American emerging seaweed farming regions, or the next generations of farmers are lacking due to more attractive employment opportunities in the cities, as the Japanese and South Korean trends show, where the farming population is aging.

3.2 Emerging Versus Established Seaweed Farming Regions

When comparing established versus emerging seaweed farming regions, we see that established areas tend to exhibit a market pull that facilitates scale. For instance, East Asia has a long-standing cultural appreciation for selected seaweeds as staple ingredients and side dishes. Similarly, Southeast Asia has experienced increased global demand for carrageenan and agar derived from cultivated seaweed. In both cases, markets were initially supplied by wild-harvested seaweed; however, demand quickly outpaced supply, justifying the development of farming. Conversely, many emerging seaweed farming regions perceive aquaculture as an opportunity for people, planet, and profit, but often lack a clear market pull to justify increased farming operations.

Stringent regulatory environments frequently compound this situation due to the novelty of the activity and limited room for error. When seaweed farming began in now-established regions, there were in most cases no regulations; now that these markets have matured, they tend to navigate existing businesses and supply chains. Facilitating the expansion of seaweed aquaculture demands the adoption of policies aimed at lowering barriers to entry for newcomers, while also prioritizing sustainability to cultivate robust sector development. This involves among others reducing licensing requirements and simplifying regulatory procedures. Yet in many emerging seaweed farming regions, seaweed farming falls under general fisheries or aquaculture acts, and policies specifically dedicated to seaweeds are rare. The status of seaweeds in various product applications is also ambiguous, complicating market entry. In the EU, seaweeds fall under the "Novel Food Catalogue," which requires adherence to safety standards that limit contaminants like heavy metals and iodine. In contrast, established seaweed-producing countries like South Korea, Japan, and China, known for their high per capita seaweed consumption, have significantly less stringent regulations regarding these levels.

Another key point to highlight is the level of mechanization in cultivation techniques between established and emerging regions (Fig. 3.4). The limited innovation in cultivation methods across tropical and temperate seaweed farms not only constrains productivity but also threatens future opportunities in many established areas. The labor-intensive nature of manual work is deterring the younger generation from carrying on their parents' business. We are already witnessing this trend in the aging farming population across East Asia, and it is likely to replicate in Southeast Asia as well. Although one might claim that labor is still abundant here, it is noteworthy that seaweed farming has enabled many children from farming families with little education to pursue university degrees. However, these children seldom return to their rural coastal communities after graduation, as they secure jobs in urban areas. Thus, it appears necessary to modernize seaweed farming practices in established tropical and temperate regions to attract the next generation. The urgency to minimize labor costs and implement automation and mechanization for scaling is most pronounced in emerging seaweed farming regions, particularly in the global North, which primarily cultivates temperate species. Here, technology-driven innovations and new farming methods from emerging regions like Europe are showcasing initial examples of technology transfer to established seaweed farming areas in South Korea.

Fig. 3.4 Mechanized harvesting of *Undaria pinnatifida* in Wando, South Korea

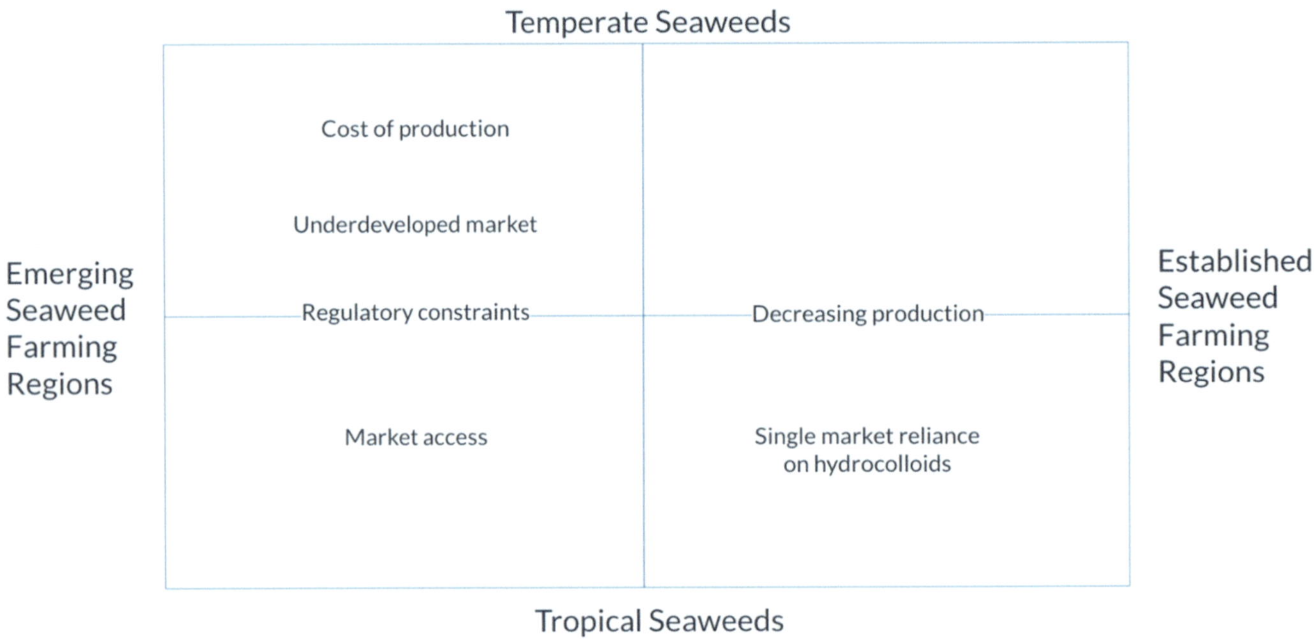

Fig. 3.5 A framework for understanding the primary challenges facing the seaweed industry

3.3 Conclusion: Mapping the Primary Challenges with this Framework

And so, we could go on and on and mention many more examples that vary significantly between temperate, tropical, emerging, and established seaweed regions.

As a result, I suggest we should go beyond talking about one generic seaweed industry. While we acknowledge the vast diversity and uniqueness of seaweed species, discussing each individually may not always be practical (Fig. 3.5). The four dimensions presented here instead may offer an approach that will elevate conversations among stakeholders who strive to address the challenges faced by their respective seaweed sectors and apply targeted solutions tailored to their unique contexts. The most significant challenges from our perspective can then be summarized in the following overview (Fig. 3.5).

As we move into the future, the next major breakthrough for seaweeds lies in a fundamental shift in approach. We must transition from merely farming what we can to strategically cultivating what the market demands. This requires greater professionalization across the seaweed aquaculture sector, developing species-specific solutions for breeding, farming, and market applications, and a renewed focus on species diversification, including the exploration of new species in both emerging and established regions. To effectively replicate the market pull that made the primary commercial seaweeds in Asia successful, we need to prioritize where seaweeds can add real value—whether as a functional ingredient or for their bioactivity—and identify which seaweeds have the most suitable composition for those applications. This market-driven approach, combined with strategic farm location assessments and diligent identification and resolution of knowledge gaps, can help us justify and unlock a new era of growth and innovation in seaweed aquaculture.

Reference

Hatch Innovation Services (2024) "World Bank (2023) Global seaweed: New and emerging markets report. http://hdl.handle.net/10986/40187 License: CC BY-NC 3.0 IGO." https://openknowledge.worldbank.org/entities/publication/4ac8c40c-cc4d-4f38-90a9-01b7bbaad3eb

Karlotta Rieve is the Latin America Portfolio Manager at Hatch Blue, a global accelerator, investment, and consultancy firm driving innovation in aquaculture. She is the author of Seaweed Insights, an open-source knowledge platform for seaweed farming worldwide. Over the past years, Karlotta has visited more than 150 seaweed and aquaculture operations across Asia, Europe, Latin America, the Caribbean, and North America. Her work centers on regenerative aquaculture development in emerging regions, with a strong focus on advancing the sustainable aquaculture sector in Latin America from her base in Baja California Sur, Mexico. Before moving to Mexico, Karlotta worked in Denmark with startups, corporates, and investors as a venture professional supporting food system innovation. She holds an MSc in Organisational Innovation and Entrepreneurship from Copenhagen Business School.

Trends in Supply and Demand for Temperate Seaweeds

4

Steven Hermans

Abstract

This chapter examines the evolving dynamics of temperate seaweed markets, challenging the prevailing narrative of uniform sector growth. While demand for key temperate seaweed species continues to rise, the most recent production data reveals a concerning decline since 2020. This supply-demand imbalance has triggered significant price increases, particularly in food markets. The study analyzes market trends across multiple segments, finding that South Korean gim exports, biostimulants, and abalone feed have demonstrated robust growth rates between 2016 and 2023, while the alginates market grew more modestly.

Production challenges for major species like *Saccharina japonica* in China and *Pyropia yezoensis* in Japan stem from climate change impacts, labor shortages, and breeding limitations. Meanwhile, Western markets face contrasting challenges with oversupply of certain species like *Saccharina latissima* but undersupply of others such as *Palmaria palmata.*

The research identified four emerging trends: geographic diversification beyond Asia, increased onshore cultivation, a continued shift from wild harvesting to cultivation, and the rising importance of biostimulants as the second-largest seaweed market segment. These findings provide critical insights for industry stakeholders addressing cultivation and application challenges in global temperate seaweed markets.

Keywords

Seaweed market · Seaweed market trends · Seaweed production data · Seaweed price dynamics · Temperate seaweeds

Literature on the development of the global seaweed market tends to characterize the sector as growing quickly, citing increasing production numbers and pointing to signs of growing demand. This article aims to complicate that narrative.

While growth in seaweed markets has been uneven, overall a strong trend toward continued demand growth for the most popular temperate seaweed species can be observed. However, latest production data points to decline since 2020. The declining production trend combined with continued strong growth in demand is leading to price hikes in food markets. Beyond food, this article examines trends in feed, biostimulant, alginate, and ecosystem services markets.

4.1 A look back: trendlines 2000–2024

4.1.1 Demand

Three temperate seaweed markets have seen strong growth in recent years: South Korean gim (*Neopyropia*/*Pyropia*), biostimulants, and abalone feed. The alginates market grew as well, but at a lower pace.

	Market size (2023)	Yearly growth rate (2016–2023)
Gim exports (S. Korea)	$770 million	11.8%
Biostimulants	$1 billion*	9.3%[a]
Abalone feed	N/A	11.6%[b]
Alginates	N/A	1.6–3%[a]

[a]Estimated
[b]2016–2021

S. Hermans (✉)
Phyconomy, Mechelen, Belgium
e-mail: steven@phyconomy.net

A. T. Critchley et al. (eds.), *Advances in Temperate Phyconomy: Algal Harvest and Cultivation in Globally Distributed Temperate Waters*, Developments in Applied Phycology 15, https://doi.org/10.1007/978-3-032-02955-3_4

4.1.2 Gim Exports

In 2016, the value of South Korea's exports of gim, dried sheets of *Pyropia yezoensis* also known as nori in Japan and laver in English, stood at 353 million (KBS World 2018). In 2023, the nation's gim exports totaled USD$771 million, representing an average 11.8% year-on-year growth in the seven-year period. Largest markets were the USA at USD$169 million, China at USD$144 million, Japan at USD$97 million, Thailand at USD$66 million, and Russia at USD$57 million (Lee 2024).

4.1.3 Biostimulants

The global seaweed extracts biostimulant market was estimated at USD$535 million for 2016 (North Sea Farm Foundation 2018). According to The Nature Conservancy and Bain and Company (2023), the seaweed biostimulants market is worth around USD$1 billion in 2023, implying a yearly growth rate of 9.3% in the past seven years. While tropical seaweeds have taken a share of this market in recent years, the vast majority of products sold remain based on temperate seaweeds.

4.1.4 Abalone Feed

Farmed abalone production grew from 156,875 tons FW in 2016–2017 to 243,506 tons FW in 2020–2021. Abalone are generally fed a mixture of temperate-water seaweeds, much of which is grown on farms at nearby locations, while some is sourced from wild stands or algal blooms (Cook 2023). As such, abalone aquaculture has boosted seaweed production. However, due to the self-contained nature of the trade, with many farms cultivating both abalone and the seaweeds to feed them, an estimate on market value is not available.

4.1.5 Alginates

In 2015, global alginate production totaled 48,144 tons (Porse and Rudolph 2017). In 2023, global alginate production was estimated at 55,000 tons (Mackey D pers. comm), implying a compound annual growth rate of 1.6%. This is lower than the estimate of the hydrocolloids research firm IMR, which puts the growth rate at 2.5–3% (Seisun 2024). Sausage casings have been a strong growth market in recent years. Due to improved processing technology, alginate sausage casings now offer considerable cost and efficiency savings compared to incumbent hog, sheep, collagen, or cellulose casings. The growth of plant-based and halal markets are additional drivers of demand (Reiser 2021).

The market for alginate-based acid reflux medicine is also growing. Previously confined to Europe, pioneering remedy Gaviscon has seen increased uptake in the United States. Competitors have sprung up and found success too (Gilbert 2023). The market for alginate-based dental molds is likely in decline due to the advent of digital imaging (Seisun 2024). For the size and growth trajectory of other alginate markets like textile printing paste, wound dressings, and various food applications, no reliable numbers were found.

4.1.6 Food

Trends in food markets for temperate seaweeds vary by country. Japanese seaweed consumption has nearly halved in 20 years, from 4.0 g per person per day in 2002 to 2.1 g per person per day in 2022, an annual decline of 3.2% (Ministry of Agriculture, Forestry and Fisheries of Japan 2024). This is in line with the trend in Japanese fish and shellfish consumption, which also saw a decline close to 50% this century, from 40.2 kg per capita in 2001 to 22 kg in 2023 (Loew 2024). High prices, a desire for convenience, and a diversification of the Japanese diet toward more meat and foreign cuisines are the main reasons cited for the long-term decline (Ministry of Agriculture, Forestry and Fisheries of Japan 2021).

In Korea, daily seaweed consumption remained stable: in 2014, 3.3 g of seaweeds were consumed per person per day. In 2022, this had risen slightly to 3.4 g per person per day (KOSIS 2024). In China, increasing urbanization and consumers' growing nutritional knowledge drove growth in seaweed consumption in the first decade of the century, when total seaweed consumption increased from approximately 59 g per month, per person in 2004 to 94 g per month per person in 2009 (Peng et al. 2021). Insights into trends in recent years are hampered due to a lack of reliable data.

In Europe and North America, the consumption of algae remains limited but is experiencing a growing trend due to increases in product availability and positive media attention (Avelin 2024). However, a lack of data regarding availability and consumption prevents accurate statements about market size and growth trends (Ficheux et al. 2022; Bouga and Combet 2015).

One notable feature of Western seaweed food markets is the contrast between high demand and insufficient supply of *Palmaria palmata* and low demand and oversupply for *Saccharina latissima*. The increasing demand for *P. palmata*, primarily from food manufacturers, exceeds the current production based primarily on wild-harvested biomass. This can explain the relatively high bulk retail value (business-to-business) of dried *P. palmata,* which was reported to range from 16–19 € kg^{-1} approximately 10 years ago and is now estimated between 40 and 75 € kg^{-1}. As a comparison, sales value for the cultivated kelps *S. latissima* and *A. esculenta* was reported to be 2–4 € kg^{-1} in Norway between 2019 and 2021. The current median consumer retail price of dried dulse is estimated to be 13.40 € 100 g^{-1} DW (Stévant et al. 2023).

4.1.7 Cosmetics and Personal Care

Seaweed extracts for use in cosmetics and personal care products could be a source of significant value. However, extracts tend to be sourced from small, specialized firms before formulation by bigger brands, which makes it difficult to gauge market size. Indicators of market growth are not consistently pointing in the same direction: Setalg, a subsidiary of agribusiness conglomerate Groupe Roullier specializing in seaweed-based cosmetics, closed its doors in 2024 due to a lack of profitability (Desevedavy 2024). In the same year, Australia's Marinova, a provider of fucoidans for personal care brands, expanded its production capacity (Lim 2024).

Emerging markets, including nutraceuticals, biomaterials, feed and pet food supplements, methane-reducing feed additives, specialty chemicals, and bioremediation, are still nascent and for now too small to estimate.

4.1.8 Supply

Harvests of three main temperate seaweed species in top producing countries in wet tons (FAO 2024)

		2000	2020	2022
Saccharina japonica	China	4,952,921	11,164,633	9,670,687
	South Korea	14,160	675,074	560,848
	Japan	53,846	30,304	29,800
Neopyropia yezoensis	China	424,913	2,220,380	2,176,580
	South Korea	130,488	536,127	550,221
	Japan	391,681	289,396	232,400
Undaria pinnatifida	China	1,511,135	2,256,040	2,061,290
	South Korea	212,429	501,501	585,955
	Japan	66,676	53,809	47,200

Yearly rate of growth or decline of harvest of the three main temperate seaweed species in top producing countries

		Yearly growth rate 2000–2020	Yearly growth rate 2020–2022
Saccharina japonica	China	4.1%	−6.9%
	South Korea	21.3%	−8.9%
	Japan	−2.8%	−0.8%
Neopyropia yezoensis	China	8.6%	−1%
	South Korea	7.3%	1.3%
	Japan	−1.5%	−10.4%
Undaria pinnatifida	China	2.4%[a]	−4.4%
	South Korea	4.4%	8%
	Japan	−1%	−6.3%

[a]Calculated from 2003 when statistics started

4.1.9 *Saccharina* spp.

According to FAO (2024), cultivation of *Saccharina japonica* in China grew at a yearly rate of 4.1% between 2000 and 2020, from 4.9 million to 11.2 million tons wet weight.

While the exact size of China's production is sometimes called into question (Porse and Rudolph 2017), the fact that the industry underwent strong growth is undeniable. However, recent events indicate that this rapid expansion has now hit a bottleneck.

In 2022, around 440,000 tons (dry weight) of *S. japonica* in Shandong bleached and decayed owing to environmental fluctuations, most likely due to a breeding-induced, biased allocation of energy resources in terms of the growth-defense trade-off of the kelp species. Kelp breeding systems in China have encountered problems relating to germplasm diversity, regulation, and technological innovation, as well as polluted environments and changing ocean conditions, all of which have led to production decline in combination with the pressures of climate change (Hu et al. 2021, 2023).

In South Korea, harvest numbers exhibit a similar pattern. Cultivation of *S. japonica* expanded massively for two decades but declined after 2020. In Japan, the decline started earlier and has been less pronounced. Japanese production of *S. japonica* declined at a yearly rate of 3.4 percent between 2001 and 2021, from 63,200 to 31,691 tons. The price of kombu reached a high of 1620 yen (USD $11.4) per kg in 2024 after another bad harvest blamed on rising ocean temperatures (Sasaki 2024).

In Europe and North America, the cultivation of *Saccharina latissima* was initiated around 15 years ago (Sæther et al. 2024). While kelp farming is now established with dozens of small-scale farms operating across North America and northwestern Europe, volumes have remained limited, with a lack of demand seen as the major impediment to greater production volumes. In the United States, around 600 tons of seaweeds were harvested in Maine in 2024 (The Fish Site 2024a, b). In Alaska, the seaweeds harvest decreased from nearly 400 tons in 2022 to around 100 tons in 2024 (Robidoux and Good 2023; Friedman 2024). Norway notched up 600 tons in 2023 (Tangen Haldorsen et al. 2023). Leading companies in France and the Faroe Islands harvested 200 and 240 tons of seaweeds, respectively, in 2023 (Maignant 2024; The Fish Site 2024b). In each location, the majority of seaweeds were *S. latissima*.

To sum up: in Asia, the cultivation of *Saccharina* spp. is in decline due to changing ocean conditions, labor constraints, lack of investment, and past kelp breeding strategies. The production shortfall is leading to price spikes. In the meantime, the opposite is true in Europe and North America: production is at overcapacity, but expansion is hampered by a lack of processing infrastructure and product-market fit.

4.1.10 *Neopyropia yezoensis* and *Undaria pinnatifida*

While production numbers for *N. yezoensis* show a similar pattern to those for *S. japonica* for the first two decades of the new millennium (Table 1,2), the picture is less marked beyond 2020. In sum, the production of *N. yezoensis* by the three biggest producers has seen a slow decline of 1.4% annually from 2020 to 2022, from 3,045,903 to 2,959,201 tons. In combination with the strong increase in demand from international markets, this has led to dramatic price spikes (Jae-hyuk 2024).

Harvest numbers for *U. pinnatifida* exhibited the same trend as *N. yezoensis*: after 2020, growth turned to decline for China, decline accelerated in Japan, while South Korea managed to continue to expand production. Although the total decline is less pronounced compared to *N. yezoensis* and *S. japonica*, here too Japanese consumers have been hit by price hikes, with the average price of boiled wakame reaching 2479 yen ($16) per kg in 2024, up 54% from last year (Sasaki 2024).

4.2 A look forward: emerging trends

4.2.1 South Korean Gim Exports

Demand for South Korean gim seems set to continue to rise strongly due to a combination of factors: on the one hand, declining competition from Chinese and Japanese production is lowering supply; on the other hand, the still-growing popularity of Korean cuisine as part of the *hallyu* Korean wave is increasing demand (Jin et al. 2024; Raman and Bok 2024).

The recent price hikes outlined above suggest that current production growth of around 1 percent per year is not enough to address the growing demand. The two key challenges for South Korea are labor and seed stock. Large labor shortages exist in rural areas and are set to rise with the average age of the South Korean farmer at 66 years old (Cho 2021; Suk 2023). However, only 19% of Koreans support an increase in immigration (Kuhn 2021). With this in mind, seaweed producers have strong incentives to increase investments in mechanization to address growing labor shortages.

When it comes to seed stock, the South Korean government's Golden Seed Program is bearing fruit. More cultivars are being registered in an effort to balance resilience, productivity, and taste. The focus is on *Neopyropia/Pyropia/Porphyra* spp., where the amount of registered cultivars went from 1 in 2014 to 13 in 2019 (Hwang et al. 2019; Hwang and Park 2020).

In conclusion, while demand for Korean gim is set to continue to rise, recent price movements indicate that the work on cultivars and automation is not moving quick enough to offset the negative effects of climate change and labor shortages in the near term, which leads us to conclude prices will continue to rise in the near future. This in turn is likely to spur innovation and a search for new geographies (Sea4Earth 2024).

4.2.2 Biostimulants

According to The Nature Conservancy and Bain and Company (2023), the seaweed biostimulants market is expected to grow at 13% per year. That would make it roughly a USD$2.5 billion market in 2030, covering about 1% of all global farmland. The World Bank Group (2023) estimated the market to grow to USD$1.87 billion by 2030, implying a lower growth rate of 9.3%.

Macro trends, including the push toward lower fertilizer usage, a growing focus on soil health, and the need for improved crop resilience in the face of changing climatic conditions, reveal a large market opportunity, bolstered by increased attention from regulators. However, seaweed-based biostimulants face competition from other types of biostimulants based on acids, microorganisms, and other sources. Other challenges include the long time to market, the diversity of seaweed biostimulants, and the lack of scientific understanding of their varying modes of action and the complexity of plant-soil interactions with biologicals (Hermans 2024).

Nonetheless, the profusion of companies producing seaweed biostimulants across the globe and biostimulant science focused on seaweeds (Biostimulant.com 2022; Corsi et al. 2022; Wazeer et al. 2024) makes it seem likely that the sector will be able to gradually overcome its issues and post strong growth numbers until 2030 and beyond. Recent market trends indicate the start of a gradual transition from wild harvested to cultivated seaweed sources for biostimulants, while the current proliferation of products is setting the stage for future consolidation of the sector.

4.2.3 High-G alginates

In 2010, a review of the hydrocolloid industry concluded: "Within the last 10 years, a marked shift has taken place in the seaweed species being used to extract alginates. In 1999, *Ascophyllum* and *Macrocystis* accounted for 58% of the wild harvest, but today *Laminaria* and *Lessonia* are dominant, accounting for 81% of the harvest. This reflects a shift in market demand. Today's most important and profitable applications require the high-G extracts from *Laminaria* and *Lessonia*, not the low-G extracts from *Ascophyllum* and *Macrocystis* (Bixler and Porse 2010)."

In 2022, that picture has largely remained the same, with 84% of alginate-bearing wild-harvested seaweeds coming from members of the Lessoniaceae (314,000 tons FW) and Laminariaceae (180,000 tons FW). The Fucalean *Ascophyllum nodosum* harvests have rebounded from a low of 56,000 tons FW in 2015 to 78,000 tons FW in 2022; however, the majority of this harvest now goes to biostimulants (FAO 2024).

Advances in scientific research of applications for alginates in food and medical fields (Xing et al. 2020; Adamiak and Sionkowska 2023; Zhang et al. 2024) suggest the demand for wild-harvested *Lessonia* spp. and *Laminaria* spp. will continue to grow as they are the main sources of the high-G alginates in demand by these industries.

To fill growing demand, the harvesting capacity of *Laminaria hyperborea* from Norway is expanding (Alginor 2024). However, harvest numbers of *Lessonia* spp. from South America seem to suggest that this supply has plateaued. While harvest numbers quickly grew from 100,000 tons in 1999 to 330,000 tons FW in 2014, they have not grown further since, stalling at 315,000 tons FW in 2022 (Vásquez 2016; FAO 2024). At the same time, prices went from USD$350 in 2000 to USD$950 in 2009, peaking at USD$2500 in 2023 before coming down again to USD$1250 in 2024 (Van Haaren C pers. comm.). The price surge suggests that demand is not being met.

However, no high-G alginate is currently under commercial cultivation. In India, a high-G *Sargassum* species are harvested at some 35,000 tons FW in 2024, but supply is tightly regulated and controlled by a limited set of actors (Nehemiah K pers. comm.). Alginate from bacteria is an avenue that has so far not been seriously explored by science and industry (Hay et al. 2013).

With growing demand and constrained supply, we expect a continued price rise for high-G alginate-bearing species, which is likely to spur a more concerted effort to supplement wild harvests with cultivated alternatives (Salazar et al. 2024).

4.2.4 Mixed-G alginates

A second source of predicted demand growth for alginates is biomaterials. Over the past decade, startups have developed biomaterials made in part from alginates for the commercial market. This includes textiles, foams, coatings, and plastics. With varied approaches and material requirements, these applications are less in need of high-G alginates and can instead tap into a variety of species with different M:G ratios.

However, commercial rollout has been hampered by a lack of demand for products that are seen as not competitive by the market in terms of cost and functionality compared to incumbent materials, despite their ecological credentials. Projections of the future evolution of the industry vary widely. Financial institutions predict the biomaterials market to grow enormously: the World Bank Group (2023) values seaweed-based textiles and bioplastics collectively at USD$1.6 billion by 2030, while Standard Chartered predicts the value of seaweed bioplastics alone to reach USD$1.8 billion by 2030 and USD$45 billion (Klerk 2023). On the other hand, registrar DNV sees little significance for the biomaterials market in seaweeds (Aarnes et al. 2023).

At the time of writing, the total market for seaweed-derived biomaterials is estimated by market participants at no more than a few million dollars in total. While it is possible that the seaweed biomaterials market manages to overcome challenges of cost competitiveness and functionality in the mid to long term, the evidence suggests that it will not reach a significant size in the near term.

New sources of low to medium-G alginates under development include *Sargassum* spp. from harmful algal blooms and cultivated *Macrocystis pyrifera* from Namibia, Alaska, New Zealand, and Chile.

4.2.5 Food

In response to very high prices for *Palmaria* spp., cultivation efforts are increasing. In Canada, 50–100 tons FW are wild harvested annually (Chopin and Ugarte 2006); onshore cultivation amounts to 350 tons FW per year (Hafting 2024). Onshore cultivation is also ramping up in Belgium, the Netherlands, and the USA (Tielemans 2024; Seaweedland 2024; McDarris 2024), while companies in Ireland and Denmark are piloting commercial cultivation of *P. palmata* at sea (Howell 2022; Techsavvy Media 2024).

Saccharina spp. markets are in the curious situation of under-supply in Japan and possibly in China, while European and North American farms have excess capacity and must regularly dump unsold crops or keep them in storage. While companies' efforts to increase consumption of *Saccharina* spp. in Europe and North America are likely to continue to grow the market, the demand side gap is considered to be large, and it is thus reasonable to assume it will take years to fill (Van Exel H pers. comm.).

The obvious answer here would be to sell Western *Saccharina* spp. to Eastern consumers, either directly or through joint ventures, with Asian companies making use of European and American farm waters and automation technology, while their counterparts could avail of the strong product development and marketing skills. As an example, Canadian company Acadian Seaplants produces 1000 tons FW of onshore cultivated *Chondrus crispus* annually for the Japanese salad market (Hafting 2024).

The fact that this approach has not been pursued more often points to the challenges in building that bridge: different taste and sensory profiles of the species and a lack of size and sophistication among most Western producers are some of the barriers holding back possible rapprochement between markets.

4.2.6 Animal Feed Supplements for Methane Reduction

Between 2020 and 2023, the cultivation of *Asparagopsis* spp. as a feed supplement for livestock to reduce their methane emissions was funded by USD$143 million of venture capital investments (Phyconomy 2024). These startup companies, which began operations between 2018 and 2022, have now reached the pilot stage. Although producers of *Asparagopsis* spp. started cultivation at sea, all have moved onshore now to increase growth rates and production of the key ingredient bromoform (Watson 2024). Only in Vietnam, trial production of *A. taxiformis.* remains at sea (Kessler 2023).

While much science remains to be done, the fact that some companies have secured offtake agreements and are moving into the market shows that the sector is becoming commercially viable. What is still unclear is to what extent *Asparagopsis* extracts can grab market share in the emerging livestock methane reduction market amid a wide range of competitors. Regulation, efficacy, cost, and supply constraints will all play a role in determining the outcome (Watson 2024; World Bank Group 2023).

4.2.7 Ecosystem Services

Bioremediation of wastewater from aquaculture, households, and desalination plants is an area garnering more interest from commercial actors, as some seaweeds can act as biofilters in a cost-effective way. Post-treatment, the seaweed biomass can be sold into different products to increase profits. Additionally, pollutant reduction credits can be obtained (Burdekin Shire Council 2024; Fletcher 2024; Eco-Markets Australia 2024).

In 2017, influential environmentalist Tim Flannery started advocating for the large-scale cultivation of seaweeds for carbon dioxide capture (Flannery 2017). Since then, his ideas have come under scrutiny from phycologists and marine biologists. The verification, reporting, and monitoring of possible carbon sequestration is considered to be very challenging, with many gaps in the science remaining and a general consensus that the benefits of "blue carbon" are oversold (Williamson and Gattuso 2022; Boyd et al. 2024). Alongside, carbon credit markets saw a collapse in credibility in 2023 (Greenfield 2024).

With carbon markets discredited and the largest startup dedicated to sinking seaweeds folding in 2024 (Velev 2024), the narrative focus of startups, environmentalists, and academics has shifted to mitigating carbon emissions through displacing or reducing carbon emissions of livestock, biomaterials, or chemicals, or through the timely pickup and processing of harmful algal blooms (Ross et al. 2023; Bullen et al. 2024; Canvin et al. 2024). The lack of life cycle assessments (LCAs) and the wide variety of end results within existing LCAs are key impediments to being able to state these arguments as fact (Hasselström and Thomas 2022). Those still looking toward credits aim for the new and not-yet tainted biodiversity and nitrogen credit markets.

While the idea of large-scale carbon capture through seaweeds has ultimately proven to be a dud, Tim Flannery's advocacy should be credited for attuning a new generation to the power of seaweeds, who have since gone on to found innovative seaweed companies (Crerar 2024; Envirostrat 2022).

4.3 Conclusion

To conclude, a wide-angled look at temperate seaweed trends in supply and demand reveals several macro shifts that are likely to shape market formation in the years to come.

1. *A gradual shift away from Asia, both in consumption and production*

   Japan, South Korea, and China are no longer the near-exclusive producers and consumers of temperate seaweeds. Although the overwhelming majority of production and consumption will remain rooted in Asia for the foreseeable future, consumption of temperate seaweeds as food is growing globally, and cultivation is now established across Europe and North America, as well as in Chile and Namibia.

2. *Growing role of onshore cultivation*

   While temperate seaweed cultivation started and developed almost exclusively in the past 70 years in the nearshore environment, the need for controlled environments for lower volume, higher value products has seen a greater share of projects focusing on land-based operations, among others, the cultivation of *Palmaria* for food, *Asparagopsis* for feed supplements, and *Ulva* for food and abalone feed.

3. *Shift toward cultivation continues*

   Seaweed cultivation development is market-driven. If demand is low and natural resources are adequate, cultivation is unnecessary. As demand increases, natural populations frequently become inadequate, and attempts are made to increase production. Cultivation commenced in Asia, first for food markets and later for aquafeed and hydrocolloid markets. Now market movements are pointing to the next phase with the cultivation of dulse for food, various species for bioremediation, and *Saccharina* spp. and *Macrocystis* spp. for biostimulants and alginates.

   Wild harvesting, however, still plays an important role in the supply of high-G alginates, abalone feed, biostimulants, food, cosmetics, lambda carrageenans, and high-grade agar that is unlikely to disappear soon.

4. *Biostimulant importance grows*

   The food market remains the most important market for seaweeds, both in terms of value and volume. However, biostimulants are the fastest growing segment of the seaweed market and have likely already outflanked the agar, carrageenan, and alginate markets. Projected to double in size by 2030, the biostimulant market is now second to food in size and importance for seaweeds globally.

References

Aarnes Ø, Vold Bjordal M, McConnell E (2023) Seaweed forecast: ocean's future to 2050. DNV

Adamiak K, Sionkowska A (2023) State of innovation in alginate-based materials. Mar Drugs 21:353. https://doi.org/10.3390/md21060353

Alginor (2024) Hypomar Ocean 1 delivered — Alginor. In: Alginor. https://alginor.no/2024/05/hypomar-ocean-1-delivered/. Accessed 9 Oct 2024

Avelin C (2024) Évaluation et suivi de la contribution sociale et économique de la filière des macroalgues - Synthèse. FranceAgriMer

Biostimulant.com (2022) Biostimulants research global trends in 2021. In: Biostimulantcom. https://www.biostimulant.com/blog-29-biostimulants-research-global-trends-in-2021/. Accessed 9 Oct 2024

Bixler HJ, Porse H (2010) A decade of change in the seaweed hydrocolloids industry. J Appl Phycol 23:321–335. https://doi.org/10.1007/s10811-010-9529-3

Bouga M, Combet E (2015) Emergence of seaweed and seaweed-containing foods in the UK: focus on labeling, iodine content, toxicity and nutrition. Foods 4:240–253. https://doi.org/10.3390/foods4020240

Boyd PW, Gattuso J-P, Hurd CL, Williamson P (2024) Limited understanding of basic ocean processes is hindering progress in marine carbon dioxide removal. Environ Res Lett 19:061002. https://doi.org/10.1088/1748-9326/ad502f

Bullen CD, Driscoll J, Burt J et al (2024) The potential climate benefits of seaweed farming in temperate waters. Sci Rep 14:15021. https://doi.org/10.1038/s41598-024-65408-3

Burdekin Shire Council (2024) World first wastewater facility to be commissioned in September—Burdekin Shire Council. In: Burdekin Shire Council. https://www.burdekin.qld.gov.au/news/article/2292/world-first-wastewater-facility-to-be-commissioned-in-september. Accessed 9 Oct 2024

Canvin MC, Moore PJ, Smale DA (2024) Quantifying growth, erosion and dislodgement rates of farmed kelp (*Saccharina latissima*) to examine the carbon sequestration potential of temperate seaweed farming. J Appl Phycol 36:3091–3102. https://doi.org/10.1007/s10811-024-03323-w

Cho H-S (2021) Lack of foreign labor leaves aging farmers without options. In: Korea JoongAng Daily. https://koreajoongangdaily.joins.com/2021/05/05/business/economy/agriculture-farming-labor/20210505133100321.html. Accessed 9 Oct 2024

Chopin T, Ugarte R (2006) The seaweed resources of Eastern Canada. In: Critchley AT, Ohno M (eds) World Seaweed Resources. An authoritative reference system. BioInformatics Publishers, pp 1–46

Cook PA (2023) Worldwide abalone production: an update. NZ J Mar Freshw Res 59:4–10. https://doi.org/10.1080/00288330.2023.2261869

Corsi S, Ruggeri G, Zamboni A et al (2022) A bibliometric analysis of the scientific literature on biostimulants. Agronomy 12:1257. https://doi.org/10.3390/agronomy12061257

Crerar S (2024) Sea Forest lick block wins Good Design Award Gold. Smart Company

Desevedavy M (2024) « On est un peu dans la détresse » : la fermeture de Nuwen Setalg inquiète des goémoniers bretons. Ouest-France

Eco-Markets Australia (2024) Celebrating a milestone: the world's first RegenAqua wastewater treatment facility at Burdekin joins the reef credit scheme. In: Eco-Markets Australia. https://eco-markets.org.au/2024/07/01/celebrating-a-milestone-the-worlds-first-regenaqua-wastewater-treatment-facility-at-burdekin-joins-the-reef-credit-scheme/. Accessed 9 Oct 2024

Envirostrat (2022) Aotearoa New Zealand's seaweed sector framework. In: Sustainable Seas Challenge. https://www.sustainableseaschallenge.co.nz/assets/dms/Guidance/Seaweed-Sector-Framework/Kelp-Blue-Case-Study-FINAL.pdf

FAO (2024) Global Capture Production. Fisheries and Aquaculture

Ficheux A-S, Pierre O, Le Garrec R, Roudot A-C (2022) Seaweed consumption in France: key data for exposure and risk assessment. Food Chem Toxicol 159:112757. https://doi.org/10.1016/j.fct.2021.112757

Flannery T (2017) Sunlight and seaweed: an argument for how to feed, power and clean up the world. Text Publishing

Fletcher R (2024) How seaweed can prevent marine destruction around desalination plants. The Fish Site. https://thefishsite.com/articles/how-seaweed-can-prevent-marine-destruction-around-desalination-plants

Friedman S (2024) Alaska seaweed industry review https://alaskamariculturecluster.org/wp-content/uploads/2024/12/Alaska-Mariculture-Industry-Update-Fall-2024-12.12.24.pdf

Gilbert M (2023) Are Alginates All That? AAO-HNS Bulletin. https://bulletin.entnet.org/clinical-patient-care/article/22879164/are-alginates-all-that

Greenfield P (2024) Market value of carbon offsets drops 61%, report finds. The Guardian

Hafting J (2024) Question on cultivation statistics. Acadian Seaplants Ltd
Hasselström L, Thomas J-BE (2022) A critical review of the life cycle climate impact in seaweed value chains to support carbon accounting and blue carbon financing. Cleaner Environ Syst 6:100093. https://doi.org/10.1016/j.cesys.2022.100093
Hay ID, Rehman ZU, Moradali MF et al (2013) Microbial alginate production, modification and its applications. Microb Biotech 6:637–650. https://doi.org/10.1111/1751-7915.12076
Hermans S (2024) How to win at biostim. Phyconomy. https://phyconomy.substack.com/p/how-to-win-at-biostim
Howell M (2022) Jumpstarting Ireland's seaweed success. The Fish Site. https://thefishsite.com/articles/jumpstarting-irelands-seaweed-success-dulse-pure-ocean-algae
Hu Z, Shan T, Zhang J et al (2021) Kelp aquaculture in China: a retrospective and future prospects. Rev Aquac 13:1324–1351. https://doi.org/10.1111/raq.12524
Hu Z, Shan T, Zhang Q et al (2023) Kelp breeding in China: challenges and opportunities for solutions. Rev Aquac 16:855–871. https://doi.org/10.1111/raq.12871
Hwang EK, Park CS (2020) Seaweed cultivation and utilization of Korea. Algae 35:107–121. https://doi.org/10.4490/algae.2020.35.5.15
Hwang EK, Yotsukura N, Pang SJ et al (2019) Seaweed breeding programs and progress in eastern Asian countries. Phycologia 58:484–495. https://doi.org/10.1080/00318884.2019.1639436
Jae-hyuk P (2024) Oceans minister promises to stabilize local seaweed supply as gimbap prices soar. The Korea Times. https://www.koreatimes.co.kr/www/tech/2025/03/129_373466.html
Jin F, Kim S-H, Choi Y-K, Yoo B-K (2024) A study on the impact of Hallyu (Korean wave) on Korea's consumer goods exports to China: panel analysis using big data and provincial-level data. Sustainability 16:4083. https://doi.org/10.3390/su16104083
KBS World (2018) K-food exports to surpass 10 trillion won this year. In: KBS World. https://world.kbs.co.kr/service/contents_view.htm?lang=e&menu_cate=&id=&board_seq=347555. Accessed 8 Sept 2024
Kessler R (2023) An 'aquatic moonshot' in Vietnam aims to fight livestock methane with seaweed. In: Mongabay Environmental News. https://news.mongabay.com/2023/10/an-aquatic-moonshot-in-vietnam-aims-to-fight-livestock-methane-with-seaweed/. Accessed 9 Oct 2024
Klerk E (2023) Seaweed: kelp is on the way. Standard Chartered. https://phyconomy.net/wp-content/uploads/2023/12/standard-chartered-seaweed-investment.pdf
Korean Statistical Information Service (KOSIS) (2024) Daily intake trend by food group (Standardized). https://kosis.kr/eng/
Kuhn A (2021) As workforce ages, South Korea increasingly depends on migrant labor. NPR. https://www.npr.org/2021/06/02/1001194446/as-workforce-ages-south-korea-increasingly-depends-on-migrant-labor
Lee S-J (2024) Gim exports hit record 1.032 trillion won as K-culture continues global rise. In: Korea JoongAng Daily. https://koreajoongangdaily.joins.com/news/2024-02-05/business/economy/Gim-exports-hit-record-1032-trillion-won-as-Kculture-continues-global-rise/1974379. Accessed 8 Sept 2024
Lim A (2024) Fucoidan's promise: Marinova eyes opportunities in wound healing and inflammation. In: cosmeticsdesign-asia.com. https://www.cosmeticsdesign-asia.com/Article/2024/06/25/marinova-eyes-opportunities-in-wound-healing-and-inflammation. Accessed 9 Sept 2024
Loew C (2024) Japan's seafood consumption continues to drop, according to industry white paper. In: SeafoodSource. https://www.seafoodsource.com/news/supply-trade/japan-s-per-capita-seafood-consumption-continues-to-drop-according-to-industry-white-paper. Accessed 17 Sept 2024
Maignant V (2024) Production et transformation d'algues. Agolesko (29) lance une levée de fonds de 5 millions d'euros. In: Bretagne Économique. https://www.bretagne-economique.com/actualites/production-et-transformation-dalgues-agolesko-29-lance-une-levee-de-fonds-de-5-millions-deuros/. Accessed 9 Oct 2024
McDarris A (2024) Bye, kale: marine scientists want you to eat more seaweed. In: Popular Science. https://www.popsci.com/environment/eat-more-seaweed/. Accessed 9 Oct 2024
Ministry of Agriculture, Forestry and Fisheries of Japan (2021) Status of fishery product consumption 2021. In: Fisheries Agency of Japan. https://www.jfa.maff.go.jp/j/kikaku/wpaper/r03_h/trend/1/t1_1_2.html. Accessed 17 Sept 2024
Ministry of Agriculture, Forestry and Fisheries of Japan (2024) Food supply and demand table final report food supply and demand table for FY 2022. Government Statistics Portal of Japan. https://www.maff.go.jp/e/data/stat/98th/attach/pdf/index-4.pdf
North Sea Farm Foundation (2018) Identification of the seaweed biostimulant market. https://www.northseafarmers.org/public/documents/Bio4safe_WP1_D112_Seaweed-biostimulant-market_North-Sea-Farm-Foundation_December-2018.pdf
Peng J, Min S, Qing P, Yang M (2021) The impacts of urbanization and dietary knowledge on seaweed consumption in China. Foods 10:1373. https://doi.org/10.3390/foods10061373
Phyconomy (2024) Investments in seaweed companies 2020–2024. https://phyconomy.net/database/
Porse H, Rudolph B (2017) The seaweed hydrocolloid industry: 2016 updates, requirements, and outlook. J Appl Phycol 29:2187–2200. https://doi.org/10.1007/s10811-017-1144-0
Raman Z, Bok GI (2024) Beyond kimchi and K-dramas: the Korean food phenomenon among university students in Malaysia. EDUCATUM J Soc Sci 10:61–76 https://doi.org/10.37134/ejoss.vol10.1.6.2024
Reiser (2021) Why alginate has become the sausage casing of the future. In: Food Manufacture. https://www.foodmanufacture.co.uk/Latest/Promotional-Features/Why-alginate-has-become-the-sausage-casing-of-the-future. Accessed 9 Sep 2024
Robidoux J, Good M (2023) State of the states: status of U.S. Seaweed Aquaculture Sea Grant. https://seaweedhub.extension.uconn.edu/wp-content/uploads/sites/3646/2023/04/2023-State-of-the-States_For-Posting_Dec2023.pdf
Ross FWR, Boyd PW, Filbee-Dexter K et al (2023) Potential role of seaweeds in climate change mitigation. Sci Total Environ 885:163699. https://doi.org/10.1016/j.scitotenv.2023.163699
Sæther M, Diehl N, Monteiro C et al (2024) The sugar kelp *Saccharina latissima* II: recent advances in farming and applications. J Appl Phycol 36:1953–1985. https://doi.org/10.1007/s10811-024-03213-1
Salazar L, Dresdner J, Figueroa Y et al (2024) Sustainability of seaweed supply and price dynamics in the Chinese hydrocolloid industry. Aquac Econ Manage 28:537–551. https://doi.org/10.1080/13657305.2024.2368783
Sasaki T (2024) Japan's poor seaweed harvest drives up prices, hits restaurants. Nikkei Asia. https://asia.nikkei.com/Business/Food-Beverage/Japan-s-poor-seaweed-harvest-drives-up-prices-hits-restaurants
Sea4Earth (2024) Sea4Earth. In: Sea4Earth. https://www.sea4earth.com/. Accessed 9 Oct 2024
Seaweedland (2024) Seaweedland. In: Seaweedland. https://seaweedland.com/palmaria. Accessed 9 Oct 2024
Seisun D (2024) Hydrocolleague tidbits. In: IMR International. https://myemail-api.constantcontact.com/Hydrocolleague-Tidbits%2D%2D.html?soid=1101349586918&aid=aAVh71_mn6k. Accessed 9 Sep 2024
Stévant P, Schmedes PS, Gall LL et al (2023) Concise review of the red macroalga dulse, *Palmaria palmata* (L.) Weber & Mohr. J Appl Phycol 35:523–550. https://doi.org/10.1007/s10811-022-02899-5
Suk LY (2023) South Korea to bring in 110,000 migrant workers this year, but some businesses want more. In: CNA. https://www.

channelnewsasia.com/asia/south-korea-migrant-workers-labour-shortage-3232771. Accessed 9 Oct 2024

Tangen Haldorsen S, Borgersen G, Duinker A, et al. (2023) The kelp industry: opportunities, barriers and next steps. In: Norwegian Blue Forests Network. https://nbfn.no/kelp-industry-brief/. Accessed 9 Oct 2024

Techsavvy Media (2024) Dansk startup har knækket koden til bæredygtig tang: "USA er meget interessant for os" - TechSavvy. In: TechSavvy - Klogere på fremtiden i dag. https://techsavvy.media/dansk-startup-har-knaekket-koden-til-baeredygtig-tang-usa-er-meget-interessant-for-os/. Accessed 9 Oct 2024

The Fish Site (2024a) Atlantic Sea farms reports record breaking kelp harvest. The Fish Site. https://thefishsite.com/articles/atlantic-sea-farms-reports-record-breaking-kelp-harvest

The Fish Site (2024b) Ocean Rainforest nets €2.5 million investment. The Fish Site. https://thefishsite.com/articles/ocean-rainforest-nets-2-5-million-investment

The Nature Conservancy, Bain & Company (2023) Analysis of farmed seaweed carbon crediting and novel markets to help decarbonize supply chains. https://www.nature.org/content/dam/tnc/nature/en/documents/SeaweedMarketsAnalysis.pdf

Tielemans J (2024) Sander (38) en Hendrik (36) kweken zeewier… aan land: "Dat van ons is veel toegankelijker." In: Made in. https://www.made-in.be/oost-vlaanderen/sander-38-en-hendrik-36-kweken-zeewier-aan-land-dat-van-ons-is-veel-toegankelijker/. Accessed 9 Oct 2024

Vásquez JA (2016) The Brown seaweeds fishery in Chile. In: Fisheries and aquaculture in the modern world. InTech

Velev V (2024) Running tide shuts down citing lack of demand from the voluntary market. In: Carbon Herald. https://carbonherald.com/running-tide-shuts-down-citing-lack-of-demand-from-the-voluntary-market/. Accessed 9 Oct 2024

Watson E (2024) Enteric methane reduction deep dive: from seaweed to gene editing. In: AgFunderNews. https://agfundernews.com/methane-busting-feed-supplements-are-beginning-to-scale-but-who-will-foot-the-bill-and-what-will-drive-widespread-adoption. Accessed 9 Oct 2024

Wazeer H, Shridhar Gaonkar S, Doria E et al (2024) Plant-based biostimulants for seeds in the context of circular economy and sustainability. Plants 13:1004. https://doi.org/10.3390/plants13071004

Williamson P, Gattuso J-P (2022) Carbon removal using coastal blue carbon ecosystems is uncertain and unreliable, with questionable climatic cost-effectiveness. Front Clim 4:853666. https://doi.org/10.3389/fclim.2022.853666

World Bank Group (2023) Global seaweed new and emerging markets report 2023. World Bank Group. https://www.worldbank.org/en/topic/environment/publication/global-seaweed-new-and-emerging-markets-report-2023

Xing M, Cao Q, Wang Y et al (2020) Advances in research on the bioactivity of alginate oligosaccharides. Mar Drugs 18:144. https://doi.org/10.3390/md18030144

Zhang Z, Liu H, Yu D-G, Bligh S-WA (2024) Alginate-based electrospun nanofibers and the enabled drug controlled release profiles: a review. Biomolecules 14:789. https://doi.org/10.3390/biom14070789

Temperate Rhodophyta in the Tropics: Cultivated and Wild Cold-Water Red Genera in Hawai'i

5

Karla J. McDermid, Cecilia G. Viljoen, Satoshi Yoshida, and Maria C. Haws

Abstract

The native marine flora of the Hawaiian Islands is a mixture of tropical species and cold-temperate species. Since 1997, Pacific dulse (*Devaleraea mollis),* an "arctic-cold temperate" red algal species, has been successfully cultured in a large-scale land-based aquaculture farm on the west coast of the island of Hawai'i and used as feed for abalone production at the Big Island Abalone Company. The success with Pacific dulse in land-based culture suggests that other native macroalgae with temperate affinities may also be candidates for culture in Hawai'i.

Keywords

Aquaculture · *Chondrus* · Cold-temperate macroalgae · *Devaleraea mollis* · *Kallymenia* · Pacific dulse · *Palmaria* · *Pyropia*

5.1 Introduction

The marine flora of the Hawaiian Islands currently stands at 661 species, including 450 red algae (Rhodophyta), 137 green algae (Chlorophyta), 71 brown algae (Phaeophyceae), and three seagrasses (Tracheophyta), with 13.2% endemism (Sherwood and Guiry 2023). The marine flora of the Hawaiian Islands is a mixture of typical tropical species, plus cold-temperate species, and species with disjunct distributions in primarily subtidal depths in the Main Hawaiian Islands (MHI), Northwestern Hawaiian Islands (NWHI), and temperate locations (Abbott and Huisman 2003; McDermid and Abbott 2006; McDermid et al. 2019; Kittle et al. 2024).

The Hawaiian archipelago, which stretches 2493 km from Kure Atoll (28.417°N 178.333°W) to Hawai'i Island (19.6°N 155.5°W), is considered part of the tropical Indo-West Pacific phytogeographic region (van den Hoek 1984). The waters surrounding the MHI (from south to north: Hawai'i Island, Maui, Koho'olawe, Lana'i, Moloka'i, O'ahu, Kaua'i, and Ni'ihau) have sea surface temperatures that typically range from 25 to 28 °C (Juvik et al. 1998); however, in the NWHI (south to north: Nihoa, Necker, French Frigate Shoals, Gardner Pinnacles, Maro Reef, Laysan, Lisianski, Pearl and Hermes Atoll, Midway Atoll, and Kure Atoll), sea surface temperatures range from 19 to 26 °C (Veillerobe 2004). The MHI and NWHI perch atop giant mountain peaks that rise from the 4000 to 5000-m-deep seafloor and were built up by volcanic eruptions, which continue today on and offshore of Hawai'i Island. The steep geomorphology of the Hawaiian Islands translates into deep water close to shore. Water temperatures in the Hawaiian subtidal zone (10–30 m deep) to the mesophotic zone (30–150 m deep) are significantly lower than on shallow reefs and can be as chilly as 16 °C. All of the NWHI, except Nihoa, are north of the Tropic of Cancer and thus in temperate waters, defined as waters between the Tropic of Cancer (23.5° N) and the Arctic Circle (66.5° N) and between the Tropic of Capricorn (23.5°S) and the Antarctic Circle (66.5° S) (International Seafood Sustainability Foundation 2025). Temperate climates of the Earth are characterized by relatively moderate mean annual temperatures, with average monthly temperatures above 10 °C in their warmest months and above −3 °C in their colder months. In addition to latitude, water temperatures are influenced by sea currents, prevailing wind direction, and water depth.

K. J. McDermid (✉)
Marine Science Department, University of Hawai'i at Hilo, Hilo, HI, USA
e-mail: mcdermid@hawaii.edu

C. G. Viljoen · S. Yoshida
Big Island Abalone Company, Kowa Premium Food Hawai'i Corporation, Kailua-Kona, HI, USA

M. C. Haws
Pacific Aquaculture and Coastal Resources Center, College of Agriculture, Forestry and Natural Resource Management, University of Hawai'i at Hilo, Hilo, HI, USA

A. T. Critchley et al. (eds.), *Advances in Temperate Phyconomy: Algal Harvest and Cultivation in Globally Distributed Temperate Waters*, Developments in Applied Phycology 15, https://doi.org/10.1007/978-3-032-02955-3_5

5.2 Wild Cold-water Macroalgae in the Hawaiian Islands

Several red marine macroalgae with cold-temperate biogeographic affinities grow in the wild in the MHI and NWHI. *Kallymenia sessilis* Okamura with large lobed rosy red blades can be found subtidally on Hawai'i Island and in the NWHI (Abbott and McDermid 2002), and *Kallymenia thompsonii* I.A.Abbott & McDermid with perforated red blades occurs in the NWHI (Abbott and McDermid 2002). *Kallymenia* species are "unusual occurrences in the tropics" (Abbott 1999), since most species in this genus are cool-temperate species of North and South America and Japan. In 2024, large storm swells brought significant quantities of *K. sessilis* onshore in Hilo on the windward coast of Hawai'i Island (Figs. 5.1 and 5.2).

The healthiest thalli were placed in culture in flasks (Fig. 5.3) and in tumble culture in large 550 L tanks at the University of Hawai'i at Hilo's Pacific Aquaculture and Coastal Resources Center (PACRC).

All *Kallymenia* thalli died within a week probably because of the water temperature pumped from the 140-m-deep seawater well onshore at the PACRC was 20–23 °C—too warm for this subtidal, cool-water species normally found at 10–15 m depths. Multiple, deeper wells around the island were drilled in 1993 for geological exploration (Thomas et al. 1996) and trace element measurements (Bienfang et al. 2009). A 365 m deep well that was drilled at the PACRC, at one time, produced 4 °C seawater (Haws personal communication) and could be re-sleeved and used for aquaculture, agriculture, or even air-conditioning.

A small red *Chondrus* species grows in the low intertidal zone in Hilo, Hawai'i Island (Fig. 5.4).

Fig. 5.1 Fragments of red, ruffled *Kallymenia sessilis* thalli washed ashore on the basalt shoreline after a large storm. Oct. 15, 2024, at Puhi Bay, Hilo, Hawai'i Island

Fig. 5.2 Drift *Kallymenia sessilis* from Puhi Bay, Hilo, was used to initiate cultures in flasks and tanks.

Fig. 5.3 *Kallymenia sessilis* in culture in flasks at the Pacific Aquaculture and Coastal Resources Center, Hilo, HI

William Setchell collected the first specimens from Keokea Point in 1900 and labeled them as *Chondrus,* which is a temperate red genus of the North Atlantic and North Pacific. Abbott (1999) tentatively identified the specimens as *C. ocellatus.* Recently, old herbarium material and recent collections were identified as *Chondrus retortus* (Kittle et al.

Fig. 5.4 Wild *Chondrus retortus* from Hilo, HI

Fig. 5.5 A handful of *Chondrus retortus* grown in 19 L containers with running seawater (23 °C) and aeration at the Pacific Aquaculture and Coastal Resources Center, Hilo, HI

Fig. 5.6 A spring bloom of *Pyropia vietnamensis* in the high intertidal zone on basalt above the band of green *Ulva fasciata* in Kona, HI. Patches of sun-bleached *Ahnfeltiopsis cocinna* (Rhodophyta) can be seen in the background

2024), based on morphology, reproductive anatomy, and DNA sequences from five different genes. *Chondrus retortus* K.Matsumoto & S.Shimada is a cold-water species known from Japan and Korea that may have arrived via currents or boat bottoms, but now is naturalized in areas of cold, freshwater seeps along the coast on east Hawai'i Island. *Chondrus retortus* has been cultured at the Pacific Aquaculture and Coastal Resources Center (Fig. 5.5) and studied experimentally to determine the effects of irradiance, nutrient enrichment, and CO_2 on its photosynthetic efficiency (unpublished C. Whetstone and K. McDermid). Other experiments have tested the water temperature and salinity required to maintain *C. retortus* in culture (K. McDermid, unpublished). This species has potential as a cultured and edible species in Hawai'i. In Japan, *Chondrus* species are known as "*tsunomata*", and in Chiba and Ibaraki Prefectures, *tsunomata* is used to make a jelly-like food, *kaisō konnyaku*, or eaten fresh in salads (Kaneryo Media 2025).

Another well-known temperate genus, *Neopyropia/Pyropia* (formerly *Porphyra*), is a member of the Hawaiian marine flora. Most species of *Neopyropia/Pyropia* are native to temperate oceans, typically inhabiting the intertidal zone and preferring cooler temperatures. However, *Pyropia vietnamensis* Tak.Tanaka & P.H.Hô) J.E.Sutherland & Monotilla, with a tropical distribution in South Asia, the Middle East, Brazil, and throughout the Hawaiian Islands (Guiry 2024), is known as *limu pahe'e* (slippery seaweed) in Hawaiian, and exhibits a biphasic life history with its macroscopic, edible gametophyte stage appearing on high intertidal basalt rocks in the MHI and NWHI in February to April. Some work has been done at Big Island Abalone Company on *P. vietnamensis*. *Pyropia* was collected from the high intertidal in Kona (Fig. 5.6) and grown in tumble culture in abalone effluent water with an initial stocking biomass of 1 kg. Biomass increased to 5.8 kg in a week and nearly doubled in the second week (9.7 kg) (Fig. 5.7). Spores were trapped on abalone

Fig. 5.7 *Pyropia vietnamensis* in tumble tank culture in abalone effluent water at Big Island Abalone Company, Kailua-Kona, HI

Fig. 5.8 *Conchocelis* stage (filamentous 2 N sporophyte stage) of *P. vietnamensis* growing on tank walls and abalone shells at Big Island Abalone Company

and oyster shells, and the filamentous *Conchocelis* (2 N sporophyte stage) was cultured (Figs. 5.8 and 5.9).

5.3 Japanese *Nori* and Giant Kelp in Hawai'i

In the 1980s, non-native, temperate, Japanese *Pyropia tenera* (formerly *Porphyra tenera)* (Kjellman) N.Kikuchi, M. Miyata, M.S.Hwang & H.G.Choi was cultured in seven 1-m^3 tank compartments with a mixture of deep (pumped from 600 m deep) and surface water at the Natural Energy Laboratory in Kona, HI (Mencher et al. 1983; Mencher and Katase 1988). This species of *Pyropia* is the highest quality species in Japan for making dried sheets of edible *nori* (Mencher and Katase 1988). This cold-water macroalga could be maintained in land-based tanks in Hawai'i at 15 °C water temperature, 30 l/min flow rate, 26–29 μg/L influent N concentration, 2–2.5 μg/L influent P concentration, and an 8-h photoperiod. The initial stocking density was 20 g wet weight (4.1 g dry wt) per compartment. After 39 days, the "*nori*" averaged 4700 ± 1525 g wet wt (622 ± 121 g dry wt) per compartment. Maximum short-term production rate was 27.9 ± 7.7 g dry wt m^{-2} day^{-1} (Mencher et al. 1983). *Nori* fronds reached a harvestable size of 30–40 cm in length from the initial size of 3 cm or less in 3–5 weeks. These successful experiments with *P. tenera* marked the first attempt to culture temperate red macroalgae in Hawai'i.

Other red genera in the Hawaiian marine flora that are considered temperate to sub-tropical include *Gelidium*, *Rhodymenia*, and *Pterocladiella*. Another temperate red genus primarily of the North Pacific, *Prionitis*, has been found in Hawaiian waters (Abbott 1999). Many newly described species from the mesophotic zone (30–150 m deep) in the Hawaiian Islands (Cabrera et al. 2021; Sherwood and Guiry 2023) will no doubt expand the list of Hawaiian red macroalgae with temperate biogeographic affinities.

In addition to temperate Rhodophyta species in Hawai'i, a handful of temperate brown, phaeophycean macroalgal genera also occur in the Hawaiian Islands with disjunct distributions and affinities with distant cold-water floras. For instance, *Distromium flabellatum* Womersley occurs in the NWHI, MHI, and southern Australia. All other species of this genus are restricted to Japan and the Juan Fernandez Islands near Chile (McDermid and Abbott 2006). *Sporochnus moorei* Harvey is known from Necker Island, NWHI, at 38–72 m deep (Abbott and Huisman 2003) and Midway Atoll, NWHI, at 20 m deep (McDermid and Abbott 2006), as well as southern Australia and New Zealand. Most *Sporochnus* species are cool water species from Japan, China, Australia, Scandinavia, California, and the Galapagos Islands (McDermid and Abbott 2006). *Nereia intricata* Yamada was reported from 32 to 94 m deep at Maro Reef,

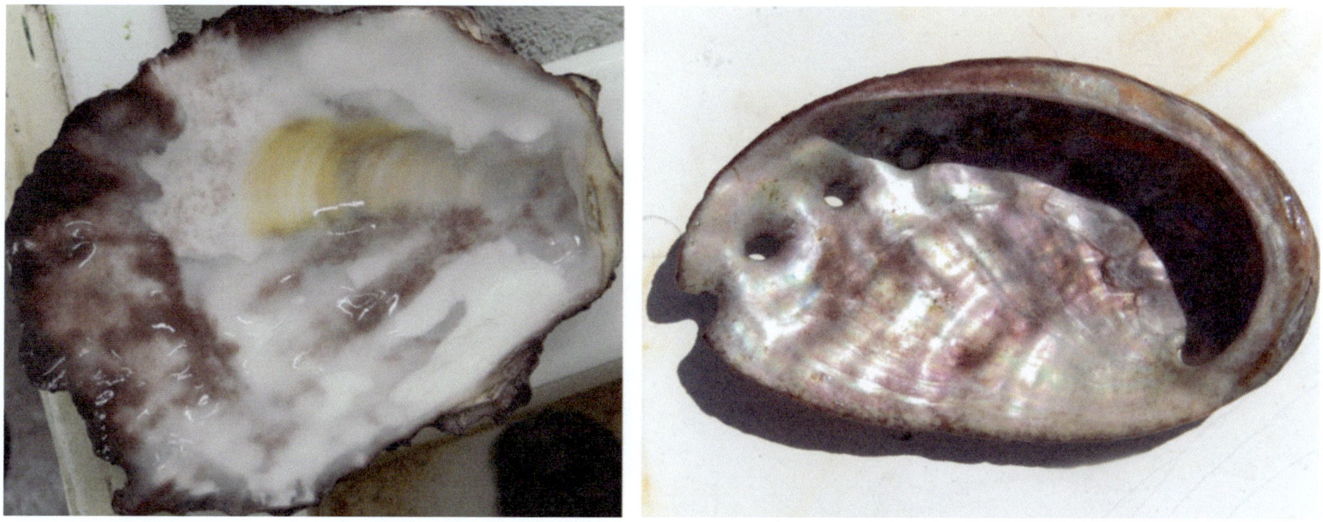

Fig. 5.9 *Conchocelis* stage of *P. vietnamensis* growing on the insides of oyster and abalone shells.

NWHI (Abbott and Huisman 2003), and Japan. A brown macroalga common to cool temperate and polar regions is *Desmarestia ligulata* (Stackhouse) J.V.Lamouroux (acid kelp in the order Desmarestiales), which has been collected 30–56 m deep off of Necker Island, NWHI (Abbott and Huisman 2003; Abbott and Huisman 2004). *Desmarestia ligulata* frequently occurs with kelps (order Laminariales) from Alaska to Antarctica and often in California (Abbott and Huisman 2004).

No Laminariales (kelps), no *Fucus*, no *Ascophyllum*, nor other typical temperate members of the Fucales occur naturally in the Hawaiian Islands. In 1984, the Monterey Abalone Farm (founded in Monterey, CA, in 1972) moved to Kailua-Kona to use the cold deep ocean water at the Natural Energy Lab, and it became Ocean Farms of Hawaii (Kresnak 1984). Ocean Farms of Hawaii cultured the brown seaweed, *Macrocystis pyrifera* (Linnaeus) C.Agardh, giant kelp, from California, as the primary food for cultured abalones. On their 21-acre farm, Ocean Farms of Hawaii also produced salmon, temperate sea urchins, and oysters in a novel system using deep ocean water pumped from 600 m deep into onshore tanks and ponds (4 acres in area and 5 m deep) (Halloran 1990). The endeavor became uneconomical in the 1990s (Hawaii Final Environmental Assessment 2003). The company no longer operates in Hawai'i.

5.4 Pacific Dulse in Hawai'i

Pacific dulse or *Devaleraea mollis* (Figs. 5.10 and 5.11) (Setchell & N.L. Gardner) G.W. Saunders, C.J. Jackson, & Salomaki previously known as *Palmaria mollis* (Setchell & N.L. Gardner) van der Meer & C.J. Bird and *Rhodymenia palmata* f. *mollis* Setchell & N.L. Gardner is considered a temperate species, specifically classified as an "arctic-cold temperate" species, because it thrives in cold temperate regions of the North Pacific Ocean.

Fig. 5.10 Pacific dulse, *Devaleraea mollis* at Big Island Abalone Company

The type locality of *D. mollis* is Whidbey Island (48.19790, −122.69291), Washington State, USA. Pacific dulse is known north to south from the Aleutian Islands, coastal Alaska, British Columbia, Washington, Oregon, and California, as well as from Kamchatka, the Sea of Japan, Russia, and the Commander Islands (Guiry 2023).

Twenty years ago, researchers focused on the production of Pacific dulse and abalone in land-based systems (Demetropoulos and Langdon 2004a, b, c, d; Langdon et al.

Fig. 5.11 A handful of red Pacific dulse, *Devaleraea mollis* from a starter tank

Fig. 5.12 The Pacific dulse is grown in a tumble culture in 32 large algal production tanks, and several small and medium starter tanks at Big Island Abalone Company

2004) in Kona, HI. Now, Pacific dulse is in large-scale culture at Big Island Abalone Company at the Natural Energy Lab aquaculture park in Kona on the west coast of Hawai'i Island.

5.5 Big Island Abalone Company

Research started at Big Island Abalone Company in 1997, and originally the plan was to co-culture Pacific dulse and red abalone, *Haliotis rufescens*. Then, driven by a better market demand and price, the decision was made to switch to the Japanese abalone (*ezo*), *Haliotis discus subsp. hannai*. Sales of this species started in 2001, and the *ezo* abalone were fed solely Pacific dulse. In 2010, the farm expanded to an abalone production capacity of 100 MT per year FW, and at that time the dulse production was not sufficient to feed all the abalone.

After various feed trials, a decision was made to feed a combination of Pacific dulse and pelleted feed. Abalone can eat 5% of their body weight per day, so for a 100MT farm, up to 5 MT of wet feed would be required per day. Increased growth was achieved with the combination feed, and the dark red shell pigmentation could be maintained. The nursery abalone feed on diatoms and dulse, and then after that, the dulse/pellet combination is used, with dulse as a polishing feed before market.

Big Island Abalone Company, a 10-acre farm, has 32 large algal production tanks and several small and medium starter tanks and large production tanks (Figs. 5.12, 5.13, 5.14, and 5.15) at the state's Hawai'i Ocean Science and Technology Park (HOST Park), part of the Natural Energy Lab), in Kailua-Kona on the west or leeward coast of Hawai'i Island.

A 1.4-m-diameter pipeline delivers deep seawater from 1000 m deep as well as surface seawater to the farm. Commercial fertilizer and micronutrients are added at night, when the seawater is off. The design is to run 3190 GPM over 13 daytime hours to maintain production tank temperatures between 12 and 15 °C. Only nutrient-rich and cold (5–8 °C incoming) deep seawater is used in the algal tanks. The starter tanks run a little cooler, around 10 °C, to keep bacteria and epiphyte growth down.

The Pacific dulse is grown in tumble culture, in production tanks less than 1 m deep (Figs. 5.15 and 5.16).

These lined cement production tanks are 42 m long and 4.88 m or 6 m wide. The densities are kept at levels to control self-shading, and the abrasion in the tumble culture reduces the epiphyte growth. The production tanks were designed such that the harvesting rotation will allow harvest from a tank every four weeks (Figs. 5.17).

Production averages about 250 g m^{-2} day^{-1}, and the maximum dulse feed production was 149,676 kg FW $year^{-1}$ (Fig. 5.18).

Fig. 5.13 Small starter tanks at Big Island Abalone Company full of Pacific dulse in tumble culture with full sunlight

Fig. 5.14 Medium starter tanks at Big Island Abalone Company with Pacific dulse in tumble culture and full sunlight

Fig. 5.15 The lined cement production tanks are 42 m long and 4.88 m or 6 m wide

Fig. 5.16 The production tanks are less than 1 m deep

Fig. 5.17 Production tanks are harvested and cleaned every four weeks. A worker is hosing out the empty tank

Fig. 5.18 Production averages about 250 g m^{-2} day^{-1} m-^{2}, and the maximum dulse feed production was 149,676 kg FW year^{-1}. Workers are unloading large containers of freshly harvested Pacific dulse

5.6 Conclusion

The successful large-scale cultivation of Pacific dulse in land-based tanks on Hawai'i Island demonstrates the potential that exists for the culture of other native Hawaiian macroalgal species, including cold-water subtidal species. Cultivation of non-native macroalgae presents many biosecurity issues, and stringent precautions are required to prevent the spread of introduced temperate species into cold-water subtidal areas of the Hawaiian Islands. More work needs to be accomplished to determine the optimal water temperature, light levels, and nutrient regimens required to grow cold-water native seaweeds with economic or ecological value, such as Hawaiian *Kallymenia*, *Chondrus,* and *Pyropia* species. Access to cold, nutrient-rich seawater through deep ocean pipelines and deep seawater wells (Bienfang et al. 2009) will be key to future cultivation of temperate seaweeds in the tropics.

Literature Cited

Abbott IA (1999) Marine red algae of the Hawaiian islands. Bishop Museum Press Honolulu Hawai'i, p 477

Abbott IA, Huisman JM (2003) New species, observations, and a list of new records of brown algae (Phaeophyceae) from the Hawaiian Islands. Phycol Res 51:173–185

Abbott IA, Huisman JM (2004) Marine green and brown algae of the Hawaiian Islands. Bishop Museum Press Honolulu, p 259

Abbott IA, McDermid KJ (2002) On two species of *Kallymenia* (Rhodophyta: Gigartinales: Kallymeniaceae) from the Hawaiian Islands, Central Pacific. Pac Sci 56(2):149–162

Bienfang P, De Carlo EH, Christopher S, DeFelice S, Moeller P (2009) Trace element concentrations in coastal Hawaiian waters. Mar Chem 113(3–4):164–171

Cabrera FP, Huisman JM, Spalding HL, Kosaki RK, Sherwood AR (2021) Diversity of Kallymeniaceae (Gigartinales, Rhodophyta) associated with Hawaiian mesophotic reefs. Eur J Phycol 57(1):68–78. https://doi.org/10.1080/09670262.2021.1891462

Demetropoulos C, Langdon C (2004a) Pacific dulse (*Palmaria mollis*) as a food and biofilter in recirculated, land-based abalone culture systems. Aquac Eng 32(1):57–75

Demetropoulos CL, Langdon CJ (2004b) Enhanced production of Pacific dulse (*Palmaria mollis*) for co-culture with abalone in a land-based system: effects of seawater exchange, pH, and inorganic carbon concentration. Aquaculture 235(1–4):457–470

Demetropoulos CL, Langdon CJ (2004c) Enhanced production of Pacific dulse (*Palmaria mollis*) for co-culture with abalone in a land-based system: effects of stocking density, light, salinity, and temperature. Aquaculture 235(1–4):471–488

Demetropoulos CL, Langdon CJ (2004d) Enhanced production of Pacific dulse (*Palmaria mollis*) for co-culture with abalone in a land-based system: nitrogen, phosphorus, and trace metal nutrition. Aquaculture 235(1–4):433–455

Guiry GM (2023, September 01) *Devaleraea mollis.* In: Guiry MD, Guiry GM AlgaeBase (eds) World-wide electronic publication, National University of Ireland, Galway. https://www.algaebase.org; searched on 13 January 2025

Guiry MD (2024) *Pyropia vietnamensis*. In: Guiry MD, Guiry GM 03 April 2024. AlgaeBase. World-wide electronic publication, National University of Ireland, Galway. https://www.algaebase.org. Accessed 19 Jan 2025

Halloran R (1990) Tapping ocean's cold for crops and energy. New York Times. May 22, 1990. https://www.nytimes.com/1990/05/22/science/tapping-ocean-s-cold-for-crops-and-energy.html

Hawaii Final Environmental Assessment (2003) Kona Blue Water. 278 pp. https://files.hawaii.gov/dbedt/erp/EA_EIS_Library/2003-07-23-HA-FEA-Kona-Blue-Water.pdf

International Seafood Sustainability Foundation (2025). https://www.iss-foundation.org/glossary/temperate-waters#:~:text=Waters%20in%20the%20region%20of,Circle%20in%20the%20southern%20hemisphere. Accessed 20 Jan 2025

Juvik SP, Juvik JO, Paradise TR (eds) (1998) Atlas of Hawai'i. University of Hawaii Press

Kaneryo Media (2025) Sea Vegetable: Tsunomata https://www.kaneryo.co.jp/media/en/seaweed-info/tsunomata-info/ accessed19 Jan. 2025

Kittle RP, Veillet A, Schmidt WE, Fredericq S, McDermid KJ (2024) *Chondrus retortus* (Gigartinales, Rhodophyta) in Hawai'i: a taxonomic and biogeographic puzzle. Bot Mar 67(1):15–30. https://doi.org/10.1515/bot-2023-0019

Kresnak WE (1984) Commercial abalone farming beginning on Hawaii. United Press International Archives. July 8, 1984. https://www.upi.com/Archives/1984/07/08/Commerical-abalone-farming-beginning-on-Hawaii/9809458107200/

Langdon C, Evans F, Demetropoulos C (2004) An environmentally-sustainable, integrated, co-culture system for dulse and abalone production. Aquac Eng 32(1):43–56

McDermid KJ, Abbott IA (2006) Deep subtidal marine plants from the Northwestern Hawaiian Islands: new perspectives on biogeography. Atoll Res Bull 543:525–532

McDermid KJ, Martin KJ, Haws MC (2019) Seaweed resources of the Hawaiian Islands. Bot Mar 62:443–462. https://doi.org/10.1515/bot-2018-0091

Mencher FM, Katase SA (1988) Growth of *nori* (*Porphyra tenera*) in an experimental ocean thermal energy conversion system at the Natural Energy Laboratory of Hawai'i. OTEC aquaculture in Hawai'i. University of Hawai'i-Sea Grant-MR-89-01, UH Sea Grant College Program Honolulu, pp 70–83

Mencher FM, Spencer RB, Woessner JW, Katase S, Barclay DK (1983) Growth of *nori* (*Porphyra tenera*) in an experimental OTEC-aquaculture system in Hawaii. J World Maric Soc 14(1–4):458–470. https://doi.org/10.1111/j.1749-7345.1983.tb00098.x

Sherwood AR, Guiry MD (2023) Inventory of the seaweeds and seagrasses of the Hawaiian Islands. Biology 12(2):215

van den Hoek C (1984) World-wide latitudinal and longitudinal seaweed distribution patterns and their possible causes, as illustrated by the distribution of Rhodophytan genera. Helgol Meeresunters 38:153–214. https://doi.org/10.1007/BF01997483

Veillerobe YJ-P (2004) Sea surface temperature variability in the Northwestern Hawaiian Islands region. Thesis Global Environmental Science, University of Hawai'i at Mānoa, p 77

Karla J. McDermid has been studying Hawaiian seaweeds for 42 years, first as a graduate student with Izzie Abbott, and now as a Professor in the Marine Science Department at the University of Hawaii at Hilo. She teaches three courses a semester, writes grant proposals, conducts research, and helps students to grow macroalgae. She is grateful for Dr. Critchley's invitation to write a chapter that allowed her to pull together lots of facts and old memories, and take a unique look at a facet of aquaculture in Hawaii that had not been examined previously: wild and cultured cold-water macroalgae in the Hawaiian Islands.

Cecilia G. Viljoen has been ascinated with the ocean since a child, when she helped her dad collect bait for fishing. This very valuable time by the sea also sparked an interest in what is edible along the shoreline. From early years she loved foraging for food along the South African shores. After a brief period as an environmental journalist at a daily newspaper and a subsequent year on a mission in Southern Africa in a 1968 Volkswagen bus, she switched to aquaculture, a field that she has worked in for 30 years: abalone, oysters, flat fish, urchins, sea cucumbers, and algae. At Big Island Abalone Corporation (BIAC) in Hawaii, she was excited to learn how to grow algae on a commercial scale. Her role at BIAC started off as grow out supervisor, but diverged to abalone and algae manager, farm manager, R&D and Health Management Director, and R&D, QC Advisor. Hawaii is a very expensive island, and it makes sense to optimize land and water use sustainably in our business. At BIA, I have looked at growing local algae in the abalone effluent water. If we could re-use the abalone water, the algae would be a valuable feed and "polish" the abalone effluent water, and the use of the nutrient rich abalone effluent water would stop the need for commercial fertilizer.

Satoshi Yoshida is currently COO at Kowa Premium Foods Hawaii. He earned his Master of Pharmaceutical Science at Nagoya City University in Japan in 2005, and has the license of Pharmacist in Japan. After completing his master's degree, Satoshi joined Kowa Company in Japan, he worked for QC and QA departments in the pharmaceutical division. After 15 years of experience in Japan, he accepted the position of COO at Kowa Premium Foods Hawaii, known as Big Island Abalone.

Maria C. Haws serves as Professor of Aquaculture at the University of Hawaii Hilo, based at the Pacific Aquaculture and Coastal Resources Center, and is the Director of the University of Hawaii Center of Excellence for Sustainable Aquaculture. She has 40 years of experience in aquaculture research and development and international aquaculture development, beginning with having served in the Peace Corps in Honduras. Her research and extension work focuses on marine fish, bivalves and seaweed, including recent trials using shellfish and seaweed for restorative aquaculture. Biosecurity, best management practices and policy development for molluscan and shrimp aquaculture have also been part of her portfolio. She teaches multiple aquaculture courses and leads the PACRC's Aquaculture Student Workforce Training Program. She also heads the largest aquaculture extension program in Hawai'i. She also specializes in climate change adaptation with an emphasis on the impacts on women and other marginalized groups.

The Development and Spread of Land-Based Aquaculture of *Ulva prolifera* in Japan

6

Masanori Hiraoka and Yoichi Sato

Abstract

Ulva is currently called SeaWheat in the EU and is considered an important biological resource for the blue bioeconomy, and its production is being promoted. In East Asia, especially in Japan, brackish *Ulva prolifera* has been used as a traditional and expensive food ingredient and has long been produced by natural collection or net culture. However, in recent years, the surface seawater temperature off the coast of western Japan, the main *U. prolifera* producing temperate region, has been warming at twice the global average rate, causing a sharp decline in production. To compensate for the lack of supply, *Ulva* land-based aquaculture has rapidly expanded. The land-based aquaculture technology was developed in Kochi, southwestern Japan, and commercial production began in 2004. In the 1990s, the natural collection of *U. prolifera* in Kochi was 10–20 t-dry y^{-1}, but production fell to zero in 2020, and there has been no harvest since. However, land-based aquaculture has made it possible to produce equal or greater quantities of *Ulva* than before. This Kochi method is spreading across the country, and it is estimated that the total production capacity will reach 100 t-dry y^{-1} as of 2025, based on the tank area of each commercial land-based aquaculture facility. In the near future, the Kochi Method will likely be used to commercialize other kinds of seaweed.

M. Hiraoka (✉)
Usa Marine Biological Institute, Kochi University, Usa, Tosa, Kochi, Japan
e-mail: mhiraoka@kochi-u.ac.jp

Y. Sato
Algal Research and Innovation Centre, Riken Food Co., Ltd., Natori, Japan

Keywords

Blue bioeconomy · Germling cluster method · Land-based aquaculture · SeaWheat · *Ulva prolifera*

6.1 Introduction

Currently, seaweeds are increasingly recognized as an important biological resource for the blue bioeconomy. Among seaweeds, *Ulva* or sea lettuce has become a particularly important key player. In the EU, *Ulva* (Chlorophyta) is called SeaWheat instead of seaweed, and a project to actively cultivate it is underway (Buck and Shpigel 2023; Hofmann et al. 2024). Key properties of *Ulva* include high morphological plasticity, diverse biochemical composition for valorization, massive proliferation, high growth rates, and the ability to thrive at high stocking density, rapid nutrient uptake potential, and wide environmental tolerance (Fig. 6.1 in Hofmann et al. 2024). These characteristics are highly valued as having the potential to play a central role in the European seaweed farming industry. However, of the world's seaweed biomass production, the production of green macroalgae, including *Ulva,* is less than 1% of the total seaweed production, which is overwhelmingly smaller than brown macroalgae and red macroalgae (Moreira et al. 2022). Given the current situation, there is ample room for the development of *Ulva* production.

In temperate East Asia, *Ulva prolifera* has long been used for food (Ohno 1993; Park and Hwang 2011; Liu et al. 2013). However, its production has been decreasing in recent years due to the effects of global warming. The attached *U. prolifera* populations, which have supported its edible use, switch between vegetative growth and spore production depending on the water temperature, producing spores at high water temperatures of 20 °C or higher (Hiraoka 2021). As a result, as seawater temperatures rise, the grown *Ulva* plants turn into spores and flow out from the tips, reducing

A. T. Critchley et al. (eds.), *Advances in Temperate Phyconomy: Algal Harvest and Cultivation in Globally Distributed Temperate Waters*, Developments in Applied Phycology 15, https://doi.org/10.1007/978-3-032-02955-3_6

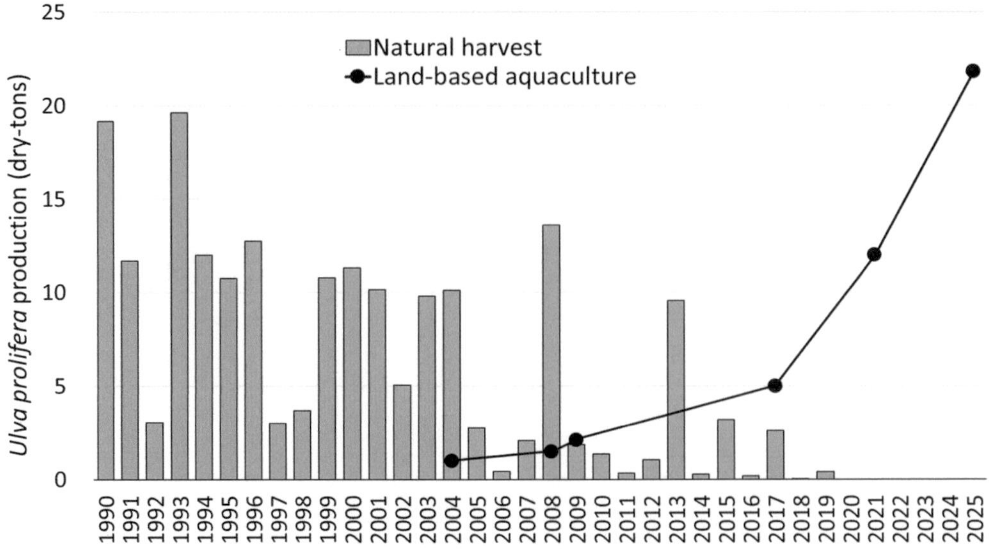

Fig. 6.1 Changes in annual production of *Ulva prolifera* in Kochi Prefecture

the yield of the plants that can be collected. In Japan, *U. prolifera* is produced by both natural collection and net culture (Ohno 1993), and the largest natural collection area is the brackish waters of the Shimanto River estuary in Kochi, southwestern Japan (Hiraoka and Higa 2016). The coast of Kochi is a region that is greatly influenced by the warm Kuroshio Current, of which the surface ocean warming rate is two to three times faster than the global mean surface ocean warming rate (Wu et al. 2012). In agreement with this, the surface seawater temperature in this region is also rising at twice the rate. As a result, temperate species of subtidal canopy-forming seaweeds (Laminariales and Fucales) in the coastal area of more than 700 km off Kochi have declined and has been replaced overwhelmingly by tropical species (Tanaka et al. 2012). The subtropical species *Ulva ohnoi*, which caused green tides in this region, has also changed to the tropical species *Ulva reticulata* (Hiraoka et al. 2020b). As for *U. prolifera*, the largest edible seaweed resource in this region, the harvest gradually decreased and finally reached zero in 2020 and has remained zero since then (Fig. 6.1). A national organization has published a correlation between environmental factors (water temperature, salinity, dissolved nitrogen, dissolved phosphorus, etc.) during the *U. prolifera* growth period in the Shimanto River estuary and the yield (Nakamura River and National Highway Office, Shikoku Regional Development Bureau, Ministry of Land, Infrastructure, Transport and Tourism 2023). According to the results, the *Ulva* yield has a strong negative correlation with water temperature and a moderately strong negative correlation with salinity. This suggests that the *Ulva* thalli will decrease if the seawater that enters the brackish area from the ocean during tidal periods is at a high temperature. In the laboratory, it has been confirmed that the *U. prolifera* in this region releases spores from the tips of the plant body at high water temperatures of 20 °C or higher, and the plants disappear, or stop growing, at about 10 cm long (Hiraoka 2021). Having only short plants is causing zero yields, as fishermen need the thalli to be long enough to harvest them. If ocean warming continues at this rate, the harvest is not expected to recover in the future.

6.2 Land-Based Aquaculture Started in Kochi

As seaweed production becomes difficult in the warming ocean, a highly efficient tank cultivation technique for *Ulva* was developed in Kochi, and commercial production began in 2004 (Hiraoka and Oka 2006). This Kochi Method has three main features: (1) the adoption of the germling cluster method (Hiraoka and Oka 2008), a floating high-density cultivation method; (2) the adoption of a multi-step tank system; and (3) the use of running deep seawater.

6.2.1 Germling Cluster Method

When cultivating seaweeds in tanks, unattached culture, as a free-floating form with agitation, allows for three-dimensional use of the tank interior and maintains high density, rather than attached culture. Schramm (1991) described some advantages of unattached culture: planting and harvesting is easy and less labor and capital intensive, the density of cultures can easily be controlled and maintained at desired optimal levels by continuous intermediate removal of biomass, and therefore yield per unit area can be considerably higher under a sufficient supply of radiant energy and nutrients. Such an unattached culture or floating culture can be carried out by vegetative propagation through fragmentation of thalli. However, repeated vegetative propagation of the same thallus causes contamination with other algae and pathogens, which risks significantly reducing productivity

(Craigie et al. 2019). In addition, many *Ulva* species are difficult to continuously propagate vegetatively due to fragmentation of the thalli, which causes sporulation (Hiraoka 2021). To solve these problems, the germling cluster method was developed (Hiraoka and Oka 2008). In this method, a large number of spores are released from a single isolated *Ulva* cultivar, plated and cultured at high density in a Petri dish, and multiple germlings are connected to each other during the developmental stage. The mat-like aggregation of germlings that is connected together is torn into smaller pieces and cultured under aeration, to create a large number of germling clusters (Fig. 6.2). More detailed methods are described in Hiraoka and Oka (2008) and Tsubaki et al. (2017). This method has the following advantages:

- Seedlings are obtained that are free from contamination by competing algae.
- By repeatedly using a selected cultivar, seedlings can be cloned.
- The risk of sporulation, which is often an issue with repeated vegetative propagation, is reduced.
- It makes it easier to adjust the density of algae in the tank.
- Stirring the tank leads to uniform growth, producing materials of uniform age.
- Easy harvesting without the need to remove the seedlings from the substrate.

The germling cluster method was also tested on several species other than *U. prolifera*, and it was demonstrated that growth rates and biomass productivities were consistently higher for clusters, as compared to non-cluster morphologies (Lawton et al. 2021; da Silva et al. 2024).

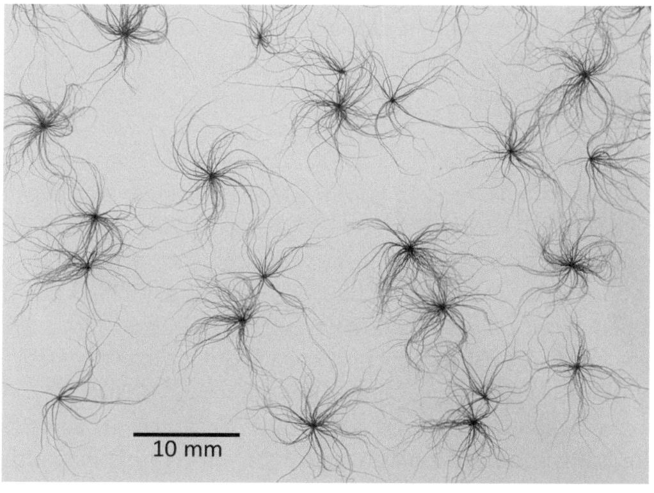

Fig. 6.2 Germling clusters of *Ulva prolifera*

6.2.2 Multi-step Tank System

When cultivating seaweeds outdoors, under sunlight, a multi-step tank system was designed to continue the exponential growth of the selected seaweed and increase production efficiency per unit light-receiving surface area or volume of the tank. A demonstration test showed that when 100 g-fresh of *Ulva* germling clusters were cultured in a cylindrical 1-ton tank, they grew 10 times to 1 kg-fresh weight within one week, and the growth rate only started to decline when they reached 1 kg-fresh weight (Hiraoka et al. 2004). This production efficiency standard of "increasing 100 g-fresh per week to 1 kg-fresh per ton of volume" was used to design the system. In the single unit of the multi-step tank system shown in Fig. 6.3, *Ulva* is continuously cultivated in four stages of tanks with a volume expanded 10 times each, and the initial amount of 10 g-fresh is increased to 100 kg-fresh in four weeks, a 10^4-fold increase, for harvest. This unit has a total tank volume of 111.1 tons and can produce 100 kg-fresh weight of *Ulva* per week, so the production efficiency is 0.90 kg-fresh t^{-1} $week^{-1}$. On the other hand, if there is only one final tank of 100 tons instead of multiple stages, the harvest is once every weeks, so the production efficiency is 100 kg-fresh/ 100 t/ 4 weeks = 0.25 kg-fresh t^{-1} $week^{-1}$. The production efficiency of the 4-step tank system is 3.6 times that of a single tank system. At the facility in Muroto City, Kochi, where the multi-step system was first introduced, six units were set up so that harvesting could be done on a six days per week cycle, since one unit in the figure was harvested every week, as mentioned above (Tsubaki et al. 2017). The annual production of six units is estimated to be 100 kg-fresh × 6 days of harvesting per week × 52 weeks = 31.2 t-fresh, and actual production has been achieved, as estimated (Hiraoka 2012).

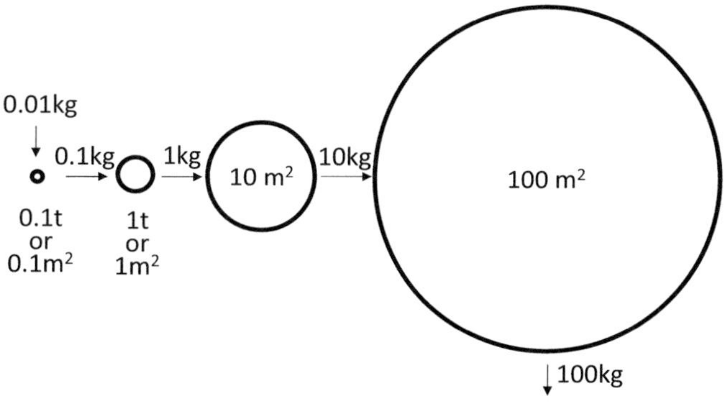

Fig. 6.3 A four-step tank system capable of producing 100 kg-fresh of *Ulva prolifera* per week

6.2.3 Deep Seawater

To produce *U. prolifera* stably year round, the water temperature needs to be kept below 20 °C. However, maintaining a constant water temperature all year round requires a large energy cost. Therefore, in the case of Muroto City, deep seawater, which is cold and stable all year round, was used. Muroto City pumps 4000 tons of deep seawater per day from a depth of 374 m off the coast through a pipeline. The temperature of the deep seawater supplied to the aquaculture facility is about 10 °C all year round. When the deep seawater is continuously supplied to the tanks, the water temperature inside the tanks fluctuates seasonally due to the influence of air temperature, but it has been shown that *U. prolifera* grows at an average daily growth rate of 37% throughout the year (Hiraoka and Oka 2008).

6.3 Recent Improvements That Led to Widespread Use

6.3.1 Saline Groundwater

The use of deep seawater is effective for land-based production of seaweeds, but the areas where it can be taken are limited, and the construction costs of a deep seawater intake facility are approximately 1 billion yen. Saline groundwater is an alternative to deep seawater, with a stable water temperature year round, naturally filtered, and usually free of contaminating organisms. Saline groundwater filtered through the strata can be pumped up from wells near the sea. Pumping up saline groundwater requires a well and a pump, but the construction costs are several million to ten million yen. Deep strata can contain many minerals that have a negative effect on seaweed growth, so water is taken from a relatively shallow depth of about 5 to 10 m, being careful to ensure that the depth is deeper than the sea level at the lowest tide. The first *Ulva* production project using saline groundwater was actually carried out in 2015 in Oshima, Yawatahama City, Ehime, and since then, the use of groundwater has become widespread (Hiraoka 2022).

6.3.2 Strain Selection

Ulva prolifera is a brackish species, first described in the Baltic Sea in 1778 (Cui et al. 2018). It is distributed in brackish waters around the world, including a subspecies that causes green tides in summer along the Chinese coast of the Yellow Sea, and there are many genetically distinct local populations (Hiraoka 2021). There are sexual variants with a sexual life cycle in which isomorphic sporophytes and male and female gametophytes alternate generations, and obligate asexual variants with an asexual life cycle that does not have a male and female gametophyte generation and repeats clonal proliferation with diploid asexual spores (Hiraoka et al. 2003). These spores specific to the asexual life cycle were termed zoosporoids (Bliding 1963). These sexual and asexual variants coexist within the population (Hiraoka and Higa 2016). In land-based aquaculture, the asexual variants are used, whose genetic characteristics do not change, even after subculture. Sexual variants have not been used so far because sexual reproduction can change the genetic combination. Among asexual strains collected from all over Japan, strains that do not form spores and can continue stable vegetative growth at high temperatures of 20 °C, or higher, have been selected (Sato et al. 2021). Such high-temperature tolerant strains, or high-temperature sterile strains, are collected and used in warm regions from Western Japan to the Okinawa Islands. The Usa Marine Biological Institute, Kochi University, holds hundreds of *Ulva* strains collected and isolated from all over Japan, as well as from all over the world, including the Baltic Sea and the Yellow Sea, and provides strains that are suitable for business facilities upon request for commercialisation.

6.3.3 Nutrient Supply

In addition to sunlight, the resources and energy required for outdoor land-based aquaculture of seaweeds are: (1) seawater as a medium, (2) agitation energy, and (3) fertilizer. When seaweeds are cultivated at high density in tanks, nitrogen and phosphorus supplies are required. The deep seawater mentioned above contains relatively high concentrations of nitrate (NO_3-N; 12.1–26.0 μM) and phosphate (PO_4-P; 1.1–2.0 μM), and if the deep seawater is supplied continuously to the tank at an exchange rate of three volumes per day, the algae will grow without the addition of the fertilizer (Hiraoka and Oka 2008). However, the iron concentration in the deep seawater is lower than that of surface seawater, and it has been reported that the addition of iron increases the yield of *Ulva* (Oka et al. 2008). As saline groundwater began to be used in various locations, the water quality varied from place to place. In Oshima, Yawatahama City, where saline groundwater was first introduced into the *Ulva* land-based aquaculture business (Hiraoka 2022), the *Ulva* production was possible without any problems and this was without the addition of nitrogen, phosphorus, or iron. Meanwhile, in other regions, these ingredients must be added to seawater for cultivation. Agricultural fertilizers for hydroponic cultivation, such as sodium nitrate or ammonium nitrate for nitrogen, potassium phosphate or ammonium phosphate for phosphorus, and chelated iron for iron, are added to the seawater. The amount added is calculated from the component content and growth rate of each seaweed species. For example, if 100 g-fresh *Ulva* in a 1-ton tank becomes 10 times that amount, 1000 g-fresh, in one week, the growth rate will be 900 g-fresh in one week. Since the nitrogen and phosphorus contents of *U. prolifera* are reported to be 0.68% and 0.09% of the fresh mass, respectively (Hiraoka et al. 2020a), 6 g of nitrogen is required to produce 900 g-fresh of *U. prolifera*, which is equivalent to 37 g of sodium nitrate (molecular weight 85). The supply amount of phosphorus can be calculated in the same way. The calculated amount is supplied over a week. To supply fertilizer more accurately without excess or deficiency, the required fertilizer is calculated and supplied daily from the daily growth rate, rather than on a one-week time scale. However, in actual outdoor tanks, the water temperature and amount of sunlight changes depending on the weather, so the growth rate of the seaweed is not constant, and the amount of growth fluctuates. Ultimately, to supply fertilizer without waste, it would be ideal to use sensors to measure the time-dependent changes in fertilizer concentration in the tank and supply only the amount that has been absorbed and reduced. The challenge for the future is to increase production efficiency by making full use of such digital technology and data.

Fig. 6.4 One tank with a paddle-type agitator in the Rikuzentakata Base

6.3.4 Tank and Agitation

Early facilities used rectangular fiber-reinforced plastic (FRP) tanks, but in recent years, circular polypropylene (PP) tanks equipped with paddle-type agitators are often used (Fig. 6.4). Paddle-type agitators have been reported to be a less energy-consuming agitation method for mass cultivation of microalgae than aeration (Matsumoto 2016). PP circular tanks are less expensive than FRP tanks and are easy to scale up. Therefore, many companies have adopted this tank and paddle agitation to open large-scale *Ulva* aquaculture facilities.

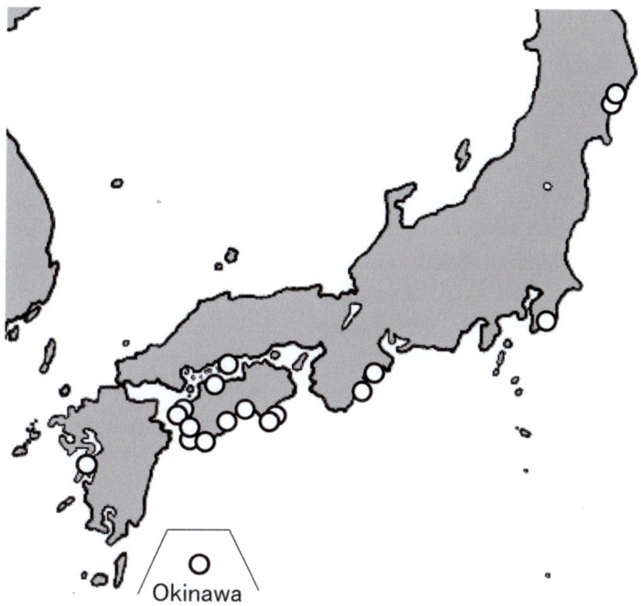

Fig. 6.5 *Ulva* land-based aquaculture facilities where the Kochi Method is adopted in Japan. 18 facilities as of 2025

6.4 Spread of the Kochi Method

6.4.1 Estimated National Production

Since its launch in 2004, the Kochi Method of land-based aquaculture spread throughout Kochi Prefecture, completely replacing the peak production of natural harvesting, which had fallen to zero in recent years in the prefecture (Fig. 6.1). It has also spread throughout Japan, and as of 2025, at least 18 aquaculture facilities are in operation for commercial production (Fig. 6.5). The Rikuzentakata Base, a representative example of them, has 5, 5, and 25 m circular PP tanks of three types with diameters of 2, 3.6, and 8 m (total tank area of 1323 m^2), and began operation in 2021 (Fig. 6.6). Although the facility was estimated to have a production capacity of 5 t-dry per year (Sato et al. 2023), actual production was 4.6 t-dry for the year from April 2023 to March 2024, confirming that estimate. The production efficiency of the Rikuzentakata Base is 5 tons/1323 m^2/year = 3.8 kg-dry m^{-2} y^{-1} = 10 g-dry m^{-2} d^{-1}. The production volume of 15 facilities in Japan that commercially produce 1 ton dry weight, or more, of *U. prolifera* per year through land-based aquaculture was compiled (Table 6.1). For factories where production volumes have not been announced, estimates were made from the tank area on Google Maps based on Rikuzentakata's production efficiency. As a result, the total land-based production, as of 2025, was estimated to be 100 t-dry per year. According to a table summarizing the global production of *Ulva* spp. (fresh mass) (Table 5 in Hofmann et al. 2024), the average annual production from

Fig. 6.6 Overall view of the Rikuzentakata Base

1950–2019 was 1882 tons DW, and the maximum value during that period was 15,440 tons. Since the dry mass of *U. prolifera* is approximately 15% of the fresh mass (Table 2 in Hiraoka et al. 2020a), the 100 dry tons produced annually by land-based aquaculture in Japan shown here is 667 tons of fresh mass, which is equivalent to 35% of the global annual average *Ulva* production in Japan.

6.4.2 The Future

According to Zertuche-González et al. (2021), most feasibility studies for the commercial cultivation of *Ulva* have strong limitations in order to be reliable. This was due to three main reasons: (a) the scale of the culture system (size and depth of tanks) is usually too small; (b) the length of the cultivation period is usually very short (few months of the year); and (c) the difference in post-harvesting methods used to report biomass increase (fresh: dry ratio and salt content). In addition, none of these studies maintained the same strain for more than a year. However, Japan overcame such problems during feasibility studies and began commercial production in 2004 and currently produces 100 t-dry per year. The reason for this commercial success is that *U. prolifera* is an expensive food ingredient in Japan, with a wholesale price of 10,000–20,000 yen per kilogram of dried product. Many seaweed species are traded in Japan, and *U. prolifera* is one of the most expensive seaweeds. It has been demonstrated that the land-based aquaculture technology introduced in this paper can be applied to other seaweeds along with *Pyropia* spp. (Rhodophyta) (Hiraoka et al. 2004) and the brown alga *Undaria* (Phaeophyceae) (Oka et al. 2004). However, if it is possible to grow various species of seaweeds using the same technology, aquaculture companies will select and produce

Table 6.1 Annual production of *Ulva prolifera* at 15 major land-based aquaculture facilities in Japan

Company name	Location (latitude and longitude)	Tank diameter φ (m) and number on the map	Total tank area (m²)	Annual production (t-dry)*
Mishima foods	Muroto City, Kochi Pref. (33°15′51.7″N 134°11′05.8″E)			**3**
Sea vegetable company	Muroto City, Kochi Pref. (33°16′06.7″N 134°09′36.5″E)	φ10 × 12, φ5 × 12	1177	4.5
Sea vegetable company	Aki City, Kochi Pref. (33°30′03.8"N 133°52′51.1″E)	φ10 × 15, φ5 × 12	1413	5.3
Sea vegetable company	Kochi City, Kochi Pref. (33°28′10.4″N 133°30′15.5″E)	φ25, φ19 × 2, φ18, φ15, φ10 × 6, φ5 × 6	2077	7.9
Umi no Kenkyusha	Sukumo City, Kochi Pref. (32°55′23.1″N 132°41′02.5″E)			**1**
Ohshima seaweed culture laboratory	Yawatahama City, Ehime Pref. (33°22′56.3″N 132°20′33.2″E)			**1**
Sato-shinju	Seiyo City, Ehime Pref. (33°19′09.9″N 132°28′06.0″E)			**1**
Sea vegetable company	Imabari City, Ehime Pref. (34°04′46.0″N 132°59′53.7″E)	φ20 × 6, φ16 × 2, φ12 × 4, φ10 × 4, φ5 × 8	3209	12.1
Sea vegetable company	Amakusa City, Kumamoto Pref. (32°32′34.0″N 130°11′22.1″E)	φ25 × 8, φ10 × 10, φ5 × 10	4906	18.5
Mishima foods	Fukuyama City, Hiroshima Pref. (34°20′54.4″N 133°25′44.7″E)	φ4 × 80, φ1.5 × 60	1111	4.2
Sea vegetable company	Owase City, Mie Pref. (34°00′55.5″N 136°14′35.0″E)	φ20 × 7, φ14 × 2, φ10 × 23, φ5 × 22	4743	17.9
Minamiise marine bio	Minamiise town, Mie Pref. (34°17′32.9″N 136°41′11.2″E)	φ7 × 24, φ3 × 16, φ1 × 10	1044	**4**
Rikuzentakata Base, Riken food	Rikuzentakata City, Iwate Pref. (39°00′23.1″N 141°38′51.1″E)	φ8 × 25, φ3.6 × 5, φ2 × 5	1323	**5**

(continued)

Table 6.1 (continued)

Company name	Location (latitude and longitude)	Tank diameter φ (m) and number on the map	Total tank area (m²)	Annual production (t-dry)*
Sea vegetable company	Rikuzentakata City, Iwate Pref. (38°56′29.2″N 141°42′19.7″E)	φ9 × 14, φ5 × 13	1145	4.3
Ie Fisheries Cooperative Association	Ie town, Okinawa Pref. (26°42′33.0″N 127°47′40.2″E)			**10**
			Total	99.8

[a]Estimated values from the map area based on the Rikuzentakata-based production efficiency and the production volume announced by the company are in bold

the most expensive seaweed. Due to ocean warming in recent years, the production of many seaweeds has decreased, causing prices to rise. In this context, it is likely that seaweeds other than *U. prolifera* will be cultivated on land in the near future.

Acknowledgements This research was supported by JST-OPERA Program (Grant Number JPMJOP1832), JST Grant Number JPMJPF2405, and a Grant-in-Aid for Core Research Project from Kochi University, Japan.

References

Bliding C (1963) A critical survey of European taxa in Ulvales. Part I. *Capsosiphon, Percursaria, Blidingia, Enteromorpha*. Opera Botan 8:1–160

Buck BH, Shpigel M (2023) *ULVA*: tomorrow's "wheat of the sea", a model for an innovative mariculture. J Appl Phycol 35:1967–1970

Craigie J, Cornish M, Deveau L (2019) Commercialization of Irish moss aquaculture: the Canadian experience. Bot Mar 62:411–432

Cui J, Monotilla AP, Zhu W, Takano Y, Shimada S, Ichihara K, Matsui T, He P, Hiraoka M (2018) Taxonomic reassessment of *Ulva prolifera* (Ulvophyceae, Chlorophyta) based on specimens from the type locality and Yellow Sea green tides. Phycologia 57:692–704

da Silva FZ, Baran TBM, Hayashi L (2024) Protocol for scaling up biomass production of the green seaweed *Ulva ohnoi* from germling clusters. J Appl Phycol 36:579–589

Hiraoka M (2012) Intensive biomass production by green seaweed *Ulva*. J Jpn Inst Energy 91:1154–1160. (In Japanese)

Hiraoka M (2021) Massive *Ulva* green tides caused by inhibition of biomass allocation to sporulation. Plants 10:2482

Hiraoka M (2022) Advances in land-based production technology for seaweed in Kochi, Japan. Foods & Food Ingredients J Jpn 227:152–156. (In Japanese)

Hiraoka M, Higa M (2016) Novel distribution pattern between coexisting sexual and obligate asexual variants of the true estuarine macroalga *Ulva prolifera*. Ecol Evol 6:3658–3671

Hiraoka M, Oka N (2006) Mariculture of seaweeds based on the new "germling cluster method" and utilizing deep seawater in Japan. In: Siew-Moi P, Critchley AT, Ang PO Jr (eds) Advances in seaweed cultivation and utilization in Asia. University of Malaya Maritime Research Centre, Kuala Lumpur, pp 53–59

Hiraoka M, Oka N (2008) Tank cultivation of *Ulva prolifera* in deep seawater using a new "germling cluster" method. J Appl Phycol 20:97–102

Hiraoka M, Dan A, Shimada S, Hagihira M, Migita M, Ohno M (2003) Different life histories of *Enteromorpha prolifera* (Ulvales, Chlorophyta) from four rivers on Shikoku Island, Japan. Phycologia 42:275–284

Hiraoka M, Ohno M, Dan A, Oka N (2004) Utilization of deep seawater for the mariculture of seaweeds in Japan. Jpn J Phycol 52(supplement):215–219

Hiraoka M, Kinoshita Y, Higa M, Tsubaki S, Monotilla AP, Onda A, Dan A (2020a) Fourfold daily growth rate in multicellular marine alga *Ulva meridionalis*. Sci Rep 10:12606

Hiraoka M, Tanaka K, Yamasaki T, Miura O (2020b) Replacement of *Ulva ohnoi* in the type locality under rapid ocean warming in southwestern Japan. J Appl Phycol 32:2489–2494

Hofmann LC, Strauss S, Shpigel M, Guttman L, Stengel DB, Rebours C, Gjorgovska N, Turan G, Balina K, Zammit G, Adams JMM, Ahsan U, Bartolo AG, Bolton JJ, Domingues R, Dürrani Ö, Eroldogan OT, Freitas A, Golberg A, Kremer KI, Marques F, Milia M, Steinhagen S, Sucu E, Vargas-Murga L, Zemah-Shamir S, Zemah-Shamir Z, Meléndez-Martínez AJ (2024) The green seaweed *Ulva*: tomorrow's "wheat of the sea" in foods, feeds, nutrition, and biomaterials. Crit Rev Food Sci Nutr:1–36

Lawton RJ, Sutherland JE, Glasson CR, Magnusson ME (2021) Selection of temperate *Ulva* species and cultivars for land-based cultivation and biomass applications. Algal Res 56:102320

Liu F, Pang S, Zhao X (2013) Molecular phylogenetic analyses of *Ulva* (Chlorophyta, Ulvophyceae) mats in the Xiangshan Bay of China using high-resolution DNA markers. J Appl Phycol 25:1287–1295

Matsumoto M (2016) What is the breakthrough technology for green oil production by microalgae toward to commercialization? Kagakutoseibutsu 54:181–190. (In Japanese)

Moreira A, Cruz S, Marques R, Cartaxana P (2022) The underexplored potential of green macroalgae in aquaculture. Rev Aquac 14:5–26

Nakamura River and National Highway Office, Shikoku Regional Development Bureau, Ministry of Land, Infrastructure, Transport and Tourism (2023) Relationship between *Ulva prolifera* harvest volume and water quality, etc. https://www.skr.mlit.go.jp/nakamura/river/shimantoshizensaisei-kentoukai/20231024_siryou3.pdf (In Japanese); accessed 17 Jan 2025

Ohno M (1993) Cultivation of the green algae, *Monostroma* and *Enteromorpha* "Aonori". In: Ohno M, Critchley AT (eds) Seaweed cultivation and marine ranching. JICA, Kanagawa, Yokosuka, pp 7–15

Oka N, Hiraoka M, Nishijima T (2004) Growth of abalone fed on *Undaria pinnatifida* (Alariaceae, Phaeophyceae) cultivated as free living form "germling cluster" in deep seawater. Jpn J Phycol 52(supplement):225–230

Oka N, Sumida T, Hiraoka M, Ohno M (2008) Effects of Fe (II) for tank cultivation of *Ulva prolifera* in deep seawater. Algal Resour 1:63–66

Park CS, Hwang EK (2011) An investigation of the relationship between sediment particles size and the development of green algal mats (*Ulva prolifera*) on the intertidal flats of Muan, Korea. J Appl Phycol 23:515–522

Sato Y, Kinoshita Y, Mogamiya M, Inomata E, Hoshino M, Hiraoka M (2021) Different growth and sporulation responses to temperature gradient among obligate apomictic strains of *Ulva prolifera*. Plants 10:2256

Sato Y, Numata Y, Kinoshita Y, Shinotsuka M, Ono K, Kawano S, Hiraoka M (2023) Beyond the valley of death for land-based aquaculture of seaweeds. Cytologia 88:1–2

Schramm W (1991) Cultivation of unattached seaweeds. In: Guiry MD, Blunden G (eds) Seaweed resources in Europe: uses and potential. John Wiley & Sons, Chichester, pp 379–408

Tanaka K, Taino S, Haraguchi H, Prendergast G, Hiraoka M (2012) Warming off southwestern Japan linked to distributional shifts of subtidal canopy-forming seaweeds. Ecol Evol 2:2854–2865

Tsubaki S, Zhu W, Hiraoka M (2017) Production and conversion of green macroalgae (*Ulva* spp.). In: Kerton FM, Yan N (eds) Fuels, chemicals and materials from the oceans and aquatic sources. John Wiley & Sons, Hoboken, New Jersey, pp 19–41

Wu L, Cai W, Zhang L, Nakamura H, Timmermann A, Joyce T, McPhaden MJ, Alexander M, Qiu B, Visbeck M, Chang P, Giese B (2012) Enhanced warming over the global subtropical western boundary currents. Nature Clim Change 2:161–166

Zertuche-González JA, Sandoval-Gil JM, Rangel-Mendoza LK, Gálvez-Palazuelos AI, Guzmán-Calderón JM, Yarish C (2021) Seasonal and interannual production of sea lettuce (*Ulva* sp.) in outdoor cultures based on commercial size ponds. J World Aquac Soc 52:1047–1058

Masanori Hiraoka, born in Osaka, Japan, in 1967. With the great support of Professor Emeritus Masao Ohno of Kochi University, he obtained his Ph.D. in 2003 from the United Graduate School of Agricultural Sciences, Ehime University, for a study on the reproductive maturation and taxonomy in *Ulva*. From 2001 to 2004, he worked as a NEDO (New Energy and Industrial Technology Development Organization) Fellow Researcher at the Kochi Prefectural Deep Seawater Laboratory, where he researched tank cultivation of useful seaweed using deep seawater. The first land-based aquaculture facility, which introduced the floating high-density cultivation method called "germling cluster method" and multi-step tank system he developed, was built in Muroto City, Kochi Prefecture, in 2004. This Kochi Method has spread throughout Japan, and 100 tons of dried *Ulva* products are expected to be produced in 2025. He became an Associate Professor of the Usa Marine Biological Institute, Kochi University, in 2004, and in 2021 became a professor of the same university. His interest has been in the growth and reproduction of seaweeds, especially *Ulva* and *Monostroma*, as well as their ecology, evolution, and speciation. Recently, he has also became interested in marine bacteria that coexist with seaweeds. He is utilizing his knowledge of seaweed growth and reproduction and applying it to aquaculture research. Understanding seaweeds means being able to draw out their potential growth ability and grow seaweed at will. In October 2024, the Shimanto Seaweed Eco-Innovation Co-Creation Platform (JST Grant Number JPMJPF2405), a national project to promote regional industrial development through seaweed production research, was launched, and he will serve as the project leader until March 2026.

Dr. Yoichi Sato has been the director of Riken Food Co., Ltd., Miyagi, Japan, since 2024, and the manager of the research institute Algal Research and Innovation Centre (ARIC) since 2021. He has worked at Riken Food since 2005. Before 2005, he was an officer in the Ministry of Agriculture, Forestry, and Fisheries (2001–2002) and an Assistant Professor at the Graduate School of Agricultural Science Faculty of Agriculture, Tohoku University (2002–2004). His main interests are the physiology and ecology of macroalgae, predominantly brown seaweeds (Wakame, Kombu, and Mozuku). His efforts have paid off for elucidating the plasticity and adaptability of seaweeds to environmental changes and for developing and improving cultivation technology. Additional fields of interest are developing the optimum processing methods of seaweeds suitable for markets and customer demands. He obtained a Ph.D. (Life Science) at the University of Tokyo in 2016.

7 Microalgal Cultivation in Temperate Regions: State of the Art, Challenges, and Future Perspectives

Léa Braud, Anne Pajot, Arianna Rizzo, Juliette Armeni-Ripari, and Vítor Verdelho

Abstract

Microalgae are a diverse group of microorganisms that grow through autotrophic, mixotrophic, and heterotrophic modes, with light availability playing a crucial role in metabolic processes. Cultivated in open ponds, closed photobioreactors, and fermenters, microalgae have promising applications in food, feed, pharmaceuticals, cosmetics, and biofertilizers, as well as potential roles in bioremediation. However, high production costs remain a major barrier to large-scale commercialization. In temperate regions, seasonal variations in temperature and light availability introduce additional challenges, necessitating adapted cultivation strategies compared to the more stable conditions found in tropical regions. This chapter examines the challenges of large-scale microalgal cultivation in temperate climates, focusing on advancements in cultivation techniques, product development, and strategies to enhance environmental, economic, and social sustainability, thereby supporting the expansion of the blue bioeconomy.

Keywords

Microalgae · Open ponds · Photobioreactors · Fermenters · Biomolecules · Sustainability · Techno-economic assessment · Life cycle assessment · Artificial intelligence · Bioeconomy

L. Braud (✉) · A. Pajot · J. Armeni-Ripari · V. Verdelho
European Algae Biomass Association, Florence, Italy
e-mail: lea.braud@eaba-association.org

A. Rizzo
European Algae Biomass Association, Florence, Italy

Blue2, Genopole, Campus 3, Évry-Courcouronnes, France

Abbreviations

AI	Artificial intelligence
C	Carbon
CAPEX	Capital expenses
DHM	Digital holographic microscopy
DW	Dry weight
EFSA	European Food Safety Authority
EU	European Union
FDA	Food and Drug Administration
FU	Functional unit
ILCD	International Reference Life Cycle Data
IoT	Internet of Things
ISO	International Standard Organization
LCA	Life cycle assessment
LCC	Life cycle costing
LCI	Life cycle inventory
LCIA	Life cycle impact assessment
ML	Machine learning
N	Nitrogen
NPV	Net present value
OPEX	Operating expenses
ORP	Open raceway pond
P	Phosphorus
PBR	Photobioreactor
PHA	Polyhydroxyalkanoates
PUFAs	Polyunsaturated fatty acids
PVC	Polyvinyl chloride
S-LCA	Social life cycle assessment
TEA	Techno-economic assessment
TLCC	Total life cycle cost
TRL	Technology readiness level
UV	Ultraviolet
WWT	Wastewater treatment

A. T. Critchley et al. (eds.), *Advances in Temperate Phyconomy: Algal Harvest and Cultivation in Globally Distributed Temperate Waters*, Developments in Applied Phycology 15, https://doi.org/10.1007/978-3-032-02955-3_7

7.1 Introduction

Microalgae are a diverse group of microorganisms found in marine, freshwater, and terrestrial environments (Raven and Giordano 2014), with significant potential as a sustainable feedstock for the bioeconomy due to their high growth rates, ability to accumulate valuable biomolecules, and minimal reliance on arable land (Acién Fernández et al. 2021; Barbosa et al. 2023). However, commercialization is hindered by high cultivation and biomass processing costs (Araújo et al. 2021; Loke Show 2022). Microalgae can grow through different trophic modes, with light availability playing a crucial role in their metabolism. Cultivation systems range from open ponds, which are cost-effective for large-scale production, to closed photobioreactors (PBRs), which provide greater environmental control for the production of high-value biomolecules but incur higher energy and material costs.

Temperate regions, located between 23.5° and 66.5° North/South of the Equator, are characterized by moderate temperatures, distinct seasons, and fluctuating precipitation patterns, all of which impact microalgal growth dynamics. The concept of "microalgal phyconomy" integrates the scientific study of algae (phycology) with agricultural principles (agronomy) to optimize cultivation, enhance biomass production, and maximize product valorization across multiple industries. Compared to tropical regions, temperate climates present unique challenges, particularly the seasonal variability in temperature and light availability. To enhance productivity and cost-efficiency, cultivation strategies must be tailored to local climatic conditions, leveraging technological innovation to scale-up production while aligning with sustainability goals.

Techno-economic assessment (TEA) and life cycle assessment (LCA) are critical tools for evaluating the sustainability of microalgae-based production systems, offering valuable insights to guide research and innovation. Current studies highlight energy consumption and mineral fertilizer use as primary contributors to the environmental impacts of microalgal cultivation. Although PBRs provide greater control over growth conditions, their significantly higher economic costs compared to open raceway ponds (ORPs) present a major challenge. Moreover, the social implications of microalgal production remain largely underexplored. Achieving economic viability while minimizing environmental and social impacts requires a holistic approach that integrates all three pillars of sustainability.

This chapter explores the scientific and practical challenges of large-scale microalgal cultivation and biomass production in temperate regions, with a particular focus on the effects of seasonal variations on cultivation strategies. It explores key factors influencing microalgal cultivation in both open and closed systems (see Sect. 7.2). The chapter then provides an overview of existing microalgae-based products (see Sect. 7.3) before assessing the environmental, economic, and social sustainability of different cultivation and biomass production systems (see Sect. 7.4). Finally, future prospects are discussed, emphasizing the role of artificial intelligence (AI) in optimizing microalgal cultivation to improve economic feasibility while reducing environmental and social impacts (see Sect. 7.5).

7.2 Current Practices and Limitations in Temperate Microalgal Cultivation

Microalgal cultivation faces significant challenges due to high production costs and technological limitations, particularly in upstream processes. Optimizing nutrient use, improving energy efficiency, and implementing resource recirculation strategies are essential to enhancing both economic and environmental sustainability. This section explores the three primary trophic modes employed in ORPs, PBRs, and fermenters, with a focus on the key challenges and recent innovations necessary for optimal cultivation in temperate climates, where seasonal variations significantly influence microalgal productivity and economic viability.

7.2.1 Microalgal Growth Requirements and Trophic Modes

Microalgae require a balanced supply of macro- and micronutrients to sustain growth and metabolic activity. Key macronutrients—nitrogen (N), phosphorus (P), and carbon (C)—are essential for the biosynthesis of proteins, lipids, and carbohydrates, while micronutrients such as iron (Fe), magnesium (Mg), zinc (Zn), and manganese (Mn) play critical roles in enzymatic functions and photosynthesis (Quigg 2016). Beyond these fundamental requirements, different species may necessitate additional nutrients tailored to their specific physiological needs. For instance, marine microalgae require salts sourced from natural or artificial seawater, while diatoms depend on silica for the formation of their frustule (Rizzo et al. 2024). Depending on their metabolic flexibility, microalgae can be cultivated under three distinct trophic modes—photoautotrophic, mixotrophic, and heterotrophic—each suited to specific cultivation systems and offering unique advantages and limitations (see Table 7.1). In addition to conventional culture media, wastewater has been investigated as an alternative source of nutrients, reducing the costs associated with external input of nutrients and enhancing the sustainability of large-scale microalgal biomass production.

Photoautotrophic cultivation relies exclusively on light as the energy source, utilizing carbon dioxide (CO_2) or bicar-

Table 7.1 Overview of the different microalgal trophic modes and associated cultivation systems

Cultivation system	Trophic mode	Advantages	Disadvantages
Open system	Photo-autotrophy	Cost-effective and scalable; utilizes sunlight as an energy source; contributes to CO_2 sequestration; suitable for WWT due to inorganic nutrient availability	Limited by light penetration and self-shading; seasonal and climatic dependence; requires large surface area; prone to contamination and evaporation losses
Open system	Mixotrophy	Higher biomass productivity than strict photo-autotrophy; can utilize both inorganic and organic carbon from wastewater; improved tolerance to light fluctuations	Risk of bacterial contamination; requires a controlled balance between organic and inorganic carbon sources; competition between microalgae and heterotrophic bacteria
Closed system (PBR)	Photo-autotrophy	Higher biomass concentration than open ponds; better control over environmental parameters; reduced contamination risk; efficient CO_2 utilization	High capital and operational costs; limited scalability; requires artificial lighting or optimized light distribution in dense cultures
Closed system (fermenters)	Heterotrophy	High biomass density and yield; independent of light conditions; ideal for large-scale production of bioactive compounds and proteins	Requires organic carbon input, increasing production costs; risk of contamination; energy-intensive oxygenation needs
Closed system (PBRs/ fermenters)	Mixotrophy	Maximizes biomass yield and lipid accumulation; flexible adaptation to varying environmental conditions; efficient nutrient recovery from wastewater	Complex regulation of carbon and energy sources; increased risk of contamination; higher operational complexity

bonate (HCO_2^-) as the carbon source. The demand for inorganic carbon is substantial, with up to 1.8 kg of CO_2 required to generate 1 kg of microalgal biomass (Cardias et al. 2023). In PBRs, compressed CO_2 is commonly used, while in ORPs, sodium bicarbonate ($NaHCO_2$) offers a cost-effective alternative. While bicarbonate supplementation increases the medium's pH, favoring alkaliphilic species such as "Spirulina", it requires pH adjustments for freshwater microalgae through CO_2 injection or buffering agents. Photoautotrophic cultivation is mainly employed in ORPs and illuminated PBRs, supporting large-scale production of oxygenic phototrophic microalgae such as *Limnospira/Arthrospira* sp. (Cyanobacteriophyta), *Chlorella* sp., and *Dunaliella* sp. (Chlorophyta). Nevertheless, light penetration constraints and self-shading effects can limit productivity, particularly in high-density cultures, while energy costs associated with artificial illumination remain a challenge (Chisti 2013).

Heterotrophic cultivation relies entirely on organic carbon sources, such as glucose and glycerol, in the absence of light. This approach enables high-density microalgal cultures and cost-effective biomass production in industrial fermenters, making it particularly suitable for large-scale applications in high-volume markets, such as protein production. In contrast, mixotrophic cultivation integrates light-driven photosynthesis with organic carbon supplementation, allowing microalgae to alternate between autotrophic and heterotrophic metabolism. This metabolic flexibility can significantly enhance biomass productivity—by up to 50 times—and promote lipid accumulation, particularly under suboptimal lighting conditions (Perez-Garcia et al. 2011; Abreu et al. 2022). However, maintaining a precise nutrient balance is crucial, as excessive organic carbon can increase bacterial contamination, potentially compromising microalgal culture stability and yield.

The need to reduce resource inputs and production costs in microalgal cultivation has driven significant interest in using wastewater as an alternative to artificial culture media. Due to their metabolic flexibility, microalgae have been extensively studied for their bioremediation potential, i.e., their ability to degrade, assimilate, and neutralize pollutants, particularly in the case of wastewater treatment (WWT). Their capacity to absorb organic contaminants and inorganic pollutants, such as phosphates and nitrates, plays a key role in mitigating eutrophication in freshwater ecosystems. Beyond reducing reliance on external nutrient inputs and lowering cultivation costs, this strategy also minimizes the need for additional wastewater treatment before discharge. Wastewaters from municipal, agricultural, and industrial sources provide an abundant supply of N, P, and organic C, making them highly suitable for microalgal growth. Photoautotrophic systems can utilize these inorganic nutrients alongside CO_2 from industrial emissions (e.g., flue gas) or microbial respiration, while heterotrophic and mixotrophic cultivation strategies efficiently process wastewater enriched with dissolved organic matter.

7.2.2 Microalgal Cultivation in Temperate Regions

The selection of a microalgal cultivation system is primarily influenced by climatic conditions and secondarily by the specific requirements of the microalgal strain. Open pond cultivation remains the predominant method for large-scale microalgal production, particularly due to the extensive cultivation of Spirulina (classified as *Limnospira/Arthrospira* sp., while Spirulina remains the collective name) in these systems. However, open ponds are most suitable for species that thrive in extreme environments—such as Spirulina (high alkalinity), *Chlorella* sp. (high nutrient concentrations), and *Dunaliella* sp. (high salinity)—as these conditions limit contamination risks and provide a competitive advantage.

Despite their commercial feasibility, open ponds face significant challenges in temperate climates, where seasonal variations in temperature and irradiance impact microalgal productivity. To mitigate photo-limitation and photoinhibition, external interventions such as temperature control and artificial lighting are often required. For example, large-scale open pond facilities, such as "Necton" in southern Portugal (200 m^3) and the University of Almería in Spain (100 m^3), benefit from relatively stable climatic conditions, enabling up to 300 production days per year. However, even in these regions, biomass productivity fluctuates between summer (34 °C) and winter (10 °C) (Rodríguez-Miranda et al. 2021). Similar seasonal limitations affect open pond systems in Brittany, France, where raceway ponds enclosed within greenhouses allow production from May to October but remain inactive in winter.

While open ponds are generally more viable in warm climates, successful implementations in cooler regions demonstrate that climate is not necessarily a limiting factor. The cultivation of Spirulina in Latvia (e.g., "SpirulinaNord") and the production of *Chlorella* sp. in Roupite, Bulgaria, have been achieved using greenhouse enclosures to compensate for lower temperatures (Stunda-Zujeva et al. 2018). Similar approaches have been observed in non-European countries within the temperate zone, considered approximately 23.5° to 66.5° North/South of the Equator, as detailed below.

In China, regional variations in Spirulina production highlight the influence of climate on cultivation strategies. In northern and central provinces (e.g., Inner Mongolia, Heilongjiang), greenhouse-covered raceway ponds are used to counteract colder temperatures, whereas southern provinces (e.g., Fujian, Yunnan, Guangdong, Hainan) can sustain open-air raceway cultivation year-round. Notably, "Inner Mongolia Rejuve Biotech Co., Ltd." is a major production hub, accounting for approximately 1300 tons of dry Spirulina biomass annually, which represents around 40% of global production.

In the United States (US), large-scale open pond facilities such as "Earthrise" (California) and "Cyanotech" (Hawaii) focus on Spirulina cultivation. Hawaii's tropical climate allows for continuous year-round production, whereas temperate regions encounter significant seasonal fluctuations. Modeling studies reveal that locations such as Phoenix, Arizona, experience extreme summer temperatures exceeding 40 °C, necessitating heat-tolerant strains. Conversely, in Lexington, Kentucky, and Colorado Springs, winter temperatures substantially reduce biomass productivity. Southern coastal regions such as Florida and Hawaii achieve higher annual biomass yields (~23 g m^2 day^{-1}), while more temperate locations, such as Nevada, display strong seasonal variability, with summer productivity levels doubling those of winter. These disparities underscore the importance of strain optimization and temperature control strategies to enhance biomass yields in temperate climates.

To support these efforts, a virtual laboratory developed by the University of Almería (Dormido et al. 2014) provides researchers with tools to assess the impact of environmental variables on microalgal growth, optimizing ORP performance (see Fig. 7.1). A related simulation tool predicts biomass productivity across various global locations, highlighting the significant influence of latitude on open pond yields (González-Hernández et al. 2024). In temperate regions between approximately 35° and 66.5° latitude, productivity can decline below 200 kg dry weight (DW) in winter while exceeding 400 kg DW in summer for *Scenedesmus/Tetradesmus* sp., with similar seasonal trends observed for *Chlorella* sp. (Chlorophyta).

7.2.3 Overview of Open Microalgal Cultivation Systems

Open cultivation systems—including raceway ponds (circular and rectangular), thin-layer cascades, and, to a lesser extent, multilayer bioreactors (Liu et al. 2013)—have been widely utilized since the 1950s, particularly for the cultivation of Spirulina and *Chlorella* sp. (see Fig. 7.2). Raceway ponds, one of the most common open system designs, can be operated outdoors or within greenhouses, with the latter being a popular practice in Southern Europe to counteract seasonal temperature fluctuations. While greenhouse enclosures increase initial capital costs, they offer better control over cultivation conditions and extend production periods, improving overall microalgal biomass production efficiency.

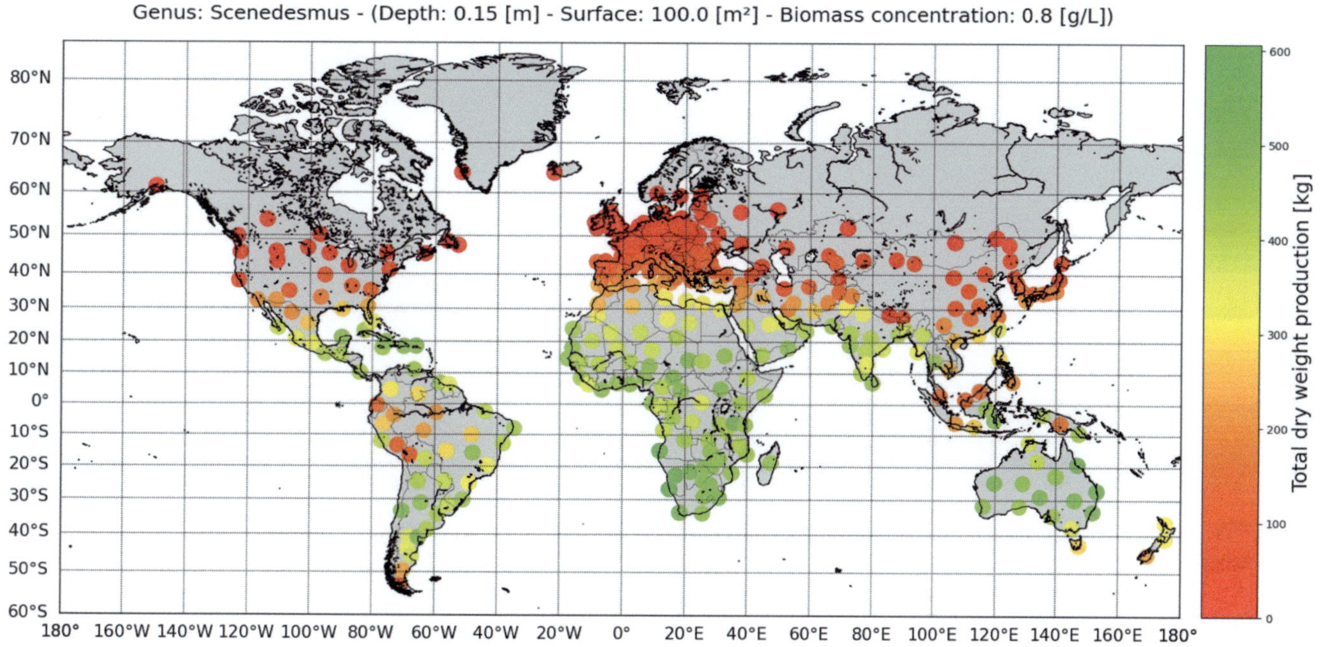

(a) September to march

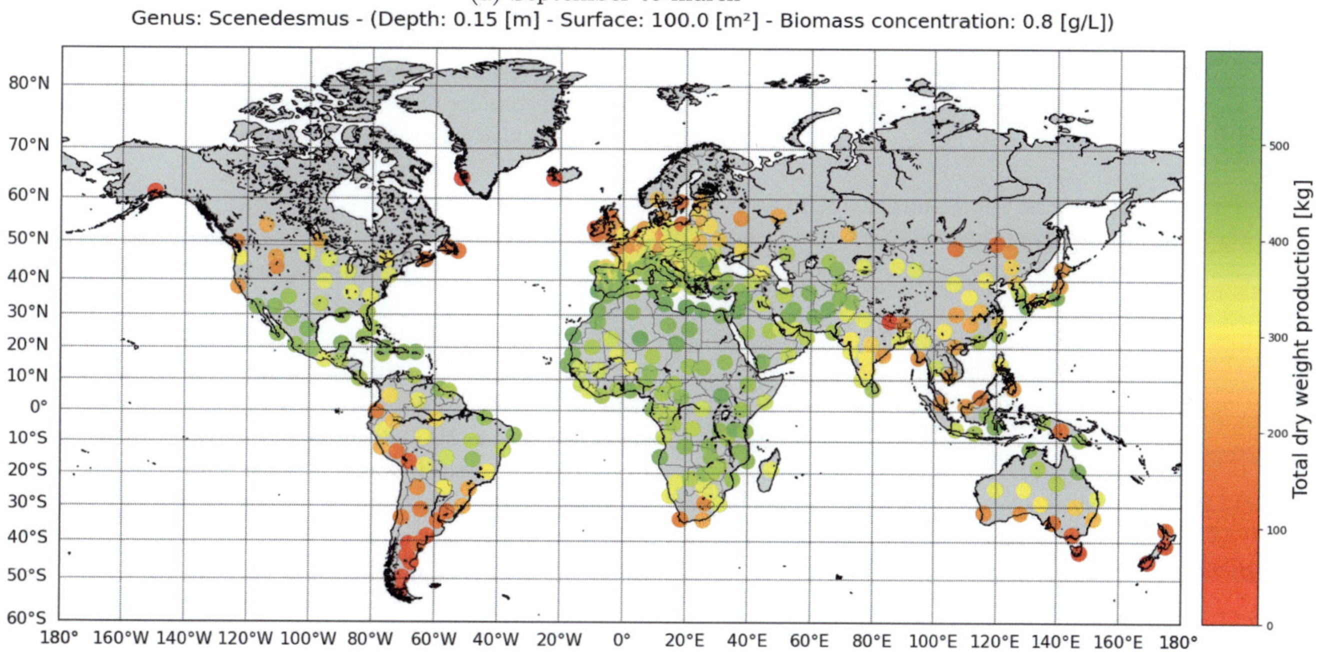

(b) March to september

Fig. 7.1 Global distribution of the microalga *Scenedesmus/Tetradesmus* sp. (Chlorophyta) in open ponds. (Reproduced from González-Hernández et al. (2024))

Raceway ponds remain the dominant choice for commercial-scale production due to their relatively low energy consumption. However, innovative designs such as thin-layer cascades and multilayer bioreactors have been developed to enhance mass transfer, mixing efficiency, and light utilization, ultimately leading to higher biomass productivity. These advancements contribute to optimizing large-scale microalgal cultivation while addressing some of the inherent limitations of traditional open pond systems. For a more comprehensive overview of open pond designs and their applications, several reviews provide extensive insights (Assunção and Malcata 2020; Chanquia et al. 2022; Sarker and Kaparaju 2023).

The growth efficiency of microalgae cultivated in open ponds depends on key factors such as light availability, mix-

Fig. 7.2 Overview of different open microalgal cultivation systems, including an open raceway pond from "Necton" (Portugal) (**a**) and a cascade open pond from "Allmicroalgae" (Portugal) (**b**). (Images: "Necton" and "Allmicroalgae")

Table 7.2 Key factors influencing the performance of open pond systems for microalgal cultivation

Factors	Main considerations
Pond depth	Optimal depth ranges from 15–30 cm in raceways and < 2 cm in thin-layer reactors (Hallmann 2015). Shallower depths improve light penetration, while deeper ponds stabilize temperature and support higher biomass concentrations
Mixing	Essential for preventing sedimentation, reducing light/dark cycles, and enhancing gas exchange. Paddle wheels, aeration, and baffles improve mixing efficiency
Gas transfer	CO_2 supplementation is crucial, as atmospheric levels are insufficient for optimal microalgal growth. Carbonation pumps and micropore diffusers enhance uptake efficiency (82–96%)
Light availability	Influenced by pond depth, microalgal concentration, and reactor orientation. Computational studies show that optimizing light/dark cycles and baffle placement avoids photo-limitation/photoinhibition
Temperature	A critical factor for productivity. Open pond temperatures fluctuate with ambient conditions, making temperate regions more susceptible to seasonal variability. Species selection and thermal regulation strategies (e.g., shadowing) are required to maintain stable microalgal growth
Microalgal species	Only microalgae with extremophile traits or high adaptability to fluctuating environmental conditions can be cultivated in open ponds. Regulatory restrictions prohibit the use of genetically modified species in outdoor open ponds

ing, gas transfer, temperature regulation, and microalgal species (see Table 7.2).

While open pond systems remain the most widely adopted method for large-scale microalgal cultivation due to their cost-effectiveness and scalability, they are inherently constrained by environmental variability. To address these limitations, technological advancements such as controlled-depth raceways, greenhouse enclosures, and microalgal strain optimization are being explored to enhance productivity, particularly in temperate regions. In addition, less common open systems, including natural lakes and offshore floating PBRs, offer alternative cultivation approaches.

Natural Lakes

Natural lakes have long served as sites for microalgal growth and harvesting, with some being adapted for commercial-scale production. In Lake Texcoco, Mexico, the spontaneous proliferation of Spirulina led to the establishment of large-scale industrial cultivation by Sosa Texcoco S.A., which operated one of the world's largest Spirulina production systems, utilizing an 800-hectare solar evaporator to generate up to five tons of dried biomass per day. Similarly, Lake Chad, Africa, has sustained traditional Spirulina harvesting by the Kanembu tribe for centuries, where biomass is processed into *dihé*, a nutrient-rich dried cake, with annual production reaching approximately 250 tons (Habib et al. 2008; Carcea et al. 2015; Batello et al. 2004). Other notable examples include Tuyn Taung Lake, Myanmar, where seasonal Spirulina harvesting yields around 150 tons per year, and the hypersaline lakes of South Australia, where *Dunaliella salina* is cultivated for high-value beta-carotene production.

Offshore Floating Photobioreactors

In an effort to overcome land constraints associated with conventional microalgal cultivation, the Offshore Membrane Enclosures for Growing Algae (OMEGA) project, led by the National Aeronautics and Space Administration (NASA) (2010–2013), explored floating PBRs as an alternative cultivation strategy. These systems utilize natural resources such as sunlight, seawater, and wave energy for mixing while benefiting from the surrounding water's temperature-regulating effect, thereby reducing thermal fluctuations.

However, despite these advantages, floating PBRs face significant challenges that have hindered widespread commercialization. The harsh marine environment exposes these systems to intense hydrodynamic forces, including storms, strong currents, and wave-induced stresses, which can compromise structural integrity. In addition, biofouling—the accumulation of microorganisms and debris on PBR surfaces—reduces light penetration and negatively impacts productivity. Economic viability remains a concern due to high installation, maintenance, and energy costs. These limitations, along with potential mitigation strategies, are extensively discussed in a review by Khor et al. (2022).

Open Ponds in Greenhouses: A Semi-Outdoor Approach
Greenhouse-enclosed open ponds, also known as semi-outdoor cultivation systems, have been increasingly adopted to mitigate some of the limitations associated with traditional open ponds (Roselet et al. 2013). By enclosing raceway ponds within glass or plastic-covered greenhouses, these systems enhance environmental control, providing greater temperature stability, protection from contaminants, and reduced water evaporation (Kurpan et al. 2024). This approach has been particularly effective in France, the leading Spirulina producer in Europe, where 94 production sites have been recorded (FSF 2025).

Despite these benefits, greenhouse enclosures do not fully eliminate certain challenges inherent to open pond systems. Limited abiotic control remains an issue, as factors such as pH and light intensity continue to fluctuate. In addition, contamination risks persist, with airborne microorganisms and competing microalgal species potentially infiltrating the system. While the enclosed structure reduces the variability in productivity due to seasonal variations, production levels still fluctuate in response to external climatic conditions. Water evaporation, although reduced, requires ongoing management strategies to maintain optimal conditions. Furthermore, self-shading effects can arise as biomass density increases, limiting light penetration and ultimately reducing cultivation efficiency.

These challenges are particularly pronounced in temperate regions and when cultivating contamination-sensitive species such as *Haematococcus lacustris* (formerly *H. pluvialis*) (Chlorophyta), which demands stringent growth conditions. To address these limitations, hybrid cultivation systems integrating ORPs with closed PBRs have been developed, offering a more controlled yet scalable approach (Olaizola 2000; Oslan et al. 2022; An et al. 2024).

Hybrid Raceway–Photobioreactor Systems
Hybrid cultivation systems combine the advantages of open ponds and closed PBRs, optimizing productivity while minimizing the drawbacks of each method. This approach is particularly beneficial for microalgal species such as *H. lacustris*, which requires controlled conditions during early growth stages before being transferred to ORPs for biomass accumulation and astaxanthin production (Adesanya et al. 2014; Skifa et al. 2024). Hybrid ORP–PBR systems offer several key advantages, including lower production costs compared to fully enclosed PBRs, a reduced risk of contamination during early-stage cultivation, and higher overall productivity due to the combination of controlled PBR conditions and the extensive surface area available in raceways. In addition, these systems provide scalability, allowing for gradual expansion of biomass production.

Despite these benefits, hybrid systems also present challenges, such as higher initial investment costs and the need for precise operational control to ensure efficiency. Nevertheless, they continue to be widely employed for astaxanthin production from *H. lacustris*, serving as an effective middle ground between cost-efficient open ponds and high-performance closed PBRs.

7.2.4 Overview of Closed Cultivation System: Photobioreactors

Unlike open ponds, closed PBRs provide a highly controlled cultivation environment, significantly reducing contamination risks while allowing precise regulation of light intensity, gas exchange, and nutrient supply. Various PBR designs have been developed, including tubular, flat-panel, airlift, and membrane-based systems, each offering specific advantages in terms of scalability, energy efficiency, and product yield (see Fig. 7.3).

At a commercial scale, horizontal tubular systems and plastic bag-based PBRs are the most widely used, whereas other designs are typically restricted to pilot-scale applications. Although the initial capital investment and operational costs of PBRs are considerably higher than those of ORPs—up to three times more expensive (Richardson et al. 2012)—these systems enable year-round production, maximize biomass density, and support applications in biofuels, pharmaceuticals, and high-value biochemicals. Their ability to maintain sterile conditions and optimize microalgal growth parameters makes them a promising alternative to open pond cultivation, particularly in regions where environmental variability poses significant challenges to outdoor production.

The growth efficiency of microalgae cultivated in PBRs depends on key factors such as the configuration of the cultivation system, mixing, gas transfer and CO_2 delivery, light availability, temperature regulation, and microalgal species (see Table 7.3).

One of the primary financial challenges associated with PBRs is infrastructure and scalability, as large-scale produc-

Fig. 7.3 Overview of different closed microalgal cultivation systems, including tubular photobioreactors from "Allmicroalgae" (Portugal) (**a**) and "Necton" (Portugal) (**b**), as well as flat-panel photobioreactors from "Necton" (Portugal) (**c** and **d**) and fermenters from "Biorea" (France) (**e**). (Images: "Allmicroalgae", "Necton", and "Biorea")

tion requires substantial capital investment and advanced processing technology. Despite ongoing advancements in PBR designs aimed at optimizing space utilization and improving scalability, maintaining consistent product quality and achieving high biomass yields remains a challenge at the commercial level. The high cost contribution of PBRs to overall production expenses has driven significant research into cost reduction strategies.

In tubular PBRs—the most commonly analyzed configuration in TEA studies—construction materials alone can account for 66–79% of total costs (Vázquez-Romero et al. 2022). However, these expenses decrease with economies of scale (Clippinger and Davis 2019). Material selection also plays a crucial role in cost reduction, with glass being the most expensive option, while alternatives such as Plexiglas, polyvinyl chloride (PVC), acrylic PVC, polyethylene, and plastic bag-based PBRs offer more cost-effective solutions.

To further address economic challenges, innovations in PBR design are enhancing both performance and financial feasibility. The integration of AI and machine learning (ML) enables real-time monitoring, automated control systems, and advanced sensor technologies, improving cultivation efficiency and reducing operating expenses (OPEX). In addition, climate adaptation strategies—such as cooling systems, heating elements, and greenhouse enclosures—are increasingly implemented to stabilize growth conditions and optimize production year-round. These advancements are

Table 7.3 Key factors influencing the performance of photobioreactors for microalgal cultivation

Factors	Main considerations
Configuration	The selection of a PBR configuration is determined by the cultivated microalgal species and the intended biomass application. Tubular PBRs are favored for the production of high-value biomolecules due to their superior gas exchange efficiency and light penetration, whereas plastic bag PBRs provide a cost-effective alternative for large-scale applications, such as biofertilizer production
Mixing	Mixing strategies include airlift circulation, mechanical stirring, and bubble column systems. Effective mixing prevents the formation of dead zones and biofilms while ensuring uniform nutrient and gas distribution. However, shear stress from excessive turbulence may negatively impact fragile microalgal species, leading to cell damage or reduced viability (Chanquia et al. 2022)
Gas transfer and CO_2 delivery	Efficient gas–liquid mass transfer can be achieved through the use of spargers, microporous membranes, and external gas recycling loops, which enhance CO_2 availability and minimize gas losses
Light availability	Light distribution is more homogeneous in flat-panel and inclined PBRs, whereas tubular PBRs are prone to self-shading effects. The integration of internal light sources or light-guiding systems can improve illumination efficiency and biomass productivity
Temperature	To maintain optimal microalgal growth conditions, PBRs can incorporate active temperature regulation strategies, including cooling and heating systems, thermal insulation, and greenhouse enclosures
Microalgal species	PBRs are suitable for cultivating a wide range of commercially relevant microalgal species, including genetically engineered strains. They are particularly advantageous for the production of polyunsaturated fatty acids (PUFAs) and other high-value biomolecules

essential for maximizing the commercial viability of PBR-based microalgal cultivation, particularly in temperate regions where environmental variability remains a significant constraint.

7.2.5 Overview of Closed Microalgal Cultivation Systems: Fermenters

Not all microalgal species exhibit facultative heterotrophy, meaning only certain strains can transition between autotrophic and heterotrophic trophic modes. The choice of cultivation strategy is influenced by multiple factors, including nutrient availability and associated costs, energy requirements for artificial lighting, and the intended application of the microalgae-based product. In addition, CO_2 emissions must be considered in heterotrophic systems, as fermentation-based cultivation does not contribute to atmospheric CO_2 absorption, unlike photoautotrophic systems, which actively sequester CO_2 during growth.

Over the past two decades, industrial-scale fermentation has emerged as an efficient alternative for microalgal cultivation, particularly for applications requiring high purity, biomass consistency, and controlled environmental conditions. The expansion of fermenter-based microalgal production has been driven by the increasing demand for sustainable protein sources, nutraceuticals, and functional bioactives. In the US, companies such as "TerraVia" (formerly "Solazyme") and "Martek" (now "DSM") focus on the heterotrophic production of omega-3 fatty acids and high-value lipids using large-scale fermentation technologies. In Asia, major producers such as "Yakult," "Nihon *Chlorella,*" "Kirin," and "Daesang" specialize in dietary supplements and functional food products using *Chlorella* sp. In Europe, firms including "Fermentalg" (France), "Lonza" (Switzerland), and "Allmicroalgae" (Portugal) employ dark fermentation either as a primary cultivation process or as part of an integrated production system, focusing on specialized microalgal biomolecules ranging from pharmaceutical precursors to cosmetic ingredients.

Fermentation-based microalgal cultivation offers several advantages over traditional PBR systems, including higher biomass concentrations, reduced contamination risks, and lower dependence on external climatic conditions. However, the high production costs associated with organic carbon substrates and bioreactor operation make other closed-system alternatives competitive, necessitating further advancements in cost-effective fermentation strategies.

7.3 Development of Microalgae-Based Products and Applications

The need to address the unprecedented challenges of the twenty-first century—such as food security and climate change—has driven significant advancements in microalgal biotechnology and led to the development of diverse microalgae-based products in the food, feed, pharmaceutical, cosmetic, biomaterial, biochemical, bioenergy, and wastewater treatment industries (Barbosa et al. 2023). Over the past decade, the scientific and industrial communities have expanded rapidly, facilitating technological advancements, scaling up production, and unlocking new opportunities for microalgal applications. This section provides an overview of microalgae-based products and applications and highlights past and ongoing European projects that have advanced the use of microalgae as a feedstock for biorefineries. Finally, it explores the journey from production to commercialization for three microalgal pigments: phycocyanin, astaxanthin, and fucoxanthin.

7.3.1 Overview of Microalgae-Based Products and Applications

Rich in proteins, lipids, pigments, and bioactive compounds, microalgae are widely used in health supplements (e.g., *Spirulina*, *Chlorella* sp.), natural food colorants (e.g., phycocyanin, astaxanthin), and functional ingredients for cosmetics. In the agricultural sector, microalgae are used as ingredients for biostimulants and biopesticides, while in the bioenergy and biomaterial sectors, applications include biofuels and bioplastics, supporting a more sustainable bioeconomy by providing alternatives to fossil-based products. Figure 7.4 presents the main sectors of activity for microalgae, along with examples of commercialized species and their main applications.

Food and Dietary Supplements

This category includes a large part of the microalgae-based products currently commercialized. *Chlorella vulgaris*, a nutrient-rich freshwater microalga, is widely used in dietary supplements aimed at detoxifying the human body and supporting the immune system. In particular, its wide range of vitamins, minerals, and antioxidants contributes to immune functions (Bito et al. 2020). The high content of chlorophyll in *Chlorella* sp. and its capacity to bind heavy metals from contaminated water—such as copper (Cu), cobalt (Co), cadmium (Cd), lead (Pb), and chromium (Cr)—make it valuable not only for detoxification but also for wastewater bioremediation (Joo et al. 2021). Among commercially available products, "SuperSmart's" supplement using *Chlorella* sp. from Luxembourg supports detoxification, liver function, and immune health. "Organika's Organic" *Chlorella* sp. Powder from Canada is recognized for its cleansing properties, facilitating heavy metal removal and supporting detoxification. The French companies "Vegalia," "nu3," and "Flamant vert," offer pure *Chlorella* sp. in powder and tablets, formulated to promote detoxification, enhance immune defences, and support digestion. Another widely commercialized microalga, *Dunaliella salina*, is known for its high beta-carotene content, a precursor of vitamin A, and is used both as a natural vitamin A supplement and a food colorant (Vol and Tran 2014). Products such as Betatene® by "BASF" (Germany) harness natural beta-carotene derived from *D. salina* to support eye and skin health. Various brands market

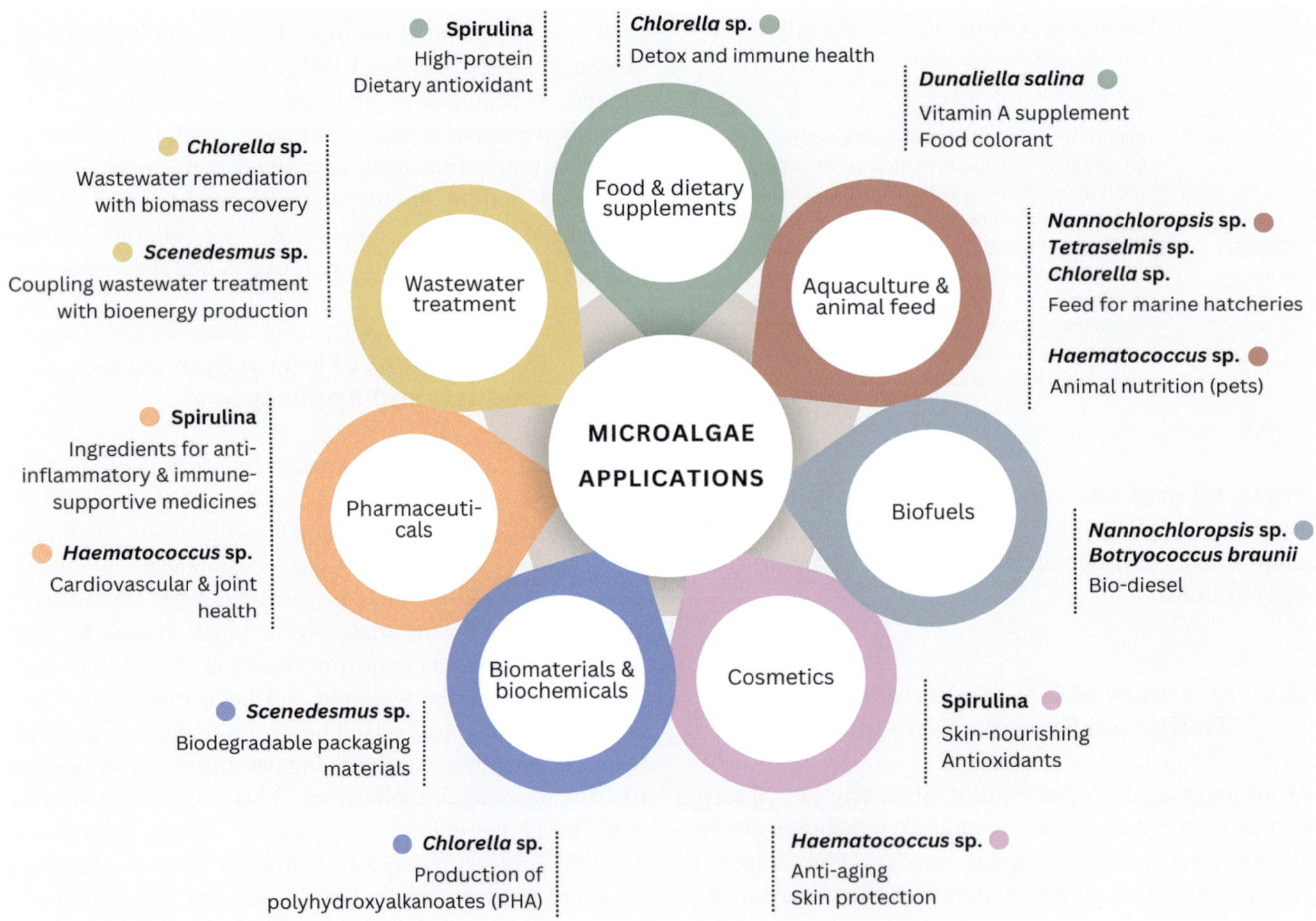

Fig. 7.4 Overview of key applications of microalgae-based products with examples of commercially available species

D. salina powder as a 100% pure, natural orange food color, emphasizing its high β- and α-carotene content.

Aquaculture and Animal Feed

The microalgal species *Nannochloropsis* sp. (Eustigmatophyceae) and *Tetraselmis* sp. (Chlorophyta) are commonly used in aquaculture hatcheries due to their high nutritional value, particularly for rotifer cultivation and as direct feed for various marine larvae (Paterson et al. 2023). Notable products incorporating these microalgae include Rotifer Diet™ HD from "Reed Mariculture" (US), which combines *Nannochloropsis* sp. and *Tetraselmis* sp., resulting in a 20% increase in rotifer production compared to pure *Nannochloropsis* sp. Nanno 3600™ and Tet 3600™ from "Planktovie" (France) are frozen whole-cell concentrates of *Nannochloropsis* sp. and *Tetraselmis* sp., used for rotifer production and enrichment, as well as for stimulating feeding in bivalves and larval shrimp.

Biofuels

It is important to note that commercial biofuel production from microalgae remains limited. Despite their high lipid content and potential as a sustainable source of biofuel, challenges such as high production costs, scalability constraints, and technological barriers have hindered widespread commercial adoption (Hoang et al. 2022). Currently, no microalgal-derived biofuels are available on a commercial scale. Ongoing research seeks to overcome these limitations, with species such as *Nannochloropsis* sp. and *Botryococcus braunii* (Chlorophyta) among the promising candidates (Moutel et al. 2015; Ma et al. 2016).

Cosmetics

Haematococcus lacustris is a highly valuable microalga rich in astaxanthin and widely used in the cosmetics industry, particularly in skincare products, for its antioxidant properties (Mularczyk et al. 2020). These properties help combat signs of aging and protect the skin. Notable examples include AstaPure® Skin from "AstaReal" (Sweden and Japan), an astaxanthin-rich ingredient stabilized with olive oil for use in skincare formulations. SUN'ALG® from "Gelyma" (France) combines *H. lacustris* extract with *Pongamia glabra* (vascular plant) seed oil, providing natural bioactive ingredients that serve as a primary shield against ultraviolet (UV) radiation.

Biomaterials and Biochemicals

The production of bioplastics from micro- and macroalgae has gained increasing attention as a sustainable alternative to fossil-based plastics (Onen Cinar et al. 2020). Research on *Scenedesmus/Tetradesmus* sp. has demonstrated its potential in bio-based composites, where microalgal-derived polymers replace conventional plastics (Garcia et al. 2021). *Chlorella vulgaris* has also been studied by companies such as "Newlight Technologies" (US) and "Bioplastics International" (US) for its ability to produce polyhydroxyalkanoates (PHA), a biodegradable plastic used in various consumer goods, including packaging. Spirulina, more specifically *Limnospira platensis* (formerly *Arthrospira platensis*) (Cyanobacteriophyta), is currently used in the development of biodegradable composites and films (Zhila et al. 2025), as well as *Dunaliella salina,* for which polysaccharides are used in the development of biodegradable materials such as films and coatings (Hadiyanto et al. 2022).

Pharmaceuticals

Microalgae have gained increasing attention in the pharmaceutical industry due to their rich nutritional profile and bioactive compounds. Spirulina is widely recognized for its high content of proteins, vitamins, and antioxidants and is commonly used in dietary supplements to support immune function, enhance energy levels, and facilitate detoxification. It has also been used to address conditions such as anemia and malnutrition. In pharmaceuticals, Spirulina extracts serve as key ingredients in anti-inflammatory and immune-supportive formulations (Wu et al. 2016). The company's "AstaReal" (Sweden and Japan) and "Nutrex" (Hawaii, US) produce astaxanthin from *Haematococcus lacustris,* which is incorporated into formulations aimed at supporting skin health and joint function due to its antioxidant and anti-aging properties (Ambati et al. 2014). *Dunaliella salina* is also valued for its strong antioxidant potential (Bezerra da Silva et al. 2021). Furthermore, certain microalgal-derived products are marketed for their potential anticancer benefits, often positioned as dietary supplements or adjunct therapies rather than primary treatments. These products emphasize antioxidant, immune-boosting, and anti-inflammatory properties, which may contribute to cancer prevention or complement conventional therapies. Notable products include astaxanthin from "AstaReal" (Sweden and Japan), Spirulina tablets from "Earthrise Nutritionals" (US), β-carotene 10% from "Algatech" (Israel) using *Dunaliella salina*, "BioAstin" from "Nutrex" (Hawaii, US) using *Haematococcus lacustris*, and "Almega PL" from "Qualitas Health" (Malaysia) using *Nannochloropsis* sp.

Wastewater Treatment

Microalgae-based wastewater treatment is an innovative and sustainable approach that leverages the natural ability of microalgae to remove pollutants from water, such as nitrogen, phosphorus, heavy metals, and organic compounds (Leong and Chang 2020). Through photosynthesis, microalgae absorb and metabolize these contaminants while simultaneously producing valuable biomass that can be used to produce biofuels, animal feed, and biofertilizers. In addition, microalgae enhance water quality by increasing oxygen levels and reducing excess nutrients, making them a valuable asset for treating industrial, agricultural, and municipal wastewater. While *Chlorella* sp. and *Scenedesmus* sp. are widely studied for wastewater treatment, their application remains largely in research and pilot phases (Gupta et al. 2016). However, some commercial entities are actively developing and integrating these species into water purification and bioremediation systems, particularly in niche markets such as aquaculture and industrial water treatment, though large-scale commercialization remains limited.

7.3.2 Microalgae as Feedstock for Biorefineries: Insights from European Research Projects

Several projects funded by the European Union (EU), in collaboration with the Bio-Based Industries Consortium through the Circular Bio-Based Europe Joint Undertaking, have demonstrated the potential of microalgae as a sustainable feedstock for biorefineries, highlighting their versatility in producing bio-based products for food, feed, cosmetics, agriculture, and energy. Initiatives such as "SABANA," "D-Factory," "MAGNIFICENT," and "ABACUS" have focused on optimizing large-scale microalgal cultivation, refining extraction techniques, and developing innovative biorefinery models. By showcasing the feasibility of microalgae-based value chains, these efforts contribute to Europe's transition toward a circular bioeconomy while reducing reliance on fossil-based resources. The following section presents projects running between 2013 and 2026.

REALM (Grant Agreement No. 101060991, 2022–2026)
This project, Reusing Effluents from Agriculture to unLock the potential of Microalgae, aims to enhance microalgal production through automation, integrating newly developed sensors and AI to reduce costs and increase productivity. It also focuses on developing novel biorefinery processes to produce biostimulants, biopesticides, and aquaculture feed from the microalgal biomass produced.

REDWine (Grant Agreement No. 101023567, 2021–2025)
This project explores the use of off-gases produced during wine fermentation (rich in CO_2) and liquid effluents from the winery for the cultivation of *Chlorella* sp. By integrating a simple microalgal biorefinery system within wineries, the project aims to produce ingredients with applications in the food, cosmetics, and agricultural sectors and demonstrate the technical, economic, and environmental feasibility of this approach.

SCALE (Grant Agreement No. 101023593, 2021–2025)
The SCALE project (Supplying bioactives Compounds from microAlgae to foster a bLue futurE) aims to establish one of the first industrial-scale facilities for producing natural ingredients from microalgae. Coordinated by "Microphyt" (France), the project focuses on developing high-value ingredients for the food, feed, and cosmetic sectors.

Nenu2PHAr (Grant Agreement No. 887474, 2020–2024)
The Nenu2PHAr project—"For a sustainable and European value chain of PHA-based materials for high-volume consumer products"—pioneered an innovative approach to produce PHA from renewable resources. This process leverages microalgal biomass, using CO_2 as a carbon source, which is then metabolized by selected bacterial strains. As renewable, biodegradable, and bio-based polymers, PHAs offer a promising alternative to fossil fuel-derived plastics for various applications.

MULTI-STR3AM (Grant Agreement No. 887227, 2020–2024)
The MULTI-STR3AM project developed a sustainable multi-strain, multi-method, multi-product microalgal biorefinery by integrating industrial side streams as feedstock. This approach supported the cultivation of microalgal species such as *Dunaliella salina* and *Nannochloropsis* sp., producing biomass for applications in the food, feed, and fragrance sectors.

SpiralG (Grant Agreement No. 792257, 2018–2023)
The SpiralG project developed a sustainable biorefinery process for Spirulina, focusing on the extraction of high-value compounds and the valorization of residual biomass. The project optimized phycocyanin extraction for applications in the food and cosmetics sectors, while the residual biomass was used to generate additional valuable products in the agricultural sector. The project explored energy-efficient open cultivation systems integrating renewable energy sources to minimize the overall environmental impacts.

ABACUS (Grant Agreement No. 745668, 2017–2021)
The ABACUS project, Algae for a Biomass applied to the production of Added value CompoUndS, focused on developing five innovative microalgal biorefinery pathways to introduce competitive products for the cosmetics and nutraceutical markets. The project also advanced the profitability and reliability of microalgal cultivation processes, contributing to the broader adoption of microalgae in high-value industries.

MAGNIFICENT (Grant Agreement No. 745754, 2017–2021)
The MAGNIFICENT project, Microalgae As a Green source for Nutritional Ingredients for Food/feed and Ingredients for Cosmetics by cost-Effective New Technologies, focused on the efficient and sustainable cultivation, recovery, and processing of microalgae. The project aimed at producing high-value chemicals and materials as well as ingredients for food, feed, and cosmetics, while demonstrating the viability of microalgae as a sustainable feedstock for various applications.

SABANA (Grant Agreement No. 727874, 2016–2020)
The SABANA project, Sustainable Algae Biorefinery for Agriculture aNd Aquaculture, developed a large scale, integrated microalgal biorefinery to produce biostimulants, biopesticides, and feed additives. By focusing on sustainable agriculture and aquaculture practices, the project aimed to enhance resource efficiency and reduce environmental impacts.

CYCLALG (2016–2018)
The CYCLALG project developed an innovative microalgal biorefinery promoting a circular economy and zero-residue production. By optimizing microalgal cultivation for biodiesel production, the project aimed to reduce costs and minimize environmental impacts, enhancing the feasibility of microalgae-based biofuels.

MAB 2.0 (2015–2017)
Supported by Climate-KIC, Europe's leading climate innovation agency, and hosted by Wageningen University and Research, the MAB 2.0 project (MicroAlgae Biorefinery 2.0) aimed to demonstrate innovative technologies and business models for large-scale microalgal biomass production. By utilizing waste streams, the project enhanced resource efficiency and promoted sustainable biorefinery solutions.

MIRACLES (Grant Agreement No. 613588, 2013–2017)
The MIRACLES project, Multi-product Integrated bioRefinery of Algae: from Carbon dioxide and Light Energy to high-value Specialties, aimed to develop an integrated, multi-product biorefinery for high-value compounds derived from microalgae. With a focus on applications in food, aquafeeds, and non-food products, the project emphasized mild cell disruption techniques and environmentally sustainable extraction processes.

D-Factory (Grant Agreement No. 613870, 2013–2017)
The D-Factory project aimed to develop a microalgal biorefinery by fractionating *Dunaliella* sp. biomass into carotenoid isomers, glycerol, and carbohydrate-rich defatted powder. The goal was to produce multiple bio-based products as sustainable alternatives to fossil-based counterparts.

PUFAChain (Grant Agreement No. 613303, 2013–2017)
The PUFAChain project aimed to establish a sustainable value chain for the production of high-value omega-3 fatty acids from microalgae. By integrating cultivation, harvesting, and processing technologies, the project sought to develop a competitive microalgal biorefinery.

Bisigodos (Grant Agreement No. 613680, 2013–2017)
This project, high-value-added chemicals and BIoreSIns from alGae biorefineries produced from CO_2 provided by industrial emissions, focused on developing a sustainable biorefinery process to convert microalgal biomass into valuable chemical products, including bioplastics and additives, demonstrating the feasibility of microalgae as a versatile feedstock.

These initiatives highlight the EU's commitment to advancing microalgal biorefineries for sustainable and diverse industrial applications.

7.3.3 From Production to Commercialization: Focus on Microalgal Pigments

The most widely utilized microalgal pigments in commercial applications include phycocyanin, a phycobiliprotein used as a natural blue food colorant and in nutraceutical and biotechnological applications; astaxanthin, a red carotenoid with antioxidant properties, extensively applied in the nutraceutical, pharmaceutical, and aquaculture sectors; fucoxanthin, a brown-gold carotenoid investigated for its potential weight loss and anti-aging effects; beta-carotene, an orange carotenoid serving as a precursor to vitamin A and a widely used natural food colorant; and lutein and zeaxanthin, yellow carotenoids recognized for their roles in ocular health and cosmetic formulations. The growing demand for these pigments is driven by their health benefits, natural coloring properties, and sustainability.

7.3.3.1 The Journey of Phycocyanin: From Discovery to Commercialization

Phycocyanin was discovered in the late nineteenth–early twentieth century. In the 1840s–1860s, scientists began identifying and characterizing natural pigments from various biological sources. The term "phycobiliproteins" emerged to describe water-soluble pigments found in cyanobacteria and red algae (Tandeau de Marsac 2003). In 1873, Cohn F. first described the blue pigment in cyanobacteria and recognized its role in photosynthesis (Cohn 1873). During the 1870s–1880s, microscopic observations of Spirulina and other cyanobacteria further hinted at the presence of unique pigments contributing to their intense blue-green color.

In 1938, Lemberg and Legge isolated and described the structure of phycocyanin by studying its spectral properties, confirming its classification as a phycobiliprotein (Lemberg and Legge 1938). Advances in spectrophotometry and chromatography in the 1950s–1960s enabled the precise characterization of its light-absorbing properties. By the 1960s, the chromophore-protein interaction in phycocyanin was elucidated, revealing its role in light absorption during photosynthesis. In 1963, Glazer, Fang, and Hixson fully characterized the phycobilisome, identifying phycocyanin as a key antenna pigment (Glazer and Fang 1973a, b; Glazer and Hixson 1977). The 1970s saw the resolution of phycocyanin's crystal structure, confirming its protein-pigment complex (Gantt et al. 1979). Interest in phycocyanin increased in the 1980s due to its potential antioxidant and anti-inflammatory properties. The food and cosmetic industries explored its use as a natural blue colorant, positioning it as an alternative to synthetic dyes (Belay et al. 1993). Regulatory milestones included the designation of Spirulina as a Generally Recognized As Safe (GRAS) ingredient in the US and Japan (Ciferri 1983), marking its first significant commercial acceptance.

By the 2000s, patented extraction techniques facilitated commercial production of phycocyanin for food and nutraceutical applications, including anti-aging, immune-boosting, and neuroprotective products (Zhang 2003). Then, phycocyanin found applications as a fluorescent probe in biotechnology for biomedical imaging and flow cytometry (Glazer 1992). The increasing demand for natural colorants led to the industrial-scale cultivation of Spirulina for phycocyanin extraction. In 2013, the Food and Drug Administration (FDA) approved phycocyanin as a natural food colorant (E18) in the EU (FDA 2013). Between 2017 and 2020, biotechnology companies developed high-purity phycocyanin (>99%) for pharmaceuticals, nutraceuticals, and high-end cosmetics. The pigment gained popularity in beverages, dairy products, and plant-based foods. Advances in microalgal cultivation systems and extraction techniques in the 2020s have further improved phycocyanin extraction efficiency, reducing costs and increasing the purity of the pigment for pharmaceutical and nutraceutical applications. Current research continues to focus on optimizing extraction processes, enhancing stability, and expanding therapeutic uses in cancer treatment, neuroprotection, and anti-inflammatory medicines.

Phycocyanin has evolved from a scientific curiosity in the nineteenth century to a high-value commercial product with applications in food, cosmetics, nutraceuticals, and pharmaceuticals. Its commercialization reflects a broader industry shift toward natural, sustainable, and health-promoting ingredients, with ongoing research driving further innovations. Several products currently available on the market are illustrated in Fig. 7.5 (non-exhaustive list).

7.3.3.2 The Journey of Astaxanthin: From Discovery to Commercialization

Astaxanthin, a red-orange carotenoid pigment, has undergone a remarkable journey from its initial discovery to widespread commercialization across various industries. This progression encompasses its identification, understanding of its natural sources, synthetic production, and diverse applications.

In 1938, German chemist Richard Kuhn first identified and isolated astaxanthin from lobster shells, recognizing it as a carotenoid due to its intense red pigmentation. Subsequent studies established its presence in various marine organisms, including crustaceans, fish, and certain bird species such as flamingos, where it contributes to their characteristic reddish-pink coloration. This discovery spurred interest in the pigment's chemical structure and potential applications. Researchers found that astaxanthin is synthesized primarily by microalgae, such as *Haematococcus lacustris*, and accumulates in higher organisms through the food chain.

In the 1950s–1970s, detailed studies on *H. lacustris* identified this microalga as the richest natural source of astaxanthin. During this period, research also elucidated its role as an antioxidant and UV-protective compound in microalgae, sparking interest in its potential health benefits. By the 1980s, astaxanthin had gained commercial significance in aquaculture, where it was utilized as a feed additive to enhance the pigmentation of farmed salmon, trout, and shrimp. To meet the increasing demand, synthetic astaxanthin, derived from petrochemical sources, was developed and became the dominant form in aquaculture due to its cost-effectiveness and scalability.

In the 1990s, researchers began exploring astaxanthin's antioxidant properties in human health. It was identified as a potent antioxidant, exhibiting 10–100 times greater free radical scavenging capacity than other carotenoids such as beta-carotene and vitamin E. Clinical studies suggested its efficacy in promoting skin health, reducing inflammation, and supporting cardiovascular function, eye health, and muscle recovery. In the 1990s–2000s, advancements in microalgal cultivation and biomass processing technologies facilitated the commercial-scale production of natural astaxanthin. Companies such as "AstaReal" (Sweden and Japan) and "Cyanotech" (Hawaii, US) pioneered the commercial cultivation of *H. lacustris* in PBRs and ORPs, making natural astaxanthin a viable alternative to synthetic forms, especially in the nutraceutical and cosmetic industries.

Throughout the 2000s, global demand for natural astaxanthin surged as its health benefits became increasingly validated through clinical research. Nutraceutical brands such as BioAstin® and AstaReal® emerged as key players in the

Fig. 7.5 Key sectors and commercially available phycocyanin-based products (non-exhaustive list)

dietary supplement industry. Applications expanded into cosmetics, leveraging astaxanthin's photoprotective and anti-aging properties, while functional food products enriched with astaxanthin gained traction in Asia and the US. Since 2010, advances in biotechnology have introduced sustainable and scalable production methods. The adoption of closed microalgal cultivation systems such as PBRs improved production efficiency, ensuring consistent quality. In addition, research into genetically engineered microorganisms, such as yeast and bacteria, has explored alternative cost-effective methods for astaxanthin biosynthesis.

Today, astaxanthin is recognized as a high-value bioactive compound with diverse applications across multiple industries. In human health, it is widely utilized in nutraceuticals for its antioxidant and anti-inflammatory properties. In aquaculture, it remains a critical feed additive for pigmentation in farmed fish and crustaceans. The cosmetics industry incorporates astaxanthin into formulations for skin protection and anti-aging benefits. Functional food and beverage sectors have embraced astaxanthin for its health-promoting attributes, while the animal health industry applies it in supplements for pets and livestock. Several commercial astaxanthin-based products are available on the market, as illustrated in Fig. 7.6.

7.3.3.3 The Journey of Fucoxanthin: From Discovery to Commercialization

Fucoxanthin is the predominant carotenoid pigment in marine ecosystems, accounting for approximately 10% of global carotenoid production (Liaaen-Jensen 1978; D'Orazio et al. 2012). It belongs to the xanthophyll class of carotenoids, which, unlike carotenes, contain oxygen in their structure. This distinctive feature, along with the presence of an epoxide group and a conjugated carbonyl group, grants xanthophylls strong antioxidant properties, making them highly sought after in aquaculture, cosmetics, and pharmacology. Since its full structural characterization in 1990, fucoxanthin has been extensively studied for its diverse bioactive properties, including anticancer, anti-diabetic, anti-obesity, anti-inflammatory, anti-angiogenic, and anti-malarial effects (Pajot et al. 2022). The increasing demand for natural bioactive compounds has driven its commercialization, particularly in the food and health industries.

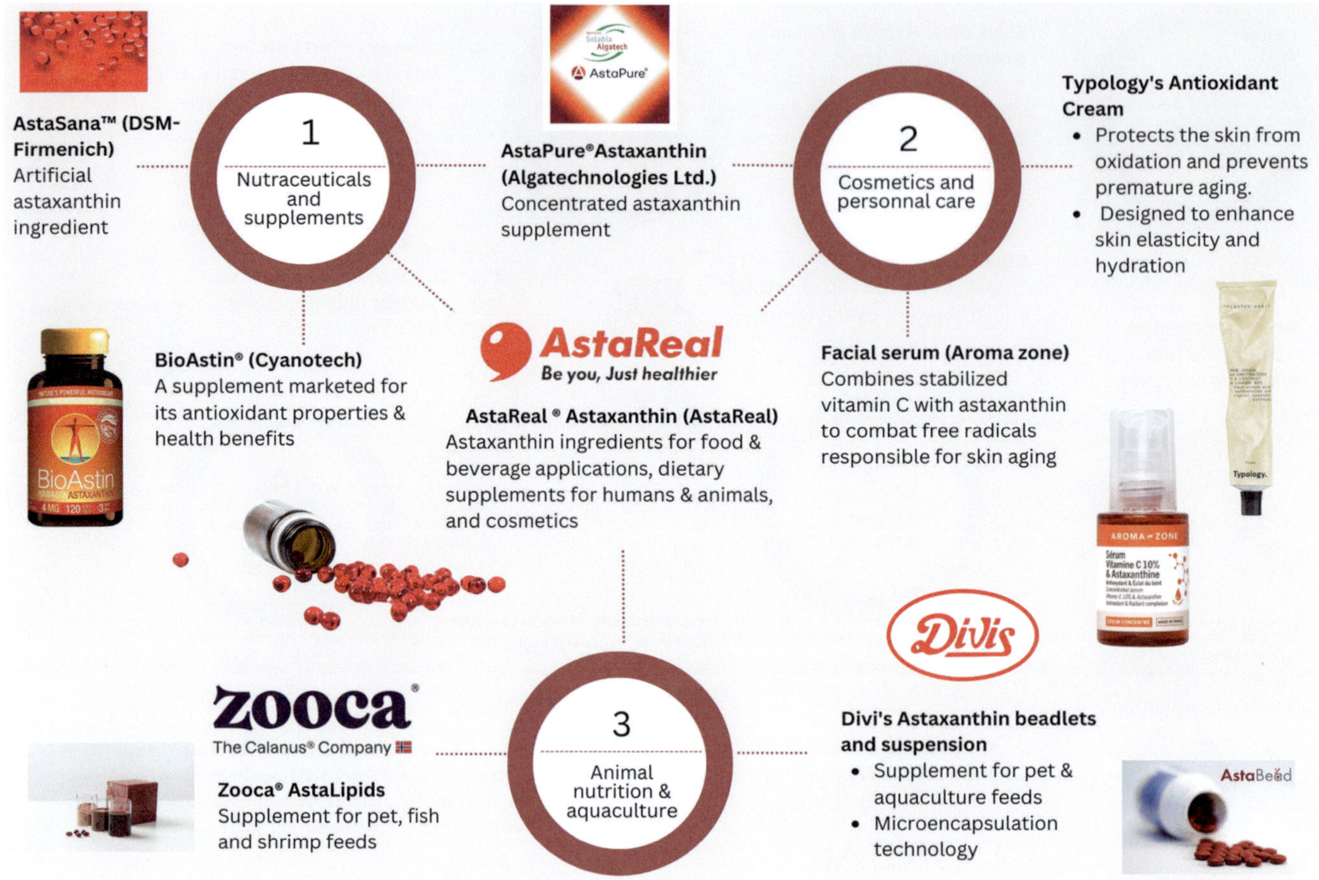

Fig. 7.6 Key sectors and commercially available astaxanthin-based products (non-exhaustive list)

The history of fucoxanthin research dates back to 1866 when Fremy M. first observed a yellow pigment in brown seaweeds, though its chemical identity remained unknown (Fremy 1866). In 1914, Willstätter and Page identified fucoxanthin as a carotenoid, distinguishing it from other algal pigments (Willstätter and Page 1914). By the 1930s, its isolation and characterization of light absorption properties confirmed its role in photosynthesis (Strain 1933). The definitive elucidation of fucoxanthin's chemical structure was achieved in 1958 by Goodwin T., establishing it as an oxygenated xanthophyll (Goodwin 1958). In the 1970s, researchers demonstrated that fucoxanthin efficiently transfers absorbed light energy to chlorophyll a, confirming its essential role in photosynthetic light harvesting (Chapman 1976). By the 1980s, studies had begun to highlight its antioxidant properties, suggesting potential health benefits (Yokoyama and Yano 1988).

In the 1990s, interest in fucoxanthin's anti-obesity effects increased, particularly its potential role in modulating fat metabolism. In 2004, Miyashita et al. demonstrated that fucoxanthin induces the expression of uncoupling protein 1 (UCP1), increasing energy expenditure and promoting weight loss, which positioned it as a promising anti-obesity agent (Miyashita et al. 2004). In recent years, research on fucoxanthin's anticancer properties has expanded significantly, with a substantial number of studies investigating its potential role in cancer prevention and therapy (Khaw et al. 2021). Between 2005 and 2010, growing awareness of its bioactivity sparked interest in its commercialization by the nutraceutical industry (Peng et al. 2011). During the 2010s, technological advancements in extraction methods, including supercritical CO_2 extraction, enhanced the efficiency and sustainability of fucoxanthin production, facilitating its entry into commercial markets (Shannon and Abu-Ghannam 2018). In the 2020s, fucoxanthin-based products, mainly enriched seaweed powders, have been introduced into the global market, targeting applications in weight management, skincare, and anti-aging supplements (Gammone and D'Orazio 2015).

Despite its growing commercial presence, fucoxanthin remains subject to regulatory limitations. As of 2025, the European Food Safety Authority (EFSA) has not yet approved fucoxanthin as a novel food ingredient (EFSA 2023). A recent EFSA assessment reviewed an ethanolic extract of the microalga *Phaeodactylum tricornutum* (Bacillariophyceae), standardized for fucoxanthin content, for use in dietary supplements. While fucoxanthin-rich extracts from *Undaria pinnatifida* (Phaeophyceae) have been legally marketed in the European Union since 1997, EFSA has not endorsed specific health claims related to fucoxanthin consumption, particularly in relation to body weight regulation (Lourenço-Lopes et al. 2023).

The fucoxanthin market is experiencing steady growth, largely driven by the increasing demand for natural weight-loss supplements and skincare formulations (Lourenço-Lopes et al. 2020; Miyashita et al. 2020). Emerging research continues to explore its anticancer, anti-inflammatory, and neuroprotective properties, expanding its potential applications in pharmaceuticals (Mohibbullah et al. 2022). Concurrently, industry efforts are focused on optimizing large scale cultivation of fucoxanthin-rich microalgae, such as *Phaeodactylum tricornutum* and *Tisochrysis lutea* (Haptophyta), to enhance commercial viability (Pereira et al. 2021). Future developments will be influenced by regulatory approvals, increasing consumer demand for natural bioactives, and improvements in cost-effective production methodologies. These will position fucoxanthin as a key ingredient in the health and wellness sector. As illustrated in Fig. 7.7, several fucoxanthin-containing products, mostly based on seaweed powder, are already available in commercial markets.

In response to current environmental and health challenges, researchers are increasingly focusing on sustainable methodologies that minimize environmental impacts while maintaining high extraction efficiency, which are expected to further drive market expansion. Traditional extraction methods often rely on organic solvents, which pose environmental and health risks. To address these concerns, alternative techniques such as supercritical CO_2 extraction (Honda et al. 2022), ultrasound-assisted extraction (Carreira-Casais et al. 2022), enzymatic-assisted extraction, and centrifugal partition chromatography (Pajot et al. 2023) are being developed. These approaches enhance yield while significantly reducing solvent consumption. In addition, the integration of membrane filtration and green chromatography techniques is improving purification efficiency, reducing waste, and facilitating scalability for industrial applications. These advancements align with the principles of green chemistry, making fucoxanthin extraction more sustainable and suitable for diverse applications in the food, pharmaceutical, and cosmetic industries.

To conclude, the journey from production to commercialization of microalgal pigments—phycocyanin, astaxanthin, and fucoxanthin—highlights the growing potential of these bioactive compounds across multiple industries. Advances in cultivation, extraction, and purification technologies have significantly improved yield, quality, and cost-effectiveness, making large-scale production increasingly viable. Phycocyanin is widely used as a natural food colorant and in cosmetic formulations. Astaxanthin is renowned for its potent antioxidant and nutraceutical benefits. Fucoxanthin is gaining attention for its metabolic and pharmaceutical applications. As demand for natural and sustainable ingredients continues to rise, further innovations in biotechnology and AI-driven process optimization will be critical for expanding market accessibility and maximizing commercial impact.

7.4 Toward Sustainable Microalgal Value Chains

Microalgae-based products have significant potential to support a sustainable bioeconomy, with applications in food, feed, chemical, pharmaceutical, cosmetic, and agricultural industries (see Sect. 7.3). Achieving sustainability requires a comprehensive evaluation across environmental, economic, and social dimensions. Life cycle-based methodologies provide structured assessment frameworks: LCA for environmental impacts, TEA and life cycle costing (LCC) for economic feasibility, and social LCA (S-LCA) for societal implications. This section reviews the environmental, economic, and social performance of microalgal value chains and discusses key methodological challenges in sustainability assessments.

7.4.1 Environmental Performance of Microalgal Value Chains

Life cycle assessment methodologies have been widely used to evaluate the environmental impacts of algae systems. A comprehensive review by Braud et al. (2023) highlights key aspects of LCA methodologies applied to the algae sector and discusses the methodological challenges related to modeling choices made by LCA practitioners. This sub-section provides a summary of key aspects of LCA methodologies. A more detailed discussion can be found in the review.

7.4.2 Life Cycle Assessment Methodologies

In the late 1990s, the International Organization for Standardization (ISO) introduced the 14040 and 14044 standards to establish a structured approach for conducting LCA studies. These standards, updated in 2006 and last reviewed in 2022, define a framework consisting of four interrelated phases: goal and scope definition, life cycle inventory (LCI), life cycle impact assessment (LCIA), and interpretation (see Fig. 7.8). The goal and scope phase determines the functional unit (FU), system boundaries, and impact assessment methods. The LCI phase involves data collection and system modeling, while the LCIA phase translates inventory flows

Fig. 7.7 Key sectors and commercially available fucoxanthin-based products (non-exhaustive list)

into environmental impacts. Finally, the interpretation phase synthesizes results and informs decision-making.

Goal definition is a fundamental step in conducting an LCA study, as it establishes its purpose, target audience, and intended application. It also specifies whether the study involves a comparative analysis or focuses on evaluating the environmental impacts of a single system. The scope outlines key elements such as the system under study, its function, FU, the life cycle stages considered within the system boundaries, the chosen LCIA method, and the selected impact categories. In addition, it includes the approach to interpretation, data quality requirements, and the need for a critical review.

The FU determines the reference flow—the amount of product required to fulfill the system's function—and can be either output or input related. According to Braud et al. (2023), most LCA studies on algae systems have used output-related FUs, such as energy (e.g., "1 megajoule of microalgal biofuel" (Zhang and Kendall 2019)), product (e.g., "1 kilogram of dried microalgal biomass" (Quintero et al. 2021), "1 kilogram of astaxanthin" (Onorato and Rösch 2020)), area (e.g., "1 hectare of cultivated area" (Seghetta et al. 2017)), distance (e.g., "1 vehicle-kilometer traveled" (Mu et al. 2014)), and time (e.g., "1 hour of marine vessel operation" (Tsang et al. 2015)). In the case of multi-product systems such as microalgal biorefineries producing several outputs, the FU determines whether a single product or a combination of products is considered. Using a combination of biorefinery outputs allows for the comparison of different configurations of the same biorefinery (Braud et al. 2025). However, such FUs hinder the assessment of the influence of changes in biomass composition on the environmental impacts of microalgal biorefineries, for instance, due to seasonal variations (ibid).

In LCA, a system refers to a unit process or a set of interconnected processes. A unit process is the smallest element for which input and output flows are quantified (Wolf et al. 2012). The foreground system comprises the processes directly modeled by the LCA practitioner, such as microalgal cultivation, harvesting, dewatering, and processing. Due to limited data on end-of-life disposal, most studies adopt a cradle-to-gate or gate-to-gate approach (Braud et al. 2023). However, some researchers advocate for cradle-to-grave system boundaries to capture the full life cycle (ibid). The background system consists of data retrieved from databases such as ecoinvent (Wernet et al. 2016). Including a system dia-

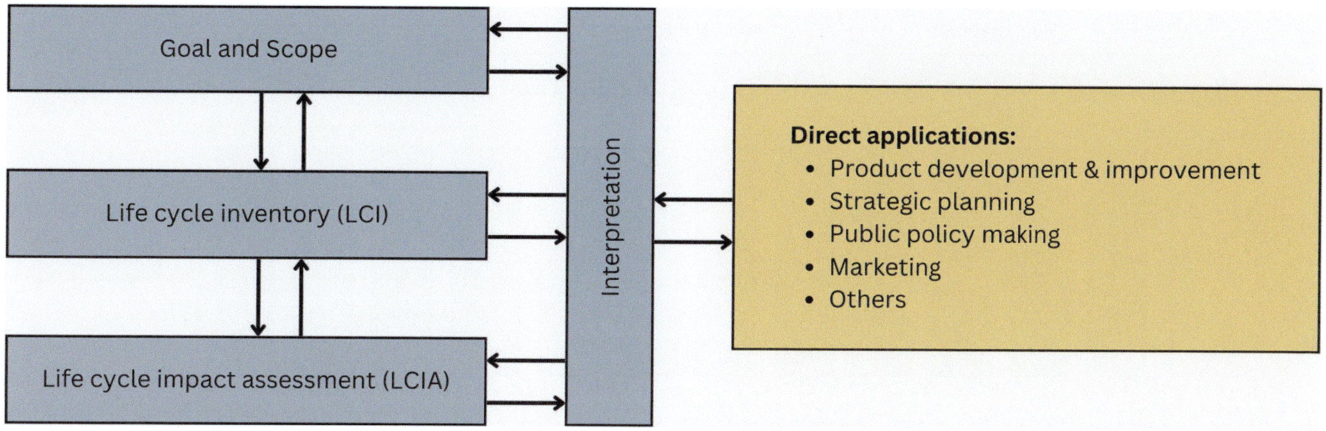

Fig. 7.8 The four interrelated phases of life cycle assessment. (Adapted from ISO 14040:2006)

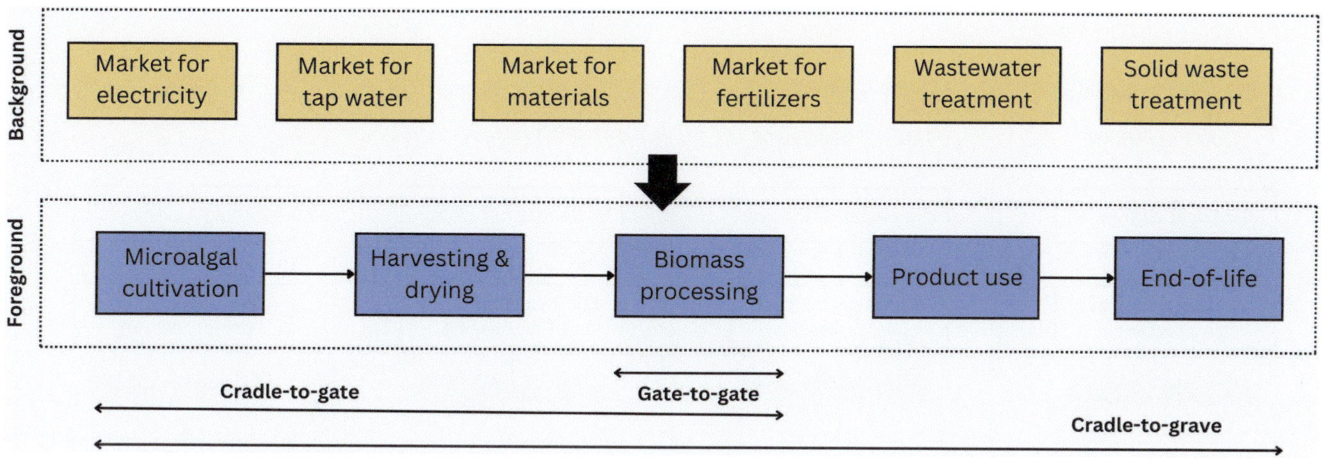

Fig. 7.9 Simplified system diagram of a microalgal system

gram in LCA studies helps clarify system design and illustrates how flows connect different processes (see Fig. 7.9).

The choice of an LCIA method and impact categories depends on the study's goal and scope. Impact assessment is conducted at two levels: midpoint and endpoint. Midpoint categories provide detailed impact scores by focusing on specific environmental effects within a cause-effect chain, whereas endpoint categories evaluate overall damage across three Areas of Protection: "Human Health," "Natural Environment" (or "Ecosystem Quality"), and "Natural Resources." Common midpoint impact categories include acidification, climate change, eutrophication, ecotoxicity, ozone depletion, ionizing radiation, and resource depletion, among others (see Fig. 7.10). In LCA studies, selected impact categories should comprehensively address the environmental concerns relevant to the system under analysis.

Despite the standardized LCA framework, methodological choices depend on practitioner expertise, data availability, and time constraints, leading to variability across studies. The International Reference Life Cycle Data System (ILCD) was developed to guide these choices by clarifying and expanding on ISO 14040 and 14044 standards (Wolf et al. 2012). The ILCD Handbook refines key aspects, including the selection of an appropriate LCI modeling approach (attributional or consequential) and the management of multifunctionality, which arises when a system generates multiple outputs. Four main methods are used to address multifunctionality: subdivision, system expansion (or enlargement), substitution, and allocation (or partitioning) (Braud et al. 2023). Figure 7.11 illustrates these approaches in the context of a microalgal biorefinery producing pigment, animal feed, and biofertilizer. The choice of method depends on the FU, which may be product-related (e.g., "1 kilogram of pigment"), area-based (e.g., "1 hectare facility"), or inclusive of all biorefinery outputs (e.g., "1 kilogram of pigment and associated co-products"). As shown by Sills et al. (2020), the choice of FU and the method used to handle multifunctionality strongly influence LCA results and should therefore be documented transparently.

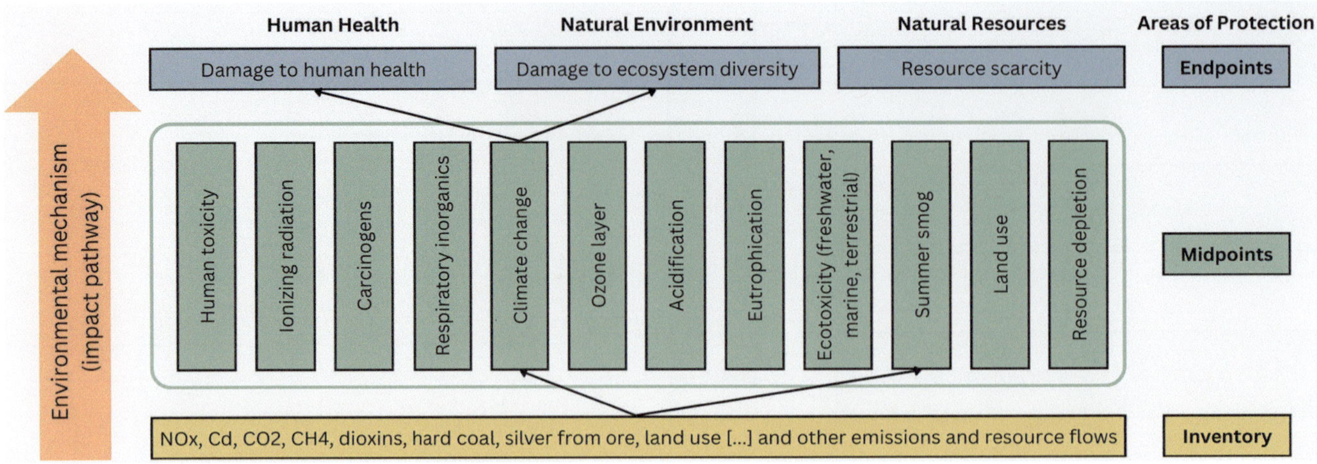

Fig. 7.10 Schematic steps from inventory to category endpoints in life cycle impact assessment. (Adapted from the ILCD Handbook (Wolf et al. 2012))

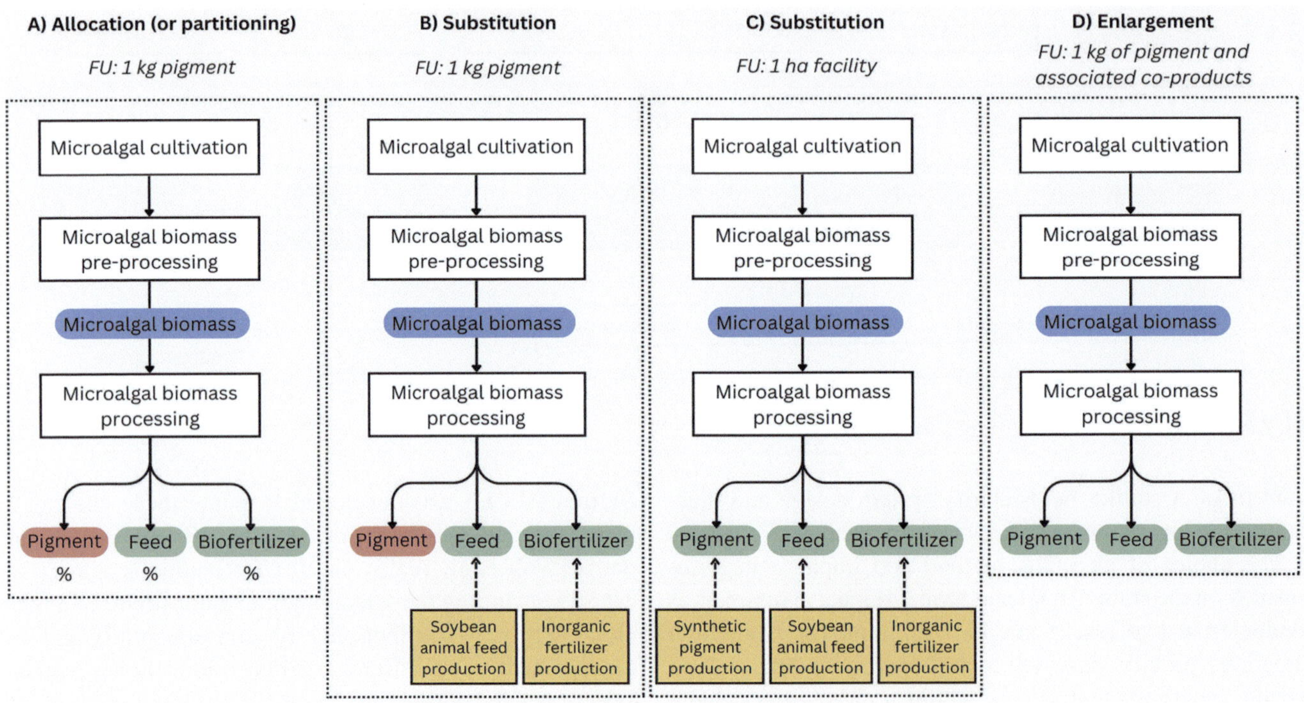

Fig. 7.11 The use of an allocation, substitution, and enlargement approach to solve or avoid multifunctionality in microalgal biorefinery systems. (Adapted from Braud et al. (2023))

7.4.3 Main Environmental Impacts and Mitigation Strategies

LCA applications in microalgal research initially focused on biofuels in the early 2000s (Collet et al. 2015). Since 2014, the scope has broadened to include high-value biomolecules such as pigments, lactic acid, and polyunsaturated fatty acids (Braud et al. 2023). This shift has expanded the range of biological, technological, and geographical considerations in algae-based LCA studies. Recent comparative assessments have examined different microalgal species (Foteinis et al. 2018; Lu et al. 2021), cultivation systems (Morales et al. 2020; Pechsiri et al. 2023), geographical locations (Beal et al. 2015), and processing technologies (Shi et al. 2019; De Souza et al. 2024). Studies have also analyzed various conversion pathways (Pankratz et al. 2020; Marangon et al. 2024) and the effects of seasonal variations in microalgal biomass production (Pérez-López et al. 2017).

Electricity consumption and mineral fertilizer use are among the most significant contributors to the environmental impacts of microalgal systems, particularly in terms of climate change, eutrophication, and acidification (Collet et al. 2015; Braud et al. 2023). The cultivation phase is often identified as a critical hotspot (Pechsiri et al. 2023), with energy consumption accounting for 40–75% of the total environmental burdens. Key contributors include temperature control, aeration, and agitation (Quintero et al. 2021; Gurreri et al. 2024). Heat pumps used to warm the culture medium in outdoor ORPs contribute up to 44% of climate impacts (Quintero et al. 2021). Artificial lighting and extended fermentation times in autotrophic and heterotrophic microalgal cultivation, respectively, exacerbate large environmental impacts (Lu et al. 2021). In addition, the use of mineral fertilizers such as diammonium phosphate and ammonium nitrate further increases environmental burdens due to high energy demands for their production and nutrient runoff in wastewater (Quintero et al. 2021; Braud et al. 2025).

Downstream processes, particularly harvesting and drying, also contribute substantially to environmental impacts. Electricity use for biomass filtration accounts for up to 31% of climate impacts, particularly when productivity is low (Duran Quintero et al. 2021; Pechsiri et al. 2023). Drying methods are another critical factor, with solar drying offering lower environmental burdens than freeze or spray drying (De Souza et al. 2024). While less significant than energy use, material selection in cultivation systems can also influence impacts. Acrylic glass in vertical tubular PBRs contributes to ozone depletion (Gurreri et al. 2024), while concrete and PVC in ORPs account for up to 36% of climate impacts (Quintero et al. 2021).

LCA studies have identified several strategies to mitigate environmental impacts. Increasing biomass productivity reduces per-kilogram impacts, as demonstrated by Zarra et al. (2024). Thermal regulation and year-round operation in high-radiation regions further enhance environmental sustainability (Quintero et al. 2021). In downstream processing, alternative energy sources, such as residual heat from anaerobic digestion, lower drying impacts compared to battery-powered methods (Braud et al. 2025). Conversion pathway optimization also plays a key role. For instance, Marangon et al. (2024) found that hydrothermal liquefaction of wastewater-grown microalgae had lower environmental burdens than gasification and Fischer-Tropsch synthesis, with catalyst use emerging as a significant hotspot.

Cultivation system design strongly influences environmental performance. Closed systems such as vertical tubular PBRs and bubble columns offer better contamination control but require higher energy inputs for mixing and lighting, which account for 40–79% of total impacts (Pechsiri et al. 2023). Conversely, heterotrophic systems have lower overall impacts when optimized with non-food carbon sources and shorter fermentation times (Lu et al. 2021). Emerging designs such as algal turf scrubbers show promise by integrating wastewater treatment with biomass production, significantly reducing fertilizer use and associated environmental impacts (De Souza et al. 2024).

The ability of microalgae to grow on waste streams (i.e., bioremediation potential) presents another key opportunity to mitigate environmental impacts. Utilizing wastewater and CO_2 from industrial sources reduces reliance on synthetic fertilizers while simultaneously treating pollutants (Mu et al. 2014; de Souza Schneider et al. 2018). Several studies have shown that integrating wastewater and flue gas within industrial symbiosis frameworks leads to substantial environmental benefits (Arashiro et al. 2018; Herrera et al. 2021). Zarra et al. (2024) found that replacing synthetic nutrients with wastewater reduced global warming potential by 27%, though biomass productivity was lower. However, transportation and injection of waste streams can contribute to increased environmental impacts (Kim et al. 2019; Collet et al. 2014). Overall, circular economy approaches—including renewable energy integration and nutrient recovery—further enhance microalgal system sustainability (Gurreri et al. 2024) (Table 7.4).

7.4.4 Economic Feasibility of Microalgal Value Chains

Evaluating the economic feasibility of microalgal value chains requires methodologies that capture both short-term costs and long-term sustainability. Techno-economic assessment and life cycle costing complement each other, providing insights into cost structures, resource allocation, technological improvements, and market potential. By identifying key cost drivers and optimization opportunities, these tools help researchers and industry stakeholders make informed decisions to support scaling and competitiveness.

7.4.4.1 Techno-Economic Assessment and Life Cycle Costing Methodologies

TEA and LCC, while complementary, provide unique perspectives suited for different stages of product and project development. TEA focuses on evaluating the technical and economic feasibility of technologies, market competitiveness, and scalability. It uses scenario modeling and sensitivity assessments to project future costs and market dynamics, helping to benchmark new technologies against existing alternatives. TEA also divides expenditures into capital expenses (CAPEX) and OPEX, pinpointing technical and economic hotspots to guide decisions about scaling and market potential. In contrast, LCC takes a cradle-to-grave approach, evaluating the total costs across the life cycle of a product or system—from raw material extraction to disposal. LCC integrates initial investments and future costs into a present value metric through discounting, and it systemati-

Table 7.4 Summary of environmental life cycle assessment studies on microalgal cultivation and biomass production based on real on-site data from pilot to industrial scale. Values were extracted from the literature as is, and no harmonization was performed

References	Species	Cultivation system	Productivity	Climate impact (kg CO_2-eq kg DW^{-1})	Main hotspots
Braud et al. (2025)	*Limnospira platensis* (formerly *Arthrospira platensis*)	Pilot scale, outdoor ORPs, natural light, 4230 m^2, Italy	28.78 kg day^{-1} (dry weight)	10.5–14.1 kg CO_2-eq kg DW^{-1} biomass	Energy for drying, nutrients
Fernández-Ríos et al. (2024)	*Limnospira platensis*	Industrial scale, ORP, Spain	N/A	3.24–3.91 kg CO_2-eq kg DW^{-1} biomass	Nutrients, pH regulators
Tzachor et al. (2022)	*Limnospira platensis*	Industrial scale, indoor flat-panel airlift PBRs, 14.4 m^3, artificial light, Iceland	150,000 kg y^{-1} (61% water)	−0.008 kg CO_2-eq kg (61% water)	Energy, materials
Pechsiri et al. (2023)	*Tetradesmus almeriensis* (formerly *Scenedesmus almeriensis*)	Pilot scale, outdoor ORP, natural light, 100 m^2, 20 m^3, Spain	1.8 kg day^{-1} (dry weight)	9.8 kg CO_2-eq kg DW^{-1} biomass	Energy for agitation
		Pilot scale, outdoor thin-layer cascade reactor, natural light, 100 m^2, 20 m^3, Spain	1.4 kg day^{-1} (dry weight)	8.8 kg CO_2-eq kg DW^{-1} biomass	Energy for agitation
		Pilot scale, outdoor vertical tubular PBR, natural light, 80 m^2, 4 m^3, Spain	1.0 kg day^{-1} (dry weight)	26.6 kg CO_2-eq kg DW^{-1} biomass	Energy for agitation
		Pilot scale, indoor light-exchange bio-column, artificial light, 0.49 m^2, 0.85 m^3, Spain	0.65 kg day^{-1} (dry weight)	24.7 kg CO_2-eq kg DW^{-1} biomass	Energy for lighting
Gurreri et al. (2024)	*Chlorella vulgaris*	Industrial scale, indoor vertically stacked horizontal PBRs, 40.4 m^3, Italy	1200 kg y^{-1} (dry weight)	153 kg CO_2-eq kg DW^{-1} biomass	Chemicals, energy for agitation & thermoregulation
Lu et al. (2021)	*Chlorella* sp.	Industrial scale, fermenters, 60 m^3	31.01 g L^{-1} (dry cells)	1.85 10^3–1.67 10^4 kg CO_2-eq t bio-oil^{-1}	Energy, carbon source, fermentation time
	Schizochytrium sp.	Industrial scale, fermenters, 7 m^3	120 g L^{-1} (dry cells)	1.85 10^3–1.67 10^4 kg CO_2-eq t bio-oil^{-1}	Energy, carbon source, fermentation time
Davis et al. (2021)	*Schizochytrium* sp.	Industrial-scale, stainless-steel agitated aerobic fermenters	N/A	4.12–4.70 kg CO_2-eq kg omega-3 fatty $acids^{-1}$	Substrate production
Herrera et al. (2021)	*Scenedesmus* sp.	Demonstration scale, ORPs, 5000 m^2 $reactor^{-1}$, Spain	20 g m^{-2} day^{-1}	0.33–1.50 kg CO_2-eq kg DW^{-1} biomass	Energy for cultivation
Pérez-López et al. (2017)	*Nannochloropsis oceanica*	Pilot scale, outdoor ORP, 4.73 m^3, 25.4 m^2, the Netherlands	2.1–10.5 g m^{-2} day^{-1}	256–4256 kg CO_2-eq kg DW^{-1} biomass	Energy for thermoregulation
		Pilot scale, outdoor horizontal tubular PBR, 0.56 m^3, 27.0 m^2, the Netherlands	1.2–12.1 g m^{-2} day^{-1}	216–2409 kg CO_2-eq kg DW^{-1} biomass	Energy for thermoregulation
		Pilot scale, outdoor vertical tubular PBR, 1.06 m^3, 31.0 m^2, the Netherlands	2.7–19.4 g m^{-2} day^{-1}	214–2665 kg CO_2-eq kg DW^{-1} biomass	Energy for thermoregulation

cally compares alternatives to identify the most cost-effective solutions. Together, these tools offer a robust framework for assessing the economic performance of microalgal value chains, with TEA providing insights for early-phase decision-making and LCC refining cost evaluations in later stages. They employ a range of distinct but complementary metrics to evaluate cost and performance, ensuring comprehensive assessments and providing insights into both short-term and long-term economic feasibility (see Table 7.5).

7.4.4.2 Key Drivers in the Economic Viability of Microalgal Value Chains

The economic viability of microalgal cultivation and biomass production systems is influenced by factors such as high capital and operating costs, scalability challenges, and market competitiveness. Microalgal production requires significant investment in infrastructure and operational expenses, resulting in a high levelized cost of production, which remains a barrier to commercialization. Achieving economies of scale and improving operational efficiency are crucial for lowering costs. While metrics such as minimum selling price and net present value should align with market prices for alternative products, technological advancements, such as optimized cultivation methods and extraction technologies, can reduce costs and improve competitiveness. In addition, government support and incentives for the development of sustainable bio-based products could help offset initial cost challenges. Continued innovation in microalgal biotechnology, advances in technology, and strategic market positioning offer promising opportunities for enhancing the economic feasibility of microalgal value chains.

Recent studies demonstrated the importance of TEA and LCC in evaluating the economic feasibility of microalgal cultivation systems, identifying key cost drivers, and outlining strategies for enhancing economic performance across various applications (see Table 7.6). For instance, Vázquez-Romero et al. (2022) used TEA to evaluate the costs of microalgal cultivation and biomass production at an industrial scale, comparing parameters such as cultivation periods and species (*Nannochloropsis oceanica* cultivated year-round vs. alternating *Phaeodactylum tricornutum* and *Tisochrysis lutea*), production scales (1 to 10 ha), harvesting methods (centrifugation vs. ultrafiltration), and drying techniques (freeze-drying vs. spray drying). The authors found that year-round cultivation of *N. oceanica* was more cost-effective than alternating species and that costs could be reduced by 18% when expanding from 1 to 10 ha and by 11% when changing nutrient sources. Centrifugation and freeze-drying accounted for 11% and 20% of costs, respectively, with savings of 7% possible by using ultrafiltration and spray drying.

Moreover, Somers et al. (2021) evaluated the costs of mixotrophic cultivation of *Galdieria sulphuraria* (Rhodophyta) using corn stover-derived cellulosic sugars, comparing two cultivation systems: glass helical PBRs and covered ORPs. Results showed that the minimum biomass selling price for the covered ORP design was lower than that for the PBR. The same conclusion was reached when including downstream processes, such as renewable diesel production via hydrothermal liquefaction. The authors discuss the co-product selling price needed to achieve economic parity with conventional fuels, depending on the amount of biomass diverted for other uses. Alternatively, Schade and Meier (2021) used TEA to assess the economic potential of microalgal omega-3 fatty acid production from *Nannochloropsis* sp. cultivated in PBRs in a so-called "cold-weather" climate in Germany. They identified major costs in PBR infrastructure, maintenance, and labor and concluded that eicosapentaenoic acid production from *Nannochloropsis* sp. resulted in

Table 7.5 Summary of key metrics for techno-economic and life cycle costing assessments of microalgal value chains

Economic metrics	Description
Key TEA metrics	
Minimum selling price	Lowest price needed to cover costs and break even. Insights: Indicates cost structure, feasibility, and market competitiveness.
Net present value (NPV)	Present value of future cash flows. Insights: Assesses profitability, long-term viability, and investment attractiveness.
Internal rate of return	Discount rate where NPV = 0. Insights: Measures return potential, efficiency, and investment comparison.
Payback period	Time to recover initial investment. Insights: Reflects risk, liquidity, and investment recovery speed.
Levelized cost of production	Average cost per unit over a project's lifetime. Insights: Evaluates cost competitiveness and financial sustainability.
Cost of goods sold	Direct production costs (materials, labor, and manufacturing). Insights: Measures production efficiency and profitability.
Break-even value	Sales volume or price needed to cover costs. Insights: Determines minimum threshold for profitability.
Key LCC metrics	
Total life cycle cost (TLCC)	Total ownership cost (operation, maintenance, disposal). Insights: Assesses long-term financial impact and cost-saving opportunities.
Net present value in LCC	NPV applied to full life cycle costs. Insights: Evaluates long-term financial feasibility.
Annualized life cycle costs	TLCC converted to equal annual payments. Insights: Standardizes costs for better long-term comparisons.
Residual value	End-of-life asset value after depreciation. Insights: Determines final costs, potential returns, or resale value.
Total cost of ownership	Comprehensive cost (purchase, operation, maintenance, disposal). Insights: Evaluates the true long-term financial impact.

Table 7.6 Summary of techno-economic and life cycle costing assessments of microalgal value chains

References	Species	Cultivation system	Productivity	Economic metrics	Main drivers
Vázquez-Romero et al. (2022)	*Nannochloropsis oceanica*	Vertically stacked tubular PBRs, 400 m^2, 38 L m^{-2}	28 t ha^{-1} y^{-1}	Biomass cost: 53 € kg DW^{-1}	PBRs, nutrients, chemicals
Vázquez-Romero et al. (2022)	*Phaeodactylum tricornutum* + *Tisochrysis lutea*	Vertically stacked tubular PBRs, 400 m^2, 38 L m^{-2}	13 t ha^{-1} y^{-1}	Biomass cost: 105 € kg DW^{-1}	PBRs, labor
Vázquez-Romero et al. (2024)	*Tetraselmis* sp. (51%) + diatoms (49%)	ORP, 10.19 ha, 30,576 m^3, use of wastewater	26.9 mg L^{-1} d^{-1}	Water treatment cost: 1.37–1.66 € m^{-3}	Equipment, labor
Watkins et al. (2024)	N/A	Rotating algal biofilm reactor, grown on wastewater	577 t y DW^{-1}	MSP of bioplastic: 3520 $ t DW^{-1}	Dewatering/drying
Schade and Meier (2021)	*Nannochoropsis* sp.	Tubular PBR, 628 m^3, 1.2 ha	53.5 t ha^{-1} y^{-1}	TIOC biomass: 9.60 € kg DW^{-1}	PBR, maintenance, labor
	Phaeodactylum tricornutum	Tubular PBR, 628 m^3, 1.2 ha	54.8 t ha^{-1} y^{-1}	TIOC biomass: 9.37 € kg DW^{-1}	PBR, maintenance, labor
Kingsley et al. (2023)	*Limnospira platensis*	ORPs, 752 m^2 $unit^{-1}$, 119 m^3 $unit^{-1}$	1 t d^{-1}	Cultivation cost: 7.54 € kg DW^{-1}	Energy for cultivation, harvesting, drying
Somers et al. (2021)	*Galdieria sulphuraria*	Helix PBR in a greenhouse, 15 m^2 m^{-3}	3 kg m^{-3}	MSP: 2869 $ t^{-1} biomass	PBR, carbon source, energy
		Covered ORP, 10 acre, 1 m^2 m^{-3}	0.5 kg m^{-3}	MSP: 921 $ t^{-1} biomass	ORP, carbon source

Values were extracted from the literature as is, and no harmonization was performed. *TIOC* Total investment and operating costs, *MSP* minimum selling price

−44% to −60% cultivation costs compared to Atlantic salmon raised in aquaculture.

Watkins et al. (2024) used TEA to assess the economic performance of microalgal biomass production with a rotating algal biofilm reactor, which is efficient for WWT but produces biomass with a high ash content (30–60% DW), impacting biomass conversion feasibility. They found that bioplastic production using whole biomass had the highest income-to-expense ratio and most cost-competitive pricing compared to bioplastic production with lipid extraction pretreatment and biocrude production. Similarly, Vázquez-Romero et al. (2024) analyzed an outdoor ORP for growing microalgae with aquaculture streams, removing 90% of N and P from wastewater. The TEA for large-scale cultivation revealed that capital costs were mainly driven by production (76–84% of equipment costs), while labor accounted for 28–36% of operational expenses. The authors concluded that if the microalgal biomass were sold as a high-value fish feed ingredient, total costs could be covered by 21–81%.

In the context of wastewater treatment, Nobre et al. (2024) found that integrating *Chlorella vulgaris* as a tertiary treatment in a municipal WWT plant was economically unfeasible due to a negative net present value and low internal rate of return. In contrast, Romagnoli et al. (2024) used LCC to demonstrate the economic viability of cultivating the same microalgal species alongside a biogas plant, utilizing digestate and flue gas to support its growth. These studies highlight the potential of industrial symbiosis and the use of the bioremediation potential of microalgae to reduce the capital costs associated with the construction of microalgal cultivation facilities.

The key drivers in the economic viability of microalgal value chains are associated with infrastructure, maintenance, and energy consumption, with labor and biomass processing also playing significant roles. To address these financial challenges, integrating microalgal cultivation within a local bioeconomy (e.g., using waste streams from WWT plants and biogas facilities) offers a promising strategy to lower capital costs while leveraging bioremediation benefits. By utilizing industrial symbiosis, resource recycling, and optimized operational models, these approaches enhance economic feasibility and contribute to more sustainable and cost-effective microalgal value chains.

7.4.5 Social Performance of Microalgal Value Chains

While the environmental and economic sustainability of microalgal value chains has been widely assessed, their social impacts remain underexplored. Factors such as labor conditions, employment generation, health and safety risks, and societal acceptance are crucial for their integration into the bioeconomy. Social LCA provides a systematic approach to evaluating these aspects, supporting data-driven decision-making to enhance social sustainability alongside the development of microalgae-based products.

7.4.5.1 Social Life Cycle Assessment Methodologies

Social LCA is a methodology used to assess the social impacts of products and services across their life cycle, from raw material extraction to final disposal (UNEP 2020). It involves the compilation and evaluation of socially relevant inputs, outputs, and potential positive or negative social impacts, focusing on key stakeholder groups such as workers, consumers, local communities, society, and value chain actors. The assessment can be quantitative, semi-quantitative, or qualitative, using both generic and site-specific data.

The ISO 14075:2024 standard on S-LCA provides specific guidance for practitioners in various sectors—including industry, government, academia, and non-profit organizations—on assessing social aspects of products in a reliable and efficient manner. The standard defines two primary functions of S-LCA: assessing the magnitude and significance of potential social impacts and forecasting these impacts throughout the product's life cycle. It emphasizes the importance of transparency for meaningful comparisons between studies and outlines four key phases: goal and scope definition, social life cycle inventory analysis, social life cycle impact assessment, and interpretation. While the standard ensures consistency in methodologies, it clarifies that it does not aim to rank products based on human rights performance but to support transparency and informed decision-making. By adhering to ISO 14075:2024, practitioners can conduct S-LCA studies that are transparent and reliable and enable the effective management of social impacts.

7.4.5.2 Social Performance of Microalgal Value Chains

Social performance in the microalgal sector has garnered increasing attention, with S-LCA emerging as a key tool to evaluate the social impacts of microalgae-based systems. For instance, Rafiaani et al. (2020) assessed social impacts at the company level within the microalgal production sector in Belgium, focusing on workers, consumers, and local communities. While the company showed positive outcomes in most social categories, a gender disparity was noted, particularly regarding equal opportunities for women. This underscores the complexity and context-dependence of social indicators in S-LCA, highlighting the need for better data availability and improvements in the methodology to enhance the reliability of social impact assessments, especially for companies in specific regional contexts.

Josa and Garfí (2023) conducted an S-LCA of microalgae-based systems for wastewater treatment, comparing three scenarios: urban WWT, industrial WWT, and using a standard growth medium. The authors found that the last scenario, with its simpler design, resulted in the lowest social impacts, minimizing health and safety risks for workers while enhancing consumer and community well-being. In contrast, the WWT scenarios, particularly industrial wastewater treatment, posed greater challenges, such as health risks and issues with community acceptance. However, urban wastewater treatment provided the benefit of local employment generation, highlighting the importance of regulatory frameworks and public awareness in facilitating the social acceptance of microalgae-based technologies.

A key contribution in this field is the approach developed by Paula Pérez-López et al. (2025), which integrates LCC outcomes with S-LCA to assess the social impacts of microalgal biorefineries. This framework prioritizes stakeholders and social impact subcategories, focusing on workers and local communities, with key areas of concern including health and safety, fair wages for workers, and employment opportunities for local communities (see Fig. 7.12). The authors applied their integrated LCC/S-LCA methodology to a case study of microalgal cultivation coupled with WWT. In this system, biomass is used to produce biomethane through anaerobic digestion, and co-products are utilized as biofertilizers for agriculture. The study's functional unit was 1 US dollar worth of biogas produced. The authors considered not only direct workers and value chain actors but also those indirectly affected by the system, such as local communities surrounding raw material extraction sites. The study used site-specific data collected from the European project PRODIGIO (developing early-warning systems for improved microalgal <u>PRO</u>duction and anaerobic <u>DIG</u>est<u>IO</u>n) to support its assessments.

Further insights into social impacts in the bioeconomy come from Rebolledo-Leiva et al. (2023), who analyzed the state of S-LCA within bioeconomy strategies. They highlighted several challenges, such as defining system boundaries, selecting social indicators, and impact assessment methods. In addition, the study discussed the complexities of multifunctionality and allocation issues in bio-based products which arise in microalgal biorefineries. Despite these challenges, they emphasized the importance of refining S-LCA to provide a comprehensive and standardized approach to assessing social impacts in the bioeconomy.

Overall, these findings highlight the importance of integrating S-LCA in the microalgal sector to better understand and manage social impacts. However, challenges persist, particularly with data availability in early-stage systems, which is common in microalgal value chains (Valdivia et al. 2021; Pérez-López et al. 2025). To address this, methodologies have been proposed to adapt S-LCA based on the technology readiness level (TRL) of a system. This approach helps identify relevant stakeholders at different stages of technological maturity, ensuring comprehensive social

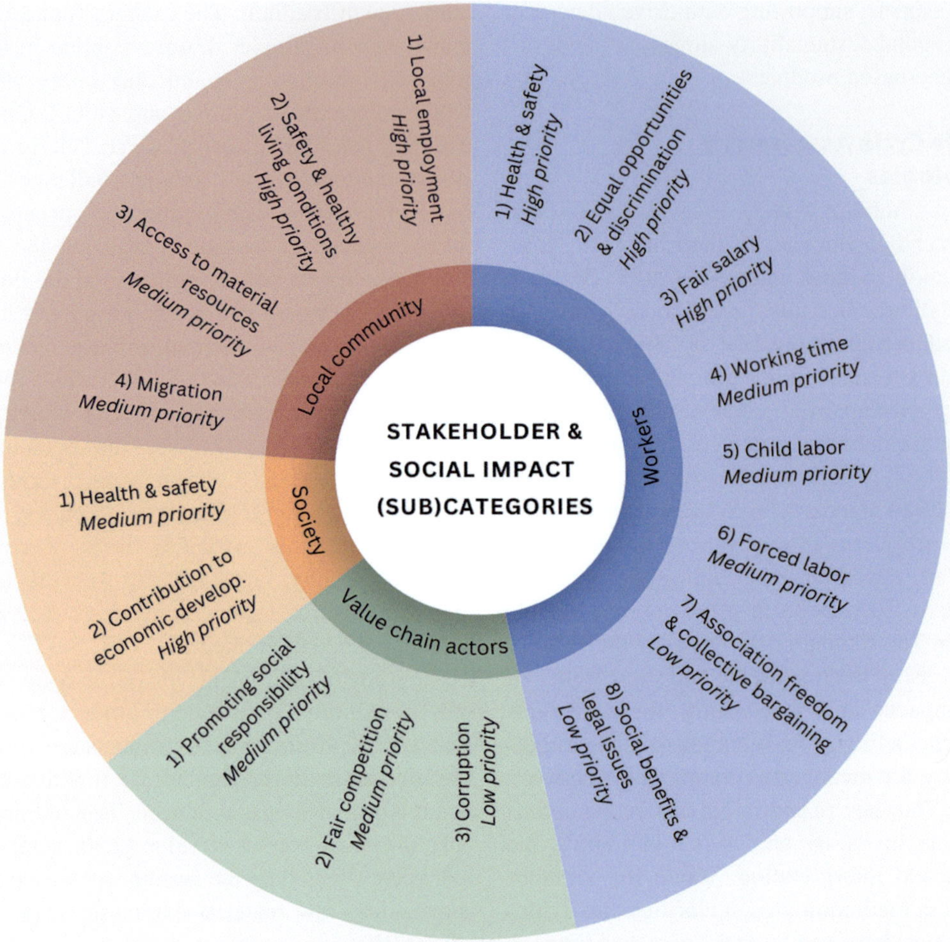

SOCIAL INDICATORS

Workers

1) Rate of fatal accidents at workplace
DALYs* due to air and water pollution
Violation of health & safety standards
Workers affected by natural disasters
Presence of sufficient safety measures
2) Women in the sectoral labor force
Men in the sectoral labor force
Gender wage gap
3) Minimum wage, per month
Living wage, per month
4) Weekly hours of work per employee
5) Children in employment
6) Frequency of forced labor
Goods produced by forced labor
Trafficking in persons
7) Right of association
Right of collective bargain
Right of strike
Trade union density
8) Evidence of violation of laws
Social security expenditure

Value chain actors

1) Membership in an initiative that promotes social responsibility along the supply chain
2) Presence of anti-competitive behavior or violation of anti-trust and monopoly legislation
3) Active involvement of enterprises in corruption and bribery
Public sector corruption

Society

1) Domestic and external health expenditure
Domestic general health expenditure
Global Peace Index
Health expenditure, external resources
Life expectancy at birth
2) Contribution of the sector to economic development
Embodied value added total
Public expenditure on education
Illiteracy rate
Youth illiteracy rate

Local community

1) Unemployment rate in the country
2) Drinking water coverage
Pollution level of the country
Sanitation coverage
3) Certified environmental management systems
Extraction of biomass
Level of industrial water use
4) Emigration rate
Immigration rate
International migrant stock
International migrant workers in the sector
Net migration rate
Number of asylum seekers in relation to total population

**DALYs: Disability adjusted life years*

Fig. 7.12 Summary of stakeholder categories and social impact subcategories relevant for microalgal value chains. (Adapted from Pérez-Lopéz et al. (2025))

assessments in both low- and high-TRL systems (Hannouf et al. 2024).

7.4.6 Challenges in Sustainability Assessments

Environmental, economic, and social sustainability assessments of microalgal value chains face key challenges that need to be addressed to improve the reliability and comparability of studies. For instance, the lack of commercial-scale data poses a major challenge in comparing microalgal value chains with conventional and well-established ones (e.g., mineral fertilizers vs. microalgae-based biofertilizers). Most LCA, LCC, and S-LCA studies rely on laboratory and pilot-scale data, which fail to capture the complexities and efficiencies of large-scale production. Scaling up these data is essential for assessing microalgal cultivation and biomass processing at different development stages, guiding the design of economically viable products with minimal environmental and social impacts.

Since most microalgae-based products are still at an early stage of development and commercialization (i.e., low TRLs), prospective LCA emerges as a promising alternative to conventional LCA for estimating the environmental impacts of microalgae-based products in the future, once deployed at a large scale (Arvidsson et al. 2018; van der Giesen 2020). This novel approach models how environmental impacts evolve as a technology matures and scales up (see Fig. 7.13). While prospective LCA has been applied since 2020 in the energy and transport sectors—and to a lesser extent in the bio-based industry—its application to macro- and microalgae remains limited. Jouannais and Pizzol (2022) developed a stochastic model to anticipate and mitigate the environmental impacts of bio-prospected microalgal biomolecules under high uncertainty. More recently, Ayala et al. (2024) conducted a prospective LCA on macroalgae-based bioplastics. The use of prospective LCA in the algae sector is expected to grow rapidly, supporting the development and commercialization of novel algae-based products with low environmental impacts.

Incorporating learning curves into TEA/LCC and LCA could be valuable for predicting how costs, efficiency, and environmental impacts improve with increasing cumulative production. Faber et al. (2022) showed that applying learning rates to both CAPEX and greenhouse gas emissions results in reductions as technology deployment progresses. Learning effects could also be applied to OPEX and other relevant metrics. This approach can help identify the scale at which cost and emissions targets can be met. Furthermore, it could assist in determining the necessary government incentives to facilitate the technology's market adoption. As microalgal cultivation and biomass production systems scale, technological advances and process optimization can lead to significant reductions in energy use, enhancing the overall sustainability of these systems. Learning curves provide a valuable tool for modeling future environmental and economic aspects, improving the accuracy of prospective studies.

Another challenge lies in harmonizing LCA studies, as substantial variability in results has been observed across the literature (Tu et al. 2017; Cruce et al. 2021). This high variability can be attributed to the unique characteristics of the microalgal value chains assessed, including parameters such as microalgal species, cultivation systems, processing techniques, and geographical locations, as well as the methodological choices made by LCA practitioners during the goal and scope definition of the studies. For instance, a comparison of environmental impacts associated with Spirulina biomass production across six distinct LCA studies revealed a broad range of outcomes, ranging from 0 to 200 kg CO_2 per kg of dry biomass produced (Braud et al. 2025). Such discrepancies complicate the interpretation of results and hinder informed decision-making by industry stakeholders, particularly those less familiar with LCA. Proposed harmonization methods, aimed at standardizing system boundaries and data assumptions, could help mitigate this variability (Handler et al. 2012). However, these approaches are often resource intensive and time consuming. Therefore, enhancing the transparency of LCA modeling, especially during the goal and scope definition phase, is a critical area requiring attention. By offering clear justifications for subjective choices made during LCA studies, transparency can enhance the comparability and reliability of results.

In conclusion, although the application of LCA to microalgae-based systems presents several challenges, it also offers substantial potential to promote more sustainable practices within the blue bioeconomy. Overcoming these challenges will require coordinated efforts among researchers, industry stakeholders, and policymakers to address the scarcity of commercial-scale data and ensure that sustainability assessments remain transparent, reproducible, and meaningful. The development of prospective life cycle methodologies presents a valuable opportunity to advance the field, while the integration of AI could further enhance the ability of sustainability assessments to generate forward-looking insights, which will be crucial in shaping the future of the microalgal sector.

7.5 Conclusion and Future Perspectives

This chapter provided an overview of the current state of microalgal cultivation and biomass production in temperate regions, emphasizing the impact of seasonal variations such as fluctuations in light intensity and temperature. It also

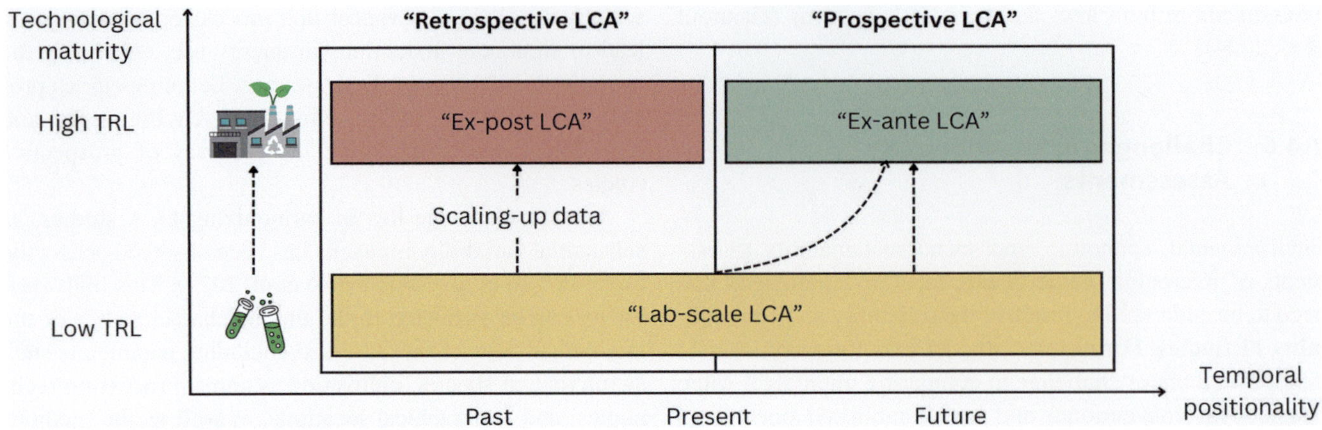

Fig. 7.13 Key terminologies for life cycle assessment are used in the literature, depending on technological maturity and temporal positionality. (Adapted from Arvidsson et al. (2024))

reviewed existing commercialized microalgae-based products and examined methodologies for assessing the environmental, economic, and social sustainability of microalgal value chains. The microalgae sector faces significant barriers to economic viability for both cultivation systems and final products, particularly when competing with conventional alternatives, even though these typically have greater environmental and social impacts (e.g., synthetic plastics vs. bioplastics, mineral fertilizers vs. biofertilizers). In temperate regions, selecting suitable microalgal species and adapting cultivation systems to fluctuating light and temperature conditions are key factors for achieving sustainability. To address these issues, digital and AI-driven technologies are increasingly being explored to optimize production efficiency and resource utilization. This final section examines how AI can improve the sustainability of microalgae-based products, maximizing economic viability while minimizing environmental and social impacts across their life cycle.

The concept of Industry 4.0, initially introduced in 2011, revolutionized manufacturing by integrating digital technologies, automation, and real-time data analytics into industrial processes (Xu et al. 2021). Building on foundational technologies such as the internet of things (IoT), cyber-physical systems, and cloud computing has led to the development of smarter factories that optimize production, enhance process efficiencies, and reduce waste generation. The evolution to Industry 5.0 and 6.0 shifts the focus toward human-centric and sustainable approaches, emphasizing environmental responsibility and ethical considerations to ensure industrial advancements align with societal well-being (Leng et al. 2022; Samuels 2025). In the microalgal sector, IoT has played a critical role in automating the monitoring of growth parameters such as temperature, pH, and biomass productivity, enhancing efficiency across both upstream and downstream processes (Wang et al. 2022). By leveraging sensor networks, real-time monitoring of biomass concentrations and biomolecule accumulation becomes more precise, significantly reducing uncertainties compared to manual data collection.

Recent breakthroughs in data storage and analytics, particularly through big data and AI, are unlocking new opportunities for process optimization and sustainability. By extracting insights from large datasets, these technologies reveal hidden patterns that improve microalgal cultivation, enhance economic viability, and reduce environmental impacts. AI-driven predictive models enable real-time adjustments, optimizing growth conditions, resource allocation, and waste reduction. In addition, AI enhances the processing of unstructured sensor data, automating critical tasks such as cell viability monitoring and feeding schedule optimization. For instance, Li et al. (2025) applied a graph learning approach using a binarized attributed network embedding model to improve the predictive performance of random forest and extreme gradient boosting in the case of *Porphyridium* sp. (Rhodophyta) cultivation. Their study identified light intensity and nutrient concentration as critical factors influencing biomass yield. Similarly, deep learning models have improved process precision, enabling more accurate predictions of pH levels and microalgal species identification. AI-driven classification systems, such as those developed by Otálora et al. (2023), have achieved 97.27% accuracy in classifying microalgae, reducing the need for manual analysis.

Other examples include "OptimALga," a digital platform developed by "DataFriends" in collaboration with "CzechGlobe," both based in the Czech Republic, for data collection and analysis in microalgal cultivation systems, such as PBRs. It centralizes data acquisition, applies AI-driven analytics, and optimizes microalgal productivity in real time. The platform supports geographically distributed data collection, integrates a centralized infrastructure for storage and processing, and maintains analytical models. "OptimALga" features an interactive web-based interface that enables monitoring, model management, and secure data transfer across system components. Moreover, "Holodetect Instruments Ltd" (Hungary) has developed sensors and instruments utilizing digital holographic microscopy (DHM) and AI for automated analysis of volumetric liquid samples. These tools can be used to identify, classify, and count microalgal cells. By digitally reconstructing holograms without mechanical focusing, this technology offers a cost-efficient approach to microalgal culture monitoring, eliminating the need for costly components and labor-intensive pre-processing. Flow-through measurements enable extended observation periods with minimal maintenance, while deep neural networks ensure precise microalgae counting and cell size assessment. The combination of DHM with fluorescence imaging supports concurrent measurement of cell morphology, chlorophyll content, and cell viability (using appropriate dyes), providing comprehensive insights into microalgal cultures.

AI enhances strain selection and genetic engineering by analyzing genetic data to identify traits linked to higher productivity and environmental resilience. AI-based tools are increasingly integrated with genetic modification techniques such as CRISPR-Cas9 to enhance genome editing precision, facilitating the development of microalgal strains optimized for improved biomolecule production (Teng et al. 2020). Beyond strain selection and cultivation, AI optimizes extraction processes and supply chain management, improving product yield and reducing waste generation across the value chains. For instance, Chong et al. (2023) demonstrated how AI models optimized the extraction of bioactive compounds such as fucoxanthin, achieving up to 99.60% prediction accuracy. AI also contributes to sustainability efforts by optimizing energy consumption and integrating waste streams into cultivation systems. Oruganti et al. (2023) showed how ML models enable precise monitoring and optimization of resource recovery and biofuel production, while Meenatchisundaram et al. (2024) demonstrated how AI outperforms traditional models in predicting biomass yield, leading to a 31.7% increase in production efficiency. These advances are critical to the sustainability of large-scale microalgal cultivation and biorefinery systems.

The integration of AI into LCA methodologies remains largely unexplored, despite its potential to fill data gaps, optimize system configurations, and incorporate real-time operational data for predictive analysis (Pérez-Almada et al. 2023; Gachkar et al. 2024). Salla et al. (2025) reviewed the use of ML in LCA, identifying artificial neural networks and random forests as the most commonly applied models for regression and classification tasks, aligning with the findings of Ghoroghi et al. (2022) and Martínez-Ramón et al. (2024). Key challenges in ML-LCA integration include data quality limitations and the static nature of conventional LCA models, underscoring the need for dynamic LCI approaches, as also highlighted by Romeiko et al. (2024). Furthermore, Martínez-Ramón et al. (2024) outlined three ML-LCA frameworks illustrating common applications of ML in LCA: (1) rapid LCI prediction using ML algorithms trained on simulation data to reduce computational demands and support sustainability assessments when primary data are unavailable (e.g., when scaling up emerging technologies), (2) surrogate models for specific process stages to facilitate the scale-up of research-stage processes based on experimental or literature data, and (3) system-level surrogate models to predict LCI data and optimize system configurations for minimal environmental impacts without detailing internal processes. While case studies on microalgal systems are lacking, research has been conducted on other bio-based systems. For instance, Li et al. (2024) employed random forest and artificial neural networks to evaluate model performance using outputs from process simulation software for bioplastic production from corn stover. They used AI-driven surrogate models of the biorefinery to compute mass and energy balances across 630 feedstock composition scenarios. More broadly, Piron et al. (2024) explored how Industry 4.0 technologies, such as cyber-physical systems and IoT, contribute to LCI development.

In the microalgal sector, digital technologies hold significant potential for enhancing economic viability while reducing environmental and social impacts. Further research is needed to integrate AI tools with sustainability assessment methods effectively. Prospective life cycle methodologies present a promising approach to supporting the development of microalgae-based products and ensuring minimized impacts once deployed at a commercial scale. In addition, integrating AI with learning curves could improve the representation of how environmental impacts evolve over time as systems scale, providing valuable insights for prospective LCA studies. AI is transforming microalgal biotechnology by enhancing real-time monitoring, optimizing processes, and enabling predictive modeling across diverse applications (see Fig. 7.14). However, unlocking the full potential of these technologies for industrial adoption requires addressing key challenges, including data quality, model accuracy, and the need for interdisciplinary collaboration between AI specialists and biotechnologists. Moving forward, strengthening industry-academic partnerships and developing standardized AI frameworks for microalgal applications will be critical to advancing the field and accelerating large-scale adoption.

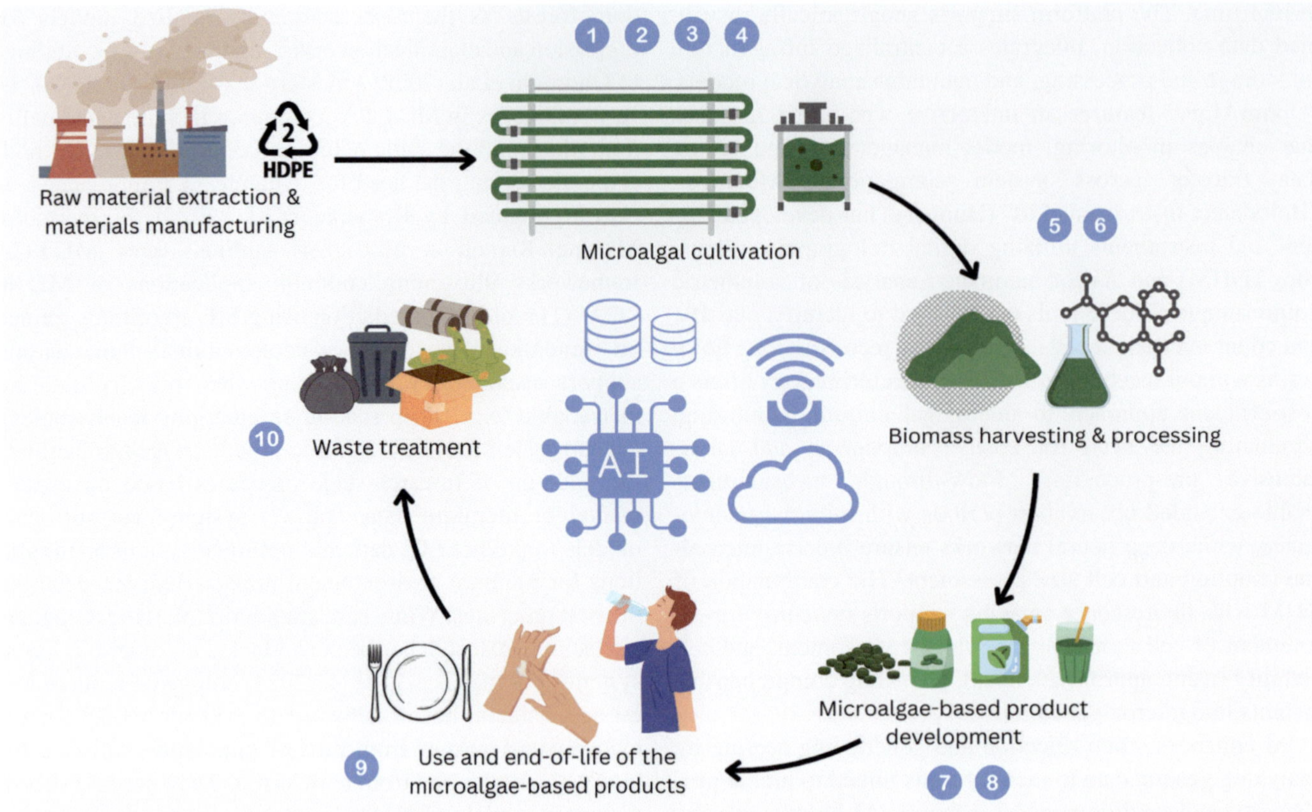

POTENTIAL USES OF DIGITAL TECHNOLOGIES

1. **Bioprospection, strain selection, and genetic engineering**
   Accelerate the discovery of high-performing microalgal strains and enhance precision in genetic modifications to optimize biomolecule production.
2. **Optimization of cultivation system design**
   Improve cultivation system design and optimize operational parameters for higher efficiency and scalability.
3. **Optimization of microalgal growth conditions**
   Enable real-time monitoring and predictive modeling to facilitate dynamic adjustments, maximizing productivity while minimizing resource use.
4. **Integration into a local bioeconomy**
   Facilitate the use of industrial waste streams for nutrient and water recovery, supporting industrial symbiosis.
5. **Optimization of biomass conversion technologies**
   Enhance process efficiency in biomass conversion, including solvent recovery and waste valorization for increased circularity.
6. **Valorization of co-products and wastes**
   Identify high-value applications for co-products and residual biomass, supporting a circular bioeconomy.
7. **Development of novel microalgae-based products**
   Accelerate the discovery of new formulations and applications for food, feed, bioplastics, biofuels etc.
8. **Optimization of supply chain logistics**
   Reduce energy use and costs, develop preservation techniques, optimize transportation routes, align supply and demand.
9. **Sustainable use and end-of-life optimization**
   Assess product durability and biodegradability, support eco-design and sustainable disposal or recycling strategies.
10. **Minimization of waste generation**
    Reduce material losses throughout production and use phases while enhancing product recyclability.

Fig. 7.14 Potential uses of digital technologies over the life cycle of microalgae-based products

Acknowledgments The authors sincerely thank Alan T. Critchley for the invitation to contribute to this book, as well as Carlos Unamunzaga, Vítor Verdelho, and Jean-Paul Cadoret for their support in facilitating and encouraging this work. We also extend our gratitude to the members of the European Algae Biomass Association for their valuable contributions in gathering knowledge for this chapter and for providing access to images of microalgal cultivation systems.

References

Abreu AP, Morais RC, Teixeira JA, Nunes J (2022) A comparison between microalgal autotrophic growth and metabolite accumulation with heterotrophic, mixotrophic and photoheterotrophic cultivation modes. Renew Sust Energ Rev 159:112247. https://doi.org/10.1016/j.rser.2022.112247

Acién Fernández FG, Reis A, Wijffels RH, Barbosa MJ, Verdelho V, Llamas B (2021) The role of microalgae in the bioeconomy. New Biotechnol 61:99–107. https://doi.org/10.1016/j.nbt.2020.11.011

Adesanya VO, Cadena E, Scott SA, Smith AG (2014) Life cycle assessment on microalgal biodiesel production using a hybrid cultivation system. Bioresour Technol 163:343–355. https://doi.org/10.1016/j.biortech.2014.04.051

Ambati RR, Phang SM, Ravi SK, Aswathanarayana RG (2014) Astaxanthin: sources, extraction, stability, biological activities and its commercial applications–a review. Mar Drugs 12(1):128–152. https://doi.org/10.3390/md12010128

An Y, Kim T, Byeon H, Rayamajhi V, Lee J, Jung S, Shin H (2024) Improved production of Astaxanthin from *Haematococcus pluvialis* using a hybrid open–closed cultivation system. Appl Sci 14(3):1104. https://doi.org/10.3390/app14031104

Arashiro LT, Montero N, Ferrer I, Acién FG, Gómez C, Garfí M (2018) Life cycle assessment of high rate algal ponds for wastewater treatment and resource recovery. Sci Total Environ 622:1118–1130. https://doi.org/10.1016/j.scitotenv.2017.12.051

Araújo R, Vázquez Calderón F, Sánchez López J, Azevedo IC, Bruhn A, Fluch S, Garcia Tasende M, Ghaderiardakani F, Ilmjärv T, Laurans M, Mac Monagail M, Mangini S, Peteiro C, Rebours C, Stefansson T, Ullmann J (2021) Current status of the algae production industry in Europe: An emerging sector of the blue bioeconomy. Front Mar Sci 7:626389. https://doi.org/10.3389/fmars.2020.626389

Arvidsson R, Tillman AM, Sandén BA, Janssen M, Nordelöf A, Kushnir D, Molander S (2018) Environmental assessment of emerging technologies: recommendations for prospective LCA. J Ind Ecol 22(6):1286–1294. https://doi.org/10.1111/jiec.12690

Arvidsson R, Svanström M, Sandén BA, Thonemann N, Steubing B, Cucurachi S (2024) Terminology for future-oriented life cycle assessment: review and recommendations. Int J Life Cycle Assess 29(4):607–613. https://doi.org/10.1007/s11367-023-02265-8

Assunção J, Malcata FX (2020) Enclosed "non-conventional" photobioreactors for microalga production: a review. Algal Res 52:102107. https://doi.org/10.1016/j.algal.2020.102107

Ayala M, Goosen N, Michalak L, Thomsen M, Pizzol M (2024) Prospective LCA of brown seaweed-based bioplastic: upscaling from pilot to industrial scale. Sustain Prod Consum 52:416–426. https://doi.org/10.1016/j.spc.2024.11.020

Barbosa MJ, Janssen M, Südfeld C, D'Adamo S, Wijffels RH (2023) Hypes, hopes, and the way forward for microalgal biotechnology. Trends Biotechnol 41(3):452–471. https://doi.org/10.1016/j.tibtech.2022.12.017

Batello C, Marzot M, Touré AH (2004) The future is an ancient lake: traditional knowledge, biodiversity and genetic resources for food and agriculture in Lake Chad Basin ecosystems. Food & Agriculture Org

Beal CM, Gerber LN, Sills DL, Huntley ME, Machesky SC, Walsh MJ, Tester JW, Archibald I, Granados J, Greene CH (2015) Algal biofuel production for fuels and feed in a 100-ha facility: a comprehensive techno-economic analysis and life cycle assessment. Algal Res 10:266–279. https://doi.org/10.1016/j.algal.2015.04.017

Belay A, Ota Y, Miyakawa K et al (1993) Current knowledge on potential health benefits of Spirulina. J Appl Phycol 5:235–241. https://doi.org/10.1007/BF00004024

Bezerra da Silva MRO, Moura YAS, Converti A, Figueiredo Porto AL, de Araújo Viana Marques D, Bezerra RP (2021) Assessment of the potential of *Dunaliella* microalgae for different biotechnological applications: a systematic review. Algal Res 58:102396. https://doi.org/10.1016/j.algal.2021.102396

Bito T, Okumura E, Fujishima M, Watanabe F (2020) Potential of *Chlorella* as a dietary supplement to promote human health. Nutrients 12(9):2524. https://doi.org/10.3390/nu12092524

Braud L, McDonnell K, Murphy F (2023) Environmental life cycle assessment of algae systems: critical review of modelling approaches. Renew Sust Energ Rev 179:113218. https://doi.org/10.1016/j.rser.2023.113218

Braud L, McDonnell K, Murphy F (2025) Beyond phycocyanin: environmental life cycle assessment of a European pilot scale Spirulina biorefinery. Bioresour Technol Rep 29:102027. https://doi.org/10.1016/j.biteb.2025.102027

Carcea M, Sorto M, Batello C, Narducci V, Aguzzi A, Azzini E, Fantauzzi P, Finotti E, Gabrielli P, Galli V, Gambelli L (2015) Nutritional characterization of traditional and improved dihé, alimentary blue-green algae from the Lake Chad region in Africa. LWT-Food Science and Technology 62(1):753–763. https://doi.org/10.1016/j.lwt.2014.10.039

Cardias BB, Barceló-Villalobos M, Lafarga T, Fernández FA, Morais MG, Costa JAV (2023) An overall analysis of CO_2 demand and utilization of microalgal cultures in pilot-scale raceway reactors. Algal Res 74:103197. https://doi.org/10.1016/j.algal.2023.103197

Carreira-Casais A, Cassani L, Soria-López A, Mansour SS, Fraga-Corral M, Perez-Gregorio R, Simal-Gandara J, Prieto MA (2022) Green extraction of fucoxanthin with promising nutraceutical applications. Biol Life Sci Forum 12(1):34. https://doi.org/10.3390/IECN2022-12380

Chanquia SN, Vernet G, Kara S (2022) Photobioreactors for cultivation and synthesis: specifications, challenges, and perspectives. Eng Life Sci 22(12):712–724. https://doi.org/10.1002/elsc.202100070

Chapman DJ (1976) Carotenoid-binding proteins in algae. J Phycol 12(1):195–202

Chisti Y (2013) Constraints to commercialization of algal fuels. J Biotechnol 167(3):201–214. https://doi.org/10.1016/j.jbiotec.2013.07.020

Chong JWR, Tang DYY, Leong HY, Khoo KS, Show PL, Chew KW (2023) Bridging artificial intelligence and fucoxanthin for the recovery and quantification from microalgae. Bioengineered 14(1):2244232. https://doi.org/10.1080/21655979.2023.2244232

Ciferri O (1983) Spirulina, the edible microorganism. Microbiol Rev 47(4):551–578. https://doi.org/10.1128/mr.47.4.551-578.1983

Clippinger JN, Davis RE (2019) Techno-economic analysis for the production of algal biomass via closed photobioreactors: future cost potential evaluated across a range of cultivation system designs. National Renewable Energy Laboratory Golden, Colorado. NREL/TP-5100-72716

Cohn F (1873) Untersuchungen über Bakterien. Beiträge zur Biologie der Pflanzen 1(3):127–224

Collet P, Lardon L, Hélias A, Bricout S, Lombaert-Valot I, Perrier B, Lépine O, Steyer JP, Bernard O (2014) Biodiesel from microalgae–life cycle assessment and recommendations for potential improvements. Renew Energy 71:525–533. https://doi.org/10.1016/j.renene.2014.06.009

Collet P, Hélias A, Lardon L, Steyer JP, Bernard O (2015) Recommendations for life cycle assessment of algal fuels. Appl Energy 154:1089–1102. https://doi.org/10.1016/j.apenergy.2015.03.056

Cruce JR, Beattie A, Chen P, Quiroz D, Somers M, Compton S, DeRose K, Beckstrom B, Quinn JC (2021) Driving toward sustainable algal fuels: a harmonization of techno-economic and life cycle assessments. Algal Res 54:102169. https://doi.org/10.1016/j.algal.2020.102169

D'Orazio N, Gemello E, Gammone MA, De Girolamo M, Ficoneri C, Riccioni G (2012) Fucoxantin: a treasure from the sea. Mar Drugs 10(3):604–616. https://doi.org/10.3390/md10030604

Davis D, Morao A, Johnson JK, Shen L (2021) Life cycle assessment of heterotrophic algae omega-3. Algal Res 60:102494. https://doi.org/10.1016/j.algal.2021.102494

De Souza Schneider RDC, De Moura Lima M, Hoeltz M, De Farias Neves F, John DK, De Azevedo A (2018) Life cycle assessment of microalgae production in a raceway pond with alternative culture media. Algal Res 32:280–292. https://doi.org/10.1016/j.algal.2018.04.012

De Souza MP, Colares GS, Wiesel PG, Rizzetti TM, Hoeltz M, Benitez LB, Machado ÊL, Schneider RC (2024) Life cycle assessment of different scenarios for pigment production from an algal turf scrubber (ATS) system. Biomass Convers Biorefinery 14(22):28163–28174. https://doi.org/10.1007/s13399-022-03530-w

Dormido R, Sánchez J, Duro N, Dormido-Canto S, Guinaldo M, Dormido S (2014) An interactive tool for outdoor computer controlled cultivation of microalgae in a tubular photobioreactor system. Sensors 14(3):4466–4483. https://doi.org/10.3390/s140304466

EFSA (2023) Safety of an ethanolic extract of the dried biomass of the microalga *Phaeodactylum tricornutum* as a novel food pursuant to Regulation (EU) 2015/2283. EFSA J. https://doi.org/10.2903/j.efsa.2023.8072

Faber G, Ruttinger A, Strunge T, Langhorst T, Zimmermann A, van der Hulst M, Bensebaa F, Moni S, Tao L (2022) Adapting technology learning curves for prospective techno-economic and life cycle assessments of emerging carbon capture and utilization pathways. Front Clim 4:820261. https://doi.org/10.3389/fclim.2022.820261

FDA (2013) FDA regulation of color additives in drug products. https://www.fda.gov/media/122829/download

Fernández-Ríos A, Laso J, Aldaco R, Margallo M (2024) Environmental implications and hidden costs of artisanal Spirulina (*Arthrospira platensis*) production and consumption. Environ Impact Assess Rev 108:107579. https://doi.org/10.1016/j.eiar.2024.107579

Foteinis S, Antoniadis-Gavriil A, Tsoutsos T (2018) Life cycle assessment of algae-to-biodiesel shallow pond production systems in the Mediterranean: influence of species, pond type, by (co)-product valorisation and electricity mix. Biofuels Bioprod Biorefin 12(4):542–558. https://doi.org/10.1002/bbb.1871

Fremy M (1866) Sur les matières colorantes des algues. C R Acad Sci 62:1126–1129

FSF (2025) Carte des producteurs de la Fédération des Spiruliniers de France. https://www.spiruliniersdefrance.fr. Accessed 25 Jan 2025

Gachkar D, Gachkar S, García Martínez A, Angulo C, Aghlmand S, Ahmadi J (2024) Artificial intelligence in building life cycle assessment. Archit Sci Rev 67(6):484–502. https://doi.org/10.1080/00038628.2024.2350491

Gammone MA, D'Orazio N (2015) Anti-obesity activity of the marine carotenoid fucoxanthin. Mar Drugs 13(4):2196–2214. https://doi.org/10.3390/md13042196

Gantt E, Lipschultz CA, Grabowski J, Zimmerman BK (1979) Phycobilisomes from blue-green and red algae: isolation criteria and dissociation characteristics. Plant Physiol 63(4):615–620. https://doi.org/10.1104/pp.63.4.615

García G, Sosa-Hernández JE, Rodas-Zuluaga LI, Castillo-Zacarías C, Iqbal H, Parra-Saldívar R (2021) Accumulation of PHA in the microalga *Scenedesmus* sp. under nutrient-deficient conditions. Polymers 13(1):131. https://doi.org/10.3390/polym13010131

Ghoroghi A, Rezgui Y, Petri I, Beach T (2022) Advances in application of machine learning to life cycle assessment: a literature review. Int J Life Cycle Assess 27(3):433–456. https://doi.org/10.1007/s11367-022-02030-3

Glazer AN (1992) Phycobiliprotein conjugates for immunofluorescence assays. U.S. Patent US5171846A U.S. Patent and Trademark Office. https://patents.google.com/patent/US5171846A/en

Glazer AN, Fang S (1973a) Chromophore content of blue-green algal phycobiliproteins. J Biol Chem 248:659–662. https://doi.org/10.1016/S0021-9258(19)44424-4

Glazer AN, Fang S (1973b) Formation of hybrid proteins from the α and β subunits of phycocyanins of unicellular and filamentous blue-green algae. J Biol Chem 248:663–671. https://doi.org/10.1016/S0021-9258(19)44425-6

Glazer AN, Hixson CS (1977) Subunit structure and chromophore composition of Rhodophytan phycoerythrins. *Porphyridium cruentum* B-phycoerythrin and b-phycoerythrin. J Biol Chem 252:32–42. https://pubmed.ncbi.nlm.nih.gov/833125/

González-Hernández J, Guzmán JL, Úbeda JCM, Acién FG (2024) A graphical simulation tool to estimate microalgae production capacity around the world. Algal Res 83:103665. https://doi.org/10.1016/j.algal.2024.103665

Goodwin TW (1958) The carotenoids of brown algae. Biochem J 69(2):251–259

Gupta SK, Ansari FA, Shriwastav A, Sahoo NK, Rawat I, Bux F (2016) Dual role of *Chlorella sorokiniana* and *Scenedesmus obliquus* for comprehensive wastewater treatment and biomass production for bio-fuels. J Clean Prod 115:255–264. https://doi.org/10.1016/j.jclepro.2015.12.040

Gurreri L, Rindina MC, Luciano A, Falqui L, Fino D, Mancini G (2024) Microalgae production in an industrial-scale photobioreactors plant: a comprehensive Life Cycle assessment. Sustain Chem Pharm 39:101598. https://doi.org/10.1016/j.scp.2024.101598

Habib MAB, Parvin M, Huntington TC, Hasan MR (2008) A review on culture, production and use of Spirulina as food for humans and feeds for domestic animals and fish. FAO Fisheries and Aquaculture CircularFAO, Rome

Hadiyanto H, Khoironi A, Dianratri I, Huda K, Suherman S, Muhammad F (2022) Biodegradation of oxidized high-density polyethylene and oxo-degradable plastic using microalgae *Dunaliella salina*. Environ Pollut Bioavailab 34(1):469–481. https://doi.org/10.1080/26395940.2022.2128884

Hallmann A (2015) Algae biotechnology–green cell-factories on the rise. Curr Biotechnol 4(4):389–415. https://doi.org/10.2174/2211550105666151107001338

Handler RM, Canter CE, Kalnes TN, Lupton FS, Kholiqov O, Shonnard DR, Blowers P (2012) Evaluation of environmental impacts from microalgae cultivation in open-air raceway ponds: analysis of the prior literature and investigation of wide variance in predicted impacts. Algal Res 1(1):83–92. https://doi.org/10.1016/j.algal.2012.02.003

Hannouf MB, Padilla-Rivera A, Assefa G, Gates I (2024) Social life cycle assessment (S-LCA) of technology systems at different stages of development. Int J Life Cycle Assess 30. https://doi.org/10.1007/s11367-024-02373-z

Herrera A, D'Imporzano G, Fernandez FGA, Adani F (2021) Sustainable production of microalgae in raceways: nutrients and water management as key factors influencing environmental impacts. J Clean Prod 287:125005. https://doi.org/10.1016/j.jclepro.2020.125005

Hoang AT, Sirohi R, Pandey A, Nižetić S, Shiung Lam S, Chen W-H, Luque R, Thomas S, Arıcı M, Viet Pham V (2022) Biofuel production from microalgae: challenges and chances. Phytochem Rev 22:1089–1126. https://doi.org/10.1007/s11101-022-09819-y

Honda M, Murakami K, Takasu S, Goto M (2022) Extraction of fucoxanthin isomers from the edible brown seaweed *Undaria pinnatifida* using supercritical CO2: effects of extraction conditions on isomerization and recovery of fucoxanthin. J Oleo Sci 71(8):1097–1106. https://doi.org/10.5650/jos.ess22077

ISO 14040 (2006) Environmental management—life cycle assessment—principles and framework. Standard. International Organization for Standardization, Geneva

ISO 14044 (2006) Environmental Management—Life Cycle Assessment—Requirements and Guidelines. Standard. International Organization for Standardization, Geneva

ISO 14075 (2024) Environmental management—Principles and framework for social life cycle assessment. Standard. International Organization for Standardization, Geneva

Joo G, Lee W, Choi Y (2021) Heavy metal adsorption capacity of powdered *Chlorella vulgaris* biosorbent: effect of chemical modification and growth media. Environ Sci Pollut Res 28:25390–25399. https://doi.org/10.1007/s11356-021-12396-w

Josa I, Garfí M (2023) Social life cycle assessment of microalgae-based systems for wastewater treatment and resource recovery. J Clean Prod 407:137121. https://doi.org/10.1016/j.jclepro.2023.137121

Jouannais P, Pizzol M (2022) Stochastic ex-ante LCA under multi-dimensional uncertainty: anticipating the production of undiscovered microalgal compounds in Europe. Environ Sci Technol 56(22):16382–16393. https://doi.org/10.1021/acs.est.2c04849

Khaw YS, Yusoff FM, Tan HT, Noor Mazli NAI, Nazarudin MF, Shaharuddin NA, Omar AR (2021) The critical studies of fucoxanthin research trends from 1928 to June 2021: a bibliometric review. Mar Drugs 19(11):606. https://doi.org/10.3390/md19110606

Khor WH, Kang HS, Lim JW, Iwamoto K, Tang CHH, Goh PS, Quen LK, Shaharuddin NMRB, Lai NYG (2022) Microalgae cultivation in offshore floating photobioreactor: State-of-the-art, opportunities and challenges. Aquac Eng 98:102269. https://doi.org/10.1016/j.aquaeng.2022.102269

Kim GY, Roh K, Han JI (2019) The use of bicarbonate for microalgae cultivation and its carbon footprint analysis. Green Chem 21(18):5053–5062. https://doi.org/10.1039/C9GC01107B

Kingsley PR, Braud L, Mediboyina MK, McDonnell K, Murphy F (2023) Prospects for commercial microalgal biorefineries: integrated pilot demonstrations and process simulations based techno-economic assessment of single and multi-product value chains. Algal Res 74:103190. https://doi.org/10.1016/j.algal.2023.103190

Kurpan D, Idà A, Körner FG, Bombelli P, da Silva Aguiar JP, Marinho LM, do Valle AF, Acién FG, Trasatti SP, Schievano A (2024) Long-term evaluation of productivity and harvesting efficiency of an industrial Spirulina (*Arthrospira platensis*) production facility. Bioresour Technol Rep 25:101741. https://doi.org/10.1016/j.biteb.2023.101741

Lemberg R, Legge JW (1938) Hematin compounds and bile pigments. Interscience Publishers

Leng J, Sha W, Wang B, Zheng P, Zhuang C, Liu Q, Wuest T, Mourtzis D, Wang L (2022) Industry 5.0: Prospect and retrospect. J Manuf Syst 65:279–295. https://doi.org/10.1016/j.jmsy.2022.09.017

Leong YK, Chang JS (2020) Bioremediation of heavy metals using microalgae: recent advances and mechanisms. Bioresour Technol 303:122886. https://doi.org/10.1016/j.biortech.2020.122886

Li J, Wang Y, Xu C, Liu S, Dai J, Lan K (2024) Bioplastic derived from corn stover: life cycle assessment and artificial intelligence-based analysis of uncertainty and variability. Sci Total Environ 946:174349. https://doi.org/10.1016/j.scitotenv.2024.174349

Li H, Chen L, Zhang F, Cai Z (2025) Graph-learning-based machine learning improves prediction and cultivation of commercial-grade marine microalgae *Porphyridium*. Bioresour Technol 416:131728. https://doi.org/10.1016/j.biortech.2024.131728

Liaaen-Jensen S (1978) Marine carotenoids. In: Scheuer PJ (ed) Marine natural products, Cambridge, MA, pp 1–73. https://doi.org/10.1016/B978-0-12-624002-3.50008-5

Liu T, Wang J, Hu Q, Cheng P, Ji B, Liu J, Chen Y, Zhang W, Chen X, Chen L, Gao L, Ji C, Wang H (2013) Attached cultivation technology of microalgae for efficient biomass feedstock production. Bioresour Technol 127:216–222. https://doi.org/10.1016/j.biortech.2012.09.100

Loke Show P (2022) Global market and economic analysis of microalgae technology: status and perspectives. Bioresour Technol 357:127329. https://doi.org/10.1016/j.biortech.2022.127329

Lourenço-Lopes C, Fraga-Corral M, Jimenez-Lopez C, Pereira AG, Garcia-Oliveira P, Carpena M, Prieto MA, Simal-Gandara J (2020) Metabolites from macroalgae and its applications in the cosmetic industry: a circular economy approach. Resources 9(9):101. https://doi.org/10.3390/resources9090101

Lourenço-Lopes C, Carreira-Casais A, Carperna M, Barral-Martinez M, Chamorro F, Jiménez-López C, Cassani L, Simal-Gandara J, Prieto MA (2023) Emerging technologies to extract fucoxanthin from *Undaria pinnatifida*: microwave vs. ultrasound assisted extractions. Mar Drugs 21(5):282. https://doi.org/10.3390/md21050282

Lu Y, Mu D, Xue Z, Xu P, Li Y, Xiang W, Burnett J, Bryant K, Zhou W (2021) Life cycle assessment of industrial production of microalgal oil from heterotrophic fermentation. Algal Res 58:102404. https://doi.org/10.1016/j.algal.2021.102404

Ma XN, Chen TP, Yang B, Liu J, Chen F (2016) Lipid production from *Nannochloropsis*. Mar Drugs 14(4):61. https://doi.org/10.3390/md14040061

Marangon BB, de Siqueira Castro J, Assemany PP, Machado NA, Calijuri ML (2024) Wastewater-grown microalgae biomass as a source of sustainable aviation fuel: life cycle assessment comparing hydrothermal routes. J Environ Manag 360:121164. https://doi.org/10.1016/j.jenvman.2024.121164

Martínez-Ramón N, Calvo-Rodríguez F, Iribarren D, Dufour J (2024) Frameworks for the application of machine learning in life cycle assessment for process modeling. Clean Environ Syst 100221. https://doi.org/10.1016/j.cesys.2024.100221

Meenatchisundaram K, Gowd SC, Lee J, Barathi S, Rajendran K (2024) Data-driven model development for prediction and optimization of biomass yield of microalgae-based wastewater treatment. Sustain Energy Technol Assess 63:103670. https://doi.org/10.1016/j.seta.2024.103670

Miyashita K, Maeda H, Hosokawa M, Sashima T (2004) Fucoxanthin as a bioactive marine carotenoid. J Mar Biotechnol 6(3):201–207

Miyashita K, Beppu F, Hosokawa M, Liu X, Wang S (2020) Nutraceutical characteristics of the brown seaweed carotenoid fucoxanthin. Arch Biochem Biophys 686:108364. https://doi.org/10.1016/j.abb.2020.108364

Mohibbullah M, Haque MN, Sohag AAM, Hossain MT, Zahan MS, Uddin MJ, Hannan MA, Moon IS, Choi JS (2022) A systematic review on marine algae-derived Fucoxanthin: An update of pharmacological insights. Mar Drugs 20(5):279. https://doi.org/10.3390/md20050279

Morales M, Bonnefond H, Bernard O (2020) Rotating algal biofilm versus planktonic cultivation: LCA perspective. J Clean Prod 257:120547. https://doi.org/10.1016/j.jclepro.2020.120547

Moutel B, André M, Kucma D, Legrand J, Grizeau D, Pruvost J, Gonçalves O (2015) Assessing the biofuel production potential of *Botryococcus braunii* strains by sensitive rapid qualitative chemotyping using chemometrically-assisted gas chromatography–mass spectrometry. Algal Res 11:33–42. https://doi.org/10.1016/j.algal.2015.05.001

Mu D, Min M, Krohn B, Mullins KA, Ruan R, Hill J (2014) Life cycle environmental impacts of wastewater-based algal biofuels. Environ Sci Technol 48(19):11696–11704. https://doi.org/10.1021/es5027689

Mularczyk M, Michalak I, Marycz K (2020) Astaxanthin and other nutrients from *Haematococcus pluvialis*—multifunctional applications. Mar Drugs 18(9):459. https://doi.org/10.3390/md18090459

Nobre ML, Tavares D, Fraga C, Oliveira B, Dias M, Mesquita S, Oliveira CM, Pires JC (2024) Techno-economic analysis of a circular microalgal approach for enhanced wastewater treatment and resource recovery in northern Portugal. J Clean Prod 434:140389. https://doi.org/10.1016/j.jclepro.2023.140389

Olaizola M (2000) Commercial production of astaxanthin from *Haematococcus pluvialis* using 25,000-liter outdoor photobioreactors. J Appl Phycol 12(3):499–506. https://doi.org/10.1023/A:1008159127672

Onen Cinar S, Chong ZK, Kucuker MA, Wieczorek N, Cengiz U, Kuchta K (2020) Bioplastic production from microalgae: a review. Int J Environ Res Public Health 17(11):3842. https://doi.org/10.3390/ijerph17113842

Onorato C, Rösch C (2020) Comparative life cycle assessment of astaxanthin production with *Haematococcus pluvialis* in different photobioreactor technologies. Algal Res 50:102005. https://doi.org/10.1016/j.algal.2020.102005

Oruganti RK, Biji AP, Lanuyanger T, Show PL, Sriariyanun M, Upadhyayula VK, Gadhamshetty V, Bhattacharyya D (2023) Artificial intelligence and machine learning tools for high-performance microalgal wastewater treatment and algal biorefinery: a critical review. Sci Total Environ 876:162797. https://doi.org/10.1016/j.scitotenv.2023.162797

Oslan SNH, Oslan SN, Mohamad R, Tan JS, Yusoff AH, Matanjun P, Mokhtar RAM, Shapawi R, Huda N (2022) Bioprocess strategy of *Haematococcus lacustris* for biomass and astaxanthin production keys to commercialization: perspective and future direction. Fermentation 8(4):179. https://doi.org/10.3390/fermentation8040179

Otálora P, Guzmán JL, Acién FG, Berenguel M, Reul A (2023) An artificial intelligence approach for identification of microalgae cultures. New Biotechnol 77:58–67. https://doi.org/10.1016/j.nbt.2023.07.003

Pajot A, Hao Huynh G, Picot L, Marchal L, Nicolau E (2022) Fucoxanthin from algae to human, an extraordinary bioresource: insights and advances in up and downstream processes. Mar Drugs 20(4):222. https://doi.org/10.3390/md20040222

Pajot A, Chollet S, Nicolau E, Marchal L (2023) Improving the extraction and the purification of fucoxanthin from *Tisochrysis lutea* using centrifugal partition chromatography. Algal Res 74:103174. https://doi.org/10.1016/j.algal.2023.103174

Pankratz S, Kumar M, Oyedun AO, Gemechu E, Kumar A (2020) Environmental performances of diluents and hydrogen production pathways from microalgae in cold climates: open raceway ponds and photobioreactors coupled with thermochemical conversion. Algal Res 47:101815. https://doi.org/10.1016/j.algal.2020.101815

Paterson S, Gómez-Cortés P, de la Fuente MA, Hernández-Ledesma B (2023) Bioactivity and digestibility of microalgae *Tetraselmis* sp. and *Nannochloropsis* sp. as basis of their potential as novel functional foods. Nutrients 15(2):477. https://doi.org/10.3390/nu15020477

Pechsiri JS, Thomas JBE, El Bahraoui N, Fernandez FGA, Chaouki J, Chidami S, Tinoco RR, Martin JP, Gomez C, Combe M, Gröndahl F (2023) Comparative life cycle assessment of conventional and novel microalgae production systems and environmental impact mitigation in urban-industrial symbiosis. Sci Total Environ 854:158445. https://doi.org/10.1016/j.scitotenv.2022.158445

Peng J, Yuan JP, Wu CF, Wang JH (2011) Fucoxanthin, a marine carotenoid present in brown seaweeds and diatoms: metabolism and bioactivities relevant to human health. Mar Drugs 9(10):1806–1828

Pereira H, Sá M, Maia I, Rodrigues A, Teles I, Wijffels RH, Navalho J, Barbosa M (2021) Fucoxanthin production from *Tisochrysis lutea* and *Phaeodactylum tricornutum* at industrial scale. Algal Res 56:2211–9264. https://doi.org/10.1016/j.algal.2021.102322

Pérez-Almada D, Galán-Martín Á, del Mar Contreras M, Castro E (2023) Integrated techno-economic and environmental assessment of biorefineries: review and future research directions. Sustain Energy Fuels 7(17):4031–4050. https://doi.org/10.1039/D3SE00405H

Perez-Garcia O, Escalante FM, De-Bashan LE, Bashan Y (2011) Heterotrophic cultures of microalgae: metabolism and potential products. Water Res 45(1):11–36. https://doi.org/10.1016/j.watres.2010.08.037

Pérez-López P, De Vree JH, Feijoo G, Bosma R, Barbosa MJ, Moreira MT, Wijffels RH, Van Boxtel AJ, Kleinegris DM (2017) Comparative life cycle assessment of real pilot reactors for microalgae cultivation in different seasons. App Energy 205:1151–1164. https://doi.org/10.1016/j.apenergy.2017.08.102

Pérez-López P, Rajaonison A, Zebian B, Bouallou C, González-Fernández C, Greses S, Gómez-Serrano C, Acién Fernández FG (2025) Exploiting outcomes of life cycle costing to conduct coherent screening social life cycle assessments of emerging systems: a case study of microalgae biorefineries. Int J Life Cycle Assess 1-22. https://doi.org/10.1007/s11367-024-02427-2

Piron M, Wu J, Fedele A, Manzardo A (2024) Industry 4.0 and life cycle assessment: evaluation of the technology applications as an asset for the life cycle inventory. Sci Total Environ 916:170263. https://doi.org/10.1016/j.scitotenv.2024.170263

Quigg A (2016) Micronutrients. In: Borowitzka M, Beardall J, Raven J (eds) The physiology of microalgae. Developments in applied phycology, vol 6. Springer, Cham

Quintero CD, Ventura A, Lepine O, Pruvost J (2021) Eco-design of Spirulina solar cultivation: key aspects to reduce environmental impacts using Life Cycle Assessment. J Clean Prod 299:126741. https://doi.org/10.1016/j.jclepro.2021.126741

Rafiaani P, Kuppens T, Thomassen G, Van Dael M, Azadi H, Lebailly P, Van Passel S (2020) A critical view on social performance assessment at company level: social life cycle analysis of an algae case. Int J Life Cycle Assess 25:363–381. https://doi.org/10.1007/s11367-019-01702-x

Raven JA, Giordano M (2014) Algae. Curr Biol 24(13):R590. https://doi.org/10.1016/j.cub.2014.05.039

Rebolledo-Leiva R, Moreira MT, González-García S (2023) Progress of social assessment in the framework of bioeconomy under a life cycle perspective. Renew Sust Energ Rev 175:113162. https://doi.org/10.1016/j.rser.2023.113162

Richardson JW, Johnson MD, Outlaw JL (2012) Economic comparison of open pond raceways to photo bio-reactors for profitable production of algae for transportation fuels in the Southwest. Algal Res 1(1):93–100. https://doi.org/10.1016/j.algal.2012.04.001

Rizzo A, Ajò A, Kang H, De Cola L, Jesus B (2024) Development of a new kappa-carrageenan hydrogel system to study benthic diatom vertical movements. PLoS One 19(4):e0297962. https://doi.org/10.1371/journal.pone.0297962

Rodríguez-Miranda E, Acién FG, Guzmán JL, Berenguel M, Visioli A (2021) A new model to analyze the temperature effect on the microalgae performance at large scale raceway reactors. Biotechnol Bioeng 118(2):877–889. https://doi.org/10.1002/bit.27617

Romagnoli F, Spaccini F, Boggia A, Paoli R, Feofilovs M, Ieviņa B, Rocchi L (2024) Microalgae cultivation in a biogas plant: environmental and economic assessment through a life cycle approach. Biomass Bioenergy 182:107116. https://doi.org/10.1016/j.biombioe.2024.107116

Romeiko XX, Zhang X, Pang Y, Gao F, Xu M, Lin S, Babbitt C (2024) A review of machine learning applications in life cycle assessment studies. Sci Total Environ 912:168969. https://doi.org/10.1016/j.scitotenv.2023.168969

Roselet F, Maica P, Martins T, Abreu PC (2013) Comparison of open-air and semi-enclosed cultivation system for massive microalgae production in sub-tropical and temperate latitudes. Biomass Bioenergy 59:418–424. https://doi.org/10.1016/j.biombioe.2013.09.014

Salla JVE, de Almeida TA, Silva DAL (2025) Integrating machine learning with life cycle assessment: a comprehensive review and guide for predicting environmental impacts. Int J Life Cycle Assess 1-23. https://doi.org/10.1007/s11367-025-02437-8

Samuels A (2025) Examining the integration of artificial intelligence in supply chain management from industry 4.0 to 6.0: a systematic literature review. Front. Artif Intell 7:1477044. https://doi.org/10.3389/frai.2024.1477044

Sarker NK, Kaparaju P (2023) A critical review on the status and progress of microalgae cultivation in outdoor photobioreactors conducted over 35 years (1986–2021). Energies 16(7):3105. https://doi.org/10.3390/en16073105

Schade S, Meier T (2021) Techno-economic assessment of microalgae cultivation in a tubular photobioreactor for food in a humid continental climate. Clean Techn Environ Policy 23:1475–1492. https://doi.org/10.1007/s10098-021-02042-x

Seghetta M, Romeo D, D'este M, Alvarado-Morales M, Angelidaki I, Bastianoni S, Thomsen M (2017) Seaweed as innovative feedstock for energy and feed–evaluating the impacts through a Life Cycle Assessment. J Clean Prod 150:1–15. https://doi.org/10.1016/j.jclepro.2017.02.022

Shannon E, Abu-Ghannam N (2018) Enzymatic extraction of fucoxanthin from brown seaweeds. Int J Food Sci Technol 53(9):2195–2204. https://doi.org/10.1111/ijfs.13808

Shi R, Handler RM, Shonnard DR (2019) Life cycle assessment of novel technologies for algae harvesting and oil extraction in the renewable diesel pathway. Algal Res 37:248–259. https://doi.org/10.1016/j.algal.2018.12.005

Sills DL, Van Doren LG, Beal C, Raynor E (2020) The effect of functional unit and co-product handling methods on life cycle assessment of an algal biorefinery. Algal Res 46:101770. https://doi.org/10.1016/j.algal.2019.101770

Skifa I, Chauchat N, Cocquet PH, Le Guer Y (2024) Microalgae cultivation in raceway ponds: advances, challenges, and hydrodynamic considerations. EFB Bioeconomy Journal 100073. https://doi.org/10.1016/j.bioeco.2024.100073

Somers MD, Chen P, Clippinger J, Cruce JR, Davis R, Lammers PJ, Quinn JC (2021) Techno-economic and life-cycle assessment of fuel production from mixotrophic *Galdieria sulphuraria* microalgae on hydrolysate. Algal Res 59:102419. https://doi.org/10.1016/j.algal.2021.102419

Strain HH (1933) Carotenoids of brown algae. J Biol Chem 103(3):501–510

Stunda-Zujeva A, Zuteris M, Rugele K (2018) Sunlight potential for microalgae cultivation in the mid-latitude region–the Baltic states. Agron Res 16(3):910–916. https://doi.org/10.15159/AR.18.126

Tandeau de Marsac N (2003) Phycobiliproteins and phycobilisomes: the early observations. Photosynth Res 76:193–205. https://doi.org/10.1023/A:1024954911473

Teng SY, Yew GY, Sukačová K, Show PL, Máša V, Chang JS (2020) Microalgae with artificial intelligence: a digitalized perspective on genetics, systems and products. Biotechnol Adv 44:107631. https://doi.org/10.1016/j.biotechadv.2020.107631

Tsang M, Fox-Lent C, Wallace S, Welp T, Bates M, Linkov I (2015) Life-cycle impacts of soybean and algae biodiesel: case study of US marine vessels. Biofuels Bioprod Biorefin 9(5):567–580. https://doi.org/10.1002/bbb.1569

Tu Q, Eckelman M, Zimmerman J (2017) Meta-analysis and harmonization of life cycle assessment studies for algae biofuels. Environ Sci Technol 51(17):9419–9432. https://doi.org/10.1021/acs.est.7b01049

Tzachor A, Smidt-Jensen A, Ramel A, Geirsdóttir M (2022) Environmental impacts of large-scale Spirulina (*Arthrospira platensis*) production in Hellisheidi geothermal park Iceland: life cycle assessment. Mar Biotechnol 24(5):991–1001. https://doi.org/10.1007/s10126-022-10162-8

UNEP (2020) In: Benoît Norris C, Traverso M, Neugebauer S, Ekener E, Schaubroeck T, Russo Garrido S, Berger M, Valdivia S, Lehmann A, Finkbeiner M, Arcese G (eds) Guidelines for social life cycle assessment of products and organizations 2020. United Nations Environment Programme (UNEP)

Valdivia S, Backes JG, Traverso M, Sonnemann G, Cucurachi S, Guinée JB, Schaubroeck T, Finkbeiner M, Leroy-Parmentier N, Ugaya C, Peña C, Zamagni A, Inaba A, Amaral M, Berger M, Dvarioniene J, Vakhitova T, Benoit-Norris C, Prox M, Foolmaun R, Goedkoop M (2021) Principles for the application of life cycle sustainability assessment. Int J Life Cycle Assess 26:1900–1905. https://doi.org/10.1007/s11367-021-01958-2

van der Giesen C, Cucurachi S, Guinée J, Kramer GJ, Tukker A (2020) A critical view on the current application of LCA for new technologies and recommendations for improved practice. J Clean Prod 259:120904. https://doi.org/10.1016/j.jclepro.2020.120904

Vázquez-Romero B, Perales JA, Pereira H, Barbosa M, Ruiz J (2022) Techno-economic assessment of microalgae production, harvesting and drying for food, feed, cosmetics, and agriculture. Sci Total Environ 837:155742. https://doi.org/10.1016/j.scitotenv.2022.155742

Vázquez-Romero B, Villar-Navarro E, Perales JA, Garrido-Pérez C, Ruiz J (2024) Techno-economic analysis of using microalgae to treat streams from fish RAS farming and replace fish meal: a case study. J Water Process Eng 59:104904. https://doi.org/10.1016/j.jwpe.2024.104904

Vol T, Tran D (2014) Carotene and antioxidant capacity of *Dunaliella salina* strains. J Nutri Health Food Eng 2(2):123–128. https://doi.org/10.12691/jnh-2-2-2

Wang K, Khoo KS, Leong HY, Nagarajan D, Chew KW, Ting HY, Selvarajoo A, Chang JS, Show PL (2022) How does the Internet of Things (IoT) help in microalgae biorefinery? Biotechnol Adv 54:107819. https://doi.org/10.1016/j.biotechadv.2021.107819

Watkins J, Zhu Y, Valdez P, Lords C, Zeller A, Bohutskyi P, Sims RC (2024) Techno-economic analysis of bioplastic and biofuel production from a high-ash microalgae biofilm cultivated in effluent from a municipal anaerobic digester. Algal Res 84:103774. https://doi.org/10.1016/j.algal.2024.103774

Wernet G, Bauer C, Steubing B, Reinhard J, Moreno-Ruiz E, Weidema B (2016) The ecoinvent database version 3 (part I): overview and methodology. Int J Life Cycle Assess 21:1218–1230. https://doi.org/10.1007/s11367-016-1087-8

Willstätter R, Page H (1914) Untersuchungen über chlorophyll. Justus Liebigs Ann Chem 401(1):147–153

Wolf M, Pant R, Chomkhamsri K, Sala S, Pennington D (2012) The International Reference Life Cycle Data System (ILCD) Handbook—Towards more sustainable production and consumption for a resource-efficient Europe. Publications Office of the European Union, Luxembourg. EUR 24982 EN JRC66506. https://data.europa.eu/doi/10.2788/85727

Wu Q, Liu L, Miron A, Klímová B, Wan D, Kuča K (2016) The antioxidant, immunomodulatory, and anti-inflammatory activities of

Spirulina: an overview. Arch Toxicol 90:1817–1840. https://doi.org/10.1007/s00204-016-1744-5

Xu X, Lu Y, Vogel-Heuser B, Wang L (2021) Industry 4.0 and industry 5.0—inception, conception and perception. J Manuf Syst 61:530–535. https://doi.org/10.1016/j.jmsy.2021.10.006

Yokoyama A, Yano K (1988) Antioxidant properties of fucoxanthin. J Agric Food Chem 36(2):300–305

Zarra T, Senatore V, Zorpas AA, Oliva G, Voukkali I, Belgiorno V, Naddeo V (2024) Advanced membrane photobioreactors in algal CO2 biofixation and valuable biomass production: integrative life cycle assessment and sustainability analysis. Sustain Chem Pharm 40:101658. https://doi.org/10.1016/j.scp.2024.101658

Zhang C (2003) Method of extracting phycocyanin from Spirulina. Chinese Patent CN1295080A. https://patents.google.com/patent/CN1295080A/en

Zhang Y, Kendall A (2019) Effects of system design and co-product treatment strategies on the life cycle performance of biofuels from microalgae. J Clean Prod 230:536–546. https://doi.org/10.1016/j.jclepro.2019.05.137

Zhila NO, Kiselev EG, Shishatskaya EI, Ghorabe FDE, Kazachenko AS, Volova TG (2025) Comparative study of the synthesis of polyhydroxyalkanoates by cyanobacteria *Spirulina platensis* and green microalga *Chlorella vulgaris*. Algal Res 85:103826. https://doi.org/10.1016/j.algal.2024.103826

Dr. Léa Braud is the Innovation Manager at the European Algae Biomass Association (EABA) and a postdoctoral researcher at ETH Zurich, Switzerland. Her research is dedicated to advancing life cycle assessment (LCA) methodologies in the algae sector, with a particular focus on prospective approaches and the integration of economic, environmental, and social dimensions to facilitate comprehensive and meaningful sustainability assessments. Dr. Braud obtained her PhD in Biosystems and Food Engineering from University College Dublin, Ireland, where she specialized in the development of open-source tools to adapt LCA methodologies for algae systems, enhancing transparency and reproducibility. She currently coordinates the EABA LCA task force, which aims to harmonize LCA methodologies within the algae sector. Beyond academia, she is passionate about science outreach for children, organizing workshops on algae species identification and the use of algae-based products in everyday life. She is dedicated to promoting sustainable algae-based solutions and actively engaging local communities in harnessing the potential of algae.

Dr. Anne Pajot is the Communication Manager at the European Algae Biomass Association (EABA) and a Project Manager at Sea4earth. At EABA, she supports the visibility and outreach of the European algae sector, fostering connections between researchers, industry, and policymakers.Trained as a biotechnology engineer at the University of Technology of Compiègne, France, Anne pursued a PhD at IFREMER Nantes, France, where she studied the impact of light and nutrients on the photosynthetic system of *Tisochrysis lutea*, focusing on fucoxanthin, a pigment of high interest for nutraceutical, pharmaceutical, and cosmetic applications. She later worked as a postdoctoral researcher at CEA Cadarache, France, exploring monoclonal antibody production in *Phaeodactylum tricornutum* within the PHAEOmAbs project. At Sea4earth, she leads projects dedicated to developing a fully integrated macroalgal production chain for human nutrition. She also teaches biology, chemistry, and biochemistry at the University of Nice, France. Passionate about algae and their potential, she is committed to advancing innovation and communication in the sector.

Dr. Arianna Rizzo is a marine biologist passionate about natural resources for human and ecosystem services, especially microalgae. Since 2017, she has worked at the intersection of academia and industry, specializing in improving microalgae cultivation and species selection to produce high-value molecules such as astaxanthin. In 2024, she defended her PhD on diatom physiology and optical properties as part of BEEP, a European program led by Cambridge University. Following her PhD, she co-founded Blue2, a startup dedicated to revalorizing industrial emissions through microalgae. At Blue2, she is responsible for species selection, scale-up processes, and the extraction of high-value molecules. Beyond her research, she organized and chaired three editions of the Young Algaeneers Symposium (YAS) under the patronage of EABA, creating a platform for young professionals in the algae sector. Through YAS, she became part of the EABA team, where she manages the EU4Algae Project, a European Commission initiative aimed at advancing the algae sector and facilitating the acceptance of algae-based products on the market.

Juliette Armeni-Ripari is the Project Innovation Manager at the European Algae Biomass Association (EABA) and the Marine Biological Resources Manager at Pôle Mer Méditerranée, France. At EABA, she helps structure the European algae industry, supporting initiatives such as the EU4Algae platform and Algaflix. She fosters collaboration between industry, academia, and policymakers to advance sustainable algae-based solutions. At Pôle Mer Méditerranée, a French competitiveness cluster dedicated to the blue economy, she is in charge of the Marine Biological Resources sector. She assists companies, research institutes, and public authorities in structuring and accelerating innovative projects, ensuring their integration within national and European funding frameworks. With a background in engineering and laboratory research, Juliette developed a strong expertise in marine biological resources, particularly through her early work on microalgae in both academic and private sectors. Passionate about marine ecosystems, she advocates for algae as a bridge between science, innovation, and public engagement.

Vítor Verdelho is a Portuguese entrepreneur and deep tech innovator with a long-standing career in marine biotechnology, entrepreneurship, and sustainable technologies. He is currently the CEO, shareholder, and co-founder of *Cell4Food*, the first Portuguese startup focused on cellular agriculture and aquaculture. Since 2018, Vítor has served as General Manager of the European Algae Biomass Association (EABA), after previously leading the organization as President from 2014 to 2018. Under his leadership, EABA has become a key European reference in algae-based biotechnologies. He co-founded *Necton* in 1987, a marine biotechnology company specializing in microalgae production and traditional sea salt. He also co-founded *A4F—Algae for Future*, a company dedicated to industrial-scale microalgae production systems. Vítor has over 30 years of experience developing and managing innovative businesses. He has worked as an advisor or shareholder in multiple startups, including *PhytoBloom*, *InnoPhage*, and *Ecotlon*, focusing on sectors ranging from health and environment to circular economy. He is actively engaged in education and mentorship. As a visiting professor at Católica Porto Business School, he mentors innovation projects and teaches in MBA and executive programs. Over the past two decades, he has supported more than 150 entrepreneurial ventures. Vítor played a key role in fostering biotech entrepreneurship at the Catholic University of Portugal, where he coordinated technology transfer and incubation programs for over a decade. He also contributed to national creative industries through ADDICT and worked with government agencies and universities on innovation strategy. He holds a degree in physics from the University of Porto and pursued postgraduate studies at the Catholic University of Portugal. His academic background blends hard science with business and innovation management. Fluent in Portuguese, English, Spanish, and French, Vítor is a frequent speaker on topics such as aquaculture, microalgae applications, and renewable bio-based industries. He is also co-author of scientific publications and holds a patent in the field of nutritional compositions. As president of the *Portuguese Association of Applied Algology*, he is furthering national engagement in marine biotechnology. With a strong European network and a deep commitment to sustainability, Vítor Verdelho is a driving force in the bioeconomy transition.

8 Integrating Traditional Knowledge into the Phyconomy of Temperate Regions

Schery Umanzor and Jan Marie Conitz

Abstract

Seaweeds have been foundational in the subsistence, culture, and economies of Indigenous coastal communities across temperate regions for millennia. In Southeast Alaska, Indigenous groups such as the Tlingit, Haida, and Tsimshian developed intricate traditional knowledge systems governing the sustainable harvest, use, and stewardship of various seaweed species, e.g., *Pyropia abbottiae*, *Devaleraea mollis*, and *Nereocystis luetkeana*. With the rise of seaweed aquaculture in Alaska and the broader global demand for seaweed-derived products, there is growing recognition of the value of integrating traditional ecological knowledge (TEK) into developing a resilient "phyconomy." This chapter explores historical and contemporary seaweed practices in Alaska, highlighting how TEK aligns with modern sustainability frameworks such as the Blue Economy. It also evaluates the ecological and economic implications of expanding aquaculture while preserving cultural heritage. Challenges remain, including fragmented policy, limited processing infrastructure, and environmental stressors affecting wild populations. Nonetheless, Alaska's efforts, such as Indigenous-led initiatives, sustainable harvest protocols, and state support for seaweed farming, illustrate a promising model for temperate regions worldwide seeking to balance ecological, cultural, and economic objectives through inclusive and sustainable seaweed industry development.

The term "phyconomy," which combines "phycology" and "agronomy," refers to commercial cultivation of seaweed.

S. Umanzor (✉) · J. M. Conitz
College of Fisheries and Ocean Science, University of Alaska Fairbanks, Juneau, Fairbanks, AK, USA
e-mail: sumanzor@alaska.edu

Keywords

Indigenous stewardship · Seaweed aquaculture · Southeast Alaska · Seaweed processing · Traditional ecological knowledge

8.1 Introduction

For thousands of years, coastal peoples across temperate regions of the world have relied on wild seaweed as a critical resource for food, materials, and trade (Abbott and Dawson 1978; MacDiarmid and Selwood 2015). Wild seaweed harvesting has played a pivotal role in the cultural traditions of temperate coastal communities worldwide (Mouritson et al. 2019; Turner and Loewen 1998). While the ancient use of seaweeds is well-documented in oral traditions and cultural practices, physical evidence of seaweed use in the archaeological record is relatively limited since seaweeds lack hard structures. However, researchers uncovered physical traces of at least nine seaweed species at one site in southern Chile, along with stone tools, mortars, and hearths (Dillehay et al. 2008). Similarly, recent archaeological research provides evidence that seaweed was consumed in Europe during the Mesolithic period, approximately 8000 years ago (Buckley et al. 2023) offering a glimpse into the deep historical relationship between coastal communities and seaweed resources (Mouritson et al. 2019; Mac Monagail et al. 2017).

Coastal Indigenous communities across the temperate regions developed various methods for harvesting, preserving, storing, and consuming seaweed (Turner 2005; Moss 2011; Garibaldi and Turner 2004), along with ingenious uses for tools, storage vessels, and other essential items. For example, Aboriginal Australians used bull kelp (*Durvillaea* spp.) to craft water carriers, demonstrating innovative use of available resources (Thurstan et al. 2018).

A. T. Critchley et al. (eds.), *Advances in Temperate Phyconomy: Algal Harvest and Cultivation in Globally Distributed Temperate Waters*, Developments in Applied Phycology 15, https://doi.org/10.1007/978-3-032-02955-3_8

Similarly, Indigenous peoples of the Pacific Northwest of North America, including the Tlingit and Haida, traditionally used bull kelp (*Nereocystis luetkeana*) to create ropes and fishing lines. The long, flexible stipes were dried, treated, and braided or spliced to produce strong cords for various marine purposes, such as anchoring and fishing (Garza 2012).

The revival of traditional knowledge and sustainable harvesting practices among coastal Indigenous communities in the temperate regions of the Pacific Ocean highlights the enduring value of ecological wisdom developed over centuries. Indigenous peoples cultivated intricate knowledge systems to manage seaweed resources sustainably, ensuring their availability for future generations. This chapter focuses on Southeast Alaska, where traditional seaweed harvesting remains a vital tradition among rural communities and Indigenous peoples, predominantly Tlingit, Haida, and Tsimshian. Seaweed species such as black seaweed (*Pyropia abbottiae*), red ribbon seaweed (*Devaleraea mollis*), and bull kelp (*Nereocystis luetkeana*) have been and remain central to diets and cultural vitality among all three cultural groups (Garza 2012 ; Williams 2022). We examine patterns of seaweed use in the region and consider the potential impacts of developing a modern seaweed aquaculture industry here as a case study to explore the opportunities and challenges of integrating traditional knowledge into the evolving phyconomy.

8.2 Cultural and Historical Context of Seaweed Harvest and Use in Southeast Alaska

Alaska boasts approximately 50,000 kilometers of tidal shoreline, supporting over 500 seaweed species, which thrive in its cold, nutrient-rich waters (Lindeberg and Lindstrom 2010; O'Clair and Lindstrom 2000). Southeast Alaska's intertidal environments vary significantly due to oceanographic conditions, glacial influence, and freshwater inputs (Hood and Berner 2009; Ham et al. 2024). The outer coast features warmer, higher salinity waters. At the same time, the inner bays and channels have less saline, colder waters, often influenced by glacial runoff to the north and warmer currents to the south. This environmental diversity shapes the distribution of seaweed species. Some have narrow ranges within specific salinity zones, temperature, shoreline type, etc., while others more broadly span the coastline from Southeast Alaska to British Columbia and Washington state (O'Clair et al. 1996).

Seaweed has been integral to the cultural, nutritional, and economic practices of rural and Indigenous communities in Southeast Alaska for centuries (Fig. 8.1). Traditional economies associated with wild seaweed harvest, preservation, and distribution contrast sharply with the global market-driven focus of modern commercial operations (Garza 2012 ; Turner 2003; Victor-Howe 2008). Traditional practices have been developed over generations around sustainability, community engagement, and ecological balance, while commercial practices generally focus on maximizing yield and market value in the near term.

Fig. 8.1 Drying seaweed on a roof in Shakan, Alaska, July 1911. Note the regularly formed, rectangular seaweed cakes (probably *Laak'ask*, black seaweed—*Pyropia abbottiae*) on woven mats. (Used with permission: University of Washington Libraries, Special Collections, John N. Cobb Photo Collection)

Traditional seaweed harvesting has always been a communal activity reinforcing social bonds and cultural values. Families and clans gather for harvests, such as the annual collection of herring roe on seaweed or hemlock branches, working together to process and share the bounty (Fig. 8.2). This collaborative approach emphasizes stewardship, cooperative ownership, and resource sharing, fostering social resilience (Kitka 1998; Victor-Howe 2008). The transmission of traditional knowledge, particularly to younger generations, continues to be integral to these practices, embedding ecological stewardship and cultural continuity in everyday life (Williams 2020).

Of the many species used by Indigenous people in Southeast Alaska, black seaweed (*Pyropia abbottiae*), known as *Laak'ask* in Tlingit, *La'axsk* in Tsimshian (*Sm'algyax*), and *Sgi'w* in Haida, was, and still is, the most highly prized for its flavor and nutritional value (Turner 2003). Red ribbon seaweed (*Devaleraea mollis*), or *X'aach* in Tlingit, is also highly edible, though less valuable in sharing and trade networks (Kitka 1998). It is more widely available throughout the Southeast Alaska region, while black seaweed requires specific conditions found only in more exposed areas near or along the outer coasts.

Many other seaweeds, including *Alaria marginata* (ribbon kelp), *Saccharina latissima* (sugar kelp), and *Nereocystis lutkeana* (bull kelp), were important in traditional diets and

Fig. 8.2 Traditionally harvested herring roe on kelp (*Macrocystis pyrifera*), 1962. (Used with permission: Alaska State Archive)

material cultures. Even the common rockweed or popweed, *Fucus distichus*, was harvested as food and prepared in various ways. Both rockweed and giant kelp (*Macrocystis pyrifera*) were used as edible substrates to collect highly valued herring roe in the spring. Kelp parts were used to craft items used in food gathering, fishing, hunting, storage, and trade. For example, bull kelp pneumatocysts were dried and crafted into storage containers to store fish oil, and its stipes were used to make fishing lines and other tools (Garza 2012 ; Naar 2020). These ancient, ongoing practices demonstrate a holistic use of seaweed, which is not isolated as a specialty or novelty item but integrated into multiple facets of life. Seaweed harvest, consumption, and trade are still a valued and highly anticipated seasonal part of life in contemporary Indigenous communities.

Indigenous communities developed intricate knowledge systems around seaweed harvesting, such as observation of environmental cues to guide harvest timing for various seaweed species. For example, phenological indicators such as the blooming of nettles, skunk cabbage, and alder, as well as increased activity of herring and humpback whales, signal that black seaweed is ready for harvest (Williams 2020; Turner 2003). Similarly, the readiness of other species, like red laver (*Porphyra/Pyropia* spp.), is marked by the appearance of migratory birds or the blooming of specific flowers, while the molting of crustaceans or fish spawning events often indicate the optimal time to collect bull kelp (*Nereocystis luetkeana*) (Lantz and Turner 2003; Turner and Clifton 2009). These observations reflect an understanding of interconnected ecological systems, resulting in harvest practices that align with natural growth cycles and ensure the regeneration of seaweed beds. For example, the *Gitga'at* in British Columbia cleared entire sections of black seaweed from rocky shores. Based on their observation, this practice promoted healthier and more robust regrowth the following year, in contrast to allowing unpicked blades to decompose (Turner 2003). For other species, harvesting is selective, where only mature fronds are often collected, while the holdfast and stipe are left intact to allow regrowth and maintain population stability (Victor-Howe 2008). These traditions and practices ensure that harvests are inherently sustainable over generations and are rooted in deep ecological knowledge. This traditional ecological knowledge (TEK) could offer valuable insights into seaweed management in the modern phyconomy.

Historically, commercial seaweed use in Alaska was initially limited to informal trade and cottage industries. Still, as commercial markets developed, industry practices often contrasted sharply with the sustainable practices of Indigenous communities. The most significant large-scale commercial seaweed harvest impacting traditional use was associated with the herring roe-on-kelp fishery. Giant kelp (*Macrocystis pyrifera*) was a preferred substrate for this commercial fishery. Herring roe attached to kelp (*kazunoko kombu*) is highly prized in Japanese markets, and the size and quality of *Macrocystis* blades align perfectly with consumer preferences (Stekoll and Else 1990). Its long, broad blades offer an extensive surface area for herring to attach their eggs, maximizing roe yield, and their size and durability allow them to support substantial egg loads without breaking apart during handling and transport (Graham et al. 2007). Initially, the roe-on-kelp fishery was a small-scale local enterprise in which wild-harvested spawn on kelp was sold to Asians who worked in Southeast Alaska canneries. However, the fishery intensified until it became unsustainable by the mid-1940s and was closed by the State of Alaska in 1967. A state-managed herring roe-on-kelp fishery opened during the 1980s and early 1990s, in which giant kelp blades were harvested and put into impoundments with spawning herring to collect the roe-on-kelp. The kelp harvest was loosely managed without adequate knowledge or consideration of its life history and regenerative capacity, resulting in ecological damage and depletion around traditional harvest areas (Victor-Howe 2008). Regulatory measures were needed to prevent further harm to the kelp forests (North 1971; McHugh 2003).

The initiative to develop giant kelp aquaculture was motivated by the need to maintain the valuable herring roe-on-kelp fishery by replacing the unsustainable wild harvest with farmed kelp. Despite decades of study and experimentation, which resulted in a more detailed understanding of the requirements for reproducing and growing giant kelp, commercial-scale production has not yet been achieved (Stekoll et al. 2021).

The ecological value of wild giant kelp forests has gained broader recognition compared to their primarily commercial use during the roe-on-kelp fisheries of the 1980s and 1990s. These forests are essential to the marine ecosystems of the northeast Pacific, providing critical spawning habitats for herring and serving as shelter and nursery areas for various marine species (Graham et al. 2007). Integrating Indigenous traditions, which emphasize sustainable practices and a deep understanding of ecosystem dynamics, can guide effective conservation strategies, ensuring the resilience of these ecosystems while respecting cultural heritage.

8.3 The Rise of Seaweed Aquaculture

The formal promotion of seaweed aquaculture in Alaska began in 2016 with the establishment of the Alaska Governor's Mariculture Task Force (ADF&G 2017), marking a pivotal moment in the state's efforts to foster a sustainable aquaculture industry.[1] The first commercial seaweed harvest under this new authority occurred in 2017, and the industry has grown steadily since then.

Seaweed aquaculture in Alaska has emerged as a response to the increasing global demand for seaweed and seaweed-derived products. Northern temperate seaweed species, predominantly kelp and fucoids, thrive in Alaska's clean, cold, and nutrient-rich waters. However, in contrast with East Asian countries such as Japan, Korea, and China, which have old traditions of seaweed farming, Alaska does not have a historically established commercial seaweed industry based on either wild harvest or aquaculture. Instead, efforts within the state focus on building an industry from the ground up, guided by principles of sustainability and economic inclusivity (State of Alaska 2016; Mariculture Task Force 2018).

The formation of the Mariculture Task Force highlighted the state's intention to "develop a viable and sustainable mariculture industry producing shellfish and aquatic plants for the long-term benefit of Alaska's economy, environment, and communities" (State of Alaska 2016). The Alaska Mariculture Development Plan further emphasized that this industry must be "environmentally sound and designed to complement rather than replace existing uses" (Mariculture Task Force 2018). This dual focus on sustainability and inclusivity seeks to preserve the cultural and ecological significance of wild seaweed harvest while capitalizing on the economic potential of aquaculture.

Seaweed consumption has risen globally due to its integration into popular Asian dishes such as sushi, health-focused natural food markets, and high-end experimental cuisines (Buschmann et al. 2017; Cikoš et al. 2019; Mouritson et al. 2019). Beyond food, seaweed extracts such as alginates and carrageenan are essential in various industries, including cosmetics, agriculture, and pharmaceuticals (Pereira and Neto 2015; Cikoš et al. 2019). Despite the growing dominance of aquaculture as a seaweed biomass supplier, wild harvest continues in temperate regions, albeit on a smaller scale, contributing less than 3% of global seaweed production as of 2019 (Cai 2021).

The rapid growth of the seaweed aquaculture industry in Alaska reflects the rising demand for "green" and sustainable products such as plant-based foods, biofuels, and natural fertilizers (Welch 2019). To date, relatively stable commercial-scale production has been achieved in two species, ribbon kelp (*Alaria marginata*) and sugar kelp (*Saccharina latissima*). Several other species, including bull kelp (*Nereocystis luetkeana*) and five-rib kelp (*Costaria costata*), are being tested at pilot scales (MRTC 2023). Processing capacity within Alaska remains limited, as the size and seasonality of seaweed harvests do not yet align with the operational requirements and timelines of large-scale processors. Smaller, year-round local fish processors, such as E.C. Phillips & Son in Ketchikan, have begun incorporating seaweed processing into their schedules. As an alternative, shipping raw seaweed out of state for processing is not cost-effective due to high energy and transport costs. Scaling seaweed production to match local processing capacities will be essential (McKinley Research Group 2021).

8.4 Management and Sustainable Wild Harvest

Wild harvest is expected to be complementary until seaweed aquaculture can reliably produce the volumes and variety of species needed for the specialty food market. Such may be the case for bull kelp (*Nereocystis luetkeana*) and giant kelp (*Macrocystis pyrifera*) to continue to meet commercial demand, particularly for uses like specialty salsas and the herring roe-on-kelp fishery (Dupuis et al. 2022; US Department of the Interior 2024).

Wild harvest management must balance ecological conservation with industry needs. Robust and regionally specific data are essential for developing management strategies that support the long-term sustainability of seaweed resources while ensuring their role in Alaska's evolving phyconomy. However, the vast and complex nature of Alaska's coastline, which is characterized by diverse habitats, varying oceanographic conditions, and seasonal dynamics, challenges the ability to establish accurate baselines and effectively monitor changes (Bennion et al. 2019).

The Alaska Department of Fish and Game (ADF&G) oversees wild seaweed harvesting through a regulatory

[1] Aquaculture in Alaska is referred to as mariculture and is focused on invertebrates and seaweeds only, as finfish farming is prohibited.

framework rooted in the Alaska Constitution. Living resources are treated as common property, managed under the "sustained yield principle" to ensure resource availability for future generations (Alaska Statute 16.05.020; ADF&G 1972).

As in other data-limited fisheries, the management of wild seaweed harvesting reflects a precautionary approach. Commercial harvest of species like black seaweed and red laver (*Pyropia* spp.), and red ribbon seaweed (*Devaleraea mollis*) is generally not allowed due to the inability to assess and monitor abundances adequately. Most permits are issued for easily identifiable kelp species such as *Saccharina*, *Alaria*, and *Nereocystis*. Harvest activities are closely monitored through annual reports submitted by harvesters, which include data on species, quantities, and regions (Victor-Howe 2008; ADF&G 2024).

In addition to human harvest activities, wild seaweed populations in many regions face other threats from human-induced and natural factors. Ecological change has variable and complex effects on seaweed populations. For example, grazing by herbivores such as sea urchins and gastropods can heavily impact seaweed abundance and recovery and change community structure, especially in areas where predator populations, such as sea otters, are declining (Estes et al. 2010). Seaweed populations in some regions of Alaska appear stable (Hollarsmith et al. 2023). Still, in other regions, changes in ocean temperatures are altering the growth, distribution, and reproductive cycles of key species like *Nereocystis luetkeana* and *Macrocystis pyrifera*. These changes can affect population resilience, potentially placing some at greater risk and reducing their economic viability for wild harvest (Graham et al. 2007; Gorra et al. 2022; Drakard et al. 2023). Additionally, anthropogenic pressures such as coastal development, pollution, and increased industrial activity further degrade critical seaweed habitats, compounding ecological stressors (Krumhansl et al. 2016). Without comprehensive studies on species abundance, productivity, and the cumulative impacts of environmental and anthropogenic factors, risks of resource depletion, ecosystem disruption, and diminished economic potential are high.

The sustainable management of wild seaweed harvests is essential for maintaining biodiversity, supporting local communities, and ensuring the long-term viability of the phyconomy of temperate regions. In Alaska, improved scientific understanding will provide the foundation for adaptive management strategies, including biologically based harvest thresholds, habitat protection, and climate resilience planning. Moreover, it will facilitate the integration of traditional ecological knowledge into modern management practices, fostering a balanced approach that aligns with ecological, cultural, and economic goals.

8.5 The Blue Economy and Revival of Traditional Practices

The resurgence of interest in seaweed as a traditional resource within Indigenous communities and its development as a global commodity have led to initiatives integrating traditional practices and modern management. In Alaska, the Mariculture Development Plan aims to balance economic growth with cultural preservation (Mariculture Task Force 2018). Collaborative projects between Indigenous communities and state agencies exemplify efforts to align traditional knowledge and industry goals (MRTC 2023). The intersection of traditional knowledge and the principles of the Blue Economy[2] offers a robust framework for sustainable seaweed industry development (Elegbede et al. 2023).

Indigenous stewardship practices are deeply aligned with the core tenets of the Blue Economy, which emphasize the sustainable use of marine resources to promote ecological, economic, and social well-being. For centuries, Indigenous communities have harvested seaweed using methods prioritizing ecological health, such as selective harvesting and aligning harvests with natural growth cycles (Turner 2003; Williams 2022). These practices ensure the renewal of resources and foster biodiversity, echoing the ecological goals of modern sustainability frameworks.

Traditional ecological knowledge provides insights into sustainable harvesting that are increasingly relevant to the modern phyconomy. For example, Indigenous practices like clearing entire sections of black seaweed (*Pyropia abbottiae*) to promote robust regrowth can inform scalable, sustainable harvest protocols for commercial operations (Turner 2003). Similarly, seaweed's capacity to absorb excess nutrients and stabilize marine habitats (Umanzor et al. 2017; Reid et al. 2013) aligns with Indigenous knowledge systems that view seaweed as integral to marine ecosystem health.

Seaweed aquaculture and the traditional wild harvesting of seaweed biomass are uniquely positioned to address global environmental challenges. As a primary producer, seaweed plays a vital role in carbon removal and nutrient cycling, potentially helping mitigate climate change drivers and reduce coastal eutrophication (Gallager et al. 2022; Stephens et al. 2024). Unlike land-based agriculture, seaweed cultivation requires no fresh water, fertilizers, or arable land, making it a relatively low-impact alternative for producing food, biofuels, and bioplastics (Buschmann et al. 2017). These ecological benefits are complemented by economic opportunities, such as producing high-value specialty food products and natural fertilizers that are increasingly sought in green markets (Welch 2019).

[2]The term "Blue Economy" encompasses the sustainable use of ocean resources for economic growth, improved livelihoods, and jobs, while preserving the health of marine ecosystems.

In Alaska, collaborative initiatives between Indigenous and rural communities, state agencies, and academia aim to bridge traditional knowledge with the growing phyconomy. For instance, Indigenous and rural communities are exploring the potential of seaweed aquaculture to generate economic benefits within their communities and traditional harvest areas. To support this vision, the Alaska Mariculture Development Plan incorporates Indigenous stewardship principles, emphasizing the need for aquaculture to complement, not replace, existing uses like subsistence and wild harvesting (Mariculture Task Force 2018). Moreover, programs that document and share traditional harvesting techniques are helping to inform best practices for sustainable aquaculture development.

While integrating traditional knowledge into phyconomy and the broader Blue Economy of temperate regions offers immense potential, challenges remain. Policies governing wild harvest and aquaculture are often fragmented, making it difficult to scale traditional harvest-informed practices across the industry (Dupuis et al. 2022). Moreover, ensuring that Indigenous communities and other non-commercial traditional harvesters can access aquaculture opportunities may require innovative approaches, such as cooperatives and limited entry systems, to prevent corporate dominance and resource exclusion (Miller 2021).

Despite challenges, integrating traditional knowledge into the modern phyconomy represents a critical step toward achieving the goals of the Blue Economy. By aligning traditional practices with sustainable aquaculture, Alaska can create an industry that generates economic value while achieving the critical goals of preserving cultural heritage and protecting marine ecosystems (Garza 2012 ; Victor-Howe 2008). The ecological benefits of seaweed, cultural significance, and market potential underscore its importance as a cornerstone of a resilient and inclusive Blue Economy (Turner 2003, 2005).

8.6 Broader Implications for Temperate Regions

The socioeconomic and cultural dynamics observed in Alaska mirror those in other temperate regions where traditional, non-commercial harvesters have long been the primary users and stewards of seaweed resources. Traditional harvest practices emphasize sustainability, ecological knowledge, and communal stewardship across the northeastern Pacific, the British Isles, and parts of South Africa and New Zealand (Turner 2003; Naar 2020). For example, workshops and programs in South Africa led by Indigenous knowledge holders emphasize sustainable harvesting techniques that foster environmental awareness and cultural pride (Nirandara 2024). Similarly, partnerships between Indigenous communities and scientific institutions in New Zealand, such as the Smart Māori Aquaculture Initiative and the He Pou Kai Āwha Project, explore sustainable aquaculture methods that respect traditional harvesting practices while addressing modern market demand. In addition, traditional seaweed practices in Ireland have been revitalized and integrated into modern applications, reflecting a blend of heritage and innovation (Zocchi and Pardo-de-Santayana 2024).

Contemporary seaweed industries can foster sustainability, cultural preservation, and ecological resilience by embracing traditional practices. This approach requires developing frameworks that integrate traditional knowledge, support community-led initiatives, and address the unique challenges of rural communities (Nalau et al. 2018; Williams 2020; Kraus-Jensen and Duarte 2022). Partnering with traditional knowledge holders is paramount in documenting best practices and developing adaptive strategies for changing ecological conditions (Turner 2003; Williams 2022). Traditional cultures promoted the teaching of sustainable harvesting techniques, raised awareness of the importance of seaweed in the natural environment, and empowered individuals within communities (Garza 2012 ; Victor-Howe 2008), all of which are critically needed in this time of continuing rapid socioeconomic and environmental change.

8.7 Conclusion

The phyconomy represents a transformative opportunity to align economic growth with ecological stewardship and cultural preservation. Alaska's experience demonstrates the potential to build an inclusive, sustainable industry that respects tradition and enhances community well-being. As other temperate regions navigate the challenges of developing their seaweed economies, Alaska's case provides a compelling example of how traditional knowledge and modern innovation can coexist to promote a resilient and equitable Blue Economy. By prioritizing sustainability and inclusivity, phyconomy can serve as a global model for the responsible use of marine resources.

References

Abbott IA, Dawson EY (1978) How to know the seaweeds. Wm. C. Brown Co. Publishers, Dubuque

ADF&G (Alaska Department of Fish and Game) (1972) Regulations, process, constitutional authority. Available at: https://www.adfg.alaska.gov/index.cfm?adfg=process.commissioner. Accessed 28 Feb 2023

ADF&G (Alaska Department of Fish and Game) (2017) Alaska Governor's Mariculture Task Force: Overview. Available at: https://www.adfg.alaska.gov/index.cfm?adfg=amtf.main. Accessed 28 Feb 2023

ADF&G (Alaska Department of Fish and Game) (2024) Aquatic farming FAQs. Available at: https://www.adfg.alaska.gov/index.cfm?adfg=aquaticfarming.faqs. Accessed 26 Dec 2024

Bennion M, Collier G, Stekoll MS (2019) Updated methodology for kelp stock assessments. J Phycol 55:725–736. https://doi.org/10.1111/jpy.12891

Buckley S, Hardy K, Hallgren F et al (2023) Human consumption of seaweed and freshwater aquatic plants in ancient Europe. Nat Commun 14:6192. https://doi.org/10.1038/s41467-023-41671-2

Buschmann AH, Camus C, Infante J et al (2017) Seaweed production: overview of the global state of exploitation, farming, and emerging research activity. Eur J Phycol 52:391–406. https://doi.org/10.1080/09670262.2017.1365175

Cai J (2021) Global status of seaweed production, trade, and utilization. FAO Fisheries and Aquaculture Circular. Available at: https://www.fao.org/3/cb5670en/cb5670en.pdf

Cikoš AM, Jerković I, Molnar M, Šubarić D, Jokić S (2019) New trends for macroalgal natural products applications. Nat Prod Res 35:1180–1191. https://doi.org/10.1080/14786419.2019.1644629

Dillehay TD, Ramirez C, Pino M et al (2008) Monte Verde: seaweed, food, medicine, and the peopling of South America. Science 320:784–786. https://doi.org/10.1126/science.1156549

Drakard VF, Hollarsmith JA, Stekoll MS (2023) High-latitude kelps and future oceans: a review of multiple stressor impacts in a changing world. Ecol Evol 13:e10277. https://doi.org/10.1002/ece3.10277

Dupuis A, Harris D, Meredith B, Salomone P (2022) Southeast Alaska herring spawn-on-kelp pound fishery management plan. Alaska Department of Fish and Game, Division of Commercial Fisheries, Regional Information Report No. 1J22-04. Available at: https://www.adfg.alaska.gov/FedAidPDFs/RIR.1J.2022.04.pdf

Elegbede IO et al (2023) Blue economy (Sustainability). In: Idowu S, Schmidpeter R, Capaldi N et al. (eds) Encyclopedia of sustainable management. Springer, Cham. https://doi.org/10.1007/978-3-030-02006-4_401-1

Estes JA, Duggins DO, Rathbun GB (2010) The ecology of extinctions in kelp forests. Conserv Biol 11:1230–1242. https://doi.org/10.1111/j.1523-1739.2010.tb00253.x

Gallagher JB, Shelamoff V, Layton C (2022) Seaweed ecosystems may not mitigate CO2 emissions. ICES J Mar Sci 79:585–592. https://doi.org/10.1093/icesjms/fsac009

Garibaldi A, Turner NJ (2004) Cultural keystone species: implications for ecological conservation and restoration. Ecol Soc 9:1. https://doi.org/10.5751/ES-00669-090301

Garza D (2012) Common edible seaweeds in the Gulf of Alaska. Fairbanks, Alaska: Alaska Sea Grant College Program, University of Alaska, Fairbanks. ISBN 978-1-56612-168-2

Gorra TR, Garcia SCR, Langhans MR et al (2022) Southeast Alaskan kelp forests: inferences of process from large-scale patterns of variation in space and time. Proc R Soc B 289:20211697. https://doi.org/10.1098/rspb.2021.1697

Graham MH, Vásquez JA, Buschmann AH (2007) Global ecology of the giant kelp *Macrocystis*: from ecotypes to ecosystems. Oceanogr Mar Biol Annu Rev 45:39–88

Ham G, Gabara S, Chase P, Konar B, Umanzor S (2024) Regional differences in the effects of watershed glacial coverage on the performance of Pacific Blue mussel, *Mytilus trossulus*. J Exp Mar Biol Ecol 580:152043. https://doi.org/10.1016/j.jembe.2024.152043

Hollarsmith JA, Cornett JC, Evenson E, Tugaw A (2023) A century of canopy kelp persistence and recovery in the Gulf of Alaska. Ann Bot 133. https://doi.org/10.1093/aob/mcad149

Hood E, Berner L (2009) Effects of changing glacial coverage on the physical and biogeochemical properties of coastal streams in southeastern Alaska. J Geophys Res Biogeosci 114:G3. https://doi.org/10.1029/2009JG000971

Kitka H (1998) Deep ties to Deep Bay: a Tlingit elder's training. Cultural Survival Q Fall:47–48

Krause-Jensen D, Duarte CM (2022) The empirical evidence for the social-ecological impacts of seaweed farming. PLOS Sustain Transform 1:e0000042. https://doi.org/10.1371/journal.pstr.0000042

Krumhansl KA et al (2016) Global patterns of kelp forest change over the past half-century. Proc Natl Acad Sci USA 113:13785–13790

Lantz TC, Turner NJ (2003) Traditional phenological knowledge of Aboriginal peoples in British Columbia. J Ethnobiol 23:263–286

Lindeberg MR, Lindstrom SC (2010) Field guide to seaweeds of Alaska. Alaska Sea Grant College Program, University of Alaska Fairbanks, Fairbanks. Publication No. 69. https://doi.org/10.4027/fgsa.2010

MacDiarmid AB, Selwood AI (2015) Seaweeds and their uses: past, present, and future. In: Sheppard G (ed) Seaweed in health and disease prevention. Academic Press, London, pp 1–26. https://doi.org/10.1016/B978-0-12-802772-1.00001-5

MacMonagail M, Cornish L, Morrison L, Araújo R, Critchley AT (2017) Sustainable harvesting of wild seaweed resources. Eur J Phycol 52:371–390. https://doi.org/10.1080/09670262.2017.1365273

Mariculture Task Force (2018) Alaska Mariculture Development Plan. Alaska Fisheries Development Foundation. Available at: https://afdf.org/research-library/alaska-mariculture-development-plan. Accessed 28 Feb 2023

McHugh DJ (2003) A guide to the seaweed industry. FAO Fisheries Technical Paper No. 441. Food and Agriculture Organization of the United Nations. Available at: https://www.fao.org/3/y4765e/y4765e00.htm

McKinley Research Group (2021) Alaska seaweed market assessment. Alaska Fisheries Development Foundation. Available at: https://afdf.org/research-library/alaska-seaweed-market-assessment. Accessed 28 Feb 2023

Miller L (2021) Legalizing local: Alaska's unique opportunity to create an equitable and sustainable seaweed farming industry. Alaska Law Review, 38(2):313–340.https://scholarship.law.duke.edu/alr/vol38/iss2/6

Moss ML (2011) Northwest coast: archaeology as deep history. University Press of Colorado, Boulder. https://doi.org/10.1353/book.112357

Mouritson OG, Rhatigan P, Perez-Llorens JL (2019) The rise of seaweed gastronomy: Phycogastronomy. Bot Mar 62:195–209. https://doi.org/10.1515/bot-2018-0041

MRTC (2023) Research and workforce development priorities. Available at: https://amrtc.org/research-section/

Naar N (2020) The cultural importance of kelp for Pacific Northwest Tribes. Appendix B in Puget Sound Kelp Conservation and Recovery Plan. Prepared for NOAA-NMFS, Seattle. Available at: https://nwstraits.org/media/2957/appendix_b_the-cultural-importance-of-kelp-for-pacific-northwest-tribes.pdf. Accessed 28 Feb 2023

Nalau J, Becken S, Schliephack J, Parsons M, Brown C, Mackey B, & Risser M (2018) The role of Indigenous and traditional knowledge in ecosystem-based adaptation: A review of the literature and case studies from the Pacific Islands. Weather, Climate, and Society, 10:851–865. https://doi.org/10.1175/WCAS-D-18-0032.1

Nirandara P (2024) Diving and foraging for wild food in Cape Town's False Bay. Condé Nast Traveler. Available at: https://www.cntraveler.com/story/coastal-foraging-cape-town. Accessed 28 Dec 2024

North WJ (1971) The biology of giant kelp beds (*Macrocystis*) in California. Nova Hedwigia 32:1–600

O'Clair RM, Lindstrom SC (2000) North Pacific seaweeds. Plant Press, Auke Bay, 159 pp

O'Clair RM, Lindstrom SC, Brodo IR (1996) Southeast Alaska's rocky shores: seaweeds and lichens. Plant Press, Auke Bay

Pereira L, Neto JM (2015) Marine algae: biodiversity, taxonomy, environmental assessment, and biotechnology. In: Tiwari A, Nigam AD (eds) Applied phycology. Springer, Cham, pp 1–36. https://doi.org/10.1007/978-3-319-26893-4_1

Reid GK, Chopin T, Robinson MC et al (2013) Weight ratios of the kelps *Alaria esculenta* and *Saccharina latissima* required to sequester dissolved inorganic nutrients and supply oxygen for Atlantic salmon (*Salmo salar*) in integrated multi-trophic aquaculture systems. Aquaculture 408–409:34–46. https://doi.org/10.1016/j.aquaculture.2013.05.044

State of Alaska (2016) Administrative Order No. 280. Available at: https://gov.alaska.gov/admin-orders/administrative-order-no-280/

Stekoll MS, Else PV (1990) Cultivation of *Macrocystis integrifolia* (Laminariales, Phaeophyta) in southeastern Alaskan waters. Hydrobiologia 204:445–451. https://doi.org/10.1007/BF00040269

Stekoll MS, Peeples TN, Raymond AET (2021) Mariculture research of *Macrocystis pyrifera* and *Saccharina latissima* in Southeast Alaska. J World Aquac Soc 52:1–16. https://doi.org/10.1111/jwas.12777

Stephens T, Li Y, Yarish C et al (2024) Does seawater nitrogen better predict the baseline farmed yield for sugar kelp (*Saccharina latissima*) rather than the final yield? Phycology 4:370–383. https://doi.org/10.3390/phycology4030020

Thurstan RH, Brittain Z, Jones DS et al (2018) Aboriginal uses of seaweeds in temperate Australia: an archival assessment. J Appl Phycol 30:1821–1832. https://doi.org/10.1007/s10811-017-1344-z

Turner NJ (2003) The ethnobotany of edible seaweed (*Porphyra abbottae* and related species; Rhodophyta: Bangiales) and its use by First Nations on the Pacific Coast of Canada. Can J Bot 81:283–293. https://doi.org/10.1139/b03-029

Turner NJ (2005) The Earth's blanket: traditional teachings for sustainable living. University of Washington Press, Seattle, p 304

Turner NJ, Clifton H (2009) "It's so different today": climate change and indigenous lifeways in British Columbia, Canada. Glob Environ Change 19:180–190. https://doi.org/10.1016/j.gloenvcha.2009.01.005

Turner NJ, Loewen DC (1998) The original "Free Trade": exchange of botanical products and associated plant knowledge in northwestern North America. Anthropologica 40:49–70. https://doi.org/10.2307/25605872

U.S. Department of the Interior (2024) Alaska kelp farming: a growing industry for sustainable food and economic development. Available at: https://www.doi.gov/sites/doi.gov/files/13-adfg-alaska-kelp-farming-508.pdf. Accessed 26 Dec 2024

Umanzor S, Ladah L, Zertuche-González JA (2017) The influence of species, density, and diversity of macroalgal aggregations on microphytobenthic settlement. J Phycol 53:1060–1071. https://doi.org/10.1111/jpy.12565

Victor-Howe L (2008) Sustainability of Indigenous roe-on-kelp harvest practices in Alaska. Am Anthropol Rev 70:267–288

Welch L (2019) Interest in growing seaweed in Alaska is gaining momentum. Anchorage Daily News. Available at: https://www.adn.com/business-economy/2019/12/10/interest-in-growing-seaweeds-in-alaska-is-gaining-momentum/

Williams SR (2020) Traditional coastal resources of the indigenous people. In: Traditional ecological knowledge and natural resource management: A practical guide Taylor & Francis. https://doi.org/10.4324/9780429273452-11

Williams T (2022) Phenological cues in traditional black seaweed harvesting: linking Indigenous knowledge and climate change. Mar Ethnobiol J 14:85–96

Zocchi DM, Pardo-de-Santayana M (2024) Gathering and cooking seaweeds in contemporary Ireland: beyond plant foraging and trendy gastronomies. Sustainability 16:3337. https://doi.org/10.3390/su16083337

Dr. Schery Umanzor serves as the mariculture faculty at the University of Alaska Fairbanks (UAF), bringing a wealth of expertise in ecology, ecophysiology, and the cultivation of seaweed and abalone. Her work extends to the development of cost-effective tools and methodologies for optimal site selection in Alaska's kelp farms, the refinement of cultivation protocols for red seaweeds, the establishment of a pinto abalone cultivation initiative tailored for Alaska, and the exploration of seaweed's potential for bio-mining rare earth metals. Dr. Umanzor earned her doctorate in marine ecology in November 2017. Shortly after, she joined the Seaweed Biotechnology Lab at UCONN, CT, where she focused on the selective breeding of sugar kelp. In 2020, she brought her invaluable insights and passion to UAF.

Jan Marie Conitz earned her Bachelor of Science degree in Cell and Molecular Biology from the University of Washington in 1983, with a break in her studies to live and travel in Southeast Alaska, working as a park ranger, ski instructor, commercial fisher, and fisheries technician. Upon returning to the University of Washington, she focused her studies on developmental biology. She credits an independent study semester with Dr. Lynn Riddiford for inspiring her interest in the interaction between the early development of living organisms and environmental change in a study on the development of tobacco hornworm *Manduca sexta* larvae. Returning to Southeast Alaska, she planned to work as a science teacher but eventually found her way back to the laboratory, working with Dr. Michael Stekoll on several projects involving the biology and mariculture of seaweeds. In between projects, she taught high school science and also fished commercially on her own 32-foot troller. Working with Dr. Stekoll and Dr. Sandra Lindstrom, she completed her Master of Science degree at the University of Alaska School of Fisheries and Ocean Science in 1999. Her thesis focused on the reproductive biology and ecological physiology of *Pyropia abbottae* and *Pyropia torta*. For inspiration and success in her research work, she is indebted to the expertise of Dr. Lindstrom in the biogeography of Alaskan seaweeds and to the deep knowledge of traditional fishermen and seaweed harvesters from several Indigenous communities. After graduation, she began a 20-year career in Alaska state and federal fisheries agencies, starting in Southeast Alaska, where her work involved conducting surveys of remote salmon spawning runs and freshwater habitat areas and working with Tribal members and leaders to identify critical subsistence needs and concerns. Later, she worked as a research supervisor in the Yukon and Kuskokwim River areas of Alaska and served as the U.S. co-chair of the Joint Technical Committee for the Yukon River Panel under the provisions of the U.S.-Canada Salmon Treaty. She retired in 2021 from fisheries work in order to return to Southeast Alaska and participate in research efforts supporting seaweed mariculture, which by this time had begun to gain a strong foothold in coastal Alaska. Her main interests are in the life history and development of seaweeds with mariculture potential and in better understanding the needs and interests of Southeast Alaska Tribal communities, who have harvested and stewarded seaweed and other marine resources since time immemorial.

Seaweed Wildcrafting and Mariculture: An Integrated Maine Phyconomy

9

Stephen D. Eddy Jr, Shep Erhart, and Seraphina Erhart

Abstract

People have wildcrafted seaweed for food, medicine, fodder, and other purposes for centuries. *Pyropia* spp. were the first macroalgae to be domesticated in the 1600s in Japan, but since then, many other species have also been cultured. By 1971, macroalgae were the first group of aquatic organisms to achieve the distinction of having more production from aquaculture than from fisheries. The growth and scale of seaweed aquaculture has coined the term phyconomy as an apt description of applying agronomic principles to seaweed mariculture. However, most phyconomic output still comes from Asian nations, while in Europe and North America, seaweed fisheries still account for most production. This means that phyconomic development in these regions takes place alongside established seaweed fisheries. In this chapter, the US state of Maine is discussed as an example of the tensions and commonalities that can arise under such circumstances. Aligned interests between the two sectors show how there is more to be gained from working together as an integrated phyconomy rather than in opposition. Collaboration can be facilitated by organizations such as the Maine Seaweed Council, originally established in 1993 to serve the wildcrafting sector but which now also works to serve growers.

Keywords

Maine phyconomy · Maine Seaweed Council · Wildcrafted seaweed · Coastal seaweeds

The original version of the chapter has been revised. A correction to this chapter can be found at https://doi.org/10.1007/978-3-032-02955-3_29

S. D. Eddy Jr (✉)
Maine Coast Sea Vegetables, Franklin, ME, USA
e-mail: steve@seaveg.com

S. Erhart · S. Erhart
Maine Coast Sea Vegetables, Hancock, ME, USA
e-mail: shep@seaveg.com; seraphina@seaveg.com

9.1 Introduction

Foraging or harvesting plants from their natural or "wild" habitat, primarily for food or medicinal purposes, is known as wildcrafting. Humans have wildcrafted seaweed for food, medicine, fodder, and other purposes for a very long time. The earliest archeological evidence of this practice in the Americas comes from Chile's famed Monte Verde site, located about 15 km from the inland marine bay of Seno de Reloncavi. The remains of nine marine algae species were found in ancient hearths there and dated to approximately 14,000 years BCE (before the current era) (Dillehay et al. 2008). In Europe, Buckley et al. (2023) found evidence, based on biomarkers extracted from dental calculus, that some humans regularly consumed marine algae during the Mesolithic period, or Middle Stone Age, between 10,000 and 8000 BCE. In Japan, Neolithic settlements from the Jōmon Era (10,500–300 BCE) yielded archeological evidence of seaweed cooked in clay pots for consumption (Nisizawa et al. 1987).

Humans have probably wildcrafted seaweed long before any of the above archeological evidence. It is, after all, a highly nutritious food, and representatives of brown, green, and red species are readily harvested from shorelines at low tide (Mouritsen 2013; Wells et al. 2017). Indeed, Cornish et al. (2017) offer the intriguing speculation that seaweed and other marine foods played key roles in the evolution of the human brain by providing brain-essential nutrients to early hominids. The kelp highway hypothesis (Erlandson et al. 2015) posits that early human settlement of the Americas, starting about 16,000 years ago, was by people following a coastal kelp highway from Asia to South America. Areas with abundant laminarian kelp beds would have been conducive to long-term habitation due to the rich assemblage of edible shellfish, fish, marine mammals, and other edible seaweeds associated with laminarian kelps. Coastal inhabitants could then disperse from these early settlements inland along large rivers.

A. T. Critchley et al. (eds.), *Advances in Temperate Phyconomy: Algal Harvest and Cultivation in Globally Distributed Temperate Waters*, Developments in Applied Phycology 15, https://doi.org/10.1007/978-3-032-02955-3_9

Compared to wildcrafting, seaweed mariculture is a relatively modern enterprise dating back to the 1600s, when Japanese fisherfolk first planted bamboo poles in shallow waters to encourage growth of naturally seeded nori *Pyropia tenera* (formerly *Porphyra tenera*) (Rhodophyta) (Baker 1965). It wasn't until the 1940s, though, that seaweed mariculture became established in Asia as a true agricultural enterprise with the life cycle under full human control from "seed to sorus." The new industry grew rapidly. By 1971, macroalgae were the first group of aquatic organisms to attain the distinction of "cultured more than caught" (Chopin 2018), and today, seaweed mariculture dwarfs wildcraft production. In 2019, farmed seaweed production was 34.7 million landed tons, while wild-harvested seaweed totaled 1.1 million landed tons (Cai et al. 2021). Asia, the birthplace of seaweed mariculture, accounted for 97.4% of global production in 2019.

Fig. 9.1 This nori farm in Eastport, Maine (ca. 1994) was one of the first attempts in the United States to farm seaweed

Unlike Asia, Europe and the Americas have only recently embraced seaweed mariculture, and in 2014, wildcrafted seaweed still accounted for 99% of domestic production in these regions (Mac Monagail et al. 2017). Chile and Norway were the top producers of wild laminarian kelps in 2014, but France, Norway, Ireland, Portugal, Spain, Iceland, Canada, and the United States all have commercially important seaweed fisheries (Mac Monagail et al. 2017). Commercial-scale seaweed mariculture, on the other hand, is still a relative novelty in these countries, with sugar kelp (*Saccharina latissima*) (Phaeophyceae) being the primary farmed species. In Europe, most of the larger seaweed mariculture companies did not establish pilot or commercial farms until after 2010 (e.g., Algolesko, France, 2014; Zeewaar, the Netherlands, 2013; Seaweed Solutions AS, Norway, 2014). Production figures for individual companies, countries, or Europe as a whole are hard to come by, but total annual European production of farmed seaweed is estimated to lie between 1000 and 3000 tons. In 2020, 376 tons of *S. latissima,* 107 tons of *Alaria esculenta* (Phaeophyceae), 50 tons of *Ulva* (Chlorophyta), and unknown quantities of various other species were farmed in Europe. Meanwhile, the European wild harvest of *Laminaria* spp. (Phaeophyceae) alone was 209,772 tons (Araújo et al. 2021).

North American attempts in the 1980s and 1990s to establish nori farms in Washington state and Eastport, Maine (Fig. 9.1), growing non-native Japanese *Porphyra* species (Mumford 1990), were unsuccessful (author Shep Erhart helped lead the effort in Maine). The longest continuously operating, commercially successful seaweed mariculture operation in North America is a shoreside facility operated by Acadian Seaplants in Charlesville, NS, Canada. Originally established in the 1970s to grow *Chondrus crispus* (Rhodophyta) in tanks for carrageenan extraction, the company now grows an edible *C. crispus* strain trademarked as Hana Tsunomata®. In the US, it wasn't until 2010 that the first commercially successful kelp (*S. latissima*) farming company (Ocean Approved, now known as Atlantic Sea Farms) had its first crop in the water. Other commercial seaweed mariculture ventures have since been established in Canada, Maine, Connecticut, and other NE US states, California, Washington state, and Alaska, though production is still insignificant compared to Asia. The annual US seaweed mariculture production from 2016 to 2022 has been about 1000 tons (Kim et al. 2019; Robidoux and Concepcion 2023). Maine and Alaska accounted for the bulk of that in 2022, harvesting 454 and 375 tons respectively (primarily sugar kelp). Cascadia Seaweed, said to be Canada's largest sugar kelp farm, produced 50 tons in 2016. Meanwhile, Maine's rockweed (*Ascophyllum nodosum*) (Phaeophyceae) fishery alone yielded almost 9000 tons in 2023 (Maine DMR), while Atlantic Canada harvested 12,655 tons of rockweed in 2019 (Cai and Giulia 2021). *A. nodosum* is the most commercially significant species in both the US and Canada, though unknown volumes of at least a dozen other species are also wildcrafted.

Although North American and European seaweed mariculture statistics show an industry in its infancy, the sector is projected to grow well beyond its current scale in both regions in the coming decades. According to NOAA, seaweed farming is the fastest growing aquaculture sector in the US. Ocean Rainforest, a California company, plans to grow one million tons per year of *Macrocystis pyrifera* (Phaeophyceae) by 2030. Alaska, with more coastline than the rest of the US combined, is aggressively promoting seaweed mariculture. Because Alaska lacks the constraints on farm size seen in states such as Maine, it may be the next frontier for industrial-scale US kelp farms. Alaska already has at least one 40-hectare kelp farm, and about a dozen applications are pending for more at that scale or larger.

In EU waters, the European Commission intends to harvest eight million tons from industrial-scale farms by 2030 (EC 2022), and the Netherlands is allocating 400 km^2 of its North Sea exclusive economic zone to seaweed farming and wind energy. Norway has seen its number of seaweed mariculture licenses increase from 54 in 2014 to 511 by 2020 (Fiskeridirektoratet 2023). Norwegian producers and trade organizations speak of millions of tons of annual production, "explosive growth," and a "billion kroner industry," though Albrecht (2023) describes these statements as "political and entrepreneurial hype" and notes that the majority of Norwegian producers are currently modest scale with annual production of less than 50 tons.

The term phyconomy has recently been introduced to describe large-scale, sustainable seaweed farming for economic benefit (Hurtado et al. 2019; Sugumaran et al. 2022). In this context, phyconomy is considered as a marine counterpart to agronomy, which is the application of science and technology from various fields to the improvement and management of the major agricultural crops of the world. Scale is a defining agronomic feature: agronomic crops occupy large acreages and form the bases of the planet's food and fiber production systems. Likewise, phyconomy has been applied to the large-scale cultivation of certain seaweed species such as the eucheumatoids that constitute major export commodities for Indonesia and the Philippines and provide raw materials for global carrageenan and agricultural value chains (Critchley et al. 2024). Like agronomy, phyconomy is the scientific application of technology to improve production efficiencies and scale, and it favors more technical approaches over simple or laborious cultivation methods. To be sustainable, however, phyconomy must take a holistic approach oriented not just toward profit, but also toward other positive outcomes such as ecosystem-based management, resilient communities, and diversified livelihoods (Hurtado et al. 2019). Phyconomy has also been defined by some as the global seaweed economy (see website https://phyconomy.net/). In this context, application of technical agronomic principles to the farming of seaweed may be a less important consideration than its contribution to regional or global economies.

The scale of production where seaweed mariculture for any one species or region develops beyond a cottage or artisanal industry with localized and relatively minor benefits to becoming a phyconomy with widespread or regionally significant benefits is not well defined. At minimum, though, phyconomy must be a well-established and economically sustainable enterprise before it can be said to meaningfully contribute to community resilience and diversified livelihoods. Based on the current early stage and small scale of seaweed mariculture in Europe and N. America, it is probably fair to state that both regions are still "phyconomically developmental." Additionally, ambitious projections to achieve phyconomic scale in Europe and N. America, whether hype or not, raise concern among some environmental organizations, fishing groups, and others. Environmental concerns include nutrient depletion, genetic pollution, marine mammal entanglement, plastic fibers, and habitat change (Campbell et al. 2019; Kross et al. 2021; Wilding et al. 2021; Gercama et al. 2022; Bhuyan 2023). Fishermen are concerned about the loss of fishing grounds and unintended ecological impacts on wild seaweed stocks and other fisheries. Phyconomy will also likely have sociological impacts (Valderrama 2012; Cerca et al. 2023; Spillias et al. 2023), some of which may affect wildcrafting. Will tensions and conflicts arise between modern agricultural and traditional wildcrafting practices when a phyconomy develops alongside an established wildcrafting sector? What lessons, if any, can seaweed farmers and harvesters learn from one another? Perhaps most importantly, how do we ensure that both sectors coexist and thrive with a commitment to sustainable practice? In this chapter, we attempt to offer some insight into these questions by focusing on what we know best—Maine's seaweed industry.

9.2 Maine's Seaweed Fishery

Maine is the 39th largest state in land area (91,633 km^2), but in coastline ranks #4 (5597 km including tidal inlets and islands). Maine's coastal waters are bounded by the Gulf of Maine, a highly productive cold-water gulf encompassing roughly 93,000 km^2. Not surprisingly, Maine has a long and proud maritime tradition that includes shipping and shipbuilding, fisheries, and recreation. Today, Maine's fisheries are best known for lobster, but over 30 other species are also commercially fished in Maine, including seaweed.

The cold waters of the Gulf of Maine support at least a dozen edible and/or commercially valuable macroalgae species, including rockweed (*A. nodosum*), bladderwrack (*Fucus vesiculosus*), sugar kelp (*S. latissima*), winged kelp (*Alaria esculenta*), horsetail kelp (*Laminaria digitata*) (Phaeophyceae), Irish moss (*Chondrus crispus*), dulse (*Palmaria palmata*), laver (*Porphyra* spp.) (Rhodophyta), sea lettuce (*Ulva* sp.) (Chlorophyta), and *Gracilaria* (*Gracilaria tikvahiae*) (Rhodophyta). Though there are no written records, and seaweed does not generally persist in archeological remains (earlier examples notwithstanding), it's likely that Maine's original coastal nations (the Maliseet, Micmac, Passamaquoddy, and Penobscot) gathered seaweed for food and medicine. Hints of evidence in support of this are found in oral tradition. For example, Sayres (1958) described the life of Sammy Louis, a Micmac member, and quotes him thusly.

Fig. 9.2 Gathering kelp (Phaeophyceae), Long Sands, York, Maine, ca. 1882. Emma Coleman

BOSTON, SUNDAY, JUNE 9, 1907. COPYRIGHT, 1907, BY THE BOSTON HERALD CO.

LUMBER KING RICHER THAN ROCKEFELLER

COLD LUNCHES AND WET TO THE SKIN

Work for 18 Hours a Day on an Average and Sleep Out of Doors.

HOW WOULD YOU LIKE TO BE A SEAWEED GATHERER?

Farmers in Maine Find the Marine Plant Very Valuable as a Fertilizer.

[Special.]

PORTLAND, Me., June 8, 1907. The calling of gathering seaweed is one of the hardest known in Maine.

The men are wet to the skin all the time. They eat cold lunches. They sleep out of doors. They average to work 18 hours a day.

There is an aged and cynical saying, common along the eastern shore, which relates how a merciful farmer always sent his hired men to pull rockweed "when it rained too hard to work out of doors." The exact wording is not nice for refined ears, though the idea conveyed requires no explanatory key.

Before the days when cheap commercial fertilizers were invented, every farmer within four miles of salt water relied upon his spring gathering of seaweed for the dressing to give his potatoes an early start, and for the production of hard and elliptical rutabaga turnips, such as are said to make

Fig. 9.3 "Cold Lunches and Wet to the Skin" describes life as a Maine seaweed harvester

Maine's first European settlers hailed from nations with long traditions of foraging seaweed, such as France, Scotland, England, and Ireland. Rockweed and laminarian kelps are both abundant in Maine, and early farmers gathered them from the wrack line for forage and fertilizer. A photograph taken in York, Maine, by the talented amateur photographer Emma L. Coleman in 1882 depicts a pair of oxen pulling a cart overflowing with brown seaweed, while the farmer poses alongside with a forkful (Fig. 9.2). In a 1907 Boston Herald article, Maine seaweed was touted as high-quality fertilizer, and gathering it was described as one of the hardest known jobs in Maine (Fig. 9.3).

Maine's seaweed industry was a cottage industry for many decades until 1936, when Algin Corporation established a plant in Rockland, Maine, at the site of a former lime factory to process local kelp (mostly rockweed) into alginate. In the 1950s the plant shifted to producing carrageenan from Irish moss locally gathered from Rockland and Orr's Island, Maine, and imported from Scituate, Massachusetts, and Maritime Canada. In the 1960s the company's name was changed to Marine Colloids, and it began sourcing seaweed from all over the world. By the 1970s most carrageenan production at the factory came from red species other than Irish moss, namely *Eucheuma* and *Kappaphycus* (Rhodophyta) sourced from Tanzania, the Philippines, China, and other countries. With that change came an end to large-scale commercial harvesting of Irish moss in Maine and Massachusetts.

That same decade and the ones following (1970s–1990s) saw the rise in Maine of several new seaweed companies founded on harvesting rockweed and other species. Some of these newcomers specialized in harvesting and processing rockweed for animal nutrition and crop fertilizer or biostimulant products, while others focused more on edible "sea vegetables" such as *Alaria*, sugar kelp, and laver for the US natural food market. Acadian Seaplants, a Canadian company operating in Maine, was also founded during this period. As of 2025, Acadian Seaplants and four other "old guard" companies from that era continued to harvest and process most of the wild seaweed landed in Maine. These companies, which are described next, were also instrumental in forming the Maine Seaweed Council, which plays a key role in promoting Maine's seaweed industry and advocating for its effective management.

9.3 SOURCE Inc

Established in 1974, the company's origin story actually began in the late 1960s when company founder Susan Domizi was competing on her eventing horse, Hull, with hopes of being on the US Equestrian Olympic Team. The horse had hoof problems that weren't resolved despite top-notch veterinary care and a scientific diet. In search of solutions, Domizi learned of the nutritional power of seaweed and eventually developed a finely ground seaweed blend that she included in her horse's feed. By 1969, Hull was named US Intermediate Horse of the Year in Combined Training. Word of his improved physical condition spread fast through the equestrian community, and Domizi was soon in business.

In 1975, SOURCE® micronutrients (Fig. 9.4) officially introduced the blend to the horse industry, where it was well received with multiple testimonials to its effectiveness at improving equine health. As of 2025, the company was manufacturing and selling a range of seaweed-based nutritional

Fig. 9.4 SOURCE logo

Fig. 9.5 NAK logo

supplements for horses, dogs, and humans. The formulae are proprietary, but Maine rockweed remains one of the main ingredients. The company does its own harvesting and also contracts with independent harvesters. (Information retrieved from SOURCE® micronutrients website 09/12/2024 https://www.4source.com/). Susan Domizi passed away on December 10, 2024, but not before passing ownership of SOURCE to operations manager Bonnie Tobey and her brother Greg Tobey, who is the general manager.

9.4 Atlantic Laboratories, Inc. DBA North American Kelp

Atlantic Laboratories, Inc., DBA North American Kelp was originally founded as the Samoset Algae Co. in 1971 by Robert Morse, Jr. The company produces high-quality kelp meals, powders, and liquid fertilizers in Maine's mid-coast region, primarily for agricultural applications. North American Kelp (Fig. 9.5) was an early pioneer in the seaweed biologicals market, joining a consortium of like-minded visionaries to demonstrate the benefits of liquid seaweed extracts on crop production. This culminated in the registration of the product Nitrozyme with the EPA in 1981 as a growth stimulant on plants. Nitrozyme's legacy lives on in North American Kelp's lineup of seaweed-based biostimulants and fertilizers for agronomic and horticultural use. The rockweed is harvested by owner-operated mechanical harvesters fishing from Eastport to Casco Bay, covering much of Maine's long coastline. After being trucked to the plant, it is conveyed to a large, rotating wood-fired drum to dry (information retrieved from North American Kelp website 09/12/2024 https://noamkelp.com/). Morse was inspired in part by the story of Luther Maddox, who operated a wood-fired rotary drum seaweed dryer at Dogfish Head in the 1860s, two miles from where North American Kelp is now located. Maddox made a dried seaweed soil conditioner that he sold to Connecticut tobacco farmers (Thayer 2016).

9.5 Ocean Organics

George Seaver worked for Robert Morse from 1977 to 1991 before striking out on his own to establish Ocean Organics Corp. (Fig. 9.6). The company, located just across the road from North American Kelp, uses proprietary methods to process rockweed into granular fertilizer and seaweed extract-based liquid products for landscaping, greenhouse operations, golf courses, and a variety of agricultural crops. Liquid extracts are mostly sold in 2.5-gallon containers and often as "private label," meaning they include additives tailored to customer specifications. Maine potato growers have reported 5% to 15% yield improvements when using Ocean Organic rockweed products, and alfalfa farmers have reported 20% or more yield improvements (information retrieved from Ocean Organics website 09/12/2024 https://oceanorganics.com/). The company purchases rockweed harvested by independent contractors (Groening 2024).

9.6 Maine Coast Sea Vegetables

Maine Coast Sea Vegetables (MCSV) (Fig. 9.7) was founded in 1971 by Shep and Linnette Erhart in their farmhouse kitchen, inspired by an especially delicious pot of miso soup. They realized that the soups' full, rich flavor came from the wild *Alaria* they had recently gathered at low tide, and they soon began sharing their Atlantic wakame and other wild-harvested sea vegetables with friends and neighbors (Fig. 9.8). At the time, there were few others harvesting and selling seaweed in Maine, and over the next two decades, the Erharts built one of the most recognized and trusted brands in the industry. In the early years, the Erharts and a few close associates did all of the harvesting for the company (Fig. 9.9), but as the company grew, it began sourcing seaweed from various independent harvesters in Maine (Fig. 9.10) and trusted suppliers in Canada and Iceland. Today, MCSV is an employee-owned company operating out of a modern processing plant in Hancock, Maine. Every year, the team of over 20 people sorts, packs, and ships several hundred thousand pounds (dry weight) of sea vegetables to US and

Ocean Organics®

Fig. 9.6 Ocean Organics logo

MAINE COAST SEA VEGETABLES

Fig. 9.7 MCSV logo

Fig. 9.8 These hand-drawn labels by Linnette Erhart adorned paper bags holding the first four sea vegetable species sold by Maine Coast Sea Vegetables

Fig. 9.9 Shep Erhart harvests sugar kelp while Seraphina takes a break from hard labor (ca. 1975)

Fig. 9.10 Independent harvesters supply Maine Coast Sea Vegetables with sugar kelp and other wildcrafted sea vegetables

Canadian customers (information retrieved from Maine Coast Sea Vegetables website 09/12/2024 https://seaveg.com/pages/about-us). The company's focus on processing a variety of culinary sea vegetables for human consumption distinguishes it from the three other companies described above, which mostly process rockweed for agricultural or animal nutrition applications.

Fig. 9.11 Acadian Seaplants logo

9.7 Acadian Seaplants

Canadian company Acadian Seaplants (Fig. 9.11) was founded in 1981 by Louis Deveau after he purchased the assets of Marine Colloids. Deveau had worked for Marine Colloids for 12 years, during which time he mastered the process of procuring and processing red seaweeds for carrageenan and helped pioneer the seaweed farming industry in the Philippines and Indonesia. Under Louis and his son Jean-Paul Deveau (now President and CEO), Acadian Seaplants expanded their products and operations beyond carrageenan and Canada. As of 2025, the company had over 400 employees and five manufacturing and production facilities focused around key harvesting areas in Canada, Scotland, and Ireland. The company is headquartered in Dartmouth, Nova Scotia, but sources seaweed globally (information retrieved from Acadian Seaplants website 09/12/2024 https://www.acadian-seaplants.com/commercial-seaweed-harvesting/). Acadian Seaplants has harvested rockweed from Maine for decades and is a key industry player there. In 2024, the company directly employed about eight Maine residents in addition to 30 harvesters who mostly utilize cutter rakes to harvest rockweed across various locations in Maine, including Jonesport, Cobscook Bay, and other surrounding areas (Acadian Seaplants 2024).

9.8 Maine Seaweed Council

Fig. 9.12 Maine Seaweed Council logo

The Maine Seaweed Council (MSC) (Fig. 9.12) was formed in 1993 by the four Maine companies described above and a coterie of independent harvesters, scientists, and consultants to address areas of common interest within Maine's growing seaweed industry. Susan Domizi, founder of SOURCE, was a driving force behind forming the council, and author Shep Erhart was a founding member. An independent harvester who also played a key role in forming the MSC was Larch Hanson (aka the Seaweed Man), founder of Maine Seaweed, LLC. (https://theseaweedman.com/about-maine-seaweed/). Hanson is emblematic of the independent and often contrary spirit of many Maine seaweed harvesters, who reject "corporate America" and embrace seaweed harvesting as a sustainable and nature-based livelihood. After helping to found the MSC, he left it, in part because of philosophical disagreements with other members.

The MSC mission statement is: "*The Maine Seaweed Council strives to protect the ecosystems of Maine's marine algae and adhere to sustainable cultivation and harvest practices, promote the use of Maine seaweeds, educate the public, regulators, and elected officials, and provide a collaborative forum for its members.*" Some of the council's earliest focus areas included how to best manage natural seaweed resources for accountability and sustainability, ways to support new harvesters and those looking to start seaweed businesses, and educating the public, lawmakers, and resource managers about the benefits of seaweeds for human and animal health and use in agriculture (retrieved from the Maine Seaweed Council website 09/12/2024 https://www.seaweedcouncil.org/). The MSC remains an all-volunteer and actively engaged seaweed advocacy organization, typically meeting four to six times per year and always welcoming new members. Membership requires paying dues, which fall within several tiers, and pledging to honor the mission statement. The MSC had 36 dues-paying members in 2024.

9.9 Management of Maine's Seaweed Fishery

Managing natural seaweed resources to ensure accountability and sustainability has always been a key mission of the Maine Seaweed Council. Prior to 1999, Maine seaweed harvesting operated essentially free of regulatory oversight, though the Maine Seaweed Council had advocated for regulations throughout the 90's as aligning with the council's sustainability mission. In 1999, a licensing requirement was enacted by the Maine Legislature in response to the growing size of the rockweed fishery, which by 2003 had surpassed 1400 tons (Maine DMR historical landings; Reiter et al. 2020). The new law authorized harvesting up to 50 pounds of seaweed daily for personal use, but anyone wishing to harvest beyond that limit or for commercial use had to obtain a Marine Seaweed Harvesting License from the Department of Marine Resources (DMR). The Maine DMR issued 120 seaweed licenses in 2000, the first year following the enactment of the new law (Feeney 2001).

The Maine Legislature enacted several other laws regulating the seaweed fishery, along with the licensing requirement. Among these was a law stating, "The (DMR) commissioner may adopt rules regulating the harvest of seaweed on a species-specific basis." Acting under this authority, the Maine DMR enacted several rules affecting the harvesting of any seaweed species. Harvester reporting requirements included specific harvest area, harvest method, species, and wet pounds of landed seaweed. However, the DMR focused most of its attention on just one species, *A. nodosum*, by far the most landed species by wet weight in Maine. Regulations affecting other seaweed species in Maine were relatively minor by comparison (and remain so as of 2025).

For about 10 years, the most impactful of the rockweed regulations was the requirement that a minimum of 16 inches (≈0.4 m) of the rockweed above the holdfast be left undisturbed by harvesting. In 2009, the Maine Legislature enacted additional rules through "An Act to Regulate the Rockweed Harvest in Cobscook Bay." Cobscook Bay lies in a northeastern region known as Downeast Maine, and at that time, about 90% of Maine rockweed landings came from its waters. The bay was divided into harvest management sectors, and certain areas deemed of ecological or other significance were closed to harvesting. Management sectors were allocated between harvesters, and the total annual biomass removed in a harvest management sector was limited to 17% of the biomass. Harvesters were required to submit an annual harvest plan to Maine DMR for approval and to make a "reasonable effort" to return marine organisms associated with the rockweed back into the sea. Finally, buyers of more than ten tons of seaweed were required to be licensed and to pay a small percentage of the purchase cost to the Maine DMR to fund management. This last rule applied statewide.

These regulations came largely in response to mounting concerns from coastal landowners and environmental groups about the perceived negative consequences of rockweed harvesting. Mechanical harvesting (Fig. 9.13), the lack of restrictions on total allowable harvest, and rockweed's value as habitat for other marine organisms and for contributing to marine food webs via detrital shedding were all frequently cited concerns (Feeney 2001; Seeley and Schlesinger 2012; Halat et al. 2015; Garbary et al. 2017). On the other side, proponents of rockweed harvesting cite studies concluding that current rockweed harvesting practices in Canada and Maine are sustainable with only short-term ecological impacts and with relatively rapid recovery (within approx. 1 year) of harvested beds, and that natural losses from storm damage and detrital shedding far exceed harvest losses (Ugarte 2011; Ugarte and Sharp 2012; Ugarte et al. 2017; Lauzon-Guay et al. 2021, 2023). Earlier studies showing negative harvesting impacts have also been criticized for drawing large conclusions from small-scale plot studies. Johnston et al. (2023) recommended that assessing bed scale impacts better captures the spatial heterogeneity of harvest impacts than the smaller scale assessments done previously. However, this publication was criticized by Seeley et al. (2024) for using inappropriate statistics and for having industry partner conflicts of interest because study sites were selected and controlled by seaweed harvesters.

Fig. 9.13 Mechanical rockweed harvester in Maine

A larger conflict of interest could be viewed as that between opposing scientific and ideological views: those who view current rockweed harvesting practices as having overall positive societal benefits with minimal environmental costs, and those who see little to no positive societal benefits with potentially significant environmental costs. This dichotomy of views is evidenced in the research community by the quick rebuttal of Seeley et al. (2024) to Johnston et al. (2023) and in earlier exchanges between Garbary et al. (2017), Ugarte et al. (2017), and Halat et al. (2015) regarding the role rockweed epidermal shedding plays in nutrient cycling and disagreement as to whether harvesting does or does not negatively impact that role.

The lack of scientific consensus regarding appropriate rockweed management could affect actual management practices, but controversy surrounding "ownership" of the rockweed resource in Maine is almost certainly more impactful at present. The controversy stems from Maine judicial interpretation that the intertidal zone (i.e., between mean high tide and mean low tide) is owned by upland property owners whose property lines extend to the shore. Under the public trust doctrine dating back to centuries of English law, that owner's title in fee is subject to a public easement for fishing, fowling, navigation, and related activities (Johnson 2019). Rockweed harvesters, who viewed their activity as fishing, believed this meant they could lawfully harvest rockweed from the intertidal zone. However, in 2019, Maine's Supreme Judicial Court (SJC) upended that understanding with a ruling in a case brought by a private property owner, Kenneth Ross. The so-called Ross decision held that rockweed was a plant owned by the upland property owner and therefore not subject to the fishing exception under the public trust doctrine (Totten 2019; Reiter et al. 2020). As such, rockweed harvesters could be prosecuted for theft if they removed rockweed without the landowner's permission.

The implications of the Ross decision continue to be debated, including even whether it is legally proper or based upon sound biological principles (rockweed is an alga, not a plant). The Maine DMR maintains that rockweed is a marine resource regardless of ownership and, as such, is subject to its authority. Some private landowners, however, maintain that they have the right to exclude harvesters from their land, regardless of whether they have deeded intertidal ownership. The Ross decision appears to have had a chilling effect on harvest levels, as some harvesters curtailed their activities to avoid conflict or potential prosecution. Bonnie Tobey of SOURCE® reports that the company now operates two shifts instead of three. Harvesters have reported being shouted at, threatened, and even shot at by irate landowners following the Ross decision. In 2018, total rockweed landings were about 10,500 tons, but in the 3 years following the 2019 Ross decision, average annual landings declined to about 6700 tons (Fig. 9.14, Maine rockweed landings). Other issues related to the Ross decision, such as intertidal ownership (landowners have been accused of inserting ownership language into deeds that previously lacked such language), taxation (intertidal land has not been subject to property taxes), and property lines (identifying underwater property lines along a changing coastline is difficult), remain unresolved. On October 10, 2024, Maine's SJC heard oral arguments in a lawsuit to return use of Maine's beaches and intertidal lands to the state's residents (Masucci et al vs Moody LLC et al). A decision on the case was pending as of this writing.

The Maine Seaweed Council has played an integral role through all of this history and controversy by advocating for sound harvesting and management practices and by educating the public about the issues at stake. In 2014–2015, the MSC published "Harvest Guidelines for Maine Seaweeds" and the accompanying "Harvester's Field Guide to Maine Seaweeds" (currently being updated and expanded for 2025 publication). These publications offer voluntary recommendations on best harvest practices for 12 species. In 2014, councilmembers also participated in writing a rockweed Fisheries Management Plan (FMP) (Maine DMR 2014) with input from regulators, industry, academia, and the environmental community. This made rockweed Maine's first commercial marine species to have such a plan. The stated goal of the FMP was to establish harvest guidelines that support the long-term health and sustainability of *Ascophyllum nodosum* and its associated biota along the Maine coast. Key recommendations included mandatory harvester training and expanding sector management and no-harvest areas beyond Cobscook Bay to the entire coastline. The proposed state-

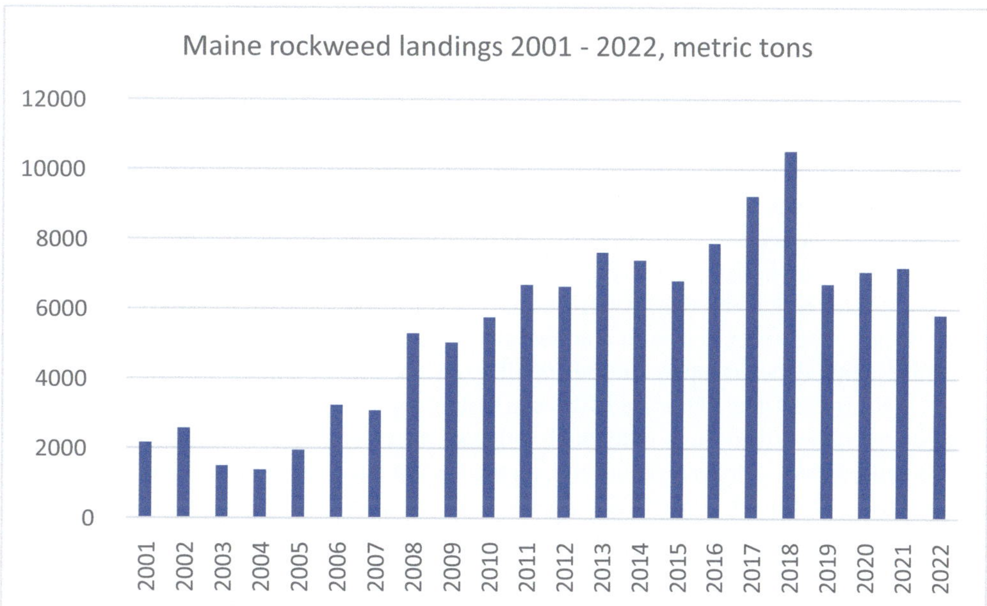

Fig. 9.14 Maine rockweed landings peaked in 2018 but then declined following the Ross decision, which declared rockweed to be privately held by upland property owners

wide sector management scheme was designed to meet the needs of both small-scale and large-scale harvesters by not restricting sector size and by allowing individuals to apply for and receive as small or as large an area as met their needs. Sector applicants had to provide certain information subject to Maine DMR review before receiving a sector, such as boundaries, training records, harvesting methods, and, crucially, biomass estimates. Landings would be limited to 17% or less of the estimated biomass per year, and the minimum 16″ cutting height under current Cobscook Bay regulations would be applied statewide.

Some of the rockweed FMP recommendations required legislative action, while others required Maine DMR rule-making. Unfortunately, neither was forthcoming, and as of 2024, the rockweed FMP remained an aspirational document with no legal authority. In part, this state of affairs may be attributed to the uncertainty regarding ownership of intertidal seaweed, which continues despite the Ross decision that seemingly laid the issue to rest (Webber et al. 2023). Another reason for inaction, however, may simply be that full implementation of the plan requires additional state management resources, enforcement resources, and funding (personal communications, Maine Seaweed Council). In 2019, the Maine DMR commissioner did go so far as to establish a Seaweed Fisheries Advisory Council (SFAC) to advise on "the health of the seaweed resource, its ecosystem, and the industry it supports" (Webber et al. 2023). However, in 5 years of existence (2019–2024), the SFAC has held just two meetings with little tangible outcomes (https://www.maine.gov/dmr/about-us/councils/seaweed-fisheries-advisory-council/meeting-archive). The SFAC is scheduled to meet in 2025 to discuss recommendations from the FMP.

The Maine Seaweed Council continues to advocate for the adoption of the rockweed FMP, while also working to improve upon it. Delayed implementation of the rockweed FMP presented an opportunity to re-evaluate it from the perspective of ecosystem-based management. The concept of ecosystem-based management (EBM) has been around for several decades and has been adopted by many international agreements and national governments (Curtin and Prellezo 2010). EBM recognizes the interdependence and interconnectedness of ecosystem components, and it is widely regarded as a more holistic approach toward managing species for sustainability, whether in terrestrial or marine settings. Because species exist in communities and the health of the community depends upon the health of each species within it, the overall goal of marine EBM is to sustain marine ecosystems rather than just individual targeted species. Traditional fisheries management criteria, on the other hand, such as maximum sustained yield, don't necessarily consider that a sustainable harvest level for a single species may adversely affect other species and fisheries or the ecological integrity necessary to sustain the fishery (Mace 2001). EBM is thought to be especially relevant for managing seaweed resources because of the vital role macroalgae play in supporting other species (Lotze et al. 2019; Clark et al. 2021). Drawing upon strategies proposed by Lotze et al. (2019), Webber et al. (2023) proposed that Maine adopt EBM for managing not just rockweed but all wildcrafted and farmed seaweed.

9.10 Maine Phyconomy: Promise and Peril

The first commercial harvest of farmed sugar kelp in Maine waters by Ocean Approved in 2010 marked a turning point for seaweed mariculture in the state and potentially for wildcrafting as well. The only other commercial activity in the Northeast US at that time was in Connecticut, where Thimble Island Oyster Co. operated a small sugar kelp farm starting in 2011 (Kim et al. 2019). Sugar kelp was also being farmed in Long Island Sound and New York coastal waters for bioremediation projects, but these are more fairly characterized as research projects rather than commercial activities.

The success of Ocean Approved was followed by an influx of state and federal funding over the next decade to promote seaweed mariculture and research. In 2012, Maine Sea Grant, a federal-state partnership program based at the University of Maine, hired its first seaweed extension agent, Sarah Redmond. Among other achievements, Redmond co-authored the *New England Seaweed Culture Handbook—Nursery* Systems (2014). This followed the Ocean Approved publication *Kelp Farming Manual: A Guide to the Processes, Techniques, and Equipment for Farming Kelp in New England Waters* (Flavin et al. 2013). Together, these two manuals were instrumental in helping new seaweed farmers become established in Maine and New England. In 2014 Maine EPSCoR (Established Program to Stimulate Competitive Research, a program sponsored by the National Science Foundation) received $20 million for a 5-year Sustainable Ecological Aquaculture Network (SEANET) program. The SEANET program addressed various aspects of Maine's diverse aquaculture sector and targeted seaweed mariculture research priorities in nutrition, processing, food safety, and marketing. In 2019, a network of ten Sea Grant programs were awarded a total of $1.1 million in federal funds to establish the National Sea Grant Seaweed Hub. This collaborative hub has since grown to include a total of 12 Sea Grant programs that serve as a central clearinghouse for science-based, non-advocacy, practical resources about seaweed aquaculture research and outreach. The Seaweed Hub has sponsored or facilitated seaweed symposiums, reports, manuals, and workgroups to spur development and innovation in the Northeast US (https://seaweedhub.extension.uconn.edu/about/).

Seaweed mariculture as a sustainable alternative to traditional terrestrial agriculture and for carbon mitigation has also attracted support from numerous non-profits, such as Food for Climate League (https://www.foodforclimateleague.org/) and Seagreen Insights (https://www.seagreeninsights.org/), and impact investors such as Builders Initiative (https://www.buildersinitiative.org/) and Coastal Enterprises, Inc. (CEI) (https://www.ceimaine.org/). These groups sponsored educational materials and marketing reports and provided investment capital to new North American seaweed mariculture companies. Some also offered hands-on training. Greenwave (https://www.greenwave.org/), co-founded in 2013 by Bren Smith and Emily Stengel, is a leading proponent of regenerative ocean farming and three-dimensional aquaculture, where seaweed, oysters, and mussels are grown vertically from the surface to the bottom. In 2015, the organization's 3D ocean farming method won the Buckminster Fuller Institute's Fuller Challenge. Greenwave offers training, education, and advocacy for seaweed growers. The Maine Seaweed Exchange (https://www.maineseaweedexchange.com/), co-founded in 2019 by Sarah Redmond and Trey Angera, is dedicated to seaweed aquaculture education, research, and advocacy, providing classroom and hands-on instruction to new growers in Maine and elsewhere.

Technology transfer from Asia was also funded via visits to South Korea and Japan. The World Wildlife Fund sponsored a South Korean tour in 2022 for seaweed companies, researchers, and others from Scotland, Wales, England, and the United States. The *South Korea—WWF Industry Exchange Visit to Seaweed Industry* report can be found at http://argyllaquaculture.co.uk/wp-content/uploads/2024/05/South-Korea-WWF-Industry-Exchange-Visit-to-Seaweed-Industry-v2.pdf. In June 2024, a delegation of 15 seaweed industry members from Maine and Alaska traveled to Hokkaido, Japan, and met with chefs, fishermen, growers, processors, retailers, restaurateurs, and scientists. The trip was sponsored in part by Coastal Enterprises Inc. and organized by Hugh Cowperthwaite of CEI and Keiichiro Hamano, CEO of Japan Fishing Machine LLC.

The support from federal, state, academic, and non-profit organizations spurred a new cohort of US growers to try their hand at farming seaweed, most notably in Maine, which offers three types of aquaculture leases: Limited Practice Aquaculture (LPA), Experimental, and Standard. These lease options vary in size, cost, lease period, and permitting. The flexible lease options encourage applicants from beginners with limited financial resources to seasoned, well-funded operators (Maine DMR Table of Standard and Experimental Aquaculture Leases 2024 https://www.maine.gov/dmr/aquaculture/maine-aquaculture-leases-and-lpas/aquaculture-lease-decisions-table).

At 400ft^2 ($\approx$37 m^2), LPAs are the smallest in area but also the most popular. LPAs are well suited for beginning growers because they have a low annual renewal fee of $100 and are relatively easy to acquire with an expedient permitting process. LPAs can be renewed indefinitely and are large enough to support commercially viable shellfish operations, and many shellfish growers in the state operate solely with one or more LPA leases. Maine had 769 active LPA leases in 2024, with 131 of these naming sugar kelp and other seaweeds as target species, usually along with shellfish.

Experimental leases are typically 3–4 acres (1–2 hectares) in size. They have a non-renewable term of 3 years (with research exceptions) and come with a $750 application fee and a $100/acre/year lease. Experimental leases help growers evaluate sites, test new species, and refine methods before committing to more costly standard leases. Maine had 22 active experimental leases in 2024, with nine of these dedicated solely to sugar kelp and other macroalgae, and another two that included shellfish along with seaweed.

Standard leases can be as large as 100 acres ($\approx$40 hectares), and the license period can be up to 20 years. They are costlier to acquire and maintain compared to the other two options, and lease applicants undergo a rigorous permitting process involving scoping sessions and public hearings. Depending on species and location, standard leases may encounter public opposition (usually when finfish or shellfish are involved, less often with seaweed). Maine had 127 active standard leases in 2024, with seven of these dedicated solely to sugar kelp and other macroalgae and another 16 that included shellfish along with seaweed, for a total of 23 seaweed standard leases.

All told, by 2024, Maine had 165 seaweed leases ranging in scale from 37m^2 LPAs to 40.34 acres ($\approx$16 hectare). The Maine DMR does not report individual harvests, making it difficult to determine whether all 165 leases regularly produce sugar kelp or other seaweed. In 2020, the Island Institute published the *Edible Seaweed Market Analysis* (Piconi et al. 2020), which projected that the seaweed mariculture sector in Maine would produce approximately 1400 tons by 2035. The *Maine Seaweed Benchmarking Report* (Brayden and Coleman 2023), which was based on grower and expert surveys, reported that approximately 454 tons were harvested in the spring of 2022, an almost 70-fold increase from the 6.6 tons reported in 2015. Sixteen seaweed growers participated in the survey.

These figures, along with continuing public support, speak to the promise of seaweed mariculture in Maine. However, the *Maine Seaweed Benchmarking Report* also identified several risk factors. Gear failure, line entanglement, and winter storms continued to cause crop losses. Biofouling and grazers were also cited as issues in the report. Some growers reported receiving poor-quality seeds or experiencing outright seed line failure, with some of that risk attributed to the practice of sourcing wild seedstock having variable genetics. Relatedly, there are few seed nurseries in the Northeast US, and selective breeding is still in its infancy.

Technical and husbandry issues such as those described above will likely be addressed through innovation and experience, but structural issues may present more significant challenges to developing a Maine phyconomy. One of these is that success currently rides on a single company and species. A large segment (about 35 growers operating on 66 leases) of Maine's seaweed mariculture sector relies upon Atlantic Sea Farms (ASF) for seed and as a guaranteed market. Many of these "partner growers" are lobster fishermen who might quickly exit seaweed farming if ASF support diminished or collapsed. Furthermore, because their business model incorporates kelp only as a supplement to their primary fishing activities, fisherfolk often favor a simplified and small-scale approach to aquaculture (Cleaver et al. 2024) whereby the entire crop can be harvested and unloaded within a short window of time and before the summer fishing season begins. This discourages experimentation, such as varying line depths, or tending activities, such as grading or thinning, because of the extra time and cost entailed. Perhaps as a consequence, experts from the wildcrafting sector have noted that farmed sugar kelp is typically paler, thinner, smaller, and less robust than wild sugar kelp (Seraphina Erhart, personal observation). This makes it less desirable as a whole-leaf product for processors and customers accustomed to wildcrafted sugar kelp, thereby limiting processing and marketing opportunities to milled and processed foods. To our knowledge, there are no published studies comparing farmed and wildcrafted sugar kelp quality attributes or consumer preference or investigating the causes for why farmed sugar kelp differs from wild sugar kelp in appearance and texture. *S. latissima* is known for its morphological plasticity (Diehl et al. 2024), and it has been speculated that the causes may be related to growing conditions, i.e., farmed sugar kelp is suspended from lines and experiences little to no depth and solar irradiance changes, whereas wild kelp grows from the bottom up and is thus subject to tidal-induced depth and irradiance changes. Farm siting (protected areas) and practices (heavy line density) may also be contributing causes. The sector's current reliance on a single species, *Saccharina latissima*, raises other obvious concerns, such as limited market opportunities, disease and parasites, and lack of innovation. These concerns are addressed in more detail in later sections.

The lack of primary processing capacity in Maine to dry or freeze freshly harvested kelp is another structural issue. Angera (2023) notes that Maine's small, distributed farm model consisting of part-time growers hinders the development of processing infrastructure. The problem is compounded by the highly seasonal nature of kelp mariculture as currently practiced. After growing through the winter, the crop reaches optimum harvest size and quality in May and June. Delayed harvest beyond this time can result in lower quality and crop losses due to biofouling. Growers tend to harvest most or all of their crops within a six-week period to make the most efficient use of vessel time and labor (Fig. 9.15). The freshly harvested kelp must then be dried or frozen within hours of harvest, and both processes are energy intensive, though solar drying has the advantages of being effective and low-cost. The weight and bulk of freshly harvested kelp add to transportation costs, so drying or freezing infrastructure is best situated in proximity to farm sites. Several such facilities would have to be located along Maine's long coastline to be accessible to all growers, and unless they could be repurposed for other crops or fisheries, they would only operate 2 months out of the year. Taken altogether, these considerations discourage investment in kelp primary processing infrastructure. However, Ocean's Balance, one of the new mariculture and processing companies to emerge in Maine, recently installed an industrial seaweed dryer able to dry up to 14,000 kg per day (Fig. 9.16). The dehydrator weighs 17 tons and was shipped from South Africa, representing a significant investment. Prior to this acquisition, the company had been blanching and freezing its farmed sugar kelp. Their new drying capacity is being offered

Fig. 9.15 Sugar kelp being harvested from a grow site in Casco Bay, Maine

Fig. 9.16 This seaweed dehydrator owned by Ocean's Balance can dry up to 14,000 kg/day (wet weight) and is offered on a fee basis to other companies

Fig. 9.17 Sugar kelp powder

on a fee basis to other companies. After the kelp is dried or frozen, it requires secondary processing, which at minimum can be packaging for sale, or more optimally can be value-added processing into food products or other goods. Maine Coast Sea Vegetables has been long established as a value-added processor for wildcrafted seaweed, but most of the state's value-added processing for farmed seaweed is currently done by new companies such as Atlantic Sea Farms, Ocean's Balance, and some smaller companies. However, these companies currently lack sufficient capacity to mill dried seaweed into the full range of flakes, granules, and powders highly sought by food and supplement manufacturers. Milled seaweed is a versatile ingredient that can be incorporated into a range of processed foods, nutritional supplements, cosmetics, and other products (Fig. 9.17). Of the state's value-added seaweed processors, as of 2024, only MCSV had capacity to mill large volumes of farmed kelp into flakes and granules, and in December 2024, they ordered another mill to produce significant volumes of powders as well.

Processing infrastructure will likely continue to develop as farmed kelp proves economically lucrative, but aquaculture leasing regulations and operational scale may stymie industry growth. Maine has developed a robust aquaculture regulatory framework over the past 50 years, but perhaps as a consequence, permitting for standard leases may require 3 years or more. Applicants complain that this is too lengthy, adds cost, and deters investment. Furthermore, Maine caps lease sizes at 100 acres (40 hectares); total acreage leased by any one entity cannot exceed 1000 acres (e.g., ten 100-acre leases or five 20-acre leases); and lease sites may not be contiguous (i.e., a leaseholder may not combine multiple leases to effectively form a 1000-acre plot). These size constraints may affect economies of scale and place Maine at a competitive disadvantage with Europe, Canada, or states such as Alaska, which permit much larger seaweed farms. Related to the matter of regulated scale is the question of appropriate scale, with differing visions of the future of Maine seaweed farming at stake. One point of view holds that Maine, with its tradition of independent lobster fishermen (4500–5000 licensed fishermen), is culturally best suited for small-scale family operated seaweed farms (between 1 and 20 acres). Others, however, believe profitability is unlikely unless seaweed farms can operate at a much larger scale (20–1000 acres), particularly when growing seaweed for agricultural or biorefinery applications.

The question of scale raises the issue of social license (Billing et al. 2021; Whitmore et al. 2025). Finfish and shellfish aquaculture has long been controversial in Maine, and several large projects involving finfish have been canceled due to public opposition. Some people find aquaculture in any form objectionable. Although seaweed mariculture has enjoyed a relatively warm public reception due to its perceived benefits, lack of input such as feed and fertilizer, and claimed ecosystem services, there are some who remain skeptical, particularly of large-scale farms (i.e., anything larger than about 8 hectares or 20 acres). Maine's lucrative and politically powerful lobster fishery becomes especially concerned when large aquaculture projects impact access to traditional fishing grounds. Development of a Maine phyconomy, which by some definition entails large-scale farming, would likely be accompanied by heightened public scrutiny of its value.

Clarity regarding food safety regulations applicable to seaweed is another challenge (Banach et al. 2022). Many of those new to the industry perceived seaweed as a novel food for Western cultures and therefore one with unknown or unquantified risks. Microbial pathogens, heavy metals and iodine content, and shellfish allergens have all been cited as warranting concern, and new growers felt regulators provided little guidance. Unlike finfish or shellfish hazards, which are extensively covered in the US Food and Drug Administration (FDA) "Fish and Fishery Products Hazards and Controls Guidance" manual, the FDA and the US Department of Agriculture (USDA) have few "seaweed-specific" regulations or guidance in place. Seaweed-specific regulations currently in place mostly regard its use as a food additive, not as a main ingredient, and the emerging industry wanted more guidance. In addition, many new growers with little to no experience marketing seafood directly to consumers lacked basic food safety knowledge and training. This lack of familiarity may have contributed to their heightened concern.

The perceived lack of federal guidance has led to different approaches between Northeast states to regulate seaweed food safety. Connecticut, which in 2024 had ten seaweed mariculture companies, took the approach of adapting shellfish regulations to seaweed. The Connecticut Department of Agriculture, Bureau of Aquaculture (DoAg), only permits seaweed cultivation in areas classified as Approved or Conditionally Approved through the state shellfish sanitation program. Connecticut was also the first state to develop and require an HAACP plan for seaweed producers, and in 2025, CT Sea Grant published what may be the nation's first Seaweed Hazards Guide. Maine, on the other hand, is far less restrictive on these matters. In large part, this may be because Maine companies were selling wildcrafted seaweed as food decades before mariculture emerged there, with no reports of food-borne illness.

While the FDA and USDA do have specific rules and guidance pertaining to certain foods, by and large, these are designed to address hazards or issues known to be frequently associated with those foods. In this context, it's important to acknowledge that edible macroalgae have been globally marketed and consumed for decades, including in the United States, with very few reported instances of food-borne illness (FAO and WHO 2022). This is not to say that edible macroalgae carry no risk of pathogens or that potential hazards shouldn't be investigated, but it does suggest that the perceived lack of seaweed-specific food safety regulations may be somewhat overstated by some as a concern. Seaweed is subject to the same regulatory standards applied to any other food, and these may be adequate to the task. To be fair, though, seaweed has traditionally been consumed in small servings (the FDA Reference Amount Customarily Consumed for dried seaweed is 5 grams). Frequent consumption of seaweed as a fresh or center-of-the-plate food at serving sizes exceeding 20–40 g may be more deserving of scrutiny for potential hazards such as excessive intake of iodine or heavy metals.

Cumulatively, the perils described above could threaten the development of a Maine phyconomy, and they are not necessarily limited to Maine or to farmed seaweed alone. Maine's wildcrafting sector developed while facing similar issues. Since wildcrafting and mariculture both produce seaweed for human consumption, they must both navigate the same food safety regulatory framework. Although seaweed mariculture has enjoyed a relatively warm public reception compared to the controversy surrounding the rockweed fishery, both sectors require social license to operate, and concerns about appropriate scale have been raised with both. Both sectors rely on the integrity of marine ecosystems for sustainability, and they face common threats such as climate change and the introduction or range expansion of exotic species. Ultimately, both sectors rise and fall on the market demand for seaweed, which depends to a large extent on the perception that seaweed is safe and sustainable. That perception could change if discussion or regulation became overly focused on theoretical hazards, or if either phyconomy or wildcrafting were perceived to cause harm to marine ecosystems. Market demand also implies competition, which in Western cultures often translates into winners and losers, but which by no means demands such. The common interests and experiences shared between seaweed mariculture and wildcrafting should instead be seen as an opportunity for each to learn from the other while creating a new Maine phyconomy.

9.11 Wildcrafting and Mariculture: Adversaries or Allies?

History shows that aquaculture and capture fisheries can become adversarial. Aquaculture is often justified as a solution to the harm fisheries cause to wild stocks and ecosystems, but fishers make similar claims about the harm caused by aquaculture. Market competition between the two sectors can also cause conflict depending upon the abundance of fisheries stocks and the aquaculture economies of scale. Cod and salmon fisheries, respectively, are examples of the interplay between natural fisheries stocks and mariculture economics. Cod farming went through a boom-and-bust cycle where the initial boom was driven by the collapse of wild cod stocks in the early 1990s, while the ensuing bust was driven in part by the doubling of wild cod catches from the Barents Sea in 2012–2013 (Nardi et al. 2021). Salmon aquaculture, on the other hand, grew to a scale where farmed output led to lower prices for both farmed and wild-caught salmon, to the detriment of Alaskan fishermen (Valderrama and Anderson 2010).

Phyconomy and seaweed wildcrafting won't be immune to similar challenges. The widespread public perception that fisheries are inherently destructive to wild resources is already being exploited by some in Maine, despite there being no evidence that this is the case with Maine's seaweed fisheries. Jake Patryn, Director of Operations at Acadian Seaplants Maine, speaks of how long-time harvesters manage their sectors with a sense of stewardship (pers. comm.). Likewise, wildcrafters who supply Maine Coast Sea Vegetables with sugar kelp, Alaria, and other species often return to the same grounds year after year. They take care not to overharvest beds because doing so for short-term gain affects long-term livelihood.

However, the Maine Seaweed Exchange claims on their website that "Kelp forests are disappearing at a rapid rate and wild harvesting seaweeds is accelerating this loss" (retrieved from https://www.maineseaweedexchange.com/ November 20, 2024). A recent mass email sent by the organization goes one step further when it exhorts recipients to "Say no to wild seaweed" and declares that "…wild seaweed is vastly overharvested" (personal communication, 2024). In touting farmed seaweed as a solution to the threat of overharvesting, though, the Maine Seaweed Exchange neglects to mention its potential impacts. The World Wildlife Fund takes a more nuanced approach when it states that while seaweed wildcrafting has been practiced by generations of indigenous and coastal communities, it can't sustain the rising global demand for seaweed without the threat of overharvesting (From Sea to Table: Wild Harvested Kelp Versus Farmed Kelp. Retrieved 11/20/2024 from https://www.worldwildlife.org/blogs/sustainability-works/posts/from-sea-to-table-wild-harvested-kelp-versus-farmed-kelp%2011/07/2024). To its credit, the WWF lists marine mammal entanglements, ecosystem changes, and invasive seaweed species as potential impacts of seaweed mariculture. Those attempting to differentiate farmed seaweed as being more sustainable than wild seaweed may regret doing so if further evidence shows otherwise. Some of the proposed benefits and ecosystem services of seaweed farming vary depending upon context, nor are they well substantiated by science (Spillias et al. 2023).

In regard to economic conflict, some Maine seaweed growers are concerned that ex-vessel prices ($/lb. wet weight) for wild-harvested sugar kelp, which as of 2024 were lower than farmed kelp production costs, make it difficult for growers to demand the higher prices they need to achieve profitability. This may explain why growers might resort to portraying the fishery as destructive and unsustainable, thus justifying farmed seaweed's price premium to consumers. Fortunately, so far both sectors have largely managed to navigate market competition, in part by expanding market opportunities and in part because many of the state's growers earn more income from growing shellfish or harvesting lobsters than from growing kelp.

Other than these examples, which largely pertain to sugar kelp, there has been minimal conflict between Maine's dominant rockweed fishery and seaweed mariculture. For the most part, rockweed harvesters and sugar kelp growers don't directly compete with one another for resources or markets, which helps minimize conflict. This could change, though, as sugar kelp growers begin targeting the same agricultural and animal feed markets targeted by rockweed harvesters. Cascadia Seaweed of Canada, for example, recently announced plans to construct a sugar kelp biorefinery to produce such products. The kelp mariculture sector has also largely avoided conflict with Maine's other fisheries. Kelp growers earned social license from Maine's lobster fishery by promoting sugar kelp as a winter crop that doesn't interfere with the summer fishery and that offers lobster harvesters an opportunity for off-season income as kelp farmers (Whitmore et al. 2025).

Ultimately, we believe cooperation and unity between mariculture and wildcrafting will benefit both sectors more than opposition and division will. President John F. Kennedy is often credited with popularizing the aphorism "a rising tide lifts all boats" to suggest that an improved economy benefits all those who partake in it. This nautically derived saying seems especially apt for the seaweed economy. In this next section, we discuss five areas of common interest between wildcrafting and phyconomy and highlight areas where the two sectors can cooperate: Organic certification, sustainability certification, integrated resource management, sector diversity, and seaweed food safety standards.

9.12 Organic Certification

Meeting organic standards and achieving certification are viewed by those in the seaweed wildcrafting sector as validation and acknowledgment of their commitment to sustainable harvesting and practices. Organic standards are also economically important. Consumers will pay more for certified organic crops because they are produced under higher standards that connote quality, sustainability, and purity (Midmore et al. 2005). These perceptions add direct and indirect value to Maine companies producing wildcrafted seaweed. For Maine Coast Sea Vegetables, the USDA organic seal reassures customers that seaweed is a healthy and sustainably produced food (Fig. 9.18). For companies such as Ocean Organics and North American Kelp, USDA approval of rockweed and other seaweeds for use in organic crop production is a market opportunity. Many of Maine's new seaweed growers appreciate the value of organic certification, and some have obtained it. However, the path to organic certification of seaweed has not been easy, and the requirements and documentation necessary to achieve and maintain it are becoming increasingly complex, maybe especially for smaller producers. In addition, some within the organic certification community, as well as those concerned with the environmental consequences of harvesting wild seaweed, are

Fig. 9.18 Certified Organic assures customers of quality, safety, and sustainability

now questioning certain aspects of certifying marine crops as organic or even of allowing their use in organic crop production.

Efforts to obtain organic certification of seaweed in the US began in the 1990s, soon after the Federal Organic Foods Production Act of 1990 was passed. The Act authorized the US Department of Agriculture (USDA) to establish the National Organics Program (NOP) to promulgate rules and standards for organic certification. The National Organics Standards Board (NOSB), a Federal Advisory Board made up of 15 public volunteers from across the organic community, was also tasked to make recommendations to the NOP, which then decides whether to initiate appropriate rulemaking. Among early recommendations were those affecting organic certification of wild crops such as mushrooms, maple syrup, and ginseng, collectively known as wild crop standards. In the early 1990s, Shep Erhart of Maine Coast Sea Vegetables and several others worked with the Organic Crop Improvement Association (OCIA) to develop organic wild crop standards for wildcrafted sea vegetables. In 1993, MCSV became the first US company, and perhaps one of the first in the world, to sell certified organic sea vegetables. Their success was followed in Europe when Seagreens® became the first European seaweed company to receive EU Organic Certification in 1998.

To be certified organic in the USA, seaweed products must be harvested sustainably to prevent the depletion of natural resources. In the Code of Federal Regulations §205.2 Terms defined, organic production is "a production system that [among other factors] promote[s] ecological balance, and conserve[s] biodiversity." In §205.207, the wild crop harvesting practice standard states that "a wild crop must be harvested in a manner that ensures that such harvesting or gathering will not be destructive to the environment and will sustain the growth and production of the wild crop" (Oakley 2024). In addition, organic certification under the NOP establishes setbacks for seaweed harvest grounds from sources of potential contamination, such as aquaculture fish pens, marinas, and municipal outfalls. Seaweed producers are certified as meeting these standards through third-party certifiers such as OCIA and the Maine Organic Farmers and Growers Association (MOFGA) Certification Services LLC. MOFGA advises that certified organic harvesters in Maine should follow the Maine Seaweed Council publication "Harvester's Field Guide to Maine Seaweeds," which describes sustainable harvesting methods for each of Maine's 11 most commonly harvested sea vegetables.

When Maine's new seaweed growers requested organic certification as well, it was relatively straightforward to adapt many of the seaweed wild crop standards to seaweed mariculture. However, certain mariculture practices, such as the use of fertility and crop nutrients (covered for terrestrial crops under NOP 205.203 Soil Fertility and crop nutrient management practice standard) and exemptions for when non-certified organic seed stock was allowed (NOP 205.204 Seed and planting stock standard), presented challenges. Common practice in sugar kelp nurseries is to use synthetic nutrients such as Guillards F/2 solution and germanium dioxide to control ciliates (Redmond et al. 2014). Technically, neither is allowed under organic standards because the only allowed substances are those listed under NOP 205.103 and NOP 205.601. Consistent production of high-quality kelp seed spools without the use of synthetic nutrients or germanium dioxide can be challenging. Though certified organic seed production is certainly possible, production failures are frequent, and there are currently just two certified organic kelp spool nurseries for the entire Northeast US and none on the West Coast. Lack of certified organic seed could limit organic production as the mariculture sector grows into a phyconomy and more producers seek certification. Growers can be granted exemption from using certified organic seed stock on the basis that they are not commercially available in the form, quality, or quantity needed, but this is widely recognized as a stopgap measure.

These issues are currently under discussion through the National Alignment of Farmed Seaweed Organic Certification Protocols and Guidelines administered through the USDA NOP (author S. Eddy serves as an advisor to the group). During group discussions, the lack of information regarding the discharge of synthetic nutrients and germanium dioxide from kelp nurseries into the environment, and residual levels on out-planted stock emerged as key considerations. The workgroup proposed a testing program to address these questions, with germanium dioxide residues on kelp seed spools being the first target for testing.

Another issue being discussed by the group is how to interpret and apply NOP 205.202 (land requirements) to kelp farms. NOP 205.202 requires that fields or land parcels being used for organic production be managed to maintain certain soil fertility and crop nutrient standards, have had no prohibited substances applied in the previous 3 years, and have distinct, defined boundaries and buffer zones to prevent unintended contact with prohibited substances used on nearby land. However, these standards are not easily applied to aquatic farming, and certifiers may interpret 205.202 differently when establishing buffer zones from potential sources of contamination such as commercial activities, harbors, and power plants. The National Alignment group is working (as of March 2025) to achieve harmony between MOFGA and Oregon Tilth Certified Organic (OTCO) "buffer zone language."

Ultimately, however, it is inherently difficult to apply organic standards developed for terrestrial crops to aquatic crops. Terrestrial crops are rooted in soil, a stationary medium that can be amended in ways that can persist for years, whereas seaweed crops grow in a fluid medium subject to vast dilution and water exchanges with every current and tidal cycle. Organic standards that promote soil and plant health may not be appropriate in determining where seaweed farms can be located and indeed may be irrelevant. Considerations such as these have led some to suggest it may be time to develop a separate set of organic standards for seaweed and other marine crops that are independent of terrestrial crop standards and that better consider their biology and the marine environment.

Moreover, USDA approval for using marine algae in organic crop production as fertilizer, soil amendment, or biostimulant is also being questioned. Wild seaweed has traditionally been allowed for use in organic farming without federal review. Synthetic seaweed products have also been allowed, provided they are on the National List of Allowed and Prohibited Substances contained within the Code of Federal Regulations. The National List permits aquatic plant extracts (other than hydrolyzed) so long as the "extraction process is limited to the use of potassium hydroxide or sodium hydroxide; solvent amount used is limited to that amount necessary for extraction." This language has allowed Maine companies that process rockweed for agricultural applications (i.e., North American Kelp and Ocean Organics) to market their products for organic crop production. However, in fall 2020 the NOSB recommended a proposed annotation to the National List for marine macroalgae used in fertilizers to provide parameters on cutting wild seaweeds (Oakley 2024). The proposal was in response to stakeholders and critics concerned about the environmental impacts of seaweed harvesting. If adopted, it would have established harvesting parameters addressing conservation areas, bottom trawling, reproduction and ecosystem functions, biomass and architecture, and bycatch. After much discussion, the NOSB voted to approve the proposal by a 13–2 vote, and it was forwarded to the NOP for review. So far, the USDA NOP has declined to put the recommendations through the rulemaking process, giving the industry a perhaps temporary reprieve (Oakley 2024). The National List is reviewed every 5 years, making it all but certain that the matter will be revisited soon. This issue is particularly problematic in Maine because of the lack of clarity regarding rockweed "ownership" and management.

Considering an activity's environmental and ecological impacts when determining whether it meets organic standards is certainly not unprecedented, and indeed is written into statute. For example, to certify organic maple syrup and sap in Vermont, producers must have a written forest management plan that preserves tree species diversity, protects ecologically sensitive areas, and addresses wildlife habitat (VOF Guidelines for Certification of Organic Maple Syrup and Sap). However, when accounting for ecological impacts, it's important to carefully weigh costs vs benefits, consider the interests of all stakeholders, and base recommendations on well-agreed science. The USDA NOP may have been trying to do just that when it declined to put the NOSB seaweed harvesting recommendations into effect. The recommendations could have made it more costly or nearly impossible to use wildcrafted seaweed as an organic crop input, which many terrestrial crop growers, certified or not, find to be of great benefit as a substitute for more environmentally damaging inputs. A large body of scientific literature shows that rockweed and other macroalgal-based extracts and fertilizers improve animal and plant health, yields, resistance to abiotic stress, and even crop nutritional value (e.g., Craigie 2011; Fan et al. 2013; Alam et al. 2014; Pereira et al. 2020). Prioritizing the ecological impacts of seaweed harvesting over these benefits by dictating harvest standards that are challenging to validate or based upon incomplete science or emotive language would be counterproductive to achieving sustainability. For rockweed harvesting in particular, it is far from clear that its ecological impacts on the Gulf of Maine are anywhere near as dire as portrayed by some (Legge 2019; Johnston et al. 2023). In this context, assertions that rockweed beds are equivalent to "old growth forests" and should be protected as such (Seeley and Schlesinger 2012) may have more emotional resonance than scientific legitimacy.

While much of the current debate around sustainability is focused on the practice of harvesting seaweed as an organic crop input, sustainability questions are just as relevant when certifying maricultured seaweed products as organic. The above discussion suggests that organic standards as applied to either wildcrafting or mariculture are likely to affect both and that stakeholders could mutually benefit from an aligned, holistic perspective when formulating or applying organic standards to seaweed. A shared understanding and agreement regarding sustainable practice and sound resource management is necessary for both sectors to not only meet organic standards but also to thrive.

9.13 Sustainability Certification

Sustainability certification has been proposed since 1994 to address the socioenvironmental and sustainability challenges of aquaculture, and about 30 aquaculture certification schemes have since emerged (Saha 2022). Certification by NGOs can effectively address issues that government policies may not, while also helping producers market products as being more environmentally, socially, and economically beneficial than non-certified products. In this sense, many of the same benefits accruing from organic certification also accrue from sustainability certification. Indeed, aquaculture sustainability certification emerged as an alternative to organic certification when it became apparent there was strong opposition to certifying organic fish production (Saha 2022).

The joint ASC-MSC seaweed standard may be one of the better-known sustainable seaweed standards. The standard is meant to be applied globally to all locations and scales of seaweed production, both wildcrafted and farmed, and it addresses five guiding principles: (1) Maintain sustainable wild seaweed populations; (2) Minimize environmental impacts; (3) Ensure effective management; (4) Promote social responsibility; (5) Strengthen community relations and interactions. Friend of the Sea, a World Sustainability Organization, also certifies seaweed production. The six core criteria of Friend of the Sea sustainable seaweed certification are: (1) No impact on critical habitat; (2) Water monitoring; (3) Prohibition on using hazardous substances; (4) Energy management; (5) Social accountability; (6) Traceability. Although the two seaweed certification schemes share similar values, such as protecting the environment and demonstrating social accountability, they differ in emphasizing other areas, such as water monitoring, energy management, and traceability (Friends of the Sea) or effective management (ASC-MSC). Regardless, producers interested in obtaining certification to these or any other standard must go through a similar process entailing fees, audits, documentation, and operational transparency, and they should be prepared to modify their operations to meet certification requirements.

The fees and process entailed in sustainability certification may be why so few companies in the US have pursued it. According to their website, only one US company is certified to the ASC-MSC seaweed standard (Veramaris, for algal oil used as an Omega-3 alternative). Nevertheless, it could be highly instructive for the industry in Maine and elsewhere to understand what is required to meet each standard and then work toward doing so, even if outright certification isn't the goal. Consumers and communities increasingly favor companies conducting business in ethical and sustainable fashions over companies that they perceive do not. Wildcrafting and phyconomy both rely upon social license to operate, and public perception that they are sustainable activities contributes to social acceptance. A unified approach toward defining and meeting sustainability standards will benefit both sectors.

9.14 Integrated Resource Management

Fisheries and aquaculture share common interests and intersect in fashions that can be negative, neutral, or positive (Clavelle et al. 2019). Capture fisheries and aquaculture both appropriate "goods and services" from marine ecosystems (Kumar 2014) and coastal communities, while facing shared threats such as pollution, climate change, and invasive species. Interactions between the two sectors may frequently lead to conflict. Among the most frequently cited concerns regarding finfish and shellfish aquaculture are those around their ecological and sociological interactions with capture fisheries. These concerns include disease, escapement, genetic interactions, pollution, and market competition, and fishers often oppose mariculture for one or more of these reasons. Although such concerns have traditionally focused on finfish and shellfish aquaculture, phyconomy comes with similar issues. For example, current Maine practice is to use wild-sourced sorus tissue for nursery production, but selective breeding will be essential for the sector to grow and compete as a phyconomy (Hwang et al. 2019; Brakel et al. 2021). The introduction of non-native species for aquaculture into Maine waters is prohibited, but there is currently no language regarding the use of selectively bred native seaweed species. Since seaweed is not contained while under culture at sea, selectively bred strains could spread into the environment through the release of reproductive spores, with unknown consequences on wild strains (Goecke et al. 2020). Preservation of local genetic diversity was identified as a key challenge by Barbier et al. (2020). Risk mitigation strategies include developing local populations for local use and using sterile sporophytes (Goecke et al. 2020), but such strategies have to be explicitly incorporated from the start into selective breeding programs to be effective. In addition, harvesting wild algae for seed stock could put pressure on the same beds harvested for wildcrafting. This could lead to overharvesting, not necessarily of the entire resource, but rather of localized wild strains that offer unique genetic value to both the wild resource and to mariculture. Bed scale or other forms of localized management might be appropriate in these cases to help avoid extinction of unique strains.

On the other hand, mariculture can reduce pressure on wild populations by filling demand that would otherwise be met by capture fisheries. Seaweed mariculture in particular is touted for having less environmental consequences than other forms of aquaculture while also providing ecosystem services. Seaweed beds, whether wild or farmed, provide multiple services to coastal ecologies, including aquatic habitat, carbon and nutrient cycling, and contributing to detrital food webs (Smale and Orr 2019). However, ecosystem services are highly dependent upon scale and context (Clark et al. 2021; Spillias et al. 2023), leading to the paradox that while large-scale seaweed production may provide meaningful ecosystem services, it could also lead to sustainability issues as industry scale increases. In addition, some of these services may be negated when seaweed is harvested for human use.

Various management schemes have been proposed that consider these and other interactions between aquaculture and fisheries. Clavelle et al. (2019) recommended that mariculture development be considered in the context of fisheries interactions to minimize conflicts and maximize positive connections between the two sectors. Newkirk (1996) advanced the concept of sustainable coastal production systems (SCPS), where a variety of trophic levels are maintained, and their interactions managed to maximize benefits from and for the whole system. As an example of SCPS, Newkirk (1996) made the prescient suggestion that seaweed could be grown as food for sea urchins, a practice currently being developed by at least one Maine seaweed grower (Spring Tide Seaweed). In Japan, seaweed farmers drop their thinned and trimmed crop on the bottom to feed wild sea urchin stocks (S. Erhart, personal communication). More recently, ecosystem-based management (EBM) has been proposed for seaweed mariculture (Langton et al. 2019) and for mariculture and wildcrafting (Webber et al. 2023). EBM fits within the SCPS framework, and an integrated approach can foster cooperation between the two sectors. In Aotearoa New Zealand, a research team recently developed a seaweed sector framework grounded in blue economy and EBM principles (*Building a seaweed sector: Developing a seaweed sector framework for Aotearoa New Zealand 2022*). The project defined EBM as a holistic and inclusive way to manage marine environments and the competing uses for, demands on, and ways New Zealanders value them. The framework was undertaken by a core research team of 20 with a budget of $560,000. The team worked for 2 years with the iwi (large Māori social units), stakeholders, industry, researchers, and government agencies that currently or plan to operate in the sector—either as farmers or as users of seaweed products.

As of 2024, the Maine Seaweed Council was working on a “Best Sustainable Practices (BSP)” document for managing Maine’s seaweed sector that includes wildcrafting and mariculture. The BSP document reflects current handling and harvesting practices for nine commercially harvested species. Although EBM is not included in the BSP document or as an official MSC policy project, EBM will be presented by members of the MSC to the DMR commissioner at an upcoming Seaweed Fisheries Advisories Council meeting. However, developing ecosystem-based-management for Maine’s seaweed sector is an ambitious undertaking that will require considerably more resources than the MSC can provide. An effort of this scope requires funding, sector-wide agreement as to its importance, and leadership from industry, regulatory bodies, and the research community. This is an area where the wildcrafting and mariculture sectors could benefit from working together to ensure a cohesive approach.

9.15 Sector Diversity

Thirty-seven Maine seaweed leases include species such as *Alaria*, horsetail kelp (*L. digitata*), dulse, laver, Irish moss, sea lettuce, and *Gracilaria tikvahiae,* but sugar kelp accounted for almost all of Maine’s 2024 production. This situation is not unique to Maine. Sugar kelp is the dominant farmed seaweed species in Alaska, Canada, and Europe. There are historical and biological reasons for this (Saether et al. 2024), but relying on a single species as the foundation for a seaweed mariculture economy raises concern. Critics note that sugar kelp can contain heavy metals and potentially harmful levels of iodine, has limited market appeal as a food, and that Maine lacks infrastructure to process large volumes of sugar kelp harvested in a short timespan (Angera 2023, 2024). Furthermore, oversupply of sugar kelp could lead to depressed prices for both farmed and wildcrafted kelp unless processing capacity and markets were developed at a pace matching production. Beyond marketing challenges, a phyconomy based on sugar kelp monocropping may be vulnerable to disease, predation, and lack of resiliency when environmental or other factors lead to crop failure.

Alaria esculenta, commonly known in Maine as winged kelp or Atlantic wakame, has been proposed as a suitable laminarian alternative. *Alaria* seeding, culture, and processing methods are similar to those used for sugar kelp, so investment in new infrastructure would be minimal. The two species share similar nutritional profiles, but *Alaria* has a much lower iodine content (Nitschke and Stengel 2015;

Roleda et al. 2018; Arlov et al. 2024), thus addressing the frequently cited concern that sugar kelp consumption could harm iodine-sensitive individuals (Blikra et al. 2024; Roleda et al. 2018). *Alaria* also possesses the desirable umami flavor characteristic of sugar kelp (Mouritsen et al. 2019). Other potential Maine seaweed mariculture species are *Laminaria digitata*, *Palmaria palmata*, *Chondrus crispus*, *Porphyra umbilicalis, Gracilaria tikvahiae* (Rhodophyta), and *Ulva* sp. (Chlorophyta). Maine growers and researchers are experimenting with all of these species, but none are yet being grown at a commercial scale, although Atlantic Sea Farms, Spring Tide Seaweed, and other growers are actively developing dulse mariculture, and limited commercial production of this species could occur by 2026. In October 2024, Acadian Seaplants announced a newly cultivated *P. palmata* strain that the company was growing at scale in tanks at a Canadian facility.

Maine's wildcrafting sector is similarly dominated by a single species. Rockweed accounts for more than 95% of Maine seaweed landings (Webber et al. 2023), but the state also has commercially viable, albeit small, fisheries for bladderwrack, rockweed, sugar kelp, winged kelp, sea lettuce, Irish moss, and dulse. These fisheries support Maine Coast Sea Vegetables and several smaller companies selling wildcrafted dried sea vegetables directly to consumers or other companies. On the other hand, wild harvest companies are well positioned to become growers themselves or to forge partnerships with growers. Although rockweed is not yet commercially farmed anywhere in the world, rockweed companies might find it advantageous to consider other farmed species such as sugar kelp or *Ulva* as alternatives or supplements to the rockweed-derived products they currently manufacture. At least one Maine kelp grower that we are aware of is considering processing farmed sugar kelp into fertilizer-type products. Wildcraft companies have longstanding knowledge of multiple species regarding marketing, optimal growing conditions, and where to source wild sorus tissue. This knowledge could be shared with growers seeking to diversify or expand beyond sugar kelp. Traditional wildcrafting companies also have processing infrastructure suitable for farmed seaweed. Maine Coast Sea Vegetables incorporated mariculture sourcing and sales into its strategic planning as a hedge against natural declines of wild beds caused by environmental factors and to fill increased demand without putting undue pressure on the resource.

Related to the issue of species diversity is the matter of product diversity. Seaweed mariculture and wildcrafting companies in Maine both initially focused on producing seaweed for human food or for agricultural applications. However, seaweed has many uses beyond nutrition and agriculture, including pharmaceutical, cosmetic, and materials applications. Maine has seen some encouraging progress on this front with the establishment of several non-traditional seaweed companies. Dulse & Rugosa, for example, is a Maine wild harvest company that manufactures a range of seaweed-based cosmetic products. Dirigo Sea Farms, based in Blue Hill, Maine, is a startup company working to replace polyvinyl alcohol with kelp bio-plastic to make sustainable and biodegradable laundry pods. Likewise, Viable Gear, another Maine-based startup, is working to replace the plastic in fishing and aquaculture equipment with a polymer derived from seaweed. Everything Seaweed intends to produce an alternative, PFAS-free coating for food containers and wrappers from refined seaweed "bio-materials." Companies outside of Maine are also looking at Maine seaweed. Soarce, which is based in Florida, is working with the University of Maine Process Development Center to produce seaweed-derived nanocellulose offering an exceptional aspect ratio that can exceed 100 and a superior water retention value of 7000%, outperforming traditional sources. Nanocellulose is a versatile, high-strength material being applied in papermaking, packaging, energy and electronics, biomedical applications, sensors, coatings, cosmetics, and more (Dhali et al. 2021). These companies, and scores of other similar ventures, will need to source large quantities of seaweed from both mariculture and wildcrafting. Their success will enable phyconomy success in Maine and elsewhere.

9.16 Seaweed Food Safety Standards

Passage of the US Food Safety Modernization Act (FSMA) in 2010 marked a shift toward a more proactive food safety approach. Some of the act's requirements are especially relevant for producers selling seaweed to other companies as an ingredient for food or supplement products. The act requires manufacturers of human and animal foods and nutritional supplements to verify the safety of foreign supply chains, whether by auditing suppliers themselves or by going through third-party auditors. Although the original intent of this requirement was to ensure the safety of imported food ingredients, many manufacturers now also apply it to domestic suppliers. Maine Coast Sea Vegetables has witnessed this trend firsthand as an increased frequency of customers requesting Global Food Safety Initiative (GFSI) or similar certification (Eddy, pers. obs.). GFSI certification is a voluntary standard, but one that is being increasingly requested as the cost of doing business.

The GFSI does not itself provide food safety certification but instead recognizes certification programs that meet GFSI Benchmarking Requirements. To achieve GFSI-recognized certification, companies must successfully undergo a third-party audit by a GFSI-recognized certification body. In turn, certification bodies must be accredited against ISO/IEC 17065 or ISO/IEC 17021, which are internationally recognized standards addressing issues such as preventing conflicts of interest, managing customer information, and qualifying personnel. In addition, Certification Bodies must be assessed by Accreditation Bodies members of the International Accreditation Forum (IAF) and signatories of the IAF Multilateral Recognition Arrangement (MLA). Needless to say, the cost of these extra layers of oversight is passed on to companies seeking GFSI-recognized food safety certification.

The cost, along with a lack of seaweed food safety expertise within US certification bodies, can make it challenging for Maine seaweed processors to obtain certification. Although many food safety principles apply regardless of food type, not all are equally relevant to every food, while some foods may warrant additional scrutiny or standards (e.g., many seafoods). Customers and certifiers unfamiliar with seaweed may not necessarily understand what are and are not appropriate or relevant standards. In such cases, they may find it convenient to apply seafood HACCP principles to seaweed despite it lacking many of the hazards associated with shellfish or finfish. Industry confusion, hesitancy, or inconsistency about such matters as the microbial safety of dried seaweed, allergen hazards, or cold chain control could lead to overly restrictive or excessive requirements and guidelines that fail to meaningfully advance food safety but do result in added costs. It is imperative that seaweed standards developed by regulators, industry, and auditors be practical, grounded in realistic concerns, and widely acceptable. Standards development is well informed by scientists but should also be driven by industry experience, including that of established wildcraft companies that have safely harvested, processed, and marketed seaweed for decades. One idea that recently emerged in Maine is for wildcrafting and mariculture companies to collectively address food safety or other regulatory concerns. Collaborative problem-solving and knowledge sharing could help the industry avoid conflicting, inconsistent, or inappropriate solutions, while helping regulators formulate appropriate guidance and rulemaking.

9.17 A Unified Phyconomy

Maine's seaweed mariculture and wildcrafting sectors share many areas of common interest, and common interests call for collaboration and cooperation. As an established organization with over 30 years of experience working on behalf of Maine's seaweed industry, the Maine Seaweed Council is perhaps in the best position to advocate for both wildcrafters and growers. This will not necessarily be an easy task. Although the MSC welcomes seaweed farmers as new members, it was founded by wild harvesters and has long worked to serve their interests. The rockweed industry continues to face challenges around appropriate management and intertidal access, and naturally, old guard companies that founded the MSC and financially support it regard these as priorities. The council's long legacy as an advocate for the rockweed industry might make it challenging to prioritize a shared agenda with mariculture companies. New growers attending MSC meetings for the first time may wonder how it benefits them when the discussion revolves mostly around rockweed harvesting.

MSC leadership is working to address this challenge through strategic planning and by coordinating with Maine Sea Grant to help organize and promote seaweed events that include harvesters and growers. In February 2024 the council held a planning session facilitated by Sea Grant Extension professionals to prioritize goals for the next 1–5 years. Seventeen members attended the meeting, though there was just one representative from Maine's mariculture sector (Mitch Lench from Ocean's Balance). A top goal identified by participants was for the MSC to be the trusted resource for reliable scientific information within the seaweed community. Relatedly, participants wanted the MSC to be a trusted source of accurate information as an effective advocate for reasonable regulations affecting both wild and farmed seaweeds. Increasing grower membership and collaboration among members were also identified as key tasks (Springuel et al. 2024). The overall theme was for the MSC to promote Maine's seaweed industry at large. Aligning with this theme, in April 2024, Maine Sea Grant organized Maine's sixth annual Seaweed Week, said to be the largest seaweed food and drink festival in North America. The event's website https://seaweedweek.org/ promoted both farmers and wild harvesters, and more than 70 wildcrafting and mariculture businesses participated.

In 2024 the MSC also formed a working subgroup named the Maine Rockweed Fisheries Alliance to help counter the Maine Rockweed Coalition, which opposes rockweed harvesting, and to help the council address rockweed fishery concerns outside of regular council meetings. In theory, this allows regular council meetings, which occur 4–6 times per year, to be less dominated by rockweed concerns and more about addressing industry-wide concerns such as those described earlier. A similar subgroup focused on the interests of growers is also possible, provided there is enough interest and leadership from within the grower community.

Encouraging more growers to join the MSC, and more importantly, to participate in meetings, remains a key challenge that may take time to resolve. Membership continues to be dominated by the traditional wildcrafting industry. Mariculture companies may regard membership with the Maine Aquaculture Association as aligning more with their interests than does MSC membership. Since most seaweed growers in the state operate on a part-time, seasonal basis, they may see little value in belonging to two organizations (or even one). This is unlike long-time MSC members for whom seaweed harvesting and processing is their primary enterprise and a full-time, year-round vocation. Seaweed mariculture is also still quite new for many, and other than for a handful of companies, it has not yet reached an economically significant scale. Rivalry between some of the new mariculture companies may discourage membership in a common organization. To that point, it's important to note that though founding members of the MSC did not always act without rivalry or disagreement, they nevertheless persevered in keeping the organization active and relevant. It should also be noted that it took almost 20 years after they founded their companies for them to join together as the Maine Seaweed Council. Industry cohesiveness can take time to develop.

We propose that there is much to be gained with cooperation between seaweed wildcrafting and mariculture companies to address their many shared challenges. Ultimately, the two sectors share stewardship for the seaweed they produce and the environment it grows in. Working on the water day after day imbues respect for marine ecosystems and a desire to protect them. Both are essential prerequisites to sustainable practice. A shared sense of stewardship could, and should, be the foundation of a unified phyconomy in Maine and elsewhere. **Note added in Proof:** In April 2025, Acadian Seaplants announced it was closing its operations in Jonesboro Maine for the 2025 harvest season. The company cited the exchange rate, transportation, and tariffs as drivers of the decision.

References

Alam MZ, Braun G, Norrie J, Hodges MD (2014) *Ascophyllum* extract application can promote plant growth and root yield in carrot associated with increased root-zone soil microbial activity. Can J Plant Sci 94(2):337–348

Albrecht M (2023) A Norwegian seaweed utopia? Governmental narratives of coastal communities, upscaling, and the industrial conquering of ocean spaces. Marit Stud 22(3):37

Angera T (2023) Business outlook: Maine seaweed farming. Retrieved from https://www.maineseaweedexchange.com/blog/business-outlook-maine-seaweed-farming. 10/08/2024

Angera T (2024) Kelp crisis: kelp supply chain challenges highlight market weakness. Retrieved from https://www.maineseaweedexchange.com/blog/kelp-crisis. Retrieved 8 October 2024

Araújo R, Vázquez Calderón F, Sánchez López J, Azevedo IC, Bruhn A, Fluch S, Ullmann J (2021) Current status of the algae production industry in Europe: an emerging sector of the blue bioeconomy. Front Mar Sci 7:626389

Arlov Ø, Nøkling-Eide K, Aarstad OA, Jacobsen SS, Langeng AM, Borrero-Santiago AR, Sæther M, Rustad T, Aachmann FL, Sletta H (2024) Variations in the chemical composition of Norwegian cultivated brown algae *Saccharina latissima* and *Alaria esculenta* based on deployment and harvest times. Alg Res 78:103421

Baker HW (1965) The story of nori in Japan. Br Phyco Bull 2:497–500

Banach JL, Koch SJ, Hoffmans Y, van den Burg SW (2022) Seaweed value chain stakeholder perspectives for food and environmental safety hazards. Foods 11:1514

Barbier M, Araújo R, Rebours C, Jacquemin B, Holdt SL, Charrier B (2020) Development and objectives of the PHYCOMORPH European Guidelines for the Sustainable Aquaculture of Seaweeds (PEGASUS). Bot Mar 63(1):5–16

Bhuyan MS (2023) Ecological risks associated with seaweed cultivation and identifying risk minimization approaches. Algal Res 1(69):102967

Billing SL, Rostan J, Tett P, Macleod A (2021) Is social license to operate relevant for seaweed cultivation in Europe? Aquaculture 534:736203

Blikra MJ, Aakre I, Rigutto-Farebrother J (2024) Consequences of acute and long-term excessive iodine intake: A literature review focusing on seaweed as a potential dietary iodine source. Compr Rev Food Sci Food Saf 23(6):70037

Brakel J, Sibonga RC, Dumilag RV, Montalescot V, Campbell I, Cottier-Cook EJ, Ward G, Le Masson V, Liu T, Msuya FE, Brodie J (2021) Exploring, harnessing and conserving marine genetic resources towards a sustainable seaweed aquaculture. Plants People Planet 3(4):337–349

Brayden C, Coleman S (2023) Maine seaweed benchmark report. Maine Aquaculture Association, Gardiner, Maine. 30pp

Buckley S, Hardy K, Hallgren F, Kubiak-Martens L, Miliauskienė Ž, Sheridan A, Sobkowiak-Tabaka I, Subirà ME (2023) Human consumption of seaweed and freshwater aquatic plants in ancient Europe. Nat Commun 14(1):6192

Cai J, Giulia G (2021) Global seaweeds and microalgae production, 1950–2019. World Aquaculture Performance Indicators (WAPI). https://openknowledge.fao.org/server/api/core/bitstreams/97409d09-2f8e-4712-b11e-60105d89959b/content. Jan 2025

Cai J, Lovatelli A, Aguilar-Manjarrez J, Cornish L, Dabbadie L, Desrochers A, Diffey S, Garrido Gamarro E, Geehan J, Hurtado A, Lucente D (2021) Seaweeds and microalgae: an overview for unlocking their potential in global aquaculture development. FAO Fisheries and Aquaculture Circular, p 1229

Campbell I, Macleod A, Sahlmann C, Neves L, Funderud J, Øverland M, Hughes AD, Stanley M (2019) The environmental risks associated with the development of seaweed farming in Europe-prioritizing key knowledge gaps. Front Mar Sci 22(6):107

Cerca M, Sosa A, Murphy F (2023) Responsible supply systems for macroalgae: upscaling seaweed cultivation in Ireland. Aquaculture 563:738996

Chopin T (2018) Seaweeds: the world's largest mariculture crop. Int Aquafeed 21(8):14–15

Clark DE, Newcombe E, Clement D, Magnusson M, Lawton RJ, CRK G, Major R, Adams S (2021) Stock take and characterization of Aotearoa New Zealand's seaweed sector: environmental effects of seaweed wild-harvest and aquaculture. In: Report for Sustainable Seas National Science Challenge project Building a seaweed sector: developing a seaweed sector framework for Aotearoa New Zealand. https://www.sustainableseaschallenge.co.nz/assets/dms/Guidance/Seaweed-Sector-Framework/Seaweed-Sector-Framework-Nov-screen.pdf. Jan 2025

Clavelle T, Lester SE, Gentry R, Froehlich HE (2019) Interactions and management for the future of marine aquaculture and capture fisheries. Fish Fisheries 20(2):368–388

Cleaver C, Johnson TR, Hanes SP (2024) Challenges and opportunities for integrating aquaculture with commercial fishing in Maine. Mar Policy 170:106407

Cornish ML, Critchley AT, Mouritsen OG (2017) Consumption of seaweeds and the human brain. J Appl Phycol 29:2377–2398

Craigie JS (2011) Seaweed extract stimuli in plant science and agriculture. J Appl Phycol 23:371–393

Critchley AT, Hurtado AQ, Neish IC (eds) (2024) Tropical Phyconomy coalition development: focus on Eucheumatoid seaweeds, vol 11. Springer

Curtin R, Prellezo R (2010) Understanding marine ecosystem based management: a literature review. Mar Poll 34(5):821–830

Dhali K, Ghasemlou M, Daver F, Cass P, Adhikari B (2021) A review of nanocellulose as a new material towards environmental sustainability. Sci Tot Environ 775:145871

Diehl N, Li H, Scheschonk L, Burgunter-Delamare B, Niedzwiedz S, Forbord S, Sæther M, Bischof K, Monteiro C (2024) The sugar kelp *Saccharina latissima* I: recent advances in a changing climate. Ann Bot 133(1):183–212

Dillehay TD, Ramírez C, Pino M, Collins MB, Rossen J, Pino-Navarro JD (2008) Monte Verde: seaweed, food, medicine, and the peopling of South America. science. Science 320:784–786

European Commission (2022) Commission communication COM/2022/592. Blue Bioeconomy – Towards a Strong and Sustainable EU Algae Sector. Available online at https://oceans--and-fisheries.ec.europa.eu/document/download/d2b345d5-9c66-4eca-aa43-6b135293bac8_en?filename=swd-2022-361_en.pdf

Erlandson JM, Braje TJ, Gill KM, Graham MH (2015) Ecology of the kelp highway: did marine resources facilitate human dispersal from Northeast Asia to the Americas? J Isl Coast Archaeol 10(3):392–411

FAO and WHO (2022) Report of the expert meeting on food safety for seaweed—current status and future perspectives. Rome, 28–29. https://wwwwhoint/publications/i/item/9789240058538. Jan 2025

Fan D, Hodges DM, Critchley AT, Prithiviraj B (2013) A commercial extract of brown macroalga (*Ascophyllum nodosum*) affects yield and the nutritional quality of spinach in vitro. Commun Soil Sci Plant Anal 44(12):1873–1884

Feeney MW (2001) Regulating seaweed harvesting in Maine: the public and private interests in an emerging marine resource industry. Ocean Coastal LJ 7:329

Fiskeridirektoratet (2023) Statistikkbanken. A06011 Alger Høsting til konsum og fôr etter art, statistikkvariabel og. https://havniaugefiskeridirno/PxWeb/pxweb/no/Fiskeridirektoratet/Fiskeridirektoratet__A%20Akvakultur__A06%20Salg/A06011px/. Accessed 19 January 2023

Flavin K, Flavin N, Flahive B (2013) Kelp farming manual: a guide to the processes, techniques, and equipment for farming kelp in New England waters. Ocean Approved. 122pp

Garbary DJ, Galway ME, Halat L (2017) Response to Ugarte et al.: *Ascophyllum* (Phaeophyceae) annually contributes over 100% of its vegetative biomass to detritus. Phycologia 56(1):116–118

Gercama I, Bertrams N, Tayler T (2022) Europe considers large-scale seaweed farming; environmental effects unknown. Retrieved from Mongabay website. https://news.mongabay.com/2022/10/europe-considers-large-scale-seaweed-farming-environmental-effects-unknown/. Dec 2024

Goecke F, Klemetsdal G, Ergon Å (2020) Cultivar development of kelps for commercial cultivation—past lessons and future prospects. Front Mar Sci 8:110

Groening T (2024) Despite legal restrictions, rockweed business thrives. Working Waterfront, The Island Institute. 09/05/2024. Retrieved from https://www.islandinstitute.org/working-waterfront/despite-legal-restrictions-rockweed-business-thrives/. Dec 2024

Halat L, Galway ME, Gitto S, Garbary DJ (2015) Epidermal shedding in *Ascophyllum nodosum* (Phaeophyceae): seasonality, productivity and relationship to harvesting. Phycologia 54(6):599–608

Hurtado AQ, Neish IC, Critchley AT (2019) Phyconomy: the extensive cultivation of seaweeds, their sustainability and economic value, with particular reference to important lessons to be learned and transferred from the practice of eucheumatoid farming. Phycologia 58(5):472–483

Hwang EK, Yotsukura N, Pang SJ, Su L, Shan TF (2019) Seaweed breeding programs and progress in eastern Asian countries. Phycologia 58:484–495

Johnson KB (2019) Unchanged Tides: the Maine Supreme Court continues to restrict public activities on navigable intertidal waters in Ross v. Acadian Seaplants, Ltd. Tul Mar LJ 44:707

Johnston EM, Mittelstaedt HN, Braun LA, Muhlin JF, Olsen BJ, Webber HM, Klemmer AJ (2023) Bed-scale impact and recovery of a commercially important intertidal seaweed. J Exp Mar Bio Ecol 561:151869

Kim J, Stekoll M, Yarish C (2019) Opportunities, challenges and future directions of open-water seaweed aquaculture in the United States. Phycologia 58(5):446–461

Kross E, Frohman G, Fong H, Luthringer O, Barden R (2021) Composite lines for kelp aquaculture. https://repository.library.noaa.gov/view/noaa/48495

Kumar MS (2014) Sustainable management of fisheries and aquaculture for food security and nutrition: policies requirements and actions. Agric Res 3:97–103

Langton R, Augyte S, Price N, Forster J, Noji T, Grebe G, St. Gelais A, Byron CJ (2019) An ecosystem approach to the culture of seaweed. NOAA Tech. Memo. NMFS-F/SPO-195, 24 p

Lauzon-Guay JS, Ugarte RA, Morse BL, Robertson CA (2021) Biomass and height of *Ascophyllum nodosum* after two decades of continuous commercial harvesting in eastern Canada. J Appl Phycol 33:1695–1708

Lauzon-Guay JS, Feibel AI, Morse BL, Ugarte RA (2023) Morphology of *Ascophyllum nodosum* in relation to commercial harvesting in New Brunswick, Canada. J Appl Phycol 35(5):2371–2381

Legge E (2019) Rockweed in Maine's intertidal zone. https://bpb-us-e1.wpmucdn.com/journeys.dartmouth.edu/dist/9/2772/files/2020/07/Policy-Brief-1.pdf. Dec 2024

Lotze HK, Milewski I, Fast J, Kay L, Worm B (2019) Ecosystem-based management of seaweed harvesting. Bot Mar 62(5):395–409

Mac Monagail M, Cornish L, Morrison L, Araújo R, Critchley AT (2017) Sustainable harvesting of wild seaweed resources. Eur J Phycol 52(4):371–390

Mace (2001) A new role for MSY in single-species and ecosystem approaches to fisheries stock assessment and management. Fish Fish 2(1):2–32

Midmore P, Naspetti S, Sherwood AM, Vairo D, Wier M, Zanoli R (2005) Consumer attitudes to quality and safety of organic and low input foods: a review. Report of EU-funded project "Improving Quality and Safety and Reduction of Cost in the European Organic and 'Low Input' Food Supply Chains." Univ. Wales, Aberystwyth, UK

Mouritsen OG (2013) Seaweeds: edible, available, and sustainable. University of Chicago Press

Mouritsen OG, Duelund L, Petersen MA, Hartmann AL, Frøst MB (2019) Umami taste, free amino acid composition, and volatile compounds of brown seaweeds. J Appl Phycol 31:1213–1232

Mumford TF (1990) Nori cultivation in North America: growth of the industry. In: Proceedings of the Thirteenth International Seaweed Symposium held in Vancouver, Canada, August 13–18, 1989. Springer, pp 89–98

Nardi G, Prickett R, van der Meeren T, Boyce D, Moir J (2021) Atlantic cod aquaculture: boom, bust, and rebirth? J World Aquac Soc 52(3):672–690

Newkirk G (1996) Sustainable coastal production systems: a model for integrating aquaculture and fisheries under community management. Ocean Coast Manage 32(2):69–83

Nisizawa K, Noda H, Kikuchi R, Watanabe T (1987) The main seaweed foods in Japan. Hydrobiologia 151:5–29

Nitschke U, Stengel DB (2015) A new HPLC method for the detection of iodine applied to natural samples of edible seaweeds and commercial seaweed food products. Food Chem 172:326–334

Oakley E (2024) Policy implications of wild seaweeds used in organic crop fertilizers. J Appl Phycol 36(1):371–383

Pereira L, Bahcevandziev K, Joshi NH (eds) (2020) Seaweeds as plant fertilizer, agricultural biostimulants and animal fodder. CRC Press, Boca Raton, FL

Piconi P, Veidenheimer R, Chase B (2020) Edible seaweed market analysis. Island Institute, Rockland Maine. 56pp

Redmond S, Green SL, Yarish C, Kim J, Neefus C (2014) New England seaweed culture handbook-nursery systems. Connecticut Sea Grant CTSG-14-01. PDF file URL: http://seagrant.uconn.edu/publications/aquaculture/handbook.pdf. 92 pp

Reiter SM, Post D, Wedding L, Strong A (2020) The future of the public trust: the muddied waters of rockweed management in Maine. Ocean & Coastal LJ 25:235

Robidoux J, Concepcion A (2023) Status and priorities for seaweed in the United States: report of the 2023 National Seaweed Symposium. https://seagrant.umaine.edu/extension/2023-national-seaweed-symposium/

Roleda MY, Skjermo J, Marfaing H, Jónsdóttir R, Rebours C, Gietl A, Stengel DB, Nitschke U (2018) Iodine content in bulk biomass of wild-harvested and cultivated edible seaweeds: inherent variations determine species-specific daily allowable consumption. Food Chem 15(254):333–339

Sæther M, Diehl N, Monteiro C, Li H, Niedzwiedz S, Burgunter-Delamare B, Scheschonk L, Bischof K, Forbord S (2024) The sugar kelp *Saccharina latissima* II: recent advances in farming and applications. J Appl Phycol 20:1–33

Saha CK (2022) Emergence and evolution of aquaculture sustainability certification schemes. Mar Poll 143:105196

Sayres WC (1958) Sammy Louis: the life history of a young MicMac. The Compass Publishing Company, New Haven, CT

Seeley RH, Schlesinger WH (2012) Sustainable seaweed cutting? The rockweed (*Ascophyllum nodosum*) industry of Maine and the Maritime Provinces. Ann N Y Acad Sci 1249(1):84–103

Seeley RH, Hardy S, Prentiss NK, Adey WH (2024) Comment: a reexamination of, bed-scale impact and recovery of a commercially important intertidal seaweed. J Exp Mar Biol Ecol 574:151984

Smale D, Orr K (2019) Ecosystem Considerations for Management of Scottish Seaweed Harvesting & Cultivation. A report commissioned by: the Sustainable Inshore Fisheries trust (SIFT). https://sift.scot/wp-content/uploads/2021/05/Seaweed-Review-Smale-Orr-SIFT-October-2019-FINAL-2.pdf. Dec 2024

Spillias S, Kelly R, Cottrell RS, O'Brien KR, Im RY, Kim JY, Lei C, Leung RW, Matsuba M, Reis JA, Sato Y (2023) The empirical evidence for the social-ecological impacts of seaweed farming. PLOS Sustain Transform 2(2):e0000042

Springuel N, Bartlett C, Robidoux J (2024) Maine Seaweed Council Priority Setting Meeting, Facilitators Report. 18pp. https://www.seaweedcouncil.org/

Sugumaran R, Padam BS, Yong WTL, Saallah S, Ahmed K, Yusof NA (2022) A retrospective review of global commercial seaweed production—current challenges, biosecurity and mitigation measures and prospects. Int J Environ Res Public Health 19(12):7087

Thayer S (2016) Back to the future—with seaweed. Boothbay register. 3/15/2016. Retrieved on-line from https://www.boothbayregister.com/article/back-future-seaweed/66958

Totten RP (2019) Defining fishing, the slippery seaweed slope, Ross v. Acadian Seaplants, LTD. Ocean Coastal LJ 24:178

Ugarte R, Sharp G (2012) Management and production of the brown algae *Ascophyllum nodosum* in the Canadian Maritimes. J Appl Phycol 24:409–416

Ugarte RA (2011) An evaluation of the mortality of the brown seaweed *Ascophyllum nodosum* (L.) Le Jol. produced by cutter rake harvests in southern New Brunswick, Canada. J Appl Phycol 23:401–407

Ugarte R, Lauzon-Guay JS, Critchley AT (2017) Comments on Halat L., Galway ME, Gitto S. & Garbary DJ 2015. Epidermal shedding in *Ascophyllum nodosum* (Phaeophyceae): seasonality, productivity and relationship to harvesting. Phycologia 54(6):599–608

Valderrama D (2012) Social and economic dimensions of seaweed farming: a global review. https://www.fao.org/4/i3344e/i3344e.pdf. Dec 2024

Valderrama D, Anderson JL (2010) Market interactions between aquaculture and common-property fisheries: recent evidence from the Bristol Bay sockeye salmon fishery in Alaska. J Environ Econ Manag 59(2):115–128

Webber HM, Claesson S, Erhart S, Schmitt CV, Muhlin JF (2023) Maine's potential to be a global leader in sustainable seaweed harvesting and management. Maine Pol Rev 32(2):70–77

Wells ML, Potin P, Craigie JS, Raven JA, Merchant SS, Helliwell KE, Smith AG, Camire ME, Brawley SH (2017) Algae as nutritional and functional food sources: revisiting our understanding. J Appl Phycol 29:949–982

Whitmore E, Davis C, Safford T (2025) Working the ground game: how Maine shellfish and seaweed farmers are building social license to operate. Aquac 15(596):741786

Wilding C, Tillin H, Corrigan SE, Stuart E, Ashton IA, Felstead P, Lubelski A, Burrows M, Smale D (2021) Seaweed aquaculture and mechanical harvesting: an evidence review to support sustainable management. Natural England Commissioned Reports. Natural England Report NECR378 https://plymsea.ac.uk/id/eprint/9590/. Dec 2024

Stephen D. Eddy Jr has been Director of the University of Maine Center for Cooperative Aquaculture Research since 2017 and Science Advisor for Maine Coast Sea Vegetables since 2018. He has worked in the aquaculture profession since 1996 and along the way developed expertise in recirculating aquaculture systems and in finfish, sea urchin, and seaweed biology and aquaculture. Prior to that he worked 10 years in commercial fisheries in the Bering Sea as a monitor on Japanese, Korean, and Soviet factory vessels. Steve grew up in Vermont and has degrees from the University of Vermont and the University of New Hampshire and a graduate certificate in food and nutrition from the University of Maine.

Shep Erhart co-founded Maine Coast Sea Vegetables in 1971 shortly after he and his wife, Linnette, moved to Maine to homestead. As the market for local, natural, and organic foods grew, so did the company, and today its simple dried sea veggie products are found nationwide. In 2017, he turned the company over to his employees in the form of an Employee Stock Ownership Plan (ESOP), which his daughter, Seraphina, is now managing. Shep was a founding member of the Maine Seaweed Council in 1993, which gives a voice to the growing and diverse seaweed community in Maine and advocates for the ecosystem of the 11 species of algae commercially harvested in Maine. He is also a member of the Seaweed Fisheries Advisory Council, a seven-member board that advises the Department of Marine Resources on all things impacting seaweed in Maine.

Seraphina Erhart is the daughter of Linnette and Shep Erhart, co-founders of Maine Coast Sea Vegetables; she jokes that she has seaweed in her blood. Many of her childhood memories revolve around time spent on rocky shores foraging for seaweed with her parents, packaging dried seaweed in the family barn, and recipe development in their farmhouse kitchen. She graduated college with a degree in the arts, soon following in the entrepreneurial spirit she grew up with by starting a fine art wet process photography print lab in Portland, Maine. Ultimately the seaweed kept calling, she recognized the value of what her family had built, and in 2010 Seraphina joined the Maine Coast Sea Vegetables team, becoming general manager in 2014 and now managing it on behalf of its employee owners. Seraphina is a member of the Maine Seaweed Council and is on the Maine Sea Grant Program Advisory Committee.

10 Economic, Technological and Regulatory Drivers in the Canadian *Ascophyllum nodosum* (L.) Le Jolis Industry

Glyn J. Sharp and R. E. Semple

Abstract

The *Ascophyllum nodosum* industry in the Maritimes stemmed from its long-standing use as fertilizer in Europe. Nova Scotia's *A. nodosum* commercialization began with the 1959 Sea Plants Harvesting Act, granting exclusive harvesting rights along coastal areas. Over 62 years, the industry evolved through five distinct periods, including changes in ownership, harvest technology, and products. The main driver was the allocation of the resource via leases, resulting in mechanization during Periods II and III, followed by demechanization in 1995, Period IV. A shortage of manual labor has necessitated the reintroduction of mechanical harvesting in Period V. Exclusive leases promoted expansion of the industry to New Brunswick and the southern and eastern shores of Nova Scotia in Periods IV and V. Control of the resource up to 15 years allowed the development of area-based annual management plans at hectare resolution enabling sustainable harvests. In Period IV, there was a significant shift in the allocation of leases in Southwestern Nova Scotia, leading to a loss in alginate production and a 67% decrease in landings. In Period V, landings recovered due to investments in biostimulant products for agriculture and continued production of traditional kelp meal products.

Keywords

Ascophyllum nodosum industry · Area-based management mechanization · Products · Nova Scotia

G. J. Sharp (✉)
KCME Inc., Halifax, NS, Canada

R. E. Semple
39 Coppermoon Court, Halifax, NS, Canada

10.1 Introduction

Ascophyllum nodosum (L.) Le Jolis plants are an assemblage of dichotomously and laterally branching shoots arising from a common holdfast floated by gas-filled vesicles. They form an extensive canopy on rising tides and homogenous mid to lower intertidal cover on stable substrates on the western Atlantic from New Jersey to Northern Quebec (Sharp 1986).

The Atlantic coast of Nova Scotia (NS), the Bay of Fundy, and southern coast of New Brunswick (NB) are in the mid-range of this distribution with abundant and accessible biomass in the intertidal. The total *Ascophyllum* biomass of NS and NB is 352,723 tons (Ugarte and Sharp 2012). The average tidal amplitude 3.5 m in the Southwestern Nova Scotia (SWNS) and 14.5 m in the Bay of Fundy on low sloping drumlin boulder and bedrock shores creates optimal habitat for *Ascophyllum* beds (Fig. 10.1).

French colonists to SWNS brought a coastal tradition of *Ascophyllum* utilization for fertilizer (Arzel 1987). Its industrial use as a fertilizer and animal fodder was well established in Europe by the 1930s (Baarsdeth 1970).

The concentration of resources on accessible islands, intertidal shoals, and coastlines in Lobster Bay (LB) attracted companies to propose investment in purchasing infrastructure and processing facilities in the industry in 1950s (Sharp 1986) (Fig. 10.2).

Ascophyllum landings in the Maritimes originated from SWNS in 1962, three years following the enactment of the NS Sea Plants Harvesting Act (SPHA) (NSDFA 1959) (Fig. 10.3) This act provided exclusive purchasing access for *Ascophyllum* to large portions of the SWNS coastline. In the following 62 years, the *Ascophyllum* industry developed in five distinctive periods in the Canadian Maritimes (Fig. 10.3) (Ugarte and Sharp 2012; Chopin and Ugarte 2006; Sharp and Semple 1997). Periods I and II were distinguished by a transition from manual harvesting to a mechanical method; period III was defined by introduction of new mechanical

A. T. Critchley et al. (eds.), *Advances in Temperate Phyconomy: Algal Harvest and Cultivation in Globally Distributed Temperate Waters*, Developments in Applied Phycology 15, https://doi.org/10.1007/978-3-032-02955-3_10

Fig. 10.1 Aerial view of intertidal concentrations of *Ascophyllum nodosum* on an island and shoals at low tide in Southwestern Nova Scotia

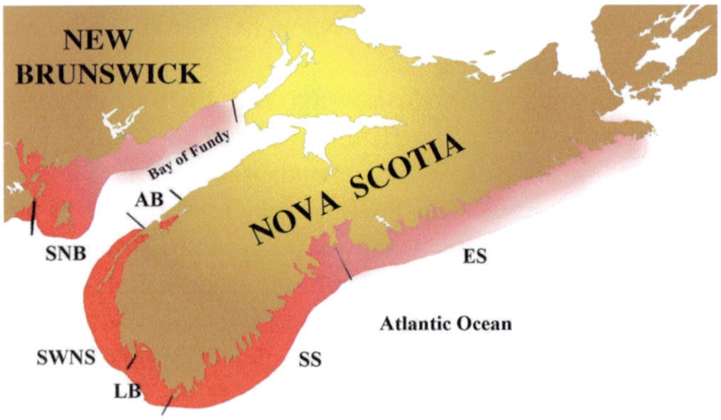

Fig. 10.2 Relative abundance of *Ascophyllum nodosum* biomass on the Atlantic and Bay of Fundy shores of Nova Scotia and New Brunswick. AB, Annapolis Basin; SNB, Southern New Brunswick; SWNS, Southwestern Nova Scotia; LB, Lobster Bay; SS, South shore; ES, Eastern Shore

harvesters; period IV brought a return to manual harvesting, geographical expansion, and changes in final products; period V to the present is marked by a decline in landings in NS and a reintroduction of mechanical harvesters.

The transitions between these five periods had three drivers: allocation of resource access via government agencies, introduction of harvesting technology by processing companies, and changes in added final products. Each driver had linkages to other factors in the local economy, the international marketplace, and resource management. This chapter examines these connections and interactions during the substantial changes that have occurred during the five periods of the *Ascophyllum* industry in the Maritimes.

10.2 Lease Allocation and Management

Access to the *Ascophyllum* resource has no provincial or federal restrictions on the number of harvesters. The NS Sea Plants Harvesting Act (SPHA) of 1959 granted the provincial minister of fisheries the authority to assign exclusive harvesting rights for *Ascophyllum* resources to companies or individuals for up to 15 years (Sharp 1986).

"No person other than the holder of the license shall willfully harvest, cut or in any way detach any sea plant from an area of the solum for which a license is in force" (NSDFA 1959). This provision allowed allocating intertidal areas in NS to individuals or companies without size limits. The act was revised under the Fisheries and Coastal Act 1996 as the Rockweed Harvesting Regulations (RHR), described as a lease (NSDFA 1997).

> A lease shall grant a person exclusive right to harvest in a particular area for processing or sale and shall be in a form prescribed by the Minister.

There are no limits to the number of individuals designated by the lessee to harvest in a lease, known as permit holders:

> A permit holder shall display the identification card issued to the permit holder pursuant to subsection 70(2) of the Act to a peace officer on demand. (NSDFA 1997)

Resource access granted by the NS SPHA license or RWR lease can determine the economic potential and viability of a company as well as the distribution of public resource value in society (Chopin and Ugarte 2006). Leases have

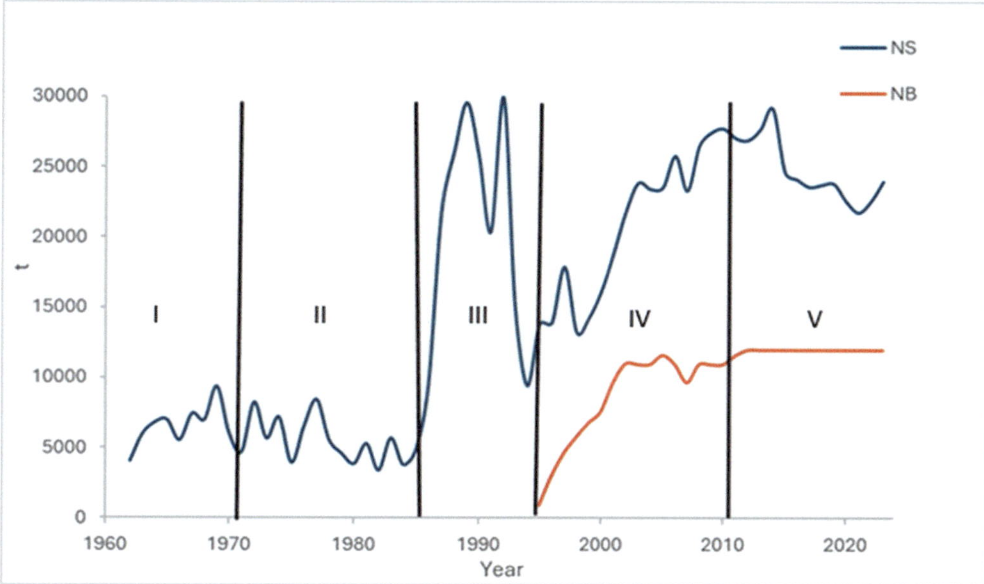

Fig. 10.3 Landings of *Ascophyllum nodosum* wet tons from 1962 to 2023 Nova Scotia and New Brunswick divided into five periods of development

strongly driven the *Ascophyllum* industry in the Maritimes over the five periods. The number of exclusive leases covering large areas of the *Ascophyllum* resource has steadily increased. The degree of responsibility for resource management in the lease conditions and costs of application and operation of lease have increased over the 62-year history of the industry in NS and NB (Ugarte and Sharp 2012; Sharp and Bodiguel 2002).

10.2.1 Periods I–II (1959–1987)

The first applicants for leases sought large sections of the coast in SWNS of *Ascophyllum* (Sharp 1986) (Fig. 10.4). Leases were granted to companies instead of individuals to foster investment in the development of infrastructure, processing facilities, and marketing. The prime area for harvestable *Ascophyllum* biomass in Southwestern Nova Scotia (SWNS) was divided between two companies: Scotia Marine Products Ltd. (SMP) and Bonda Sea Foods Ltd. (BSF)—companies purchased from independent manual harvesters operating within their lease boundaries, who landed their harvests at local wharves. The harvesters determined the locations and quantities of *Ascophyllum* to be harvested. The SPHA mandated yearly total tonnage accounting and a royalty based on tonnage purchased by the lessee. Management's responsibilities were limited to harvest methods and general conservation (Sharp 1986).

SMP's use of mechanical harvesting with salaried operators allowed direct control over daily resource management. Scotia Marine Products managed its lease in three large subsectors with 2.5 by 2.6 km grids, rotating them based on past landings records (Sharp 1986). Bonda SeaFoods continued to buy from independent manual harvesters.

The federal Department of Fisheries introduced an additional management layer with the 1978 Atlantic Coast Marine Plant Regulations (ACMPR) (DFO 2024). These regulations established 12 Maritime Marine Plant Harvesting Districts (MPHD), disregarding existing provincial lease boundaries. The Department of Fisheries and Oceans (DFO) issued marine plant licenses to individuals without entry limits in each Marine Plant Harvesting District (MPHD) or exploitation controls, except for restrictions on cutting height, harvesting gear, and whole-plant mortality limits like those found inNSDFA regulations (Chopin and Ugarte 2006). The Marine Plants Unit (MPU) of the DFO initiated a research program to assess resources, study the biology, and evaluate the impact of existing harvesting technology on all commercial algae in the Maritimes (Pringle and Mathieson 1986; Sharp 1986). The MPU provided regular resource assessment reports, evaluated harvesting impacts, and sponsored external research focused on resources (Cousens 1984; Ang et al. 1993; Lazo and Chapman 1993; Sharp 1981).

10.2.2 Period III (1988–1995)

The acquisition of SMP by Protan A/S, forming Pronova Biopolymers Ltd. (PNB), introduced highly efficient harvesting technology, expanded processing capacity, increased demand for raw materials, and heightened exploitation rates in the SWNS harvest area (Chopin and Ugarte 2006; Sharp and Semple 1997; Ugarte and Sharp 2001).

The harvest extended southward into the Bay of Fundy, encompassing both leased and open areas (Fig. 10.1) (Sharp and Semple 1991). Acadian Seaplants Ltd. (ASL), in collaboration with R&K Murphy Ltd. (RK), reported landings from the former BSF lease in Southwest Nova Scotia

Fig. 10.4 Distribution of provincial Rock Weed leases in Nova Scotia and New Brunswick 1959–2024 (NSDFA); present lease holders in 2024

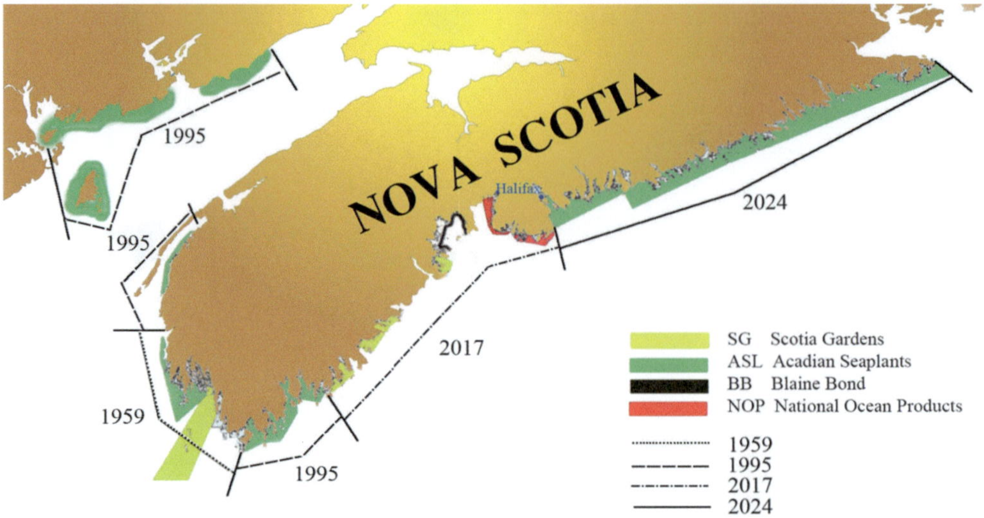

(SWNS) and expanded into the Bay of Fundy by purchasing from manual harvesters (Ugarte and Sharp 2012). In 1990, St. Mary's Bay and Digby open areas accounted for 4188 tons, representing 16.2% of the total year's landing (Sharp and Semple 1991). In response to the increased landings, the DFO biologist implemented a voluntary area-based management system (ABM). The ABM was based on 231 subsectors of leases held by PNB and RK/ASL, with a geographic resolution down to ±1 h (Sharp and Trembay 1989). The application of ABM provided the geographic resolution of landing data to subsectors ranging from 48.9 to 539 ha, permitting calculations of exploitation rates based on harvestable biomass assessments (Fig. 10.5).

10.2.3 Period IV (1996–2011)

During the transition from period III to IV, there was a significant shift in the distribution of NSDFA leases. Pronova Biopolymers Ltd. experienced a reduction of 255 hectares in its lease area, which corresponds to an annual loss of 4000–4500 tons due to mechanical harvesting. Three companies—R&K, ASL, and Scotia Garden Seafoods Ltd. (SGS)—were given parts of the relinquished leases to purchase from manual harvesters (Chopin and Ugarte 2006). This reallocation caused Protan A/S to shut its PNB processing facility in NS and leave Canada in 1995 (Ugarte and Sharp 2012). The distribution of leases between SGS and ASL remained ambiguous until NSDFA established new lease boundaries in 1997 (NSDFA 1997). The NSDFA continued to oversee the management of *Ascophyllum* leases, as well as the conditions of these leases and relevant regulations. This oversight includes enforcement, inspections, and the review of management plans and landing reports. Biologists from the Department of Fisheries Marine Plants Unit provided informal assessments, monitoring data, and information on the impacts of harvesting to the NSDFA.

Ascophyllum nodosum industry was introduced to New Brunswick in Period IV under unique terms—comanaged by DFO and the New Brunswick Department of Agriculture Aquaculture and Fisheries (NBDAF) under a memorandum of understanding (MOU) in 1990 (Chopin and Ugarte 2006; Ugarte and Sharp 2001). Companies seeking one or more of the three leases specified by NBDAF were required to submit a comprehensive business plan detailing the intended use of the resource. The ABM resource management plan required the company to assess the harvestable biomass of 96 subsectors within the three leases (Chopin and Ugarte 2006). Acadian Seaplants Ltd. received all three leases to the *Ascophyllum* resource in southern New Brunswick. The harvest was limited to a 17% exploitation rate based on assessment of the harvestable biomass in each sector (CAFSAC 1999). The company invested in a staged pilot to full-scale harvest with successive reporting, environmental reviews while building infrastructure for material handling, processing, and recruitment of harvesters (CAFSAC 1992).

The expansion of the *Ascophyllum* harvest to NB came under a wider mandate of DFO under the Oceans Act (1997). This act gave DFO responsibility beyond commercial species management to ecosystem protection in the marine environment. The Oceans Act enabled DFO to review development plans with input from NGOs, research institutions, and all stakeholders (Rangely and Davis 2000; CAFSAC 1999; Ugarte and Sharp 2001). The reviews led to lease conditions such as seasonal closures for wildfowl, closed study areas, and fishing gear protection. From 1995 to 1999, the quota was capped at 10,000 tons during the pilot phase, then gradually increased to 11,007 tons by 2002 (Fig. 10.6). The DFO provided assessment and resource biology research programs in SNB, reporting to a MOU management committee (Sharp and Bodiguel 2002).

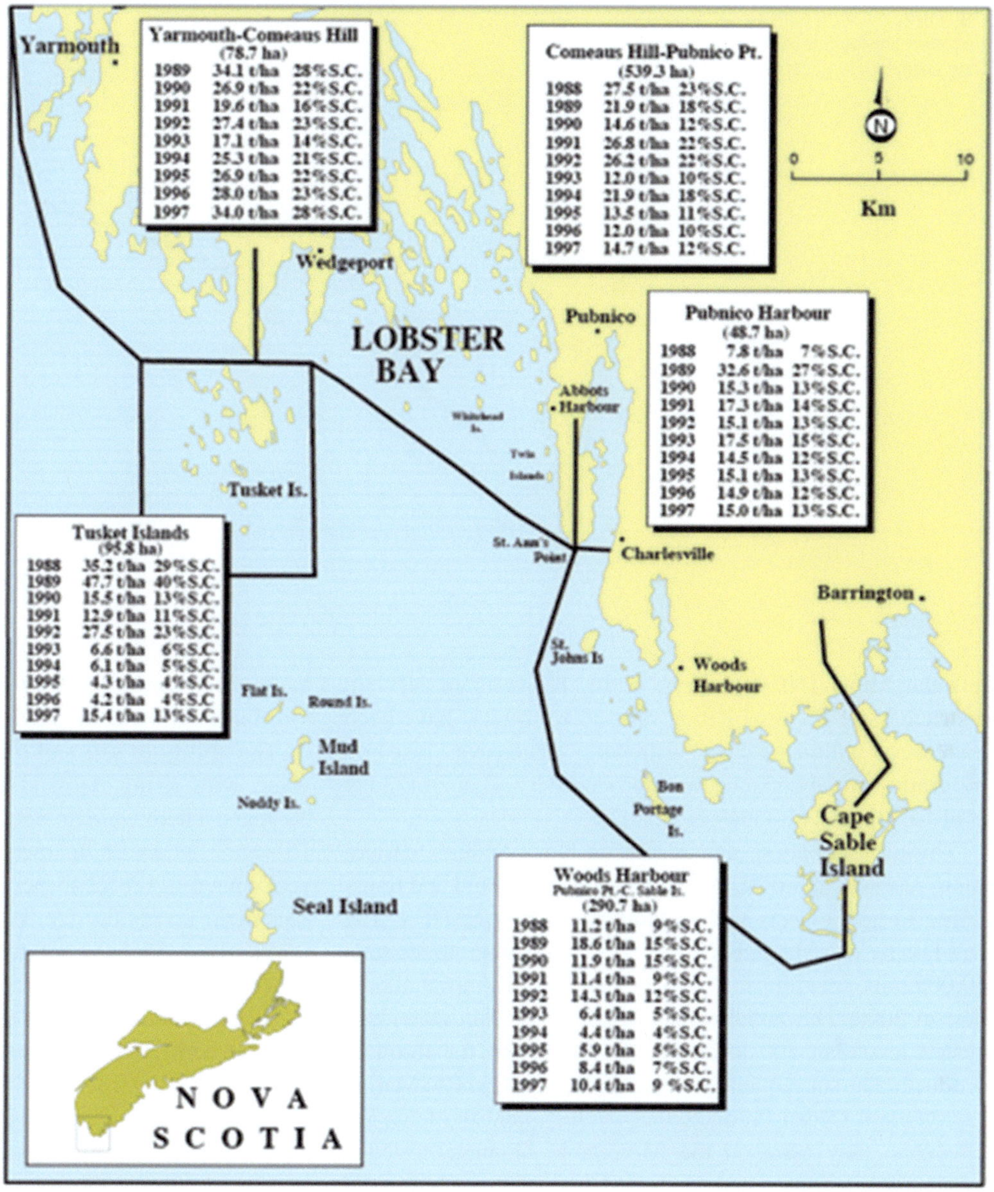

Fig. 10.5 *Ascophyllum nodosum* exploitation rates in Lobster Bay, southwestern Nova Scotia 1988–1997 based on grouping of subsectors for each of five areas (Sharp and Trembay 1989)

Leases were granted during period IV in the southern shore of Nova Scotia to ASL and SGS, in Mahone Bay to Blaine Bond (BB), and in St. Margarets Bay to Vic Bain (VB) (Fig. 10.1). Established companies made efforts to recruit and promote harvesting in open portions of the eastern shore of Nova Scotia, requiring transportation of up to 600 km to processing facilities. The 1997 NS RHR provisions for ABM require preapproval of annual harvest quotas and monthly and yearly reporting of landings within lease subsectors.

10.2.4 Period V (2012–2023)

During this period, a lease application in NS required an estimate of harvestable biomass, a harvest plan, and a business plan detailing planned use of the harvest. Fisheries and Oceans Canada aligned the ACMPR with NS lease regulations, removing the need for a marine plant license for harvesters in leases. In the absence of a lease, the harvest is still managed by DFO and requires a federal marine plant license. In 2023, DFO announced that the waters and shores of the Bay of Fundy were within provincial not federal jurisdiction and stopped issuing individual marine plant licenses for those shores. Both provinces are still developing management practices for those areas without leases in NS.

Nova Scotia landings rose over 29000 t in 2014; a peak only reached in phase III when mechanical harvesting was active. Combined with landings from NB, Maritime landings reached 40197 t in 2010, 92% of these landings were attributed to ASL (Ugarte and Sharp 2012). Management of *Ascophyllum* in NB in 2024 is solely by NBDF after DFO

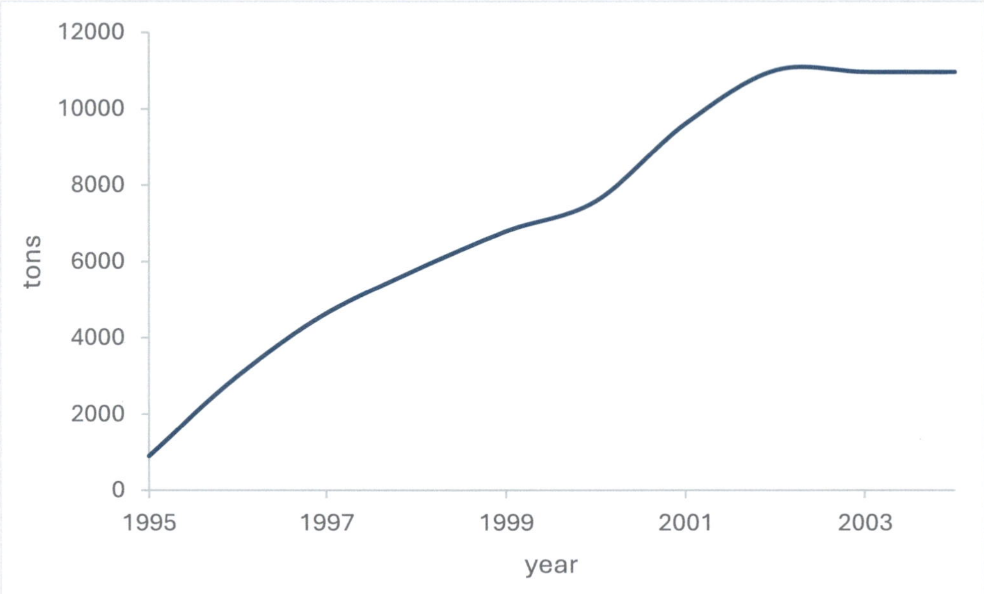

Fig. 10.6 *Ascophyllum nodosum* landings in southern New Brunswick wet tons

left the MOU (NBDFA pers. com). Harvests in NB are restricted to 17% exploitation rate; sector quotas have been increased within sectors following extensive research programs on abundance and productivity (Lauzon-Guay et al. 2021; Lauzon-Guay et al. 2023).

During this period, new leases on Nova Scotia's eastern shore received approval, and some leases were sold and transferred to new owners with authorization from NSDFA. The Halifax to St. Margarets lease area, estimated to contain 10,000 tons, was transferred twice following its initial allocation in 2000. The initial owner in 2000 Vic Bain (VB) harvested less than 500 tons annually processed for fertilizer products. The subsequent owner in 2006, EuroCan Bio Marine Ltd./Natural Ocean Products Inc. (NOPI), held the lease for 15 years; they assessed the harvestable biomass, tested a mechanical harvester, and conducted harvesting trials, but did not establish commercial harvesting or processing. The lease and NOPI were transferred to Louisburg Sea Foods Ltd. (LSF) in 2021. To date, LSF has not started commercial use of the resource in this lease. Three eastern shore leases were awarded to Maritime Rockweed in 2018 and were sold to ASL in 2019 in addition to the lease allocated to ASL in 2018.

Before 2017, lease applicants had to state their commercialization plans for their harvests, but selling raw materials to an existing processor was not excluded. Until 2017, leases required that landings be "minimally processed in NS." Revisions to these conditions in 2017 were more specific, including requirements for chopping, drying to 15% moisture, and packing for transport and sale (NSDFA pers. com). Additionally, the conditions of a lease required third-party biomass validation and an assessment of biomass prior to the first harvesting of the lease (NSDFA pers. com). This requirement was removed in 2024 and replaced with a requirement to submit all raw data to NSDFA for audit.

The ability to sell and transfer leases has increased their value with rising demand for *Ascophyllum* over the past 30 years. There were no performance criteria for leases up to 15 years. Maintaining ownership costs CAD $663.45 per year plus a CAD $2.45 per ton royalty, landings under 270 tons incur no royalty fee. Public information on lease transfer purchase prices is not available as these are private business transactions independent of NSDFA.

By 2012, the number of leases in Nova Scotia reached 19, with an additional 4 leases on the eastern shore added in 2018. With one lease discontinued, the total was 21 in 2023. The area of a lease includes all the water surface and shorelines encompassed by its boundaries. There is no direct relationship between the size of a lease and the harvestable biomass; however, the leases in Southwestern Nova Scotia and Southern New Brunswick have the greatest biomass (Ugarte and Sharp 2012). NS leases cover 490,123 hectares of water, including shorelines, islands, and *Ascophyllum*-covered intertidal shoals. Few open areas with significant commercial quantities of *Ascophyllum* remain. Distant areas from ASL and SGS processing facilities are now included in each company's annual harvest.

With NSDFA approval, leases can be transferred, resulting in a concentration of leases among a few companies or individuals over the past 25 years. In SWNS, as of 2024, two companies, ASL and SGS, hold 11 leases. Acadian Sea Plants Ltd. controls 15 out of the 21 existing leases in NS, covering 334,374.7 h of water surface and access to all the encompassed *Ascophyllum* beds, which constitutes 68.2% of the allocated area (NSDFA 2024). The area and number of the remaining leases in NS are distributed as follows: SGS 21.4%, 5; BB 6.1%, 1; NOPI 4.3%, 2.

Leases have concentrated ownership through application, allotment, renewal, and transfer. Some leases are relinquished after initial commercial assessment and available for new applications if they are not closed are managed by DFO. Certain areas in SWNS are specially closed to protect waterfowl habitats under the NS Department of Natural Resources.

Obtaining exclusive access through a lease today necessitates investment in the application process, annual fees, and the development of a management framework to comply with the lease conditions. This requires either trained company staff or third-party consultants. A company with existing facilities and product marketing has an advantage when the 15-year term of their lease is reached and an application for re-issuance must be submitted to NSDFA.

10.3 Mechanization

The transition from artisanal methods of gathering and fishing during industrialization is normally a direct and continuous transition, rarely reversing. In the Maritimes, manual techniques were initially used for marine plant harvesting, including the early *Ascophyllum* harvest (Pringle and Mathieson 1986; Sharp and Roddick 1982). The transition to mechanical from artisanal harvesting methods in the *Ascophyllum* industry began in 1973 and reversed in 1995 (Fig. 10.7) (Sharp and Semple 1997).

10.3.1 Periods I–III (1962–1995)

A manual labor force has existed to harvest *Chondrus crispus* Stackhouse in SWNS since the 1950s (MacFarlane 1956). The *Chondrus* harvest was an open fishery with no limits to entry either by federal or provincial regulations. Harvesters raked *Chondrus* along wave exposed coastline from 3 to 5 m outboard powered vessels during low tides (Fig. 10.8).

Chondrus crispus landings and participants were increased during the summer of the 1950s to 1960s, peaking in 1969 (Fig. 10.9). The industry consisted of three companies purchasing, drying, and exporting dried *Chondrus* to France, Denmark, and the USA (Pringle and Mathieson 1986).

Ascophyllum nodosum harvesters access its floating canopy on the rising or falling tide in wave sheltered to semiwave sheltered shorelines using a cutter rake, June to October (Fig. 10.10) (Ugarte and Sharp 2012).

Harvesters coordinate their harvests with tidal times and amplitude each day. A *Chondrus* harvester could potentially alternate between low tide harvesting of *Chondrus* and rising tide harvests of *Ascophyllum.*

The majority of harvestable *Ascophyllum* biomass is in wave sheltered to semiwave sheltered coastline (Sharp 1986). Safe operation of an outboard powered small vessel in a wide range of marine weather conditions particularly when loaded to capacity requires either experience or extensive training. The harvester assesses weather, wave exposure, tide level, and distance from landing sites to determine the best option for a day's harvest.

Initially, SMP bought raw material for alginate production from independent harvesters as did Bonda SeaFoods Ltd. (BSF) for kelp meal purchasing 7459 t in 1967 (Sharp 1986). Based on 156 t average seasonal landings per individual, the participation of a minimum of 48 harvesters over the harvesting season (June to October) was necessary. Although there was a large pool of harvesters with vessels, a 4:1 difference in landed value existed between *Chondrus* and *Ascophyllum* (Anderson et al. 1978; Ffrench 1971). To land equivalent value to *Chondrus* with *Ascophyllum*, the harvester must load the boat on each tide with four times the weight of *Chondrus*. The *Chondrus* vessels 3.5–5 m long had load limits of 0.5–1 t (Pringle and Mathieson 1986). A larger vessel with a more powerful outboard was needed for the *Ascophyllum* harvest. During this period, participation in the *Chondrus* harvest remained high. The number of harvesters actively harvesting *Chondrus* in Lobster Bay was 541 in June, 494 in July, and 186 in August 1978. An individual harvester could land 359 kg—605 wet kg d^{-1} between June and August (Sharp and Roddick 1982). Based on total landings and reported value of landings to harvesters, the price per wet kg was \$0.51(2024 \$CDN) kg^{-1} in 1977 (Anderson et al. 1978; Ffrench 1971). The return per day for harvesting *Chondrus* ranged from \$183 to 308 (2024 \$CDN). An equivalent landed value of *Ascophyllum* would require a daily harvest of 1000–2400 kg per day plus an investment in a larger vessel. The SWNS *Chondrus* industry reached a landing peak in1969, 18569 t.

Scotia Marine Products Ltd. needed a steady supply of raw material unrelated to season in competition for manual labor during the *Chondrus* harvesting season June through September. Kelco of California, the owner of SMP, used large vessels with reciprocating cutters to harvest *Macrocystis pyrifera* (Phaeophyceae) in California (McPeak et al. 1988). Scotia Marine Products purchased four Aquamarine[tm] harvesters equipped with reciprocating cutters mounted on a vessel propelled by paddle wheels in 1970 (Sharp 1986). Scotia Marine Products mechanically harvested all wet raw material after 1972 and purchased dried manually harvested material for processing in winter months (Fig. 10.1).

The cost of commencing a mechanical harvesting operation using the Aquamarine[tm] 475, including machines, a transport barge, and towing vessel in 1975 was CAD \$1,600,000 (2024 \$CDN) (Sharp and Neish 1976).

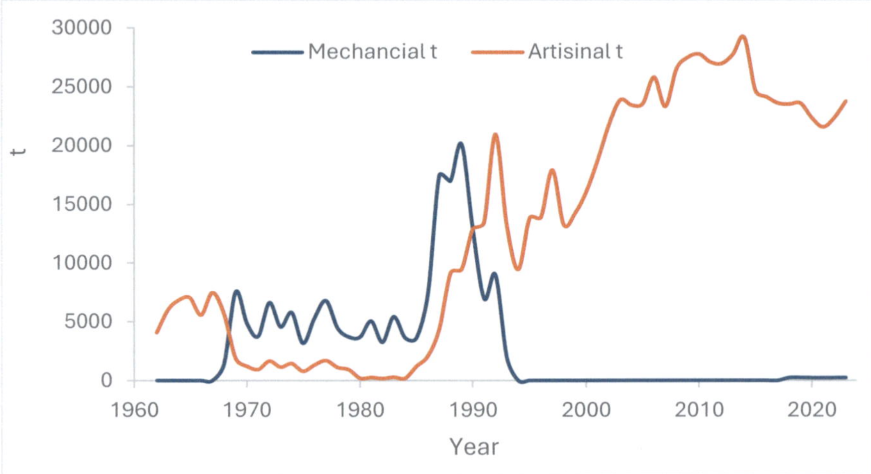

Fig. 10.7 Annual *Ascophyllum nodosum* landings wet tons in Nova Scotia 1962–2023 by mechanical and artisinal harvesting techniques

Fig. 10.8 *Chondrus crispus* harvesters raking from skiffs in Southwestern Nova Scotia at low tide, 1975

Harvesting operations required a tugboat captain, deck hand, barge man, and four machine operators. Daily operational costs over a capital amortization period of 15 years were $ 2000 per day. The estimated harvest per day is 12–18 t d^{-1} for 4 machines, with a cost of \$111–167 t^{-1} (Sharp and Neish 1976). The operators were inside a protected cab and without direct contact with seaweed; all material handling was completed by a conveyor belt system (Sharp 1986).

Manual harvesters entering the industry without a vessel or motor would require a capital expenditure of $13,000 (2024 $CDN) for a vessel capable of holding 2–3 t. Manual harvesters worked 3–6 h including travel to the harvest site, harvesting return, and unloading at the wharf to transport containers. The manual *Ascophyllum* harvester had to lift 10 kg 200–300 times with a cutter rake to earn CAD $54 to $81(2024 $CDN) per day.

10.3.2 Period III (1988–1995)

The sale of SMP to Protan A/S forming the Canadian subsidiary Pronova Biopolymers Ltd. (PNB) introduced Norwegian suction cutters to NS in 1985. These machines averaged 33.6 t per day 2–3 times more efficient than the Aquamarine^{tm} machine (Sharp 1986). Their cost in 1985 was $0.9 million (2024 $CDN) per unit and support barge and tow boat were an additional $1.8 million totaling $2.7 million (2024 $CDN) (Sharp et al. 1994).

In this period, manual harvesters were equipped with higher capacity boats (4–6 t) costing $25,000 (2024 $CDN) and harvesting 3.5 ± 1.5 t d^{-1}(Sharp et al. 1994). The cost of purchase of manual harvests and transport to the processing plant was \$126–150 t^{-1} compared to machine, \$131 t^{-1} (Sharp et al. 1994). Despite the increasing availability of former *Chondrus* harvesters due the decline in the *Chondrus* industry, PNB relied on mechanical harvesters. The assurance of raw material supply and higher mechanical exploitation rates in the traditional harvest area required control over the location of harvest, rate of harvest, and frequency of harvest not available with independent manual harvesters. Bonda Sea Foods Ltd. was sold to R&K Murphy (RK) and combined with Acadian Seaplants Ltd. (ASL) expanded their harvests at the end of period III and recruited harvesters in non-traditional open harvest areas of the Bay of Fundy and southern NS (Sharp et al. 1994). Scotia Gardens Seafood Inc. (SGS) entered the *Ascophyllum* market during this period, utilizing leases in the traditional harvest area in SWNS. Mechanical landings declined at the end of this period as PNB reduced its operations and prepared to leave Canada.

Fig. 10.9 *Chondrus crispus* landings wet tons, Nova Scotia 1948–2000

Fig. 10.10 Harvesting *Ascophyllum nodosum* on a rising tide with a cutter rake from an outboard powered skiff

10.3.3 Period IV (1996–2011)

Nova Scotia landings dropped dramatically after the departure of PNB with harvesting machines from NS to 9444 t in 1994 from 29951 in 1989. The rebuilding of the harvest was based solely on recruitment of up to 158 manual harvesters and expansion of the industry on the southern and eastern coastline of NS to new leases and open areas. The resource in these areas was not as continuous or as abundant as the traditional SWNS resource (Ugarte and Sharp 2001). The *Chondrus* industry continued to decline to below 2000 t providing a pool of experienced small boat harvesters in SWNS.

Harvesting vessels and gear were provided free or under a rental agreement to new harvesters. The defrayment or removal of the capital entry costs and the provision of landing infrastructure in nontraditional areas allowed the recovery of landings in period IV (Chopin and Ugarte 2006).

Acadian Seaplants Ltd. was granted three leases in southern NB in 1994. Recruitment of harvesters was difficult as NB did not have a history of *Ascophyllum* harvests and there was a lack of local support for a harvest (Chopin and Ugarte 2006). To reach 11612 t in 2005, 50–80 harvesters were recruited from other provinces to find (Ugarte and Sharp 2012) (Fig. 10.11).

10.3.4 Period V (2012–2023)

Landings continued to recover in NS to 29158 t in 2014 approaching the maximum landings in period III 29597 t (Fig. 10.12). A significant decline of 5724 t annually on average occurred since 2014.

Although there has been a concentration in the ownership of leases, there have been sustained efforts to purchase both in traditional areas (new leases) and open areas (Ugarte and Sharp 2012). The supply of manual labor is a potential reason for the drop in landings. Average seasonal landings of 156 t per person would require 32 harvesters to land 5000 t.

The return to the manual harvesting has not kept pace with inflation and rising value of other fisheries over the past 20 years. The price to the harvester landed at the wharf in 2004 was \$40–\$ 61 t^{-1} (2024 \$CDN); today, the price ranges from \$70 to 75 t^{-1} (Tidal Organics Inc. [TOI] and ASL, 2024 \$CDN). An *Ascophyllum* harvester will cut and load 3–5 t of seaweed per tide in 2.5–4 h. Average value landings per season for 156t at \$70 t^{-1} to \$75 t^{-1} is \$10,920 to \$11,700. These landings including travel and unloading provide a gross income of \$87 h^{-1} to \$ 98 h^{-1}. The current minimum wage in NS is \$1.52 per hour; harvesting *Ascophyllum* would appear

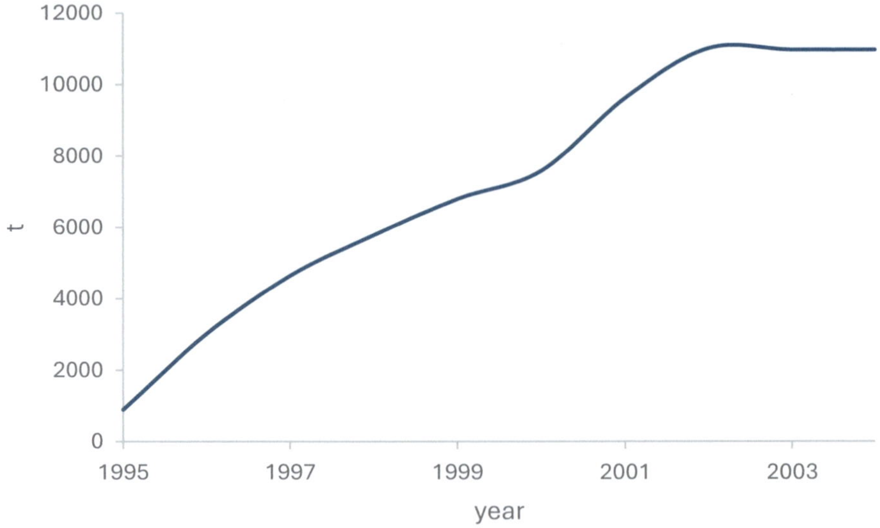

Fig. 10.11 Landings of *Ascophyllum nodosum* in New Brunswick 1995–2011 wet tons

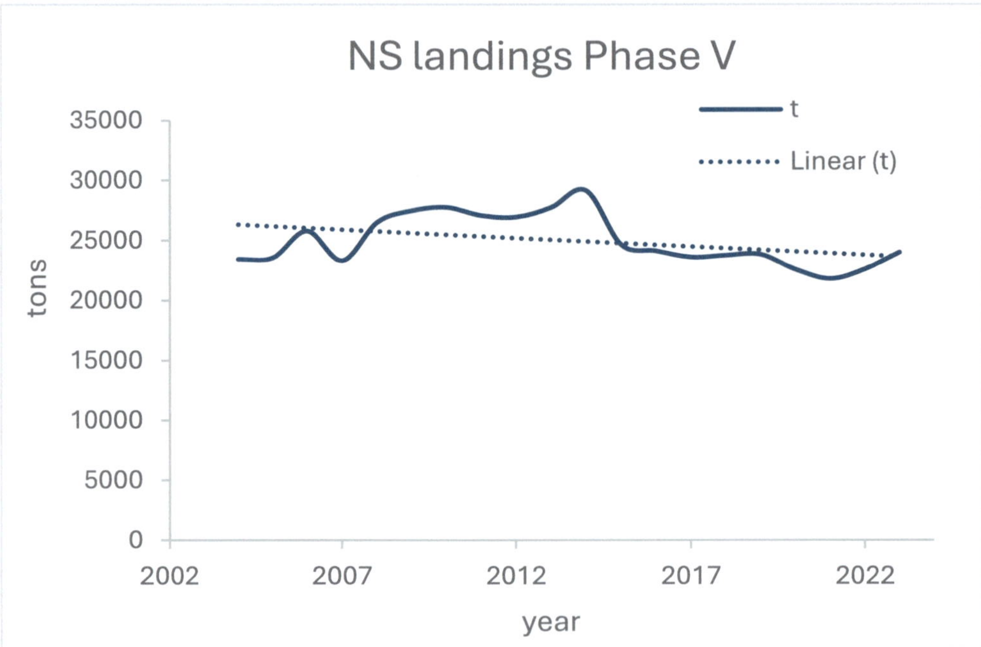

Fig. 10.12 Annual landings of *Ascophyllum nodosum* wet tons 2004–2023 and trend line

to provide an attractive wage. Harvesters range from those with a personal daily target for landings to highliners who harvest the maximum capacity of the vessel each tide June to October—less than 20% of harvesters.

Local manual *Ascophyllum* harvesters range from 14 to 55 years averaging 40 years old (TOI 2024). Except for the youngest harvesters, seasonal workers in SWNS have experience in other fisheries and associated employment. The availability of harvesters with sufficient physical skills and condition is in part related to the demographics of the local communities. In the municipality of Clare within the traditional *Ascophyllum* harvesting area, there has been a 17% drop of working-age adults since 2011 (Statistics Canada 2022).

In the primary area for *Ascophyllum* harvests LFA 34 and LFA 33, there are many employment opportunities in the fishing industry both as deckhands on licensed vessels for groundfish, herring, scallop in the summer, and lobster in the fall, winter, and spring (Fig. 10.13). Harvesters actively optimize their income from several sources; or have the option to gain employment in those fisheries as license holder, deck hands, or shore workers (Bodiguel 1996).

Fishing industries create many indirect jobs, including handling, processing, and transport of the catches. The consistent value of the fisheries and stable landings have sustained demand for labor from 2008 to 2021 (Fig. 10.14).

The success of the lobster fishing industry is critical to the choice of marine plants harvesting as an income source in

SWNS. The traditional *Ascophyllum* harvest area is encompassed by LFA 34 and 33, the highest lobster landed value in the Scotia Fundy region; $CAD 625 million 2021–2022 season. The 1656 licensed lobster fishing vessels in LFA 33–34 average 3.33 crew per vessel (DFO 2019). Although the lobster fishery is not active during the summer and early fall, its increasing value means higher returns for labor and lower requirement for supplemental income. The landings of lobsters in the Maritimes region have steadily increased since 2007–2015 with the largest share in LFA 34 and 33 (Fig. 10.15) (DFO 2019).

Recruitment of harvesters is an ongoing part of an industry dependent on independent manual harvesters with other employment options. The most difficult areas are those without any tradition of marine plant harvesting and manual fisheries. In traditional harvest areas, recruitment begin with connections to and mentoring by existing harvesters. Companies advertise through local and social media and in other provinces (Fig. 10.16).

There is a wide range of competition to an open access *Ascophyllum* harvest (Bodiguel 1996). Changes in the value of other fisheries and seasonal employment opportunities can mean participation from a few weeks to the full season (Sharp and Roddick 1982). Maintaining a steady supply of raw material to efficiently utilize a processing facility can be a challenge with a fluctuating labor force.

Increasing indirect control of harvesters to maintain both quota limits and utilization of all portions of the leased resource is at times in conflict with a harvester's goals. To deal with these challenges, the two largest lease holders, SGS and ASL, have reintroduced mechanical harvesting in period V.

Mechanical harvesting technology left NS in 1995 but still exist internationally in Norway, Iceland, and Scotland. Suction-cutter-type harvesters have been developed independently in Maine by North American Kelp and an independent harvester James Young; these harvesters have been used for over 18 years (James Young pers. comm).

Harvesting machines in Maine and NS differ significantly in design from the Aquamarine^{tm} and the Norwegian suction cutter. The seaweed is cut by a cutter mounted horizontal to the substrate in front of a wastewater pump and exhausted into a bag. Propulsion is provided by hydraulically driven thrusters. The shape of the vessels hull and outboard propulsion allows rapid movements of the harvester between *Ascophyllum* beds and bays (Fig. 10.17).

The reintroduction of mechanical harvesting to NS was not open to access but required approval by the minister of NSDFA (1997). EuroCan Ltd., a lease holder, purchased a prototype suction cutter machine from Maine and conducted harvesting trials but did not implement its commercial use. Tidal Organics Inc. chartered the James Young suction cutter

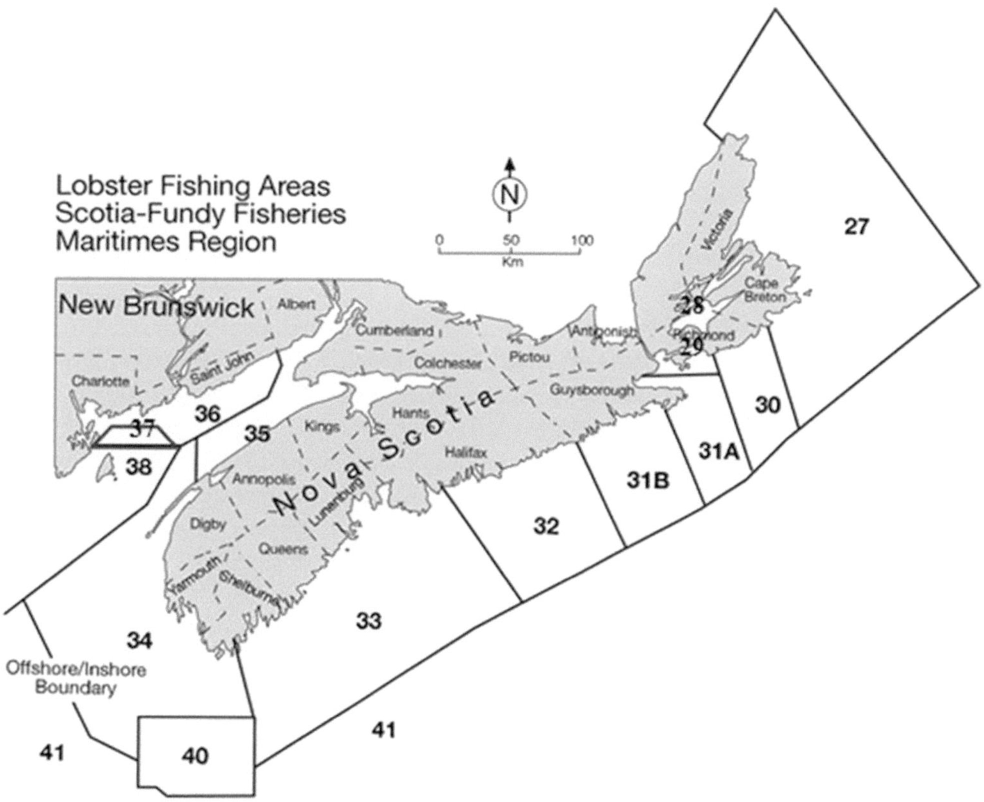

Fig. 10.13 Lobster Fishing Areas 27–41, Scotia Fundy Region, Department of Fisheries and Oceans

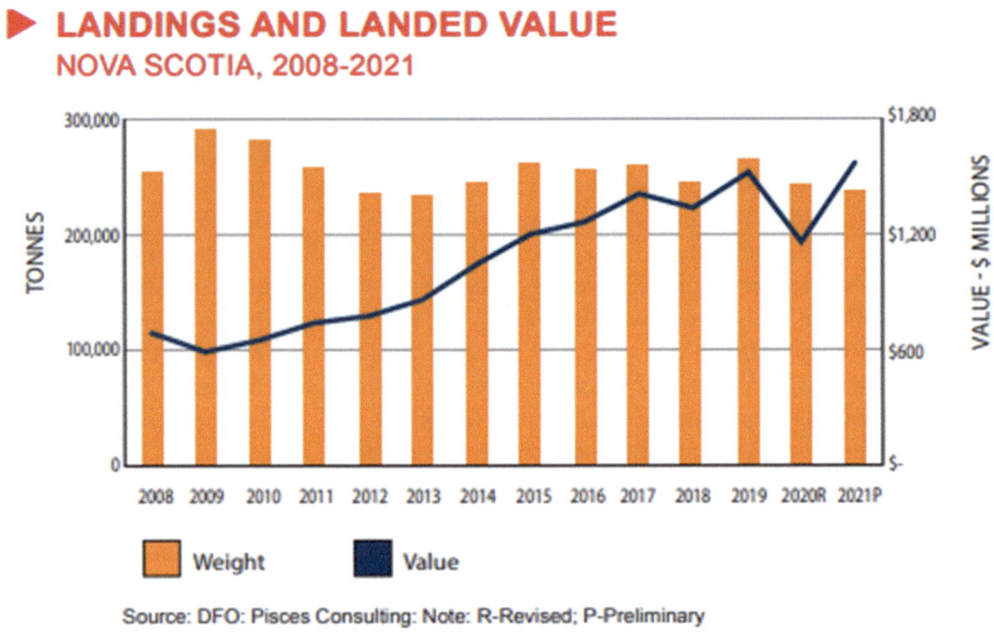

Fig. 10.14 Small boat fisheries landings and landed value 2008–2021 (Department of Fisheries and Oceans, Pices Consulting Ltd) (NS 2022)

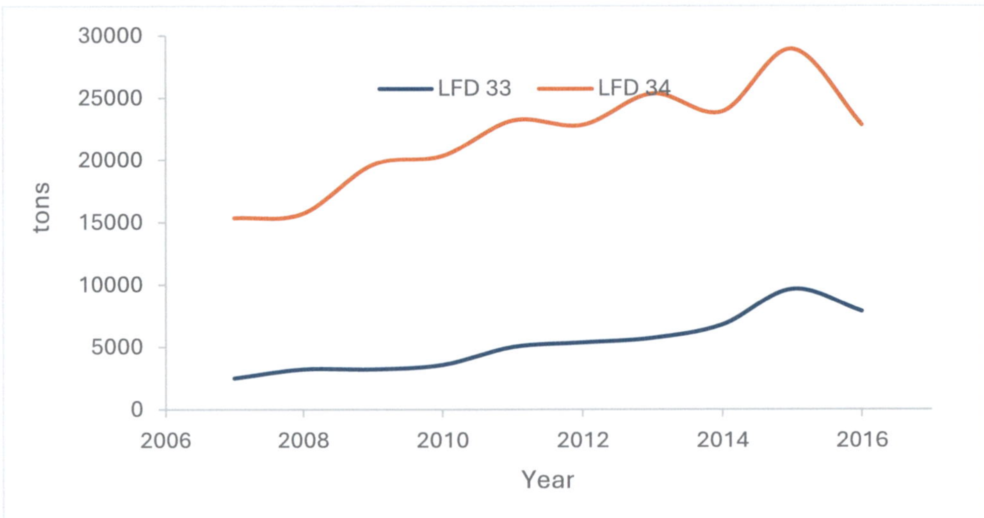

Fig. 10.15 Maritimes Region inshore lobster landed weight in Lobster Fishing Areas 33 and 34 (2006–2017)

machine under the NSDFA guidelines for experimental evaluation in NS in 2017 (Fig. 10.17). Harvesting trials followed by quota limited harvests and limited operational areas in 2020. Based on the success of the pilot harvest and impact assessment, this machine was permitted to commence commercial harvests in 2022 to the present using existing harvest regulations in traditional and new leased areas (Sharp and Sharp 2023). Mechanical landings have been from traditional SWNS and leased areas southern NS.

The cost of this harvester is low relative to previous Norwegian or Aquamarine[tm] machines. The major components of this machine are "off the shelf" accessible and do not require custom manufacture. Combined with the trailer and towing vehicle, the total costs are estimated to be between CAD $350,000 and 450,000. Material transport requires investment in combined boom truck and trailer if fixed wharf lifting gear is not available, an additional CAD $300,000–400,000.

The James Young machine is currently in complementary use to manual harvesters. The machine operates in areas considered suboptimal for manual methods, in beds distant from landing sites, and in areas where there is a shortage of reliable manual harvesters.

10.4 Products

Seaweeds are part of traditional agricultural systems in coastal regions of Europe for hundreds of years, especially in countries bordering the North Atlantic Ocean such as Ireland,

Fig. 10.16 Acadian Seaplants Ltd. advertisement for harvesters in Newfoundland 2024

Fig. 10.17 The James Young *Ascophyllum nodosum* harvesting machine 2017

Scotland, and parts of France and Scandinavia (Arzel 1987). Early coastal communities would collect drift seaweed including *Ascophyllum* and apply it directly to fields to enrich soil fertility. Reports from the 18th and 19th centuries document farmers using seaweed as mulch and as a nutritional supplement for soil and soil structure.

In the early twentieth century, interest shifted from raw seaweed use to more refined products. Industrial-scale harvesting of *Ascophyllum* began along the North Atlantic coasts for alginate extraction, a polysaccharide widely used in the food, textile, and pharmaceutical industries (McHugh 2003). By the mid-twentieth century, a growing body of empirical evidence suggested that, beyond alginates, *Ascophyllum* contained compounds capable of stimulating plant growth and resilience. Commercial seaweed extracts were subsequently developed, particularly in countries like Norway, the United Kingdom, and Canada, laying the foundation for modern seaweed-based biostimulants (Craigie 2011).

The 62-year history of the *Ascophyllum* industry in NS had a steady production of dried and milled "kelp meal" products for the fertilizer and animal feed supplement market. There has been a transition in chemical extraction from sodium alginate production to soluble extracts for plant biostimulants (Fig. 10.18). This transition was a driver influencing development in the Maritime seaweed industry.

10.4.1 Periods I–III (1962–1995)

The two companies awarded the first leases in the first 33 years of the *Ascophyllum* industry of SWNS produced two very diverse products. Bonda Seafoods Ltd. dried *Ascophyllum* by either solar or mechanical methods and produced a ground product (kelp meal) for fertilizer and animal food supplement. The first step in this process was very similar to the preparation of *Chondrus* for export existing in the Maritimes since the 1940s (MacFarlane 1952).

Scotia Marine Products Ltd extracted sodium alginate from *Ascophyllum* in a multistage, chemical, mechanical extraction process. This product was already in use and in demand for food additive and industrial uses already in Europe and the USA (Baarsdeth 1970; McHugh 2003). Kelco Ltd. of California established SMP adapting existing alginate extraction technology for *M. pyrifera* to *A. nodosum*. Differences in morphology, cell structure, alginate chemistry required investment in a research and development program prior to establishing the final processing protocol and plant design. The facility was 3000 m^2 equipped for multiple processing steps: alkali digestion, filtration, acidification, dilution, neutralization, and desiccation (McHugh 2003). The investment in this facility provided Kelco Ltd. an alginate product with unique mannuronic and glucuronic ratios with its own niche in the marketplace compared to *M. pyrifera* alginate (Borges et al. 2023). Alginate and kelp meal contributed to the first documented harvest 4073 t in 1962. Landing records averaging 5037 t y^{-1} over 22 years SMP operated in SWNS indicate SMP did not attempt to increase its production significantly (Fig. 10.19).

Kelco Ltd. sold the SMP processing facility and associated leases to Protan A/S, an established sodium alginate producer from Norway in 1984. Protan A/S brought alginate extraction expertise and markets to NS and imported harvesting machines designed and developed in Norway for its subsidiary Pronova Biopolymers Ltd. (PNB) (Sharp and Semple 1997). After 1985, landings increased five-fold due to enhanced production facilities and new mechanical harvesters, which boosted catch rates and extended the harvest-

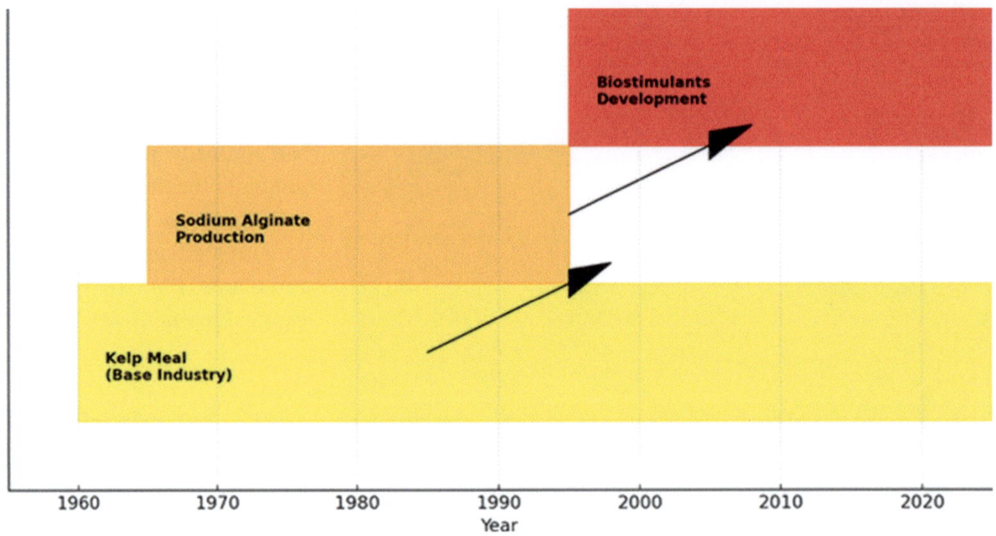

Fig. 10.18 Transitions of products from *Ascophyllum nodosum* in the Maritimes, 1962–2024

ing season (Sharp and Semple 1997). Dried material was purchased from other lease holders (RK) to maintain processing during the winter months. Total NS landings rapidly rose to 21800 t in 1987, peaking at 29951 t in 1989; PNB's were 80–90% of the landings. Sodium alginate production reached 1000 t in 1992, sixth highest in the world (Chopin and Ugarte 2006).

10.4.2 Period IV (1996–2012)

The NSDFA began to reallocate *Ascophyllum* leases of SWNS 1990–1996. ProNova Biopolymers Ltd. had anticipated renewal of its existing leases under the SPHA for 15 years. A counter argument was made for Canadian ownership and the wider distribution of harvesting jobs using independent manual harvesters. The province reallocated leases and allowed two companies to enter the industry. Acadian Seaplants Ltd. (ASL) entered the *Ascophyllum* industry using existing *Chondrus* drying and processing facilities in Yarmouth with RKM. Acadian Sea Plants expanded its harvesting force and production of dried and ground kelp meal for fertilizer and animal food supplements in 1992 (Chopin and Ugarte 2006).

Scotia Sea Garden Seafoods Inc. (SGS) Yarmouth purchased Genu Ltd., *Chondrus* drying and packing facility in Pubnico, and obtained an *Ascophyllum* lease in SWNS. Scotia Sea Garden Seafoods Ltd. invested in the Pubnico facility to produce kelp meal under Tidal Organics Inc. (TOI). The existing *Chondrus* drying and packing facilities were operating under capacity less than 2000 t due to the decline in *Chondrus* landings.

Pronova Biopolymers Ltd. did not have sufficient resources after redistribution of leases to sustain an economically successful Canadian operation and closed its operations in 1995. Nova Scotia lost production of a value-added extract from *Ascophyllum* and, for a time, a reduction in the diversity of the processing industry. Sodium alginate has a present retail value of CAD \$ 50332 t^{-1}(ingredi.com) compared to \$CDN 6299 t^{-1} (Thorvin) for kelp meal. Despite this loss of markets, the industry slowly recovered landings within 10 years (Fig. 10.19).

Acadian Sea Plants obtained access to the *Ascophyllum* resources of southern New Brunswick in 1994. It invested in a drying, grinding, and milling facility in Pennfield, NB, producing kelp meal. New Brunswick landings rose from 6800 t in 1995 to 11000 tons in 2002 (Chopin and Ugarte 2006).

Acadian Seaplants initiated research and development of a water-soluble extract from *Ascophyllum* for the biostimulant agricultural market by 1991. Acadian Seaplants Ltd. established an extraction facility in Cornwallis Nova Scotia in 1995 to produce liquid and dry soluble products.

10.4.3 Period V (2012–2023)

One of the primary reasons *Ascophyllum* is favored as a source of biostimulants is its rich composition of polysaccharides (alginate, laminarin, and fucoidan), vitamins, amino acids, phytohormones (auxins, cytokinins, and gibberellins), and minerals (K, Ca, Mg, Fe, etc.) (Khan et al. 2009). These compounds elicit various positive responses in plants, including enhanced root and shoot growth, improved nutrient uptake, and increased tolerance to abiotic stresses such as drought and salinity (Wally et al. 2013).

Increasing regulatory interest in sustainable agriculture and the demand for organic solutions have propelled the market for seaweed-based biostimulants, including those derived from *Ascophyllum* (McHugh 2003). This interest in *Ascophyllum* as a raw material for fertilizer products as

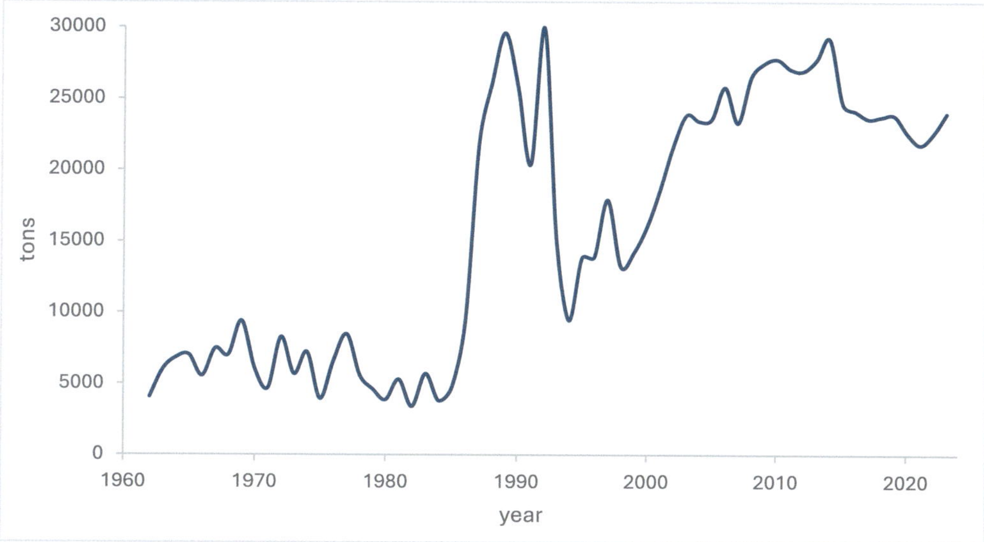

Fig. 10.19 Annual Nova Scotia landings of *Ascophyllum nodosum* wet tons, 1962–2023

reflected in the history of patent applications related to the use of *Ascophyllum* as a fertilizer rose slowly from 2000 to a dramatic rise in 2014–2015 (Fig. 10.20). *Ascophyllum nodosum* was used to produce16 of 25 plant growth and stimulant products marketed in 2009 including 4 by ASL (Khan et al., 2009). *Ascophyllum nodosum* is the seaweed most widely researched for its value as a source for industrial and commercial biostimulants (Shukla 2019). In the biostimulant marketplace, *A. nodosum* is the raw material used to manufacture extracts by 47 companies around the world (Van Oosten et al. 2017).

The Cornwallis, NS, processing facility grew on the enhanced value of a soluble extract sold in the biostimulant market. There are several proprietary modes of extraction to extract soluble biostimulants from seaweeds based chemically on water-based, acid hydrolysis, alkaline hydrolysis, and physical methods—microwave, ultrasound, supercritical fluid, and pressurized liquid (Shukla et al. 2019). The added value for the costs of extraction is reflected in the retail marketplace comparison of dried and ground *Ascophyllum*. The value of ASL kelp meal in 2024 marketed in the USA in 22 kg bags by Thorvin is \$CDN 6299 t^{-1}. Talun Canada Ottaway markets' soluble extract from *Ascophyllum* in 25 kg bags is equivalent to \$CDN 26,000 t^{-1}. Concentrate Milwaukie, the USA, is marketing soluble extract from *Ascophyllum* in 20 kg bags at \$CDN 32,170 t^{-1}. It is not possible to obtain the commercially confidential costs of production for the added retail value of *Ascophyllum* extract; however, the level of investment and growth of ASL reflects the profitability in this market.

The ASL processing facility in Cornwallis expanded in period V; a doubling was announced in 2009, and the expansion completed in 2013 (NS Business Inc. 2009). An associated Craigie Research center with 22 employees was established in Cornwallis that focused on biostimulant markets and marine plant aquaculture development with 22 employees. The production facility, the Deveau Center, employed 75 people in 2021 (NS Business Inc. 2009). A further doubling of the facility was announced in 2021 (Saltwire 2021).

The global market for seaweed extracts was valued at \$16.5 billion in 2023 and is projected to grow to \$20.9 billion by the end of 2029. The production volume of wet seaweed was estimated to be 38.5 million tons in 2023 exceeding 43.7 million tons by the end of 2029 (Global 2024).

10.5 Discussion

The management of seaweed resources has reflected the development of fisheries management in general and the assertion of national or local jurisdiction. A seaweed industry was well established in the Canadian Maritimes prior to any significant capability or interest of a government agency to control basic resource access. Seaweed harvesting was supplement to fishermen's income but rarely their sole income source in the Canadian Maritimes as it is today (Bodiguel 1996). The harvests were artisanal, a low landed value compared to other fishing activities (Sharp and Roddick 1982). Companies buying seaweeds limited their responsibilities to providing a reliable infrastructure for purchasing and processing of the raw material (Ffrench 1971). Despite the lack of initial oversight, industry has risen from open access with laissez faire management to a high-resolution area-based limited-access harvest under the provincial government legislation.

During the 62-year history of the *Ascophyllum* industry in the Maritimes, there have been strong economic, social, legislative, and technological drivers. The allocation of exclusive leases for purchasing a public *Ascophyllum* marine

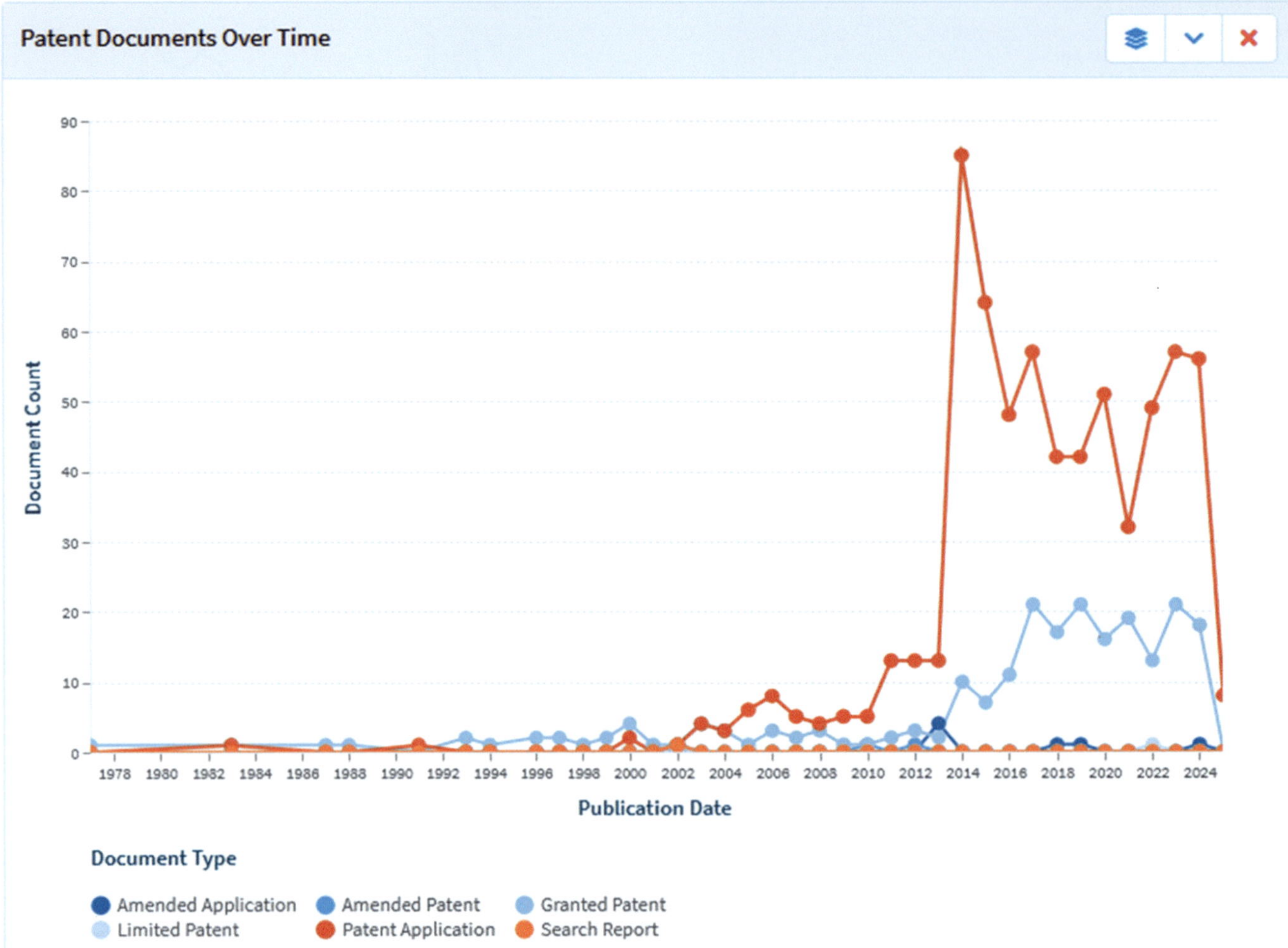

Fig. 10.20 Patent applications related to *Ascophyllum nodosum* 1978–2024 https://www.lens.org/lens/search/patent/analysis?q=Ascophyllum,%20%20fertilizer

resource has been unique. The 1867 Canadian constitution provided provincial governments jurisdiction of the "solum" interpreted to include *Ascophyllum* resources (La Forest 2008). Private ownership of the intertidal foreshore and shallow subtidal property does exist in Iceland, and the USA does include attached marine plants (Cheung 1991).

There have been similar management tools used to transfer responsibility for local resources to traditional users as an effort control. Managed Access Programs aimed at incentivizing fisher ownership and stewardship of regions to promote sustainability (Wade et al. 2019). Individual transferable quotas allocate fish resources within a defined area between fishers with the option to lease portions of a quota (Emery et al. 2014; Acheson et al. 2015; Pomeroy 1995). Leases of water bodies for aquaculture utilization are direct exclusive utilization. In comparison to the allocation of water column or water surface for aquaculture, the rocky intertidal did not have direct conflict with traditional fishery activities as has been common with aquaculture leasing (Marshall 2001).

The key driver of the *Ascophyllum* industry in NS and NB was the allocation of leases granting exclusive access to large portions of the resource. The assignment of exclusive access to a public resource provides medium-term lower risks for investment to encourage development of a processing industry. Business can provide customers with an assured constant supply of products and work to develop markets based on a known stable supply of raw material.

Reallocation of existing leases in 1990–1996 provided access to new companies to the principal concentration of harvestable resources (Chopin and Ugarte 2006). The reallocation of leases as well was the driver for changes in harvesting technology and management. These changes also led to the loss of alginate extraction technology and its replacement by biostimulant extraction.

Leases were not initially designed to limit effort and exploitation rates in an open-access harvest. Allocating resources to a single controlling entity potentially limited the number of harvesters within each lease in relationship to the processor's raw material demand. Leases provided a geographical framework for the resource to be managed by ABM in multiple subsectors (Ugarte and Sharp 2012; Sharp and Bodiguel 2002). Conditions accompanying resource access via a lease changed over the industry's history in relation to a wider view of the species in science and legislation (Chopin and Ugarte 2006). Leases are now the dominant means of managing the *Ascophyllum* resource leaving only a small portion of the least economically attractive portions of the resource in NS as open-access areas.

The evolution of leases from an economic stimulus to an ABM tool was not a planned process but a result of adaptation to higher demands for *Ascophyllum* stimulating changes in the regulatory structure. The evolution of regulation can be traced to the importance of this resource not only economically but also to the ecosystem services the *Ascophyllum* habitat and its productivity provide in the intertidal and coastal zone (Schmidt et al. 2015, Sharp et al 2008).

Area based management in fisheries management is the application of restrictions on resource exploitation within clearly defined geographic units of the coastline or ocean (Pikitch et al. 2004). The use of ABM was a component of the earliest efforts to manage fisheries (Gough 1991). Restrictions applied within geographical area allow application of limits on effort on both sessile and mobile fish stocks, as well as spreading economic value of a resource (Neher 1988; Pomeroy 1995). Sessile species, including intertidal seaweeds, can be accurately mapped at a high resolution, providing a GIS framework for monitoring and assessment at the hectare scale (Sharp et al. 2006).

Today, *Ascophyllum* leases and associated ABM have under the NS and NB government agencies shifted responsibility to industry for resource management. The present government–industry relationship fits under the label of "results-based management" (Nielsen et al. 2015). The input for stakeholders under the present conditions for lease application includes a public notice of application and an opportunity to comment on the application to the provincial government.

Companies have maximum limits set by annual resource assessment and quota limits within subsectors of a lease. Setting exploitation rate limits as quotas for resources within lease subsectors provides essential data for lessees to monitor usage and fulfill reporting obligations to management agencies. To provide this data, an assessment of the status of the resource must be completed within each subsector. It is their responsibility to oversee harvesting operations to control either the machines operated by employees or the independent manual harvesters to not exceed the sector quotas. Conditions of leases require control of cutting height, gear type, and whole-plant mortality rates (NSDFA 1997).

The increasing oversight of the lease holders with the obligation of day-to-day management of independent harvesters required development of company guidelines for harvesters and feedback control measures (Ugarte and Sharp 2012). Management of leases at the ABM sector resolution required a reliable reporting and monitoring system with independent harvesters. Each company has its own policy on enforcement of gear, reporting protocol, and quota limits for harvesters.

The lease holder in 2024 must have up-to-date knowledge of the resource base at the resolution that allows the sector to develop an annual management plan. The data used to assess harvestable biomass and annual harvests have been improved by remote sensing methods, ground truthing, and annual direct sampling programs (Ugarte and Sharp 2001; Ugarte et al. 2006; Chopin and Ugarte 2006). Harvesting impacts have been evaluated for manual and recently introduced mechanical harvesting methods (Lauzon-Guay et al. 2023; Sharp and Sharp 2023).

The allocation of most of the material supply to one company places the industries fate in the financial success of this company. The transition between PNB and ASL as the dominant company in the *Ascophyllum* industry had a period of uncertainty as new processes, products, and markets were developed by ASL. Acadian Seaplants Ltd., an established local company, was able to invest in development of biostimulant products and expand its processing capacity. The company expanded its resource base by applying for leases in open areas and purchased existing leases (Ugarte and Sharp 2012).

The processing of a seaweed ranges from simple drying or chemical preservation for export for final processing to sophisticated chemical/physical extraction or transformation to add value. The industry in the Maritimes had several transitions in products that occurred over 62 years, impacting added value and markets. Lease reallocation was the primary factor in the changes between periods III and IV. The existing alginate extraction facility lost a significant portion of its resource (Chopin and Ugarte 2006). It was replaced in period IV by biostimulant extraction and development of new markets. There was a potential for other companies to develop similar products; Argosy Seaweed Ltd. produced a liquid fertilizer from *Ascophyllum* in the 1970s but failed to reach commercial success (A McArthur pers. com.). A lease holder VB in southern NS produced liquid fertilizers in the early 2000s but ceased production when it sold its lease and business.

ASL's success in this market was also attributed to the growing demand from the agricultural sector. This demand rose from the need to produce food products with organic farming methods as well as the increasing costs of chemical fertilizers (Larramendy and Soloneski (2019). The participation of three other companies with smaller portions of the *Ascophyllum* resource available has not led to diversification of products to date. The products have remained in the classification of Kelp meal for fertilizer and animal food supplements. There is a small cottage industry utilizing bulk products from processing companies to produce cosmetic, nutraceutical, and specialized horticultural products.

A seaweed industry frequently begins with storm cast gathering or artisanal harvesting techniques (Ffrench 1971). Mechanization of fisheries has been a driver for expansion and enhancement of effort and exploitation rates particularly in artisanal fisheries in developing economies (Bhat and Bhatta 2006). The artisanal *Ascophyllum* industry has transitioned to mechanization and back over five periods of development. Between 1972 and 1995, both manual and mechanized techniques were active in the same *Ascophyllum* resource (Ugarte and Sharp 2012). The availability of labor and the need for a steady supply of raw material were primary factors leading to mechanization in SWNS. Introduction of more efficient machines enabled increased exploitation in SWNS (Sharp and Trembay 1989). High demand can lead to over exploitation without mechanization particularly in open-access areas. The most extreme example of over exploitation occurred in a portion of the resource in the Bay of Fundy by manual harvesters working in the Annapolis Basin (Sharp and Semple 1997). This was a clear example of "the tragedy of the commons"; the demand by several companies for raw material led to competitive harvesting without any controls (Hardin 1968). Machines controlled by lease holders offer the opportunity to directly manage the resource within the regulated subsectors to a resolution of individual beds (Sharp and Sharp 2023).

The loss of mechanical harvesters was eventually replaced by recruitment of manual harvesters when the competition for skilled manual harvesters was low due to the decline in the *Chondrus* industry. Investment in recruitment programs and incentives for entry to the industry lead to the success of expanded harvests in NS and NB since 2006.

Mechanical harvesting has returned to NS due to a need to fully utilize the resource available to lease holders. Harvesters have many employment options when the fisheries are increasing in value. Fewer experienced fishers opt to participate in a highly manual and physical harvest as they age. It is not in the interests of the companies to replace manual harvesters completely where they are available. Artisanal harvesting still dominates the harvest in 2023, but recently, new harvesting technology has been reintroduced to the industry period V by both TOI and ASL (Sharp and Sharp 2023). Machines are used in a complementary way to artisanal harvesters in traditional areas and to harvest resource evaluated as suboptimal for manual harvesters (TOI).

Utilization of *Ascophyllum* resources in the Maritimes has compensated for the decline of the *Chondrus* industry in Nova Scotia, increased the value of fisheries in Nova Scotia, and brought a new industry to New Brunswick. The management of this industry has steadily evolved with the recognition of its wider values beyond its direct economic return from this resource.

References

Acheson J, Apollonio S, Wilson J (2015) Individual transferable quotas and conservation: a critical assessment. Ecol Soc 20(4) http://www.jstor.org/stable/26270307

Anderson N, Frenettee E, Webster G (1978) Global village? Global pillage Irish moss from P.E.I. in the world market Social Action Commission, Charlottetown PEI p 102

Ang PO, Sharp GJ, Semple RE (1993) Changes in the population structure of *Ascophyllum nodosum* (L.) Le Jolis due to mechanical harvesting. Hydrobiologica 261:321–326

Arzel P (1987) Les goémoniers. Edition de l'estran, Douarnenez, Chasse marée, p 305

Baarsdeth E (1970) A synopsis of biological data on knobbed wrack, *Ascophyllum nodosum* (L.) LeJolis. FAO Fish Synop 36 Rev:41

Bhat M, Bhatta R (2006) Mechanization and technical interactions in multi-species Indian fisheries: implications for economic and biological sustainability. Mar Policy 30/3:237–248

Bodiguel C (1996) Des algues, des hommes et des politiques dans les Provinces Maritimes canadiennes. Vers quel mode de gestion durable? Mémoire de D.E.A., Univ. Paris I, Paris. 156 p

Borges H, Lopez-Rubio A, Martinez-Abad FMJ (2023) Overview of alginate extraction processes: impact on alginate molecular structure and techno-functional properties. Trends Food Sci Technol 140:104142

Chopin T, Ugarte R (2006) The seaweed resource of eastern Canada. In: Critchley AT, Ohno M, Largo DB (eds) World seaweed resources. 46 p. DVD -ROM. ETI BioInformatics Amsterdam ISBN: 90 75000 80 4

Cousens R (1984) Estimation of annual production by the intertidal brown alga *Ascophyllum nodosum* (L.) Le Jolis. Bot Mar 27:217–227

CAFSAC (1992) Rockweed in South-Western New Brunswick CAFSAC. Advisory Document 92/13:183–189

CAFSAC (1999) The impact of the rockweed harvest on the habitat of Southwest New Brunswick. Habitat Status Report 99/2E p 9

Cheung S (1991) Common property rights. In: Eatwell J, Milgate M, Newman P (eds) The world of economics. The New Palgrave. Palgrave Macmillan, London. https://doi.org/10.1007/978-1-349-21315-3_12

Craigie J (2011) Seaweed extract stimuli in plant science and agriculture. J Appl Phycol 23(3):371–393

DFO (2019) Lobster fishing areas 27–38 integrated fisheries management plan. https://www.dfo-mpo.gc.ca/fisheries-peches/ifmp-gmp/maritimes/2019/inshore-lobster-eng.html

DFO (2024) Fisheries act schedule 26 Atlantic fisheries regulations. Part IX Marine Plants

Emery T, Hartmann K, Green BS, Gardner C, Tisdell J (2014) Fishing for revenue: how leasing quota can be hazardous to your health. ICES J Mar Sci 71:1854–1865

Ffrench R (1971) A current appraisal of the Irish Moss industry. Industrial Development Branch Fisheries Service, Fisheries and Forestry Canada, p 230

Global (2024). https://www.businesswire.com/new/home/20241001368852/en/Seaweed-Extract

Gough J (1991) Fisheries management in Canada 1880–1910. Can Manu Rep Fish Aqua Sci 2105:96

Hardin G (1968) The tragedy of the commons. Science 162(3859):1243–1248

Khan W, Rayirath UP, Subramanian S (2009) Seaweed extracts as biostimulants of plant growth and development. J Plant Growth Reg 28(4):386–399

La Forest G (2008) Chapter 19: Rights relating to water. In: La Forest AW (ed) Anger and Honsberger law of real property, vol 2, 3rd edn. Canada Law Book, Aurora. p550

Larramendy M, Soloneski S (2019) Organic fertilizers history, production and applications. IntechOpen, London, UK. ISBN: 978-1-78985-147-2

Lauzon-Guay JS, Ugarte R, Morse BL, Robertson C (2021) Biomass and height of Ascophyllum nodosum after two decades of continuous commercial harvesting in eastern Canada. J Appl Phycol 33:1695–1708. https://doi.org/10.1007/s10811-021-02427-x

Lauzon-Guay JS, Feibel A, Morse BL, Ugarte RA (2023) Morphology of *Ascophyllum nodosum* in relation to commercial harvesting in New Brunswick, Canada. J Appl Phycol 35:2371–2381

Lazo L, Chapman ARO (1993) Components of crowding in a modular seaweed: sorting through the contradictions. Mar Ecol Prog Ser 174:257–267

MacFarlane C (1952) A survey of certain seaweeds of commercial importance in Southwest Nova Scotia. Can J Bot 30:78–97

MacFarlane C (1956) Irish Moss in the Maritime Provinces NS Research Foundation p20

Marshall J (2001) Landlords, leaseholders and sweat equity: changing property regimes in aquaculture. Mar Pol 25:335–352

McHugh D (2003) A guide to the seaweed industry. FAO Fisheries Technical Paper No. 441

McPeak R, Glantz D, Shaw R (1988) The amber Forest. Watersport Publishing, Inc, San Diego: CA, p 144

Neher PA (1988) Scientific basis for rights-based fishing. Proceedings of NATO Advanced Research Workshop Rekkjavik Iceland. App Sci Ser 169:5–8

Nielsen KN, Holm P, Aschan P (2015) Results based management in fisheries: delegating responsibility to resource users. Mar Pol 51:442–451

NS (2022) Harvesting sector: key facts and figures. https://novascotia.ca/fish/documents/harvesting-sector-facts-and-figures.pdf

NS Business Inc (2009) Acadian Seaplants Limited Builds on Success. https://news.novascotia.ca/en/2009/02/16/acadian-seaplants-limited-builds-success

NSDFA (1959) The Sea Plant Harvesting Act 1959. Chapter 9. An Act respecting the harvesting and conservation of Sea Plants 57

NSDFA (1997) Rock Weed Regulations (RWR) S.N.S. Part VI (Sea Plants Harvesting) of the Fisheries and Coastal Resources Act. Chapter 25, Acts of 1996 and (to date) the Sea Plants Harvesting Regulations of the Fisheries and Coastal Resources Act, 1989, Chapter 416

NSDFA (2024). https://novascotia.ca/fish/aquaculture/site-mapping-tool/

Pikitch EK, Santora EA, Babcock A, Bakun A, Bonfil R, Conover DO, Dayton P, Doukakis P, Fluharty D, Heheman B, Houde ED, Link J, Livingston PA, Mange M, MK MA, Pope J, Sainsbury K (2004) Ecosystem-based fishery management. Science 305:346–347

Pomeroy RS (1995) Community-based and co-management institutions for sustainable coastal fisheries management in Southeast Asia. Ocean Coast Manag 27(3):143–162

Pringle J, Mathieson A (1986) *Chondrus crispus* Stackhouse. In: Doty M (ed) Case studies of seven commercial seaweed resources. FAO Tech. Rept. 281, pp 49–122

Rangely R, Davis J (2000) Management of low trophic level fisheries in the face of uncertainty. Huntsman Marine Science Centre Occasional Report 00/1 p. 94

Saltwire (2021). https://www.saltwire.com/atlantic-canada/seaweed-plant-in-cornwallis-park-to-double-production-capacity-100586587#

Schmidt AL, Coll M, Romanuk T, Lotze HK (2015) Ecosystem structure and services in eelgrass *Zostera marina* and rockweed *Ascophyllum nodosum* habitats. Mar Ecol Prog Ser 437:51–68

Sharp GJ, Neish I (1976) Cost estimates for the harvesting holding and mechanical reduction of various seaweed products from Charlotte County New Brunswick Atlantic Mariculture Ltd to NB. Department of Fisheries 25p

Sharp GJ (1981) An assessment of *Ascophyllum nodosum* harvesting methods in southwestern Nova Scotia. Can Tech Rep Fish Aquat Sci 1012:28p

Sharp GJ, Roddick DL (1982) Catch and effort trends the Irish Moss (*Chondrus crispus* Stackhouse) fishery in southwestern Nova Scotia, 1978 to 1980. Can Tech Rep Fish Aquat Sci 1118. 43p

Sharp GJ (1986) *Ascophyllum nodosum* and its harvesting in eastern Canada. In: Case studies of seven commercial seaweed resources, FAO fisheries technical paper 281. Rome, Italy, Food and Agriculture Organization of the United Nations, pp 3–48

Sharp GJ, Trembay D (1989) An assessment of *Ascophyllum nodosum* resources in Scotia Fundy CAFSAC Res. Theatr Doc 89(1):19

Sharp GJ, Semple R (1991) An assessment of *Ascophyllum nodosum* resources in Scotia Fundy. CAFSAC Res Doc 91/52:30

Sharp GJ, Ang P, MacKinnon D (1994) Rockweed (*Ascophyllum nodosum*) (L.) Le Jolis harvesting in Nova Scotia: its socioeconomic and biological implications for coastal zone management. Proc Coastal Zone Canada 94:1632–1644

Sharp G, Semple R (1997) Rockweed (*Ascophyllum nodosum*). CASFSAC Res Doc 97/31:10 p

Sharp GJ, Bodiguel C (2002) Introducing integrated management, ecosystem and precautionary approaches in seaweed management: the *Ascophyllum nodosum* (rockweed) harvest in New Brunswick Canada and implications for the industry. In: Chapman ARO (ed) Proceedings of the 17th International Seaweed Symposium Oxford University Press, Oxford, pp 107–114

Sharp GJ, Ugarte R, Semple R (2008) The ecological impact of marine plant harvesting in the Canadian Maritimes, implications for coastal zone management science. Asia Sci Asia 32(Suppl):77–86

Sharp GJ, Sharp JT (2023) Estimating *Ascophyllum nodosum* (L.) Le Jolis harvesting impacts using GPS Tracking of mechanical harvesters in Nova Scotia, Canada. J Appl Phycol 36:605–610

Shukla PS, Mantin EG, Adil M, Bajpai S, Critchley AT, Prithiviraj B (2019) *Ascophyllum nodosum* based biostimulant: sustainable applications in agriculture for the stimulation of plant growth, stress tolerance, and disease management. Front Pl Sci 10:655. https://doi.org/10.3389/fpls.2019.00655

Statistics Canada (2022) Economic profile Clare, Nova Scotia. https://www.canada.ca/en/immigration-refugees-citizenship/campaigns/immigration-matters/local-economies/clare.html

Ugarte R, Sharp G (2001) A new approach to seaweed management in eastern Canada: the case of *Ascophyllum nodosum* Cah. Biol Mar 42:63–70

Ugarte R, Sharp G (2012) Management and production of the brown alga *Ascophyllum nodosum* in the Canadian Maritimes. J Appl Phycol 24:409–416

Ugarte R, Sharp G, Moore (2006) Changes in the brown seaweed Ascophyllum nodosum (L.) Le Jol. plant morphology and biomass produced by the cutter rake in southern New Brunswick, J Appl Phycol 18:351–359

Van Oosten MJ, Pepe O, De Pascale S, Silletti S, Maggio A (2017) The role of biostimulants and bioeffectors as alleviators of abiotic stress in crop plants. Chem Biol Technol Agric 4:5. https://doi.org/10.1186/s40538-017-0089-5

Wade E, Spalding AK, Biedenwer K (2019) Integrating property rights into fisheries management: the case of Belize's journey to managed access. Mar Pol 108:103631. https://doi.org/10.1016/j.marpol.2019.103631

Wally O, Critchley AT, Hiltz D, Craigie JS, Han X, Zaharia LI, Prithiviraj B (2013) Regulation of phytohormone biosynthesis and accumulation in *Arabidopsis* following treatment with commercial extract from the marine macroalga *Ascophyllum nodosum*. J Pl Growth Reg 32(2):324–339

Glyn J. Sharp was introduced to marine plants' world by Dr. L Hanic at Dalhousie University and Dr. C MacFarlane of the Nova Scotia Research Foundation during his undergraduate years. He was mentored by Dr. L Druehl of Simon Fraser University during his graduate studies of Macrocystis ecology.

Employment with Applied Marine Research Ltd provided him early experience in applied phycology, contributing to the development of tank culture for *Chondrus crispus* for Acadian SeaPlants Ltd. He joined the Fisheries and Oceans Marine Plants Unit under Dr. John Pringle to be part of a research team focusing on sustainable management of marine plant resources in the Canadian Maritimes, projects included the impact of harvests, demographics, and ecology of five commercial species. His work later expanded to aquaculture, the development of marine protected areas, and artificial reef systems. He collaborated with students and technicians over the years, especially his coauthor Bob Semple, a friend for 30 years. Together, they have faced many challenges in a wide variety of marine environments. While seconded to International Development Research Canada, they explored the commercial potential of Hypnea resources in Senegal with local partners. After retirement, his company KMCE Inc. has been involved in projects to develop and manage marine plant resources in Nunavik and Iceland. In Nova Scotia, KCME Inc. assisted Acadian SeaPlants Ltd., EuroCan Ltd., and Tidal Organics Inc. with research and development initiatives related to sustainable resource management. Fifty years of immersion in many aspects of applied phycology has been an exciting and satisfying ongoing learning experience.

R. E. Semple Scuba diving marked the beginning of Bob Semple's career, which he pursued from his teenage years in the 1960s. Influenced by his high school biology teacher Alvin Curley and Dr. Ian MacQuarrie, a professor of ecology at UPEI, Semple developed an interest in Marine Biology. Upon finishing high school, he commenced employment as a diver for a seaweed survey on Prince Edward Island. Subsequently, he transitioned to work as a contractor with the federal Industrial Development Branch as a research diver under H. Doug Johnson at the Miminegash Research Station on P.E.I. This role evolved into a permanent position with Fisheries & Oceans Canada, where he worked under Dr. John Pringle as a biological technician and later with Glyn Sharp in the Marine Plants Unit. Bob Semple was a key team member in various seaweed projects focusing on Irish moss and other related algae, assessing the impact of commercial harvesting on seaweed growth and biomass. In addition to seaweed research, he was involved in studies concerning artificial reefs and marine protected areas. This work extended across the Canadian Maritimes, Nunavik in Quebec, where he explored Fucus and kelp resources, and Senegal, examining Hypnea resources. He also collaborated with Dr. Raul Ugarte on seaweed aquaculture projects in Sri Lanka. Bob Semple retired after 36 years of service, continuing his passion for scuba diving, underwater photography, and traveling with a particular interest in shark species.

11 Current Global Status of Kelp Farming in Wave-Exposed Environments: Species, Sites, Methods, and Technology

Marc Gagnon, Éric Tamigneaux, and Isabelle Gendron-Lemieux

Abstract

Kelp species (Order Laminariales) are currently almost exclusively cultivated in temperate coastal waters of the North Atlantic and North Pacific and mostly in sheltered sites. There is a strong drive in several countries to expand aquaculture further offshore co-occurring with a rapid change of the conditions under which this activity will be practiced due to climate change. At the dawn of these profound changes, a global review of the current status of kelp cultivation in wave-exposed environments can serve as a benchmark for future developments. This chapter presents an inventory and description of kelp commercial farms currently in operation in temperate-exposed waters. The inventory is based on 80+ high-resolution aquaculture lease maps, most of them available as Web-GIS applications, that cover the entire study area except for a few countries. The number of exposed sites and the leased area are tallied by region, country and production method. The "longline" is the production structure used in all sites inventoried. Longline design and farm layout in exposed farms as well as structural damage and loss of cultured biomass due to hydrodynamic forces are reviewed. Information gaps, strengths and weaknesses of current practices, and recent R&D projects are discussed.

Keywords
Aquaculture · Offshore · Exposed · Temperate · Kelp · Global status

M. Gagnon (✉)
Biorex Inc., Québec, Qc, Canada
e-mail: mgagnon@biorex.com

É. Tamigneaux · I. Gendron-Lemieux
École des pêches et de l'aquaculture du Québec, Cégep de la Gaspésie et des Îles, Grande-Rivière, Qc, Canada

11.1 Introduction

Macroalgae that are classified in the Order Laminariales (Phaeophyceae) are commonly called kelp. Many kelp species grow into dense strands (kelp forests) along temperate and sub-polar rocky coastlines of the Northern and Southern Hemispheres. These forests are some of the most productive ecosystems on earth and provide a wide array of ecosystem services including habitat and food for several pelagic and benthic species, carbon sequestration, eutrophication mitigation, and coastal defense. They face serious threats from climate change, over-harvesting, and habitat destruction and are in severe decline in several regions (Bolton 2010; Eger et al. 2023; Eger et al. 2024). Wild harvest of kelp is relatively stable since the 1960s while aquaculture production has surpassed wild harvests in the 1950s and now represents roughly 97% of the global kelp production (FAO 2024). Until mid-2010, kelp farming was almost entirely practiced in the Pacific Northwest (Japan, China, North and South Korea; Buschmann et al. 2017). It has emerged in recent years in non-traditional regions (Russian Far East, northern Europe, and temperate North America) and is gaining interest in countries such as Australia, Chile, Namibia, and New Zealand (Hermans 2023).

Due to the relatively low market value and low farming density of kelp compared to high valued invertebrate and finfish species, kelp cultivation requires large-scale farms to be viable (van den Burg et al. 2016; Heasman et al. 2024a; Kim 2024). In temperate waters, most of the space potentially available for mariculture in sheltered areas (estuaries, marine lagoons, fjords, and enclosed bays) is already used for other activities. In addition, there is increasing pressure from other coastal users and regulators to move existing nearshore farms farther offshore (Wang et al. 2022; Kim 2024). In some areas where kelp farming is well developed, expansion is only possible in more exposed sites (Cai et al. 2021; St-Gelais et al. 2022). When the carrying capacity has been exceeded, as is the case in several sheltered areas, part of the production is

A. T. Critchley et al. (eds.), *Advances in Temperate Phyconomy: Algal Harvest and Cultivation in Globally Distributed Temperate Waters*, Developments in Applied Phycology 15, https://doi.org/10.1007/978-3-032-02955-3_11

moved farther offshore and in more exposed locations to reduce the density of farming operations (Komatsu et al. 2016; Liu et al. 2019; Wang et al. 2022). Another new opportunity for exposed kelp farming is its co-location inside marine renewable energy parks which are, by definition, situated in high-energy environments (Ross et al. 2025). For example, it was projected that the installed capacity of offshore wind farms will increase 15- to 24-fold between 2018 and 2040 and that these farms will occupy 47,000 to 73,000 km^2 of exposed waters, mainly in China, Europe, and northeastern USA (International Energy Agency (IEA) 2019). Marine renewable energy parks are considered promising locations for kelp farms since, among several benefits, they allow for cost sharing and are navigation-restricted areas (Buck et al. 2017).

As defined by Buck et al. (2024), exposed sites are unprotected from strong currents, large waves, and strong winds and they may be near to land or far offshore. Temperate waters where kelp is farmed are characterized by the most frequent and intense wave storms in the world (Lobeto et al. 2024). Compared to sheltered sites, exposed sites have several potential advantages for kelp cultivation including more space (larger leases), fewer user conflicts, more stable water characteristics, lower level of biofouling, better kelp growth and quality, and fewer environmental impacts. However, they also have several potential disadvantages including a limited number of potential species, nutrient limitation, the need for larger and more powerful vessels, shortened operating window, increased risks of structural failure and cultured biomass loss caused by storms, increased health and safety risks for farm employees, and increased likelihood of marine mammal entanglement (ICES 2012; Lovatelli et al. 2013; Buck and Langan 2017; Mizuta and Wikfors 2019; Yang et al. 2020; Mascorda Cabre et al. 2021; Bath et al. 2023; Paine et al. 2023).

Co-occurring with the current strong interest toward exposed waters, climate change will have a significant impact on the conditions in which kelp farming will evolve in the coming decades (Hu et al. 2021; Kim et al. 2021; Liu et al. 2024a, b; Veenhof et al. 2024). More specifically, the Intergovernmental Panel on Climate Change (Cooley et al. 2022) predicts an increase in the average sea temperature and frequency, duration, and intensity of marine heatwaves in all regions, as well as an increase in the mean wave energy and extreme wave heights in several regions during the twenty-first century. At the dawn of these profound changes, a global review of current status of commercial kelp farming in high-energy environments can be useful for the industry and the R&D community and serve as a benchmark for future development.

Several reviews of the technological aspects of kelp farming have been published recently (Fernand et al. 2017; Buck et al. 2017, 2018; Buschmann et al. 2017; Goseberg et al. 2017; Bak et al. 2020; García-Poza et al. 2020; Tullberg et al. 2022; Heasman et al. 2024a; Saether et al. 2024; Lian et al. 2024). Most of these reviews focus on practices at sheltered sites or experimental/pilot technology as opposed to current commercial practices at exposed sites. In this chapter, we first carried out a review of the current global kelp aquaculture production in temperate waters based on the best information available. We then expand and complement the global inventory of extractive species farms in exposed temperate waters conducted by Gagnon (2024a) with additional information made available over the past year and a focus on kelp farming. Longline (LL) design and farm layout, as well as the structural damages and biomass losses caused by storms and tsunamis, are examined for some exposed farms for which detailed information is available. Finally, information gaps, strengths, and weaknesses of current practices and recent R&D projects are discussed.

11.2 Study Area

The study area is limited to marine waters where the mean annual sea surface temperature (SST) is between 5 and 20 °C. These limits correspond roughly to the historical global distribution of wild kelp (Steneck et al. 2002). The study area was subdivided into eight large regions (Fig. 11.1): Atlantic Northeast (ANE), Atlantic Northwest (ANW), Mediterranean and Black seas (MBS), Pacific Northeast (PNE), Pacific Northwest (PNW), Temperate South America (TSAM), Temperate South Africa (TSAF), and Temperate Australasia (TAA). The list of countries and country subdivisions included in each region and the SST climatologies used to delimit the study area are presented in Gagnon (2024a).

The pole-ward part of all regions except MBS is situated in the global extra-tropical storm belts where wave energy is at its maximum (Lobeto et al. 2024). The coasts of the study area that are exposed directly to the open ocean are characterized by high or very high wave storminess (a variable that combines storm frequency and wave energy), while the coasts of marginal seas (e.g., Gulf of St. Lawrence, Irish Sea, North Sea, Baltic Sea, Mediterranean Sea, Okhotsk Sea, Sea of Japan, Yellow Sea) have an intermediate storminess between the above and that of sub-tropical and tropical regions (Lobeto et al. 2024). All coastal areas around the Pacific Ocean are vulnerable to large tsunamis generated by earthquakes (Reid and Mooney 2022). Also, of particular importance for kelp farming, the four major global coastal upwelling systems (CUS) are situated in the study area: along the southern coast of the PNE (California Current) and ANE (Canary-Iberia CUS), the Pacific coast of the TSAM Region (Peruvian-Chilean CUS), and in the TSAF Region (Benguela Current). These systems are the most productive

Fig. 11.1 Limits of the eight large regions (shaded areas) which make up the study area and position of the 7 kelp farming wave-exposed sites for which detailed information on longline design and farm layout is available. ANE Atlantic Northeast; ANW Atlantic Northwest; MBS Mediterranean and Black seas; PNE Pacific Northeast; PNW Pacific Northwest; TAA temperate Australasia; TSAF temperate South Africa; TSAM temperate South America

areas of the world ocean in terms of primary and secondary production (Kämpf and Chapman 2016).

11.3 Current Global Kelp Production

The only global account of kelp production is provided by the FAO (2024) who collates official figures reported by national marine agencies. These figures are questioned by industry experts (Araújo et al. 2021; Hermans 2024a; Hatch Innovation Services (HIS) 2024) as they do not reflect the current status of the kelp farming industry. Table 11.1 presents the kelp species as well as the total kelp volume produced per country in the study area in 2022, based on the best information available. When needed, the production outside the study area was subtracted from the FAO (2024) figures. After consultation of the European Commission's Algae Industry Database (Vazquez Calderon et al. 2022) and the Phyconomy Database of seaweed organizations (Hermans 2024b) to identify active farming companies and the company websites to verify the species farmed, it was concluded that the countries that are not listed in Table 11.1 did not produce kelp commercially in 2022.

More than 99% of the world kelp production comes from PNW countries, namely China, Japan, North Korea, Russia, and South Korea. The main kelp species cultivated in this region are the Japanese kelp or kombu (*Saccharina japonica*) and the sea mustard or wakame (*Undaria pinnatifida*). China is by far the largest kelp producer with an estimated production of 6.78 Mt. for the provinces that are part of the study area. There are however questions regarding the validity of the production volumes declared for this country. Kelp production in China is reported to the FAO in dry weight and the FAO applies a conversion factor to transform these into wet weights. According to industry experts, FAO figures for China are not realistic and are likely overestimated (World Bank 2023; HIS 2024). North Korea, officially the second largest producer of kombu in the world, does not report to the FAO. The value of 600 kt of kombu for this country is an FAO estimate that has been attributed each year since 2018. South Korea is the second most important producer of wakame and third producer of kombu, whereas Japan is the third producer of wakame and the fourth producer of kombu. Very little is known about the Russian kelp production in the PNW. The FAO values reported for Russia are for unspecified "brown algae (Phaeophyceae)." It is likely that the 28.6

Table 11.1 Kelp production in the study area in 2022

Country	Region	Production (2022) t (wet weight)	Main species	Source
China	PNW	6,779,314[a]	*S. japonica, U. pinnatifida*	FAO (2024)
South Korea	PNW	1,146,803	*U. pinnatifida, S. japonica*	FAO (2024)
North Korea	PNW	600,000[b]	*S. japonica*	FAO (2024), Park (2015)
Japan	PNW	64,403[c]	*U. pinnatifida, S. japonica*	FAO (2024)
Russia	PNW, ANE	28,654[d]	*S. japonica*	FAO (2024)
USA	ANW, PNE	862[e]	*S. latissima, A. marginata, N. luetkeana*	Robidoux and Good (2023)
Norway	ANE	525	*A. esculenta, S. latissima, Laminaria* sp.	Hermans (2023)
Ireland	ANE	493	*A. esculenta, L. digitata, S. latissima.*	Evans et al. (2023), BIM (2024)
France	ANE	325	*U. pinnatifida, S. latissima, A. esculenta*	Hermans (2023)
UK	ANE	<300[f]	*S. latissima, A. esculenta*	Capuzzo and MacMillan (2024), Peyton Jones F. pers. comm. (2024)
Faroe Islands (Denmark)	ANE	185[h]	*S. latissima, A. esculenta, L. digitata*	ICES (2023)
Canada	ANW, PNE	<50[fg]	*S. latissima, A. esculenta, A. marginata*	Autef S., Couturier C., Leavitt C., Vieira J. pers. comm. (2024)
Sweden	ANE	<30[f]	*S. latissima*	CEVA (2024)
The Netherlands	ANE	<20[f]	*S. latissima, A. esculenta*	CEVA (2024)
Chile	TSAM	11	*M. pyrifera*	FAO (2024)
Spain	ANE	10.8	*S. latissima*	JACUMAR (2022)
Denmark[i]	ANE	7.5	*S. latissima*	FAO (2024), CEVA (2024)
Greenland (Denmark)	ANE	?[f]	*S. latissima, A. esculenta*	Vazquez Calderon et al. (2022)
Mexico	PNE	?[f]	*M. pyrifera*	Conapesca Acuacesor (2024)
Others	–	0[j]	–	CEVA (2024), FAO (2024), Hermans (2024a, b), Vazquez Calderon et al. (2022)

[a]Production in provinces outside the study area (4953 kt) was subtracted from the FAO figure. We assumed that the additional 2051 kt reported as "brown algae (Phaeophyceae)" are not for kelp species
[b]FAO estimate
[c]Production outside the study area in southern Honshu (13.5 kt) was subtracted from the FAO figure
[d]Reported as "brown algae (Phaeophyceae)" and assumed to be 100% kelp. ANE: 12 t; PNW: 28,642 t
[e]ANW: 467 t; PNE: 395 t
[f]Known active commercial kelp farms but no official production data available for the country
[g]Estimate for 2023
[h]Value for 2021
[i]Excluding the Faroe Islands and Greenland
[j]No known active commercial kelp farm in 2022

kt of brown algae reported for the Russian Far East is entirely composed of kelp (Kawai et al. 2015). According to an industry expert, this figure is considerably overestimated (Konkurent 2021). As depicted in Fig. 11.2, kombu is composed of a short stipe and a single blade of about 3–5 m in length (maximum 10 m). Wakame has a single blade with a distinct central rib and sporophylls. It normally reaches a size of about 1–2 m. Both species have a mass density slightly higher than that of sea water and hang down from the supporting lines in still water (Gagnon 2024b).

Kelp production in the Western Hemisphere is in its infancy; it was virtually nil before 2015 (FAO 2024). The FAO figures for this part of the world are inconsistent and have many gaps; they were not retained except for Chile and Denmark. Other sources (see Table 11.1) are more up-to-date and complete for some countries (the Faroe Islands, France, Ireland, Norway, Spain, and the USA). However, for others with documented commercial kelp farming (Canada, Greenland, Mexico, the Netherlands, Sweden, and the UK), national aquaculture kelp production statistics are not available, and only rough estimates are possible for some of them. The sugar kelp (*Saccharina latissima*) followed by the winged kelps (*Alaria esculenta* and *Alaria marginata*) are the main cultivated species in the Western Hemisphere. As depicted in Fig. 11.2, sugar kelp is morphologically similar to kombu but usually has a shorter blade (1–2 m long, maximum 7 m). The winged kelps are in the same family as wakame with which they share some morphological features such as the central rib and the sporophylls.

Other species farmed commercially in 2022 in the PNW include the shreaded kombu (*Saccharina angustata*) in Japan (Yotsukura 2010) and the five ribbed kelp (*Costaria costata*)

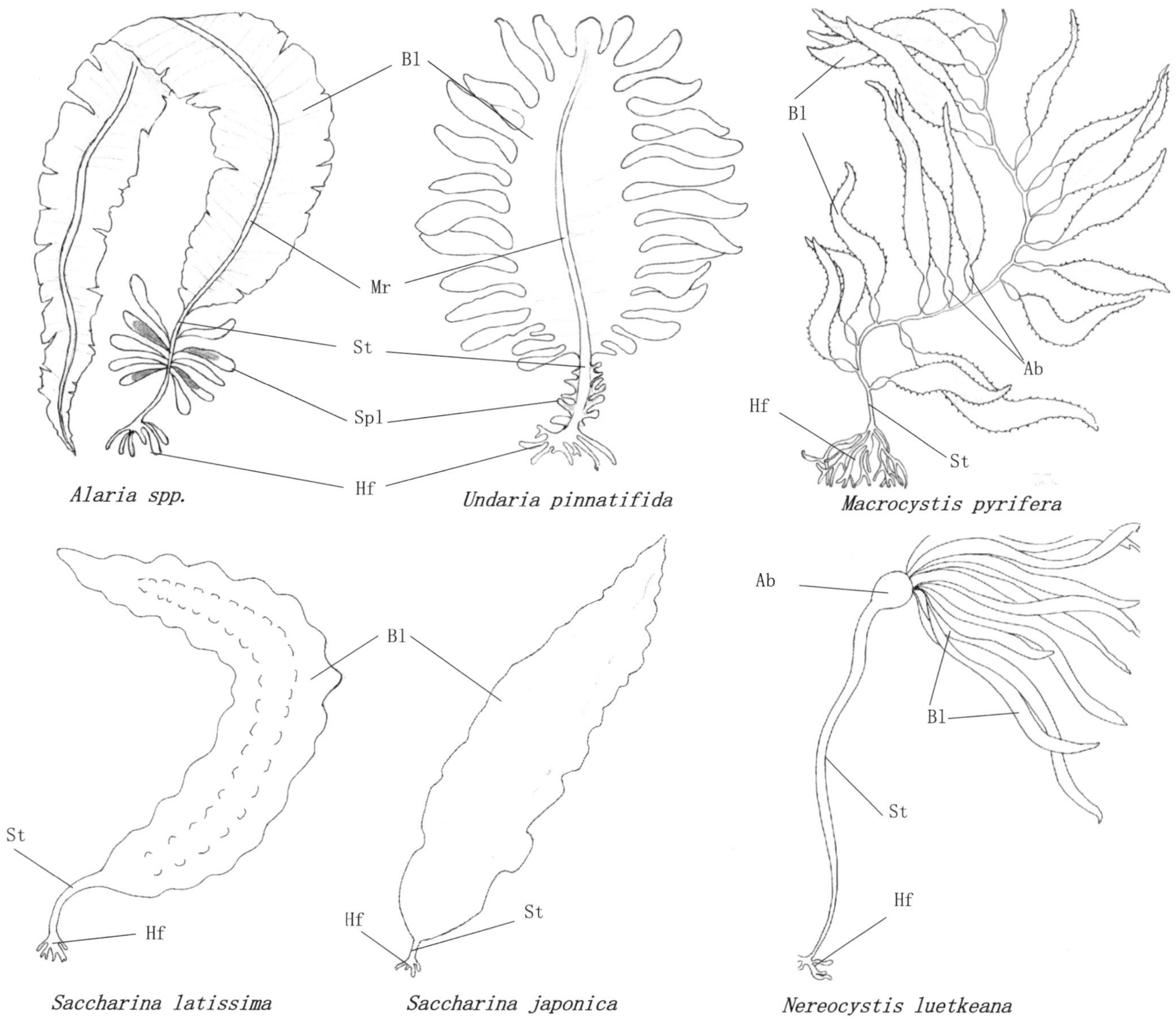

Fig. 11.2 General morphology of the main kelp species cultivated in the world. *Alaria*, *Saccharina* and *Undaria* are slightly denser than sea water and sink in still water, while the air bladders on *Macrocystis* and *Nereocystis* provide buoyancy that lifts the blades toward the surface. Ab air bladder; Bl blade; Hf holdfast; Mr. Midrib; Spl Sporophyll; St stipe. Not to scale

in South Korea (Hwang and Park 2020). In the Western Hemisphere, small quantities (< 20 t) of oarweed (*Laminaria digitata*) in the Faroe Islands and Scotland, bull kelp (*Nereocystis luetkeana)* in Alaska (USA), skinny kelp (*Saccharina angustissima*) in Maine (USA), and giant kelp (*Macrocystis pyrifera*) in Chile and Mexico were produced (see references in Table 11.1). In the case of Chili, production is primarily for the repopulation of Chilean kelp forest. *M. pyrifera* and *N. luetkeana* are very large kelp that can grow to more than 30 m long in the wild (Fig. 11.2). Unlike other kelp, they have air bladders at the base of the blades that keep the plants erect above the supporting lines. Finally, economically important species of the genus *Lessonia* and *Laminaria hyperborea* are not farmed commercially (FAO 2024).

11.4 Inventory of Exposed Kelp Farms

Wave-exposed sites where kelp is commercially cultivated were identified by Gagnon (2024a) using a systematic approach based mainly on high-resolution aquaculture lease maps (HRALMs). The criteria used to screen the thousands of aquaculture leases appearing on these HRALMs are (1) the type of lease (commercial and active), (2) the cultured species (kelp; order Laminariales), and (3) wave exposure (exposed sites). Inactive, proposed, under review, experimental, and pilot leases were not retained. Wave exposure was the only criterion used to distinguish exposed sites from sheltered sites (Table 11.2). Gagnon's (2024a) inventory was completed by consulting the European Commission's Algae

Table 11.2 Wave exposure classification. A site was assigned to the highest wave exposure class for which at least one of the four criteria was met (Gagnon 2024a)

Class	Mean annual SWH (m)	99th percentile SWH (m)	50-y-return-period SWH (m)	Mean annual WPD (kW/m)[a]
Sheltered (S)	< 0.5	< 2.2	< 4	< 1.5
Moderately exposed (ME)	0.5–1.0	2.2–3.8	4–7	1.5–8
Fully exposed (FE)	1.0–2.0	3.8–6.0	7–14	8–20
Very exposed (VE)	>2.0	>6.0	>14	>20

[a]WPD: wave power density (energy flux)

Industry Database (Vazquez Calderon et al. 2022) and the Phyconomy Database (Hermans 2024b) that provide an up-to-date listing of kelp farming companies in the Western Hemisphere and the websites of new companies. A few additional exposed sites were identified in Europe.

The results of the inventory are summarized in Table 11.3 per country or territory in each of the eight regions of the study area using three metrics: number of sites, total leased area (ha), and percentage of sites that use longlines. A "site" refers to a single isolated lease or a group of several active leases in a given site. The leased area includes the actual space occupied by the production structures, navigational channels, and buffer zones around the structures and any undeveloped part of the lease. All the sites for which the information is available use longlines (LLs). These consist of long horizontal ropes supported by buoys (floats) individually anchored to the seabed or arranged in a network of several parallel horizontal ropes anchored by a mooring grid (arrays; see next section).

Information is missing for North Korea and the Russian Far East and only a rough estimate of the exposed farmed area was possible for China. Excluding these three countries and country subdivisions, a total of 396 sites were inventoried. At least 97% of the number of sites and of the total farming area inventoried are situated in the PNW region. Because of this overwhelming importance, each country in this region is examined separately below.

11.4.1 Pacific Northwest

In the case of China, HRALMs at the national or provincial levels are not available and the official Statistical Fishery Yearbook does not distinguish between sheltered and exposed sites (Zheng et al. 2019; Wang et al. 2022). Recent nation-wide and regional estimations of the areal extent of mariculture from satellite remote sensing (Liu et al. 2022a; Wang et al. 2022; Jin et al. 2023) do not distinguish the farmed species and exposed vs sheltered sites. For these reasons, it was not possible to obtain a precise estimate of the total area of exposed farms in temperate China. Gagnon (2024a) used an indirect approach for the northern provinces included in the study area. The estimated 200 km^2 of exposed kelp sites are all situated at the northeastern end of the Shandong Peninsula, in Rongcheng County. From fall to following spring, this area is used solely for kombu farming with LLs.

In Japan, the total area of wave-exposed sites amounts to 424 km^2 which is by far the largest area of any country. Most sites are situated on the Pacific coast of the Iwate and Myagi Prefectures (northeastern Honshu) and are used to produce wakame and kombu. From the information available on the HRALMs, it is likely that 100% of these sites use LLs.

In South Korea, exposed sites are concentrated along the southeastern coast in Busan and Ulsan Provinces (Sea of Japan). The average size of the leases is quite small (< 5 ha). They are used to farm wakame and/or to a lesser extent kombu using LLs.

HRALMs are not available for North Korea and other information on the location of aquaculture farms is very scarce. For these reasons, it was not possible to determine the number and area of exposed sites in this country.

In the Russian Far East, according to the aquaculture leasing web application AquaVostok (2023), over 700 km^2 of exposed aquaculture space have been leased and are "in use." Most of this space was leased after 2015 and is located along the coasts of the Primorsky Territory (Sea of Japan). As a result, the total production of kelp in the Russian Far East increased 14-fold between 2015 and 2022 (FAO 2024). Unfortunately, information on the cultured species and production methods are not available for individual sites. Since these sites can be used to grow kelp, finfish, mollusks, and other invertebrates, it was not possible to obtain a reliable estimate of the number and extent of exposed kelp sites in this territory.

11.4.2 Other Regions

Outside Asia, only 14 exposed sites were inventoried, most of them in northern Europe. In these small- to intermediate-sized sites, the sugar kelp (*Saccharina latissima*) and winged kelp (*Alaria esculenta*) are usually the two main species cultivated. Only one of the several kelp farms in Canada and the USA is classified as wave-exposed. Two sites, one in northern Chile (TSAM) and the other in Baja California, Mexico (PNE), are used to farm the giant kelp (*M. pyrifera*). No sites were inventoried in the MBS, TAA, and TSAF regions. This is also the case in parts of other regions where several kelp

Table 11.3 Number and areal extent of wave-exposed kelp farms in the study area

Region	Country	Number of sites	Leased area (ha)	% LL	Main species
ANE	Faroe Islands	2	220	100	*S. latissima*
	France	3	341	100	*S. latissima*
	Spain	2	6	100	*S. latissima*
	UK	4	235	100	*S. latissima*
	Other countries	0	0	–	–
ANW	Canada	1	84	100	*S. latissima*
	USA	0	0	–	–
	Other countries	0	0	–	–
MBS	All countries	0	0	–	–
PNE	Mexico	1	5	?	*M. pyrifera*
	Other countries	0	0	–	–
PNW	China	?	20,000[a]	100	*S. japonica*
	Japan	218	42,378	100	*U. pinnatifida*, *S. japonica*
	North Korea	?	?	?	*S. japonica*
	Russia	?	?	?	*S. japonica*
	South Korea	164	1147	100	*U. pinnafitida*, *S. japonica*
TSAM	Chile	1	4	?	*M. pyrifera*
	Other countries	0	0	–	–
TSAF	All countries	0	0	–	–
TAA	All countries	0	0	–	–

% LL percentage of the sites that use longlines
Estimate made by Gagnon (2024a)

farms are located in sheltered areas (e.g., Alaska and Maine (USA), British Columbia (Canada), Norway, and Scotland (UK)).

11.5 Culture Method, Longline Design, and Farm Layout

Detailed information on LL design and farm layout was found for only seven sites (Table 11.4 and Fig. 11.1). Gagnon (2024a) reviewed the information for four of these sites (# 3, 4, 6 and 7). Detailed information on two additional sites (# 2 and 5) inventoried by Gagnon (2024a) was found in ASC and MSC (2022) and Canvin et al. (2024). Finally, Site # 1 was added to the list because detailed information was available for this major kombu farming site in Japan. The information on Sites # 6 and 7 was completed and validated by the operator's response to a questionnaire and by direct contacts between the operators and the authors.

The location and environmental setting of the seven sites are summarized in Table 11.4. Although limited in spatial extent (3 of the 8 regions), the sites cover a wide range of years of establishment (1960–2018), size (2–2415 ha), water depth (8–70 m), distance to coast (0.4–8 km), water temperature (0.9–14.7 °C), and water transparency (3.8–16.8 m Secchi disk depth (SSD)). The first three Asian sites regroup several leases and have been active for more than 30 years while those in the Western Hemisphere are individual farms that were established after 2010. The distance to the nearest coast does not exceed 8 km and is less than 4 km in most sites and the water depth does not exceed 70 m and is less than 25 m in most sites. The mean water temperature from December to April is relatively high (10.9–14.7 °C) at three sites, intermediate (5.5–6.9 °C) at three other sites, and very low (0.9 °C) at the last site. Four sites have a high water transparency (SDD > 9 m) while the three others are moderately turbid (SDD between 3.8 and 6.5 m). Finally, five sites are moderately exposed to waves and the two others are fully exposed. Note that no very exposed site, as defined in Table 11.2, has been inventoried.

Table 11.5 provides a summary of the cultivation method. In the Asian sites, the main species produced are kombu and wakame, while in the Western Hemisphere farms, the main species is the sugar kelp in all cases. In all sites, except in the Faroe Islands, kelp-ropes are seeded in autumn and are fully harvested in the following spring or summer. The seeding method refers to how the kelp is attached to the kelp-line at the start of the at-sea grow-out phase. Four different methods are used (Kawashima 1984; Kerrison et al. 2017; HIS 2024): (1) twine with attached seedlings (length < 2 cm) wound spirally around the kelp-line (twine winding) or (2) cut into short segments and inserted through the lay of the kelp-line at regular intervals (twine insertion), (3) juveniles (length between 20 and 40 cm) with holdfast inserted through the lay of the kelp-line at regular intervals (holdfast insertion), and (4) direct application of seedlings suspended in a binder solution on the kelp-line (direct seeding). Currently, seeding is done manually at all sites except the Faroe Islands Site #4. The harvesting method refers to the method used to remove the kelp biomass from the kelp lines. Currently, this is also done manually at all sites.

The LL design and site layout for each site are summarized in Tables 11.6 and 11.7. These tables refer to three main types of LLs (Fig. 11.3), four different arrangements of the kelp-lines (Fig. 11.4), four main types of mooring lines (Fig. 11.5), and two main types of farm layout (Fig. 11.6). There is currently no standard LL design and farm layout. Each site has its own particularities which are described below.

Table 11.4 Location and environmental setting of the seven exposed kelp farming sites for which detailed information on longline design and farm layout are available

Site #	Location	Water Body	Year established[a]	Nb of leases	Leased Area(ha)	Distance to the nearest coast (km)	Water Depth (m)	Water temperature (°C)[c]	Secchi disk depth (m)[c]	Storminess[d]	Wave exposure[e]
1	Minamikayabe, Hokkaido, Japan	Pacific Ocean	1960s	Several	2415	0.5–1.5	15–25	6.8	13.3	Intermediate	ME
2	Jusanhama, Miyagi, Japan	Pacific Ocean	1960s and 2012[b]	19	1990	0.7–3	10–50	14.7	14.8	Intermediate	FE
3	Sanggou Bay, Shandong, China	Yellow Sea	1990s	Less than a dozen	Up to 1500 ha per lease	5–8	10–30	5.5	3.8	Intermediate	ME
4	Funningsfjord, Faroe Islands	Funnigsfjord	2014	1	6	0.4–1.1	25–70	6.9	16.8	Very high	ME
5	Porthallow, Cornwall, UK	English Channel	2017	1	2	0.5	8–15	10.9	9,7	Intermediate	ME
6	Lesconil, Brittany, France	Atlantic Ocean	2013	1	150	1–2.5	15–25	12.0	5.6	High	FE
7	Paspébiac, Québec, Canada	Gulf of St. Lawrence	2018	1	84	2–3	15–20	0.9	6.5	No data	ME

[a]First year of commercial operation
[b]Farms were reconstructed after their destruction by the 2011 Great Tohoku Tsunami
[c]Average from December 1 to April 30, 2018–2022 (Copernicus Viewer https://marine.copernicus.eu/access-data/myocean-viewer; accessed 11 December 2024)
[d]Lobeto et al. (2024)
[e]*ME* moderately exposed; *FE* fully exposed. See Table 11.2 for the wave exposure classification

11.5.1 Site 1. Minamikayabe, Hokkaido, Japan (Figs. 11.3b, 11.4d, 11.5a, b, 11.6c)

Minamikayabe is the oldest and the most important kombu farming site in Japan. It is located along a peninsula at the southern end of Hokkaido between the Tsugaru Strait and Uchiura Bay and is moderately exposed to Pacific Ocean storms, swells, and tsunamis. It includes an unknown number of leases where LL arrays are used. Each array has an area of roughly 1.6 ha with 8–10 parallel semi-submerged kelp-lines placed 5 m apart. Kombu is farmed from October to the following summer mainly for the food market. To ensure a high-quality product, twine insertion is used to seed the lines and the vertical sinking kelp-lines are manually thinned twice to reduce plant density. During winter, the kelp-lines are vertical and sinking (Fig. 11.4d) and in the spring, their bottom free end is raised to make them horizontal, and the depth of the ML is reduced from 2 m to 0.5 m to ensure optimal lighting. The farms are operated by individual families that use traditional labor-intensive seeding, thinning, and harvesting methods (Olson 1987; Kawashima 1984, 1997; Liu et al. 2013; Yotsukura 2010; HIS 2024).

11.5.2 Site 2. Jusanhama, Ishinomaki, Myagi, Japan (Figs. 11.3b, 11.4a, 11.5a, 11.6a)

The Jusanhama site is situated in the main area of wakame farming in Japan. It includes 19 family-owned leases that are fully exposed to storms, swells, and tsunamis from the Pacific Ocean and have the highest mean water temperature of the seven sites. The LLs used in this site are individually anchored and have two parallel vertically separated horizontal MLs, a first one to grow wakame at a depth of 1–2 m and a second one to grow kombu at a depth of 5–10 m. Kelp is seeded using the twine winding method and is thinned in late December/early January. The wakame is harvested in late winter and the kombu later in spring (ASC and MSC 2022; HIS 2024; MPFCA 2024; Saijo 2024).

11.5.3 Site 3. Sanggou Bay, Rongcheng County, Shandong, China (Figs. 11.3a, 11.4b, 11.5a, 11.6c)

The outer half of Sanggou Bay and the offshore waters facing the bay constitute the epicenter of Japanese kelp (*S. japonica*) farming in China (Zheng et al. 2019). The high-density longline fields in and outside this semi-enclosed bay are clearly visible on Google Earth and extend up to 14.5 km from the inner bay shore into the Yellow Sea (Fig. 11.7). Remote sensing-based mapping shows that the exposed kelp culture area outside the limits of the bay increased approximately eightfold between 1990 and 2018 (Wang et al. 2022).

This site is moderately exposed to storms and swells from the Yellow Sea. Its waters are the least transparent of the seven sites. Each array is composed of 4 or 5, 70–100 m long surface LLs individually anchored 4–5 m apart. The kelp-lines are attached horizontally between two adjacent LLs at 1–2 m intervals. The farming density in this site, in terms of kelp-line length per ha, is the highest of the seven sites (6000–15,000 m/ha); it is currently too high and there are efforts to reduce it. The main lines are parallel to the swells, but the kelp-lines are perpendicular to them because of this

Table 11.5 Culture method used in the seven exposed sites listed in Table 11.4

Site #	Location	Main species	Seeding		Thinning	Harvesting		
			Method[a]	Timing	Timing	Method[b]	Timing	Biomass density (fronds per m of kelp-line)
1	Hokkaido, Japan	*S. japonica*	Twine insertion	Oct	Jan and Mar	Manual	July and August	12–15
2	Myagi, Japan	*U. pinnatifida*, *S. japonica*	Twine winding	Nov and Dec	Late Dec to early Jan	Manual	*Undaria*: Feb to April *Saccharina*: May–June	?
3	Shandong, China	*S. japonica*	Holdfast insertion	Oct to Jan	No	Manual	May and June	7–13
4	Faroe Islands	*S. latissima*	Direct seeding	Nov	No	Manual	Multiple	Hundreds
5	Cornwall, UK	*S. latissima*	Twine winding	Nov	No	Manual	May	139
6	Brittany, France	*S. latissima*	Twine winding	Oct	No	Manual	March and April	Hundreds
7	Québec, Canada	*S. latissima*	Twine winding	Oct	No	Manual	May and June	Hundreds

[a]Method used to attach seedlings/juveniles to kelp-lines (see text)
[b]Method used to detach kelp biomass from kelp-lines

Table 11.6 Longline design in the seven exposed sites of Table 11.4

Site #	Location	LL type[a]	Mainline length (m)	Mainline depth (m)	Kelp-line type[b]	Mooring line type[c]	Anchor type
1	Hokkaido, Japan	SS	120	2 at beginning and 0.5 at harvest	Vertical-sinking during winter and horizontal-mainline during spring	Single rope or tensioner buoy	Concrete blocks and sand bags
2	Myagi, Japan	SS	130 x 2 (vertically superposed)	Undaria: 1–2 Saccharina: 5–10	Horizontal-mainline	Single rope	Concrete blocks
3	Shandong, China	S	70–100	0–1	Horizontal- perpendicular	Single rope	Pile (stake) or concrete blocks
4	Faroe Islands	FS	500	10	Vertical-floating	Single ropes (two at each end)	Drag embedment anchor
5	Cornwall, UK	S	200	1–3	Vertical-sinking	Single rope	Concrete blocks
6	Brittany, France[d]	SS	1000	1	Horizontal-mainline	Tensioner buoy and sinker	Concrete blocks
7	Québec, Canada	FS	100	3 in spring and 6–7 in autumn and winter	Horizontal below mainline	Single rope	Concrete blocks

[a]*S* surface, *SS* semi-submerged, *FS* fully submerged. See Fig. 11.3 for a representation of the different types of LLs
[b]See Fig. 11.4 for a representation of the different types of kelp-lines
[c]See Fig. 11.5 for a representation of the different types of mooring lines
[d]Design used in 2013–2014

Table 11.7 Farm layout in the seven exposed sites of Table 11.4

Site #	Location	Layout type[a]	Distance between MLs (m)	Farm density (m ha^{-1})[b]	Mainline orientationrelative to:	
					Tidal currents	Swells
1	Hokkaido, Japan	Array	5	1660–1850	Parallel	Perpendicular
2	Myagi, Japan	Individually anchored	25–50	?	Parallel	Perpendicular
3	Shandong, China	Array	4–5	6000–15,000	Perpendicular	Parallel
4	Faroe Islands	Individually anchored	20	2500–5000	?	?
5	Cornwall, UK	Individually anchored	40	200	Parallel	Perpendicular
6	Brittany, France[c]	Individually anchored	10	980	Parallel	Parallel
7	Québec, Canada	Individually anchored	50	110	Parallel	Perpendicular

[a]See Fig. 11.6 for the different types of layouts
[b]Meters of kelp-lines per hectare
[c]Layout used in 2013–2014

particular arrangement. It is the only site where kelp-lines are seeded by inserting the holdfast of juveniles (30–50 cm long) through the lay of the kelp-line. At harvest, the kelp-lines are detached from the main lines and transported on shore where the kelp is manually cut off the lines. The farms are operated by a small number of large companies that use labor-intensive seeding and harvesting methods. (Zhang et al. 2017; Liu et al. 2019; Fang et al. 2020; Liu and Zhang 2022; Li et al. 2023a; HIS 2024).

11.5.4 Site 4. Funningsfjord, Faroe Islands (Figs. 11.3c, 11.4c, 11.5a, 11.6a)

The farm in the Faroe Islands as described in Tables 11.4, 11.5, 11.6, and 11.7 is the first of three sites used by the same operator; it was used from 2014 to 2020 and had several particularities. Firstly, although it has a maximum wind fetch of only 10 km, the local topography refracts the north Atlantic swells into the outer part of the fjord where the farm is located. Secondly, its waters are characterized by a low seasonal variability with relatively cool waters during the summer and a high transparency which allow the sugar kelp to grow down to a maximum depth of 20 m. Biofouling during the summer is minimal which allows multiple harvests (coppicing) of the perennial sugar kelp (*S. latissima*) over a two- or three-year growing cycle without reseeding. Thirdly, the ML is maintained at a depth of 10 m and vertical kelp-lines with a small float at their free end are attached along the main line at 1–2 m intervals (Fig. 11.4c). With this particular design, the kelp-lines are pushed out of the surface layer during storms. Finally, it is the only site where direct seeding and drag embedment anchors are used (Bak et al. 2018, 2020; Broch et al. 2019; Ciravegna et al. 2023).

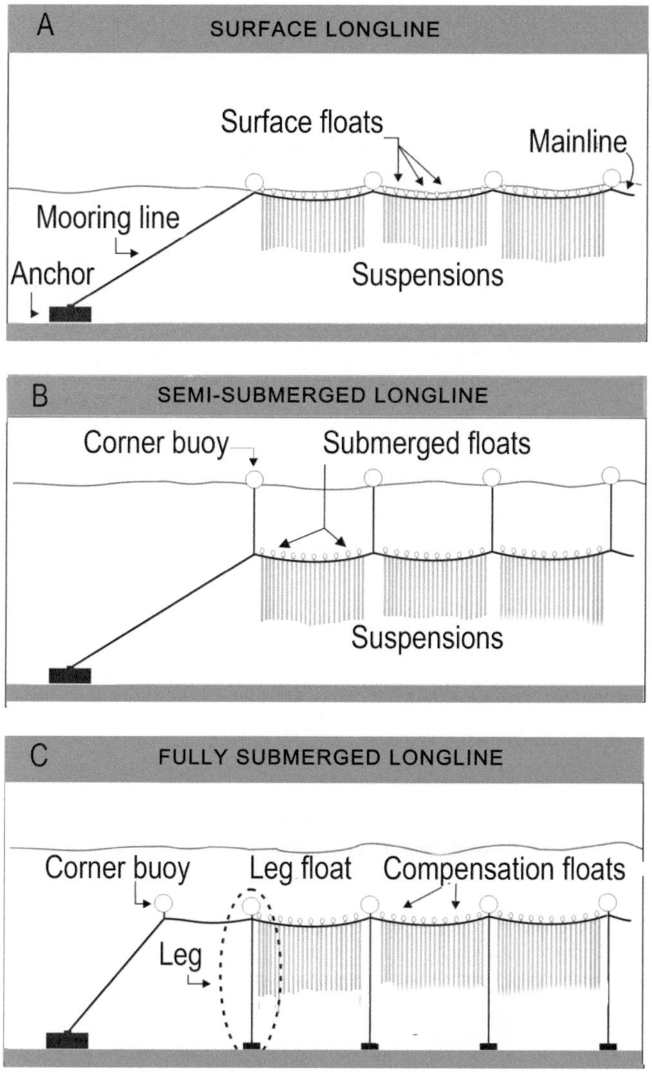

Fig. 11.3 Schematic representation of the three main types of longlines and their components. (**a**) surface longline; (**b**) semi-submerged longline; (**c**) fully submerged longline. Suspensions may be kelp fronds, vertical kelp-lines, mussel droppers, lantern nets, and other types of enclosures. Not to scale

In 2020, the operator acquired a new 50 ha lease in the same part of the fjord and changed the farm layout from individually anchored LLs to arrays of eight parallel 500 m mainlines, 15 m apart with a grid of anchors. Mechanized seeding and harvesting were recently tested at this site. In 2021, the operator acquired an additional 155 ha lease at a fully exposed site in Gotuvik, 20 km southeast of the first lease (Gregersen 2020, 2021; Gregersen et al. 2021; Aquavitae 2022; á Norði et al. 2023).

11.5.5 Site 5. Porthallow, Cornwall, England, UK (Figs. 11.3a, 11.4d, 11.5a, 11.6a)

The Porthallow farm is the smallest, shallowest, and closest to shore of the seven sites. It is moderately exposed to storms, wind waves, and swells from the English Channel. It is part of a larger (16 ha) integrated multi-trophic farm in combination with blue mussel (*Mytilus edulis*). Individually anchored surface LLs with vertical sinking 6 m long droppers attached to the main line 1.5–3 m apart are used (Fig. 11.4d). The typical manual twine winding seeding technique is used in November with a manual harvest carried out in May (Corrigan et al. 2023, 2024; Walker et al. 2023; Canvin et al. 2024; Wilding et al. 2024).

11.5.6 Site 6. Lesconil, Brittany, France (Figs. 11.3b, 11.4a, 11.5c, 11.6a)

The Lesconil site harbors one of the largest kelp farms in Europe (150 ha). It is fully exposed to storms and swells from the Atlantic Ocean (Fig. 11.8). Harvesting is relatively early (March) compared to other Western Hemisphere sites and this is related to the relatively high water temperature in this site from December to April (12 °C). The longline design and farm layout described in Tables 11.6 and 11.7 are those used in the first year(s) of operation in 2013–2014. Individually anchored semi-submerged 700 m long LLs with legs every 100 m and moorings with a tensioner buoy and sinker were used. The MLs were oriented parallel to the prevailing tidal currents and oceanic swells. Since then, LL design and farm layout were changed but detailed information on these changes is confidential. The farm operator owns another 180-ha lease 50 km east of this site but it is not in operation yet (CEVA 2013; Kholler et al. 2022; Gagnon 2024a; Ino-Rope 2024; Lemoine M. pers. comm. 2024).

11.5.7 Site 7. Paspébiac, Québec, Canada (Figs. 11.3c, 11.4a, 11.5a, 11.6a)

The Paspébiac farm is likely the only exposed commercial kelp farm in North America. It is moderately exposed to storms and swells from the Gulf of St. Lawrence. Its waters are the coldest of the seven sites with a mean temperature from December to April close to 1 °C and the only one exposed to drifting sea ice in winter. Individually anchored fully submerged LLs with legs are used (Fig. 11.9). The ML

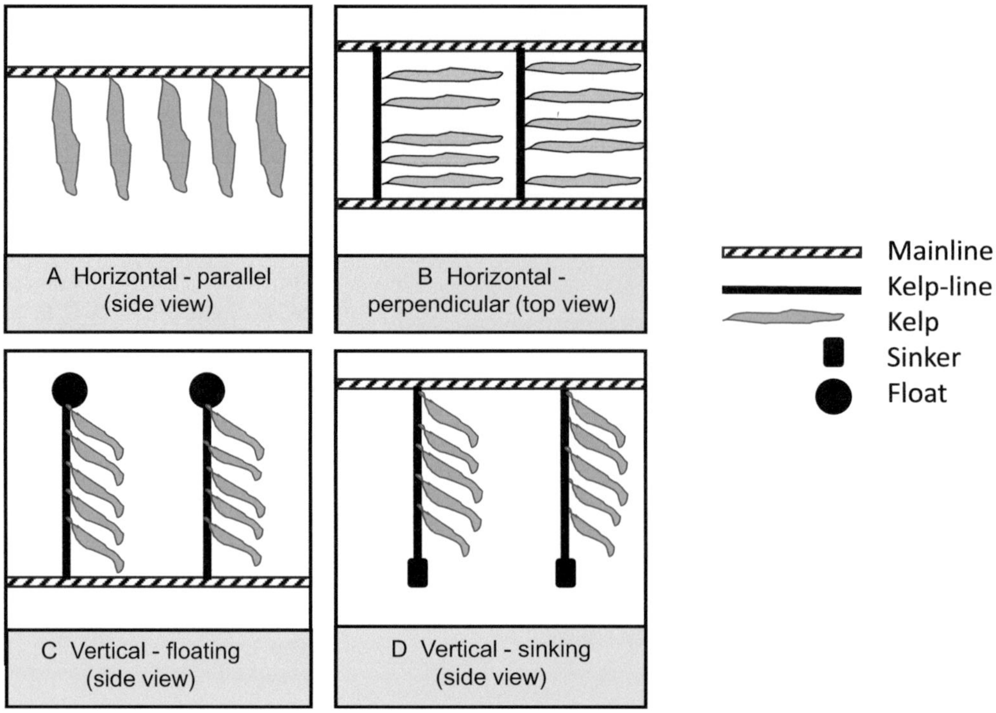

Fig. 11.4 Schematic representation of the different arrangements of kelp-lines on commercial longlines. (**a**) horizontal and parallel to mainline; (**b**) horizontal and perpendicular to mainlines; (**c**) vertical and floating above mainline; (**d**) vertical and sinking below mainline. Not to scale

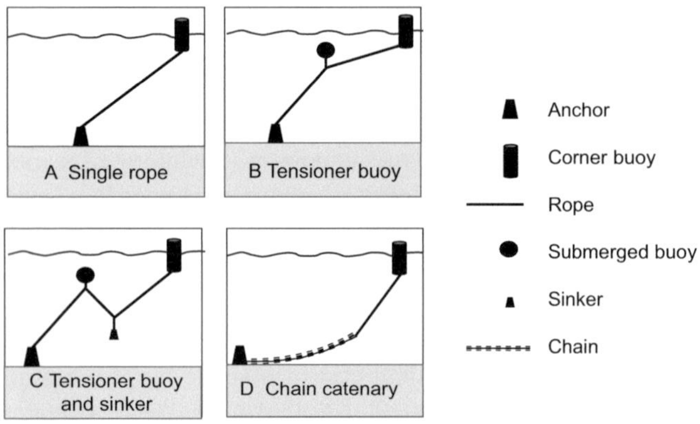

Fig. 11.5 Schematic representation of the main types of mooring lines used on longlines. (**a**) single rope; (**b**) two ropes with a submerged float between the anchor and the corner buoy; (**c**) three ropes with a submerged float and a sinker between the anchor and the corner buoy. (**d**) chain at bottom end and rope at top end. Not to scale. (Adapted from Bompais (1991))

is maintained 3 m below the sea surface between September and October, lowered to 6–7 m from November to March to avoid drifting ice and is raised again to a depth of 3 m between April and June (harvest) to allow for better access to light while protecting the kelp from the spring surface freshet. The kelp-line is attached parallel to the ML and 1 m below it to avoid crushing the kelp during husbandry operations (Tamigneaux et al. 2012; Rondeau and Bossé-Côte 2021; Autef S. pers. comm. 2024).

11.6 Structural Damage and Loss of Cultured Biomass

The severity of the damage caused by hydrodynamic forcing to LL farms may vary from a partial loss of the cultured biomass to the complete dislocation of the farm (Gagnon 2024a). Kelp biomass loss can be the result of the continuous erosion of the oldest tissue at the tip of the blade (erosion), the breakage of the blade (breakage), and the dislodgement of the complete blade due to the breakage of the stipe or detachment of the holdfast (dislodgement; Zhang et al. 2012). Additionally, biomass loss can be caused by friction of the kelp-lines with each other or with other LL components (abrasion). Structural damages include the loss of buoys, line breakage, line entanglement, and anchor dislodgement. Here, we review the information available for each of the seven sites described in the previous section.

11.6.1 Site 1. Minamikayabe, Japan

The information available for the Minamikayabe site is scarce and anecdotal. This site was hit by a large extratropical storm in January 2016 that caused severe damage to kelp arrays. In April 2017, a violent storm caused less severe damage (Ameba 2017).

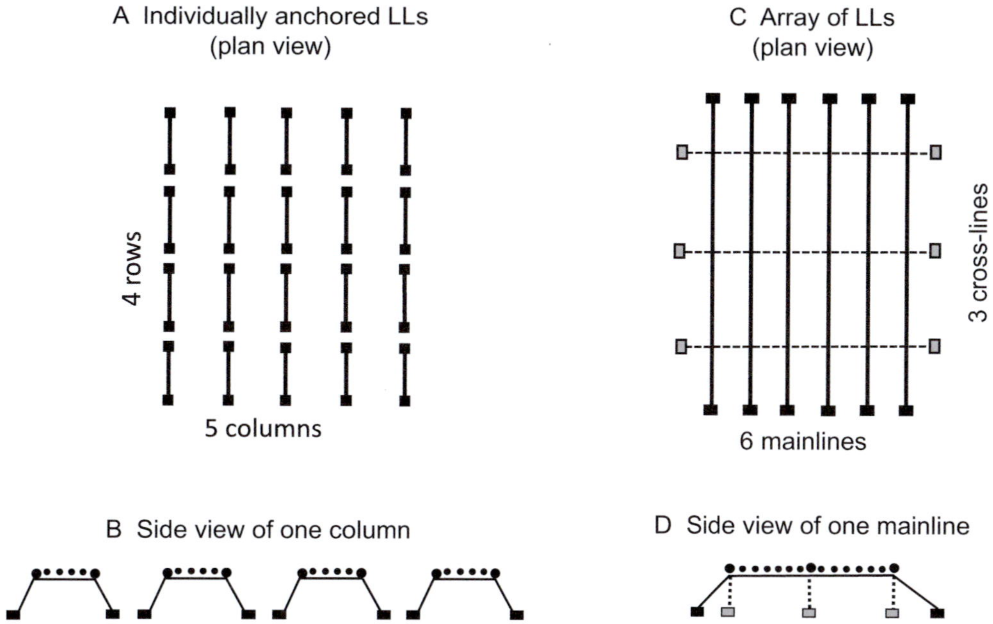

Fig. 11.6 The two main types of farm layout. (**a**) Plan view of a block of individually anchored longlines arranged in five columns and four rows. Large farms may be composed of several such blocks. (**b**) Side view of a column of individually anchored LLs. (**c**) Plan view of an array of longlines. Such arrays may or may not have crosslines. In large farms, several arrays may be arranged in columns and rows. (**d**) Side view of the LL in the middle of the array. Not to scale

Fig. 11.7 Satellite photo (4 November 2016) of the high-density longline fields in Sanggou Bay, Rongcheng County, Shandong, China (Site # 3). The kelp longlines occupy the outer part of the bay and the adjacent open waters. (Google Earth ©; Image © 2024 Maxar Technologies; Image © 2024 CINES/Airbus)

Fig. 11.8 Oblique aerial view of the kelp farm off Lesconil, southern Brittany, France (Site # 6). (Photo credit: Algolesko)

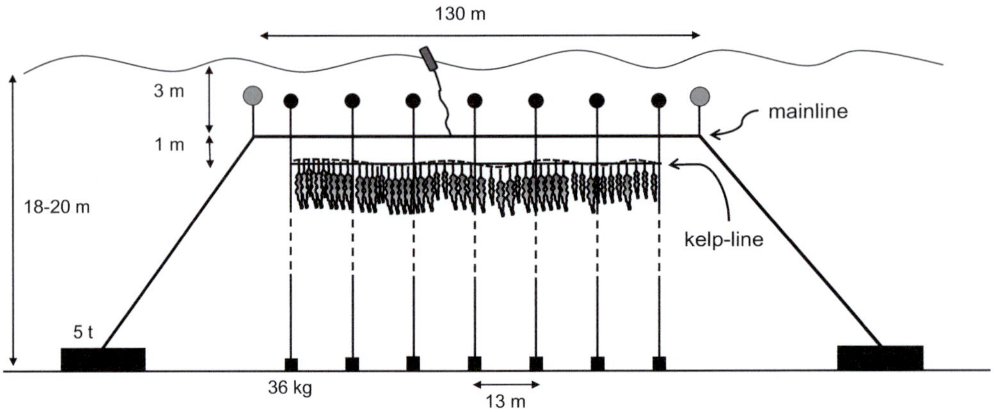

Fig. 11.9 Side view of the fully submerged longline used in Paspébiac, Québec, Canada (Site # 7). The main line is kept at 3 m between September and October and it is lowered to 6–7 m depth between November and March and thereafter at 3 m depth between April and June. Depth adjustments are made by changing the length of the legs. Not to scale

11.6.2 Site 2. Jusanhama, Japan

Kelp farms in the Jusanhama site were completely destroyed by the Great Tohoku Tsunami of March 2011 (ASC and MSC 2022). The Tohoku disaster reduced the national production of wakame by 64% and that of kombu by 42% in 2011 compared to the previous year (FAO 2024). The farms in Jusanhama were rebuilt by 2014. In 2024, the site was hit by an unprecedented series of three consecutive severe extra-tropical storms between the end of January and the end of March. These storms caused a 66% reduction of the production of wakame in the Myagi Prefecture, while the kombu that was cultivated on lines below the wakame was not destroyed (Ishinomaki Kahoku 2024; Terashima 2024).

11.6.3 Site 3. Sanggou Bay, China

In a controlled grow-out experiment carried out in the middle of the LL field offshore of Sanggou Bay, Zhang et al. (2012) report that 16% of the kelp plants were dislodged mostly in winter (juveniles), 4% of the blades were broken mostly at the end of the grow-out cycle due to a weakening of the kelp tissues, and 53% of the gross production was lost during the grow-out cycle due to the constant erosion of the distal part of the blade. In another controlled grow-out experiment just north of Sanggou Bay, in the outermost area, Liu et al. (2022b) report juvenile fall-off rates between 25% and 40% depending on the type of rope used. No information on the wave climate during both of the above experiments is reported. In a review of the problems currently faced by the offshore expansion of kelp farming in this site, Liu et al. (2019) report a high level of juvenile fall-off (up to 50%) and blade breakage (up to 70%) in the outermost exposed areas because the cultivars used were not developed for high-energy environments. Lost juveniles are replaced by new ones during husbandry operations.

11.6.4 Site 4. Funningfjord, Faroe Islands

According to Bak et al. (2018, 2020), there was no loss of equipment during the years the individually anchored LLs were used in the Faroe Islands site where extreme waves of

7–8 m have been reported. They attribute this to the fact that the vertical kelp-lines with a float at their free end are bent out of the surface layer during rough weather. Contrary to most other sites in Europe and North America, fouling during the summer is very low and this considerably reduces blade breakage. Losses due to the erosion of old tissue at the tip of the blade are also reduced by multiple harvesting.

11.6.5 Site 5. Porthallow, UK

Canvin et al. (2024) determined that the loss of biomass in the Porthallow farm was due to blade erosion (21% of total gross production) and to dislodgement (17%) mostly at the beginning of the monitoring period (January). Blade breakage was not important because the kelp was harvested before fouling started.

11.6.6 Site 6. Lesconil, France

In the fully exposed Lesconil site, nearly half of the LLs were destroyed by a large storm during the first winter of operation in February 2014 (Télégramme (Le) 2014). Eyrolles et al. (2018) report that the rough oceanic conditions and passing fishing and recreational vessels caused severe damage to the structures and biomass loss but these are not quantified. Since then, LL design and farm layout were changed, and all mooring lines were replaced by stronger ropes. Two years after this change, there had been no structural damage to the farm (Ino-Rope 2024). Biomass losses caused by storms are difficult to estimate and vary greatly from year to year but appear to be greater in January–February (Lemoine M. pers. comm. 2024).

11.6.7 Site 7. Paspébiac, Québec, Canada

The operator of the Paspébiac farm (Autef S. pers. comm. 2024) reports that blade breaking is a problem only in July and August when kelp harvest is delayed, and the blade gets progressively covered by the bryozoan *Membranipora membranacea*. Fall storms occasionally cause minor structural damage, mainly the loss of surface marker buoys (Fig. 11.9). Spring storms may cause the kelp-line to roll over the ML and cause kelp abrasion.

11.7 Discussion and Conclusion

11.7.1 Global Kelp Production

The FAO FishstatJ database (FAO 2024) is the only comprehensive global database on kelp production currently available. Although it is undeniable that eastern Asia largely dominates current kelp production, production figures for China and the Russian Far East are likely overestimated and there are many gaps in the production figures for the Western Hemisphere. These gaps are associated with the nascent nature of the industry: they are due to a lack of data collection/collation processes, omissions due to confidentiality (when there is a limited number of producers), or exclusion of non-animal species from aquaculture databases. The lack of accurate figures on seaweed aquaculture production has been identified by Araújo et al. (2021) and Wilding et al. (2021) as a constraint to the development of the farming industry in Europe.

After an annual 7.8% increase of the combined kelp (kombu and wakame) production in China, Japan, and South Korea from 2000 to 2020, this production decreased by 12% from 2020 to 2022 (FAO 2024; Hermans 2024a). This, in part due to a catastrophic kombu bleaching episode in 9300 ha of the Rongcheng County farming area (including Sanggou Bay Site #3) in 2022, is caused by red tides that generated phosphorus limitation, abnormal water transparency, and the development of pathogenic bacteria (Li et al. 2023b). As we will see below, the decreased production is also attributed to a lack of innovation in farming practices. In the Western Hemisphere, there was an overcapacity in kelp production in 2022 and expansion was hampered by a lack of processing infrastructure and product-market fit (Hermans 2024a). Since the inception of the industry in mid-2010s, profitability is elusive for most farmers except in Maine (USA) in recent years (Brayden and Coleman 2023).

11.7.2 Inventory of Exposed Sites

The criteria used to identify exposed sites for the systematic inventory carried out in this paper (Gagnon 2024a) are based on wave exposure. This is in line with the argument of Buck et al. (2024) and Heasman et al. (2024b) to the effect that the distance to shore (port) should be decoupled from hydrodynamic exposure when evaluating an aquaculture site. Indeed, as shown in Table 11.4, wave-exposed sites may be close or far from the coast. The exposure criteria used for this present review are however different from those developed by Lojek et al. (2024) and used by Heasman et al. (2024b) to characterize the level of hydrodynamic exposure of some selected kelp, shellfish, and finfish commercial and experimental/pilot exposed sites. The former authors proposed six hydrodynamic exposure indices related to the structural design and operation of aquaculture systems that are based not only on the wave energy but also on steady current velocity. One of these indices, the specific exposure energy (SEE), is proportional to the squared maximum horizontal water velocity generated by steady currents and waves and to the maximum drag force exerted on a fixed bluff body. As an example,

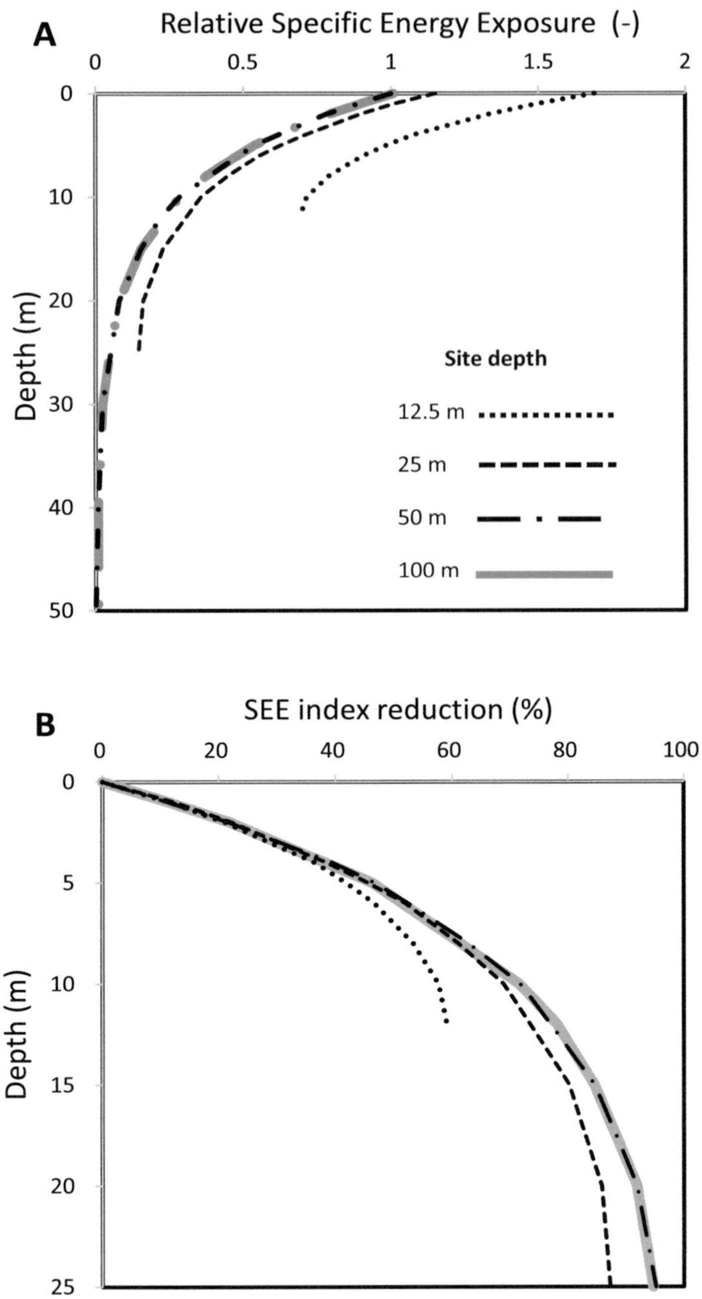

Fig. 11.10 Effect of site depth and distance from the sea surface on wave exposure. (**a**) Vertical profile of the relative Specific Exposure Energy (SEE) index developed by Lojek et al. (2024) for a 4 m and 8 s wave and no steady current (0 m s^{-1}) at four sites having a depth between 12.5 and 100 m. All values are relative to the surface (0 m) value of the index at the deepest site (100 m). (**b**) Reduction (%) of the SEE index as a function of the position of the structure in the water column at the four sites. All values are relative to the surface (0 m) value of the index at each site. (Produced using Kelson Marine's Exposure Index Calculator (https://www.kelsonmarine.com/resources; accessed December 2024))

Fig. 11.10 shows, for a 4 m and 8 s period wave without steady current (0 m s^{-1}), how the SEE index varies as a function of site depth (12.5, 25, 50 and 100 m) and distance from the sea surface (0–50 m). The SEE at the sea surface decreases with increasing site depth up to 37 m, which is equal to half the deep-water wavelength (74 m). According to the SEE index, exposure to extreme waves is maximal where storm waves interact with the sea bottom (where the water depth is less than half the wavelength) as shown by Lojek et al. (2024) for the southeastern North Sea.

Although the exposure indices developed by Lojek et al. (2024) provide a more precise and complete evaluation of the hydrodynamic exposure of aquaculture sites, they are difficult to apply in the context of the current global inventory, with thousands of sites to classify as sheltered or exposed. This is because the data needed to calculate the indices for each individual site is not readily available or do not have an adequate spatial resolution. Moreover, Lojek et al. (2024) and Heasman et al. (2024b) do not propose a site classification system similar to the one we used (Table 11.2). For example, they do not propose a value of the SEE index that separates sheltered sites from moderately exposed ones. In a recent model used to map site suitability for kelp farming at the global and European scales (van Oort et al. 2023; Macias et al. 2025), wave exposure is one of the variables retained. However, the former authors highlight the high uncertainty of the relationship they modeled between site suitability and wave exposure compared to that of the other model variables (temperature, salinity, irradiance, nitrogen, and phosphorus), this uncertainty being due to a lack of studies on seaweed cultivation in very exposed sites. Results of several at-sea trials in very exposed sites (see below) will become available in the next few years and these will increase considerably the knowledge base supporting the classification of exposed sites.

The fact that 99% of the wave-exposed kelp farms we inventoried are located in the PNW (Table 11.3) is not surprising considering that more than 99% of the global kelp production comes from that region (Table 11.1). Exposed farms in the Western Hemisphere are very scarce because kelp farms are mostly operated by startups (Brayden and Coleman 2023; Hermans 2023) that likely minimize their risks by farming in sheltered sites when available. What is more outstanding is that two-thirds of the PNW exposed area are located in Japan (Table 11.3), while this country's share of the PNW kelp production is only 1% (Table 11.1; if the FAO figures for China, North Korea, and Russia are considered accurate). This is tied to Japan's particular geographic and market settings. Firstly, a large proportion of Japan's temperate coasts are directly exposed to the Pacific Ocean (Eastern Hokkaido, Iwate, and Myagi Prefectures; Lobeto et al. 2024). Secondly, kelp is almost entirely destined for the food market in Japan (HIS 2024) and the market value of kelp varies considerately with the quality of individual blades. It is recognized that farming in exposed sites produces higher quality kombu and wakame (i.e., long, narrow, thick, and flat blades) than in sheltered sites (Nanba et al. 2011; Liu et al. 2022b; Peteiro et al. 2016; Broch et al. 2019; Liu et al. 2019; Sato et al. 2023; Skjermo et al. 2020; Sun et al. 2020). By analyzing the HRALMs in the PNW Region, a strong pattern in the distribution of the different types of aquaculture farms was found. Kelp leases are usually located in more exposed sites while those for shellfish are usually located in more sheltered ones. Examples of this can be found in Nanba et al. (2011), Liu et al. (2013), Fang et al. (2016), and Yokoyama (2021). We will see below that the likely reason for such a pattern is the greater ability of kelp to resist rough weather than other types of cultured biomass.

Given that the number of active farms and farming technology are currently changing quite rapidly, this inventory should be considered a snapshot of the early 2020s. It can eventually be used later as a benchmark to measure what has actually been gained or lost and how technology has changed. Climate change will have a significant effect on the extent of the temperate regions and exposed areas as defined in this paper. More specifically, Cooley et al. (2022) predict a poleward migration of the 5 and 20 °C SST isotherms used in this paper to delimit the study area and it is likely that, by the end of this century, some of the sites not included in this inventory will exceed the wave height and power density thresholds used to identify exposed sites and that others that were included will not exceed these thresholds (Aarnes et al. 2017; Casas-Prat et al. 2018; Lobeto et al. 2021).

11.7.3 Culture Method

In PNW, the main markets for kelp are food and animal feed (HIS 2024) and these determine the cultivation method. When it is for food, the quality and value of kombu and wakame depend on the morphological and organoleptic characteristics of individual blades (Liu et al. 2024b). Long, narrow, thick, and flat blades command a premium price on the markets. Complex and labor-intensive seeding, thinning, and harvesting operations are required to ensure this high quality. Kelp is grown at a low density (plants per m of kelp-line), is thinned once or twice, and its depth is adjusted to optimize lighting during the grow-out period. When kelp is cultivated for abalone or sea urchin feed, optimizing biomass yield is the main objective and the cultivation method is simpler (high density, no thinning, etc.; E. Tamigneaux pers. observ.). The above is illustrated in Fig. 11.11 which compares a low-density kelp-line at harvest typical of Asian cultivation for the food market with a high-density kelp-line typical of Asian cultivation for the abalone feed market and of Western Hemisphere kelp cultivation.

In all regions, seeding, thinning, and harvesting operations are currently essentially manual with the exception of twine insertion seeding that has been mechanized in South Korea (HIS 2024) and direct seeding in the Faroe Islands (Aquavitae 2022). Due to the lack of innovation, kelp farming in PNW faces serious challenges. Farming is very hard work and labor-intensive, the farmers are getting old, and there are few newcomers to the industry. This is the main reason why kelp production in Japan decreased by 60% between 1986 and 2022 (FAO 2024). The lack of succession is also beginning to be a problem in South Korea. In China, current practice is becoming less profitable because of rising labor costs. Automation and mechanization is urgently needed in all countries (Hasegawa and Suzuki 2005; FAO 2009; Liu et al. 2019; Hu et al. 2021; Watanabe and Sakami 2021; HIS 2024). There is also a need for new cultivars adapted to global warming and high-energy environments (Liu et al. 2019; Hu et al. 2023).

In the Western Hemisphere, wages are high, and farming operations are still mostly manual. Kelp farms are currently operated by start-up companies that are in the process of scaling up their operations to attain profitability (Brayden and Coleman 2023; HIS 2024). In several regions, scaling-up will likely be possible by expanding into more exposed waters where adequate space is available and this will require

Fig. 11.11 Examples of two kelp densities at harvest. (**a**) Winged kelp (*A. esculenta*) longline at the Paspébiac, Québec, Canada farm. The relatively high density in A compared to B results from the winding of the seedling twine around the horizontal kelp-line and the absence of thinning operations during the grow-out period. The mainline is the line without kelp while the biomass is attached to the kelp-line installed 1 m below the mainline (see Fig. 11.9). Photo credit: Sandra Autef (AGHAMW). (**b**) Japanese kelp (*S. japonica*) longline in Wando County, South Jeolla, South Korea. The relatively low kelp density on the longline of B compared to A reflects different culture techniques. In A, the twine carrying A. esculenta seedlings was rolled around the mainline and no thinning occured during the cultivation. S. japonica clumps in B reflect a method where 3–5 cm twine segments with seedlings are inserted through the lay of the mainline at 30 cm intervals and, when the kelp grows, the clumps are regularly thinned before harvesting. (Photo credit: E. Tamigneaux)

the automation and mechanization of seeding and harvesting operations combined with remote sensing of kelp, LL, and farm condition.

11.7.4 Longline Design and Farm Layout

The number of wave-exposed farms with detailed information on LL design and farm layout is quite limited. This is mostly because farm owners and operators are reluctant to release sensitive business information. Peer-reviewed papers, company websites, and ASC-MSC documents usually provide only summary and incomplete information on the subject, while governmental/industrial lease/permit application/review documents usually provide complete information but are published only by a few administrations (e.g., British Columbia (Canada), Maine (USA), UK).

The most striking feature of the group of farms for which information is available is the very large range of LL designs and farm layouts currently used to farm kelp. There are no standards in that matter except at the local scale in PNW as a result of more than 30 years of experience in adapting to local conditions. We discuss below four main questions concerning LL design and farm layout in exposed sites: (1) mainline depth, (2) kelp-line types, (3) layout type, and (4) orientation relative to currents and waves.

The best way to reduce wave forces on a LL at a given exposed site is to maintain the mainline at the largest depth possible because wave-induced forces decrease with increasing depth in the water column. This is illustrated in Fig. 11.10b where the reduction of the specific exposure energy (SEE) index from its value at the sea surface is given as a function of the distance from the sea surface. At the two deep sites (50 and 100 m), the drag force decreases exponentially with increasing distance from the surface. The reduction rate is smaller for the two shallower sites (12.5 and 25 m) and the difference between the shallow and deep sites is significant only when the distance from the surface is larger than 5 m. This is why Heasman et al. (2024a) indicates that the best exposed sites for farming are the deeper ones, where the waves do not interact with the sea bottom and, we should add, when the structure can be lowered at a depth of more than 5 m.

However, unlike shellfish LLs (Mizuta and Wikfors 2019; Gagnon 2024a), the maximum possible mainline depth for kelp is limited by the necessity of keeping the fronds in the illuminated surface water layer (photic zone), at least during daytime. This maximum depth depends on water transparency. Kelp species have a high photosynthetic efficiency (Gattuso et al. 2006; Kerrison et al. 2015; Peteiro et al. 2016; Stekoll et al. 2021). For example, in offshore clear waters of the North Atlantic (SDD between 15 and 30 m), *S. latissima* could grow significantly down to a depth of 50 m and maximum production would be at a depth around 20 m (Strong-Wright and Taylor 2022). In the relatively turbid coastal waters (SDD < 6 m) off China (e.g., Site #3, Sanggou Bay) and southern Brittany (Site #6, Lesconil), kelp is maintained within a few meters of the sea surface, while in the relatively clear (SDD > 14 m) coastal waters off northeastern Honshu (e.g., Site #2, Juanhama) and in the Faroe Islands (e.g., Site #4, Funningsfjord), it is cultivated at depths up to 10 m. The Paspébiac (Canada) site (#7) is a special case. Although the water is relatively turbid (Table 11.4), the mainline is maintained at a depth of 7 m in winter to protect the structure from drift ice and 3 m in spring because of the strong surface freshet at that time. The design adopted in the clear waters of Farm # 4 in the Faroe Islands, where the negatively buoyant kelp is grown on vertical lines with a small float at the free end, allows lowering the ML to 10 m while maintaining the kelp biomass in the illuminated layer. This approach is also used to grow positively buoyant kelp (*M. pyrifera* and *N. luetkeana*) but this time by having the kelp attached directly to the ML (Camus et al. 2018; Stekoll et al. 2024).

Depending on the site, the kelp lines may be horizontal or vertical (Table 11.3, Fig. 11.4). The likely reasons for the choice of the kelp-line type are the following: firstly, in turbid sites (# 3, 6 and 7) the optimally illuminated layer is vertically limited and a horizontal kelp-line allows a better illumination of all blades than a vertical one by keeping the blades at the optimal depth and minimizing blade self-shading (Peteiro et al. 2016). Secondly, in fully exposed sites (Sites # 2 and 6), there would be an increased risk of kelp abrasion if vertical kelp-lines were used because these can entangle with one another or with the mainline (Buck and Buchholz 2004).

Longlines may be individually anchored 10 to 50 m apart or arranged in arrays of closely spaced kelp-lines (Table 11.7 and Fig. 11.6). In the PNW, the first type is mostly used in fully exposed sites in order to reduce abrasion caused by the interactions between longlines during storms (Perez et al. 1992; Hasegawa and Suzuki 2005; HIS 2024; MPFCA 2024). This comes with the disadvantage of lower kelp-line density in the leased area.

LL orientation relative to currents and waves varies considerably between the farms (Table 11.7). The question of which orientation is best for the survivability of the structure and the retention, growth and quality of the cultured biomass in exposed sites is unresolved. Some researchers believe that the kelp-lines should be parallel to the prevailing swell direction, while others recommend that they be oriented parallel to the prevailing currents. However, the effect of LL orientation on the forces exerted by waves on the structures and the kelp biomass is unknown and the question cannot be answered with current knowledge (Gagnon 2024b).

11.7.5 Structural Damage

Structural damage to commercial and pilot shellfish and kelp LL farms caused by storms and tsumanis was reviewed by Gagnon (2024a). In the case of storms, information on structural damage is scarce and mainly anecdotal. For that reason, it is not possible to correlate the severity of damage with LL design, farm layout, current velocity, and wave characteristics. LL farms in wave-exposed environments are vulnerable to extreme storms (i.e., with a large return period). Depending on the location and season, the frequency of occurrence and severity of such storms will increase or decrease due to climate change (Lobeto et al. 2021). Compared to shellfish LLs with the same basic design and weight of the cultured biomass (5–15 kg wet weight per meter), kelp LLs are less vulnerable to strong currents because the mass density of kelp (around 1100 kg m^{-3}) is much less than that of shellfish (over 1250 kg m^{-3}). This has two main effects: firstly, kelp LLs require less floats for support and consequently the drag and inertia forces on the floats and the risk of snap loads in the lines are reduced. Secondly, kelp fronds incline more in the current than, for example, mussel droppers and scallop lantern nets. At a current velocity of more than 0.2 m s^{-1}, they adopt a horizontal posture and reconfigure into a more streamlined shape that reduces the drag forces on the kelp biomass and on the longline as a whole, when compared to that on partially inclined rigid mussel droppers and lantern nets (Gagnon 2024b).

In the case of tsunamis, although large ones have a return period of several decades (Reid and Mooney 2022), some of them have been responsible for severe or catastrophic damage to longline farms around the Pacific (Gagnon 2024a) including those in the Jusanhama site (# 2). After the large tsunamis of 1960, 2010, and 2011 on the Pacific coast of Japan, the level of damage to LLs was much higher in sheltered areas than in open ocean sites and was not related to sea surface elevation but rather to current velocity; LLs were destroyed when it exceeded 1.0–1.2 ms^{-1} (Kato et al. 2010; Suppasri et al. 2018). On the open coast, this velocity is rarely attained during a tsunami in areas where the water depth is >65 m and consequently open deep waters are a refuge from tsunamis for surface structures (Lynett et al. 2014).

11.7.6 Biomass Loss

The rate of erosion of the oldest tissue at the tip of the blade is a continuous process that usually increases during the grow-out cycle and may be accelerated by rough sea conditions (Yoshikawa et al. 2001; Dolliver and O'Connor 2022). In late spring to mid-summer depending on latitude, heavy fouling of the kelp blades starts, kelp tissue deteriorates, and blade breakage increases rapidly. Biomass loss can reach 50% by late summer and almost 100% in late fall leaving only holdfasts and stipes attached to the lines (Gendron and Tamigneaux 2008; Krumhansl et al. 2011; Skjermo et al. 2020; Fieler et al. 2021). This is why kelp is usually harvested in late spring or early summer before heavy fouling starts (Table 11.5). Fouling (and presumably breakage) is usually less important or delayed in exposed sites (Broch et al. 2019) with high transparency and cool summer waters such as those found at the Faroe Islands site (# 4).

Kelp dislodgement caused by breakage of the stipe or detachment of the holdfast is usually important only at the start of the grow-out cycle. In the case of *S. japonica* in

China, which is seeded with juveniles at low densities, fall-off can be measured and it occurs mainly in winter likely due to the increased storminess and higher vulnerability of juvenile plants (Liu et al. 2022b). In the case of *S. latissima* which is seeded at high densities, fall-off due to hydrodynamic forces is difficult to quantify because it is partly masked by the natural self-thinning caused by intraspecific competition for space and/or nutrients that considerably reduces plant density during the grow-out cycle (Kerrison et al. 2017; Dolliver and O'Connor 2022). Laboratory testing concluded that the forces exerted on sugar kelp biomass attached to LLs are not likely to cause dislodgement of medium- and large-sized kelp even in very exposed sites (Buck and Buchholz 2005; Chen et al. 2023). This is in large part because kelp species in exposed sites develop narrow, flat, and smooth blades that reduce the drag force exerted by waves and currents and develop strong holdfasts (Kawamata 2001; Buck and Buchholz 2005; Koehl et al. 2008; Nanba et al. 2011; Peteiro et al. 2016).

Friction between adjacent vertical droppers on LLs because their linear weight varies along the mainline may cause abrasion of mussel and kelp biomass (Bompais 1991; Buck and Buchholz 2004). Furthermore, when the distance between the kelp-lines is small (< 10 m) and the pretension is not the same in all lines, friction between the lines causes kelp abrasion. In Maine, where the LLs are closely spaced (average of 3.5 m), line entanglement is the main cause of biomass loss; 67% to 100% of crop losses on the entangled lines are reported (Brayden and Coleman 2023). Increasing the spacing of the lines and the pretension in the lines and orientating the lines parallel to the wave direction can reduce this abrasion (Flavin et al. 2013; Brayden and Coleman 2023; Moscicki et al. 2024). Finally, in large waves perpendicular or oblique to the mainline, kelp blades attached directly to the mainline (Zhu 2020; Chen et al. 2023), vertical droppers (Fredheim and Lien 2001; Buck and Buchholz 2004), and horizontal kelp-lines attached below the mainline (Gendron-Lemieux I. pers. observ.) may roll over this line. This may cause kelp abrasion, dislodgement, and/or breakage.

The fact that kelp farms are located in exposed sites while shellfish farms are located in sheltered sites in the PNW is likely explained by the superior ability of kelp to resist current and wave force. For example, Wu et al. (2024) report severe mussel sloughing (fall off) near harvest time on surface LLs in a semi-sheltered site in Shengsi County, China, due to passing tropical storms. In the case of scallops, they are very sensitive to the acceleration and inclination of their enclosure by currents and waves (Ventilla 1982; Oshino 2010; Natsuike et al. 2022).

11.7.7 Innovation

The exposed kelp farms in the present inventory are located in environments that are less energetic than where most offshore wind and wave farms are or will be established in the next decade (4C Offshore Ltd. 2023). For example, the average annual wave power density in the North Sea wind farms varies between 4 kW m^{-1} off Belgium and 33 kW m^{-1} off Norway (Beels et al. 2007; Rusu and Rusu 2021), while the maximum wave power density in the inventoried exposed kelp sites is less than 10 kW m^{-1} (Dong et al. 2020; Eum et al. 2020; Webb et al. 2020). Therefore, the question arises as to whether the current technology is suitable for very exposed environments where it is planned to be colocated with marine renewable energy farms. In the past five years, several R&D programs have been terminated or been initiated to determine the feasibility of kelp farming in wind and wave farms and other very exposed sites (Table 11.8). Currently used designs, new designs based on the basic longline, and other technologies have or will be tested (Heasman et al. 2024a, b). Here we review some results of the at-sea testing of innovations that are relevant to the above discussion.

One particular design tested is an array with a catenary system used to distribute forces evenly across closely spaced kelp-lines (less than 5 m apart) in order to reduce abrasion. Testing of this new design in moderately exposed sites in Scotland (UK) and Maine (USA) gave disappointing results (Bond and Elbourne 2024; Moscicki et al. 2024). In the former case, the commercial operator concluded that catenary arrays are not commercially viable. In the second case, the synthetic ropes used for the mooring lines and kelp-lines were replaced by fiberglass rebar to reduce the risk of marine mammal entanglements. The resulting low yield was likely due to the reduced attachment strength of the kelp on the rebar compared to that on ropes and/or to abrasion of the kelp biomass due to friction between the kelp-lines (Moscicki et al. 2024).

Another design proposed for the PNE offshore areas is a very long (5 km) free-floating horizontal LL with *N. luetkeana* floating above the kelp-line and *S. latissima* growing downward. Numerical simulations show that this design could give higher yields than conventional farms due to higher water transparency and nutrient availability caused by wave mixing and lower biofouling (Whiting et al. 2020). However, they also show that the surface buoys may be damaged and result in the whole system sinking to the bottom and that kelp-kelp and kelp-structure interactions may cause important biomass loss (Chen et al. 2023).

Table 11.8 List of R&D programs focusing on kelp farming in high-energy environments (terminated or initiated between 2014 and 2024)

Program	Country	Main objective	Web site/Reference
Aquavitae	Faroe Islands	Growing kelp offshore	https://aquavitaeproject.eu/
AOOF[a]	New Zealand	Growing kelp in high energy environments	https://openocean.cawthron.org.nz/
Blue Economy CRC	Australia	Growing kelp offshore	https://blueeconomycrc.com.au/
Genialg	Norway	Growing kelp in exposed sites	https://genialgproject.eu/
Macrocascade	Faroe Islands	Mechanical kelp seeding and harvesting	https://www.macrocascade.eu/
Macrosea	Norway	Industrial kelp production	https://www.sintef.no/projectweb/macrosea/
Mariner	USA	Industrial kelp production	https://arpa-e.energy.gov/technologies/programs/mariner
Madoc	UK	Kelp production in Wales	https://projectmadoc.cymru/home/
North Sea Farm #1	The Netherlands	Growing kelp in wind farms	https://www.northseafarmers.org/projects/nsf1
Olamur	Denmark, Germany	Growing kelp in wind farms	https://olamur.eu/
Seaweed Carbon Solutions	Norway	Growing kelp in offshore exposed sites	https://www.sintef.no/en/projects/2021/seaweed-carbon-solutions-jip/
United and Ultfarms	Germany, The Netherlands, Belgium	Growing kelp in wind farms	https://www.h2020united.eu/; https://ultfarms.eu/
Weir & Wind	The Netherlands	Growing kelp in high energy environments	https://www.northseafarmers.org/sector/wier-en-wind

[a]Anchoring our Open Ocean Future (Ngā Punga o te Moana)

Direct (binder) seeding is an innovation that has been tested in the past ten years in Europe. This method can significantly reduce the cost and time of hatchery cultivation and seeding operations and was adopted by one of the seven commercial farms described (Site #4 Faroe Islands). However, compared with twine seeding, binder seeding may result in delayed blade growth because of the early energy investment in the development of the holdfast and poor yields due to the dislodgement of the young plants before they have attached securely. Reports on the more or less successful use of direct seeding are from trials in sheltered sites (as defined in Table 11.2) except in the case of the Wilding et al. (2024) study in Site #5 (Porthallow Bay, Cornwall, UK), a moderately exposed site. In that study, twine seeding outperformed direct seeding fourfold in terms of biomass weight per m of kelp-line likely because of dislodgement of directly seeded seedlings in the days or weeks following deployment. The better performance of direct seeding in the Faroese farm compared to the Cornish farm could be related to the different LL design and wave climate. Indeed, the surface LLs of the Cornish farm are more prone to snap loads in the kelp-lines caused by wind seas acting on the surface buoys (Gagnon 2024b).

Several seeding and harvesting machines have been tested recently (Chung et al. 2015; Zhang et al. 2017; Merinov 2019, 2021; Choi 2020; Hwang and Park 2020; Solvang et al. 2021; Wilding et al. 2021; Aquavitae 2022; Jiang et al. 2022; WHOI 2022; North Sea Farmers (NSF) 2022, 2023; Zhu et al. 2023; Seawiser 2024). Each machine was adapted to the specific cultivation method and LL design used in the region of development. To our knowledge these machines are not used on commercial farms yet except in the case of the one used in South Korea to insert seeded twine segments in the lay of the kelp-rope (Hwang and Park 2020; HIS 2024) and in the Faroe Islands for direct seeding (Aquavitae 2022).

None of the seven farms described in this chapter use screw (helical) anchors. When the bottom sediment type and thickness are suitable for their installation, this type of anchor have a much higher holding capacity than conventional deadweights, drag embedment anchors, and piles (Gunnlaugsdottir et al. 2013; Heasman et al. 2024b).

Some of the new LL designs being tested are based on depth cycling using variable buoyancy components (submersible buoys). This cycling can be diurnal (in nutrient-rich deep water during the night), occasional (lowering at the approach of a storm, lifting for seeding and harvesting operations), and seasonal (lowering during winter). It is likely that this approach can eventually become economically feasible only for large LL arrays (Bale 2017; Goseberg et al. 2017; Capron et al. 2018; Godsiff 2020; Navarrete et al. 2021; OOS 2022; Kite-Powell et al. 2022; Lian et al. 2023, 2024).

Acknowledgments The authors wish to thank one anonymous reviewer for his helpful comments and suggestions. Thank you to Sandra Autef (AGHAMW), Cyr Couturier (Memorial University of Newfoundland), Cory Leavitt (Department of Agriculture, Aquaculture and Fisheries, New Brunswick), Mathide Lemoine (Algolesko), Florence Peyton Jones (Crown State, UK), Rhianna Rees (Scottish Seaweed Industry Association), Jonathon Vieira (Government of British Columbia), and Wendy Vissers (Department of Fisheries and Aquaculture, Nova Scotia) who provided information on kelp production and farming practices.

Author Contributions MG—Conceptualization, literature review, analysis, writing and editing. ET—data collection, literature review, graphics and editing. IGL—data collection, graphics, review and editing.

Conflict of Interest (Competing Interests) The authors declare that the research was conducted in the absence of any commercial or financial relationships that could be construed as a potential conflict of interest. M. Gagnon has worked for Biorex inc. É. Tamigneaux receives funding from two FRQNT strategic clusters of researchers: Ressources Aquatiques Québec and Institut sur la nutrition et les aliments fonctionnels.

Availability of Data and Material Access to data is available upon reasonable request from the lead author.

Code Availability not applicable.

Funding This research did not receive any specific grant from funding agencies in the public, commercial, or not-for-profit sectors.

References

4C Offshore (2023) Global offshore windfarms map. https://map.4coffshore.com/offshorewind/. Accessed 15 Mar 2024

á Norði G, Mols-Mortensen A, Hancke K, Borgersen G, Wegeberg S, Andreasen B (2023) Workshop: macroalgal cultivation on Faroese fjords - the need for environmental assessment and monitoring. Fiskaaling rit 2023-12. Hvalvík, Faroe Islands. https://doi.org/10.5281/zenodo.10185198

Aarnes OJ, Reistad M, Breivik Ø, Bitner-Gregersen E, Ingolf Eide L, Gramstad O, Vanem E (2017) Projected changes in significant wave height toward the end of the 21st century: Northeast Atlantic. J Geophys Res Oceans 122:3394–3403. https://doi.org/10.1002/2016JC012521

Ameba (2017) The season for "quick-boiled kelp" cultivated in Minamikayabe, Hakodate, is drawing to an end. https://ameblo.jp/dodenkyo/entry-12269760399.html. Accessed 3 Nov 2024

AquaVitae (2022) New species, processes and products contributing to increased production and improved sustainability in emerging low trophic, and existing low and high trophic aquaculture value chains in the Atlantic. Deliverable D1.4 Report on second development phase for hatchery/seedling production in CSs, with user acceptance testing feedback, pp 1–781

Aquavostok (2023). https://aquavostok.ru/#project. Accessed 12 Dec 2023

Araújo R, Vázquez Calderón F, Sánchez López J, Azevedo IC, Bruhn A, Fluch S, Garcia Tasende M, Ghaderiardakani F, Ilmjärv T, Laurans M et al (2021) Current status of the algae production industry in Europe: an emerging sector of the Blue Bioeconomy. Front Mar Sci 7:626389

ASC and MSC (2022) ASC-MSC seaweed (algae) audit group reporting. Junsanhama Gyokyo, Youth Association, 108 p

Bak UG, Mols-Mortensen A, Gregersen O (2018) Production method and cost of commercial-scale offshore cultivation of kelp in The Faroe Islands using multiple partial harvesting. Algal Res 33:36–47. https://doi.org/10.1016/j.algal.2018.05.001

Bak UG, Gregersen Ó, Infante J (2020) Technical challenges for offshore cultivation of kelp species: lessons learned and future directions. Bot Mar 63(4):341–353. https://doi.org/10.1515/bot-2019-0005

Bale ES (2017) Development of area-efficient and standardized structures for large-scale macroalgae cultivation. SINTEF Rep OC2017A171, pp 1–24

Bath GE, Price CA, Riley KL, Morris JA (2023) A global review of protected species interactions with marine aquaculture. Rev Aquac 15:1686–1719. https://doi.org/10.1111/raq.12811

Beels C, Henriques JCC, De Rouck J, Pontes MT, De Backer G, Verhaeghe H (2007) Wave energy resource in the North Sea. In: Proceedings of the European Wave and Tidal Energy Conference (EWTEC), Porto, Portugal, 11–13 September 2007. https://doi.org/10.13140/2.1.4640.6403

Bolton JJ (2010) The biogeography of kelps (Laminariales, Phaeophyceae): a global analysis with new insights from recent advances in molecular phylogenetics. Helgol Mar Res 64:263–279. https://doi.org/10.1007/s10152-010-0211-6

Bompais X (1991) Les filières pour l'élevage des moules: guide pratique. Institut français de recherche pour l'exploitation de la mer (IFREMER), pp 1–247

Bond I, Elbourne P (2024) Reflections on our seaweed farming operations 2016–2023. New Wave Foods Ltd. Horizon Seaweed, pp 1–80

Bord Lascaigh Mhara (BIM) (2024) Annual aquaculture report. Findings of the national seafood survey 2024

Brayden C, Coleman S (2023) Maine seaweed benchmarking: economically assessing the growth of an emerging sector. Aquac Econ Manag 28:491–514. https://doi.org/10.1080/13657305.2024.2319806

Broch OJ, Alver MO, Bekkby T, Gundersen H, Forbord S, Handå A, Skjermo J, Hancke K (2019) The kelp cultivation potential in coastal and offshore regions of Norway. Front Mar Sci 5:529. https://doi.org/10.3389/fmars.2018.00529

Buck BH, Buchholz CM (2004) The offshore-ring: a new system design for the open ocean aquaculture of macroalgae. J Appl Phycol 16:355–368. https://doi.org/10.1023/B:JAPH.0000047947.96231.ea

Buck BH, Buchholz CM (2005) Response of offshore cultivated *Laminaria saccharina* to hydrodynamic forcing in the North Sea. Aquaculture 250:674–691. https://doi.org/10.1016/j.aquaculture.2005.04.062

Buck BH, Langan R (eds) (2017) Aquaculture perspective of multi-use sites in the open ocean: the untapped potential for marine resources in the Anthropocene. Springer, Cham

Buck BH, Nevejan N, Wille M, Chambers M, Chopin T (2017) Offshore and multi-use aquaculture with extractive species: seaweeds and bivalves. In: Buck BH, Langan R (eds) Aquaculture perspective of multi-use sites in the open ocean: the untapped potential for marine resources in the Anthropocene. Springer, Cham, pp 23–69. https://doi.org/10.1007/978-3-319-51159-7_2

Buck BH, Troell MF, Krause G, Angel DL, Grote B, Chopin T (2018) State of the art and challenges for offshore integrated multitrophic aquaculture (IMTA). Front Mar Sci 5:165. https://doi.org/10.3389/fmars.2018.00165

Buck BH, Bjelland H, Bockus A, Chambers M, Costa-Pierce BA, Dewhurst T et al (2024) Resolving the term "offshore aquaculture": the importance of decoupling it from "exposed" and "distance from the coast". Front Aquac 3:1428056. https://doi.org/10.3389/faquc.2024.1428056

Buschmann AH, Camus C, Infante J, Neori A, Israel A, Hernández-González MC, Pereda SV, Gomez-Pinchetti JL, Golberg A, Tadmor-Shalev N, Critchley AT (2017) Seaweed production: overview of the global state of exploitation, farming and emerging research activity. Eur J Phycol 52:391–406. https://doi.org/10.1080/09670262.2017.1365175

Cai J, Lovatelli A, Aguilar-Manjarrez J, Cornish L, Dabbadie L, Desrocher, A, Diffey S, Garrido Gamarro E, Geehan J, Hurtado A, Lucente D, Mair G, Miao W, Potin P, Przybyla C, Reantaso M, Roubach R, Tauati M, Yuan X (2021) Seaweeds and microalgae: an overview for unlocking their potential in global aquaculture development. FAO Fisheries and Aquaculture Circular No. 1229. FAO, Rome. https://doi.org/10.4060/cb5670en

Camus C, Infante J, Buschmann AH (2018) Overview of 3 year pre-commercial sea farming of *Macrocystis pyrifer*a along the Chilean coast. Rev Aquac 10:543–559

Canvin MC, Moore PJ, Smale DA (2024). Quantifying growth, erosion and dislodgement rates of farmed kelp (*Saccharina latissima*) to examine the carbon sequestration potential of temperate seaweed. J Appl Phycol 36:3091–3102

Capron ME, Blaylock R, Lucas K, Chambers MD, Stewart JR, DiMarco SF et al (2018) Ocean forests: breakthrough yields for macroalgae. In: Proceedings of the OCEANS 2018 MTS/IEEE Charleston, Charleston, SC, USA, 22–25 October 2018, pp. 1–6. https://doi.org/10.1109/OCEANS.2018.8604586

Capuzzo E, MacMillan I (2024) Review of species and farming methods. Seaweed in East Anglia (SEA) Project. CEFAS, pp 1–52

Casas-Prat M, Wang XL, Swart N (2018) CMIP5-based global wave climate projections including the entire Arctic Ocean. Ocean Model 123:66–85. https://doi.org/10.1016/j.ocemod.2017.12.003

Centre d'Étude et de Valorisation des Algues (CEVA) (2013) Algolesko. Tome 1. Déclaration au titre de l'article R214-32 du code de l'Environnement. https://docplayer.fr/112426611-Algolesko-tome-1-declaration-au-titre-de-l-article-r-du-code-de-l-environnement.html. Accessed 15 Oct 2023

CEVA (2024) Report on the current algae industry in Europe. CIRCALGAE Deliverable 1.1, pp 1–148

Chen M, Yim SC, Cox DT, Yang Z, Huesemann MH, Mumford TF et al (2023) Modeling and analysis of a novel offshore binary species free-floating longline macroalgal farming system. J Offshore Mech Arct Eng 145:021301. https://doi.org/10.1115/1.4055803

Choi KJ (2020) Development of the automation system for seaweed biomass mass production. J Korean Soc Ind Converg 23:351–359. https://doi.org/10.21289/KSIC.2020.23.2.351

Chung H, Kim NG, Choi KJ, Woo HC (2015) Novel ocean system for high density mass production of seaweed biomass in Korea. In: 2015 World Congress on Advances in Structural Engineering and Mechanics (ASEM15), Incheon, Korea, August 25–29, 2015

Ciravegna E, Koch S, van den Burg S (2023) The business case for seaweed aquaculture in the North Sea: learning from international experiences. Wageningen U Res Rep 2023-018, pp 1–37.

Conapesca Acuasesor (2024) Portal informativo acuicola. Comision national de acuacultura y pesca. Mexico. https://acuasesor.conapesca.gob.mx/. Accessed 20 Nov 2024

Cooley S, Schoeman Bopp L, Boyd P, Donner S, Ghebrehiwet DY, Ito S-I, Kiessling W, Martinetto P, Ojea E, Racault MF, Rost B, Skern-Mauritzen M (2022) Oceans and coastal ecosystems and their services. In: Pörtner HO, Roberts DC, Tignor M, PoloczanskaES, K. MintenbeckK, Alegría A, Craig M, Langsdorf S, Löschke S, Möller V, Okem A, Rama B (eds) Climate change 2022: impacts, adaptation and vulnerability. Contribution of Working Group II to the Sixth Assessment Report of the Intergovernmental Panel on Climate Change. Cambridge University Press, Cambridge, UK and New York, pp 379–550. https://doi.org/10.1017/9781009325844.005

Corrigan S, Brown AR, Tyler CR, Wilding C, Daniels C, Ashton IGC, Smale DA (2023) Development and diversity of epibiont assemblages on cultivated sugar kelp (*Saccharina latissima*) in relation to farming schedules and harvesting techniques. Life 13:209. https://doi.org/10.3390/life13010209

Corrigan S, Smale DA, Tyler C, Brown AR (2024) Quantification of finfish assemblages associated with mussel and seaweed farms in southwest UK provides evidence of potential benefits to fisheries. Aquacult Environ Interact 16:145–162. https://doi.org/10.3354/aei0047

Dolliver J, O'Connor NE (2022) Estimating growth, loss and potential carbon sequestration of farmed kelp: a case study of *Saccharina Latissima* at Strangford Lough, Northern Ireland. Appl Phycol 3:324–339

Dong J, Shi J, Zhao JC, Zhang C, Haiyan X (2020) Wave energy assessment in the Bohai Sea and the Yellow Sea based on a 40-year hindcast. Water 12:2087. https://doi.org/10.3390/w12082087

Eger AM, McHugh TA, Eddy N, Vergés A (eds) (2024) State of the world's kelp forests 2024 report. Kelp Forest Alliance, Sydney

Eger AM, Marzinelli EM, Beas-Luna R, Blain CO, Blamey LK et al (2023) The value of ecosystem services in global marine kelp forests. Nat Commun 14:1894. https://doi.org/10.1038/s41467-023-37385-0

Eum HS, Jeong WM, Chang YS, Oh SH, Park JJ (2020) Wave energy in Korean seas from 12-year wave hindcasting. J Mar Sci Eng 8:161. https://doi.org/10.3390/jmse8030161

Evans J, Hotchkiss S, Hurst D (2023) A strategic review of Irish macroalgae cultivation. Bord lascaigh Mhara (BIM), pp 1–182

Eyrolles P, Lesueur M, Le Bris H, Laidin C, Brunner L, Hugues A et al (2018) Définir un cadre pour le développement de l'AMTI sur l'Arc Atlantique: Lot n°6 - Rapport d'action N°2: exemples d'implantations de systèmes AMTI sur l'Arc Atlantique. Délivrable 6–2. Agrocampus Ouest, pp 1–30

Fang J, Zhang J, Xiao T, Huang D, Liu S (2016) Integrated multi-trophic aquaculture (IMTA) in Sanggou Bay, China. Aquac Environ Interact 8:201–205

Fang JH, Jiang ZJ, Lin F, Gao YP, Fang JG, Zhang P, Guo ZL, Du MR, Meng S (2020) Analysis on the advantages of standard kelp long line culture in Sanggou Bay. Progr Fish Sci 41:134–140

FAO (2009) Laminaria japonica. Cultured aquatic species fact sheets. https://www.fao.org/fishery/docs/DOCUMENT/aquaculture/CulturedSpecies/file/en/en_japanesekelp.htm. Accessed 12 Nov 2024

FAO (2024) FishStatJ - Software for fishery and aquaculture statistical time series. https://www.fao.org/fishery/en/statistics/software/fishstatj. Accessed 24 Oct 2024

Fernand F, Israel A, Skjermo J, Wichard T, Timmermans KR, Golberg A (2017) Offshore macroalgae biomass for bioenergy production: environmental aspects, technological achievements and challenges. Renew Sust Energ Rev 75:35–45. https://doi.org/10.1016/j.rser.2016.10.04

Fieler R, Greenacre M, Matsson S, Neves L, Forbord S, Hancke K (2021) Erosion dynamics of cultivated kelp, *Saccharina latissima*, and implications for environmental management and carbon sequestration. Front Mar Sci 8:632725. https://doi.org/10.3389/fmars.2021.632725

Flavin K, Flavin N, Flahive B (2013) Kelp farming manual: a guide to the processes, techniques, and equipment for farming kelp in New England waters. Ocean Approved, pp 1–123

Fredheim A, Lien E (2001) General analysis of long-line constructions used for the cultivation of blue mussels *(Mytilus edulis*). SINTEF Fish Aquac N-7465, pp 1–16

Gagnon M (2024a) Status of off-bottom mariculture in wave-exposed environments. Part 1. Global inventory of extractive species commercial farms in temperate waters. Front Aquac 3:1411749. https://doi.org/10.3389/faquc.2024.1411749

Gagnon M (2024b) Status of off-bottom mariculture in wave-exposed environments. Part 2. Comparative loading and motion of longline designs currently used in exposed commercial farms. Front Aquacult 3:1422173. https://doi.org/10.3389/faquc.2024.1422173

García-Poza S, Leandro A, Cotas C, Cotas J, Marques JC, Pereira L et al (2020) The evolution road of seaweed aquaculture: cultivation technologies and the industry 4.0. Int J Environ Res Public Health 17:6528. https://doi.org/10.3390/ijerph17186528

Gattuso JP, Gentili B, Duarte CM, Kleypas JA, Middelburg JJ, Antoine D (2006) Light availability in the coastal ocean: impact on the distribution of benthic photosynthetic organisms and their contribution to primary production. Biogeosciences 3:489–513. https://doi.org/10.5194/bg-3-489-2006

Gendron L, Tamigneaux L (2008) Expériences de culture de l'algue brune *Saccharina longicruris* en 2007: essais en bassin et en mer au large de Paspébiac et de Grande-Rivière (Québec). Can Tech Rep Fish Aquat Sci 2820:1–48

Godsiff S (2020). Autonomous variable buoyancy system for marine farming. MSc. Thesis, U. Canterbury, Christchuch, New Zealand, pp 1–75

Goseberg N, Chambers MD, Heasman K, Fredriksson D, Fredheim A, Schlurmann T (2017) Technological approaches to longline- and cage based aquaculture in open ocean environments. In: Buck B, Langan R (eds) Aquaculture perspective of multi-use sites in the open ocean: the untapped potential for marine resources in the Anthropocene. Springer, Cham, pp 71–95

Gregersen K (2020) Ocean Rainforest secures permit to cultivate seaweed through 2032. Ocean Rainforest Blog 3 August 2020. https://www.oceanrainforest.com/blog-en/. Accessed 10 Dec 2024

Gregersen O (2021) New cultivation license in The Faroe Islands. Ocean Rainforest Blog, 6 June 2021. https://www.oceanrainforest.com/blog-en/. Accessed 10 Dec 2024

Gregersen O, Bak UG, Marsmanrg E, Schepper J (2021) Mechanical harvest and seeding. MacroCascade Final Conference, 2-03-2021. https://www.macrocascade.eu/outcomes/reports/macrocascade-final-conference/. Accessed 10 Dec 2024

Gunnlaugsdottir H, Jensen S, Pordarson G (2013) Offshore aquaculture: development, building and testing of a deep water mooring system. Skýrsla Matís 04-13, pp 11–16

Hasegawa K, Suzuki S (2005) Work analysis of harvesting and salt-preserved wakame seaweed (*Undaria pinnatifida*) processing. Tech Rep Nat Res Inst Fish Eng 27:61–80

Hatch Innovation Services (HIS) (2024) Seaweed insights: a practical guide to global seaweed farming. https://seaweedinsights.com/. Accessed 22 Oct 2024

Heasman K, Scott N, Sclodnick T, Chambers M, Costa-Pierce BA, Dewhurst T et al (2024a) Variations of aquaculture structures, operations and maintenance with increasing ocean energy: trying to avoid evolution and aim for revolution. Front Aquac 3:1444186

Heasman K, Sclodnick T, Goseberg N, Scott N, Chambers M, Dewhurst T et al (2024b) Utilization of the site assessment index for aquaculture in exposed waters: biology, technology, and operation and maintenance. Front Aquac 3:1427168. https://doi.org/10.3389/faquc.2024.1444186

Hermans S (2023) 2023 seaweed state of the industry. Phyconomy Newsletter, 7 January 2023. https://phyconomy.net/articles/2022-seaweed-review/. Accessed 14 Oct 2024

Hermans S (2024a) Temperate seaweed markets, an overview (part 1: a look back). Phyconomy Newsletter 12 November 2024. https://phyconomy.substack.com/p/temperate-seaweed-markets-an-overview. Accessed 14 Oct 2024

Hermans S (2024b) A database of seaweed organizations. https://phyconomy.net/database/. Accessed 10 Oct 2024

Hu ZM, Shan TF, Zhang J, Zhang QS, Critchley AT, Choi HG et al (2021) Kelp aquaculture in China: a retrospective and future prospects. Rev Aquac 13:1324–1351. https://doi.org/10.1111/raq.12585

Hu ZM, Shan TF, Zhang QS, Liu FL, Jueterbock A, Wang G et al (2023) Kelp breeding in China: challenges and opportunities for solutions. Rev Aquac 16:855–871. https://doi.org/10.1111/raq.12871

Hwang EK, Park CS (2020) Seaweed cultivation and utilization of Korea. Algae 35:107–121

ICES (2023) Aquaculture overviews. Faroe ecoregion, pp 1–23

Ino-Rope (2024) Lignes de mouillage innovantes pour le parc d'aquaculture Algolesko. https://inorope.com/. Accessed 6 Nov 2024

International Council for the Exploration of the Sea (ICES) (2012) Report of the Working Group on Marine Shellfish Culture (WGMASC). CM2012/SSGHE:15, Copenhagen, pp 1–117

International Energy Agency (IEA) (2019) Offshore wind outlook 2019, pp 1–97

Ishinomaki Kahoku Cat's Eye (2024) Nori farming hit hard by storms and high waves in January and March: fishermen ask city for assistance with disposal costs. Article 4 July 2024 (in Japanese)

JACUMAR (2024) Produccion de acuicultura. Historico de produccion de engorde 2022. Accessed 4 Dec 2024

Jiang T, Hong Y, Lu L, Zhu Y, Chen Z, Yang M (2022) Design and experiment of a new mode of mechanized harvesting of raft aquacultured kelp. Aquac Eng 99:102289

Jin R, Ye Z, Chen S, Gu J, He J, Huang L, Christakos G, Agusti S, Duarte CM, Wu J (2023) Accurate mapping of seaweed farms with high-resolution imagery in China. Geocarto Int 38:2203114. https://doi.org/10.1080/10106049.2023.2203114

Kämpf J, Chapman P (2016) Upwelling systems of the world: a scientific journey to the most productive marine ecosystems. Springer, Berlin

Kato H, Tanji Y, Fujima K, Sighihara Y (2010) Study on measures against drifting of cultivation rafts by tsunami: report on the results of 2010 Chile earthquake tsunami. Jap Inst Fish Infrastruct Commun 21:111–120. (in Japanese)

Kawai T, Galanin D, Krupnova T, Yotsukura N (2015) Harvest and cultivation of *Saccharina japonica* in Northern Hokkaido, Japan, and South Sakhalin and Primorye, Russia: a review. Algal Res 8:155–163

Kawamata S (2001) Adaptive mechanical tolerance and dislodgement velocity of the kelp *Laminaria japonica* in wave induced water motion. Mar Ecol Prog Ser 211:89–104. https://doi.org/10.3354/meps211089

Kawashima S (1984) Kombu cu1tivation in Japan for human foodstuff. Jap J Phycol 32:379–394

Kawashima S (1997) Cultivation of the brown alga, *Laminaria* «kombu». In: Ohno M, Critchley, AT (eds) Seaweed cultivation and marine ranching. Kanagawa International Fisheries Training Centre. Japan International Cooperation Agency, pp 25–40

Kerrison PD, Stanley MS, Edwards MD, Black KD, Hughes AD (2015) The cultivation of European kelp for bioenergy: site and species selection. Biomass Bioenergy 80:229–242. https://doi.org/10.1016/j.biombioe.2015.04.035

Kerrison PD, Stanley MS, Black KD, Hughes AD (2017) Assessing the suitability of twelve polymer substrates for the cultivation of macroalgae *Laminaria digitata* and *Saccharina latissima* (Laminariales). Algal Res 22:127–134. https://doi.org/10.1016/j.algal.2016.10.001

Kholler D, Lavarde F, Chaigneau G (2022) Présentation et développement de l'algoculture en France. CGAAER Report 21125, pp 1–109

Kim KY (2024) Harnessing seaweed farming for climate mitigation in South Korea: evaluating carbon dioxide removal potential and future research directions. Algae 39:329–347

Kim T, Han S, Lee S, Kim H (2021) Impact of seawater temperature on Korean aquaculture under 362 representative concentration pathways (RCPs) scenarios. Aquaculture 542:736893. https://doi.org/10.1016/j.aquaculture.2021.736893

Kite-Powell H, Erick A, Augyte S, Bailey D, Decker J, Goudey C et al (2022) Estimating production cost for large-scale seaweed farms.

Appl Phycol 3:435–445. https://doi.org/10.1080/26388081.2022.2111271

Koehl MAR, Silk WK, Liang H, Mehadevan L (2008) How kelps produce blade shapes suited to different flow regimes: a new wrinkle. Integr Comp Biol 48:834–851

Komatsu T, Sasa S, Montani S, Nishimur O, Sakamaki T, Yoshimura C et al (2016) Coast hit by a huge tsunami on 11 March 2011. Satoumi approach for realizing sustainable coastal use in rias-type bay: a case of Shizugawa Bay in Sanriku. MECS 11—Sea Coasts XXVI, 22–27 August 2016, St Petersburg, Russia

Konkuret (2021) A maricultural disaster: the scallop business is paralyzed. Article 3 November 2021. (in Russian)

Krumhansl KA, Lee JM, Scheibling RE (2011) Grazing damage and encrustation by an invasive bryozoan reduce the ability of kelps to withstand breakage by waves. J Exp Mar Biol Ecol 407:12–18. https://doi.org/10.1016/j.jembe.2011.06.033

Li J, Bergman K, Thomas JBE, Gao Y, Grondahl F (2023a) Life cycle assessment of a large commercial kelp farm in Shandong, China. Sci Total Environ 903:166861. https://doi.org/10.1016/j.scitotenv.2023.166861

Li X, Su L, Li X, Li J, Xu Y, Chang R, Yu R, Yang D, Pang S (2023b) Comprehensive analysis of large-scale *Saccharina japonica* damage in the principal farming area of Rongcheng in Shandong Province from 2021 to 2022. J Agric Sci Technol 25:206–222

Lian Y, Shen S, Zheng J, Boamah S, Yim Solomon C (2023) A design and numerical study on new kelp culture facility. In: Proceedings of the ASME 2023 42nd International Conference on Ocean, Offshore and Arctic Engineering, Melbourne.

Lian Y, Boamah SO, Pan Z, Zheng J, Chen W, Ma G, Yim SC (2024) Engineering design and economic analysis of offshore seaweed farm. Front Mar Sci 11:1276552. https://doi.org/10.3389/fmars.2024.1276552

Liu X, Zhang X (2022) Impacts of high-density suspended aquaculture on water currents: observation and modeling. J Mar Sci Eng 10:1151. https://doi.org/10.3390/jmse10081151

Liu Y, Saitoh SI, Radiarta IN, Isada T, Hirawake T, Mizuta H (2013) Improvement of an aquaculture site-selection model for Japanese kelp (*Saccharina japonica*) in southern Hokkaido, Japan: an application for the impacts of climate events. ICES J Mar Sci 70:1460–1470

Liu FL, Liang ZR, Zhang PY, Wang WJ, Sun XT, Wang FJ et al (2019) Preliminary discussion on the development of *Saccharina japonica* offshore aquaculture in China. Progr Fish Sci 40:161–166. (in Chinese)

Liu X, Wang Z, Yang X, Liu Y, Liu B, Zhang J et al (2022a) Mapping China's offshore mariculture based on dense time-series optical and radar data. Int J Digit Earth 15:1326–1349. https://doi.org/10.1080/17538947.2022.2108923

Liu Y, Liang ZR, Yu WW, Chang LR, Lu LF, Xiao LY et al (2022b) Effects of cultivation area of different offshore distances on the agronomic traits and breeding rope aging of *Saccharina japonica*. Progr Fish Sci 43:156–164. https://doi.org/10.19663/j.issn2095-9869.20210317002

Liu X, Liu Y, Jiang Z, Cao L (2024a) Spatiotemporal variation of China's mariculture potential under climate change. Rev Fish Biol Fish 34:315–335. https://doi.org/10.1007/s11160-023-09814-2

Liu X, Wu H, Xin M, Chi W, Xu M, Song J, Sui H, Lv F, Zhan D (2024b) Comprehensive analysis of the integrated quality of cultivated kelp (*Saccharina japonica*) in three major producing provinces of China. Aquacult Rep 39:102484

Lobeto H, Menendez M, Losada IJ (2021) Future behavior of wind wave extremes due to climate change. Sci Rep 11:1–12. https://doi.org/10.1038/s41598-021-86524-4

Lobeto H, Semedo A, Lemos G, Dastgheib A, Menendez M, Ranasinghe R, Bidlot JR (2024) Global coastal wave storminess. Sci Rep 14:3726. https://doi.org/10.1038/s41598-024-51420-0

Lojek O, Goseberg N, Moe Fore H, Dewhurst T, Bölker T, Heasman K et al (2024) Hydrodynamic exposure—on the quest to deriving quantitative metrics for mariculture sites. Front Aquacult 3:1388280. https://doi.org/10.3389/faquc.2024.1388280

Lovatelli A, Manjarrez J, Soto D (eds) (2013) Expanding mariculture farther offshore. Technical, environmental, spatial and governance challenges. FAO Fish Aquacult Proc 24. FAO, pp 1–73

Lynett PJ, Borrero J, Son S, Wilson R, Miller K (2014) Assessment of the tsunami-induced current hazard. Geophys Res Letters 41:2048–2055. https://doi.org/10.1002/2013GL058680

Macias D, Guillen J, Duteil O, Garcia-Gorriz E, Ferreira-Cordeiro N, Miladinova S, Parn O, Piroddi C, Polimene L, Serpetti N, Stips A (2025) Assessing the potential for seaweed cultivation in EU seas through an integrated modelling approach. Aquaculture 594:741353

Mascorda Cabre L, Hosegood P, Attrill MJ, Bridger D, Sheehan EV (2021) Offshore longline mussel farms: a review of oceanographic and ecological interactions to inform future research needs, policy and management. Rev Aquac 13:1864–1887. https://doi.org/10.1111/raq.12549

Merinov (2019) Mécanisation des opérations de culture en filière de la laminaire. https://merinov.ca/projets/mecanisation-des-operations-de-culture-en-filiere-de-la-laminaire/. Accessed 18 Dec 2024

Merinov (2021) Amélioration et mécanisation du procédé de mise à l'eau des plantules d'algues https://merinov.ca/projets/amelioration-et-mecanisation-du-procede-de-mise-a-leau-des-plantules-dalgues/. Accessed 18 Dec 2024

Mizuta DD, Wikfors GH (2019) Depth selection and in-situ validation for offshore mussel aquaculture in Northeast United States federal waters. J Mar Sci Eng 7:293. https://doi.org/10.3390/jmse7090293

Moscicki Z, Swift MR, Dewhurst T, MacNicoll M, Chambers M, Tsukrov I et al (2024) Design, deployment, and operation of an experimental offshore seaweed cultivation structure. Aquac Eng 305:102413. https://doi.org/10.1016/j.aquaeng.2024.102413

Myagi Prefectural Fisheries Cooperative Association (MPFCA) (2024) Wakame cultivation methods and facilities https://www.jf-miyagi.com/omotenashi/html/zukan03b.html. Accessed 6 Nov 2024

Nanba N, Fujiwara T, Kuwano K, Ishikawa Y, Ogawa H, Kado R (2011) Effect of water flow velocity on growth and morphology of cultured *Undaria pinnatifida* sporophytes (Laminariales, Phaeophyceae) in Okirai Bay on the Sanriku coast, Northeast Japan. J Appl Phycol 23:1023–1030

Natsuike M, Nishida Y, Kanamori M, Sato M, Honke K (2022) Field observations of the physical environment and behavior of culture cages during juvenile Yesso scallop (*Mizuhopecten yessoensis*) culture in Funka Bay, Hokkaido. Jap Sci Rep Hokkaido Fish Res Inst 102:23–30. (in Japanese)

Navarrete IA, Kim DY, Wilcox C, Reed DC, Ginsburg DW, Dutton JM et al (2021) Effects of depth-cycling on nutrient uptake and biomass production in the giant kelp *Macrocystis pyrifera*. Renew Sust Energ Rev 141:110747. https://doi.org/10.1016/j.rser.2021.110747

North Sea Farmers (NSF) (2022) Successful seaweed harvest operation in the North Sea. https://www.northseafarmers.org/news/220627wwharvest; accessed 12 December 2024

NSF (2023) Optimizing large-scale sowing with the project ZeewierSeeder. https://www.northseafarmers.org/news/20231221-zeewierseeder. Accessed 4 Dec 2024

Olson WM (1987) Seaweed cultivation in Minamikayabe, Hokkaido, Japan: potential for similar mariculture in southeastern Alaska. Alaska Sea Grant College Program, U Alaska. Mar Adv Bull 27:1–23

OOS International (2022) OOS Group introduces the Semi-submersible Mussel Farm (SMF). https://www.oosinternational.com/oos-group-introduces-the-semi-submersible-mussel-farm-smf/. Accessed 5 Dec 2024

Oshino A (2010) Macroscopic relationship between scallop (*Patinopecten yessoensis*) on-growing and buoyancy caused by wave action in Miyagi coast. Myagi Pref Rep Fish Sci 10:45–62

Paine ER, Boyd PW, Strzepek RF, Ellwood M, Brewer EA, Diaz-Pulido G et al (2023) Iron limitation of kelp growth may prevent ocean afforestation. Commun Biol 6:607. https://doi.org/10.1038/s42003-023-04962-4

Park SK (2015) North Korean fisheries status and cooperation through international organizations. J Fish Business Admin 46:83–101

Perez R, Kaas R, Campello F, Arbault S, Barbaroux O (1992) La culture des algues marines dans le monde. Institut Francais de Recherche pour l'Exploitation de la Mer.(IFREMER) Plouzané, France.

Peteiro C, Sánchez N, Martínez B (2016) Mariculture of the Asian kelp *Undaria pinnatifida* and the native kelp *Saccharina lattisima* along the Atlantic coast of southern Europe: an overview. Algal Res 15:9–23. https://doi.org/10.1016/j.algal.2016.01.012

Reid JA, Mooney WD (2022) Tsunami occurrence 1900–2020: a global review, with examples from Indonesia. Pure Appl Geophys 180:1549–1571. https://doi.org/10.1007/s00024-022-03057-1

Robidoux J, Good M (2023) State of the States: status of US seaweed aquaculture. 2023 National Seaweed Symposium. https://seaweed-hub.extension.uconn.edu/. Accessed 4 Oct 2024

Rondeau MH, Bossé-Côte F (2021) Salaweg: an update on kelp farming and processing. APC 22nd Annual Fis Conf, January 2021

Ross FWR, Malerba ME, Macreadie PI (2025) Global potential for seaweed aquaculture on existing offshore infrastructure. Heliyon 11:e41248

Rusu E, Rusu L (2021) An evaluation of the wave energy resources in the proximity of the wind farms operating in the North Sea. Energy Rep 7:19–27. https://doi.org/10.1016/j.egyr.2021.05.058

Saether S, Diehl N, Monteiro C, Niedzwiedz S, Burgunter-Delamare B, Bischof K et al (2024) The sugar kelp *Saccharina latissima*. II: recent advances in farming and applications. J Appl Phycol 36:1953–1985. https://doi.org/10.1007/s10811-024-03213-1

Saijo K (2024) Research on seaweed farming methods to increase carbon dioxide absorption. MSc thesis Tokyo U MMar Sci Technol Grad School, pp 1–43

Sato Y, Fujiwara T, Endo H (2023) Density regulation of aquaculture production and its effects on commercial profit and quality as food in the cosmopolitan edible seaweed *Undaria pinnatifida*. Front Mar Sci 10:1085054. https://doi.org/10.3389/fmars.2023.1085054

Seawiser (2024) Precision seeding of seaweed with the Seaseeder. https://www.seawiser.com/technology/. Accessed 10 Dec 2024

Skjermo J, Broch OJ, Endresen PC, Forbord S, Lona E (2020) Analysis of growth of cultivated macroalgae at an exposed and a sheltered site in Møre og Romsdal - Tareal 2. SINTEF Rep 2020:01053 (in Norwegian)

Solvang T, Bale ES, Broch, O.J, Handå A, Alver MO (2021) Automation concepts for industrial-scale production of seaweed. Front Mar Sci 8:613093. doi: https://doi.org/10.3389/fmars.2021.613093

Stekoll MS, Peeples TN, Raymond AET (2021) Mariculture research of *Macrocystis pyrifera* and *Saccharina latissima* in Southeast Alaska. J World Aquacult Soc 52:1031–1046. https://doi.org/10.1111/jwas.12765

Stekoll M, Pryor A, Meyer A, Kite-Powell HL, Bailey D, Barbery K, Goudey CA, Lindell S (2024) Optimizing seaweed biomass production-a two kelp solution. J Appl Phycol 36:2757–2767. https://doi.org/10.1007/s10811-024-03296-w

Steneck RS, Graham MH, Bourque BJ, Corbett D, Erlandson JM, Estes JA, Tegner MJ (2002) Kelp forest ecosystems: biodiversity, stability, resilience and future. Environ Conserv 29:436–459. https://doi.org/10.1017/S0376892902000322

St-Gelais AT, Fredriksson DW, Dewhurst T, Miller-Hope ZS, Costa-Pierce BA, Johndrow K (2022) Engineering a low-cost kelp aquaculture system for community-scale seaweed farming at nearshore exposed sites via user-focused design process. Front Sustain Food Sys 6:848035. https://doi.org/10.3389/fsufs.2022.848035

Strong-Wright J, Taylor JR (2022) Modeling the growth potential of the kelp *Saccharina latissima* in the North Atlantic. Front Mar Sci 8:793977. https://doi.org/10.3389/fmars.2021.793977

Sun Q, Liu H, Shang W, Yu L, Jiang X, Cai B, Chang L, Xiao L (2020) GIS-based assessment for culture suitability of kelp (*Saccharina japonica*) in Sanggou Bay and surrounding waters. Progr Fish Sci 41:41–50

Suppasri A, Fukui K, Yamashita K, Leelawat N, Hiroyuki O, Imamura F (2018) Developing fragility functions for aquaculture rafts and eelgrass in the case of the 2011 Great East Japan tsunami. Nat Hazards Earth Syst Sci 18:145–155. https://doi.org/10.5194/nhess-18-145-2018

Tamigneaux E, Licois A, Hersant G (2012) An overview of applied research on kelp aquaculture in Québec, Canada. The Seaweed Scene, University of Maine, 30 August 2012.

Télégramme (Le) (2014) Lesconil. Première récolte pour Algolesko 7 May 2014. https://www.letelegramme.fr/finistere/pont-l-abbe-29120/spanlesconilspan-premiere-recolte-pour-algolesko-2128516.php. Accessed 10 Jan 2025

Terashima H (2024) Ishinomaki's specialty, Jusanhama wakame seaweed, has been reduced to one-third of its original catch due to abnormally high sea temperatures and heavy storms. Article Tohoku360, 26 April 2024 (in Japanese)

Tullberg RM, Nguyen H, Wang CM (2022) Review of the status and developments in seaweed farming infrastructure. J Mar Sci Eng 10:1447. https://doi.org/10.3390/jmse10101447

van den Burg SWK, van Duijn AP, Bartelings H, van Krimpen MM, Poelman M (2016) The economic feasibility of seaweed production in the North Sea. Aquacult Econ Manag 20:235–252. https://doi.org/10.1080/13657305.2016.1177859

van Oort PAJ, Verhagen A, van der Werf K (2023) Can seaweeds feed the world? Modelling world offshore seaweed production potential. Ecol Model 484:110486

Vazquez Calderon F, Guillén Garcia J, Araujo R, Sanchez Lopez J (2022) Algae producing industry. European Commission, Joint Research Centre (JRC). http://data.europa.eu/89h/fa59f544-bf77-4812-8869-f34d9b096638. Accessed 24 Oct 2024

Veenhof RJ, Burrows MT, Hughes AD, Michalek K, Ross ME, Thomson AI, Fedenko J, Stanley MS (2024) Sustainable seaweed aquaculture and climate change in the North Atlantic: challenges and opportunities. Front Mar Sci 11:1483330. https://doi.org/10.3389/fmars.2024.1483330

Ventilla RF (1982) The scallop industry in Japan. Adv Mar Biol 20:309–382. https://doi.org/10.1016/S0065-2881(08)60142-X

Walker C, Corrigan S, Daniels C, Wilding C, Woodward EMS, Widdicombe CE, Smale DA, Ashton IGC, Brown AR (2023) Field assessment of the potential for small scale co-cultivation of seaweed and shellfish to regulate nutrients and plankton dynamics. Aquacult Rep 33:101789

Wang J, Yang X, Wang Z, Ge D, Kang J (2022) Monitoring marine aquaculture and implications for marine spatial planning: an example from Shandong Province, China. Remote Sens 14:732. https://doi.org/10.3390/rs14030732

Watanabe S, Sakami T (2021) Problems and challenges of aquaculture in Japan. In: Proceedings of the International Workshop on the Promotion of Sustainable Aquaculture, Aquatic Animal Health, and Resource Enhancement in Southeast Asia, Aquaculture Department, Southeast Asian Fisheries Development Center. Iloilo City, Philippines, 25–27 June 2019, pp 64–69

Webb A, Waseda T, Kiyomatsu K (2020) A high-resolution, long-term wave resource assessment of Japan with wave-current effects. Renew Energy 161:1341–1358. https://doi.org/10.1016/j.renene.2020.05.030

Whiting JM, Wang T, Yang Z, Huesemann MH, Wolfram PJ, Mumford TF et al (2020) Simulating the trajectory and biomass growth of free-floating macroalgal cultivation platforms along the US west coast. J Mar Sci Eng 8:938. https://doi.org/10.3390/jmse8110938

Wilding C, Tillin H, Corriga SE, Stuart E, Ashton IA, Felstead P, Lubelski A, Burrows M, Smale D (2021) Seaweed aquaculture and mechanical harvesting: an evidence review to support sustainable management. Natural England Res Rep NECR378, pp 1–109

Wilding CM, Smith CE, Daniels CL, Knoop J, Smale DA (2024) The influence of seeding method and water depth on the morphology and biomass yield of farmed sugar kelp (*Saccharina latissima*) at a small-scale cultivation site in the northeast Atlantic. J Appl Phycol 37. https://doi.org/10.1007/s10811-024-03394-9

Woods Hole Oceanographic Institution (WHOI) (2022) Rapid seeding spool. https://whoi-technologytransfer.technologypublisher.com/tech/Rapid_Seeding_Spool. Accessed 12 Dec 2024

World Bank (2023) Global seaweed: new and emerging markets report, 2023. Washington, DC

Wu Z, Xiong M, Shi Y, Song Y, Dai Y, Zhang S (2024) Characteristics and distribution model of wind and wave in Shengsi mussel culture area under the impacts of tropical cyclones. J Water Climate Change 15:518. https://doi.org/10.2166/wcc.2024.448

Yang X, Ramezani R, Utne IB, Mosleh A, Lader PF (2020) Operational limits for aquaculture operations from a risk and safety perspective. Reliab Eng Syst Saf 20:107208. https://doi.org/10.1016/j.ress.2020.107208

Yokoyama T (2021) Adaptive capacity in fishery restoration process at tsunami-affected areas: assets, flexibility, and social organization. In: Ikeguchi A, Yokoyama T, Sakita S (eds) Adaptive fisheries governance in changing coastal regions in Japan. Springer, Berlin/Heidelberg, pp 109–130

Yoshikawa T, Takeuchi I, Furuya K (2001) Active erosion of *Undaria pinnatifida* suringar (Laminariales, Phaeophyceae) mass-cultured in Otsuchi Bay in northeastern Japan. J Exp Mar Biol Ecol 266:51–65. https://doi.org/10.1016/S0022-0981(01)00346-X

Yotsukura N (2010) Production of kelp in Japan: various natural resources and the established aquaculture technique. In: Seaweeds for Human Consumption, Bioactive Compounds, and Combating of Diseases: An international interdisciplinary symposium, Carlsberg Academy, Copenhagen, August 26–27, 2010. http://www.tangbog.dk/wp-content/uploads/youtsukura-2.pdf. Accessed 5 Oct 2024

Zhang J, Fang J, Wang W, Du M, Gao Y, Zhang M (2012) Growth and loss of mariculture kelp *Saccharina japonica* in Sungo Bay, China. J Appl Phycol 24:1209–1216. https://doi.org/10.1007/s10811-011-9762-4

Zhang Y, Chang Z, Zheng Z, Yang J (2017) Harvesting machine for kelp culture in floating raft. Aquac Eng 78:173–179. https://doi.org/10.1016/j.aquaeng.2017.07.005

Zheng Y, Jin R, Zhang X, Wang Q, Wu J (2019) The considerable environmental benefits of seaweed aquaculture in China. Stoch Environ Res Risk Assess 33:1203–1221

Zhu L (2020) Wave attenuation capacity of suspended aquaculture structures with sugar kelp and mussels. PhD Thesis, University of Maine. https://digitalcommons.library.umaine.edu/etd/3222. Accessed 5 Nov 2023

Zhu Y, Hong Y, Jiang T, Yang M, Lu L, Yu Y (2023) Design and test of an efficient automatic clip seedling system for raft aquaculture kelp. J Mar Sci Eng 11:2301

Marc Gagnon, MSc. oceanography, He holds a BSc. in biology and a MSc. in oceanography from Université Laval, Québec, Canada. He is co-founder and was from 1987 to 2021 the general manager of Biorex Inc., a marine environmental consulting firm. He has been the principal investigator for several R&D projects in fisheries and aquaculture funded by governments and industry in Eastern Canada. In the former field, he conducted experimental trials on novel fishing gear including traps/pots, longlines, dredges and trawls and inventories of underutilized fisheries resources. In the field of aquaculture, his work focused on the characterization, hydrodynamics and modeling of at-sea grow-out structures and feasibility studies of various types of finfish, crustacean and mollusc farming.

Éric Tamigneaux, M.Sc. oceanology, Ph.D. biology, After studying biology at Université Libre de Bruxelles and a master's degree in oceanology at Université de Liège in Belgium, the author obtained a Ph.D. in biology from Université Laval in Québec. Since 1999, he has been working at Cégep de la Gaspésie et des Îles, where he combines teaching in the Techniques d'aquaculture program at École des pêches et de l'aquaculture du Québec, and applied research with collaborators from Ressources Aquatiques Québec, Institut sur la Nutrition et les Aliments and Centre collégial de transfert de technologie des pêches. Initiator of the Centre d'étude et de valorisation des algues marines du Québec in partnership with Université Laval, he then held an industrial research chair from Natural Science and Engineering Council of Canada on marine macroalgae. As senior project manager, he worked closely with aquaculture companies to adapt seaweed cultivation and processing techniques to the regulatory and climatic context of Eastern Canada, resulting in the creation of the first collegial training program on macroalgal culture and the production of several guides and teaching tools on the subject. In 2003, the author received an honorary mention from the Association québécoise de pédagogie collégiale and, in 2018, he was awarded the ACFAS Denise-Barbeau prize for college research and the Fonds de recherche québécois nature et technologies prize for excellence. Since 2020, he has been diversifying his professional experience by participating in rural oyster farming projects in Senegal and working on hydroponic cultivation of saltwort from Eastern Canada.

Isabelle Gendron-Lemieux, During her bachelor's degree in marine biology (Halifax, Nova Scotia) and her master's degree in environmental management (Sherbrooke, Quebec), Isabelle Gendron-Lemieux developed a keen interest in aquaculture. After completing her training, she moved to Gaspesie (Quebec) to work as a project manager at the Fisheries and Aquaculture college research centre (Merinov). After one or two years working on various aquaculture projects, she concentrated her research on seaweed cultivation techniques. She worked on optimizing and scaling up cultivation for sugar kelp and developing techniques for other seaweed in Quebec. Isabelle is currently a teacher-researcher in the aquaculture program at the École des pêches et de l'aquaculture du Québec in Gaspesie. Her research has recently been focused on the culture of the Atlantic nori species *Wildemania amplissima*. She also participates in projects aiming at using macro and microalgae for bioremediation purposes.

Temperate Seaweed Farms May Be Better at Reducing Emissions than Removing Carbon

12

Albert Pessarrodona, Thomas Wernberg, and Karen Filbee-Dexter

Abstract

Seaweed farming has garnered significant attention as a rapidly expanding sector with the potential to contribute to climate change mitigation, food security, and environmental sustainability. This chapter critically examines the potential of temperate seaweed farming as a solution to mitigate climate change, focusing on its capacity to reduce greenhouse gas emissions and facilitate atmospheric carbon dioxide removal (CDR). We synthesize recent evidence on carbon fluxes within seaweed farms, including emissions generated during farming operations at sea and the carbon sequestration achieved through biomass burial under farms. While seaweed farming is often considered within the ocean CDR portfolio, our analysis highlights that the net climate impacts of seaweed farming are not always positive. Emissions from farm operations can often exceed the carbon removed during farming at sea, especially in temperate seaweed farms where energy and material inputs are higher. Furthermore, current methodologies for assessing CDR capacity of seaweed farms are limited, and the contribution of seaweed farming to climate change mitigation remains small relative to protecting or restoring natural coastal ecosystems like mangroves and seagrasses. Still, it is evident that seaweed farming is one of the most sustainable production systems around, with modest carbon emissions and low environmental impacts compared to other production systems. This suggests that, although seaweed farming may not always be a carbon sink, it remains a promising avenue for climate change mitigation through net emission reductions. This net benefit will further increase where products are replaced by low-carbon seaweed alternatives and supply chains are decarbonized.

Keywords

Carbon sequestration · Carbon capture · Blue carbon · Macroalgae · kelp · Aquaculture · Mariculture · CO_2 removal

12.1 Introduction

Seaweed farming is one of the fastest-growing production sectors worldwide, being expected to contribute significantly toward future food and energy security while providing a range of socioeconomic benefits (Ritschard 1992; Duarte et al. 2022; Spillias et al. 2023a). Due to its low requirements for energy, water, and fertilizer, seaweed farming can sustainably generate products with low carbon footprints for use in a wide variety of applications (e.g., seafood, hydrocolloids, biofuel, fertilizer, livestock feed, cosmetics, and pharmaceuticals) (Buschmann et al. 2017). Seaweed farming can also improve social well-being, particularly for women, who often take care of the farms (Periyasamy et al. 2014; Msuya and Hurtado 2017), and produce a range of environmental benefits, such as improving water quality (He et al. 2008; Kim et al. 2015; Xiao et al. 2017). The growing appreciation for seaweed and their benefits has galvanized phycological research during the last decade and opened a new era of phycological appreciation: the "Phycocene." While such enthusiasm is very much welcome and often well-founded, hype and overpromise of seaweed benefits could lead to unsustainable outcomes and negative impacts on developing

A. Pessarrodona (✉)
UWA Oceans Institute and School of Biological Sciences, University of Western Australia, Crawley, WA, Australia

Conservation International, Arlington, VA, USA

International Blue Carbon Institute, Singapore, Singapore
e-mail: albert.pessarrodona@uwa.edu.au

T. Wernberg · K. Filbee-Dexter
UWA Oceans Institute and School of Biological Sciences, University of Western Australia, Crawley, WA, Australia

Institute of Marine Research, Flødevigen Research Station, His, Norway

A. T. Critchley et al. (eds.), *Advances in Temperate Phyconomy: Algal Harvest and Cultivation in Globally Distributed Temperate Waters*, Developments in Applied Phycology 15, https://doi.org/10.1007/978-3-032-02955-3_12

coastal economies (Boettcher et al. 2021) and ecosystems. An empirical grounding of the benefits and risks of seaweed farming is thus critical to ensure farming is scaled responsibly.

Climate change mitigation is one of the commonly attributed benefits of seaweeds and seaweed farming (Chung et al. 2013; Liu et al. 2015; Zhang et al. 2017; Li et al. 2022; Pessarrodona et al. 2023). In principle, seaweed farms can contribute to climate change mitigation by (i) contributing to atmospheric carbon dioxide removal (CDR) or (ii) reducing human greenhouse gas (GHG) emissions. During growth, seaweeds take up dissolved CO_2 from the surrounding seawater and convert it to organic carbon, a portion of which can be exported and stored in long-term carbon sinks, leading to a removal of atmospheric carbon (Pessarrodona et al. 2024; Filbee-Dexter et al. 2024). In order to contribute to GHG emission reductions, existing products (e.g., fuel, plastics, and food) must be substituted for seaweed-based alternatives whose emissions during the farming process and the value chain are lower than those of the substituted product—and ideally as close to zero as possible (Fig. 12.1). In the case of CDR, the net emission balance must be negative, i.e., the emissions generated throughout the value chain must be exceeded by the amount of CO_2 or GHGs removed from the atmosphere during the farming process (Fig. 12.1).

Seaweed farming is often highlighted for its climate mitigation benefits (United Nations Global Compact 2021; Duarte et al. 2022; Hurd et al. 2022). However, quantitative evidence on its ability to remove carbon or reduce emissions remains scarce and scattered through the literature. This chapter synthesizes recent findings on the carbon removal capacity of seaweed farms and provides a comprehensive review of their emissions, offering a more nuanced perspective on the potential for seaweed farming to address climate change.

12.2 Seaweed Farming and CDR

A comprehensive understanding of the CDR capacity of a seaweed farm requires quantifying both the atmospheric removal of carbon as well as the storage of this carbon away from the atmosphere in any of several possible long-term reservoirs (Fig. 12.1; see also Pessarrodona et al. 2024 for a review of potential reservoirs and both of these processes). Yet, carbon storage is challenging to verify in some reservoirs; not all of them are included in carbon accounting methodologies or frameworks, and only a few are likely to be important in the context of seaweed farming. For instance, long-term storage in dissolved organic carbon is, at present, challenging to reliably quantify (Paine et al. 2021; Pessarrodona et al. 2023) and is currently not included in climate change mitigation frameworks. Although the actual CDR potential (and therefore the real climate impact) of seaweed farming may be larger than accounted for, currently, most accountable CDR is likely to happen as a result of storage in long-life seaweed products (e.g., biochar, non-degradable biopolymers) or in sediments below seaweed farms (Pessarrodona et al. 2024). Overall, seaweed farming will only contribute to CDR if the activity results in greater GHG removal compared to what was occurring in the natural ecosystem where the farm is placed, i.e., if there is CDR additionality. That is a critical—yet often overlooked—distinction, because farms may modify natural CDR processes. For instance, farms anchored on top of seagrass beds may reduce seagrass productivity and carbon sequestration capacity due to shading and trampling by farmers (Unsworth et al. 2018; Kelly et al. 2020), or farms in some ocean areas may modify natural carbon removal by phytoplankton due to nutrient reallocation ("nutrient robbing"; Bach et al. 2021).

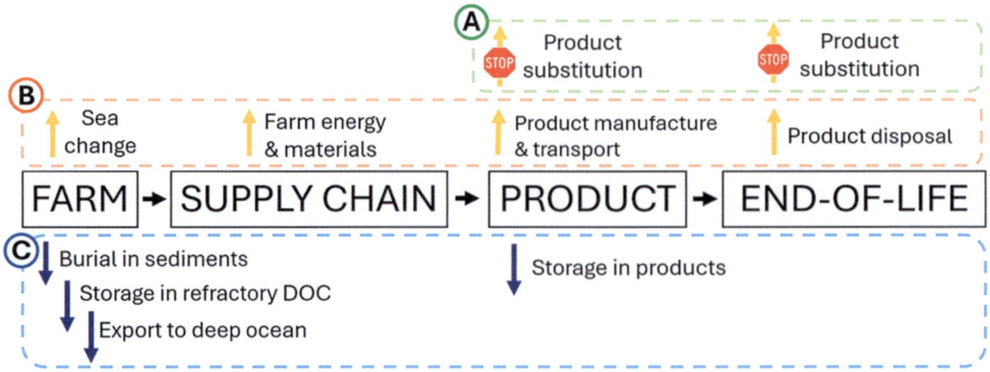

Fig. 12.1 Greenhouse gas emissions and removals associated during different phases of seaweed farming and their products. Within each phase, blue arrows indicate flows leading to removals, and orange arrows indicate flows leading to emissions. In order for seaweed farming to reduce GHG emissions, emissions of products substituted by seaweed (A) must be higher than the emissions incurred during the seaweed product's life cycle (B). In order to contribute to net carbon dioxide removal, removals of greenhouse gases (C) must be greater than emissions during the seaweed product's life-cycle (B), thus leading to net negative emissions. DOC = Dissolved Organic Carbon

12.3 Evidence for Carbon Sequestration: Burial

Evidence to date suggests that seaweed farming, and aquaculture in general, enhances the burial of organic carbon in the sediments below farms compared to unfarmed conditions (Pan et al. 2021; Wang et al. 2023; Yang et al. 2024; Duarte et al. 2025; but see Liu et al. 2015). Burial rates are relatively modest (median burial increase ± SD: 2.5 ± 62.2 gC m^{-2} yr^{-1}, mean: 25.2 gC m^{-2} yr^{-1}; 11 studies) and particularly small when compared to the burial capacity of natural coastal habitats, such as mangrove forests, seagrass meadows, or wild seaweed forests (Fig. 12.2a). This increase is often followed by a drawdown of atmospheric carbon (Han et al. 2021). At present, determining carbon burial and long-term sequestration in sediments is logistically challenging and expensive, as it requires coring and dating of underwater sediments. Building models that predict burial based on commonly available data such as total harvested biomass or sediment characteristics is a promising avenue to solve this issue. Yet, burial rates appear to be widely variable between farms (Liu et al. 2015; Pan et al. 2021) and have weak relationships with farm yields or sediment type (data not shown), making accurate predictions difficult. For instance, out of 20 farms examined by burial was negligible (e.g., due to net sediment erosion) or could not be determined (e.g., due to sediment mixing) in 40% of the farms. This suggests that most of the CDR potential may depend on the fate of the harvested crop, which represents the majority of the total carbon assimilated by seaweeds throughout the farming process (Fig. 12.2b).

12.4 Evidence for Carbon Sequestration: Long-term Products

Currently, the majority (ca. 80%) of harvested seaweed production is used for human consumption (Ferdouse et al. 2018; Hwang et al. 2019) and therefore does not contribute to CDR, as its carbon will eventually return to the atmosphere through respiration or combustion (Hurd et al. 2022). Leaving the biomass on farms or purposefully sinking it for CDR purposes appears to be a poor use of a high-quality product that yields low economic returns (Coleman et al. 2022; DeAngelo et al. 2023; Spillias et al. 2024), as well as having numerous potentially negative environmental risks in the deep sea and legal and regulatory challenges (Webb et al. 2021; Boyd et al. 2022; Ricart et al. 2022; Ross et al. 2022). Directing some of the production toward long-term carbon storage products is a much more promising way to enhance seaweed farming CDR potential. For instance, seaweed biochar—a fertilizer produced from the pyrolysis of seaweed biomass—contains stable and recalcitrant carbon that may be stored for hundreds of years when applied to the soil (Sun et al. 2022). Similarly, long-term storage in some types of bioplastics or their end products (compost) may constitute a carbon sink (Bishop et al. 2021).

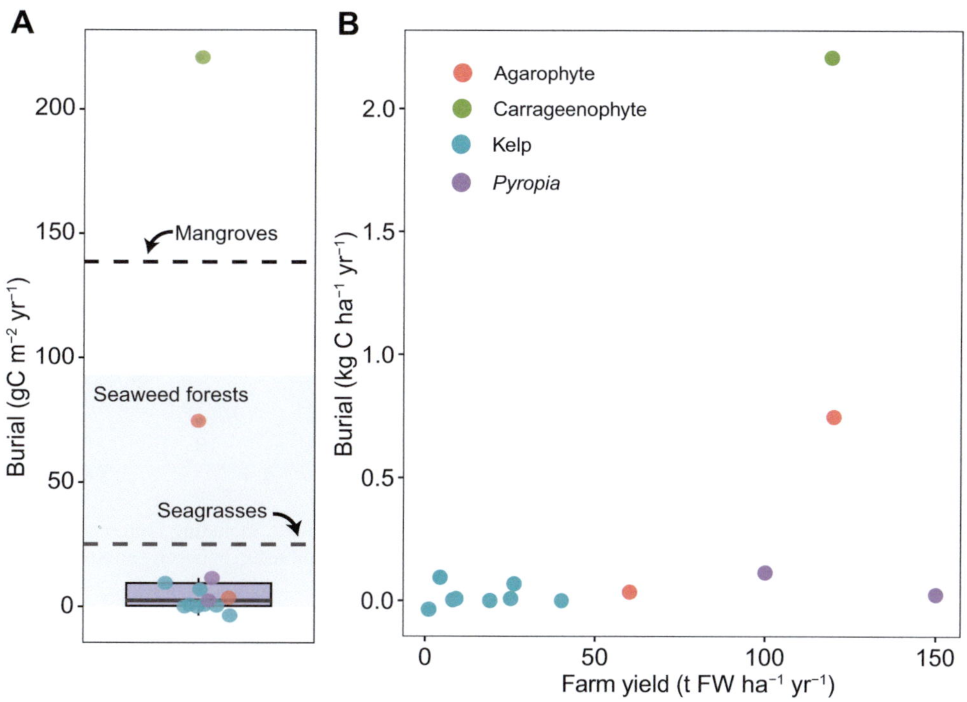

Fig. 12.2 (**a**) Estimates of increased organic carbon burial beneath seaweed farms, considering only excess burial relative unfarmed sites or a prefarm period. Dashed lines represent the average carbon burial rates for seagrass meadows (Arias-Ortiz et al. 2025) and mangrove forests (Breithaupt and Steinmuller 2022), while the gray box shows the range of carbon sequestration by wild seaweed forests (Filbee-Dexter et al. 2024). (**b**) Relationship between organic carbon burial and farm yield. Yields were converted from fresh weight to carbon weight using species- or group-specific wet-to-dry weight and dry-to-carbon weight ratios provided in Brey et al. (2010)

12.5 Seaweed Farming Emissions

Regardless of the long-term fate of seaweed carbon, it is becoming increasingly clear that the GHGs emitted across the seaweed value chain (i.e., from the seaweed hatchery to a seaweed product's end-of-life) play a key role in determining the overall net mitigation potential of seaweed farming. That is because the net GHG drawdown capacity of farms can be reduced or entirely offset by carbon emitted throughout the seaweed value chain (Coleman et al. 2022; Lian et al. 2023; Pessarrodona et al. 2024). Yet, emissions are often ignored in assessments of CDR capacity of seaweed farms (e.g., Froehlich et al. 2019; Duarte et al. 2025), probably due to the scarcity of data. Here, we synthesized emissions data from the literature (Pessarrodona et al. 2024; Thomas et al. 2024; Waqas et al. 2024), yielding a total of 27 entries. Overall, emissions of species cultivated in temperate areas (kelps, 15–287 kg CO_2 tonne seaweed^{-1}) appear to be greater than those typically cultivated in tropical areas (carrageenophytes, agarophytes, 3.9–5.3 kg CO_2 tonne seaweed^{-1}, Fig. 12.3a). This difference is most likely due to temperate kelp species being cultivated on floating farms that require large material (e.g., plastic lines and buoys, moorings) and energy (e.g., boat fuel) inputs, whereas cultivation in tropical areas predominantly uses natural products (e.g., bamboo) and can be maintained by hand from the shore (Ganesan et al. 2011; Ghosh et al. 2015; Ferdouse et al. 2018; Li et al. 2023). Additionally, temperate seedlings often require raising in hatcheries, which involves energy consumption for aquaria, whereas tropical species are often propagated vegetatively from existing seedstock (Ghosh et al. 2015; Ferdouse et al. 2018). There was no significant relationship between emissions and farm yield (Fig. 12.3b), with average emissions being 92.9 ± 67.5 (SD) kg CO_2 per tonne of cultivated seaweed.

Crucially, most of the emissions literature focused on kelp species (Fig. 12.4a). That is partly because most studies were conducted in Europe (Fig. 12.4b), where kelps are the most commonly cultivated species (Buschmann et al. 2017). However, Europe contributes to <1% of the global seaweed production, further evidencing previously identified disparities between the distribution of life cycle assessments and the world's leading producing regions (Li et al. 2023; Waqas et al. 2024). Asian seaweed farms ($n = 5$) had lower emissions per kg of product than farms in other regions. For instance, emissions in Asian farms ranged from 15.0 to 80.2 kg CO_2 per tonne of cultivated kelp compared to 22 to 287 kg CO_2 in European farms, suggesting that streamlined large-scale production in Asian farms led to lower electricity and fuel consumption (Li et al. 2023). Per tonne of fresh seaweed, cultivation on land facilities appears to have larger carbon footprints (3330 kg CO_2 t FW^{-1}, Dias et al. 2021; 9200 kg CO_2 t FW^{-1}, Nilsson and Martin 2022) compared to cultivation at sea (92.9 ± 67.5 kg CO_2 t FW^{-1}, mean ± SD).

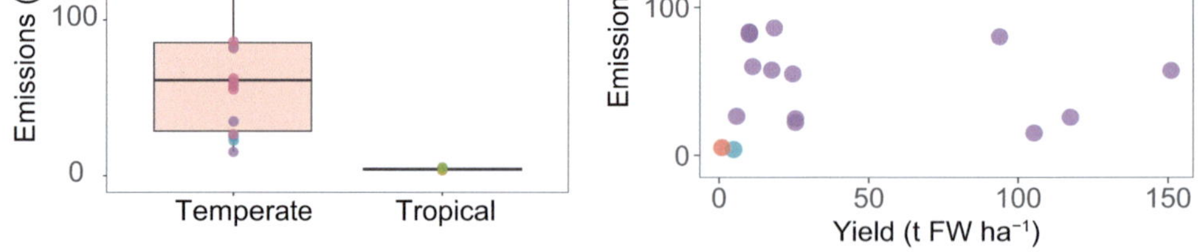

Fig. 12.3 CO_2 emissions incurred during the cultivation stage of seaweed, including both hatchery and at-sea operations. (**a**) Emissions from farms cultivating temperate ($n = 23$) and tropical ($n = 2$) species. (**b**) Relationship between emissions and farm yield in tonnes of fresh weight (FW) per hectare.

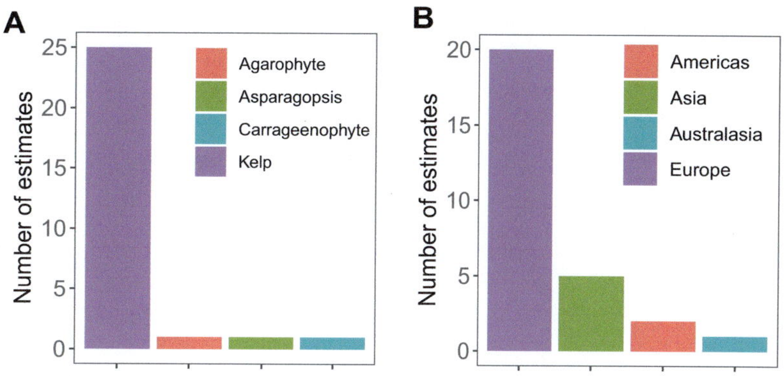

Fig. 12.4 (**a**) Number of emission estimates across different seaweed groups. (**b**) Number of emission estimates across different continents. Emission estimates account for regional variations in the electricity mix used at each location

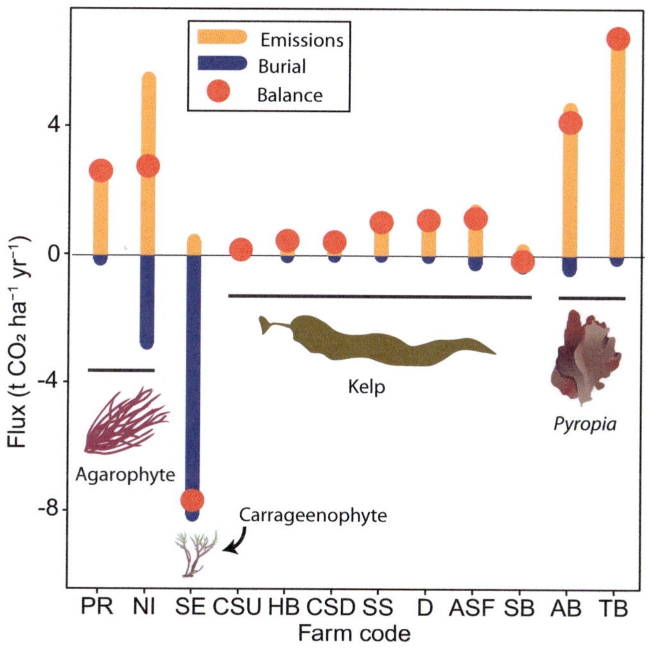

Fig. 12.5 Comparison of measured carbon burial (blue lines) and estimated emissions (orange lines) across various seaweed farms. The primary seaweed species cultivated at each farm is noted above the corresponding data. Red dots represent the net carbon balance between burial and emissions for each farm. See Supplementary Data for details of farm codes

12.6 Seaweed Farming Emissions Likely Offset Most Burial

Despite having clear gaps in spatial and taxonomic coverage, data on seaweed farm emissions are becoming increasingly available with the rise of seaweed product life cycle assessments, where the climate impact of processes across the seaweed value chain is calculated in a systematic manner. These assessments however often do not use empirically based carbon removal rates to calculate CDR, instead either relying on assumed removal rates or not considering carbon removal pathways altogether (Rose 2021). This often precludes understanding the actual net CDR capacity of seaweed farms and how it varies across farm types. We compiled available data on seaweed carbon burial rates and estimated their farm emissions using data from farms in the same region or farms elsewhere with the same cultivated species (when available), assuming their regional electricity sources and farm material requirements would be similar, and that emissions scale up linearly with yield (i.e., a fixed ratio of emissions/kg of cultivated seaweed, *sensu* Fig. 12.3b; see Supplementary Data for details). Emissions entirely offset carbon burial in all but two of the 11 farms where burial data were available, even when applying the lowest emissions reported in the literature for a given type of farm (Fig. 12.5). Considering burial is likely the primary pathway for CDR in most seaweed farms (Lian et al. 2023; Pessarrodona et al. 2024; Duarte et al. 2025), this suggests that the overall CDR capacity of most farms examined to date may be either very small or nonexistent. Other farm-derived carbon fluxes (e.g., carbon export to deep ocean, production of refractory dissolved organic carbon) not considered here may increase the total CDR capacity of seaweed farms, although those are typically more challenging to measure and account for in existing climate change mitigation frameworks (Pessarrodona et al. 2024).

We then explored how the amount of carbon burial necessary to offset farms' emissions varied with yield across different seaweed farm types, considering either the highest or lowest per-kg emissions recorded for a given farm type (hereafter low and high emissions scenarios, respectively) and their recorded maximum burial rates. We assumed that farm emissions would linearly increase as more biomass of seaweed was produced per hectare of farm (*sensu* Fig. 12.3b). As expected, the burial required to offset emissions differed between farm types, with temperate kelp farms being more emission intensive (and requiring higher burial rates) than the other two types (Fig. 12.6). Importantly, the carbon burial rates necessary to offset emissions from kelp farms was almost always higher than what has been recorded in the

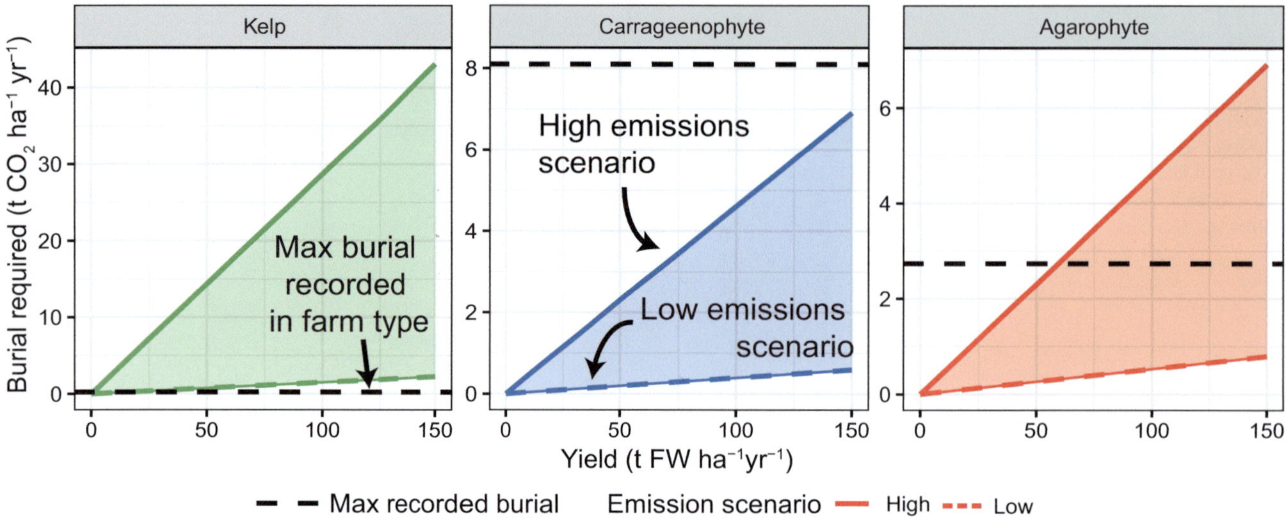

Fig. 12.6 Carbon burial required to offset either high (continuous line) or low (dashed line) emissions for a given farm yield. The black dashed line denotes actual carbon burial rates for that type of farm recorded in the literature. Note that total emissions (i.e., t CO_2 yr^{-1}) scale linearly with farm yield. Also note differences in scale of required burial rates across the three farm types for the same farm yields

literature (dashed black line), with farms with small yields being the only exception (Fig. 12.6). This suggests that kelp farming may have a more limited capacity to contribute to net CDR than other types of seaweed farms (Hasselström and Thomas 2022; Pessarrodona et al. 2024). In contrast, carbon burial rates required to offset emissions in tropical carrageenophyte farms were always lower than those recorded in the literature, regardless of the yield and emission scenario considered (Fig. 12.6). Finally, burial could only offset emissions in an agarophyte farm when per-kg emissions were low (e.g., in tropical systems) or when farm yields were lower than 50 t FW ha^{-1} yr^{-1} (Fig. 12.6). Further data on farm burial rates and emissions are crucial to reduce current uncertainties and refine our understanding of the CDR capacity of seaweeds—particularly since only two burial rates are available for agarophyte farms and one for carrageenophyte farms. However, our analysis reveals important trends: farming activities at sea will likely contribute to net CDR only under specific scenarios, such as high burial rates combined with low emissions. As such, farmed seaweeds are more likely to contribute to emission reductions than carbon removal per se.

12.7 Emissions Reductions from Seaweed Farming

Replacing products with high carbon footprints with low-carbon seaweed products is likely the primary way that seaweed farms can help reduce GHG emissions and mitigate climate change. It is quite clear that seaweed farming is one of the most sustainable production systems available at the moment, with metanalyses consistently finding modest carbon emissions and low environmental impacts compared to other systems (Gephart et al. 2021; Jones et al. 2022; Spillias et al. 2023a). For instance, emissions generated to produce a tonne of edible seaweed ranged from 299 to 2,403 kg CO_2, while emissions to produce a tonne of edible chicken or farmed tilapia ranged from 2,818 to 8,973 and 9,654 to 11,752 kg CO_2, respectively (Gephart et al. 2021). There are always exceptions (e.g., Norwegian seaweed protein had four times higher emissions than soy protein from Brazil, Koesling et al. 2021); so conducting robust life cycle assessments for different products and farm operations will still be the key. There is also clear potential to further decarbonize seaweed farming and associated supply chains by powering facilities and operations with renewable energy and using farming materials with low GHG footprints. These represent key avenues to cut emissions further and approximate net zero or net CDR states, regardless of the farm size, setup, or location (Coleman et al. 2022; Li et al. 2023). Although there can be issues with life cycle assessments and their comparability, it is still clear that seaweed farming could be an important approach to reduce net GHG emissions associated with food production. About 20–25% of global anthropogenic GHG emissions are associated with farming on land (16 Gt CO_2eq yr^{-1}; IPCC 2019), highlighting the critical need for innovation in food production to decarbonization.

12.8 Conclusions

Although it is clear that scaling seaweed farming has the potential to contribute towards a lower emission economy while displacing some of the negative environmental impacts of agriculture, seaweeds and seaweed farming are no panacea, and we must be realistic about their limitations. At locations where farms are currently placed, increases in carbon burial due to farming seem to be relatively small, which limits the ability of farms to be used as CDR tools, particularly when compared to interventions on natural ecosystems. Strategically placing farms over areas with high sedimentation might increase carbon burial, but still unlikely to be enough to offset the full emissions associated with the complete production life cycle. In that sense, decarbonization of the production chain will be the key. Additionally, the areas currently farmed are still very small (~4000 km^2), limiting seaweed farming's potential to make global transformative changes in the near- and medium-term future (Pessarrodona et al. 2024). For instance, the areas currently occupied by seaweed farming are two to three orders of magnitude lower than what would be required to reduce the environmental impacts of agriculture or to have climate-relevant CDR effect (Froehlich et al. 2019; Spillias et al. 2023b). Moreover, modeling studies often use highly optimistic targets or miss important parameters (e.g., emissions). For example, although significant reductions in agriculture's emissions, water, and fertilizer use are predicted if seaweed provides 10% of global diets (Spillias et al. 2023b), current maximum seaweed diet consumption is ca. 2% in Asian countries. Studies often assume farm production scales linearly with size, but production in many areas is still volatile due to cultivation, growth, and drying conditions being affected by changing ocean conditions and rainfall (Langford et al. 2022), all of which are poorly reflected in current models. The future of seaweed farming is promising, but to fully harness its benefits as farming expands to new geographies, thorough consideration of the social, economic, and ecological barriers that prevent their current development and evidence-based management will be essential.

Acknowledgments The authors were supported by Conservation International, The Norwegian Research Council (BlueArc project no. 334760) and The Australian Research Council (LP220200004, DE250100125, FT230100214).

References

Arias-Ortiz et al. (2025). https://www.aconf.org/conf_198615/abstract/1483.html

Bach LT, Tamsitt V, Gower J et al (2021) Testing the climate intervention potential of ocean afforestation using the Great Atlantic Sargassum belt. Nat Commun 12:2556. https://doi.org/10.1038/s41467-021-22837-2

Bishop G, Styles D, Lens PNL (2021) Environmental performance comparison of bioplastics and petrochemical plastics: a review of life cycle assessment (LCA) methodological decisions. Resour Conserv Recycl 168:105451. https://doi.org/10.1016/j.resconrec.2021.105451

Boettcher M, Brent K, Buck HJ et al (2021) Navigating potential hype and opportunity in governing marine carbon removal. Front Clim 3. https://doi.org/10.3389/fclim.2021.664456

Boyd PW, Bach LT, Hurd CL et al (2022) Potential negative effects of ocean afforestation on offshore ecosystems. Nat Ecol Evol. https://doi.org/10.1038/s41559-022-01722-1

Breithaupt JL, Steinmuller HE (2022) Refining the global estimate of mangrove carbon burial rates using sedimentary and geomorphic settings. Geophys Res Lett 49:e2022GL100177. https://doi.org/10.1029/2022GL100177

Brey T, Müller-Wiegmann C, Zittier ZMC, Hagen W (2010) Body composition in aquatic organisms – a global data bank of relationships between mass, elemental composition and energy content. J Sea Res 64:334–340. https://doi.org/10.1016/j.seares.2010.05.002

Buschmann AH, Camus C, Infante J et al (2017) Seaweed production: overview of the global state of exploitation, farming and emerging research activity. Eur J Phycol 52:391–406. https://doi.org/10.1080/09670262.2017.1365175

Chung IK, Oak JH, Lee JA et al (2013) Installing kelp forests/seaweed beds for mitigation and adaptation against global warming: Korean Project Overview. ICES J Mar Sci 70:1038–1044. https://doi.org/10.1093/icesjms/fss206

Coleman S, Dewhurst T, Fredriksson DW et al (2022) Quantifying baseline costs and cataloging potential optimization strategies for kelp aquaculture carbon dioxide removal. Front Mar Sci 9:966304. https://doi.org/10.3389/fmars.2022.966304

DeAngelo J, Saenz BT, Arzeno-Soltero IB et al (2023) Economic and biophysical limits to seaweed farming for climate change mitigation. Nat Plants 9:45–57. https://doi.org/10.1038/s41477-022-01305-9

Dias A, Raluy RG, Helmes R, van den Burg SWK (2021) GENIALG Deliverable 5.6. Report on final LCA. In: GENIALG. https://genialgproject.eu/results/deliverables/. Accessed 20 Sept 2024

Duarte CM, Bruhn A, Krause-Jensen D (2022) A seaweed aquaculture imperative to meet global sustainability targets. Nature Sustainability 5:185–193. https://doi.org/10.1038/s41893-021-00773-9

Duarte CM, Delgado-Huertas A, Marti E et al (2025) Carbon burial in sediments below seaweed farms matches that of Blue Carbon habitats. Nat Clim Chang 15:180–187. https://doi.org/10.1038/s41558-024-02238-1

Ferdouse F, Løvstad Holdt S, Smith R, et al (2018) The global status of seaweed production, trade and utilization

Filbee-Dexter K, Pessarrodona A, Pedersen MF et al (2024) Carbon export from seaweed forests to deep ocean sinks. Nat Geosci 1–8. https://doi.org/10.1038/s41561-024-01449-7

Froehlich HE, Afflerbach JC, Frazier M, Halpern BS (2019) Blue growth potential to mitigate climate change through seaweed offsetting report blue growth potential to mitigate climate change through seaweed offsetting. Curr Biol 1–7. https://doi.org/10.1016/j.cub.2019.07.041

Ganesan M, Sahu N, Eswaran K (2011) Raft culture of *Gracilaria edulis* in open sea along the south-eastern coast of India. Aquaculture 321:145–151. https://doi.org/10.1016/j.aquaculture.2011.08.040

Gephart JA, Henriksson PJG, Parker RWR et al (2021) Environmental performance of blue foods. Nature 597:360–365. https://doi.org/10.1038/s41586-021-03889-2

Ghosh A, Vijay Anand KG, Seth A (2015) Life cycle impact assessment of seaweed based biostimulant production from onshore cultivated Kappaphycus alvarezii (Doty) Doty ex Silva-is it environmentally sustainable? Algal Res 12:513–521. https://doi.org/10.1016/j.algal.2015.10.015

Han T, Shi R, Qi Z et al (2021) Impacts of large-scale aquaculture activities on the seawater carbonate system and air-sea CO2 flux in a subtropical mariculture bay, southern China. Aquac Environ Interact 13:199–210. https://doi.org/10.3354/AEI00400

Hasselström L, Thomas JBE (2022) A critical review of the life cycle climate impact in seaweed value chains to support carbon accounting and blue carbon financing. Clean Environ Syst 6. https://doi.org/10.1016/j.cesys.2022.100093

He P, Xu S, Zhang H et al (2008) Bioremediation efficiency in the removal of dissolved inorganic nutrients by the red seaweed, *Porphyra yezoensis*, cultivated in the open sea. Water Res 42:1281–1289. https://doi.org/10.1016/j.watres.2007.09.023

Hurd CL, Law CS, Bach LT et al (2022) Forensic carbon accounting: assessing the role of seaweeds for carbon sequestration. J Phycol 58:347–363. https://doi.org/10.1111/jpy.13249

Hwang EK, Yotsukura N, Pang SJ, Su L & Shan TF (2019) Seaweed breeding programs and progress in eastern Asian countries, Phycologia, 58:5, 484–495. https://doi.org/10.1080/00318884.2019.1639436

Jones AR, Alleway HK, Mcafee D, et al (2022) Climate-friendly seafood: the potential for emissions reduction and carbon capture in marine aquaculture. XX:1–21

Kelly ELA, Cannon AL, Smith JE (2020) Environmental impacts and implications of tropical carrageenophyte seaweed farming. Conserv Biol 34:326–337. https://doi.org/10.1111/cobi.13462

Kim JK, Kraemer GP, Yarish C (2015) Use of sugar kelp aquaculture in Long Island Sound and the Bronx River Estuary for nutrient extraction. Mar Ecol Prog Ser 531:155–166. https://doi.org/10.3354/meps11331

Koesling M, Kvadsheim NP, Halfdanarson J et al (2021) Environmental impacts of protein-production from farmed seaweed: comparison of possible scenarios in Norway. J Clean Prod 307:127301. https://doi.org/10.1016/j.jclepro.2021.127301

Langford A, Zhang J, Waldron S et al (2022) Price analysis of the Indonesian carrageenan seaweed industry. Aquaculture 550:737828. https://doi.org/10.1016/j.aquaculture.2021.737828

Li H, Zhang Z, Xiong T et al (2022) Carbon sequestration in the form of recalcitrant dissolved organic carbon in a seaweed (kelp) farming environment. Environ Sci Technol 56:9112–9122. https://doi.org/10.1021/acs.est.2c01535

Li J, Bergman K, Thomas JBE et al (2023) Life cycle assessment of a large commercial kelp farm in Shandong, China. Sci Total Environ 903. https://doi.org/10.1016/j.scitotenv.2023.166861

Lian Y, Wang R, Zheng J et al (2023) Carbon sequestration assessment and analysis in the whole life cycle of seaweed. Environ Res Lett 18:074013. https://doi.org/10.1088/1748-9326/acdae9

Liu S, Huang J, Yang Q et al (2015) Burial fluxes and source apportionment of carbon in culture areas of Sanggou Bay over the past 200 years. Acta Oceanol Sin 34:23–30. https://doi.org/10.1007/s13131-015-0724-6

Msuya FE, Hurtado AQ (2017) The role of women in seaweed aquaculture in the Western Indian Ocean and South-East Asia. Eur J Phycol 52:482–494. https://doi.org/10.1080/09670262.2017.1357084

Nilsson J, Martin M (2022) Exploratory environmental assessment of large-scale cultivation of seaweed used to reduce enteric methane emissions. Sustain Prod Consum 30:413–423. https://doi.org/10.1016/j.spc.2021.12.006

Paine ER, Schmid M, Boyd PW et al (2021) Rate and fate of dissolved organic carbon release by seaweeds: a missing link in the coastal ocean carbon cycle. J Phycol 57:1375–1391. https://doi.org/10.1111/jpy.13198

Pan Z, Tan YM, Gao QF et al (2021) A 120-year record of burial fluxes and source apportionment of sedimentary organic carbon in Alian Bay, China: implication for the influence of mariculture activities, and regional environment changes. Aquaculture 535:736421. https://doi.org/10.1016/j.aquaculture.2021.736421

Periyasamy C, Anantharaman P, Balasubramanian T (2014) Social upliftment of coastal fisher women through seaweed (*Kappaphycus alvarezii* (Doty) Doty) farming in Tamil Nadu, India. J Appl Phycol 26:775–781. https://doi.org/10.1007/s10811-013-0228-8

Pessarrodona A, Franco-santos RM, Wright LS et al (2023) Carbon sequestration and climate change mitigation using macroalgae: a state of knowledge review. Biol Rev. https://doi.org/10.1111/brv.12990

Pessarrodona A, Howard J, Pidgeon E et al (2024) Carbon removal and climate change mitigation by seaweed farming: a state of knowledge review. Sci Total Environ 918:170525. https://doi.org/10.1016/j.scitotenv.2024.170525

Ricart AM, Krause-Jensen D, Hancke K et al (2022) Sinking seaweed in the deep ocean for carbon neutrality is ahead of science and beyond the ethics. Environ Res Lett 17:081003. https://doi.org/10.1088/1748-9326/ac82ff

Ritschard RL (1992) Marine algae as a CO2 sink. Water Air Soil Pollut 64:289–303. https://doi.org/10.1007/BF00477107

Rose DJ (2021) Life cycle of carbon in macroalgae for various products. Pacific Northwest National Laboratory

Ross F, Tarbuck P, Macreadie PI (2022) Seaweed afforestation at large-scales exclusively for carbon sequestration: critical assessment of risks, viability and the state of knowledge. Front Mar Sci 9. https://doi.org/10.3389/fmars.2022.1015612

Spillias S, Kelly R, Cottrell RS et al (2023a) The empirical evidence for the social-ecological impacts of seaweed farming. PLOS Sustain Transformation 2:e0000042. https://doi.org/10.1371/journal.pstr.0000042

Spillias S, Valin H, Batka M et al (2023b) Reducing global land-use pressures with seaweed farming. Nat Sustain 6:380–390. https://doi.org/10.1038/s41893-022-01043-y

Spillias S, Cottrell RS, Layton C et al (2024) Having our kelp and eating it too: minimizing trade-offs from seaweed farming. J Clean Prod 448:141150. https://doi.org/10.1016/j.jclepro.2024.141150

Sun et al. (2022). https://doi.org/10.1016/j.biortech.2021.126258

Thomas J-BE, Ahlgren E, Hornborg S, Ziegler F (2024) Life cycle environmental impacts of kelp aquaculture through harmonized recalculation of inventory data. J Clean Prod 450:141987. https://doi.org/10.1016/j.jclepro.2024.141987

United Nations Global Compact (2021) Seaweed as a nature-based climate solution. Vision Statement

Unsworth RKF, Ambo-Rappe R, Jones BL et al (2018) Indonesia's globally significant seagrass meadows are under widespread threat. Sci Total Environ 634:279–286. https://doi.org/10.1016/j.scitotenv.2018.03.315

Wang Y, Yang W, Zhao X et al (2023) Changes in the carbon source and storage in a cultivation area of macro-algae in Southeast China. Mar Pollut Bull 188. https://doi.org/10.1016/j.marpolbul.2023.114680

Waqas MA, Hashemi F, Mogensen L, Knudsen MT (2024) Environmental performance of seaweed cultivation and use in different industries: a systematic review. Sustain Prod Consum 48:123–142. https://doi.org/10.1016/j.spc.2024.05.001

Webb RM, Silverman-Roati K, Gerrard MB (2021) Removing carbon dioxide through ocean alkalinity enhancement and seaweed cultivation: legal challenges and opportunities. Sabin Center for Climate Change Law, Columbia Law School The

Xiao X, Agusti S, Lin F et al (2017) Nutrient removal from Chinese coastal waters by large-scale seaweed aquaculture. Sci Rep 7:1–6. https://doi.org/10.1038/srep46613

Yang S, Yang Q, Song X et al (2024) Blue carbon in sediment from Sanggou Bay: composition, burial flux and its response to human activities. Front Mar Sci 10. https://doi.org/10.3389/fmars.2023.1342750

Zhang YY, Zhang JH, Liang YT et al (2017) Carbon sequestration processes and mechanisms in coastal mariculture environments in China. Sci China Earth Sci 60:2097–2107. https://doi.org/10.1007/s11430-017-9148-7

Dr. Albert Pessarrodona is an Australian Research Council DECRA Fellow with a deep interest in understanding how ecosystems work, and how stressors alter their functioning. He is particularly fascinated by underwater seaweed forests and his current research broadly focuses on how we can predict and adapt to climate-driven changes in their functioning. Recently he has become very interested in the role seaweeds play in the coastal carbon cycle at the global level, and whether seaweeds can provide solutions to climate change mitigation.

Professor Thomas Wernberg is an Australian Research Council Laureate Fellow and one of Australia's leading temperate marine ecologists. His research covers a broad spectrum of topics, but is largely focused on the ecological interactions in and around shallow sub-tidal habitats. His research has provided understanding of how coastal habitats might respond to stressors such as climate change, marine heatwaves, invasive species and eutrophication. Recently, he has also increased his focus on solutions to habitat decline through restoration and increasing our appreciation of the diversity of services our marine ecosystems provide, such as their potential role in climate mitigation.

Dr. Karen Filbee-Dexter is an Australian Research Council Future Fellow studying the drivers and patterns of long-term change in coastal ecosystems. Dr. Filbee-Dexter's expertise centers on climate-driven changes to kelp forests, including shifts to alternate ecosystems such as turf reefs and sea urchin barrens, and expansion along arctic coasts. Her recent work has focused on restoration approaches in kelp forests through the new Green Gravel tool. Her research also tackles interdisciplinary questions such as links between social and ecological systems during regime shifts, and ethical questions associated with emerging use of genetic restoration techniques.

13 The African Seaforest: A Socio-Cultural and Ecological Tapestry

Akshata Mehta

Abstract

The African Seaforest represents a vibrant intersection between ecology and culture in South Africa's coastal regions. While kelp has long been acknowledged for its environmental importance—such as providing habitat for marine life and offering coastal protection—recent scholarship highlights its broader socio-cultural significance. Drawing on the concept of phyconomy, this chapter investigates how local communities, artisanal harvesters, and emerging kelp-based industries are navigating changing ocean conditions and structural inequalities. Through ethnographic interviews, policy analysis, and case studies—ranging from small-scale foragers and women-led salting cooperatives to artisanal chefs experimenting with kelp-inspired cuisine—this chapter elucidates the complex social and ecological webs that kelp inhabits, arguing for an inclusive and equitable vision of kelp forest management.

Keywords

African Seaforest · Kelp · Relational value · Marine resource management · Cultural value · Decolonial governance

Kelps are large seaweeds in the order Laminariales, which form extensive underwater forests that are critical to temperate marine ecosystems across the globe (Krumhansl et al. 2016). Within South Africa, the African Seaforest, sometimes referred to as the Great African Seaforest (GASF), is increasingly recognized not only for its ecological contributions—biodiversity support, carbon sequestration, and coastal protection—but also for its profound cultural and relational significance (Prew et al. 2024). This duality of value, encompassing both ecological and socio-cultural dimensions, underscores the need for a holistic approach to kelp forest management.

Recent work suggests that kelp forests in South Africa may be expanding along sections of the Benguela Current Large Marine Ecosystem as climate-driven upwelling intensifies (Bolton and Blamey 2017). This shift highlights the importance of understanding not just where kelp forests are going, but also who has access to them and how these ecosystems are valued.

Thus far, the valuations around kelp forests usually take the form of market-based approaches. While these economic valuations can be useful, they often obscure the myriad of intangible connections people have with the African Seaforest. These elusive, non-material facets are better unearthed through plural valuation approaches that encompass the more ethereal contributions of nature (and in this regard, kelp forests). To contextualize the various dimensions of value expounded in this chapter, a brief understanding of how values are classified is necessary. A useful way to classify the different kinds of value assigned to ecosystems is by distinguishing intrinsic from instrumental values (Farber et al. 2002). Intrinsic value denotes the inherent worth of an ecosystem, existing independently of whether it provides benefits to humans (Chan et al. 2016; Pascual et al. 2017). This notion can be further separated into subjective and objective forms (Sandler 2012). In the subjective sense, humans confer worth on an entity because of what it *is*—as illustrated by personal objects or ceremonies that are valued for their symbolic meaning rather than for any utility. By contrast, objective intrinsic value refers to worth that is not created by humans but simply recognized by them. In this perspective, the ecosystem's value is seen as unconditional and based on qualities inherent to nature itself (Sandler 2012).

A. Mehta (✉)
Department of Biodiversity & Conservation Biology, University of the Western Cape, Bellville, South Africa

A. T. Critchley et al. (eds.), *Advances in Temperate Phyconomy: Algal Harvest and Cultivation in Globally Distributed Temperate Waters*, Developments in Applied Phycology 15, https://doi.org/10.1007/978-3-032-02955-3_13

Instrumental values, on the other hand, arise from an entity's capacity to produce, or help produce, desired outcomes (Chan et al. 2012). Such values are inherently conditional: the worth of an object depends on the extent to which it serves a specific purpose. For instance, a fishing line is highly valuable to a fisher who needs it for catching fish but loses that value if a more effective net is introduced (Sandler 2012). These instrumental values can take many forms and can shift over time. A third category now gaining attention is relational values, which stem from the relationships and responsibilities people hold toward nature, rather than from nature's inherent properties (Chan et al. 2016). At the individual level, this might manifest as a strong sense of identity linked to a particular environment (e.g., a fisher's bond with the ocean) or spiritual connections to specific natural features. At broader scales, relational values can foster social cohesion through communal stewardship practices, reinforcing cultural identity and moral bonds within groups (Sanborn and Jung 2021). In effect, relational values can intersect with both social and economic domains, highlighting how people's interactions with and obligations toward nature shape what they value.

Although much research focuses on kelp's instrumental roles—such as raw material for alginate and abalone feed (Rothman et al. 2020)—an emerging wave of scholarship and advocacy emphasizes the relational and cultural dimensions of kelp forests (Mehta et al. 2023; Pratson et al. 2023). An overreliance on market-based metrics alone can obscure the intangible, yet deeply felt, values that kelp holds for coastal communities (Hill et al. 2020). This is especially vital for the concept of phyconomy—the study of seaweed's societal and economic roles.

Using the African Seaforest as a case study, in this chapter, we explore how kelp forests are woven into the socio-cultural fabric of the Western Cape, shaped by colonial histories, identity formations, and everyday interactions. This is explored by drawing upon a series of ongoing interviews and storytelling initiatives around the kelp forests across South Africa. By drawing on environmental research, ethnographic interviews, and local narratives, we uncover a more inclusive view of kelp—one that transcends a purely ecological or economic lens and emphasizes the importance of historical context, community knowledge, and spiritual traditions.

13.1 Socio-Economic Impacts and Community Integration in the Kelp Value Chain

Building on the concept of phyconomy—the study of algal life in socio-economic contexts—phyconometrics aims to quantitatively assess kelp's market potential through its provision of ecosystem services. This approach can illuminate untapped opportunities in the "blue economy," such as kelp-based biofuels or carbon credits (Hynes et al. 2021). Yet, an overfocus on metrics can overlook power asymmetries and exclude cultural dimensions. By integrating participatory methods—interviews, community workshops, and iterative feedback—phyconometrics can reflect diverse knowledge systems, ensuring that valuation encompasses cultural ecosystem services as well as resources (Bennett et al. 2022).

This is especially pertinent in the case of the African Seaforest, where ongoing tensions in the Western Cape highlight the need for inclusive governance. South Africa's marine legislation, particularly the Marine Living Resources Act (MLRA), aimed to remedy historical injustices by incorporating small-scale fishers into the concession system (DFFE 2013). Yet, governance gaps persist. For instance, a 15-year concession validity can favor well-funded commercial actors while hamstringing community-based fishers who lack the capital and bureaucratic influence to maintain consistent kelp operations (Green and de Villiers 2018). This top-down approach perpetuates existing inequities—commercial operators easily absorb permit fees, legal stipulations, and market fluctuations, whereas small-scale harvesters face infrastructural and bureaucratic barriers (Rothman et al. 2020). In "Assessing Values and Perceptions of Kelp in the Western Cape Region of South Africa," Mehta notes that smaller artisanal harvesters remain on the fringes of the industry, navigating cumbersome application processes and lacking essential processing infrastructure (Mehta et al. 2023).

Addressing these structural challenges means co-creating policy by establishing local committees that include fishers, conservationists, indigenous groups, and government officials. Such multi-stakeholder dialogue can expand phyconometrics to include social justice indicators (Díaz et al. 2015). If ecological data on kelp biomass is paired with metrics on livelihoods, cultural continuity, and equity, then conservation strategies can be both ecologically robust and socially fair.

As such, research indicates widespread agreement on kelp's instrumental worth, but value dissonances appear when discussing issues of property rights, harvest permits, and tourism. The tourism sector, for example, may promote high-end scuba tours in pristine kelp forests, even as local fishers find themselves restricted by conservation regulations that limit customary use (Mehta et al. 2024). Understanding these conflicting perspectives is crucial for informed policymaking that balances biodiversity conservation with community livelihoods.

The socio-economic impacts and resultant (lack of) community integration are further assessed in the 2023 Kelp Value Chain Analyses by the Bivalve Shellfish Farmers Association of South Africa (BSASA) and the Department of Forestry, Fisheries and the Environment (DFFE). The analyses indicate that the kelp industry in South Africa currently supports roughly 350 direct jobs, focused on harvesting, processing, value addition, and distribution. It is noteworthy that 47% of these positions are held by women, with around 38% belonging to individuals under the age of 35 (BSASA 2023). While these figures reflect the sector's capacity to foster inclusive growth, several systemic issues constrain both profitability and the equitable distribution of socio-economic benefits. One such obstacle is the absence of food-grade processing infrastructure in the Western Cape, which escalates transport and handling costs, particularly for businesses seeking to market kelp-based food products and nutraceuticals. Establishing a local facility could significantly lower operational expenses, encourage value addition, and bolster the accessibility of processed kelp goods.

Another challenge lies in the limited supply of fresh kelp for high-end restaurants. Despite an increasing interest in kelp-centric dishes, inconsistent access and tenuous permitting processes hinder formal sales, leaving some dining establishments reliant on relatively informal and restricted channels. A structured framework—one that allows smaller concession holders, newly established kelp farms, or small-scale fishers to legally provide top-quality kelp—would not only diversify culinary offerings but also stimulate local livelihoods. Supply inconsistency has further repercussions for feed producers, small processors, and local entrepreneurs, since large and often exclusive purchasing agreements between major concessionaires and processing facilities can absorb most of the available stock. Inadequate logistics and unpredictable harvesting conditions also compound this challenge, resulting in fluctuating costs and making it difficult for new entrants or smaller actors to maintain steady operations (BSASA 2023).

Box 13.1: From Shore to Plate: Navigating *Kelp Foraging* and *Cultural Heritage* in a *Regulatory Void*

Journey into the Kelp Shack, the bustling kitchen of Chef Phil Mansergh, whose passion for coastal ingredients leads him regularly to the shoreline in search of fresh kelp. Unlike other chefs who rely on suppliers or local markets, Phil has taken it upon himself to explore the intertidal zone, gathering kelp by hand. Following him barefoot along rocky outcrops, he sifts through the thick brown seaweed cast ashore or still swaying in the shallows. He speaks candidly about the lack of formal regulations regarding the use of kelp in restaurants, revealing both a sense of freedom and an undercurrent of uncertainty.

On one hand, he cherishes this open access—being able to harvest what the sea offers each morning and incorporate it into his signature dishes by afternoon. In the film, "Kelp: South Africa's Golden Forests", his excitement is captured as he hurries back to his kitchen, carrying a basket of freshly gathered kelp. Once there, he dunks these thick, glossy fronds in a large basin, rinsing away the sand and salt. His enthusiasm is palpable; he marvels at the slick texture and faint aroma of brine, describing how the seaweed can add an understated sweetness and a soft crunch to everything from seafood stews to sautéed side dishes. In that same breath, Phil acknowledges the regulatory grey area around his practice. He points out that although recreational kelp collection is permitted in modest amounts, nothing formally addresses the use of wild-harvested kelp for catered meals. As the film lingers on his face, a thoughtful tension emerges. He doesn't want to undermine conservation efforts or disregard sustainability principles; yet he feels that the current legal vacuum leaves conscientious foragers like him without guidance or a framework for responsibly incorporating wild seaweed into a restaurant's offerings. (Fig. 13.1)

Nevertheless, he sees the artistry in bridging ocean and plate. In a voiceover, he talks about how foraging fosters a more intimate connection with marine ecosystems—there is a reverence in kneeling at the water's edge, selecting only what can be spared, leaving the rest to continue growing or feed other creatures. Scenes of him plating a kelp-inspired dish, garnished with the bright hues of coastal flowers, show how he attempts to pay homage to the environment that gifts him these ingredients. By inviting diners to see kelp not as an obscure novelty but as an integral part of local gastronomy, Phil hopes to expand the culinary narrative around seaweed. In doing so, he also subtly challenges authorities and policymakers to reconsider how and why—for the sake of both sustainability and cultural food heritage—responsible kelp foraging by local chefs might be recognized, supported, and brought out of its current regulatory limbo.

Later in the film, attention shifts to a small group of women harvesters from Weskusmadjie, a coastal community where the tradition of collecting and drying kelp has been passed down for generations. Their operation is modest yet vibrant: at low tide, they can be seen moving deftly across the rocks, selecting choice pieces of kelp that have washed ashore. Although they salt and dry these fronds to preserve them for later sale, their enterprise is similarly shaped by a regulatory gap—much like Chef Phil, they must negotiate a landscape in which there are no formal guidelines that address kelp gathering as a source of local income. Despite the legal grey area, the women have developed a meticulous routine, rinsing away sand and residual salt water in large outdoor barrels before laying the strips of kelp across wooden frames to dry under the sun. The film's close-ups emphasize each step of the process, from the crisp, sun-baked texture of the seaweed to the camaraderie that arises from working together. (Fig. 13.2)

As they work, the women chat about the challenges of sustaining a livelihood in a region where employment is scarce. Some describe this drying and salting tradition as part of their inherited knowledge, recalling older family members who once collected bundles of dried kelp for various uses. Now, they hope to expand their market, offering these salted fronds to anyone—local or tourist—willing to appreciate the delicacy's briny flavours. Yet, they explain that the absence of clear regulations renders them vulnerable, uncertain whether an official crackdown might one day halt their activities. Much of the passion behind their work lies in the sense of cultural continuity it provides by harvesting in a respectful manner that avoids over-collecting; they believe they are both preserving a lifeline for their own households and carrying forward a piece of their coastal heritage. Their endeavours are emblematic of how traditional knowledge can support micro-economies, even as these women remain, for now, at the fringes of formal legal recognition.

This dynamic perpetuates the asymmetrical distribution of benefits: large companies export dried kelp for alginate or abalone feed, capturing most economic gains, while local harvesters struggle to compete in markets that favor bulk processing (Rothman et al. 2020). Such inequities necessitate a more democratic, decolonial approach to kelp governance—one that leverages indigenous and local knowledge, invests in capacity building, and recognizes coastal communities as co-stewards rather than passive observers.

At the community level, integrating coastal populations more directly into the kelp value chain will require reducing barriers inherent in the concession and permitting processes, which can currently discourage small-scale fishers from formally engaging in harvesting and selling. Strengthening local communities' participation in the kelp sector demands more than simplified or redesigned permitting systems, especially when it comes to effectively farming this seaweed. It involves direct capacity-building initiatives in which prospective participants receive training in kelp cultivation, processing, and effective business operations, ensuring they have the knowledge and tools to maximize economic and social gains. Partnerships are also possible between larger concessionaires and small-scale harvesters or local fisher cooperatives, wherein a defined portion of beach-cast or fresh kelp is allocated to local communities under equitable terms. This model of shared value creation would position corporate actors and local communities as collaborators in maintaining resource sustainability while jointly benefiting from kelp's rising economic value (Isaacs et al. 2022).

As outlined in the Kelp Value Chain Roadmap, an overarching goal is to coordinate each of these initiatives so that they collectively improve the sector's resilience and inclusivity (BSASA 2023). By uniting government bodies, industry stakeholders, and local communities in a shared pursuit of sustainable expansion, it becomes feasible to tap into broader market prospects, including farmed kelp for human consumption, pharmaceutical and nutraceutical industries, and high-end gastronomy. Although the kelp sector can offer substantial socio-economic advantages, particularly for women and youth, strategic interventions in processing, supply-chain logistics, policy structure, and community engagement are essential to unlocking its full potential. In this manner, adopting a collaborative approach will not only enhance livelihoods in coastal regions but also position South Africa's kelp industry to become a critical component of the nation's blue economy (BSASA 2023).

13.2 Decolonizing Ecological Narratives

While economic valuations often dominate discussions, the case of the African Seaforest perfectly highlights kelp's less tangible but equally vital roles. Mental and emotional well-being emerges repeatedly in user surveys (Mehta et al. 2023). The gentle sway of kelp fronds, the rustling of sea bamboo (*Ecklonia maxima*), and the interplay of sunlight and water create "a natural balm for the mind," reported by recreational divers and coastal dwellers alike (Lucrezi 2021; Mehta et al. 2023). This punctuates how relational and non-material contributions to human well-being must feature in any kelp management plan (Bennett and Roth 2019). Popular media often portrays kelp forests through idyllic, decontextualized imagery, epitomized by films like *My Octopus Teacher* (Steingo 2021). While such representations can spark public interest in marine conservation, they can sidestep the colonial legacies and systemic inequalities that determine who benefits from these resources. During apartheid, coastal property and kelp concessions were largely restricted to white elites, leaving historically marginalized communities with minimal access and representation in formal management structures (Sowman and Sunde 2021).

Fig. 13.1 Left: Chef Phil Mansergh processing freshly harvested kelp. Right: The finished product—kelp ravioli on a bed of tomato confit, topped with dune spinach. (Photos by Akshata Mehta)

Fig. 13.2 The women of Weskusmandjie collecting seaweeds off the coast of St Helena Bay, South Africa. (Photo by Akshata Mehta)

In this context, romantic narratives risk perpetuating a white, middle-class ecological gaze (Steingo 2021). By overlooking the spatial reorganization of apartheid—where prime beachfronts remain privately owned and townships stand on the periphery—such narratives reinforce a partial truth: one that focuses on nature as a personal sanctuary rather than a public commons steeped in unequal power dynamics (Childs and Hicks 2019).

Counter-narratives such as “Kelp: South Africa’s Golden Forests” or productions like “Phefumla” supported by Nature, Environment, Wildlife Filmmaking (NEWF) highlight local wisdom, foraging traditions, and non-Western science perspectives. These works reveal that kelp forests carry relational values—cultural identity, sense of place, spiritual depth—that surpass mere utility (Chan et al. 2016). For Xhosa and Afrikaans-speaking communities, kelp may function as a protective talisman or a symbol of heritage, echoing beliefs passed down through generations (Mehta 2022). When these voices occupy the forefront, kelp management begins to resemble a co-created tapestry of scientific data, cultural memory, and ecological stewardship (Boswell 2022).

Moving beyond the anthropocentric or purely intrinsic paradigms, relational ethics promote the recognition of local customs and spiritual beliefs alongside scientific knowledge (Chan et al. 2016). This perspective, rooted in ecological justice, compels us to ask: Who gains or loses when kelp is harvested at scale? Whose narratives are privileged in policy-making? How do we ensure that future generations inherit not just the resource but also the relational ties that make kelp forests culturally significant?

Box 13.2: Relational Values as Explored through Art

In the film *Kelp: South Africa's Golden Forests*, we encounter artist Jeremy Revett (Jero, the "Wild Wanderer") who frames kelp not merely as a botanical curiosity but as an ever-present embodiment of the ocean's rhythms, carrying with it the pulse of coastal life. Early on, he is shown walking along the shore, gently sifting through pieces of dried and fresh kelp that have washed up from the African Seaforest. This process of gathering is deeply mindful, as he talks about the kelp's delicate textures and shapes—curling blades, spongy pneumatocysts, and elongated fronds. For him, each specimen is more than raw material; it represents a unique aspect of the intertidal zone, shaped by the currents and tides to which humans seldom pay close attention. The artist thus begins his craft from a place of reverence, acknowledging that what he is collecting is both a piece of nature's artistry and a tangible reminder of how the ocean sculpts marine life over time. (Fig. 13.3)

Either in his studio or at the beach itself, if weather conditions allow for it, Jero translates these kelp fragments into cyanotype prints, a technique that uses a light-sensitive chemical solution brushed onto paper, cloth, or other surfaces. The film captures his carefully orchestrated procedure: he arranges the kelp on the treated surface, positions them to reveal their natural curves and edges, and then exposes the composition to sunlight. By placing the kelp directly onto the paper, he fuses process and medium, so that the cyan-blue tones highlight the kelp's intricate patterns and textures. He explains how even the slightest shift in the blades can alter the resulting image—a nod to the fluid, often unpredictable rhythms of the ocean itself. Through this analogue printing approach, the artist is effectively mapping the shapes of kelp onto a new, two-dimensional plane, preserving forms that would otherwise decompose, dry out, or be carried back to sea.

The film emphasizes the contemplative nature of this practice. The artist's narrative weaves together recollections of childhood beach walks, family gatherings by the coast, and his subsequent fascination with transforming ephemeral organic material into lasting visual records. He sees cyanotypes not simply as a tribute to marine biology but as a way of merging photography, printmaking, and environmental awareness. His voiceover describes how each print is a trace of the kelp's existence—an index of a moment in which he and the ocean intersect. This concept also informs his view that kelp, in its daily dance with the tides, invites humans to re-evaluate their relationship with the shoreline: how easily we discard it, how rarely we notice its resilience. Here, the notion of relational value becomes especially clear in his reflections on the ocean's constant renewal of its kelp as an "offering." By capturing the kelp's forms in stark, crisp silhouettes of Prussian blue, the artist strives to remind viewers that these seaweeds shape entire ecosystems—binding wave-swept coastlines, feeding marine species, and fuelling local imaginations. (Fig. 13.4)

In further discussing reflections on his creative motivations (not captured in the film), Jero explains that cyanotype's historical roots in scientific illustration resonate with his quest to document nature in a sincere, direct manner. He draws a parallel between Victorian-era botanists who catalogued marine algae to better understand coastal environments, and his own modern ambition to spark ecological mindfulness. He notes that although kelp might appear commonplace or even inconvenient—particularly as decaying heaps on the sand—its presence in these prints offers a window into the intricate structures hidden in plain sight. The exposure process itself, with the intense blue tones of cyanotypes, metaphorically then, places marine life centre stage. Through this alchemy of light, water, and photosensitive solutions, the artist reveals swirling patterns and organic silhouettes, each a quiet testament to the ocean's creative energies. After a morning's work, lingering shots of the completed cyanotypes as captured on camera illustrate how a single piece of kelp, pressed against a photographic surface, can conjure narratives of oceanic wonder, cyclical regeneration, and the often unseen human dependency upon maritime worlds.

As scientists, we are often trained to view knowledge as something that emerges from controlled studies, measurable outcomes, and empirical data. But wisdom—a deeper, often experiential understanding of the world—can take many forms, and it is frequently found in the lived experiences of people and their relationships with nature. While science is undoubtedly a powerful tool, it is only one way of knowing, and it can be limiting when used as the sole lens through which we understand complex ecosystems.

Local communities, particularly those who have relied on and coexisted with kelp forests for generations, possess forms of wisdom that go beyond scientific metrics (Hill et al. 2020). This wisdom is embedded in their cultural practices, stories, and interactions with the land and sea. Recognizing this knowledge as valuable and complementary to scientific understanding allows us to develop a more holistic view of kelp forests, one that honors both ecological integrity and human dignity.

Fig. 13.4 Washed cyanotype on cloth by Jero, "the wild wanderer". (Photo by Akshata Mehta)

Fig. 13.3 Artist Jeremy Revett (Jero) identifying an ideal kelp specimen for his cyanotype artworks, off the coast of Standford, South Africa. (Photo by Akshata Mehta)

13.3 Unveiling Hidden Entanglements: Future Imaginaries in Phyconometrics

As we work toward transition to a just marine future, it is worth mentioning Artificial Intelligence (AI) and its potential uses for furthering this field of phyconomy. Given that AI can be inherently biased yet will inevitably be used, it is necessary to consider an inclusive framework so as to not further marginalize those dependent on the kelp as a resource and/or ecosystem.

Recent innovations in AI image generation—like DALL·E or MidJourney—allow for the visualization of complex socio-ecological relationships. Melding scientific data, local narratives, and artistic motifs, these AI-driven visuals can unearth alternative framings of kelp forests (and nature at large) (Braidotti 2020). For instance, many default images might depict kelp forests as empty, pristine habitats without human presence. Co-creation with local communities ensures depictions also show small-scale harvesters, historical markers of colonial violence, or alternative futures where kelp forests are equitably managed (Fig. 13.5).

This AI-generated cyanotype-style artwork explores the deep human connection to South Africa's temperate kelp forests, specifically featuring *Ecklonia maxima* (giant kelp). Created using artificial intelligence, the image blends realism with symbolic abstraction to illustrate the ecological, cultural, and governance dynamics surrounding kelp harvesting. In the foreground, small-scale fishers and divers sustainably harvest kelp while children learn from elders, emphasizing generational knowledge transfer. Ghostly ancestral handprints and ancient tools embedded in the seabed highlight the long-standing relationship between coastal communities and the ocean. A glowing seaweed thread visually links humans and marine life, symbolizing interdependence. The underwater scene is rich with South African marine species, including Cape fur seals, galjoen fish, West

Fig. 13.5 Threads of the Kelp Forest: Ancestry, Ecology, and Governance

Coast rock lobsters, and shysharks weaving through the thick, floating fronds of *Ecklonia maxima*. Above the water, fragmented bureaucratic structures loom in the distance, representing the governance struggles over kelp access and use. Rendered in a monochromatic blue-and-white palette, this AI-assisted cyanotype evokes a dreamlike yet critical reflection on conservation, cultural heritage, and power imbalances in marine resource management. The composition highlights the contrast between harmonious coexistence and contested access, underscoring the importance of sustainable stewardship of South Africa's golden forests.

If used ethically, AI-generated imagery can be used in community workshops to spark conversations about how kelp might shape—and be shaped by—climate change, new economic ventures, or decolonial reparations. Participants critique initial images, share alternative visions, and refine them in real time, forging collective foresight (Strand et al. 2022). This process foregrounds local voices, particularly those historically marginalized, fostering decolonial science communication. By visualizing speculative futures, communities can better articulate their aspirations—like co-op kelp farms, spiritual tourism, or climate-resilient shoreline protection—within formal governance debates (Strand and Khupe 2024).

13.4 Conclusion

The African Seaforest highlights how kelp extends far beyond a mere ecological resource, functioning instead as a nexus of cultural identity, economic opportunity, and communal heritage. Whether experienced through foraged dishes in a small coastal restaurant, salted and dried by women harvesters seeking household income, or rendered in cyanotype prints that exalt its ephemeral beauty, kelp weaves human narratives into the tapestry of the marine environment. These myriad connections only further highlight the reality that transforming kelp governance and management calls for more than scientific expertise or market-driven incentives; it also demands collective efforts to uplift historically marginalized voices, integrate local wisdom, and ensure equitable sharing of benefits.

Under South Africa's current legislative structures, however, top-down permit systems tend to privilege large-scale operations, leaving artisanal harvesters, coastal communities, and emerging kelp farmers at the margins. By embracing a more inclusive vision—one that respects the relational ties between people and the African Seaforest—policymakers and industry stakeholders can move toward a more just and sustainable stewardship model. The proposals set forth below provide a framework for actualizing such a shift:

1. Shorter, renewable concession terms would introduce the flexibility needed to adapt to changing kelp distributions and community priorities.
2. Community-based monitoring, carried out by trained local fishers, could merge scientific metrics with traditional knowledge and thereby deepen ecological insights.
3. Transparent benefit-sharing mechanisms would ensure that coastal communities receive a tangible return for their roles in cultivation, research, and responsible harvesting.
4. Finally, trans/interdisciplinary research funding—with a mandate to incorporate social sciences and cultural studies—would highlight kelp's relational and intrinsic values alongside its tangible economic contributions.

In embracing these recommendations, stakeholders—ranging from government agencies and commercial actors to local cooperatives and conservation NGOs—can jointly create a future in which kelp is not merely a commodity but a shared inheritance. Such an approach positions kelp forests as living entities that sustain ecosystems, shape cultural identities, and fuel livelihoods. By weaving ecological science with socio-cultural wisdom, policy with practice, and community engagement with political will, the African Seaforest can evolve into a cornerstone of an inclusive blue economy that honors both the ocean's regenerative capacity and the needs of those who rely on it.

References

Bennett NJ, Roth R (2019) Realizing the transformative potential of conservation through the social sciences, arts and humanities. Biol Conserv 229:A6–A8. https://doi.org/10.1016/j.biocon.2018.07.023

Bennett NJ et al (2022) "Social science for conservation in working landscapes and seascapes," Front Conserv Sci 3. https://doi.org/10.3389/fcosc.2022.954930

Bolton J, Blamey L (2017) Kelps in southern Africa are thriving, but some key inhabitants of kelp forests are not. The Conversation. https://theconversation.com/kelps-in-southern-africa-are-thriving-but-some-key-inhabitants-of-kelp-forests-are-not-73069

Boswell R (2022) Salted identities: biocultural heritage for a rehumanized ocean management in South Africa. Anthropol Hum 47:363–380. https://doi.org/10.1111/anhu.12402

Braidotti R (2020) "We" are in this together, but we are not one and the same. J Bioethical Inq 17:465–469. https://doi.org/10.1007/s11673-020-10017-8

Chan KMA et al (2016) Why protect nature ? Rethinking values and the environment. Defining relational values. PNAS 113:1462–1465

Chan KMA, Satterfield T, Goldstein J (2012) Rethinking ecosystem services to better address and navigate cultural values. Ecol Econ 74:8–18. https://doi.org/10.1016/j.ecolecon.2011.11.011

Childs J and Hicks CC (2019) "Securing the blue: political ecologies of the blue economy in Africa," J Polit Ecol 26(1). Available at: https://doi.org/10.2458/v26i1.23162

Department of Forestry, Fisheries and the Environment (DFFE) (2013) Description of the commercial seaweed sector. https://www.dffe.gov.za/sites/default/files/reports/descriptionofcommercial_seaweedsector.pdf

Díaz S, Demissew S, Carabias J, Carlos J, Lonsdale M et al (2015) The IPBES conceptual framework — connecting nature and people. Curr Opin Environ Sustain 14:1–16. https://doi.org/10.1016/j.cosust.2014.11.002

Farber SC, Costanza R, Wilson MA (2002) Economic and ecological concepts for valuing ecosystem services. Ecol Econ 41:375–392. https://doi.org/10.1016/S0921-8009(02)00088-5

Green N, de Villiers L (2018) Information document for kelp management in Overberg District. (see Rothman et al. 2020)

Hill R et al (2020) Working with indigenous, local and scientific knowledge in assessments of nature and nature's linkages with people. Curr Opin Environ Sustain 43:8–20. https://doi.org/10.1016/j.cosust.2019.12.006

Hynes S et al (2021) Valuing the ecosystem service benefits from kelp forest restoration: a choice experiment from Norway. Ecol Econ 179:106833. https://doi.org/10.1016/j.ecolecon.2020.106833

Isaacs M, Hara M, Dennis T, Rouhani Q, Mannarino C, Jaffer N (2022) A situational analysis of small-scale fisheries in South Africa: From vulnerability to viability. https://www.masifundise.org/wp-content/uploads/2022/10/A-Situational-Analysis-of-SSF-in-South-Africa_V2V-Working-Paper-No.-2022-9.pdf

Krumhansl KA et al (2016) Global patterns of kelp forest change over the past half-century. Proc Natl Acad Sci USA 113:13785–13790. https://doi.org/10.1073/pnas.1606102113

Lucrezi S (2021) Characterising potential participants in kelp monitoring in the recreational diving community: a comparative study of South Africa and New Zealand. Glob Ecol Conserv 28:e01649. https://doi.org/10.1016/j.gecco.2021.e01649

Mehta A (2022) 'Golden forests' of the sea: assessing values and perceptions of kelp in the Western Cape region of South Africa. Master Thesis. Faculty of Science. Available at: https://open.uct.ac.za/handle/11427/36490 (Accessed: 28 July 2023)

Mehta A, Wynberg R, Ramcharan-Kotze C, Smit AJ (2023) Exploring local perceptions around the value of marine biodiversity: the case of kelp in the Western Cape, South Africa. Ecosyst People 19. https://doi.org/10.1080/26395916.2023.2234499

Mehta A, Wynberg R, Smit AJ (2024) Embracing diversity: enhancing the management of South Africa's kelp forests in an era of change. Bot Mar 67:443–447. https://doi.org/10.1515/bot-2023-0058

Pascual U et al (2017) Valuing nature's contributions to people: the IPBES approach. Curr Opin Environ Sustain 26–27:7–16. https://doi.org/10.1016/j.cosust.2016.12.006

Pratson DF, Adams N, Gould RK (2023) Relational values of nature in empirical research: a systematic review. People Nat 5:1464–1479. https://doi.org/10.1002/pan3.10512

Prew Z et al (2024) The African seaforest: a review. Bot Mar 67. https://doi.org/10.1515/bot-2023-0060

Rothman M, Anderson R, Kandjengo L, Bolton J (2020) Trends in seaweed resource use and aquaculture in South Africa and Namibia over the last 30 years. Bot Mar 63:315–325. https://doi.org/10.1515/bot-2019-0074

Sanborn T, Jung J (2021) Intersecting social science and conservation. Front Mar Sci 8:676394. https://doi.org/10.3389/fmars.2021.676394

Sandler R (2012) Intrinsic value, ecology, and conservation. Nat Educ Knowledge 3:4. https://www.nature.com/scitable/knowledge/library/intrinsic-value-ecology-and-conservation-25815400/ (Accessed: 7 Aug 2021)

Sowman M, Sunde J (2021) A just transition? Navigating the process of policy implementation in small-scale fisheries in South Africa. Mar Policy 132:104683. https://doi.org/10.1016/j.marpol.2021.104683

Steingo S (2021) Learning from Alexis (Review of Alexis Pauline Gumbs's undrowned: Black feminist lessons from marine mammals). https://www.boundary2.org/2021/05/gavin-steingo-learning-from-alexis-review-of-alexis-pauline-gumbss-undrowned-black-feminist-lessons-from-marine-mammals/

Strand M, Khupe C (2024) How to be inclusive of community in research. Sage Research Methods: Diversifying and Decolonizing Research. https://doi.org/10.4135/9781529690729

Strand M, Rivers N, Snow B (2022) Reimagining Ocean stewardship: arts-based methods to "hear" and "see" indigenous and local knowledge in ocean management. Front Mar Sci 9:886632. https://doi.org/10.3389/fmars.2022.886632

Akshata Mehta is currently based at the Department of Biodiversity & Conservation Biology at the University of the Western Cape. Her research areas of interest include human dimensions of marine and coastal ecosystems (specifically, seaweeds), the marine bioeconomy, marine conservation and management, and climate change and resilience.

14 Phyconomy in Europe

Joël Fleurence

Abstract

The algae has been traditionally used in Europe for animal feeding and sometimes in human nutrition. Since the twentieth century, the seaweed resource, especially brown algae, are the main source for the production of phycocolloids such as alginates. The main countries concerned by this production are France and Norway. For four decades, another use of algae has been developing in Europe and more particularly in France. This application concerns the use of seaweeds as sea vegetables or ingredients. This approach is also developed in Ireland and the Netherlands for the production of new food products such as the "Seameat burger"©. Besides this, algae, particularly brown algae, are used by the European industry for the manufacture of products for agriculture (e.g., biostimulants). This chapter describes the historical and current applications of algae in Europe, the species concerned, the production methods, the regulatory constraints, and the possible prospects for the use of this resource in a temperate European environment.

Keywords
Seaweeds · Marine resource · Food · Agriculture · Companies · Europe

14.1 Introduction

In some European countries such as Iceland, Norway, Ireland, United Kingdom, and France (Brittany), seaweeds have been often consumed by the coastal populations. The red seaweed, *Palmaria palmata* (Fig. 14.1), well known under the name of Dulse, is used as food since the Viking period in Norway (Delaney et al. 2016). This species, together with the brown algae *Alaria esculenta* (Atlantic Wakame, Irish Wakame), was once part of the normal diet of Icelander (Sauvageau 1920). *Laminaria digitata*, known under the common name of Tangle, was frequently sold as vegetable on the streets of Edinburgh and Belfast in the early twentieth century (Sauvageau 1920).

Other algae have been traditionally used as ingredients in culinary preparations. This is the case of the red algae *Chondrus crispus* (Fig. 14.2) (Irish moss, Pioca, Sea Lichen, Cairgean, and Carreegan), which was used for its gelling properties for the preparation of dairy desserts.

Therefore, the valorization of algae in human food is a traditional practice also in Europe. It is less developed than in Asia, but it is based on a historical reality. In the field of traditional food applications such as the production of phycocolloids (e.g., alginates), Europe remains an active producer. The use of algae as fertilizers for agriculture or as feed for livestock breeding (cows, pigs, and sheep) is also a historical practice in Europe (Delaney et al. 2016; Sauvageau 1920). This type of application is still being developed today.

The production methods of algae in Europe, the species concerned, the different applications (direct and indirect human food, agriculture), and the regulatory constraints applied to algae are the main subjects of this chapter.

14.2 Algae Production in Europe

Total production of algae, seaweeds, and microalgae was 287,386 t wet weight in 2019 in Europe (FAO 2021). Microalgae production (cyanobacteria, green microalgae) represents 353 t or 0.12% of European algae production.

Global algae production in the world was 35.82×10^6 t. Therefore, European production represents just under 1% of global algae production. In Europe, algae produced are mainly obtained by harvesting (96%) (Fig. 14.3).

J. Fleurence (✉)
Nantes Université, Isomer, 2 chemin de la Houssinière, Nantes Cedex, France
e-mail: joel.fleurence@univ-nantes.fr

A. T. Critchley et al. (eds.), *Advances in Temperate Phyconomy: Algal Harvest and Cultivation in Globally Distributed Temperate Waters*, Developments in Applied Phycology 15, https://doi.org/10.1007/978-3-032-02955-3_14

Fig. 14.1 *Palmaria palmata* (formerly *Rhodymenia palmata*) or dulse. (Photo credit ©Fleurence J. 2021)

Fig. 14.2 *Chondrus crispus,* red seaweed known under the name of "Pioca" or "Sea Lichen" in France. (Photo credit © Fleurence J 2024)

Algae collected are mainly macroalgae (275,908 t wet weight). The number of seaweed species available in the North Atlantic area is estimated to be 1,205 species. The distribution by European country is as follows: 689 for France, 577 for the United Kingdom, 501 for Ireland, and 415 for Scotland (Guiry and Hession 1998).

Algae cultivation in Europe mainly concerns Cyanobacteriophyta such as "Spirulina" (*Arthrospira, Limnospira*) or green microalgae (*Chlorella*) (353 t of wet weight).

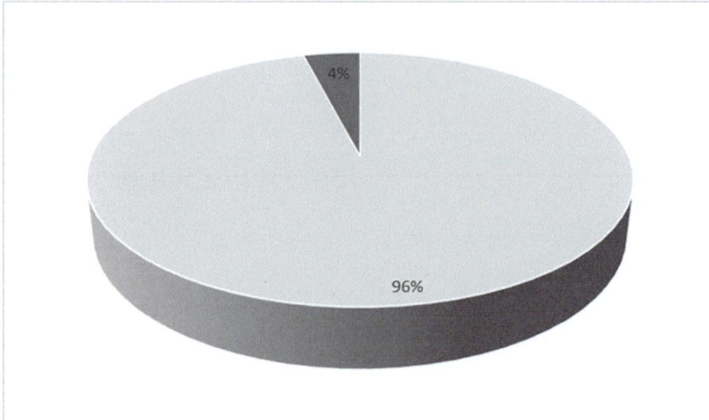

Fig. 14.3 Distribution of algal production methods (harvesting, cultivation) in Europe in 2019 (FAO 2021)

However, some seaweeds are also cultivated in Europe. There are brown algae belonging to the genera *Laminaria* (*Saccharina*) or *Undaria* (Table 14.1).

In Europe, the main countries involved in the production and valorization of seaweeds are Norway, France, and Ireland (Fig. 14.4).

In 2019, the total Norwich production of seaweeds by collection and cultivation was 163,197 t expressed in wet weight (FAO 2021). This production is mainly focused on the exploitation of the brown algae (162,941 t) like *Laminaria hyperborea*, *Saccharina latissima*, and others such as *Alaria esculenta*, *Fucus* spp., and *Ascophyllum nodosum* (Phaeophyceae).

Laminaria and *Saccharina* represent 36,844 t or 22% of the tonnage of brown seaweeds exploited. Most of the brown seaweed resource is wild harvested (Table 14.2).

However, some species are also produced by cultivation. These are *Saccharina latissima* and *Alaria esculenta*. In 2019, the production level was 117 t wet weight. The main species cultivated is the brown seaweed *Saccharina latissima,* well known under the name of sea belt. Seventy-seven tons of this alga are produced by cultivation in Norway. The production of *Alaria esculenta* (Atlantic or Irish wakame), the second species cultivated, is 44 t (FAO 2021).

In France, the total production of seaweeds is 51,476 t in wet weight (FAO 2021). The French production is focused on the harvest of two species, *Laminaria digitata* (Kombu Breton) (38,202 t) and *Laminaria hyperborea* (12,939 t). In France, two brown seaweeds are cultivated; these are *Saccharina latissima* (Royal kombu), mainly produced in Brittany by the French company "Algolesko." The second species, cultivated by several companies also located in Brittany, is *Undaria pinnatifida* or Wakame. The total tonnage of Wakame produced in France is 105 t expressed in wet weight.

Table 14.1 Separation of European production of seaweeds between harvesting and cultivation in 2019 (FAO 2021)

	Collected (t of wet weight)	Cultivated (t of wet weight)	Total (t of wet weight)
Brown seaweeds			
Laminaria/*Saccharina*	91,724	229	91,953
Undaria pinnatifida	209	105	314
Miscellaneous brown seaweeds	178,248	10,678	
Red seaweeds			
Carrageenan species	–	–	–
Gelidium, Gracilaria	390	–	390
Porphyra (*Pyropia*, Neopyropia)	8	–	8
Miscellaneous red seaweeds	1,803	–	
Green seaweeds			
Ulva lactuca, U. rigida	128	–	
Miscellaneous green seaweeds	800	36	

Table 14.2 Seaweed species collected or cultivated in main European producers

Country	Harvesting	Cultivation
Norway	*Laminaria hyperborea* (European kelp) *S. latissima* (Royal kombu) *Alaria esculenta* (Irish wakame) *Ascophyllum nodosum* *Fucus* spp.	*Saccharina lattissima* (Royal kombu) *A. esculenta* (Irish wakame)
France	*Laminaria digitata* (Kombu Breton) *L. hyperborea* (European kelp) *Palmaria palmata* (Dulse) *Ulva rigida, U. lactuca* (Sea lettuce)	*Undaria pinnatifida* (Wakame) *S. latissima* (Royal kombu)
Ireland	*A. nodosum* *L. hyperborea* (European kelp) *A. nodosum*	*A. esculenta* (Irish wakame)

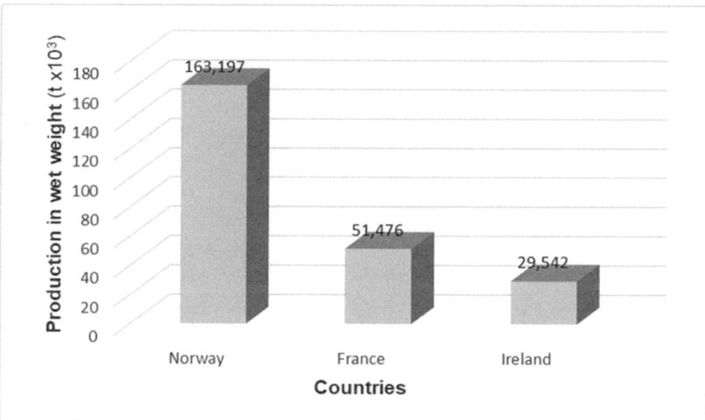

Fig. 14.4 Main seaweed producers in Europe (FAO 2021)

Paradoxically, France is the leader in Europe for the cultivation of *Spirulina* sp., with a production of 201 t (wet weight).

Ireland is the third European country involved in algae production. The total production is 29,542 t. The main volume produced is the brown seaweeds (29,442 t). *Alaria esculenta* is the main seaweed produced by cultivation (Table 14.2). The production is to 42 t expressed in wet weight.

In the 1970s, Portugal was a major producer in the world of agar from *Gelidium corneum* and *Pterocladiella capillacea* (Rhodophyta) (Santos and Melo 2018). These species have been harvested at the continental coast and in the Azores Islands, respectively. However, the harvest of these agarophytes has declined sharply between 1950 and 2014 (Fig. 14.5). Similarly, agar production decreased strongly during this period. For example, between 1970 and 1990, agar production fell by 75%.

Spain is another European country that has been involved in *Gelidium* harvesting and agar production. Since the 1990s, the harvest of this agarophyte has become very low, 300 t instead of 6,000 t in the 1980s. Morocco, a neighboring country of Spain and Portugal, has today become the main producer of agar, replacing these two European countries on the phycocolloid market.

As in Asia, seaweed production in Europe is mainly focused on various food applications (phycocolloid extraction, sea vegetables, or food ingredient manufacturing).

14.3 Food Applications

14.3.1 Sea Vegetables, Ingredients, or Condiments

The use of seaweeds as sea vegetables, ingredients, or condiments is a traditional practice for some coastal populations. The brown seaweed *Fucus spiralis* is consumed fresh by Azoreans. The red seaweed *Porphyra* can also be used as an ingredient or condiment by these populations. It is incorporated into a soup or an omelet. Likewise, after marinating in vinegar and onions, seaweeds *Laurencia* sp. or *Osmundia* sp. can be consumed as condiments. In some cases, these marinated algae are combined with more traditional dishes, such as as fried fish (Neto et al. 2005). On the continental Atlantic coast, the seaweeds are sometimes consumed in the traditional way (e.g., sea vegetable or ingredient) (Table 14.3). The fresh seaweeds are also usually eaten in Mediterranean

Fig. 14.5 Evolution of harvest red agarophyte algae at the Portuguese coasts between 1950 and 2014 (Santos and Melo 2018)

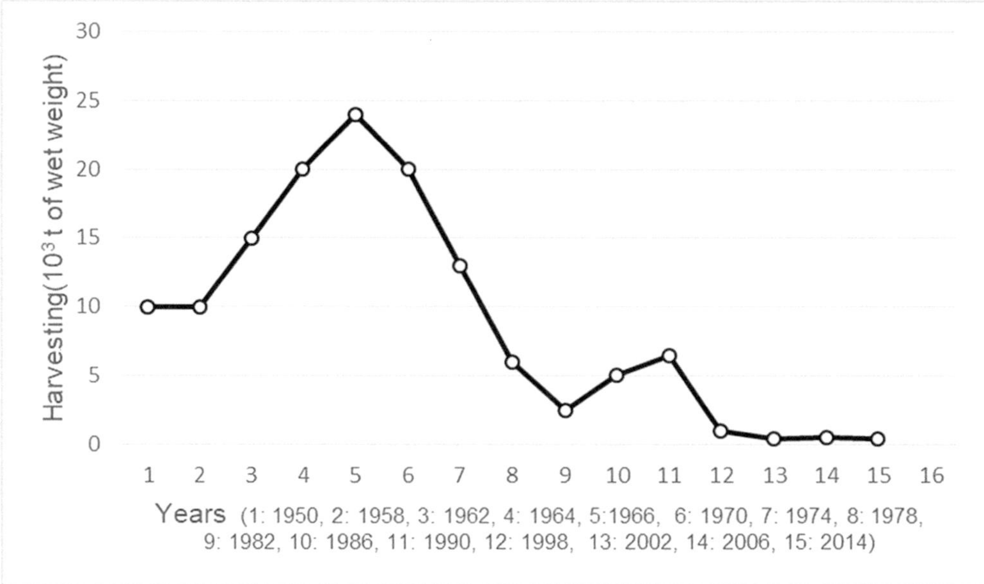

Table 14.3 Examples of traditional uses of seaweeds in Europe (O'Connor 2017)

Seaweed species	Azores	Ireland	England (Wales)	France (Brittany)	Scandinavia countries	Scotland
Palmaria palmata (dulse)	–	Sea vegetable	–	Sea vegetable	Sea vegetable Ingredient (Salt source)	Sea vegetable Ingredient (Thickener, colorant)
Chondrus crispus (Irish moss)	–	Ingredient (Thickener)	–	–	–	–
Porphyra sp.	Ingredient Condiment	–	Sea vegetable	–	–	–
Laurencia sp.	Condiment	–	–		–	–
Osmundea sp.	Condiment	–	–		–	–
Fucus spiralis	Sea vegetable	–	–		–	–
Alaria esculenta (Irish Wakame)	–	Sea vegetable	–	–	–	Sea vegetable
Laminaria digitata	–	–	–	Sea vegetable	–	Sea vegetable Condiment

coasts. The green seaweed *Ulva lactuca* and the red alga *Chondracanthus teedei,* called "Mauru," are consumed with squid by the fishermen in Sicily (Mendes et al. 2022).

However, this type of traditional practice is based on the consumption of fresh or dried seaweed. Today, fresh or dried products are always available. However, the fresh seaweed is previously salted to microbiologically stabilize the food. Other more or less complex processes can also be applied. The processes are, however, those usually developed by the food industry. These include salting, canning, fermentation, or maturation (Stévant et al. 2020, Fleurence, 2024). Among these processes, canning is very widespread. It is mainly applied to three edible seaweeds: *Undaria pinnatifida* (Wakame), *Laminaria digitata* (Kombu), and *Himanthalia elongata* (sea spaghetti) (Table 14.4). Fermentation is a process classically used in the food industry. In Europe, this type of treatment is mainly applied to the brown algae *Laminaria digitata* (Kombu) and *Himanthalia elongata* (sea spaghetti) (Table 14.4). However, fermentation processes have been tested on other seaweed species but on a laboratory scale (Table 14.5). These processes mainly involve lactic bacteria and rarely fungi or yeasts. However, there is a marine fungal flora associated with certain macroalgae. The use of this flora for the transformation of edible algae appears to be an interesting opportunity, especially in Europe, which is a leader in this type of research (Le Strat et al. 2022).

Maturation is also a classical process in the food sector. It is an endogenous fermentation process based on the activation of the bacterial or fungal flora present in the algae or on its surface (Fleurence 2024). It has been experimentally

Table 14.4 Examples of processes applied on the seaweed resource in some European countries ([a]experimental process no industrial application)

Type of product	France	Spain	Norway
Drying	Sea lettuce Wakame Royal kombu Dulse Sea spaghetti Nori	–	Sea lettuce
Salting (Fresh)	Sea lettuce Dulse Wakame Royal kombu	Wakame Sea lettuce *Codium*	–
Canned	Sea spaghetti	Wakame Kombu Sea spaghetti *Codium*	Sea spaghetti
Fermented	Sea spaghetti Royal Kombu	–	–
Matured	–	–	Dulse[a]
Enzymatic treatment	Dulse[a]	–	–

Table 14.5 Examples of fermentation processes applied at laboratory scale on different seaweed species in Europe[a] and other countries (Monteiro et al. 2021)

Seaweed species	Bacteria or fungi strains used for fermentation process
Palmaria palmata[a]	*Trichoderma pseudokoningii, Microscopus rhizopus, Aspergillus oryzae*
Kappaphycus alvarezii	*Bifidobacterium* sp.
Gracilaria sp.	*Lactobacillus acidophilus, Lactobacillus plantarum*
Ulva sp.	*Lactobacillus brevis, L. plantarum, L. casei, L. rhamnosus*
Saccharina japonica	*Monascus purpurus*

studied in Norway on red seaweed *P. palmata* (Stévant et al. 2020). This process is applied to dehydrated (6% moisture) or semidehydrated (20% moisture) algae stored at 12 °C in the darkness for 126 days. According to dehydration level, the organoleptic qualities of the products (texture, flavors, and odors) are different at the end of the maturation process (Fig. 14.6). The maturation process applied to seaweeds with 20% of moisture gives a product with a soft texture in regard to the final product obtained from dried seaweed (6% moisture). In addition, the fish skin odor is lower with semidry products compared to those reported for dehydrated seaweed.

The maturation process developed for *Palmaria palmata* appears to be a low-cost approach and should allow the development of new products on the European market.

In addition to conventional industrial processes, other processes have been tested to improve the nutritional value of algae, particularly as a source of food proteins. These processes developed on an experimental scale have not yet been the subject of industrial development. These enzymatic treatment processes of the algal raw material aim to degrade polysaccharides, anti-nutritional factors limiting the digestibility of algal proteins (Fig. 14.7). This type of process was developed on the algae *Palmaria palmata* (dulse) and has been protected by a patent (Fleurence et al. 2000). Unfortunately, this patent has not yet been applied on an industrial scale.

The use of seaweed as an ingredient in burger formulation is a new way to valorize algae in the European market. A first approach is to promote the seaweed incorporation in the vegan burger (Fernandes et al. 2023). The other way is to use seaweed in a classical burger to replace a fraction of beef proteins.

A vegan burger incorporating the red seaweed *P. palmata*, in complement to different plants such as common pea (*Pisum sativum*), chickpea (*Cicer arietinum*), and other ingredients, has been developed on a laboratory scale (Table 14.6). The general composition of the vegan burger is available, but the proportion of different compounds or ingredients is not published.

A study performed on a panel of 89 people aged between 16 and 64 years old shows a good acceptance of this type of product by the consumers. Sixty-four percent of the panel members consider the vegan burger as pleasant and twelve percent as very pleasant (Fernandes et al. 2023).

These results suggest that a part of the population could be open to the consumption of seaweeds through this approach. Probably, the young consumers frequently eaten burger, and the vegan people represent a market target for the development of this type of product.

However, this type of burger already exists in Europe. It is offered by the company Dutch Weed burger (https://dutchweedburger.com/). It includes 10% of seaweed in the final product. The species concerned is the brown seaweed *Saccharina latissima* (Royal kombu) cultivated in the Zeeland region (North Sea).

The development of burgers including seaweed and beef as major compounds is also a reality in Europe. This burger called "SeaMeat burger" was developed and traded by the Seaweed Company (Fig. 14.8) (https://www.theseaweedcompany.com/our-products/seameat). This company is a Dutch startup specialized in the cultivation of algae and the processing of this resource for food purposes. It has cultivation sites in Europe (Ireland) but also on the Asian continent (India).

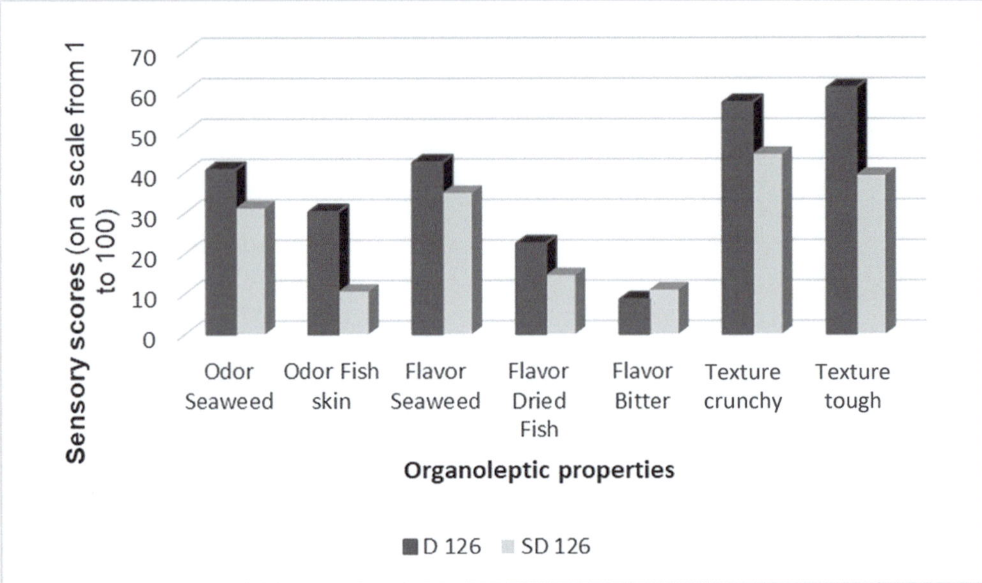

Fig. 14.6 Effect of maturation process of *Palmaria palmata* (dulse) dehydrated (D 126) and semidehydrated (SD 126) samples stored at 12 °C in darkness during 126 days (Stévant et al. 2020)

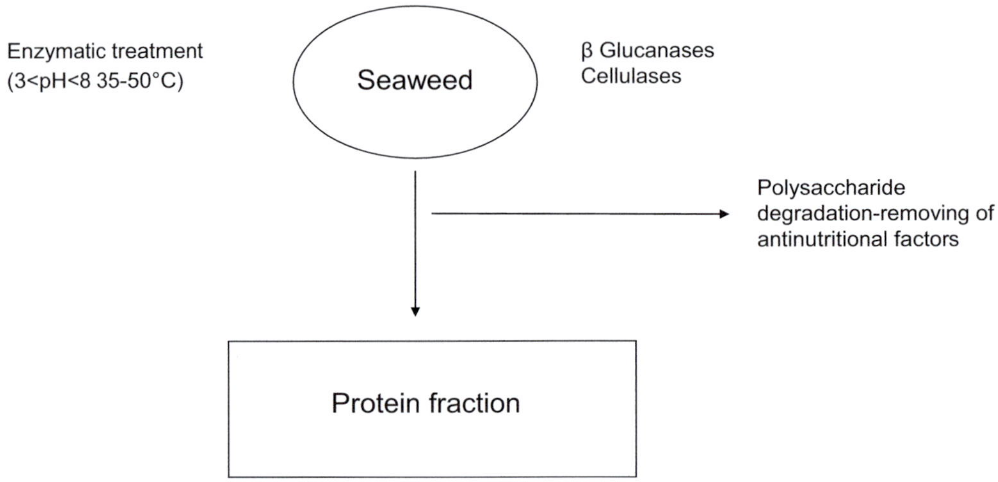

Fig. 14.7 Process principle of enzymatic treatment of the red seaweed *Palmaria palmata* (Fleurence et al. 2000)

Fig. 14.8 SeaMeat burger developed by the Seaweed Company. (Photo credit © Kraan S, Seaweed Company 2024)

The SeaMeat ingredient is composed of a blend of dried seaweed, probably brown seaweeds cultivated such as *Alaria esculenta* or *Saccharina latissima,* which are produced in Irish sites. The level of seaweed incorporation is to 25% for 1 kg of beef. The ecological impact of the SeaMeat burger is lower than that of the classical beef burger (Table 14.7).

The development of algae consumption in Europe will probably involve the inclusion of these marine plants in frequently consumed products. This could be burgers that are very popular with young consumers or vegans who are looking for products based on plant proteins. In both cases, algae-based burgers should interest these consumer populations.

14.3.2 Phycocolloids

Phycocolloids are mainly used by industry for their thickening or gelling properties. There are alginates extracted from brown seaweeds (*Laminaria digitata*, *L. hyperborea*), agar mainly obtained from red seaweeds *Gracilariopsis longissima* (formerly *Gracilaria verrucosa*), or *Gelidium corneum* (formerly *Gelidium sesquipedale*). Carrageenan, another family of phycocolloids, are extracted from *Chondrus*

Table 14.6 Partial composition of the vegan burger made with *P. palmata* as an ingredient (nc: noncommunicated) (Fernandes et al. 2023)

Vegan burger composition	% of final product
Grass pea	nc
Chickpea	nc
Palmaria palmata (powder)	nc
Palmaria palmata (diced)	nc
Raw carrot	nc
Diced onion	nc
Oatmeal flakes	nc
Lupin flour	nc
Apple cider vinegar	nc
Olive oil	nc
Mixture of herbs and spices	nc

Table 14.7 Comparison of ecological impact between classical beef burger and SeaMeat burger (for 1 kg of burger meat) (https://www.theseaweedcompany.com/our-products/seameat)

Beef burger	Ecological criteria	SeaMeat burger
33.2 kg CO_2 eq	Warning impact	25.1 kg CO_2 eq
89 m^2	Land use	66.8 m^2
6,850 L	Freshwater use	5,210 L
2 g PO_4 eq	Freshwater chemical contamination	1.5 g PO_4 eq

crispus or *Mastocarpus stellatus,* which are also red algae. Several European countries have been involved in the harvesting of the algae for the industrial extraction of these substances. There are mainly Norway, France, Ireland, Portugal, and Spain. Unfortunately, information on the quantities of phycocolloids produced from the raw material is not easily accessible in Europe. These data are often considered as confidential by the producing companies.

Generally, the only information published gives the quantity of algae collected for extraction of alginates, agar, or carrageenan. For instance, between 50,000 and 60,000 t (wet weight) of *Laminaria* (*L. digitata, L. hyperborea*) per year are collected for the extraction of alginates in France. Similarly, each year 1000 t of *Gelidium corneum* are collected on the French Basque coast for the production of agar.

However, the French health agency (ANSES) reported a global production of alginates to 13,500 t in 2015 for Europe (https://www.anses.fr/fr/system/files/VSR2014SA0236Ra-4.pdf).

The European industry for the production or trading of phycocolloids is mainly located in Norway, France, Spain, or Italy (Table 14.8).

Phycocolloid production remains an important sector for valorization of seaweeds in Europe. However, it appears much lower in comparison to Asian production (China, Philippines, Indonesia).

Table 14.8 Examples of some European companies involved in phycocolloid production

Companies	Alginates	Agar	Carrageenan
JRS Marine Product Landerneau SAS (France)	+	–	–
CEAMSA (Spain)	+	–	+
FMC Biopolymer (Norway)	+	–	–
ROKO (Spain)	+	+	+
B&V "The Agar Company" (Italy)	+	–	–

14.4 Agricultural Uses

14.4.1 Biostimulant Application

In Europe, seaweeds are traditionally used for amendment (e.g., Maërl) or fertilization of soils. Nowadays, another application is mainly developed for the use of algae in agriculture. This new way of valorization is focused on the development of biostimulant products for the growth of plants with an agronomic interest.

The biostimulant effect of brown seaweed extracts is well known (De Saeger et al. 2020). The stimulation activity concerns not only the plant growth but also the resistance of pathogens (bacteria, fungi, and viruses). Generally, the main species involved to these mechanisms belong to the brown seaweed group, Laminariales, or Fucales families. In Europe, the species concerned are *Laminaria digitata* and *Ascophyllum nodosum* (Fleurence 2022). The liquid products obtained from these seaweeds are rich in degraded polysaccharides such as algin or fucoidan (Fig. 14.9).

Degraded fucoidans are good stimulators of soil microorganism flora and root systems (Craigie 2011). In addition, seaweed extracts, particularly those of *A. nodosum*, also contain phytohormones such as 3-indole acetic acid (auxin) or abscisic acid (Fig. 14.10) (Chatzissavvidis and Therios 2014). These hormones are known to contribute to plant growth and development.

The simultaneous presence of auxins such as 3-indole acetic acid and abscisic acid already has been detected in most commercial products (Wally et al. 2013). Several European companies produce seaweed biostimulants for plant growing (Table 14.9).

These companies are mainly Norwich, Spanish, French, Irish, British, Dutch, and German. Some of these companies are grouped together within a European professional council; the European Biostimulants Industry Council (EBIC). It is located in Brussels and represents the plant biostimulants industry in the capital of Europe. The Irish companies BioAtlantis and Yara are members of EBIC (https://biostimulants.eu/membership/our-members/).

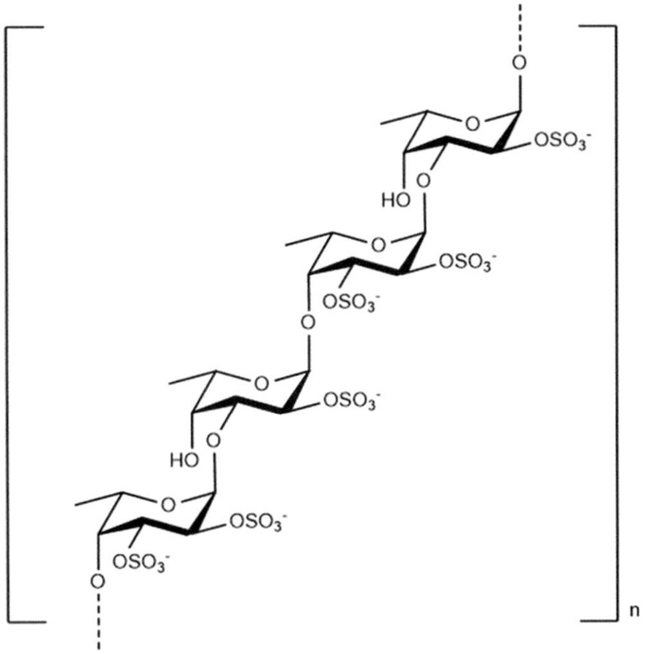

Fig. 14.9 Chemical structure of fucoidan from the brown seaweed *Ascophyllum nodosum*

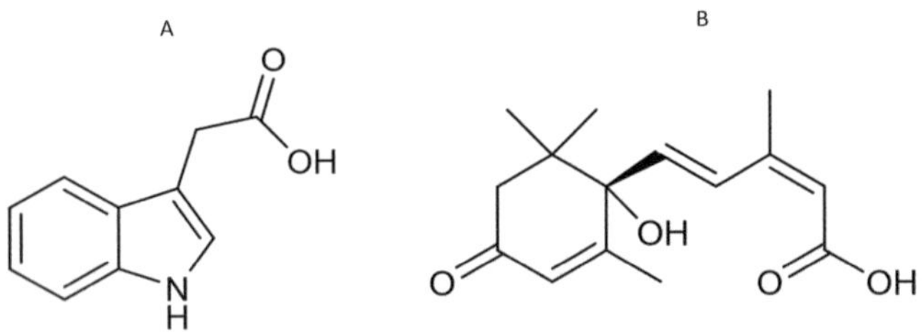

Fig. 14.10 Phytohormones characterized in *Ascophyllum nodosum* and *Laminaria* sp. extracts (**a**: 3-indole acetic acid **b**: abscisic acid)

Table 14.9 Examples of European companies involved to the production and the trading of seaweed biostimulants

Country	Company	Product	Seaweed species
Germany	Neomed Pharma Gmbh	Algifol	*Ascophyllum nodosum*
Netherlands	Van Iperen	FoliaStim	*A. nodosum*
France	Germineo	Ultrag B	Brown seaweed
	Setalg	Algovert	*Laminaria digitata* *A. nodosum*
	Goëmar Laboratories	Goëmar	*A. nodosum*
Spain	Fertival (Grupo Atlantica) Artal Smart Agriculture	Algival Algoton	*A. nodosum* *A. nodosum*
Ireland	BioAtlantis Ltd C-BIO Yara	Ecolicitor C-BioKelp YaraAmplix Optitrac	*A. nodosum* *L. digitata* *A. nodosum*
United Kingdom	Micromix Plant Health Ltd	GS35	*A. nodosum*
Norway	Kelpinor Polar Algae	Kelpinor Polar Ascophyllum	*Saccharina latissima* *A. nodosum*

14.4.2 Animal Feeding and Health Care

Algae have been often used for the feeding of livestock in Europe. There are many examples of this type of application in Scandinavian countries and in Irish, Scottish, and French regions. Currently, algae are sometimes used to feed livestock and for the preservation of animal health. Indeed, some products are manufactured to improve the microbiota of pigs, especially. The brown algae *Ascophyllum nodosum* is the main species used in the formulation of these products (Table 14.10). This species is an abundant resource on the European Atlantic coasts and already highly valued for the production of extracts stimulating plant development. Other products are made from a mixture of brown, red, and green algae without giving the name of the species concerned. This type of information is considered confidential by the manufacturers and is therefore not accessible. The effects of seaweed diets on the health care of livestock are well known (Fleurence 2023). For instance, the consumption of *A. nodosum* by weaned piglets has a positive influence of the intestinal flora of the animal (Dierick et al. 2009). The algal diet impact also concerns the quality of products such as color, organoleptic properties, and nutritional value. For these reasons, animal food products, including seaweeds, are well developed in Europe.

14.4.3 Cosmetic and Body Care Applications

Algae are often used in cosmetic preparations such as additive, excipient, or active substance (Couteau and Coiffard 2016). Brown seaweeds are the main algal resource used for this type of application. Indeed, they contain alginates, which possess gelling properties, Fucoidan, and laminarin (laminaran) with anti-inflammatory, anti-oxidant, and anti-aging activities (Fig. 14.11) (Morais et al. 2021).

Red and green seaweeds can also be applied in cosmetic formulations. Carrageenan and agar present in the red algae are suitable additives with their thickening and emulsifying properties. For the green seaweed, Ulvans with their gelling and anti-oxidant properties are good candidates for the elaboration of cosmetic preparations (Morais et al. 2021). In addition, seaweeds are rich in phenolic compounds (phlorotannins, bromophenols, and flavonoids) and various pigments (Dumay and Morançais 2016), presenting an interest for the cosmetic industry.

Table 14.10 Examples of some companies developing the use of seaweeds in animal feeds and health care

Country	Company	Product	Animal target	Seaweed species
United Kingdom	Ocean Harvest Technology	Oceanfeed Poultry Oceanfeed Swine Oceanfeed Bovine Oceanfeed Equine Oceanfeed pet Oceanfeed Aqua	Poultry Swine Bovine Horse Dogs cats Shrimp, Finfish	Green, red, brown seaweed
France	Olmix	Algimun MBiota	Swine, Poultry, Horse Ruminants	Green seaweed
Ireland	Sealac	Sealyte for Pigs Seaweed Flake	Pigs Cattle Sheep Calves	*Ascophyllum nodosum*
		Sealac 60	Horse	
	Titan	Titan® for Animals	Horse Dog	*Ascophyllum nodosum*
Norway	Algea "The Artic Company"	AlgeaFeed	Poultry Horse Fish Shrimp Pig Dog Cat	*Ascophyllum nodosum*

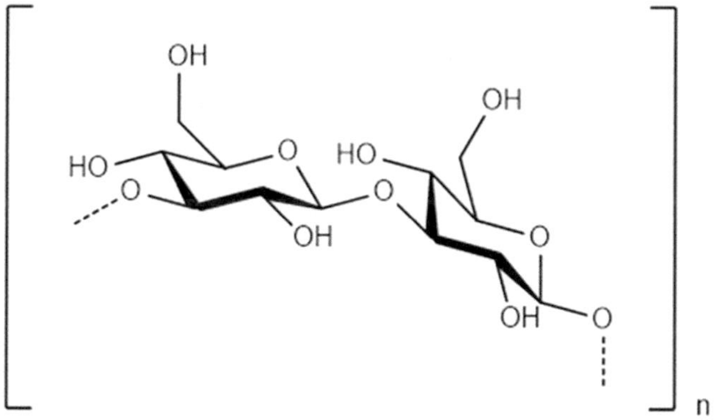

Fig. 14.11 Structure of laminarin from *Ascophyllum nodosum*

Most of the companies developing seaweed cosmetic products are present on European Atlantic coasts (Table 14.11). However, the Italian company Guam proposes a lot of body care manufactured from algae absent or little present in the Mediterranean (Armeli Minicante et al. 2022).

As in the case of many applications previously described, brown algae remain the most exploited algal resource in Europe for various applications and economic valorization.

14.5 Seaweed Sector Market in Europe and Economic Value

Seaweed market information is generally quite limited. It is collected by administrations, professional unions, researchers in the maritime economy, and private experts. Data collected by private expertise are not always easily available because they are often paid for.

Much information, especially that given by professionals, is strictly declarative and cannot always be verified. The most accessible and verifiable information concerns the edible seaweed market. Indeed, this market is based on trade data, such as the import, export, and production of seaweed for food purposes. For this type of application, there is easily analyzable data. In Europe, the market for edible algae is mainly supported by the import of algae, particularly from China. European production, although increasing, does not yet meet all needs. The United Kingdom is the main importer of algae from the European continent (Fig. 14.12) (https://www.cbi.eu/market-information/natural-ingredients-health-products/seaweed/market-potential). It is followed by France. If we take into consideration Brexit and political Europe, France then becomes the leading European importer.

In 2018, France imported 3034 t of seaweed mainly in dry form and 1800 t in 2019. This decrease of almost 42% is probably linked to the increase in French production capacity. During the same year, the volume of algae produced by cultivation for the French market was 125 t in fresh weight, or approximately 625 t of dried algae (https://sensalg.fr/la-production-dalgues-en-europe-et-la-culture-en-france/). The main companies involved in the cultivation of seaweed produce for the market of sea vegetables and ingredients are shown in Table 14.12. The species cultivated are *Undaria pinnatifida* (Wakame) and *Saccharina latissima* (Royal kombu).

Imported seaweed is mainly used by certain processors, if local production does not meet demand in Europe. The main

Table 14.11 Examples of European companies involved to the use of seaweed as active cosmetic

Country	Company	Product	Species
France	Algotherm	Ocean life	4 seaweeds
		Algoclear	*Fucus spiralis Tetraselmis chui* (green microalga)
		Ocean spa	*Laminaria* sp.
	Phytomer	Expert Jeunesse	Brown seaweed
	Dulsée	Balm	*Chondrus crispus*
Ireland	Algaran	Skin Detox Drop	*Laminaria* sp., *Ascophyllum nodosum, Ulva lactuca*
		Cream for the hand	*Asparagopsis armata*
	Wasi Seaweed	Cream Soap	Irish seaweeds
Norway	Lofoten Seaweed	Seaweed soap Artic Travel Essentials (skin and hair lotions)	*Alaria esculenta Fucus serratus*
Italy	Guam	Fanghi d'alga Guam	*Laminaria digitata, Fucus* sp. *Undaria pinnatifida*
Spain	Algae	Ulvanean®Posidonia	*Ulva* sp. *Posidonia Oceanica* (seagrass)

importing countries on the European continent (United Kingdom, France, and Germany) have a relatively well-developed network of producers, processors, and sellers (Table 14.13).

In 2019, the value of importation of seaweed was to 11.4 M Euros and 7.8 M Euros for the UK and France, respectively (Fig. 14.13).

Paradoxically, the importation of algae into Europe is done through intracommunity trade. China and Chile are only the two main extra-community exporters (Fig. 14.13; Table 14.14).

In general, for all applications, the economic value of the European seaweed sector is estimated to be between 500 and 600 M Euros by year (Mendes et al. 2022; see Fig. 14.13). In 2019, the world seaweed market was 10.9 B Euros. The specific value of the seaweed market depends on the production mode (e.g., harvesting or culture), species concerned, and type of application. The leading markets are the supply of biostimulants for agriculture, feed for animals, phycocolloids for the industry, and sea vegetables or ingredients for culinary uses. However, the market for edible algae, although growing in Europe, remains limited.

For all applications previously cited, the seaweeds mainly traded are brown algae belonging to the following species, *Ascophyllum nodosum, Alaria esculenta, Laminaria hyperborea, L. digitata, Saccharina latissima,* and *Undaria pinnatifida.*

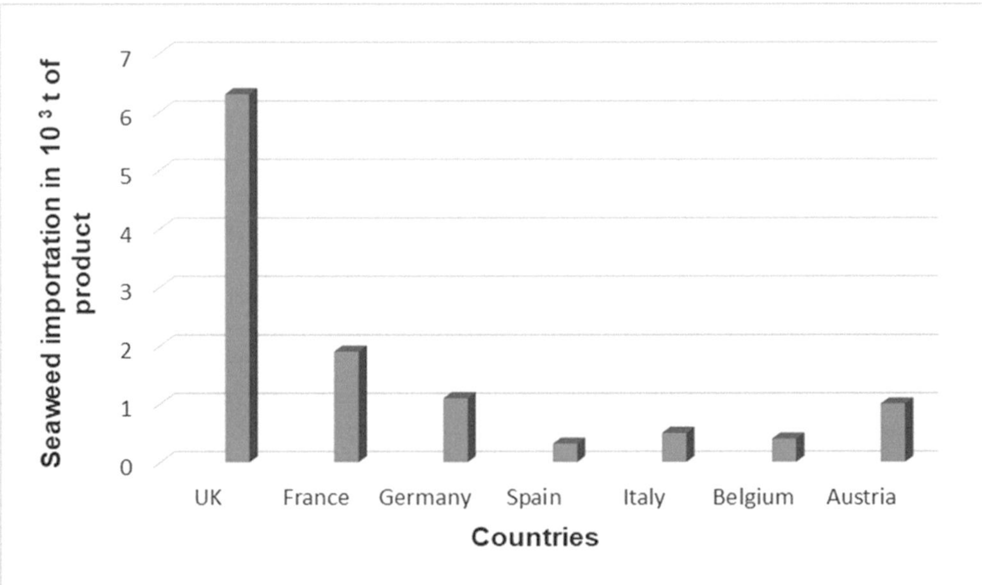

Fig. 14.12 European importers of seaweeds for food application in 2019 (https://www.cbi.eu/market-information/natural-ingredients-health-products/seaweed/market-potential)

Table 14.12 Main French companies producing seaweed by cultivation for the market of edible seaweed in Europe

Companies	Pond culture	Offshore Culture
Algolesko		+
Meristeme	+	
Nividic		+
C-weed aquaculture		+

Table 14.13 Main processors and importers in the three countries mainly involved in the market of edible seaweeds

Country	Processors	Importers
United Kingdom	The Cornish Seaweed Company Seagrown Shore Beachfood Company	
France	Algoplus Roscoff Bord à Bord GlobeXplore Marinoë	Angkor-Store Paris Store Tang Frère Asie Caraibes
Germany		Edeka Nanuko Go Asia

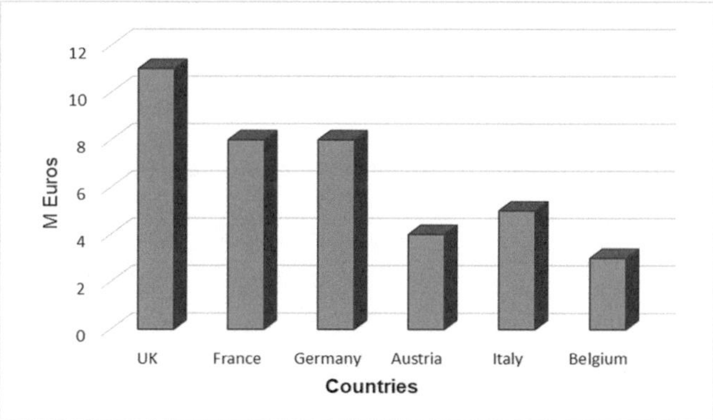

Fig. 14.13 Economic value of the edible seaweed market in Europe in 2019

For a few years now, the European seaweed market has shown rapid annual growth (7–10%) (Mendes et al. 2022). The species concerned by this progression are *Alaria esculenta*, *Saccharina latissima*, *Pyropia* sp., *Palmaria palmata,* and *Ulva* sp.

14.6 Regulation Constraints Applied to Algae

European regulations concerning the use of algae in human food are relatively lax. They are based on the "Novel Food" directive published in 1997. This document has been changed in 2015, and the use of seaweed as food now depends on regulation text 2015/2283 (https://eur-lex.europa.eu/eli/reg/2015/2283/2021-03-27).

In Europe, France has implemented specific regulations on the use of algae in human food.

This regulation provides a list of algae authorized for use as sea vegetables, condiments, or ingredients. This list includes seaweeds, microalgae, and a cyanobacterium, Spirulina (Table 14.15) (Fleurence 2016) (https://www.ceva-algues.com/wp-content/uploads/2019/10/V14-synthese-reglementaire2024-1.pdf).

In addition to the positive list of authorized species, French regulations provide toxicological criteria on heavy metals (Table 14.16). It also gives recommendations on the bacterial quality of seaweeds consumed (Table 14.17).

The French regulation is unique in Europe. Algae and seaweeds in particular are considered specific foods compared to others. This situation is probably the consequence of the interest shown in France for the use of algae in human food.

Table 14.14 Geographical distribution of importer and exporter on the European market of edible seaweeds

Importer country	Exporter country	% imported
United Kingdom	Spain	83.1
France	Chile Ireland	47.1 16.1
Germany	Netherlands Ireland China	32.0 19.6 10.3
Austria	Germany Italy	79.8 10.0
Italy	Ireland Germany China	35.7 13.7 13.1
Belgium	Netherlands France China	48.7 21.4 17.0

14.7 Conclusion

Phyconomy in Europe is based on centuries-old traditions. The oldest are based on the use of seaweed to fertilize soils or feed livestock. These historical uses are being taken up again in Europe today through the development of biostimulants for crops and feed that promote the growth and health of livestock. The consumption of seaweed by coastal populations in Europe is also part of the tradition, although less developed than in Asia. France, a country with a culinary tradition accustomed to preparing seafood and consuming seaweeds through Brittany, has become a leading country in the development of the food seaweed sector.

The development of seaweed consumption in Europe is mainly linked to the establishment of Japanese cuisine. In 2023, Europe had 14,870 Japanese restaurants. In this sector,

Table 14.15 Algae authorized in France as sea vegetables, condiments, or ingredients

Taxonomic family	Species scientific names	Common name
Brown seaweeds (Phaeophyceae)	*Ascophyllum nodosum*	Goemon
	Fucus vesiculosus, F. serratus	Varech, Black goemon
	Himanthalia elongata	Sea spaghetti
	Undaria pinnatifida	Wakame
	Laminaria digitata, Saccharina latissima, S. japonica	Breton kombu, Royal kombu, Kombu
	Alaria esculenta	Atlantic wakame or Irish wakame
Red seaweeds (Rhodophyceae)	*Palmaria palmata*	Dulse
	Porphyra umbilicalis, Pyropia tenera, Neopyropia yezoensis, Porphyra dioica, Porphyra purpurea, Erythroglossum laciniatum, Neopyropia leucosticta	Nori
	Chondrus crispus	Pioca, Lichen
	Gracilariopsis longissima	Ogonori
	Phymatolithon calcareum	Mäerl
Green seaweeds (Chlorophyceae)	*Ulva* sp. (formerly *Enteromorpha* sp.)	Sea Lettuce Aonori
Microalgae	*Odontella aurita* (Diatom) *Chlorella* sp. (Chlorophyta)	
Cyanobacteria	Spirulina spp. (*Arthrospira* sp. *Limnospira* sp.)	

Table 14.16 Maximal levels in Iodine and heavy metals for algae authorized in human nutrition in France (Fleurence 2016)

Mineral element	Maximal level (mg/kg of dry weight)
Mineral Arsenic (inorganic)	3
Cadmium	0.5
Mercury	0.1
Lead	5
Tin	5
Iodine	2,000

France appears to be a leader with the presence of 4,680 Japanese restaurants on its territory (https://www.statista.com/statistics/1479828/japan-number-japanese-restaurants-europe-by-country/). Italy, with a presence of 2,460 restaurants, comes in second place. In Europe, countries with a renowned culinary tradition seem to be countries favorable to the use of algae in consumption.

Table 14.17 Microbiological criteria applied to edible algae in France (Fleurence 2016)

Microbiological contaminants	Maximal levels (bacteria by g of dry weight)
Total aerobic flora	100,000
Fecal coliforms	10
Anaerobic flora	100
Staphylococcus aureus	100
Clostridium perfringens	1
Salmonella	0

The future of the phyconomy in Europe should be able to rely on the development of the food algae sector. The rise of algae cultivation and a better knowledge of these foodstuffs by the European consumer is probably the key to this development. The demand for environmentally friendly products in agricultural practice (biopesticides, biostimulants) also represents a promising challenge for the valorization of algae in Europe.

References

Armeli Minicante S, Bongiorni L, De Lazzari A (2022) Bio-based products from Mediterranean seaweeds: Italian opportunities and challenges for a sustainable blue economy. Sustainability 14(9):5634

Chatzissavvidis C, Therios I (2014) Role of algae in agriculture. In: Pomin VH (ed) Seaweeds. Nova Sciences Publishers, pp 1–37

Couteau C, Coiffard L (2016) Seaweed application in cosmetics. In: Fleurence J, Levine I (eds) Seaweed in health and disease prevention. Academic Press, London, pp 423–441

Craigie JS (2011) Seaweed extract stimuli in plant science and agriculture. J Appl Phycol 23:371–393

De Saeger J, Van Praet S, Vereecke D, Park J, Jacques S, Han T, Depuydt S (2020) Towards the molecular understanding of the mechanism of *Ascophyllum nodosum* extracts on plants. J Appl Phycol 32:573–597

Delaney A, Frangoudes K, Ii SA (2016) Society and seaweed: understanding the past and present. In: Fleurence J, Levine I (eds) Seaweed in health and disease prevention. Academic Press, London, pp 7–40

Dierick N, Ovyn A, De Smet S (2009) Effect of feeding intact brown seaweed *Ascophyllum nodosum* on some digestive parameters and on iodine content in edible tissues in pigs. J Sci Food Agric 89(4):584–594

Dumay J, Morançais M (2016) Proteins and pigments. In: Fleurence J, Levine I (eds) Seaweed in health and disease prevention. Academic Press, London, pp 275–318

FAO (2021) Global seaweeds and microalgae production, 1950–2019. https://openknowledge.fao.org/server/api/core/bitstreams/97409d09-2f8e-4712-b11e-60105d89959b/content

Fernandes WD, Pinto FR, Barroso S, Gil MM (2023) Development, characterisation, and consumer acceptance of an innovative vegan burger with seaweed. Sustainability 15(14):10869. https://doi.org/10.3390/su151410869

Fleurence J (2016) Seaweed as food. In: Fleurence J, Levine I (eds) Seaweed in health and disease prevention. Academic Press, London, pp 149–168

Fleurence J (2022) Biostimulant potential of seaweed extracts derived from *Laminaria* and *Ascophyllum nodosum*. In: Ramawat N, Bhardwaj V (eds) Biostimulants: exploring sources and applications. Springer Nature Singapore, Singapore, pp 31–49

Fleurence J (2023) Algae in agrobiology: realities and perspectives. John Wiley & Sons, London

Fleurence J (2024) Biotechnological processing methods. In: Legrand J, Trystam G (eds) Aquatic biotechnologies: from genetic engineering to enzymatic or fermentation engineering. John Wiley & Sons, London, pp 83–131

Fleurence J, Antoine E, Luçon M (2000) Procédé d'extraction et d'amélioration de la digestibilité des protéines de *Palmaria palmata*. France Patent 2811866

Guiry MD, Hession CC (1998) The seaweed resources of Ireland. In: Crtitchley AT, Ohno M (eds) Seaweed resources of the world. Japanese International Cooperation Agency, Yokusuka, pp 210–216

Le Strat Y, Ruiz N, Fleurence J, Pouchus YF, Déléris P, Dumay J (2022) Marine fungal abilities to enzymatically degrade algal polysaccharides, proteins and lipids: a review. J Appl Phycol 34(3):1131–1162

Mendes MC, Navalho S, Ferreira A, Paulino C, Figueiredo D, Silva D, Gao F, Gama F, Bombo G, Jacinto R, Aveiro SS, Schulze PSC, Gonçalves AT, Pereira H, Gouveia L, Patarra RF, Abreu MH, Silva JL, Navalho J, Varela JCS, Speranza LG (2022, 1871) Algae as food in Europe: an overview of species diversity and their application. Foods 11(13)

Monteiro P, Lomartire S, Cotas J, Pacheco D, Marques JC, Pereira L, Gonçalves AM (2021) Seaweeds as a fermentation substrate: a challenge for the food processing industry. Processes 9(11):1953. https://doi.org/10.3390/pr9111953

Morais T, Cotas J, Pacheco D, Pereira L (2021) Seaweeds compounds: an ecosustainable source of cosmetic ingredients? Cosmetics 8(1):8. https://doi.org/10.3390/cosmetics8010008

Neto AI, Tittley I, Raposeiro PM (2005) Flora marinha do litoral dos Açores, Secretaria Regional do Ambiente e do Mar

O'Connor K (2017) Seaweed. A global History. Reaktion books, London

Santos R, Melo RA (2018) Global shortage of technical agars: back to basics (resource management). J Appl Phycol 30:2463–2473

Sauvageau C (1920) Utilisation des algues marines. Gaston Doin, Paris

Stévant P, Ólafsdóttir A, Déléris P, Dumay J, Fleurence J, Ingadóttir B, Jónsdóttir R, Rebours C, Rustad T (2020) Semi-dry storage as a maturation process for improving the sensory characteristics of the edible red seaweed dulse (*Palmaria palmata*). Algal Res 51:102048. https://doi.org/10.1016/j.algal.2020.102048

Wally OSD, Critchley AT, Hiltz D, Craigie JS, Han X, Zaharia LI, Abrams SR, Prithiviraj B (2013) Regulation of phytohormone biosynthesis and accumulation in *Arabidopsis* following treatment with commercial extract from marine macroalga *Ascophyllum nodosum*. J Plant Growth Regul 32:324–339

Dr. Joel Fleurence, PhD is a Professor at the University of Nantes. He is specialized in the study of seaweed proteins. The fields of application of his research concern the valorization of algae in human and animal food. He has contributed to more 100 international publications, several books and 3 patents. In addition to scientific books, Joël Fleurence has also contributed to a children's book (Grateloupia) to introduce algae to younger generations.

15 Harvesting, Growing and Cooking with Seaweeds in the Bay of Cádiz (Spain)

J. Lucas Pérez-Lloréns, Alba Yamuza-Magdaleno, Juan J. Sánchez-Marabot, Fernando G. Brun, and Juan Juan Vergara

Abstract

The Bay of Cádiz (southern Spain) is a shallow embayment where, due to tidal movement, light, and optimal temperature, some seaweeds such as *Ulva* spp. or *Gracilaria* spp. grow abundantly in the earthen ponds. These species are harvested in the wild and sold fresh, dried, powdered, or salted to (mostly) local restaurants. In addition to the wild harvesting, several pilot research projects have been carried out in these earthen ponds to farm the species mentioned above, as well as others with a high culinary potential, such as *Gracilariopsis longissima*, *Chondracanthus tedeei*, or *Codium taylorii*. Probably the most limiting step in the use of seaweeds for culinary purposes in Cádiz (and in Spain as a whole) is not the availability of raw material, but the lack of knowledge about where to buy seaweeds and use them in the kitchen. Some high-end and mid-range restaurants have included local seaweeds in their menus, but the challenge is to get people to use this "blue food" in their home menus. In this sense, outreach activities such as seaweed tastings, show-cookings (demonstrations), television programs, radio interviews, or public talks on phycogastronomy (i.e., the art and science of cooking with seaweeds) are being carried out by the authors to increase the acceptance of seaweeds in home cooking.

J. L. Pérez-Lloréns (✉) · A. Yamuza-Magdaleno · J. J. Sánchez-Marabot · F. G. Brun · J. J. Vergara
Instituto Universitario de Investigación Marina (INMAR), Campus de Excelencia Internacional/Global del Mar (CEI·MAR), Departamento de Biología, Facultad de Ciencias del Mar y Ambientales Universidad de Cádiz, Campus Universitario de Puerto Real, Cádiz, Spain
e-mail: joselucas.perez@uca.es

Keywords

Blue Food · Cádiz Bay · Cultivation · *Gracilaria* · Harvesting · phycogastronomy · *Ulva*

15.1 Introduction

Population growth, rising incomes and changing food preferences will significantly increase global demand for nutritious food in the coming decades. Food from the sea, known as "blue food" (Bank et al. 2022; Tigchelaar et al. 2022), will play an important role in the global food supply through harvesting of noncommercial species and discards, as well as increased mariculture production, including seaweed farming (i.e., "phyconomy," Hurtado et al. 2019) (Costello et al. 2019). Historical records (from 1951 to the present) show no large variations in seaweed harvest, except for green algae. The highest harvest values (in tonnes of fresh weight) are for brown seaweeds (up to 800 K), followed by reds (up to 325 K), and greens (up to 60 K) (Fig. 15.1a). In contrast, in recent years, there has been a significant increase in the cultivation of brown (i.e., *Saccharina*, *Undaria*) and red seaweeds (i.e., Eucheumatoids, *Porphyra/Pyropia*, *Gracilaria*) (Fig. 15.1b) in Asian countries, mainly for human consumption (i.e., phycophagy), fresh feed in aquaculture or as hydrocolloids, and as new and emerging markets (e.g., biostimulants, animal feed, pet food, nutraceuticals, fabrics, bioplastics, construction) that have the potential to grow up to $11.8 billion by 2030 (World Bank 2023).

In contrast to the use of seaweed as a traditional food in many Asian or South American countries, its use in Western societies is more anecdotal (Dillehay et al. 2008; Mouritsen et al. 2018, 2024; Pérez-Lloréns 2019a), although recent archaeological research has shown that, contrary to popular belief, seaweeds and other aquatic plants were consumed extensively in many European coastal communities throughout the Neolithic transition to agriculture and into the early Middle Ages (Buckley et al. 2023).

A. T. Critchley et al. (eds.), *Advances in Temperate Phyconomy: Algal Harvest and Cultivation in Globally Distributed Temperate Waters*, Developments in Applied Phycology 15, https://doi.org/10.1007/978-3-032-02955-3_15

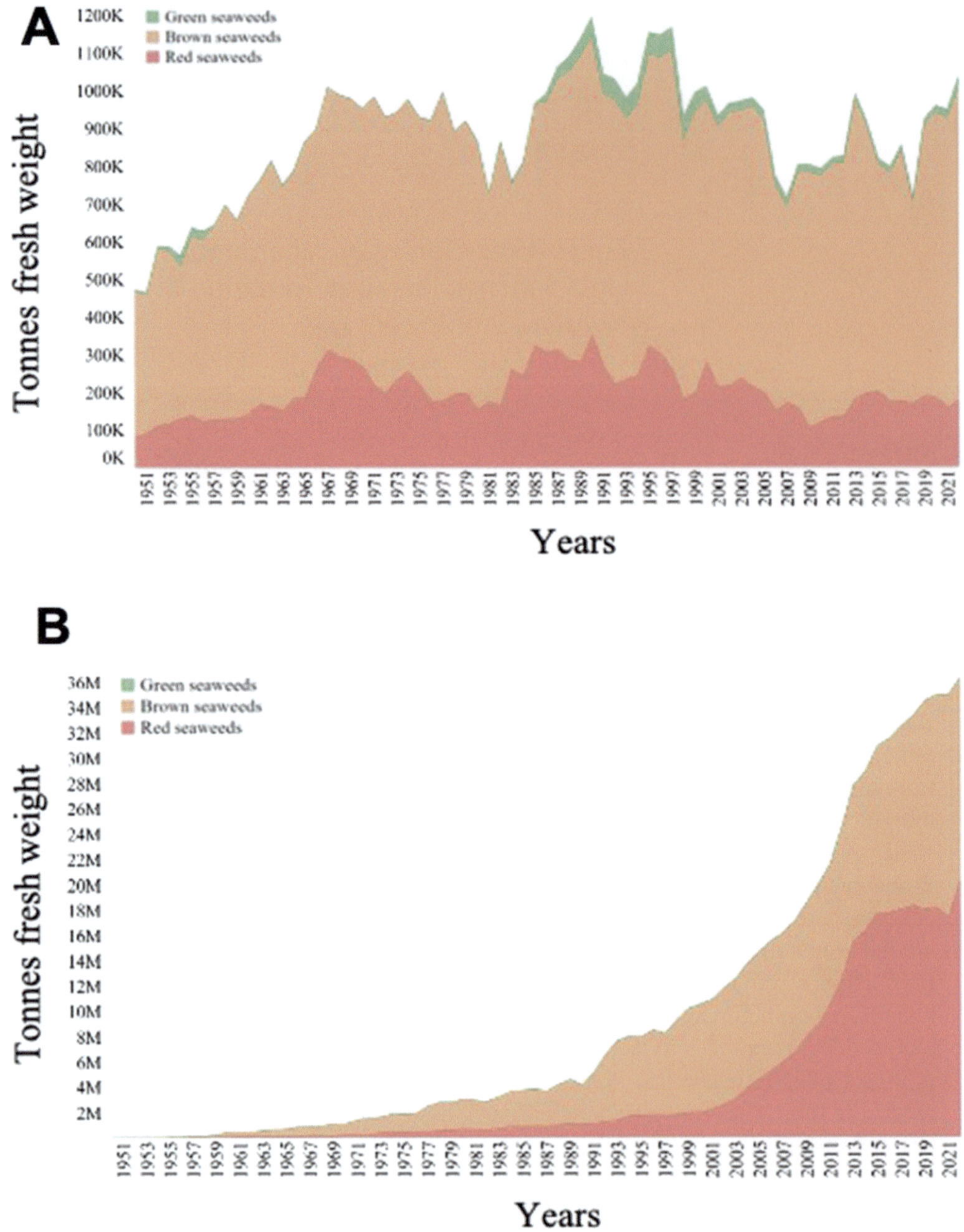

Fig. 15.1 Time series (1950–2022) of global wild seaweed harvest (**a**); global aquaculture seaweed production (**b**). Color code: green (green seaweeds), red (red seaweeds), brown (brown seaweeds). Please note the different scales of the y-axes. (Source: FishStat, FAO Fisheries & Aquaculture)

In northern Europe, seaweed consumption has traditionally been associated with hard times of poverty and famine. In southern Europe, neither the ancient Greeks nor the Romans seemed to be too keen on seaweeds, judging by some of the opinions expressed by famous personalities of the time: *immo ego Sardoniis videar tibi amarior herbis, horridior rusco, proiecta vilior algâ* (you would think me more bitter than the grass of Sardinia, rougher than the butcher's broom, and more vile than the seaweed thrown up by the sea) (Bucolic VII, Virgil, 70 BC); *et genus et virtus, nisi* cum *re, vilioralgâ est* (high birth and meritorious deeds, unless they are accompanied by wealth, are as useless as seaweed) (Satires II, v8 by Horace, 65 BC); *or refunditur algâ* (the sea detests seaweed and throws it ashore) (anonymous). In fact, there is no mention of seaweed as an ingredient in ancient culinary books or treatises such as *Hedypatheia* (*The Life of Luxury*) by the Greek writer and philosopher Archestratus of Gela (?–330 BC), *De Re Coquinaria* (*The Art of Cooking*) attributed to the Roman merchant and epicurean Apicius (first-century AD) or *Deipnosophistae* (*The Banquet of the Learned*) by the Greek Athenaeus of Naucratis (170–223 AD) (Pérez-Lloréns et al. 2018).

Nowadays, however, it is increasingly common to find opinions in the Western media promoting the use of this marine ingredient as another element of the so-called green transition (Mouritsen 2022) and a healthy and sustainable diet, although it is not listed as such in the EAT-Lancet Commission Report (Willett et al. 2019). Consider the following examples: "In 20 years kelp will rank as one of the top most consumed seafoods in America. If I could buy kelp futures, I would" (Greenberg 2014); "Food of the future: What will feed seven billion people? Insects, fake meat, seaweed and 3D-printed food all have the potential to address malnutrition, but could you stomach it?" (McEachran 2014); "Ants, seaweeds, chocolate beer and (maybe) less meat: The future of food" (King 2017).

There are two main reasons why seaweed is under-eaten in Western countries. One is neophobia (in this case phycophobia) which is the reluctance to try unfamiliar foods as a result of an evolutionary adaptation to avoid the potential risks of eating novel foods (Rozin and Fallon 1987; Alley and Potter 2011). Scientific research has shown that the acceptance of a particular food can be influenced by a variety of reasons (psychological, social, religious, anthropological, etc.) in addition to evolutionary ones (Rozin and Fallon 1985; Spence 2017). Another factor would be what we could call "phycoagnosis," i.e., a lack of knowledge about these organisms, in particular, their nutritional properties, how and where to obtain different species and how to cook them. Overcoming phycophobia and phycoagnosis is therefore the key to enjoying what is known as phycogastronomy (i.e., the art of cooking and enjoying with seaweeds, exploiting their nutritional and organoleptic properties; Mouritsen et al. 2019). This should be done by informing the general public (especially children), through educational talks and tastings, about the nutritional and organoleptic virtues of seaweeds and how they can be easily incorporated into their daily menus.

The aim of this chapter is to describe the species of seaweeds that are harvested and/or cultivated for human consumption in the Bay of Cádiz, as well as to present a selected series of events and promotional activities promoted by the University of Cádiz (UCA) and the Spanish Society of Phycology (SEF) in favor of the culinary use of seaweed.

15.2 Bay of Cádiz: "Eden" for Seaweeds and Phycologists

The province of Cádiz is a biodiversity hotspot, as it straddles two continents (Europe and Africa) and two seas (Mediterranean and Atlantic) (Pérez-Lloréns et al. 2023). The first phycological studies in Spain were carried out in Cádiz in the eighteenth and nineteenth centuries by famous personalities of the time, some of them not necessarily botanists but with great naturalistic interest. Some of them were the Valencian botanist Simón de Rojas Clemente y Rubio (1777–1827), the senior deacon of Cádiz Antonio Cabrera (1762–1827), and the Cádiz pharmacist Julián Heredia (18th–19th?) (Pérez-Lloréns et al. 2012). The Seville paediatrician and amateur naturalist Ramualdo González-Fragoso (1862–1928) wrote:

> Few regions are as worthy of detailed botanical study as the province of Cádiz. The advanced sentinel of the Iberian Peninsula, almost touching the nearby African coasts, its flora has a special character that cannot be confused with any other; and if this is true of the terrestrial flora, it is even more so of the marine flora. On its coasts, the waters of the Mediterranean and the Atlantic merge, and seaweed from both seas is tossed onto its beaches by storms. It is a land of promise for the phycologist, who can, in a short time, make a very rich collection of those beautiful plants that populate the bosom of the seas, delighting the most indifferent eye with their brilliant colours and making one admire nature, which is always admirable in its details as well as in its grandeur. It is a great pity that the study of our marine flora and fauna is so neglected, and may this note, as modest in its pretensions as its author is poor in science, be the sign of new and more important works on the flora of our coasts … (González-Fragoso 1886).

Cádiz is the type locality for more than 20 species whose botanical authorities are famous phycologists such as Clemente, Kützing, Agardh, or Greville, among others (Table 15.1).

Within the Province of Cádiz, the Bay of Cádiz is an important socio-ecosystem, since it is surrounded by cities with more than half a million inhabitants and a strong tourist pressure in the summer. It is a paradise for phycologists due to the diversity ot its marine flora. It is a shallow (average depth of 3 m with respect to the mean low water level) and luminous tidal (average tidal amplitude of 2.3 m) system, which is sometimes turbid due to strong easterly winds. Due to its shallowness, it is also subject to strong thermal oscillations, reaching greater thermal extremes than the open Atlantic waters, both in summer and in winter: It is a system with a great heterogeneity of habitats (Fig. 15.2), from the subtidal, where three of the four European seagrasses (*Zostera noltei*, *Z. marina*, and *Cymodocea nodosa*) (Tracheophyta) thrive to a variety of algal species. A total of 67 marine macrophyte species have been described in the bay, and more than a hundred if nearby ecosystems are included (Hernández et al. 2010, 2019). Species such as *Sporobolus maritimus*, *Sarcocornia* spp., and *Salicornia* spp. (Tracheophyta) inhabit in the high intertidal zone and in the low and medium marshes and earthen ponds, where fish aquaculture is currently being developed. The Bay of Cádiz therefore offers ideal conditions for the wild growth, harvesting, and cultivation of some species of edible seaweeds which have been exploited in some pilot projects carried out in this area (see next sections).

Table 15.1 Seaweed species with Cádiz as type locality, according to Algaebase (https://www.algaebase (https://www.algaebase.org), a global database of algal taxonomy, nomenclature, and distribution (Guiry and Guiry 2024).

Species	Phylum	Order
Cladophora albida https://www.algaebase.org/search/species/detail/? species_id=224 (Nees) Kützing 1843 (formerly *Cladophora humilis* Kützing 1849)	Chlorophyta	Cladophorales
Codium lineare https://www.algaebase.org/search/species/detail/? species_id=66182 C. Agardh 1822	Chlorophyta	Bryopsidales
Ulva intricata Clemente 1807 https://www.algaebase.org/search/species/detail/?species_id=48872	Chlorophyta	Ulvales
Ulva clathrata https://www.algaebase.org/search/species/detail/? species_id=W92dbf9511a17e069 (Roth) C. Agardh 1811 (formerly *Ulva muscoides* Clemente 1807)	Chlorophyta	Ulvales
Ulva rigida C. Agardh 1823 https://www.algaebase.org/search/species/detail/?species_id=40	Chlorophyta	Ulvales
Dictyota dichotoma var. intricata(C. Agardh) Greville 1830 https://www.algaebase.org/search/species/detail/?species_id=1308	Heterokontophyta	Dictyotales
Phyllariopsis brevipes (C. Agardh) E.C. Henry & G.R. South 1987 https://www.algaebase.org/search/species/detail/?species_id=1370	Heterokontophyta	Tilopteridales
Colpomenia sinuosa(Mertens ex Roth) Derbès & Solier 1851 https://www.algaebase.org/search/species/detail/?species_id=831	Heterokontophyta	Ectocarpales
Gracilaria multipartita (Clemente) Harvey 1846 https://www.algaebase.org/search/species/detail/?species_id=439	Rhodophyta	Gracilariales
Champia parvula (C. Agardh) Harvey 1853 https://www.algaebase.org/search/species/detail/?species_id=134	Rhodophyta	Rhodymeniales
Gloiocladia repens (C. Agardh) N. Sánchez & Rodríguez-Prieto 2007 https://www.algaebase.org/search/species/detail/?species_id=132993	Rhodophyta	Rhodymeniales
Halymenia elongata https://www.algaebase.org/search/species/detail/? species_id=654 C. Agardh 1822	Rhodophyta	Halymeniales
Halymenia floresii (Clemente) C. Agardh 1817 https://www.algaebase.org/search/species/detail/?species_id=1459	Rhodophyta	Halymeniales
Cryptonemia seminervis (C. Agardh) J. Agardh 1846 https://www.algaebase.org/search/species/detail/?species_id=650	Rhodophyta	Halymeniales
Caulacanthus ustulatus (Turner) Kützing 1843 https://www.algaebase.org/search/species/detail/?species_id=689	Rhodophyta	Gigartinales
Solieria chordalis (C. Agardh) J. Agardh 1842 https://www.algaebase.org/search/species/detail/?species_id=437	Rhodophyta	Gigartinales
Stenogramma interruptum(C. Agardh) Montagne 1846 https://www.algaebase.org/search/species/detail/?species_id=126	Rhodophyta	Gigartinales
Ceramium gaditanum(Clemente) Cremades 1990 https://www.algaebase.org/search/species/detail/?species_id=468	Rhodophyta	Ceramiales
Heterosiphonia crispella(C. Agardh) M.J. Wynne 1985 https://www.algaebase.org/search/species/detail/?species_id=772	Rhodophyta	Ceramiales
Lophosiphonia obscura(C. Agardh) Falkenberg 1897 https://www.algaebase.org/search/species/detail/?species_id=1481	Rhodophyta	Ceramiales
Melanothamnus collabens(C. Agardh) Díaz-Tapia & Maggs 2017 https://www.algaebase.org/search/species/detail/?species_id=162875	Rhodophyta	Ceramiales
Platysiphonia delicata (Clemente) Cremades 1990 https://www.algaebase.org/search/species/detail/?species_id=3294	Rhodophyta	Ceramiales
Pterosiphonia complanata (Clemente) Falkenberg 1897 https://www.algaebase.org/search/species/detail/?species_id=512	Rhodophyta	Ceramiales
Rytiphlaea tinctoria (Clemente) C. Agardh 1824 https://www.algaebase.org/search/species/detail/?species_id=809	Rhodophyta	Ceramiales
Gelidium microdon Kützing 1849	Rhodophyta	Gelidiales
Helminthora divaricata(C. Agardh) J. Agardh 1852	Rhodophyta	Nemaliales

Fig. 15.2 Aerial view of the marshes and intertidal mudflats of the Bay of Cádiz, with the city in the background (**a**); detail of the tidal creeks and earthen ponds. The different types of vegetation can be easily identified by the changes in color (**b**); detail of an earthen pond with floating masses of *Ulva* spp. (**c**). (Photos by the authors)

15.3 Edible Seaweed Species Harvested from the Wild

Despite the richness of the seaweeds in the Bay of Cádiz, only a limited number of species are currently harvested by hand for culinary purposes (Fig. 15.3). The main local company that harvests, processes, and markets seaweeds to specialized shops and restaurants is Suralgas https://suralgas.com/, a company founded 15 years ago by Consuelo Guerra, a former student from the Faculty of Marine and Environmental Sciences (University of Cádiz).

The species harvested and commercialized are *Ulva* spp. (Chlorophyta) (both the laminar type, sea lettuce, and the tubular type, green aonori), *Gracilaria* spp., and *Gracilariopsis longissima* (red ogonori) (Rhodophyta) and, rarely, *Chaetomorpha linum* (Chlorophyta) for some local tasting events or restaurants (on request) who want to test different seaweeds (Fig. 15.4). There is also some harvesting of *Codium* spp. (Chlorophyta) growing on hard substrata (either on the bay or on the coast), although its abundance is not sufficient to satisfy the gastronomic demand and it is mainly imported and commercialized from the north of Spain (Galicia). There is seasonality in the abundance of seaweeds. For example, *Ulva* spp. (both laminar and tubular forms) have maximum values in winter and spring, and minimal values in summer due to light and temperature stress. *Gracilaria* spp. and *Gracilariopsis longissima* occur mainly in spring-summer in earthen ponds (Hernández et al. 2006, 2008).

15.4 Processing Harvested Seaweeds for Culinary Use

Once seaweeds have been carefully harvested by hand, they are processed in the companys' facilities to ensure food safety in different formats. In the case of Suralgas, seaweeds are sold fresh, salted, or dehydrated (Fig. 15.5). Salted and especially dehydrated seaweeds have a longer shelf life. Other product formats marketed are "marine spices" (a dehydrated and coarsely ground mixture of *Ulva* and *Gracilaria*), "seaweed salt" (a mixture of flower of salt and dehydrated mixture of *Ulva* and *Gracilaria*), and "seaweed essence" (a concentrated mixture of dehydrated and finely ground mixture of *Ulva* and *Gracilaria*), which add an intense aroma and flavor of the sea to all types of dishes, from traditional to avant-garde cuisine (Fig. 15.6).

15.5 Edible Seaweed Cultivation in Cádiz Bay: Pilot Experiences

There have been experiences with seaweed cultivation in the Bay of Cádiz since the 2000s. However, most of these crops have been the result of various pilot research projects, mainly carried out by the University of Cádiz (Table 15.2). There are no local companies dedicated to seaweed production, and it could be a good niche market for the creation of companies related to the blue economy.

Fig. 15.3 Main species of seaweed growing on the intertidal mudflats and earthen ponds of Cádiz Bay, harvested for culinary use. *Ulva* spp. (**a**); *Ulva* spp. (formerly *Enteromorpha* spp.) (**b**); *Gracilaria* spp./*Gracilariopsis longissima* (**c**); *Codium* spp. (**d**). (Photos by the authors) (Scale bar =??? cm. All whole plants of seaweeds must have a scale bar)

The first of these projects was carried out in the early 2000s in collaboration with an aquaculture company, Acuinova S.L. The aim was to analyze the capacity of seaweeds growing in earthen ponds to biomitigate nutrient pollution caused by fish aquaculture effluents, and to see the possible natural growth and use of seaweed biomass, especially of green Ulvaceans (laminar and tubular species) and the red alga *Gracilariopsis longissima*. This project included studies from the microscale (laboratory) to the mesoscale (outdoor culture tanks) and macroscale (earthen ponds) (Hernandez et al. 2002, 2005, 2006, 2008).

In the field, maximum growth rates of 8% day^{-1} for *Gracilariopsis longissima* and 12.7% day^{-1} for *Ulva intestinalis* (Chlorophyta) were achieved in late summer, with losses in summer due to extreme light and temperature conditions. Correspondingly, maximum values for photosynthesis were found in autumn and minimum in summer for both species (Pérez-Lloréns et al. 2004).

In the Acuinova earthen ponds, the maximun densities of *Ulva lactuca* (formerly *Ulva rotundata*) (Chlorophyta) were as high as 600 gDW·m^{-2}, with a production about 20.5 gC·m^{-2}·day^{-1}, an enormous amount of biomass that grows by itself and is exploitable; *Gracilariopsis longissima* was cultivated in ropes (Fig. 15.7a), with growth rates oscillating between 5% and 6% day^{-1}, and maximum biomasses varying between 639 g FW·m^{-1} rope in earthen ponds with incoming water, and 1560 g FW·m^{-1} in open reservoirs (enriched with nutrients from the fish farm) (Hernández et al. 2006).

Another gracilarioid species, *Gracilaria gracilis*, has recently been monitored in the bay, with maximum growth rates in spring (3.6% day^{-1}) and negative values in summer, like *Gracilariopsis longissima*. Another possible target species for cultivation is *Chondracanthus teedei* (Rhodophyta) (Figs. 15.7b and 15.8a), with maximum growth rates of 4.7% day^{-1} in autumn (López-Campos et al. 2022). Natural populations of the latter species are scarce and widely dispersed, but *C. teedei* has promising gastronomic potential if it can be cultivated. Other studies have grown this species and other gracilarioid species in ropes and obtained growth rates in the same order of magnitude (Bermejo et al. 2022). These authors reinforce the idea that the best period for seaweed cultivation in the Bay of Cádiz is from mid-winter to early summer, avoiding extreme temperature conditions.

In a recent EU project (Novafoodies), the Andalusian Aquaculture Technology Centre (CTQUA) in Cádiz is developing cultures of three interesting species, the red alga *Chondracanthus teedei*—already mentioned—which is

Fig. 15.4 Harvesting wild seaweeds: *Ulva* spp. (**a**) and *Chaetomorpha linum* (**b**). (Photos by the authors)

interesting for its phycocolloids and for its culinary potential, and the greens *Codium taylorii* and *Ulva ohnoi* (Chlorophyta).

In the future, there are a number of species in the Cádiz area that are neither collected nor exploited and that could be a priority target for cultivation and human consumption for their culinary potential, such as the red alga *Nemalion elminthoides* (Rhodophyta) (sea noodle or tsukomo nori, in Japanese), or the edible brown alga rich in proteins *Scytosiphon lomentaria* (leather tube or *sugara*, in Japanese) (Fig. 15.8), which is cultivable and of interest as an alternative resource to other edible species (Zhang et al. 2023). A third species, for which there is an ongoing project in the bay is the brown alga *Cladosiphon* sp. (Table 15.2), a seasonal and rare species in Cádiz but may be of interest for cultivation and promising for a variety of applications, including human consumption.

15.6 Not Seaweed ... But a "Roommate": Seagrass

The few populations of *Zostera marina* (eelgrass) (Tracheophyta) in the Bay of Cádiz are very patchy and sparse, as this species is at its southern limit. A recent pilot research project led by the University of Cádiz has, for the first time, managed to domesticate this species in the natural earthen ponds of the Bay of Cádiz on a small scale in order to obtain a seed bank for restoration projects (Pérez-Lloréns and Brun 2023) (Fig. 15.9). It has opened up a new paradigm

Fig. 15.5 Different commercial formats of seaweed from the Bay of Cádiz that can be purchased in local establishments. Fresh seaweed (**a**); salted seaweed (**b**); dehydrated seaweed (**c**). (Photos by the authors)

Fig. 15.6 Seaweed spices, seaweed salt or seaweed essence in various formats. (Photos by the authors)

Table 15.2 Projects developed at the University of Cádiz related to applied phycology (bioproducts, aquaculture, consumption, bio-mitigation, etc.) with different sources of funding (European, national, regional, private companies, etc.)

Project	Years
Integrated Aquaculture Pilot Project: Reducing the environmental impact of nutrient pollution from an aquaculture farm by growing local marine macroalgae for human consumption	2000–2001
Assessment for the study of marine flora of commercial interest in the Barbate fishing grounds	2009–2010
Optimization of the collection and outdoor cultivation of macroalgae for the food industry in earthen ponds in the bay of Cádiz. Nutritional/gastronomic potential and environmental impact	2014–2019
Macroalgae in integrated aquaculture: an innovative application for production systems in Andalusia	2015
La Caleta (Cádiz) and associated rocky intertidal areas: a permanent window on biodiversity	2018
Technical assistance for the development of cultivation in earthen ponds of the seagrass *Zostera marina*[a]	2018–2020
Promotion of the emerging and sustainable micro- and macroalgae sector in Spain: diversification of species and applications with high economic and environmental value	2021–2023
Toward the domestication of seagrasses in the earthen ponds of the Bay of Cádiz[a]	2021–2024
Innovation in the development of functional algae foods for a more sustainable and healthier society	2022–2024
Obtaining seaweeds of different colors for phycogastronomy	2023
Obtaining seaweeds of different colors: biomass quality	2023
Research on the cultivation and applications of a native macroalga: *Cladosiphon* sp.	2023–2025
Phytomariculture: toward the creation of new opportunities for (associated with) the cultivation of seagrasses in earthen ponds[a]	2023–2025

[a]Seagrass projects

that could economically and ecologically revitalize currently degraded areas by cultivating seagrasses for blue carbon restoration projects. In addition, part of the seed production, as a high value-added by-product of seagrass restoration, could be redirected toward new culinary possibilities and/or new products (bread, drinks, cosmetics, etc.). Indeed, the world's media announced this pioneering domestication and the intention of Aponiente (a local 3 Michelin star restaurant) to include it (as "sea cereal" or "sea rice") on its menu in the short to medium term (Goulding 2021; Kassam 2021). However, there is not enough seed production to offer it on the menu of this fine dining restaurant (Cano 2023; Galisteo 2023). More research and investment is needed to develop intensive seagrass cultivation in the Bay of Cádiz, as seagrasses are usually protected species and their potential ecological and gastronomic use in a given location cannot depend on plants or seeds imported from other regions (Pérez-Lloréns and Brun 2023).

15.7 Seaweeds as Ingredients in Local Gourmet Products

In addition to the marketing of fresh, salted (brine) or dried seaweed (see previous section), mainly for restaurants and home use, there is an increasing number of gourmet products that use seaweed from the Bay of Cádiz as an ingredient. The bakery "La Cremita" (Chiclana, Cádiz), for example, stands out for the creativity and originality of its creations, which are based on quality local raw materials (i.e., "terroir"). Its star products are bread, little breadsticks ("picos"), little crackers ("regañas") (Fig. 15.10a, b), croissants, and tarts ("cocas") made with seaweed. One of the most innovative proposals presented at InterSICOP 2024 (International Bakery, Patisserie and Coffee Show) was the seaweed bread, made with sea lettuce harvested in Sancti Petri (Bay of Cádiz), shrimp (*Palaemon serratus*), and seawater (González 2022). This bakery works with the most prestigious chefs in the province, looking for unique breads to accompany the dishes on their tasting menus. "Picos" and "regañas" are used to accompany Spanish tapas (especially in southern Spain). Another much sought-after product is cheese with seaweeds. This is the case of El Gazul, a cheese dairy in the Cádiz mountains (Sierra de Cádiz) which produces a semi-hard and dense artisanal "payoyo" cheese (made with pasteurized milk from the rare Payoya goat, native to the mountain pastures of the Sierra de Grazalema, Cádiz), to which is added *Ulva* collected in the Bay of Cádiz (Fig. 15.10b). The company Oleum Viride from Zahara de la Sierra (Cádiz) has also added dried seaweeds (*Ulva* and *Gracilaria*) to one of its extra virgin oils. It is considered the

Fig. 15.7 Seaweed species of culinary interest cultivated using the rope technique in the estuaries of the Bay of Cádiz: *Gracilariopsis longissima* (**a**) and *Chondracanthus teedei* (**b**). (Photos by the authors)

ideal complement to salads or oriental cuisine. "Suralgas" company has also marketed elaborated products such as "Seaweed Tartar" (containing 80% raw *Ulva*) or "Seaweed Pesto" (containing 70% of raw *Ulva* and *Gracilaria*). Seaweeds (*Gracilaria* and *Ulva*) collected in the Bay of Cádiz are also used by a local confectioner (Los tres Martínez) to make delicious chocolates.

15.8 Seaweed: A Rising Star in Cádiz's Restaurants

Cádiz, surrounded by the Atlantic Ocean and the oldest city in Europe (3000 years old), has a deep-rooted and profound tradition of seafood cuisine (Pérez-Lloréns 2019b; Pérez-

Fig. 15.8 Some species of seaweed found along the coasts of Cádiz with gastronomic potential, but not currently exploited: *Chondracanthus teedei* (with cystocarps) (**a**); *Scytosiphon lomentaria* (**b**); *Nemalion elminthoides* (**c**). (Photos by the authors)

Lloréns et al. 2021). However, until a few years ago, the use of seaweed in restaurants (i.e., phycogastronomy, Pérez-Lloréns et al. 2018; Mouritsen et al. 2019) and, especially, at home was rather marginal, as it is often still considered an exotic component of menus. Seaweed can now be found on the menus of some fine restaurants whose gastronomic discourse focuses on sustainability and the search for new local ingredients ("terroir") that are rarely used and that can

Fig. 15.9 *Zostera marina* planting in natural earthen ponds in the Bay of Cádiz (**a**); flowering shoot with unripe seeds (**b**); mature seeds (**c**); Kaiseki-inspired salad dish, with a makizushi made with *Zostera marina* seeds (instead of Japonica rice), garnished with an assortment of local pickled seaweeds on daikon slices (dish by J. L. Pérez-Lloréns). (Photos by the authors)

Fig. 15.10 Seaweed from the Bay of Cádiz in some gourmet products: Crackers ("regañas") (**a**); breadsticks ("picos") and cheese (**b**). (Photos by the authors)

surprise diners with their novelty, exoticism, or nutritional qualities. Local seaweeds (*Gracilaria*, *Ulva* and *Codium*, mostly) and saltmarsh plants (*Salicornia*) are used in a wide variety of preparations and culinary techniques in high-end restaurants: in "crème fraiche" (Fig. 15.11a) to accompany grilled fish; in vinegar to accompany "tataki" (Fig. 15.11b); floured and fried to an intense crisp to accompany fried egg (Fig. 15.11c); to make earth or sand (Fig. 15.11d); sauces (Fig. 15.11e); or even as desserts (covered in dark or white chocolate) (Fig. 15.11f).

Mid-range restaurants have also incorporated local seaweed in their menus (Fig. 15.12) due to the good acceptance of seaweed in high-end restaurants. This makes it more accessible to a less affluent public that cannot afford the high bills typical of high-end restaurants, but is eager to try new ingredients. In these cases, they are often used as an additional ingredient or as a substitute for an ingredient in traditional dishes. For example, they can be added raw or lightly poached to small sandwiches of salmon or smoked cod (Fig. 15.12a), as vinaigrettes to accompany prawn or fish skewers (Fig. 15.12b), to replace parsley in chickpea shrimp fritters ("tortitas de camarones"), perhaps the most popular dish in Cádiz cuisine (Fig. 15.12c), mixed (dried seaweed powder) with flour for deep-frying fish (Fig. 15.12d), in various salads (Fig. 15.12e), or in soups (Fig. 15.12f).

Recently, in the Bay of Cádiz, in the middle of the salt marshes, restaurants have sprung up in old restored salt flat

Fig. 15.11 Seaweed from the Bay of Cádiz on some of the menus of local high-end restaurants. Grilled red mullet fillet with *Ulva* spp. crème fraiche (**a**); bluefin tuna tataki with *Ulva* spp. (**b**); deep-fried *Gracilaria* nest with fried egg and *Salicornia* (**c**); seaweed earth and foam (**d**); low-temperature cooked hake with light seaweed and saffron sauce (**e**); *Codium* coated in white chocolate and sprinkled with seaweed essence (**f**). (Photos by the authors)

Fig. 15.12 Seaweed from the Bay of Cádiz on some menus of local mid-range restaurants. *Gracilaria* spp. and smoked cod sandwich (**a**); prawn skewer dressed with *Ulva* spp. vinaigrette (**b**); shrimp fritters ("tortitas de camarones") with *Ulva* spp. (**c**); piece of fried fish in batter with *Ulva* spp. (**d**); lettuce hearts salad with *Gracilaria* spp. (**e**); pumpkin soup with lightly poached *Ulva* spp. (**f**). (Photos by the authors)

Fig. 15.13 Aerial view of Marambay (**a**) and Salina de Santa Teresa (**b**) restaurants in the heart of the Cádiz Bay salt marshes, where local specialities such as seaweed can be tasted. (Photos by the authors)

houses, offering their guests gastronomic experiences with very fresh local products, as well as other activities such as fishing, guided tours to observe birds and flora of the marshes, and thalassotherapy. Among these restaurants, "Marambay" https://marambay.com/ and "Salinas de Santa Teresa https://salinasantateresa.es/" deserve special mention (Fig. 15.13).

15.9 Seaweed: The Untapped Ingredient in Cádiz Home Cooking

Unlike in restaurants, seaweed is rarely used as a culinary ingredient in home cooking. Seaweed can be used in traditional dishes such as hot or cold soups (e.g., "vichyssoise," "gazpacho") (Fig. 15.14a), omelets (Fig. 15.14b), stews (Fig. 15.14c), rice dishes (Fig. 15.14d), fish balls (Fig. 15.14e), in purées or hummus (Fig. 15.14f), or in any other home dish. The possibilities are endless, all it takes is imagination and a desire to try something new, nutritious, and tasty!

Before the culinary use of seaweed, some local restaurants started to use microalgae in their menus. Specifically, *Tetraselmis chui* (Chlorophyta), which is marketed under the brand name "plancton marino" by the local company Fitoplancton Marino S.L. Once dried, this micro-alga turns into a fine green powder. It can be easily rehydrated and used in many home (Fig. 15.15) or avant-garde recipes: from risotto to mayonnaise, to flavor marinades and sous-vide cooking, or to enrich a savory ice cream (Pérez-Lloréns et al. 2018). Other ingredients with which it can be combined include sushi, extra virgin olive oil, aromatic white wines, and late harvest wines. The only drawback to its use in home cooking is its excessive price, although it can be replaced by a concentrated mixture of dehydrated and finely ground *Ulva* and *Gracilaria* (Fig. 15.6), sold by the local company Suralgas, which is much more affordable (Pérez-Lloréns 2020; Pérez-Lloréns and Vergara 2023).

15.10 Popularizing the Nutritional Qualities and Culinary Potential of Seaweed

The main reason why seaweed is not a regular part of household menus in Cádiz is a lack of knowledge about where to buy it, how to prepare it, and how to include it in daily menus, rather than phycophobia itself. This, at least, is the conclusion we have reached after several years dedicated to spreading the message of nutritional and culinary benefits to a public hungry for new knowledge.

Below are some selected examples of the work we have carried out at the University of Cádiz, INMAR, CEI·Mar, and in some cases with the participation of the Spanish Phycological Society (SEF). They range from monographic and specialization courses for students to a variety of scientific outreach activities such as lectures on the culinary potential of seaweed, often combined with tastings or showcookings prepared with seaweeds (Fig. 15.16) (Pérez-Lloréns 2021a, 2023; Pérez-Lloréns and Bartual 2022).

During the summers of 2014 and 2015, the course "The Amazing World of Seaweeds: An Unforgettable Journey from Biology to Gastronomy" was held in Cádiz (Spain), as part of the International Summer School training program jointly organized by the International Campus of Excellence in Marine Science (CEI-Mar) and the University of Cádiz. On both occasions, Prof. Ole G. Mouritsen and Dr. Prannie Rhatigan were invited as speakers of recognized prestige in the field of phycogastronomy. On these occasions, the students had the opportunity to delve into the theoretical and practical aspects of the biology, cultivation and gastronomic potential of seaweeds.

A workshop on "phyco-cuisine" was also held at the University of Cádiz (2016), aimed at high school students to increase their interest in science, technology and innovation. On this occasion, they had the opportunity to taste a seaweed "fideuá" and "canapés" with plankton spherification (Fig. 15.16a) (Bartual and Pérez-Lloréns 2016).

Fig. 15.14 Some homemade dishes prepared with seaweed. *Vichyssoise* with crunchy seaweed flakes (**a**); spongy omelet with *Ulva* spp. puree (**b**); chickpea, octopus, and seaweed stew (**c**); rice with seafood and seaweed (**d**); cuttlefish meatballs in a light *Ulva* spp. sauce (**e**); hummus with octopus and crunchy seaweed (dishes by J.L. Pérez-Lloréns). (Photos by the authors)

Fig. 15.15 Some homemade dishes prepared with plankton (*Tetraselmis chui*). Courgette bisque, smoked sardine fillet and plankton pearls (**a**); potato, cuttlefish, and plankton stew (**b**); grilled scallop with plankton butter and black Venere rice (**c**); plankton and squid risotto (**d**); bluefin tuna and prawn curry balls in coconut milk sauce and plankton couscous (**e**); potato soufflé with low-temperature hake and seaweed tempura with smoked paprika and plankton sauces (**f**) (dishes by J. L. Pérez-Lloréns). (Photos by the authors)

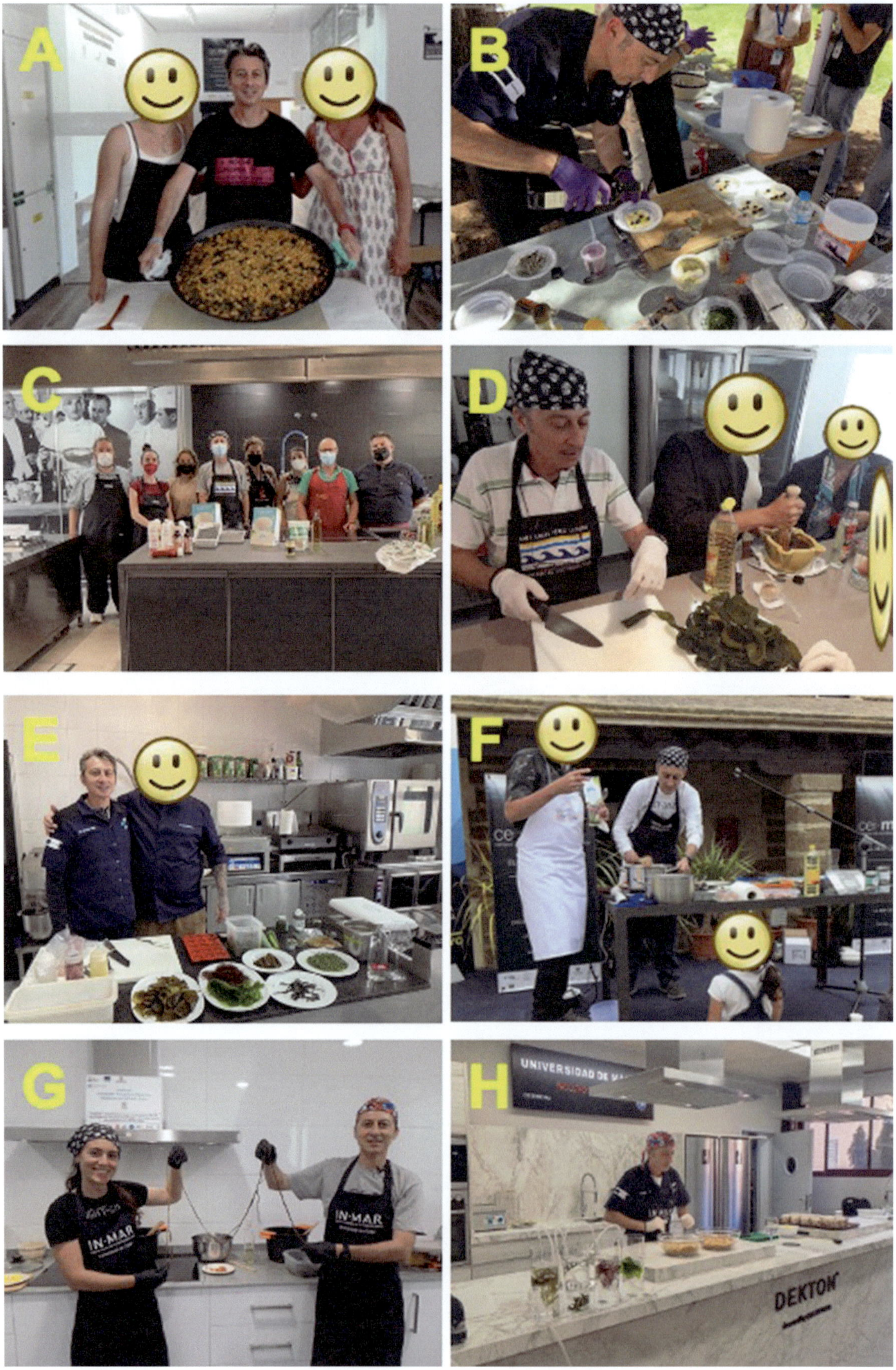

Fig. 15.16 Demonstration cooking and tasting menus with seaweed in several outreach events. Summer workshop at the University of Cádiz with high school students (**a**); annual meeting of the Spanish Society of Phycology (SEF) (**b**); Congress of Young Marine Scientists (**c**); students of the Master's in Innovation and Culinary Culture (**d**); show-cooking at the Culinary School of Cádiz with chef Mauro Barreiro (**e**); "tuna skin" event in the city of Cádiz (**f**); event at the University of Cádiz (**g**); show-cooking at "soludable" meeting (Faculty of Gastronomy and Tourism, University of Malaga) (H). (Photos by the authors)

The University of Cádiz's Master "Culinary Innovation and Culture (Masterñam)" includes a course on the biology and gastronomy of seaweed, in which students take part in the preparation (and tasting) of a menu made with local seaweeds (Fig. 15.16d).

Some of the seaweed tastings took place at important events. One of them was "Innovazul" (2022), where a tasting of a six-course menu was held: sea spaghetti tempura (Fig. 15.17a); sea lettuce and prawns stew (Fig. 15.17b); chickpea salad with seaweed (Fig. 15.17c); Korean omelet (Fig. 15.17d); squid tagliatelle ("trompe-l'œil") with seaweed (*Ulva*) "pesto" (Fig. 15.17e); and pineapple topped with crunchy seaweed flakes (Fig. 15.17f).

Another important event was the European Researchers' Night (2024), where participants were able to taste the menu created by the executive chef Juan José Sánchez-Marabot (Fig. 15.18a), which consisted of various dishes: beetroot ("salmorejo," Andalusian thick tomato soup) with topping of apple crumbles and crunchy seaweed flakes (Fig. 15.18b); local bluefin tuna tartare with seaweed and passion fruit spheres on seaweed biscote (Fig. 15.18c); prawn "Takoyaki," "katsuobushi," and *Gracilaria* spp. (Fig. 15.18d); prawn and seaweed "risotto" with crunchy *Ulva* (Fig. 15.18e); and "false truffle" made of carob flour with *Codium* and sprinkled with powdered *Ulva* (Fig. 15.18f).

Another relevant activity took place at the Cádiz Culinary School, where a lecture on the culinary potential of seaweed was given to young cookery students and a demonstration cooking session was held by the renowned local chef Mauro Barreiro (Fig. 15.16e). The menu prepared for the occasion was: seaweed butter (Fig. 15.19a); prawns in "dashi" broth and *Codium* semigel (Fig. 15.19b); monkfish *Ulva* roll with seaweed hollandaise sauce (Fig. 15.19c); and semi-cured monkfish "sashimi" with seaweed powder and extra virgin olive oil (Fig. 15.19d) (Pérez-Lloréns 2021b).

Other outreach activities include participating in local television programs such as "Sabor Gourmet" (Gourmet Taste), a program in which the guest is interviewed while preparing a series of dishes. On this occasion, we had the opportunity to prepare seaweed dishes in an emblematic setting of Cádiz city, such as the beach of La Caleta.

Last but not least, in terms of dissemination, the publication of the books "Ũ Las algas se comen?" (Pérez-Lloréns et al. 2016) and "Those Curious and Delicious Seaweeds" (Pérez-Lloréns et al. 2018), which have received numerous national and international awards (e.g., "Best in the World" at the Gourmand World Cookbook Awards, 2017 and 2019, respectively and "Best of the Best" in 2021), have undoubtedly contributed to a better understanding of these organisms by the general public eager to expand their culinary boundaries (Fig. 15.20).

15.11 Conclusion

The province of Cádiz, and the Bay of Cádiz in particular, is a man-altered ecosystem (i.e., earthen ponds) where the ideal conditions of light, temperature, and hydrodynamics (due to the effect of the tides) exist for the growth of a large number of species of seaweed, some of which are of great culinary interest. Some local restaurants of high and medium rank include seaweeds in their menus, but it is still necessary for researchers (and chefs) to develop outreach activities (conferences, tastings, interviews, and TV programs) so that this marine ingredient gradually becomes part of the daily diet.

Fig. 15.17 Six-course seaweed-based menu offered at the Innovazul event (2022). Sea spaghetti tempura with smoked chili sauce (**a**); sea lettuce and prawns stew (**b**); chickpea salad, smoked sardine fillets and seaweed with black garlic emulsion (**c**); Korean rolled omelet ("Gyeran Mari") with seaweed (**d**); false fettuccini (squid strips) with *Ulva* spp. pesto sauce (**e**), natural pineapple sprinkled with crunchy seaweed flakes (**f**) (dishes by J. L. Pérez-Lloréns). (Photos by the authors)

Fig. 15.18 Five-course seaweed-based menu served at European Researchers' Night (2024) at the University of Cádiz. Chef (J. J. Sánchez-Marabot) and scientist (J. L. Pérez-Lloréns) in charge of the activity (a). Beetroot "salmorejo" (Andalusian thick tomato soup) with topping of apple crumbles and crunchy seaweed flakes (b); local bluefin tuna tartare with seaweed and passion fruit spheres on seaweed biscote (c); prawn takoyaki, katsuobushi and Gracilaria spp. (d); prawn and seaweed risotto with crunchy Ulva spp. (e); false truffle made of carob flour with Codium spp. and sprinkled with powdered Ulva spp. (f) (dishes by J. J. Sánchez-Marabot) (photos by the authors).

Fig. 15.19 Dishes from the demonstration menu prepared by Chef Mauro Barreiro at the Cádiz Culinary School. Seaweed (*Ulva*) butter (**a**); prawns in *dashi* broth and *Codium* semi-gel (**b**); monkfish *Ulva* roll with seaweed hollandaise sauce (**c**); semi-cured monkfish *sashimi* with seaweed powder and extra virgin olive oil (**d**). (Photos by the authors)

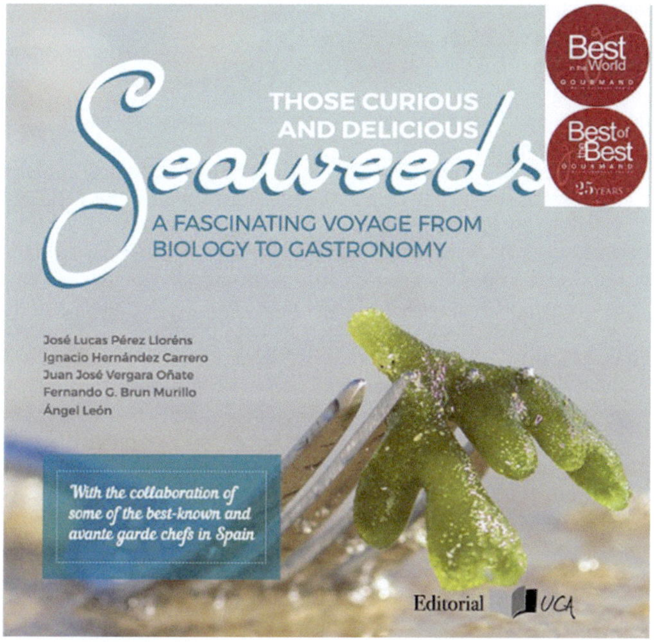

Fig. 15.20 The book "Those Curious and Delicious Seaweeds" (Foreword by O.G. Mouritsen and A. T. Critchley), which has been awarded "Best in the World" and "Best of the Best" at the Gourmand World Cookbook Awards in 2019 and 2021, respectively. (Photo by the authors)

Acknowledgments This study was partially supported by the projects DAME [PDC 2021-120792-100], funded by the Ministry of Science and Innovation, the National Agency of Research, and the European Union through the Next Generation EU funds of the Recovery, Transformation and Resilience Plan and FINOCAME [PCM_00104.C17.I03], project co-financed by the Department of University, Research and Innovation of the Regional Government of Andalusia and by the European Union through the Next Generation EU funds of the Recovery, Transformation and Resilience Plan. The authors would also like to thank Consuelo Guerra from Suralgas for the information provided.

Competing Interests The authors declare that they have no known competing financial interests or personal relationships that could have appeared to influence the work reported in this paper.

Availability of Data and Material Data sharing is not applicable to this article as no new data were created or analyzed in this study.

Authors' Contributions J. Lucas Pérez-Lloréns: Conceptualization, Bibliographic Investigation, Writing—Original Draft, Writing-review and editing, Cooking, Supervision.

Alba Yamuza-Magdaleno: Formal Analysis, Writing—Original Draft, Cooking.

Juan J. Sánchez-Marabot: Cooking.

Fernando G. Brun: Bibliographic Investigation, Writing—Original Draft.

Juan J. Vergara: Bibliographic Investigation, Writing—Original Draft.

Funding This research has been partially funded by projects DAME [PDC 2021–120792-100] and FINOCAME [PCM_00104.C17.I03].

References

Alley TR, Potter KA (2011) Food neophobia and sensation seeking. In: Preedy VR, Ross Watson R, Martin CR (eds) Handbook of behavior, food and nutrition. Springer, Dordrecht, pp 707–724

Bank MS, Duarte CM, Sonn C (2022) Intergovernmental panel on blue foods in support of sustainable development and nutritional security. Environ Sci Technol 56:5302–5305

Bartual A, Pérez-Lloréns JL (2016) Reseña del curso: taller de ficococina en los campus científicos de verano (CCV)-edición 2016. Algas 52:52–53

Bermejo R, Cara CL, Macías M, Sánchez-García J, Hernández I (2022) Growth rates of *Gracilariopsis longissima*, *Gracilaria bursa-pastoris* and *Chondracanthus teedei* (Rhodophyta) cultured in ropes: implication for N biomitigation in Cádiz Bay (southern Spain). J Appl Phycol 32:1879–1891

Buckley S, Hardy K, Hallgren F, Kubiak-Martens L, Miliauskienė Ž, Sobkowiak-Tabaka I, Subirà ME, Sheridan A (2023) Human consumption of seaweed and freshwater aquatic plants in ancient Europe. Nat Commun 14:6192

Cano CG (2023) El cocinero Ángel León denuncia el engaño de los "lobbies" del cereal y acusa a René Redzepi de haber cometido "un acto de egoísmo exagerado". https://cadenaser.com/nacional/2023/02/11/el-cocinero-angel-leon-denuncia-el-engano-de-los-lobbies-del-cereal-y-acusa-a-rene-redzepi-de-haber-cometido-un-acto-de-egoismo-exagerado-cadena-ser/. Accessed 12 Mar 2023

Costello C, Cao L, Gelcich S et al (2019) The future of food from the sea. World Resources Institute, Washington, DC. www.oceanpanel.org/future-food-sea. Accessed 30 Sept 2024

Dillehay TD, Ramírez C, Pino M, Collins MB, Rossen J, Pino-Navarro JD (2008) Monte Verde: seaweed, food, medicine, and the peopling of South America. Science 320:784–786

Galisteo A (2023) Ángel León: "A base de perder clientes aprendí que no puedo ser ta radical". https://www.esquire.com/es/donde-comer-beber/a43634808/angel-leon-chef-mar-aponiente-entrevista/. Accessed 1 May 2023

González S (2022) Del cruasán de ortiguillas al pan de algas y camarones: el universo de La Cremita (Chiclana, Cádiz). https://www.huleymantel.com/barras-estrellas/cruasan-ortiguillas-pan-algas-camarones-universo-creativo-cremita-chiclana_100227_102.html. Accessed 20 Sept 2024

González-Fragoso R (1886) Plantas marinas de la Provincia de Cádiz. Anales de la Sociedad Española de Historia Natural 15:117–130

Goulding M (2021) Seeding the Ocean: inside a Michelin-Starred Chef's Revolutionary Quest to Harvest Rice from the Sea. https://time.com/5926780/chef-angel-leon-sea-rice/. Accessed 12 Feb 2021

Greenberg P (2014) American catch: the fight for our local seafood. Penguin Press, London

Guiry MD, Guiry GM (2024) AlgaeBase. World-wide electronic publication, University of Galway. https://www.algaebase.org. Accessed 7 Oct 2024

Hernández I, Martínez-Aragón JF, Tovar A, Pérez-Lloréns JL, Vergara JJ (2002) Biofiltering efficiency in removal of dissolved nutrients by three species of estuarine macroalgae cultivated with sea bass (*Dicentrarchus labrax*) waste waters. 2. Ammonium. J Appl Phycol 14:375–384

Hernández I, Fernández-Engo MA, Pérez-Lloréns JL, Vergara JJ (2005) Integrated outdoor culture of two estuarine macroalgae as a biofilters for dissolved nutrients from *Sparus auratus* L. waste waters. J Appl Phycol 17:557–567

Hernández I, Pérez-Pastor A, Vergara JJ, Martínez-Aragón FJ, Fernández-Engo MA, Pérez-Lloréns JL (2006) Studies on the biofiltration capacity of *Gracilariopsis longissima*: from microscale to macroscale. Aquaculture 252:43–53

Hernández I, Pérez Pastor A, Mateo JJ, Megina C, Vergara JJ (2008) Growth dynamics of *Ulva rotundata* (Chlorophyta) in a fish farm: implications for biomitigation at a large scale. J Phycol 44:1080–1089

Hernández I, Bermejo R, Pérez-Lloréns JL, Vergara JJ (2010) Contribución al conocimiento de los macrófitos marinos del saco interno y caños adyacentes de la bahía de Cádiz. Algas 43:11–16

Hernández I, Sánchez R, Pérez-Lloréns JL, Vergara JJ, Cara C, Bermejo R (2019) Macroalgas del entorno de La Caleta. Resultados del proyecto "La Caleta (Cádiz) e intermareales rocosos asociados: una ventana permanente al conocimiento de la biodiversidad marina (BioCaleta)". Algas 55:7–20

Hurtado AQ, Neish I, Critchley AT (2019) Phyconomy: the extensive cultivation of seaweeds, their sustainability and economic value, with particular reference to important lessons to be learned and transferred from the practice of eucheumatoid farming. Phycologia 58(5):472–483

Kassam A (2021) The rice of the sea: how a tiny grain could change the way humanity eats. https://www.theguardian.com/environment/2021/apr/09/sea-rice-eelgrass-marine-grain-chef-angel-leon-marsh-climate-crisis. Accessed 8 Apr 2023

King B (2017) Ants, seaweed, chocolate beer and (maybe) less meat: the future of food. https://www.npr.org/sections/13.7/2017/08/31/547550216/ants-seaweed-chocolate-beer-and-maybe-less-meat-the-future-of-food. Accessed 28 Sept 2024

López Campos M, Pérez-Lloréns JL, Barrena F, Pérez-González CM, Hernández I (2022) Culture of *Gracilaria gracilis* and *Chondracanthus teedei* from vegetative fragments in the field and carpospores in the laboratory. J Mar Sci Eng 10:1041

McEachran R (2014) Food for the future: what will feed 7 billion people? https://www.theguardian.com/global-development-professionals-network/2014/aug/12/insects-algae-lab-meat-food. Accessed 28 Sept 2024

Mouritsen OG (2022) Taste for sustainability and a green transition. Food Stud Interdiscip J 12(2):9–18

Mouritsen OG, Rhatigan P, Pérez-Lloréns JL (2018) World cuisine of seaweeds: science meets gastronomy. Int J Gastr Food Sci 14:55–65

Mouritsen OG, Rhatigan P, Pérez-Lloréns JL (2019) The rise of seaweed gastronomy: phycogastronomy. Bot Mar 62:195–209

Mouritsen OG, Cornish ML, Critchley AT, Pérez-Lloréns JL (2024) History of seaweeds as foods. In: Heftt D, Adetunji C (eds) Applications of seaweeds in food and nutrition. Amsterdam Elsevier pp. 1–17

Pérez Lloréns JL, Brun FG, Hernández I, Bermejo R, Vergara JJ (2023) Macrófitos marinos (algas y angiospermas) de las costas de Cádiz. In: Pérez Rubín J, Ramírez T (eds), El Estrecho de Gibraltar: llave natural entre dos mares y dos continentes. Memorias SEHN 2nd period, vol 16, pp 133–149

Pérez-Lloréns JL (2019a) Seaweed consumption in the Americas. Gastronomica 19(4):49–59

Pérez-Lloréns JL (2019b) Bluefin tuna and Cádiz: a pinch of history and gastronomy. Int J Gastron Food Sci 17:100170

Pérez-Lloréns JL (2020) Microalgae: from staple foodstuff to avant-garde cuisine. Int J Gastron Food Sci 21:100221

Pérez-Lloréns JL (2021a) Las algas y angiospermas marinas presentes en el evento gastronómico 'Sabor en Cantabria'. Algas 57:55–56

Pérez-Lloréns JL (2021b) Jornada ficogastronómica en la Escuela de Hostelería de Cádiz. Algas 57:50–53

Pérez-Lloréns JL (2023) Jornada de ficogastronomía en el restaurante 'Casa Juan'. Algas 59:69–72

Pérez-Lloréns JL, Bartual A (2022) Las algas: protagonistas en el bar 'La Casapuerta'. Algas 58:95–99

Pérez-Lloréns JL, Brun FG (2023) "Sea rice": from traditional culinary customs to sustainable crop for high-end gastronomy? Int J Gastron Food Sci 34:100814

Pérez-Lloréns JL, Vergara JJ (2023) Microalgae in phycogastronomy. In: Jabob-Lopes E, Queiroz MI, Manzoni M, Queiroz L (eds) Handbook of food and feed from microalgae. Academic, London, pp 349–355

Pérez-Lloréns JL, Brun FG, Andría JR, Vergara JJ (2004) Seasonal and tidal variability of environmental carbon related physico-chemical variables and inorganic C acquisition in *Gracilariopsis longissima* and *Enteromorpha intestinalis* from Los Toruños salt marsh (Cádiz Bay, Spain). J Exp Mar Biol Ecol 304:183–201

Pérez-Lloréns JL, Hernández I, Bermejo R, Peralta G, Brun FG, Vergara JJ (2012) Flora marina del litoral gaditano. Biología, ecología, usos y guía de identificación. Monografías Ciencias de la Naturaleza. Servicio de Publicaciones de la Universidad de Cádiz, Cádiz

Pérez-Lloréns JL, Hernández I, Vergara JJ, Brun FG, León A (2016) ≚Las algas se comen? Un periplo por la biología, la historia, las curiosidades y la gastronomía. Servicio de Publicaciones de la Universidad de Cádiz, Cádiz

Pérez-Lloréns JL, Hernández I, Vergara JJ, Brun FG, León A (2018) Those Curious and Delicious Seaweeds. A Fascinating Voyage from Biology to Gastronomy. Servicio de Publicaciones de la Universidad de Cádiz, Cádiz

Pérez-Lloréns JL, Acosta Y, Brun FG (2021) Seafood in Mediterranean countries: a culinary journey through history. Int J Gastron Food Sci 26:100437

Rozin P, Fallon AE (1985) That's disgusting. Psychol Today 19(7):60–63

Rozin P, Fallon AE (1987) A perspective on disgust. Psychol Rev 94(1):23–41

Spence C (2017) Gastrophysics. The new science of eating. Penguin Books, London

Tigchelaar M, Leape J, Micheli F, Allison EH, Basurto X, Bennett A et al (2022) The vital roles of blue foods in the global food system. Glob Food Sec 33:100637

Willett W, Rockström J, Loken B, Springmann M, Lang T, Jonell M, Clark M, Gordon LJ, Fanzo J, Hawkes C, Zurayk R, Rivera JA, De Vries W, Sibanda LM et al (2019) Food in the anthropocene: the EAT-lancet commission on healthy diets from sustainable food systems. Lancet 393:447–492

World Bank (2023) Global seaweed. New and emerging markets report. https://documents1.worldbank.org/curated/en/099081423104548226/pdf/P175786073c14c01609fe409c202ddf12d0.pdf. Accessed 21 Oct 2024

Zhang D, Luo QY, Zhong MY, Hu ZM, Zhang QS (2023) Efficient methods of seedling culture and cultivation of the economic macroalga *Scytosiphon lomentaria* (Phaeophyta): a systematic review. Aquaculture 567:739253

J. Lucas Pérez-Lloréns, Ph.D., is a biologist and Professor of Marine Ecology at the Faculty of Marine and Environmental Sciences (University of Cádiz, Spain) and a member of IN·MAR (Institute for Marine Research). He is President of the Spanish Phycological Society. He is the author of four scientific books and more than 100 scientific papers. His research interests range from the ecology and ecophysiology of seagrasses and seaweeds to aquaculture. He has been ranked among "World's top 2% of Scientists 2023" by Stanford University in the fields of Marine Biology and Food Science. He is passionate about the culinary ethnology of marine products (especially seaweeds), including their use in avant-garde cuisine. He loves to cook and has collaborated with several chefs and cooking schools. He has given numerous radio and television interviews on the world of seaweed, particularly its use in food and haute cuisine. He has been awarded prizes in several culinary competitions. Two of his books: *¿Las algas se comen?* (2016) and *Those Curious and Delicious Seaweeds* (2018) were awarded as "Best of the World" in the categories of Nutrition and Health Institutions and Translation, respectively at the "Gourmand World Cookbook Awards (2016 and 2018)". He won the first prize in the VII Spanish National Contest "GastroUNED: Literature and Gastronomy" (2020) for the recipe: "Galician-style hake…reinterpreted." Together with some colleagues from the University of Cádiz, he has advised the Aponiente restaurant (three Michelin stars) on the experimental cultivation of "marine rice" (eelgrass seed), which has enabled the owner of Aponiente to win the National Gastronomic Innovation Award 2021 from the Royal Spanish Academy of Gastronomy.

Alba Yamuza-Magdaleno is a Spanish predoctoral researcher specializing in marine ecology. She holds a double degree in Marine Sciences and Environmental Sciences from the University of Cadiz (UCA) (2020), and a Master in Conservation and Management of the Natural Environment (2021). She is pursuing her Ph.D. at the Department of Biology (UCA) on the remineralization potential of sedimentary organic matter in coastal benthic communities—particularly seagrass meadows—the impact of global change on the carbon cycle, and the restoration of seagrass meadows in the Bay of Cadiz. She actively participates in science outreach activities and educational events in primary and secondary schools, helping to raise environmental awareness among students and the general public. In addition, Alba has a great passion for cooking, inspired by her grandmother, who taught her traditional recipes and passed on her culinary knowledge from a very young age. This passion has led her to participate in several "phycogastronomic" events in collaboration with Professor J. Lucas Pérez-Lloréns, where science and gastronomy merge to appreciate sustainable marine products and their responsible consumption.

Juan J. Sánchez-Marabot trained at the Escuela de Hostelería Fernando Quiñones de Cádiz, worked as a sous-chef in the teams of Michelin-starred chefs Ferran Adrià and Martín Berasategui, and made a name for himself as head chef at the Sopranis restaurant, where he combined innovation with traditional Cadiz cuisine. He is currently the head chef and gastronomic consultant for several restaurants in the province of Cádiz. He is the gastronomic director of the Shaokao Grill, the first restaurant in the province of Cádiz dedicated to Korean cuisine. Representative of the association of former students of "Masterñam alumni et al" and the association Arte Culinario Andaluz (ARTCUA), founded in 2017 and made up of more than 120 Andalusian chefs, whose main objective is to defend the Andalusian product.

Fernando G. Brun Murillo is a Professor of Ecology and is co-responsible for the Spanish Group of Experts on Blue Carbon Ecosystems and the Blue Carbon Laboratory at the University of Cádiz. His research focuses on the ecology of coastal ecosystems, with a special emphasis on vegetated ecosystems such as seagrass meadows and salt marshes, the effects of human activities on them, and how to mitigate these impacts through management, conservation, and restoration. The search for high-value products derived from these habitats with gastronomic, industrial, or pharmaceutical interest is also one of his research lines, aimed at improving the quality of life of local communities through environmental enhancement. He has received various awards, such as the Gourmand World Cookbook Awards in 2017 and 2019 and the National University Publishing Awards in 2013 and 2016.

Juan Juan Vergara earned a Biology degree from the University of Málaga in 1988 and completed a Ph.D. in 1993, focusing on nitrogen metabolism and the ecophysiology of red algae. He joined the University of Cádiz's Faculty of Marine and Environmental Sciences, securing a permanent position in the Department of Ecology in 1998 and a professorship in 2010. There, he led research on the ecology and physiology of seaweeds and seagrasses, exploring topics from photobiology and nitrogen metabolism to green tide dynamics and eutrophication effects. His work also encompassed seagrass community structure, herbivory, biogeochemistry, integrated aquaculture projects, and applying the Water Framework Directive to macroalgae and seagrasses. He pursued short-term Ph.D. studies at the University of North Carolina Wilmington (USA) and a postdoctoral fellowship with P. Falkowski at Brookhaven National Lab (USA) in 1995. His career includes leadership roles such as heading the EDEA research group (2010–2016), directing the Marine Research Laboratory (Labimar) at the University of Cádiz (2014), serving as General Director of Scientific Projection (2016), and coordinating the International Campus of Excellence of the Sea (CEI.Mar) from March 2017 to July 2019. From late 2021 to 2023, he directed the Marine Research Institute (INMAR) at the University of Cádiz. Since 2016, he has represented the University of Cádiz in Euromarine, a European network of marine laboratories, and has been a Steering Committee member since 2019, with his term extending through 2025. These roles have provided him with a comprehensive, multidisciplinary perspective on marine sciences, enriched by interactions with private enterprises and public administrations at various levels.

Portuguese Examples of Temperate Seaweed Use

16

Leonel Pereira

Abstract

This chapter explores the multifaceted roles and benefits of seaweed in temperate regions, with a particular focus on Portugal. Seaweed are highlighted as crucial ecological assets, contributing to marine biodiversity, coastal stabilization, and carbon sequestration. Their ability to absorb excess nutrients from agricultural runoff helps mitigate eutrophication, thereby maintaining coastal ecosystem health. In agriculture, seaweed are presented as sustainable resources that enhance soil fertility, stimulate plant growth, and improve crop resilience through the use of bio-fertilizers, reducing reliance on synthetic alternatives. The chapter also delves into the biotechnological and industrial applications of seaweed, emphasizing their adaptability to temperate climates and their potential for large-scale cultivation. Seaweed provide renewable biomass for various industries, including food, cosmetics, and biotechnology, supporting both environmental conservation and the economic viability of coastal communities. A significant portion of the chapter is dedicated to the culinary and nutritional value of seaweed. Workshops like "Algas à Mesa" have promoted seaweed as a versatile ingredient in Portuguese cuisine, showcasing its health benefits and culinary potential. Recipes such as "Seaweed and Cod Fritters" and "Seaweed Risotto" illustrate how seaweed can enhance traditional dishes while aligning with sustainable gastronomic practices. By integrating seaweed into sustainability initiatives, the chapter underscores their importance in supporting environmental conservation and economic resilience. Through Portuguese examples, the chapter highlights the diverse applications and benefits of temperate seaweed, advocating for their broader use in sustainable practices.

Keywords

Temperate region · Ecosystem services · Sustainable agriculture · Bio-fertilizers · Carbon assimilation · Renewable biomass · Portuguese cuisine · Biotechnology

16.1 Introduction

Various seaweeds play a critical role in temperate, coastal regions, offering a diverse range of marine ecosystem services and applications that extend across environmental, agricultural, and biotechnological domains. The ecological and economic importance of these seaweeds underscores their value as sustainable resources with multifaceted benefits (Cotas et al. 2023).

Seaweeds contribute significantly to ecosystem services, acting as primary producers that support marine biodiversity, stabilize coastal environments, and play a role in carbon sequestration. Their ability to absorb nutrients, such as nitrogen and phosphorus, from surrounding waters is particularly relevant in temperate regions, where nutrient runoff from agriculture can lead to eutrophication (Khan et al. 2024). By mitigating such impacts, various seaweeds contribute to maintaining the health of coastal ecosystems (Cotas et al. 2023).

In agriculture, selected seaweed species serve as a promising resource for making practices more sustainable and eco-friendlier. They are rich in bioactive compounds and nutrients that enhance soil fertility, stimulate plant growth, and improve crop resilience. The application of seaweed-based bio-fertilizers not only reduces dependence on synthetic fertilizers but also mitigates the environmental footprint of agricultural practices. This aligns with the growing demand for eco-conscious solutions in temperate

L. Pereira (✉)
Department of Life Sciences, CFE—Centre for Functional Ecology: Science for People & Planet, Marine Resources, Conservation and Technology—Marine Algae Lab, University of Coimbra, Coimbra, Portugal
e-mail: leonel.pereira@uc.pt

A. T. Critchley et al. (eds.), *Advances in Temperate Phyconomy: Algal Harvest and Cultivation in Globally Distributed Temperate Waters*, Developments in Applied Phycology 15, https://doi.org/10.1007/978-3-032-02955-3_16

regions, where agricultural expansion often intersects with fragile marine ecosystems (Mamede et al. 2023).

Furthermore, seaweeds possess a remarkable adaptability to the climatic conditions of temperate regions, making them ideal candidates for large-scale cultivation. Their seasonal availability and fast growth cycles align with the multiple demands for renewable biomass, which can be utilized in various industries, including food, cosmetics, and biotechnology (Loureiro et al. 2015).

By emphasizing their ecological significance and diverse applications, it is important to integrate specific types of seaweeds into the broader framework of sustainability initiatives in temperate regions. These efforts not only support environmental conservation but also promote the economic viability of coastal communities, where seaweed harvesting and aquaculture can serve as critical livelihood sources (Khan et al. 2024; Pereira and Cotas 2024).

This chapter will further explore these contributions, focusing on Portuguese examples to illustrate the versatile and sustainable use of temperate seaweeds.

16.2 History and Tradition of Seaweed Uses in Portugal

Portugal has a rich history of utilizing seaweed, deeply rooted in traditional practices that have evolved to address contemporary needs. With its extensive coastline and temperate marine environment, the country has long relied on selected seaweed resources to support various sectors, from agriculture to gastronomy and industry (Pereira and Cotas 2019). Historically, seaweeds were primarily used as natural fertilizers in agriculture (de Oliveira et al. 1990). Coastal communities, particularly in the northwest of Portugal, harvested seaweeds to enrich sandy and nutrient-poor soils. Known locally as "Moliço" or "Sargaço," these seaweeds were spread on fields to improve crop yields. "Sargaço" (also called "Argaço and Limos" since the time of the first kings) is a mixture composed exclusively of diverse marine macroalgae (e.g., *Saccorhiza, Laminaria/Saccharina, Fucus*—Phaeophyceae, *Codium*—Chlorophyta, *Palmaria, Gelidium,* and *Chondrus*—Rhodophyta), which grow in rocky areas of the coastal area. This practice, which dates back to at least the fourteenth century, highlights the early recognition of the agronomic value of marine biomass. Additionally, seaweeds held cultural and medicinal importance. Coastal populations leveraged their mineral-rich composition in traditional remedies, although these uses were often localized and less documented than their agricultural applications (Pereira 2024a, b).

In recent decades, Portugal has seen a shift toward sustainable and innovative seaweed utilization, spurred by advances in research and an increasing emphasis on the bioeconomy. Today, selected seaweeds are cultivated and harvested for diverse purposes, encompassing food production, biotechnology, cosmetics, and environmental management. In the culinary sector, Portuguese chefs and food producers have begun to incorporate seaweeds such as *Ulva* spp., *Gracilaria* spp., and *Porphyra* (*Pyropia*) spp. into gourmet dishes, traditional recipes, and health-focused products. This represents a growing appreciation for their nutritional value and gastronomic versatility (Gaspar et al. 2019). Building on historical practices, seaweed-based bio-fertilizers are now widely used in sustainable farming, reducing dependency on synthetic fertilizers and aligning with eco-friendly agricultural policies (Pereira and Cotas 2019).

Industrially, Portugal has become a hub for seaweed-related innovation. Companies and research institutions are exploring the potential of seaweed-derived compounds, such as polysaccharides and antioxidants, in bioplastics, pharmaceuticals, and cosmetics (Lomartire et al. 2022). Furthermore, seaweeds play a crucial role in environmental management. Their capacity to absorb excess nutrients and carbon positions them as key tools in bioremediation projects, helping to combat coastal pollution and mitigation of eutrophication (Khan et al. 2024).

This seamless integration of traditional knowledge with modern innovation highlights the enduring significance of seaweeds in Portuguese culture and economy. Portugal's example underscores how temperate seaweeds can serve as a cornerstone for sustainable practices, contributing to ecological resilience and economic vitality (Soares et al.2017).

The harvesting of algae in the regions of "Entre-os-Rios" and "Douro" (Northern Portugal) was, from the Middle Ages until the middle of the twentieth century, an economically and socially important activity in mainland Portugal, as is clearly demonstrated by the Dionysiac and Manueline provincial laws of Póvoa de Varzim and Maia. The provincial law published on March 9, 1308, granted to the former regiment of Varzim de Jusão and its mandates, King D. Dinis determined that the harvesting of "Sargaço" was an economic activity of great importance at that time for fertilization, and the land belonged to its residents. These provisions were later confirmed in the Manueline Order of 1514, but, realizing that it was a natural resource capable of generating revenue, it also established a real tax on the "Sargaço" trade (Pereira 2024a, b).

Returning to the last point that unequivocally shows the close connection between algae and the way of life of a people, it is convenient to distinguish the two main mixtures of seaweed traditionally used as fertilizer, that is, "Moliço" from "Sargaço." As already mentioned, "Moliço" is a mixture of green (Chlorophyta) and red (Rhodophyta) algae and also some marine plants (such as seagrasses). "Moliço" is composed of specimens of the genera *Enteromorpha* (algae currently belonging to the genus *Ulva*), *Chaetomorpha/*

Rhizoclonium (Chlorophyta), *Gracilaria* and *Ceramium* (Rhodophyta), and seagrasses (angiosperms) belonging to the genera *Zostera*, *Ruppia,* and *Potamogeton*, all collected in the Ria de Aveiro (Aveiro, Portugal) (Pereira and Cotas 2019; Pereira 2024a, b).

The traditional "Sargaço" harvest consisted of collecting, on the beach or by the sea, algae that came from the rocks (Fig. 16.1), through the action of wave movement, as already mentioned previously. After being collected, it was spread on the beach sand to dry in the air and in the sun. The traditional utensils used to collect "Sargaço" are (Fig. 16.2): the cart (used to transport the raft), the raft (used to reach the rocks furthest from the beach where the algae settle), the cutting and collecting instruments ("sickle," "croque," "engaceira"), and also the baskets (made of wooden splints to pack them and transport them to land) to take to the place where they would dry; another of the utensils used is the round, wide-mesh bag used to collect "Sargaço" loose in the water. Transporting the "Sargaço," after drying, to the haystacks, was also done using the same wooden splint baskets. These haystacks (Fig. 16.3) were thus designated as being covered with straw (to waterproof and protect from rain and sun) (Pereira and Cotas 2019; Pereira 2024a, b).

"Sargaço" harvesting may be historically related, to some extent, to smallholding and intensive agricultural crops. It was in fact in these cases that the use of algae as fertilizer was especially effective and viable. These conditions were particularly prevalent along the northern coastal areas, where this activity held greater significance. Moreover, the collection of "Sargaço" and the subsequent agricultural work were predominantly carried out under a family-based exploitation system (Pereira and Correia 2015).

In North of the Douro River, this activity exhibited distinct and unique characteristics, shaped by significant sociological and economic factors. These factors were often tied to local patterns, such as the distribution of labor by gender and the hybrid nature of the activity itself. Over time, it evolved from a form of family labor, initially subsidiary to agriculture, into an autonomous enterprise with direct commercial objectives. This transformation was progressively driven by individuals from different social categories who began to specialize in it. In the vast dune areas of "Aguçadoura," this shift was closely linked to a well-defined movement of sand exploration and valorization that began in the late eighteenth century. This movement, aimed at uncovering arable land for intensive horticulture—particularly the region's characteristic apple fields—was both enabled by and a catalyst for the activity (Pereira and Cotas 2019).

In some cases, this phenomenon played a critical role in coastal settlement, becoming one of the most consequential aspects of "Sargaço" collection and use. It contributed to the demographic processes that originally established certain population centers, some of which have since grown into sig-

Fig. 16.1 Recreation of the "Sargaço" harvest by the "Sargaceiros"

Fig. 16.2 Traditional utensils used to collect "Sargaço"

Fig. 16.3 "Sargaço" haystacks

nificant urban areas. This interplay between *Sargassum* harvesting and regional development underscores its importance in shaping both the economy and settlement patterns of Portugal's northern coast (Pereira 2024a, b).

In traditional medicine, various seaweeds were valued for their healing properties and used as remedies for various ailments. Use of certain species in treating skin conditions, wounds, and even gastrointestinal issues. The high iodine content in some seaweeds, for example, made them a natural remedy for iodine deficiency disorders, which were more prevalent in earlier centuries. Additionally, the mucilaginous extracts of certain species were applied topically to soothe burns or irritations, a practice passed down through generations in coastal communities (Lomartire and Gonçalves 2022).

These traditional uses of selected seaweeds in Portugal reflect a deep understanding of their bioactive properties long before modern science validated their potential. Today, there is a renewed interest in exploring seaweed-derived compounds for pharmaceuticals and nutraceuticals, bridging the gap between historical knowledge and contemporary scientific innovation. This historical perspective enriches our understanding of how seaweeds have shaped Portuguese cultural and practical life, providing a foundation for their ongoing relevance in modern applications (Gaspar et al. 2019; Pereira 2019; Cotas et al. 2024a).

16.3 Temperate Species of Seaweeds Used in Portugal

Mainland Portugal and the Islands, located in the westernmost part of Europe, are characterized by having an impressively long coastline. The Portuguese coast has an approximate length of 1736 km—distributed over a continental area of 833 km, plus 691 km of the Azores archipelago and 212 km of the Madeira archipelago. They can be classified into four main types: beaches, cliffs, wetlands, and highly modified coasts (Gaspar et al. 2019).

Ardré (1970) recognized the existence of 246 species of Rhodophyta, 98 Phaeophyceae, and 60 Chlorophyta, and these numbers have not changed substantially since then. There is a sharp gradient in algal flora along the Portuguese coast. The mid-coastal flora in the north of Portugal is similar to that of Central Europe (Brittany, Southern British Isles). The macroalgal flora of southern Portugal is very different, suffering a marked influence from Mediterranean and African species (Gaspar et al. 2019).

The Portuguese coast has some macroalgae from cold temperate waters, such as *Laminaria hyperborea* and *Saccharina latissima* (formerly *Laminaria saccharina*) (Phaeophyceae). The presence of species from temperate cold waters is decreasing along the coasts toward the south, and *Sacchoriza polyschides* (lower midlittoral) and *Laminaria ochroleuca* (midlittoral) are now more common. Other algae from cold and temperate waters have their southern limits of European distribution on the Portuguese coast, e.g., *Chorda filum, Fucus serratus,* and *Himanthalia elongata* (Phaeophyceae). Some other macroalgae, from cold and temperate waters, have their southern European limits on the Portuguese coast, e.g., *Delesseria sanguinea, Ahnfeltia plicata,* and *Palmaria palmata* (Rhodophyta). From North to South, there is an increase in the number of species of red algae (Rhodophyta) and a decrease in species of brown algae (Phaeophyceae) (Gaspar et al. 2019; Pacheco et al. 2021).

The complex oceanographic features influencing the Iberian Peninsula, such as upwelling systems and currents, play a crucial role in shaping the distribution and abundance of both native and non-native algae (García-Huidobro et al. 2023). Changes in these patterns, whether due to natural processes or human activities, can affect the success of non-native species. Additionally, the presence of non-native macroalgae alongside native species creates opportunities for hybridization. Hybrid offspring may inherit traits that enhance their fitness and adaptability, potentially giving them a competitive advantage within the ecosystem (Cacabelos et al. 2013).

While these factors facilitate the introduction of non-native algae, their establishment and ecological impact are influenced by species-specific traits, local environmental conditions, and interactions with native flora and fauna. Vigilant monitoring, early intervention, and proactive management are critical to controlling and mitigating the spread of non-native macroalgae in the region (Blanco et al. 2021).

In the European Union, over 41% of marine invasive species of special concern are macroalgae. These invasive seaweeds, like invasive species in general, significantly alter the structure and functioning of invaded communities and ecosystems, thereby impacting ecological processes and ecosystem services. On the west coast of the Iberian Peninsula (Portugal and Spain), the composition and structure of seaweed communities have undergone dramatic changes over the past two decades, largely due to climate change and biological invasions (Blanco et al. 2021; Pereira 2024a).

The spread and establishment of invasive species pose increasing challenges for environmental managers, as eradicating established invaders is often costly, difficult, or even impossible. Consequently, contingency plans must prioritize prevention over eradication. However, effective management policies rely on a comprehensive understanding of the factors influencing the distribution of non-indigenous marine species (NIMS), which include biotic, abiotic, and anthropogenic variables. The main non-native, seaweed species present in Portugal are: *Grateloupia turuturu, Asparagopsis armata* (Rhodophyta), *Colpomenia peregrina, Sargassum muticum, Undaria pinnatifida, Rugulopteryx okamurae*

(Phaeophyceae), and *Codium fragile* (Chlorophyta) (Pereira 2024a).

Invasive seaweeds, while posing ecological challenges, are rich sources of bioactive compounds with substantial industrial potential. These compounds, including polysaccharides like phycocolloids (e.g., agars and carrageenans from red seaweeds) and alginates (from brown seaweeds), have diverse applications in food, pharmaceuticals, cosmetics, and biotechnology. The gelling and encapsulation properties of alginates make them particularly valuable in wound dressings and drug delivery systems (Lomartire and Gonçalves 2023; Pereira 2024a).

Invasive seaweeds also contain bioactive molecules such as polyphenols, flavonoids, and terpenes, known for their antioxidant, antimicrobial, and anti-inflammatory properties. These compounds are promising for functional foods, dietary supplements, and skincare products. Additionally, their high carbohydrate content and rapid growth make them attractive feedstock for biofuels and bioplastics (Pereira and Cotas 2023).

Various seaweeds have long been used as sustainable feed in aquaculture and livestock, and their potential to reduce environmental impacts is under study (Morais et al. 2020). Their ability to absorb heavy metals and pollutants is being explored for bioremediation in contaminated coastal areas (Chugh et al. 2022). In pharmaceuticals, seaweed-derived compounds are being researched for applications in treating cancer, inflammation, and infections. Similarly, in cosmetics, seaweed extracts are prized for moisturizing, anti-aging, and soothing properties (Kalasariya et al. 2024).

While their industrial potential is significant, sustainable harvesting and minimizing ecosystem impacts are crucial. Ongoing research aims to optimize extraction processes and explore new applications in sectors such as bio-fertilizers, bioactivities, and food, highlighting the economic and ecological value of these species (Pereira and Cotas 2024).

16.4 Traditional and Modern Dishes Incorporating Seaweeds

Seaweeds have long been integral to the livelihoods of coastal communities in countries like Portugal, Spain, France, the UK, Ireland, Denmark, Norway, and Iceland. Traditionally gathered by hand without mechanization or specialized equipment, these particular seaweeds were incorporated into diets and also used to fertilize agricultural fields and supplement cattle feed. For centuries, storm-cast, mineral-rich seaweeds have played an indirect yet vital role in food production (Mendes et al. 2022).

Despite the abundance of various seaweeds on the Portuguese coast, their use in food has no great tradition in Portugal, except for some coastal communities in the Azores. In these communities, *Pyropia leucosticta* (formerly *Porphyra leucosticta*), known as "erva patinha," is harvested and consumed fried or incorporated into soups, pies, and omelets. *Osmundea pinnatifida*, known as "erva malagueta," is preserved in vinegar and consumed throughout the year, accompanying fried fish (Palminha 1971; Neto et al. 2005).

The Azores archipelago, with its rich history and unique biodiversity, has long offered sustenance through its land and seas (Santos et al. 2020). Amongst the natural treasures of the islands are seaweeds, historically foraged and consumed as part of the Azorean diet. Yet, despite their ecological and nutritional benefits, these seaweeds have largely fallen into disuse. Factors contributing to this decline include labor-intensive foraging processes, seasonal availability, and the global shift toward processed and imported foods. However, revitalizing seaweed consumption could address critical issues of sustainability, health, and cultural preservation in the region (Paiva et al. 2014).

16.4.1 Seaweeds in the Azorean Culinary Tradition

In the past, Azoreans relied on their surrounding environment to meet their dietary needs. Seaweeds like *Porphyra* spp., locally known as "erva-do-calhau" or "erva patinha," were essential components of the Azorean diet. Recipes such as "tortas de erva-do-calhau" (seaweed patties) were passed down through generations, particularly on islands such as Flores and Pico. These patties, often fried with corn flour and seasoned with local ingredients, exemplified the resourcefulness of islanders who crafted meals from the limited resources available (Matos 2013; Paiva et al. 2014).

Despite their historical importance, seaweeds are now rarely consumed. Today, "tortas-do-calhau" are considered exotic, primarily appealing to tourists rather than locals. This decline parallels broader global trends in which traditional, sustainable foods are replaced by often highly processed alternatives. Yet, the nutritional profile of seaweeds—rich in iron, calcium, proteins, and vitamins—positions them as a potential "superfood" capable of addressing modern dietary deficiencies (Matos 2013).

16.4.2 Challenges to the Seaweed Revival

Efforts to reintroduce seaweeds into the Azorean diet face several challenges. Conversations with local stakeholders, such as Mrs. Rosa, a restaurant owner on the island of Flores, highlight perceptions of seaweeds as outdated, or unappealing. Mrs. Rosa, for instance, initially viewed seaweeds as unconventional food items but eventually embraced them due to their unique flavors and cultural significance.

Similarly, the Dias family shared insights into traditional foraging methods and recipes, emphasizing both the ecological value of sustainable harvesting and the cultural stigma that frames seaweed as a: "hunger food" from a bygone era (Neto et al. 2005; Matos 2013).

16.4.3 Culinary Innovation: Workshops and Recipes

To address these challenges, workshops were conducted with culinary students on São Miguel and Terceira Islands. These sessions introduced young chefs to foraging practices and the culinary potential of various seaweed species, including *Porphyra* spp. (Rhodophyta), *Ulva rigida* (sea lettuce) (Chlorophyta), *Fucus spiralis* (spiral wrack) (Phaeophyceae), and *Osmundea pinnatifida* (pepper-dulse) (Rhodophyta). The students were encouraged to experiment, leading to a creative array of recipes such as Taro Root Omelet with Sea Lettuce, Pork Cheeks with Spiral Wrack Vinaigrette, and Creamy Rice with Limpets and Laver (Neto et al. 2005; Matos 2013).

Notably, some students broke from tradition by using various seaweeds in desserts, showcasing their adaptability and potential appeal to broader audiences. These workshops demonstrated that exposure and hands-on experience could shift perceptions, turning scepticism into curiosity and appreciation for seaweeds as culinary ingredients.

16.4.4 Toward a Sustainable Food Culture

Revitalizing the consumption of a variety of various seaweeds in the Azores requires a multifaceted approach. Collaborative efforts between biologists, chefs, and local communities are essential to dispel misconceptions and foster a renewed appreciation for these marine resources. Educational initiatives and cookbooks featuring modernized recipes can bridge the gap between tradition and innovation, making seaweeds more accessible and appealing to contemporary palates.

Moreover, promoting seaweeds aligns with global sustainability goals. As low-impact crops that require no arable land, fresh water, or fertilizers, selected seaweed offer a viable alternative to resource-intensive foods. By integrating selected seaweeds into local diets and culinary practices, the Azores can serve as a model for sustainable food cultures, blending ecological stewardship with cultural preservation.

The revival of seaweed consumption in the Azores represents an opportunity to reconnect with the region's culinary heritage while addressing modern challenges of health and sustainability. Through education, collaboration, and innovation, seaweeds can reclaim their place as a vital component of the Azorean diet and a symbol of the islands' rich and resilient food culture.

16.4.5 Growing Interest in Sustainable Gastronomy

Portugal, renowned for its rich culinary traditions and strong ties to the sea, is experiencing a wave of innovation in sustainable gastronomy. Among the most exciting developments is the growing interest in edible seaweeds, a resource that combines health benefits, ecological sustainability, and culinary versatility. Workshops like "Algas à Mesa" (Seaweeds at the Table) (Pereira 2021) have played a pivotal role in promoting seaweeds as a viable ingredient in Portuguese cuisine, inspiring chefs and home cooks to embrace these "vegetables of the sea" (Pereira 2009).

Seaweeds are an extraordinary resource. They are packed with essential nutrients such as iodine, calcium, and magnesium, along with antioxidants and unique bioactive compounds. Unlike land-based crops, they grow rapidly, require no freshwater or fertilizers, and help assimilate carbon dioxide, making them one of the most sustainable food sources available. This aligns perfectly with Portugal's focus on preserving its culinary heritage while embracing modern, environmentally conscious practices (Pereira et al. 2024).

The "Algas à Mesa" workshop introduced participants to the diversity of edible seaweed available along Portugal's extensive coastline, such as Nori (*Porphyra, Neopyropia)*, Wakame (*Undaria pinnatifida*), and Royal Kombu (*Saccharina latissima*). Practical sessions demonstrated how these seaweeds could be incorporated into traditional and contemporary dishes, turning an often-overlooked resource into a gastronomic delight (Pereira 2009; Mendes et al. 2022).

One of the standout recipes from the workshop was "Seaweed Stuffed with Mussels." A twist on the classic "pastéis de bacalhau" (cod pastries), this recipe blends rehydrated wakame with mashed potatoes, shredded cod, garlic, and parsley (Fig. 16.4). The mixture is shaped into small fritters and fried until golden. The result is a savory snack that is crispy on the outside, soft on the inside, and bursting with "umami" flavors from the seaweed (Pereira 2009, 2016).

Another highlight was the "Seaweed Risotto," an elegant and nutritious dish. To prepare it, chefs sautéed onions and garlic in olive oil before adding Arborio rice and a splash of white wine. Rehydrated kombu was finely chopped and incorporated into the mix, along with vegetable stock. As the rice absorbed the liquid, the kombu imparted a subtle oceanic flavor. The dish was finished with a drizzle of lemon juice and a sprinkle of grated Parmesan, offering a perfect

Fig. 16.4 Seaweed Stuffed with Mussels

Fig. 16.5 Cold Lemon-Mousse of Agar with Honey

balance of creamy texture and marine freshness (Pereira 2009, 2016).

Integrating seaweeds into Portuguese cuisine does more than expand culinary horizons; it resonates deeply with the country's maritime identity. Traditional dishes such as "caldeirada" (fish stew) can be elevated with the addition of carefully selected seaweeds, thereby enhancing their flavor profile while reducing the need for salt. For dessert, seaweeds can even be used to create innovative recipes like "Agar-Agar Gelatin with Citrus and Honey" (Fig. 16.5), combining the gelling properties of red algae with fresh Portuguese oranges and a touch of regional honey (Pereira 2009, 2016).

Despite their benefits, seaweeds are not yet a staple in Portuguese households, largely due to unfamiliarity and limited commercial availability. However, initiatives like "Algas à Mesa" are helping to address these challenges by raising awareness and fostering a deeper appreciation for this versatile ingredient. As consumers become more aware of the environmental and nutritional advantages of seaweed, their inclusion in everyday meals is likely to grow (Pereira 2009, 2016).

By embracing edible seaweeds, Portugal has the opportunity to lead by example in sustainable gastronomy. These marine treasures can help reduce the ecological footprint of food production, enrich traditional recipes, and pave the way for a more sustainable and delicious future. With workshops, innovative recipes, and a commitment to preserving the ocean's bounty, Portugal stands poised to inspire the world with its culinary ingenuity and environmental stewardship (Pereira 2009; Pereira et al. 2024).

16.4.6 Studies on Nutritional Value and Health Benefits

Various seaweeds have long been recognized for nutritional richness and potential health benefits. In Portugal, recent research has delved into the composition and applications of native seaweed species, particularly focusing on their suitability as a sustainable food source and their role in promoting human health (Cotas et al. 2024b).

Portuguese kelps, as detailed in the work "Portuguese Kelps: Feedstock Assessment for the Food Industry," offer a rich source of macronutrients, vitamins, and minerals. These kelps are particularly high in iodine, calcium, and magnesium, making them valuable for addressing dietary deficiencies (Pacheco et al. 2021). Additionally, the carbohydrate content, predominantly alginates, is a promising resource for food industry applications, including gelling agents and thickeners. The article "Seaweeds as Valuable Sources of Essential Fatty Acids for Human Nutrition" highlights the abundance of omega-3 and omega-6 fatty acids in Portuguese seaweed species. These essential fatty acids are crucial for cardiovascular health, brain function, and inflammatory response modulation, underscoring the nutritional significance of incorporating seaweed into the diet (Rocha et al. 2021).

As reviewed in: "An Overview to the Health Benefits of Seaweeds Consumption," regular intake of seaweeds is associated with a range of health advantages, including improved gut health, enhanced immune response, and anti-inflammatory effects. The polysaccharides found in seaweeds, such as fucoidans and laminarins, exhibit prebiotic properties, fostering a healthy gut microbiota. Seaweeds also show potential in cancer prevention and therapy (Lomartire et al. 2021). According to "Seaweeds' Nutraceutical and Biomedical Potential in Cancer Therapy: A Concise Review," certain compounds, including phlorotannins and carotenoids, possess antioxidant and antiproliferative properties. These bioactive molecules are being explored for their ability to inhibit cancer cell growth and induce apoptosis (Cotas et al. 2021a).

The dual nature of the many health effects of various seaweeds is addressed in "On the Health Benefits *vs.* Risks of Seaweeds and Their Constituents: The Curious Case of the Polymer Paradigm." While seaweed constituents such as carrageenans and alginates have proven benefits in weight management and cholesterol reduction, excessive consumption may pose risks due to low molecular weight fractions and the bioaccumulation of heavy metals such as arsenic and cadmium. This necessitates careful sourcing and processing to ensure safety and maximize benefits (Cotas et al. 2021b).

The potential for integrating seaweeds into various food products is explored in the: "Valorisation and Use of Seaweeds in the Food Sector" project. Seaweed are increasingly incorporated into snacks, seasonings, and functional foods, appealing to health-conscious consumers. Portuguese culinary traditions, including soups and stews, provide an excellent platform for the inclusion of seaweeds, enhancing both flavor and nutritional content (Moreira-Leite et al. 2023).

The studies reviewed highlight the significant nutritional and therapeutic potential of seaweeds in Portugal. As the global demand for sustainable and health-promoting foods grows, Portuguese seaweeds stand out as a versatile resource. Continued research and innovation in cultivation, processing, and application will be key to fully harnessing their benefits while mitigating potential risks (Pereira et al. 2024).

16.5 Algae Aquaculture and Harvesting

Seaweed cultivation has been an active area of research for over two decades (Sousa-Pinto 1998; Pereira et al. 2004, 2006; Matos et al. 2006; Abreu et al. 2011), laying the groundwork for its development in Portugal. The concept of Integrated Multi-Trophic Aquaculture (IMTA) has been successfully applied in Ílhavo (Aveiro region) at a fish farm producing seabass and seabream, managed by the company ALGAplus. Established in 2012, ALGAplus has introduced three brands: ALGA + (bulk seaweed—https://www.algaplus.pt/marcas-algaplus), Tok de Mar (food products—https://www.algaplus.pt/marcas-algaplus#Tok-de-Mar), and SeaOriginals (well-being products—https://www.algaplus.pt/marcas-algaplus#Sea-Originals).

Currently, the company is significantly expanding its operations with support from MAR 2020, demonstration project GENIALG, and more recently SeaMark and i3-4 Seaweed (https://i3-4-seaweed.eu/about-us/). The project is led by the University of Coimbra (www.uc.pt). ALGAplus cultivates species such as *Ulva* spp. and *Porphyra* spp., overseeing all production stages from hatchery to packaging to ensure complete traceability and quality (https://www.algaplus.pt/). Other species marketed by the company include *Codium tomentosum* (Chlorophyta) (Fig. 16.6), *Chondrus crispus*, *Gracilaria* spp., *Palmaria palmata*, *Porphyra dioica*, *P. umbilicalis* (Rhodophyta), and *Fucus vesiculosus* (Phaeophyceae).

Fig. 16.6 *Codium tomentosum* from aquaculture at ALGAplus

In addition to the pioneering company in Portugal, ALGAplus, artisanal *Spirulina*/*Limnospira* (Cyanobacteria), and other producers of algae also deserve to be mentioned here, such as: Tomar Natural (https://tomarnatural.pt/), 5essentia *Spirulina* Azores (https://5essentia.com/) and Spiralgae (https://www.facebook.com/spiralgae/), as well as companies dedicated to development, research, innovation and production in the algal sector (mainly with microalgae), such as A4F—Algae For Future, which has the satellite companies Algafarm SECIL (https://a4f.pt/en/projects/algafarm), Allmicroalgae (https://www.allmicroalgae.com/en/) (Fig. 16.7), Necton (https://necton.pt/en/home-necton-en/), among others, GreenAqua (https://greenaqua.pt/), Ínclita Seaweed Solutions (https://www.inclitaseaweedsolutions.com/), Arborea (https://arborea.io/), SeaExpert (https://seaexpert-azores.com/pt_pt/biotecnologia-algas/), Umimare (https://umimare.com/), Alga2O (https://www.alga2o.pt/en/the-company/who-we-are) Algicel—Biotechnology and Research (https://algicel.pt/en/), Stellarialga (https://www.stellarialga.com/pt/copy-of-blue-biotechnology) and Phytoalgae (https://phytoalgae.pt/en/). This diversity of actors in the algal sector contributes to the vitality and continued expansion of the sector in Portugal, generating not only economic value but also boosting research and knowledge in the field of algae and promoting sustainable production practices (AICEP 2024).

The Azorean company SeaExpert, established in 2004, specializes in harvesting specific seaweeds for the pharmaceutical, cosmetics, and biotechnology industries, as well as for the restaurant sector, with a focus on edible seaweeds. Seaweed collection for biotechnological applications involves species such as: *Asparagopsis armata*, *A. taxiformis*, *Gelidium spinosum*, *Pterocladiella capillacea* (Rhodophyta), *Gongolaria abies-marina*, *Cystoseira humi-*

Fig. 16.7 One of the Allmicroalgae raceways

lis, *Halopteris scoparia*, *Sargassum vulgare*, and *Zonaria tournefortii* (Phaeophyceae). This process is conducted by scuba divers who hold the required legal permits and are trained in specialized harvesting techniques.

For edible seaweeds such as: *Asparagopsis taxiformis*, *Porphyra* spp., *Nemalion elminthoides*, *Osmundea pinnatifida* (Rhodophyta), *Petalonia binghamiae*, *Fucus spiralis* (Phaeophyceae), *Ulva intestinalis*, and *U. rigida* (Chlorophyta), harvesting takes place in rocky intertidal zones. Qualified harvesters cut these seaweeds carefully, ensuring sustainable practices by targeting only the upper-base zone to allow regrowth.

Harvesting is carried out at appropriate developmental stages, following a scattered and selective approach to prevent habitat destruction and maintain the stability of marine ecosystems (Gaspar et al. 2019; https://www.seaexpert-azores.com).

Portugal is in sixth place in terms of average turnover generated by algal companies in Europe, with the first places in this ranking being occupied by France, Ireland, Norway, and Spain (annual average data, 2016–2020, published in the study "The future of the EU algae sector") (Parliament et al. 2023). The European Algae Biomass Association (EABA) estimated that the turnover of algal-based products in European Economic Area (EEA) countries was more than 350 million Euros in 2018 (including transformation), taking into account the companies and jobs. This estimate reached more than 400 million Euros when equipment and research and development companies are also included. The majority of sales refer to macroalgae (68%), followed by microalgae (17%) and *Spirulina* (Cyanobacteria) (11%). Turnover of the sector is reflected by employment figures, with France, Ireland, Spain, and Portugal recording the highest percentage of workers in the seaweed sector. Half of workers in the European algal sector are involved in the macroalgal value chain, as compared to 25% for microalgae (average annual employment data, 2016–2020) (AICEP 2024; https://www.eaba-association.org/en).

16.6 The Phycocolloid Industry in Portugal

The macroalgae in greatest demand, due to their agar content, are *Gelidium corneum* (Fig. 16.8), collected in the Central and Southern areas of the Portuguese coast, and *Pterocladiella capillacea* (Fig. 16.9), which is collected in the Azores (Palminha 1971; Sousa-Pinto 1998; Gaspar et al. 2019). When agar from Asia, especially Japan, became scarce due to World War II, the Portuguese phycocolloid

Fig. 16.8 *Gelidium corneum* specimen

Fig. 16.9 *Pterocladiella capillacea* specimen

industry began. This industry became important in terms of global agar production due to the abundance and quality of Portuguese algae (Vieira and Santos 1995; Sousa-Pinto 1998; Gaspar et al. 2019).

The first agar factory in Portugal was built in 1947, and their number increased to six factories by 1971. At that time, Portugal had become one of the world's largest producers of agar (Santelices 1999), producing around 1620 dry t per year (Palminha 1971). However, the decline or disappearance even of some agarophyte populations, the inability to diversify, the focus on quality, and the unfavorable international situation led to the disappearance of this industry, with only one company remaining today, i.e., Iberagar. Established in 1964 in Portugal, originally called Unialgas, Iberagar was initially focused on the production of agar for the food and pharmaceutical industries, exporting a large part of its production to Japan, the USA, and Great Britain (http://www.iberagar.com). From 1980 onwards, Iberagar became the only Portuguese company to produce carrageenan, supplying a significant portion of the internal market, mainly for the food industry (desserts and meat products) and also for expanding areas such as molecular biology and microbiology (Iberagar—Portuguese-Spanish Society of Marine Colloids) (Vieira and Santos 1995).

16.7 Research and Biotechnological Applications

Presently, nearly all higher education institutions in Portugal host research groups dedicated to studies of various algae, comprising macro- and microalgae, spanning ecological and environmental assessments to algal production, utilization, and applications. This concerted academic effort paves the way for innovative and sustainable practices in the development of novel value-added products, revitalizing this sector. A total of 47 organizations, comprising 13 universities, six polytechnics, eight collaborative laboratories (COLABS), 11 other organizations, and 10 culture collections, actively contribute to the advancement of algal-related research and applications. Building on the knowledge generated in the academic domain over the past two decades, numerous innovative companies have emerged, reshaping the paradigm within the sector. In today's context, the close collaboration between academia, collaborative laboratories, and companies has created a unique ecosystem in Portugal aimed at fostering innovation in the algal sector, not only related to biomass production but also to the development of novel value-added products (Vasconcelos et al. 2019; https://www.greencolab.com/).

16.7.1 Scientific Associations

Portugal is home to several notable scientific associations dedicated to the study and promotion of algae, including PROALGA and the Portuguese Association of Applied Algology (APAA) (https://www.facebook.com/associacaoalgologia/).

PROALGA (Portuguese Association of Algae Producers) is a private, non-profit association that represents the interests of algae producers in Portugal. Its primary goals are to promote, defend, and stimulate the algae production sector, encompassing both microalgae and macroalgae. Prolaga works closely with public and private entities, both nationally and internationally, to support the growth and quality of algae production. The association also organizes events and initiatives to foster collaboration and innovation within the industry (https://proalga.pt/).

The Portuguese APAA focuses on the applied aspects of algology, bridging the gap between research and practical applications. APAA is involved in various activities, including organizing conferences, workshops, and seminars to disseminate knowledge and advancements in algology. The association also collaborates with international organizations and networks to promote the sustainable use of all types of algae in biotechnology, environmental management, and other fields.

These associations play a crucial role in advancing the field of algology in Portugal, supporting both scientific research and industrial applications (https://www.linkedin.com/company/associa%C3%A7%C3%A3o-portuguesa-de-algologia-aplicada/).

16.7.2 Biobanks

Portugal is home to several prominent biobanks dedicated to the study and preservation of algae. These biobanks play a crucial role in advancing research in biodiversity, marine biology, and blue biotechnology.

The UC-ACOI (Coimbra Collection of Algae) is a scientific resource center at the University of Coimbra that promotes interdisciplinary research between biodiversity, fishing, and blue biotechnology. It is made up of a team of experts with a range of skills and knowledge in various domains of phycology and marine biology, this allows for access to technological platforms with cutting-edge equipment supported by specialized personnel. It provides privileged access to a unique collection of live microalgae and cyanobacteria (https://www.uc.pt/en/coimbra-collection-of-algae/).

The Bank of Algae and Cyanobacteria of the Azores (BACA) was created in 2018 and is located at the University of the Azores, Portugal. BACA is also a member of the World Federation for Culture Collections (WFCC—WDCM #1242), a Pt-mBRCN member, and an ECCO member (https://cibio.uac.pt/sites/default/files/2021-09/Catalogue_BACA_2021_final.pdf).

The LEGE-CC is a biological resource center located at the Interdisciplinary Centre of Marine and Environmental Research (CIIMAR) and is connected to all research going on at the Blue Biotechnology and Ecotoxicology Laboratory (https://lege.ciimar.up.pt/introduction/).

The CCMAR biobank (Faro, Portugal), created as part of the Portuguese Blue Biobank network, maintains and distributes samples of marine macroalgae and microorganisms associated with marine organisms like seagrasses, macroalgae, and corals. The biological samples were collected from Portuguese ecosystems, and also from other countries worldwide (https://ccmarbiobank.pt/).

16.7.3 Use of Seaweeds as Modern Fertilizers and Biostimulants

Certain seaweeds have emerged as a vital resource in sustainable agriculture, offering dual benefits as fertilizers and biostimulants. The application of seaweed-derived products in crop cultivation not only enhances soil fertility but also improves plant growth, resilience, and productivity, as evidenced by numerous scientific studies (Mamede et al. 2023).

The use of seaweed-derived polysaccharides as biostimulants is highlighted in: "Seaweed polysaccharides as potential biostimulants in turnip greens production." This study demonstrated that polysaccharides extracted from selected seaweeds enhanced nutrient uptake, promoted root development, and boosted overall biomass yield in turnip greens. Such effects are attributed to the bioactive compounds present in those seaweeds, which stimulated the engogenous hormonal activity and metabolic processes of the treated plants (Mamede et al. 2024).

The article: "Seaweed polysaccharides in agriculture: A next step towards sustainability" emphasized the role of seaweed-based products in fostering sustainable agricultural practices. Seaweed polysaccharides such as: alginates, carrageenans, and fucoidans act as natural soil conditioners, improving water retention and nutrient availability. These properties make certain seaweeds a valuable addition to sustainable farming systems, reducing dependency on chemical fertilizers (Mamede et al. 2023).

Research in: "Seaweed as plant biostimulants" underscords the ability of those seaweed extracts tested to enhance plant resilience to abiotic stresses, including drought, salinity, and temperature fluctuations. Bioactive compounds such as phenolics and phytohormones in seaweeds were found to bolster plant immune systems, enabling crops to withstand adverse environmental conditions. This makes seaweed-derived biostimulants particularly valuable in the face of climate change (Pereira et al. 2019; Pinto 2019).

The production of bio-fertilizers from specific seaweed species is detailed in: "Production of bio-fertilizer from *Ascophyllum nodosum* and *Sargassum muticum* (Phaeophyceae)." These brown algae are processed into bio-fertilizers that provide slow-release nutrients, reduced soil erosion, which minimize environmental impact. Such innovations highlighted the potential for integrating seaweed-derived fertilizers into modern agricultural practices (Silva et al. 2019).

The collective insights from these studies underscore the transformative potential of seaweed in agriculture. As fertilizers, they enhance soil health and nutrient content, while as biostimulants, they promote plant growth and resilience. The adoption of seaweed-based products represents a sustainable approach to improving agricultural productivity and environmental stewardship (Mamede et al. 2023).

16.8 What Next... the Future of Portuguese Seaweed Utilization and Cultivation

Seaweeds are invaluable assets in temperate regions, offering a wide array of ecosystem services and applications across environmental, agricultural, and biotechnological fields. Their ecological and economic significance underscores their role as sustainable resources with numerous benefits.

16.8.1 Ecological Contributions

Seaweeds play a crucial role in ecosystem services by acting as primary producers that support marine biodiversity, stabilizing coastal environments, and contributing to carbon assimilation and sequestration. Their ability to absorb nutrients, such as nitrogen and phosphorus, from surrounding waters is particularly relevant in temperate regions, where nutrient runoff from agriculture can lead to eutrophication. By mitigating such impacts, seaweeds help maintain the health of coastal ecosystems.

16.8.2 Agricultural Benefits

In agriculture, certain seaweeds serve as a promising resource for making practices more sustainable and eco-friendly. They are rich in bioactive compounds and nutrients that enhance soil fertility, stimulate plant growth, and improve crop resilience. The use of seaweed-based bio-fertilizers reduces dependence on synthetic fertilizers and mitigates the environmental footprint of agricultural practices. This aligns with the growing demand for eco-conscious solutions in temperate regions, where agricultural expansion often intersects with fragile marine ecosystems.

16.8.3 Biotechnological and Industrial Applications

Seaweeds possess remarkable adaptability to the climatic conditions of temperate regions, making them ideal candidates for large-scale cultivation. Their seasonal availability and fast growth cycles meet the demand for renewable biomass, which can be utilized in various industries, including food, cosmetics, and biotechnology. Integrating seaweeds into these industries supports environmental conservation and promotes the economic viability of coastal communities, where seaweed harvesting and aquaculture can serve as critical livelihood sources.

16.8.4 Culinary and Nutritional Value

Seaweed are gaining recognition in the culinary world for their nutritional benefits and versatility. Rich in essential nutrients such as iodine, calcium, and magnesium, along with antioxidants and unique bioactive compounds, seaweeds are a sustainable food source. They grow rapidly, require no freshwater or fertilizers, and help assimilate carbon dioxide dissolved in seawater. Workshops like: "Algas à Mesa" have promoted seaweeds as viable ingredients in Portuguese cuisine, inspiring chefs and home cooks to incorporate these "vegetables of the sea" into traditional and contemporary dishes. Recipes such as: "Seaweed and Cod Fritters" and "Seaweed Risotto" showcase how seaweeds can enhance flavor profiles and nutritional value, aligning with Portugal's focus on sustainable gastronomy.

16.8.5 Sustainability and Economic Viability

Emphasizing their ecological significance and diverse applications, it is important to integrate seaweeds into the broader framework of sustainability initiatives in temperate regions. These efforts support environmental conservation and promote the economic viability of coastal communities. Seaweed harvesting and aquaculture can serve as critical livelihood sources, enhancing the resilience and sustainability of these communities.

Acknowledgements FCT supported this work through the Centre for Functional Ecology Strategic Project (UIDB/04004/2025, UIDP/04004/2025), and by TERRA Associate Laboratory (LA/P/0092/2020).

Conflict of Interest The author declares no conflict of interest.

References

Abreu MH, Pereira R, Yarish C, Buschmann AH, Sousa-Pinto I (2011) IMTA with *Gracilaria vermiculophylla*: Productivity and nutrient removal performance of the seaweed in a land-based pilot scale system. Aquaculture, 312(1):77–87. https://doi.org/10.1016/j.aquaculture.2010.12.036

AICEP (2024) Empresas portuguesas lideram o caminho da aquacultura de algas. Available online at: https://www.portugalglobal.pt/pt/noticias/2024/maio/empresas-portuguesas-lideram-o-caminho-da-aquacultura-de-algas/. Consulted on 25 Dec 2024

Blanco A, Larrinaga AR, Neto JM, Troncoso J, Méndez G, Domínguez-Lapido P et al (2021) Spotting intruders: species distribution models for managing invasive intertidal macroalgae. JEM 281:111861. https://doi.org/10.1016/j.jenvman.2020.111861

Cacabelos E, Olabarria C, Viejo RM, Rubal M, Veiga P, Incera M et al (2013) Invasion of *Sargassum muticum* in intertidal rockpools: patterns along the Atlantic Iberian Peninsula. Mar Environ Res 90:18–26. https://doi.org/10.1016/j.marenvres.2013.05.008

Chugh M, Kumar L, Shah MP, Bharadvaja N (2022) Algal bioremediation of heavy metals: an insight into removal mechanisms, recovery of by-products, challenges, and future opportunities. Energy Nexus 7:100129. https://doi.org/10.1016/j.nexus.2022.100129

Cotas J, Pacheco D, Gonçalves AMM, Silva P, Carvalho L, Pereira L (2021a) Seaweeds' nutraceutical and biomedical potential in cancer therapy: a concise review. J Cancer Metastasis Treat 7:13. https://doi.org/10.20517/2394-4722.2020.134

Cotas J, Pacheco D, Araujo GS, Valado A, Critchley AT, Pereira L (2021b) On the health benefits vs. risks of seaweeds and their constituents: the curious case of the polymer paradigm. Mar Drugs 19(3):164. https://doi.org/10.3390/md19030164

Cotas J, Gomes L, Pacheco D, Pereira L (2023) Ecosystem services provided by seaweeds. Hydrobiology 2(1):75–96. https://doi.org/10.3390/hydrobiology2010006

Cotas J, Tavares JO, Silva R, Pereira L (2024a) Seaweed as a safe nutraceutical food: how to increase human welfare? Nutraceuticals 4(3):323–362. https://doi.org/10.3390/nutraceuticals4030020

Cotas J, Lomartire S, Gonçalves AMM, Pereira L (2024b) From ocean to medicine: harnessing seaweed's potential for drug development. Int J Mol Sci 25(2):797. https://doi.org/10.3390/ijms25020797

de Oliveira EV, Galhano F, Pereira B (1990) Actividades agro-marítimas em Portugal (1). Etnográfica Press, Lisboa. https://doi.org/10.4000/books.etnograficapress.6338

García-Huidobro MR, Aldana M, Varela P, García-Herrera C, Monteiro JN, Leitão F et al (2023) The influence of upwelling on key bivalves from the Humboldt and Iberian current systems. Mar Environ Res 189:106031. https://doi.org/10.1016/j.marenvres.2023.106031

Gaspar R, Pereira L, Sousa-Pinto I (2019) The seaweed resources of Portugal. Bot Mar 62(5):499–525. https://doi.org/10.1515/bot-2019-0012

Kalasariya HS, Maya-Ramírez CE, Cotas J, Pereira L (2024) Cosmeceutical significance of seaweed: a focus on carbohydrates and peptides in skin applications. Phycology 4(2):276–313. https://doi.org/10.3390/phycology4020015

Khan N, Sudhakar K, Mamat R (2024) Macroalgae farming for sustainable future: navigating opportunities and driving innovation. Heliyon 10(7):e28208. https://doi.org/10.1016/j.heliyon.2024.e28208

Lomartire S, Gonçalves AMM (2022) An overview of potential seaweed-derived bioactive compounds for pharmaceutical applications. Mar Drugs 20(2):141. https://doi.org/10.3390/md20020141

Lomartire S, Gonçalves AMM (2023) Algal phycocolloids: bioactivities and pharmaceutical applications. Mar Drugs 21(7):384. https://doi.org/10.3390/md21070384

Loureiro R, Gachon CMM and Rebours C (2015) Seaweed cultivation: potential and challenges of crop domestication at an unprecedented pace. New Phytol 206:489–492. https://doi.org/10.1111/nph.13278

Lomartire S, Marques JC, Gonçalves AMM (2021) An overview to the health benefits of seaweeds consumption. Mar Drugs 19(6):341. https://doi.org/10.3390/md19060341

Mamede M, Cotas J, Bahcevandziev K, Pereira L (2023) Seaweed polysaccharides in agriculture: a next step towards sustainability. Appl Sci 13(11):6594. https://doi.org/10.3390/app13116594

Mamede M, Cotas J, Pereira L, Bahcevandziev K (2024) Seaweed polysaccharides as potential biostimulants in turnip greens production. Horticulturae 10(2):130. https://doi.org/10.3390/horticulturae10020130

Matos S (2013) Designing food cultures: propagating the consumption of seaweed in the Azores Islands through recipes. Iridescent 2(3):24–33. https://doi.org/10.1080/19235003.2012.11428512

Mendes MC, Navalho S, Ferreira A, Paulino C, Figueiredo D, Silva D, Gao F, Gama F, Bombo G, Jacinto R et al (2022) Algae as food in Europe: an overview of species diversity and their application. Foods 11(13):1871. https://doi.org/10.3390/foods11131871

Morais T, Inácio A, Coutinho T, Ministro M, Cotas J, Pereira L, Bahcevandziev K (2020) Seaweed potential in animal feed: a review. J Mar Sci Eng 8(8):559. https://doi.org/10.3390/jmse8080559

Moreira-Leite B, Antunes R, Cotas J, Martins N, Costa N, Noronha JP, Mata P, Diniz M (2023) Modified atmosphere packaging (MAP) for seaweed conservation: impact on physicochemical characteristics and microbiological activity. Foods 12(14):2736. https://doi.org/10.3390/foods12142736

Neto AI, Tittley I, Raposeiro PM (2005) Rocky shore marine flora of the Azores. Secretaria regional do Ambiente e do Mar, Horta:156 pp. ISBN: 972-99884-0-4

Pacheco D, Miranda G, Rocha CP, Pato RL, Cotas J, Gonçalves AMM, Santos SMD, Bahcevandziev K, Pereira L (2021) Portuguese kelps: feedstock assessment for the food industry. Appl Sci 11(22):10681. https://doi.org/10.3390/app112210681

Paiva L, Lima E, Patarra RF, Neto AI, Baptista J (2014) Edible Azorean macroalgae as source of rich nutrients with impact on human health. Food Chem 164:128–135. https://doi.org/10.1016/j.foodchem.2014.04.119

Palminha FP (1971) Exploração e utilização de algas marinhas na plataforma portuguesa e nas ilhas do Arquipélago dos Açores. Junta Nacional do Fomento das Pescas 7:25–36

Parliament E, Union D.-G. f. I. P. o. t, Kuech A, Breuer M, Popescu I (2023) The future of the EU algae sector. European Parliament. https://www.europarl.europa.eu/RegData/etudes/STUD/2023/733114/IPOL_STU(2023)733114_EN.pdf

Pereira L (2009) Algas à Mesa – workshop sobre algas comestíveis. Polis Litoral Norte, Castelo do Neiva, 16 pp. Available online at: https://www.researchgate.net/publication/340815919_Algas_a_Mesa_-_Workshop_sobre_algas_comestiveis. Accessed on 2 Jan 2025

Pereira L (2004) Estudos em macroalgas carragenófitas (Gigartinales, Rhodophyceae) da costa portuguesa: aspectos ecológicos, bioquímicos e citológicos. Estudo Geral - Universidade de Coimbra, Coimbra, 293. https://estudogeral.sib.uc.pt/handle/10316/10017

Pereira L (2006) Identification of phycocolloids by vibrational spectroscopy (Chapter 11.10), pp. 1-22. In Alan T. Critchley, Masao Ohno & Danilo B Largo (eds.) World Seaweed Resources - An authoritative reference system. ETI Information Services Ltd. Hybrid Windows and Mac DVD-ROM. ISBN 90-75000-80-4

Pereira L (2016) Edible seaweeds of the world, 1st edn. CRC Press. https://doi.org/10.1201/b19970

Pereira L (2019) Traditional use of seaweed in agriculture in mainland Portugal. The Silk-Road Universities Network's WEB Magazine 1:49–54. https://doi.org/10.13140/RG.2.2.33849.31843

Pereira L (2021) As Algas na Alimentação. Rev Ciência Elem V9(01):006. https://doi.org/10.24927/rce2021.006. Available online at: https://rce.casadasciencias.org/rceapp/pdfr/9/1/. Accessed on 17 Dec 2024

Pereira L (2024a) Non-indigenous seaweeds in the Iberian Peninsula, Macaronesia Islands (Madeira, Azores, Canary Islands) and Balearic Islands: biodiversity, ecological impact, invasion dynamics, and potential industrial applications. Algal Res 78:103407. https://doi.org/10.1016/j.algal.2024.103407

Pereira L (2024b) Algas na Agricultura – Moliço e Sargaço. Voz do Campo. Available online at: https://www.researchgate.net/publication/381126926_Algas_na_Agricultura_-_Molico_e_Sargaco. Accessed on 15 Dec 2024

Pereira L, Correia F (2015) Algas Marinhas da Costa Portuguesa – Ecologia, Biodiversidade e Utilizações. Nota de Rodapé Editores, Paris. ISBN 978-989-20-5754-5

Pereira L, Cotas J (2019) Historical use of seaweed as an agricultural fertilizer in the European Atlantic area (Chapter 1). In: Pereira L, Bahcevandziev K, Joshi NH (eds) Seaweeds as plant fertilizer, agricultural biostimulants and animal fodder. CRC Press, Taylor & Francis Group, Boca Raton. 224 pp. ISBN: 978-1-13-859706-8 https://doi.org/10.1201/9780429487156

Pereira L, Cotas J (2023) Therapeutic potential of polyphenols and other micronutrients of marine origin. Mar Drugs 21(6):323. https://doi.org/10.3390/md21060323

Pereira L, Cotas J (2024) Seaweed: a sustainable solution for greening drug manufacturing in the pursuit of sustainable healthcare. Explor Drug Sci 2:50–58. https://doi.org/10.37349/eds.2024.00036

Pereira L, Bahcevandziev K, Joshi NH (eds) (2019) Seaweeds as plant fertilizer, agricultural biostimulants and animal fodder, 1st edn. CRC Press. https://doi.org/10.1201/9780429487156

Pereira L, Cotas J, Gonçalves AM (2024) Seaweed proteins: a step towards sustainability? Nutrients 16(8):1123. https://doi.org/10.3390/nu16081123

Pinto C (2019) Making agriculture more eco-friendly and sustainable with the help of seaweeds. UC News. Available online at: https://www.uc.pt/en/uc-news/articles/making-agriculture-more-eco-friendly-and-sustainable-with-the-help-of-seaweeds/. Accessed on 2 Jan 2025

Rocha CP, Pacheco D, Cotas J, Marques JC, Pereira L, Gonçalves AMM (2021) Seaweeds as valuable sources of essential fatty acids for human nutrition. Int J Environ Res Public Health 18(9):4968. https://doi.org/10.3390/ijerph18094968

Santelices B (1999) A conceptual framework for marine agronomy. In: Kain JM, Brown MT, Lahaye M (eds) Sixteenth international seaweed symposium. Developments in hydrobiology, vol 137. Springer, Dordrecht. https://doi.org/10.1007/978-94-011-4449-0_3

Santos R, Novoa-Pabon A, Silva H, Pinho M (2020) Elasmobranch species richness, fisheries, abundance and size composition in the Azores archipelago (NE Atlantic). Mar Biol Res 16(2):103–116. https://doi.org/10.1080/17451000.2020.1718713

Soares C, Machado S, Vieira EF, Morais S, Teles MT, Correia M, Carvalho MA, Domingues VF, Ramalhosa MJ, Delerue-Matos C, Antunes F (2017) Seaweeds from the Portuguese coast: A potential food resource? IOP Conference Series: Mater Sci Eng 231:012126. https://doi.org/10.1088/1757-899X/231/1/012126

Silva LD, Bahcevandziev K, Pereira L (2019) Production of bio-fertilizer from *Ascophyllum nodosum* and *Sargassum muticum* (Phaeophyceae). J Ocean Limnol 37:918–927. https://doi.org/10.1007/s00343-019-8109-x

Sousa-Pinto I (1998) The seaweed resources of Portugal. In: Critchley AT, Ohno M (eds) Seaweed resources of the world. Japan International Cooperation Agency, Yokosuka, pp 176–184

Vasconcelos V, Moreira-Silva J, Moreira S (eds) (2019) Portugal blue bioeconomy roadmap – BLUEandGREEN. CIIMAR, Matosinhos, (pub), 68 pp. Available online at: https://www2.ciimar.up.pt/pdfs/resources/roadmap_digital_hGBit_.pdf. Accessed on 26 Dec 2024

Vieira VV, Santos M (1995) Directório de aquacultura e biotecnologia marinha. Escola Superior de Biotecnologia da Universidade Católica Portuguesa, Porto, 113 pp

Leonel Pereira has a degree in Biology (Scientific branch), a PhD in Biology (specialty Cell Biology), and a Habilitation on Biosciences, specialty in Biotechnology, from the Faculty of Science and Technology of the University of Coimbra, where he is currently Associate Professor with habilitation, Master's Degree in Biodiversity and Plant Biotechnology Coordinator, and ACOI - Coimbra Collection of Algae coordinator. In addition to teaching at this university, he is an integrated researcher at CFE—Centre for Functional Ecology: Science for People & Planet. His interests focus mainly on the areas of Marine Biodiversity (algae), Marine Biotechnology (algae bioactive compounds), and Marine Ecology (environmental assessment). He is, since 2008, the author and editor of the electronic publication MACOI—Portuguese Seaweeds Website (http://www.flordeutopia.pt/macoi/), since 2018 he is a member of the editorial board of the scientific journals Marine Drugs an Applied Sciences, he is editor of several special issues of scientific journals (Journal of Applied Phycology, Botanica Marina, Marine Drugs, Applied Sciences) has been editor or author of more than 20 books, author of more than 200 indexed articles, and more than 100 book chapters. He has delivered over a hundred lectures and oral communications at various national and international scientific events. He is the coordinator of several national and international research projects. In 1998, he received the Francisco de Holanda Award (Honourable Mention) and, more recently, the Mar Rei D. Carlos Award (18th edition), and winner of the CHOICE Award—outstanding academic title award 2016—Book title: Edible Seaweeds of the World, CRC Press. He is included in the The World's Top 2% Scientists 2021, 2022, 2023, and 2024 list released by the University of Stanford and Elsevier, and holds an h-index of 43 WoS, h 51 Scopus, and h 53 ResearchGate.

The Sustainability of the *Saccharina japonica* Cultivation Industry in China

17

Xiuliang Wang, Jie Zhang, Jianting Yao, and Delin Duan

Abstract

The cultivation of *Saccharina japonica* (formerly *Laminaria japonica*) has been a cornerstone in China's marine aquaculture industry for nearly seven decades, contributing significantly to food security, economic development, and environmental sustainability. This article examines the historical development, technological innovations, and challenges facing the *S. japonica* cultivation industry, with a focus on sustainability in the context of climate change. The key achievements, such as the summer seedling method, floating raft cultivation systems, and the development of high-yield cultivars, have enabled large-scale production, with yields increasing from 22 t (dry weight) in 1952 to 170,000 t in 2021. However, the kelp industry was affected by ocean warming, acidification, and other marine influences. These changes threaten kelp growth and natural seaweed-derived ecosystems and their stability. Offshore cultivation and mechanization are emerging as potential solutions to these challenges, though both are in their infancy. The article also explores the role of *S. japonica* cultivation in carbon sequestration and its potential for mitigation during climate change, while acknowledging the limitations of current practices. Overall, the sustainability of the *S. japonica* industry will depend on continuous technical innovation, its adaptation to environmental changes, and the development of advanced technologies and strains for the offshore and mechanized cultivation.

Keywords

Kelp · Cultivation · *Saccharina japonica* · Sustainability · Climate change · Breeding · Offshore cultivation · Mechanization

X. Wang · J. Zhang · J. Yao · D. Duan (✉)
Key laboratory of Experimental Marine Biology, Institute of Oceanology, Chinese Academy of Sciences, Qingdao, China
e-mail: yaojianting@qdio.ac.cn; dlduan@qdio.ac.cn

17.1 Introduction

Saccharina japonica, previously named *Laminaria japonica* (Phaeophyceae), is valued as a nutritious sea vegetable and as a traditional remedy for goiter in many Asian countries, particularly Japan, Korea, and China. The genera *Saccharina* and *Laminaria* comprise approximately 50 species, most of which are native to the Northwestern and eastern Pacific (Bolton 2010). Only a few species, such as *Laminaria digitata*, *L. hyperbora*, and *L. solidungula,* exist along the European and Arctic coasts, while *L. abyssalis* inhabits deeper waters in the South Atlantic (Bartsch et al. 2008). Among these, *Saccharina japonica* is the only species that is commercially cultivated on a large scale in China, South Korea, North Korea, and Japan (Tseng 1981, 1987). More recently, *Saccharina latissima* was explored for cultivation in Europe and North America, primarily as a feedstock for biofuel production and health food (Huang et al. 2023). Currently, kelp aquaculture is regarded as a critical component of the "Blue Carbon" initiatives, aimed at sequestering excess CO_2, which contributes to ocean warming and acidification. This role mirrors the natural function of kelp forests in marine ecosystems (Zhang et al. 2017b).

The cultivation of *S. japonica* in China began in the 1920s, following its accidental discovery along the coast of Dalian. After several decades of trial and error, the floating raft cultivation system, also known as the long-line system, was established in 1952, laying the foundation for large-scale production. In 1958, the first kelp seedling hatchery was successfully constructed in Qingdao to culture sporophytic seedlings, enabling mass cultivation. Subsequent innovations, such as the use of nutrient-rich suspended plastic bags and blade-tip cutting to enhance yield, further advanced the industry (Tseng and Wu 1962; Duan et al. 2015). These breakthroughs not only established *S. japonica* as one of the world's first marine crops but also ushered in a "Golden Age" for marine culture in China. It was accompanied by the rise of shellfish farming in the 1960s–70s, shrimp

A. T. Critchley et al. (eds.), *Advances in Temperate Phyconomy: Algal Harvest and Cultivation in Globally Distributed Temperate Waters*, Developments in Applied Phycology 15, https://doi.org/10.1007/978-3-032-02955-3_17

farming in the 1980s–90s, fish farming in the 1990s–2000s, and a surge in abalone (*Haliotis* sp.) and sea cucumber (*Apostichopus* sp.) cultivation after 2000 (Wang et al. 2018a).

Since the initiation of large-scale cultivation of *S. japonica* in 1958, the industry operated sustainably for nearly seven decades. Production increased from 6,253 t (dwt) in 1958 to an estimated 1,770,000 t in 2023, with cultivation areas expanding from 18,000 to 50,795 hectares (Tseng 1981; FABPRC 2024). This expansion resulted in large biomass production by marine aquaculture in terms of fresh weight. In China, 70% of *S. japonica* biomass is currently used for human consumption, while 30% is utilized as feed for sea cucumber and abalone (https://submariner-network.eu/wp-content/uploads/2024/01/Seaweed_Industry_in_China_compressed.pdf). An increasing proportion of *S. japonica* biomass is expected to support the rapidly growing abalone farming sector in the future (Hwang et al. 2019). The sustainability of the *S. japonica* cultivation industry was attributed to several factors, including the summer sporophytic seedling technique, the floating raft culture system (comprising floating lines, suspended culture ropes, anchor ropes, and floats), eco-physiological studies, and the development of new climate-resistant cultivars.

Given the projected increase in ocean warming, acidification, and the frequency and intensity of marine heatwaves and storms (Wang et al. 2023a), global climate change is expected to have a significant impact on the kelp cultivation industry in China. This chapter outlines kelp adaptation strategies and addresses the changes and challenges required during developments in cultivation techniques. Moreover, offshore cultivation of *S. japonica* is proposed as a potential solution to include mechanization of the production processes to mitigate the local labor shortages in this industry.

17.2 Summer Seedling Production Under Hatcheries or Greenhouse Conditions

The cultivation of *S. japonica* sporophytic seedlings during the low-temperature seawater periods, during the summer, represents an innovative and transformative concept of kelp cultivation. This method was developed to address the issue of fouling organisms (including epiphytic seaweeds), which rapidly colonize kelp seedling culture ropes. These pests can significantly impede the growth of young sporophytes when the ropes, freshly inoculated with kelp zoospores, are introduced into the seawater in October or November. Tseng et al. (1955) demonstrated that summer seedlings not only effectively avoided competition for light from fouling seaweeds but also enhanced kelp production by extending the growing period by two months. In 1958, the first hatchery dedicated to culturing summer seedlings was established in Qingdao (Duan et al. 2015). Since then, the number of summer-produced seedling hatcheries has increased, with approximately 20 hatchery factories now producing 30–40 billion seedlings annually to support 50,000 h of China's bay areas (FABPRC 2024).

The "summering seedling" method primarily consisted of three components, namely: (a) a seawater refrigeration system, (b) a water circulation system, and (c) a hatchery equipped with concrete tanks, light shields, and zoospore collectors. This method has evolved into two distinct models, the Shandong Model, used in northern China, and the Fujian Model, employed in the south. The Shandong Model involves the direct use of mature parent sporophytes with visible sori on the kelp blades, collected from floating rafts in the open water cultivation areas. These are used to release and collect the zoospores in the hatchery when the seawater temperatures reach 20–21 °C in early August. In contrast, the Fujian Model transfers parent sporophytes without sori to hatcheries in late July, when the temperatures in the East China Sea reach 25–26 °C. These sporophytes are then cultured in tanks with circulating low-temperature seawater (8–15 °C) until their sori develop in September. At this point, zoospores are collected (Wang and Xue 1959; Cai 1982). Prior to the establishment of the Fujian model, seedlings required for *S. japonica* production in Fujian were transferred from northern hatcheries, which incurred significant time and financial costs. Today, Fujian hatcheries not only supply seedlings to local cultivation families but also offer high-quality seedlings to the kelp cultivation companies in northern China. In 2023, Fujian hatcheries produced 24.6 billion seedlings, as compared to 8.9 billion from Shandong (FABPRC 2024).

Since its inception, the summer seedling method has undergone minimal modification (Su et al. 2017). However, a major drawback of this method is its high energy consumption, as substantial amounts of electricity are required to cool large volumes of seawater to 8–10 °C during the summer seedling cultivation process (Li et al. 1999). To address this issue, alternative approaches, such as using vegetative gametophytes to produce seedlings, have been proposed to reduce both the time and energy costs (Wu et al. 1994; Li et al. 1999; Xu et al. 2009). Nevertheless, this alternative method faces significant challenges, including the clustering and overcrowding of juvenile sporophytes and their detachment from attachment ropes (Xu et al. 2009). As a result, complete replacement of the summer seedling method with vegetative gametophyte cultivation is not feasible until the technical obstacles are resolved.

17.3 Breeding of New Kelp Cultivars for the Cultivation Industry

So far, approximately 20 cultivars of *S. japonica* have been developed and utilized in China's kelp cultivation industry (Wang et al. 2020; Hu et al. 2021). These cultivars have significantly contributed to enhancing both the quantity and quality of *S. japonica* production (FAO 2014). For instance, the No.860 and No.1170 cultivars, which are characterized by their high iodine content, were developed through X-ray-induced mutagenesis of cultivated populations over a period of 5–15 years in the 1970s. These cultivars were specifically bred to meet the demands of the iodine extraction industry and are frequently cited as exemplary models in China's kelp cultivation sector (IOIMF 1976). In 1991, the National Certification Committee for Aquatic Stocks and Varieties (NCCASV) was established to advance genetic breeding and the development of new varieties (Zhou and Gui 2018). To date, 283 new aquatic varieties have been approved in China, including 11 *S. japonica* cultivars (Table 17.1). Among these, the "901" cultivar, the first approved by NCCASV, exhibits a growth rate that is 60% faster than that of control populations (Zhang et al. 2007). Other strains, such as Dongfang No.6 and No.3, were bred for traits like high seawater temperature tolerance and resistance to high levels of light (Li et al. 2008, 2016a).

China is amongst the pioneering countries in the breeding of marine crops (van der Meer 1983; Hwang et al. 2019). To date, nearly all *S. japonica* cultivars have been developed using traditional methods, primarily selection and hybridization. The first cultivar, Haiqing No.1, was bred in 1962 through inbreeding and selection, with the aim of achieving faster growth, larger blade size, and high-temperature tolerance. This approach was based on the hypothesis that the natural cultivation populations exhibited a high degree of heterozygosity (Fang et al. 1963). This strategy has continued to be employed in the development of more recent cultivars, such as "Zaohoucheng No.1" (Tian and Yuan 1989) and "Huangguan No.1" (Liu et al. 2014b). The development of "Huangguan No.1" involved selecting 150 sporophytes with desirable traits, such as large blade size, as the founding breeding population. Through five rounds of recurrent mass selection from a mixed population of 1200 individuals annually, the final cultivar demonstrated a 30% increase in yield and a 25% improvement in growth rate, as compared to control varieties (Liu et al. 2014a). The life cycle of *S. japonica* consists of two stages: the macroscopic diploid sporophytic generation and the microscopic haploid gametophyte generation (Roe 1916; Tseng 1981, 1987). The isolation and vegetative culture of male and female gametophytes enabled the hybridization of gametophytes from different strains or species, proving to be an efficient breeding method for *S. japonica* (Fang et al. 1983, 1985). Furthermore, heterosis in yield-related blade length has been found to exhibit one significant linear relationship with the genetic distance between parental gametophytes (Li et al. 2007). Several cultivars, including "Dongfang No.2" (Li et al. 2007), "Dongfang No.3" (Li et al. 2008), and "Dongfang No.6"(Li et al. 2016a), have been developed as intraspecific or interspecific hybrids. However, only the first filial generation of these hybrids is used in cultivation, as subsequent generations exhibited significant trait segregation (Li et al. 2007).

Molecular breeding has been proposed as a promising approach to the future of kelp breeding (Robinson et al. 2013; Hwang et al. 2019; Goecke et al. 2020; Wang et al. 2020; Hu et al. 2021, 2024). Genetic linkage mapping and quantitative trait locus (QTL) mapping for traits such as blade length and width have been conducted for *S. japonica* (Liu et al. 2009, 2010; Yang et al. 2009; Wang et al. 2018b, 2023b) and other economically important seaweeds (Huang and Yan 2019; Xu et al. 2015b; Shan et al. 2015). The draft genomic sequence of *S. japonica* has also been deciphered (Ye et al. 2015; Liu et al. 2019b). Wang et al. (2023b) constructed a genetic linkage map for *S. japonica* comprising 33 linkage groups with 3397 SSR and SNP markers, identifying marker M1895 as closely associated with major QTLs for blade length and width, with a LOD threshold greater than 5.0. Despite these advances, the application of molecular marker-assisted selection in *S. japonica* breeding strategies remains to be adopted. The genomic selection, with the genomic expected breeding value (GEBV) for selecting the candidate loci based on linkage disequilibrium (LD), was recognized as an efficient breeding strategy for over a decade in agriculture and, more recently, in marine culture (Meuwissen et al. 2001; Zhou and Gui 2018; D'Agaro et al. 2021). At the same time in China, genomic selection was applied to marine culture in the breeding of the scallop variety "Penglaihong No.2," using Best Linear Unbiased Prediction (BLUP) analysis (Zhou and Gui 2018). For kelp, genomic selection has been tested in *S. latissima*, an economically and ecologically important species in the eastern North Pacific and the North Atlantic Oceans, to evaluate the accuracy of the Genomic Best Linear Unbiased Prediction (GBLUP) model in a species with a biphasic life cycle (Huang et al. 2023).

In addition, the emerging technology of CRISPR-Cas9 gene editing has shown promise as a breeding tool (Xu et al. 2024). In land crops, these techniques have been used to knock out the *KRN2* gene in maize, increasing kernel row number, and to delete 54 base pairs of regulatory regions of the *IPA1* gene in rice, enhancing grain yield while reducing tillering (Chen et al. 2022; Song et al. 2022). Similarly, CRISPR-Cas9 has also been preliminarily applied to macroalgae, including *Ectocarpus* (Phaeophyceae) species 7 (Badis et al. 2021), *Ulva prolifera* (Chlorophyta) (Ichihara et al. 2022), *Gracilariopsis lemaneiformis* (Rhodophyta)

Table 17.1 New cultivars developed for *S. japonica*

Cultivar	Parent	Characters	Method for breeding	Reference
Dongfang No.7	Femal gametophyte: Kuanbao (*S. japonica*) Male gametophyte: Korea kelp (*S. japonica*)	Wide, thick, smooth, dark, brown and wrinkle less blade, strong holdfast, flat-cylindrical stripe, oblate base, non-obvious vertical central groove, and less margin	Hybridizing gametophytes and selection	Li et al. (2016b)
205	Female gametophyte: Rongcheng cultivar (*S. japonica*) Male gametophyte: Korea (*S. japonica*)	Deeply brown, wide, good yield, high-temperature resistance, light stress resistance, dry blade with heavy green color	Hybridizing gametophytes and selection	Pang et al. (2015)
Dongfang No.3	Female gametophyte: Line 7 (*S. japonica*) Male gametophyte: Line LZZ Introgressed genetically with (*S. longissima*)	Dark brown, wide, uniform, ripple less, vary tough blade, deep vertical groove and a half-round blade base, robust holdfase, tolerance of strong irradiance and high seawater temperature, resistance to the tissue rottenness of blade tip and edge, mature later,	Hybridizing gametophytes	Li et al. (2008)
Sanhai	Fujian (*S. japonica*) and Rongfu Introgressed genetically with *S. latissima*	With obvious growth advantage	Hybridization of zoospores combined backcrossing with self-crossing	Zhang et al. (2018)
Rongfu	Female gametophytes: Fujian (*S. japonica*) Male gametophytes: Yuanza No.10 Introgressed genetically with *S. latissima*	High yield and high seawater temperature tolerance	Hybridization of gametophytes, continuous self-crossing and targeted selction	Zhang et al. (2011)
Dongfang No.2	Female gametophyte: *S. japonica* Male gametophyte: *S. longissima*	High yield, wide long deep-brown blade with distinct middle groove, robust holdfast	Hybridizing gametophytes	Li et al. (2007)
90-1	Female gametophyte: *S. longissima* Male gametophyte: Zaohoucheng (*S. japonica*)	Grow faster, loss of less tissue, higher yield, high-temperature resistance, with desirable amounts of iodine, mannitol and algin	Hybridizing gametophytes in combination with successive self-crosing and selection	Zhang et al. (2007)
Huangguan No.1	Fujian sporophyte population	Much longer, wider, thicker blade, faster growth rate, good yield, high-temperature resistance	Selection and continuous inbreeding for 5 years	Liu et al. (2014b)
Dongfang No.6	Female gametophyte: Korea ecotype (*S. japonica*) Male gametophyte: Lianza No.1 (*S. japonica*)	Fast in growth, tolerant to high seawater temperature, strong irradiation resistance, mature later, high in processed product yield	Hybridizing gametophytes	Li et al. (2016a)
Ailunwan	Sporophyte population (*S. japonica*)	High yield	Selection and continuously inbreeding	Zhang et al. (2016)
Zhongbao No.1	Female gametophyte: Korea ecotype (*S. japonica*) Male gametophyte: Rongcheng cultivar (*S. japonica*)	High yield	Hybridizing gametophytes	Su et al. (2021)

(Zhang et al. 2023), *Neopyropia yezoensis* (formerly termed as *Pyropia yezoensis*) (Rhodophyta) (Wang et al. 2024b), and *Saccharina japonica* (Shen et al. 2023). Whereas in *S. japonica*, the adenine phosphoribosyl transferase (*APT*) gene in gametophytes was successfully edited using CRISPR-Cas9 ribonucleoprotein (RNP) complexes delivered via microinjection, with mutants selected using 2-Fluoroadenine (2-FA) (Shen et al. 2023). Despite these advances, gene editing technology has not yet been applied to *S. japonica* breeding programs, as most economically important traits are quantitative and controlled by multiple minor genes (Wang et al. 2023b). Further technologies for simultaneous editing of multiple genes will be necessary to fully realize the potential of CRISPR-Cas9 in kelp breeding (Shen et al. 2023).

17.4 Climate Change and Kelp Cultivation Industry

Climate change is one of the most challenging issues facing present society (IPCC 2021). Recent studies indicated that atmospheric CO_2 concentrations were primarily driven by human activities and have risen from approximately 278 ppm in 1750 to the current level of 420 ppm (State of the Climate 2024 Update for COP29). The increase in atmospheric CO_2 is the primary driver of ocean warming, acidification, marine heatwaves (MHWs), sea level rise, and other oceanic anomalies (Blunden and Boyer 2024). Ocean warming and acidification have profound impacts on kelp forests, affecting individuals, populations, and entire kelp community ecosystems (Harley et al. 2012; Steneck and Johnson 2013; Smale 2020; Wernberg et al. 2019, 2024). In the seas surrounding China, the average surface temperature increased by 0.10–0.14 °C per decade from 1950 to 2021, while the annual frequency and average intensity of MHWs have risen by 1–2 per decade and 0.1–0.3 °C per decade, respectively, since the 1980s (Wang et al. 2023a). These environmental changes pose significant challenges to the development of the *Saccharina japonica* cultivation industry, which is predominantly located in coastal bay areas sheltered from strong waves and typhoons but remains vulnerable to the inevitable effects of warming and acidification.

In addition, seawater temperature is a critical factor influencing cellular chemical reaction rate, growth rate, timing of meiosporogenesis and gametogenesis, and the latitudinal distribution of seaweeds (Hurd et al. 2014). Each seaweed species, or different ecotypes within the same species, has been seen to exhibit a bell-shaped temperature range for optimal growth and development (Harley et al. 2012). When ocean temperatures exceed the upper thermal tolerance limits of a seaweed species, its growth is threatened, leading to shifts in population distribution to the most suitable niches or, in extreme cases, local extinction (Harley et al. 2012; Wernberg et al. 2024).

17.4.1 Ocean Warming and Its Impacts on Kelp Forests

Ocean warming encompasses both gradual increases in sea temperature and acute events such as MHWs. An analysis of a global database of kelp forest changes over the past half-century revealed that 38% of eco-regions experienced declines, 27% showed increases, of eco-regions, and 35% exhibited no detectable change; these shifts have been attributed to a combination of human impacts, including overfishing, direct harvesting, and climate change (Krumhansl et al. 2016). In southwestern Japan, field observations demonstrated that the decline and local extinction of a canopy-forming kelp (*Ecklonia*) were driven by gradual sea surface temperature increases over the past 40 years, largely influenced by the warming Kuroshio Current (Tanaka et al. 2012). Similarly, Starko et al. (2019) documented declines in the abundance of kelp species such as *Hedophyllum sessile* (formerly termed *Saccharina sessilis*) and *Egregia menziesii* (Phaeophyceae) in the Northeast Pacific, following decades of gradual warming in British Columbia. MHWs could also lead to significant declines in kelp forest abundance and even trigger ecosystem phase shifts. For example, in 2011, sea temperatures along more than 2000 km of the southeast Australian coast surged to unprecedented anomalies of 2–4 °C, persisting for over 2 months. This serious MHW resulted in a reduction of habitat-forming seaweeds and a shift in ecosystem structure, which was characterized by increased dominance of turf-forming algae and a tropicalization of fish communities (Wernberg et al. 2012, 2016). Other MHWs occurred in Pile Bay, South Island of New Zealand, during the exceptionally hot summer of 2017/18, leading to the complete extinction of local *Durvillaea* populations (Thomsen et al. 2019). Lima et al. (2007) compared recent surveys of seaweed abundance along the Portuguese coast with historical records from 60 to 70 years ago. They reassessed trends in coastal air and sea temperatures over the last 50 years. They found that cold-water species shifted either northward or southward, while warm-water species expanded their ranges northwards. These findings align with the statistical analysis by Krumhansl et al. (2016), which highlighted that the regional shifts in seaweed species, due to rising temperatures, varied spatio-temporally and were species-specific in nature.

17.4.2 Ocean Acidification and Its Effects on Seaweeds

The pH of ocean surface water has decreased by approximately 0.1 units since the beginning of the industrial era, primarily due to the dissolution of fossil fuel-derived CO_2 into seawater. Projections suggested that by the year 2100, surface ocean pH could drop by an additional 0.5 units, with an annual decline of about 0.002 units (IPCC 2013; Raven et al. 2005; Ji and Gao 2021). The physiological responses of seaweeds to ocean acidification (OA) are highly species-specific. For example, some subtidal red algae, which lack effective carbon-concentrating mechanisms (CCMs), cannot utilize bicarbonate (HCO_3^-) and may exhibit enhanced photosynthesis and growth under elevated CO_2 conditions (Harley et al. 2012; Koch et al. 2013; Hurd et al. 2014). In contrast, seaweed species with CCMs may show positive, neutral, or negative responses depending on the species or life stage examined (Leal et al. 2017; van der Loos et al. 2019; Hengjie et al. 2023). Laboratory studies have extensively investigated the effects of OA on various physiological and biochemical processes in seaweeds, including growth, germination, photosynthetic activity, respiration, nutrient absorption, pigment composition, and enzyme activity (Ji and Gao 2021; Hengjie et al. 2023). However, few studies have explored the impacts of OA on the range shifts or ecosystem structure of seaweed communities (Harley et al. 2012; Wernberg et al. 2024). It is likely that OA will act as a multi-stressor, interacting with other factors such as warming, pollution, and eutrophication to influence seaweed communities. Brodie et al. (2014) demonstrated that the diverse seaweed canopies in the northeast Atlantic could be replaced by calcified and non-calcified turf-forming algae under the combined pressures of seawater warming, OA, and increased storm intensity.

17.4.3 Impacts of Ocean Warming and Acidification on the Kelp Cultivation Industry

Limited research has been conducted on the effects of ocean warming and OA on the kelp cultivation industry, particularly on the *S. japonica* cultivation (Chung et al. 2017; Clements and Chopin 2017; Largo et al. 2017; Ji and Gao 2021; Hengjie et al. 2023). An early case study demonstrated that when the young sporophytes (1–3 cm) from seven different populations of *S. japonia* were exposed to temperatures of 10, 21, and 23 °C for seven days, the offspring from the Qingdao wild population exhibited the highest capacity to resist high temperatures. However, the blade elongation rate of young sporophytes from all populations was significantly reduced at 21 and 23 °C (Liu and Pang 2010). Similar findings have been reported for other kelp species, such as *S. latissima* and *Laminaria digitata*, where tissue loss in sporophytes with the blade length of 0.5–1 m was observed to be at least twice as high at 21 °C as compared to 11 °C within the first two weeks of exposure (Simonson et al. 2015). These results suggested that rising temperatures could contribute to the decline of natural kelp populations and reduce productivity in the kelp cultivation industry.

In China, the *S. japonica* cultivation industry typically harvests kelp populations in June and July in northern regions and in March and May in southern regions, before the sea temperatures exceed 20 °C (Duan et al. 2015). The optimal temperature range for sporogenesis in *S. japonica* has been extensively studied, with temperatures between 6 and 15 °C, even up to 17 °C, reported as suitable for this process (Liu et al. 2023; Mizuta et al. 1999b). Notably, at 15 °C, sorus formation occurs earlier than at 5 °C and 10 °C, and the sorus area on blade discs is also larger (Mizuta et al. 1999a). Similar studies on *L. digitata* identified an optimal sporogenesis temperature range from 5 to 12 °C (Buchholz and Lüning 1999; Bartsch et al. 2013). In China, *S. japonica* seedlings are produced using the summer seedling method in the hatcheries, where sporogenesis occurs under controlled conditions of temperature, light, and nutrients, particularly in the Fujian area. This approach allows seedling production to avoid the adverse effects of ocean warming and OA (Duan et al. 2015; Su et al. 2017).

Additionally, molecular studies on *S. japonica* have investigated the role of the *HSP70* gene, transcriptome, and microRNA in acclimation to temperature stress (Fu et al. 2009; Liu et al. 2014a, 2015). These studies indicated that the HSP70 gene could serve as a molecular marker for monitoring thermal acclimation, although the molecular responses to temperature stress are highly complex.

17.4.4 Interactions Between Seaweed Cultivation and Ocean Acidification

Seaweed cultivation and OA are intricately linked, with OA impacting the productivity of cultivated seaweed, while seaweed farming is increasingly regarded as a potential method for carbon dioxide removal (CDR) to mitigate climate change (Chung et al. 2017; Duarte et al. 2017; Troell et al. 2023; Pessarrodona et al. 2024). A study by Xiao et al. (2021) demonstrated that *S. japonica* farms exhibited the highest capacity to buffer acidification, increasing local pH by 0.10 units, compared to farms cultivating *Gracilariopsis lemaneiformis* and *Pyropia haitanensis* (Rhodophyta). Additionally, Zhang et al. (2012) found that *S. japonica* farms acted as a source of dissolved organic carbon (DOC) and particulate organic carbon (POC), with a portion of these carbon forms exported and ultimately sequestered in sediments or buried in the deep sea. This supports the role of seaweed farms as a carbon sink in climate change mitigation. Furthermore, Zhang et al. (2017b) estimated that seaweed cultivation in China released approximately 0.6 million t of recalcitrant, dissolved organic carbon into seawater annually, contributing significantly to the oceanic carbon sink.

Nevertheless, the efficacy of seaweed cultivation as a carbon sequestration strategy remains debatable (Troell et al. 2023; Pessarrodona et al. 2024). The total area dedicated to seaweed aquaculture (e.g., 1500–2000 km^2) is relatively small compared to the 3.5 million km^2 occupied by wild seaweed populations. So far, global seaweed cultivation has been estimated to capture about 2–3 million t of CO_2 annually, whereas wild seaweed populations are estimated to capture 173 million t (Krause-Jensen and Duarte 2016; Froehlich et al. 2019). Moreover, most seaweed aquaculture products are used for daily human consumption, animal feed, or chemical extraction, representing short-term carbon storage, rather than long-term carbon removal, such as carbon burial in sediments or the deep sea (Krause-Jensen and Duarte 2016; Troell et al. 2023). Additionally, the energy and materials used in seaweed cultivation often have significant carbon footprints, particularly the energy-intensive processes involved in hatchery-based kelp seedling production, which typically rely on non-renewable energy sources (Pessarrodona et al. 2024).

17.4.5 Carbon Concentration Mechanisms in *S. japonica*

The carbon concentration mechanism (CCM) in *S. japonica* has been confirmed through the identification of several carbonic anhydrase genes (Ye et al. 2014; Wang et al. 2024a), as well as C_4-related genes involved in bicarbonate concentration, identified via genome-wide analysis (Shao et al. 2019). Recent studies showed that elevated pCO_2 levels (400 and 1000 μatm) promoted sporophyte growth and enhanced photosynthetic performance in three examined cultivars. This was likely due to energy savings from the down-regulation of CCM. However, elevated CO_2 concentrations were also found to reduce the light utilization efficiency, suggesting that the projected level of OA may have some negative effects on the growth of cultivated *S. japonica* (Xu et al. 2015a; Zhang et al. 2020). Similarly, Gao et al. (2019) reported that the increased CO_2 concentrations stimulated the photochemical efficiency of *S. japonica* sporophytes but inhibited gametophytic growth.

To date, there are no definitive conclusions on whether OA has a net positive or negative impact on the *S. japonica* cultivation. However, aside from the effects of ocean warming and acidification, other factors such as disease outbreaks and unpredictable anomalies in marine environments may also significantly affect *S. japonica* production.

17.4.6 The 2021–2022 Kelp Farming Disaster in Rongcheng City

From November 2021 to April 2022, Rongcheng City, located on the eastern coast of the Shandong Peninsula and home to the largest aquaculture area and the highest yield of *S. japonica* cultivation industry in northern China, experienced an unprecedented loss of kelp. This event was accompanied by a severe, harmful red algal bloom (Geng et al. 2024). It was termed the "2021–2022 Kelp Farming Disaster" (Li et al. 2023, 2024b). Investigations revealed that this disaster was likely caused by a combination of factors, including the depletion of phosphate nutrients due to red tides, low salinity levels (30 PSU), and high seawater transparency resulting from a significant influx of freshwater from the Yellow River. This freshwater influx was driven by the winter coastal current along the Shandong Peninsula.

Further studies indicated that extreme rainfall events in northern China during the summer of 2021, located thousands of kilometers inland, led to unusually high nutrient levels in the seasonal runoff of the Yellow River. The nutrient-rich run-off contributed to the rare occurrence of a winter red tide. The extreme rainfall was linked to significant anomalies of the western Pacific sub-tropical high (WPSH), which were influenced by global climate change (Zhang et al. 2022; Geng et al. 2024; Li et al. 2024a). This case highlights how unpredictable anomalies in marine environments can pose significant threats to the kelp cultivation industry, particularly in the context of increasing ocean warming, MHWs, and OA in the future.

17.5 Offshore Cultivation

The concept of offshore mariculture was introduced to address the growing demand for food and feed driven by the world's rapidly increasing population, the over-exploitation of wild fish stocks, and the mounting pressure to expand the inshore mariculture in increasingly crowded coastal zones (Duarte et al. 2009; Lovatelli et al. 2013; Liu and Su 2017). More recently, the cultivation of seaweeds in open sea or offshore environments has been recognized as a strategic approach for producing large quantities of biomass, which can be utilized for biofuel production or CDR to mitigate climate change (Notoya 2010; Wu et al. 2023).

The definition of offshore marine culture remains operational and subject to debate, with no universally accepted standard. Generally, offshore mariculture refers to aquaculture activities conducted in areas located more than 2 km

Fig. 17.1 Macroalgal Cultivation Rig in the Faroe Islands. (Buck et al. 2017)

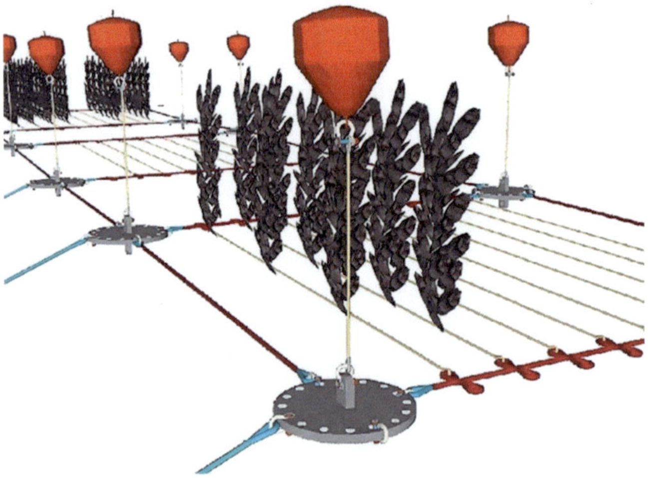

Fig. 17.2 Design of a kelp cultivation grid in Chile and modification. (Camus et al. 2018)

from the coastline, with water depths exceeding 50 m, and exposed to strong waves, ocean currents, and winds (Lovatelli et al. 2013). Buck et al. (2018) defined offshore aquaculture as the relocation of farm installations from sheltered environments to more exposed locations, as well as the establishment of new aquaculture operations in such exposed sites. An example of offshore cultivation is found around Zhangzidao Island in the Yellow Sea, approximately 8 km offshore, with water depths of up to 40 m and strong ocean currents reaching speeds of 100 cm/s. Here, the Zhangzidao Fishery Group Co. Ltd. operates a 40,000-ha farm cultivating multiple species, including scallops, sea urchins, abalone, and sea cucumbers (Buck et al. 2018). Another example is Sanggou Bay, located east of the Shandong Peninsula, which spans 13,000 ha, extends 8 km from the coastline, and has depths of 20–30 m with currents up to 60 cm/s. This site cultivates species such as scallop *Chlamys farreri*, kelp *S. japonica,* and the abalone *Haliotis discus* subsp. *hannai* (Buck et al. 2017).

Offshore cultivation typically requires advanced aquaculture technologies and costly infrastructure due to the harsh environmental conditions, including strong winds and waves. Additionally, the species cultivated must either have a high market value or be produced in large quantities to justify the significant initial investment. Despite the concept of offshore cultivation being proposed decades ago, the number of offshore aquaculture farms worldwide remains limited. Only a few offshore finfish farms exist, such as those cultivating yellowtail (*Seriola rivoliana*) off the coasts of Kona, Hawai'i, and Bahia de la Paz, Mexico, as well as gilthead seabream (*Sparus aurata*) off Ashdod, Israel. Experimental systems for offshore mussel and oyster cultivation have also been developed and tested (Buck et al. 2018).

Since the introduction of the ring system for offshore cultivation of *Saccharina latissima*, designed for integration with offshore wind farms (Buck and Buchholz 2004), various offshore cultivation systems for kelps have been developed (Bak et al. 2020; Tullberg et al. 2022). Examples include the Macroalgal Cultivation Rig in the Faroe Islands (2010–2019) (Bak et al. 2018) (Fig. 17.1), the Bio Architecture Lab's cultivation grid in Chile (Camus et al. 2018) (Fig. 17.2), the Seaweed Carrier in Norway (2010) (Bak et al. 2020), the Tension-Leg Platform (TLP) in Korea (2016) (Chung et al. 2015), and SPAR buoys and H-frames in the Netherlands (2011–2013) (Buck et al. 2017) (Fig. 17.3). Recently, a new offshore kelp cultivation system was designed and deployed in the Gulf of Maine, USA, for 3 months. After removal, the system exhibited only minor corrosion and damage (Moscicki et al. 2024). However, commercial-scale offshore seaweed cultivation has yet to be realized, and the offshore cultivation of selected seaweeds, including kelps, remains in its early stages.

Fig. 17.3 H-frame seaweed farm offshore off Texel, the Netherlands and modified by (Bak et al. 2020)

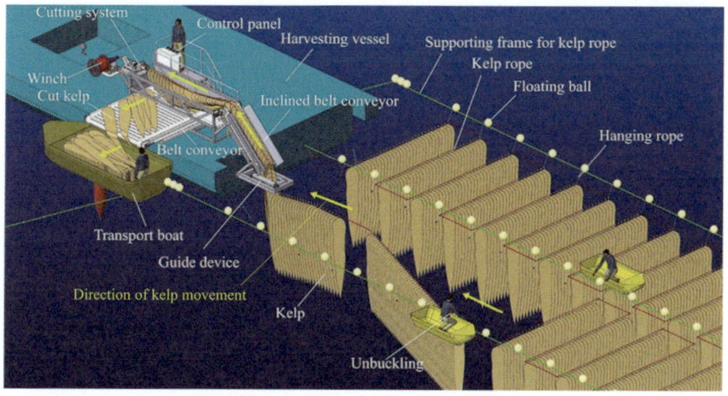

Fig. 17.4 One case of mechanization design for kelp harvest cited from (Jiang et al. 2022)

For the *S. japonica* cultivation industry in China, offshore cultivation has been proposed, and preliminary testing of cultivars and facilities has been conducted. Results indicated that certain cultivars are suitable for offshore cultivation, and specialized ropes can extend the operational lifespan of cultivation systems while reducing seedling loss (Liu et al. 2019a; Liu et al. 2022). Nevertheless, significant advancements are still required before offshore cultivation of *S. japonica* can be fully realized in China.

17.6 Mechanization in the Kelp Industry

Mechanical methods for harvesting wild kelp populations have proven to be more efficient than manual harvesting techniques. To meet the demand for large quantities of kelp biomass, particularly for the alginate industry, various specialized vessels have been developed. Examples include dredging vessels in Norway (Vea and Ask 2011) and the "*scoubidou*" system introduced in France (Sander and Karisson 2023).

As early as 1962, mechanization was identified as a critical issue requiring advancing the *S. japonica* cultivation industry (Tseng and Wu 1962). Despite the introduction of mechanization, the majority of cultivated kelp, particularly in northern China, is still harvested manually, which is labor-intensive, inefficient, and physically demanding. In some regions of Fujian, however, shipboard electric winches are used to raise or hoist the cultivating ropes laden with kelp, offering a partial solution to the challenge of manual harvesting (Li et al. 2022). In recent years, several mechanized harvesting systems have been designed and tested for the kelp cultivation industry (Zhang et al. 2017a; Jiang et al. 2022) (Fig. 17.4). While these systems have demonstrated significant improvements in harvesting efficiency, certain tasks, such as releasing and detaching the kelp ropes, still require manual intervention. These newly designed machines remain in the conceptual or experimental stages and have yet to be widely implemented in the kelp cultivation industry. Consequently, the mechanization of the *S. japonica* cultivation industry remains a pressing and unresolved issue that must be addressed in the future (Zhu et al. 2023).

17.7 Conclusions

Over the past 70 years, the *S. japonica* cultivation industry in China has achieved remarkable successes, driven by technological innovations such as the summer seedling method, floating raft systems, and the breeding of high-yield cultivars. These advancements have not only established *S. japonica* as the world's largest commercially cultivated marine crop but also positioned China as a global leader in kelp aquaculture. However, the industry now faces significant challenges, including ocean warming, acidification, and the increasing frequency of marine heatwaves, which threaten both natural kelp forests and cultivated populations.

To ensure the long-term sustainability of the industry, several strategies must be pursued. First, the development of new cultivars with enhanced resistance to high temperatures and other environmental stressors is essential. Secondly, offshore cultivation offers a promising avenue for expanding production while reducing pressure on the coastal ecosystem, though it requires significant technological and infrastructural advancements. Thirdly, the mechanization of harvesting and cultivation processes is critical to improve efficiency and address labor shortages, but current systems remain experimental and require further refinement. Additionally, the role of *S. japonica* cultivation in carbon sequestration and climate change mitigation should be further explored, though its effectiveness as a long-term carbon sink is limited by the current use of kelp biomass for food, feed, and industrial purposes. Finally, the integration of renewable energy sources into hatchery operations and other energy-intensive processes could reduce the carbon footprint of the industry.

In conclusion, the *S. japonica* cultivation industry in China has demonstrated remarkable resilience and adaptability, and its future sustainability will depend on continued innovations, proactive adaptation to environmental changes, and the development of advanced technologies for offshore and mechanized cultivation.

Acknowledgments We thank Dr. Xiaoqi Yang and Xiaohan Song for manuscript improvements and the pictures cited in the chapter. This research was supported by the project (2023YFE0106200) supported by the National Key R&D Program of China and the National Natural Science Foundation of China (32270389, 31772848).

References

Badis Y, Scornet D, Harada M, Caillard C, Godfroy O, Raphalen M, Gachon CMM, Coehlho SM, Motomura T, Nagasato C, Cock JM (2021) Targeted CRISPR-Cas9-based gene knockouts in the model brown alga *Ectocarpus*. New Phytol 231:2077–2091

Bak UG, Mols-Mortensen A, Gregersen O (2018) Production method and cost of commercial-scale offshore cultivation of kelp in The Faroe Islands using multiple partial harvesting. Algal Res 33:36–47

Bak UG, Gregersen Ó, Infante J (2020) Technical challenges for offshore cultivation of kelp species: lessons learned and future directions. Bot Mar 63(4):341–353

Bartsch I, Wiencke C, Bischof K, Buchholz CM, Buck BH, Eggert A, Feuerpfeil P, Hanelt D, Jacobsen S, Karez R, Karsten U, Molis M, Roleda MY, Schubert H, Schumann R, Valentin K, Weinberger F, Wiese J (2008) The genus *Laminaria sensus lato*: recent insights and developments. Eur J Phycol 43(1):1–86

Bartsch I, Vogt J, Pehlke C, Hanelt D (2013) Prevailing sea surface temperature inhibit summer reproduction of the kelp *Lamianria digitate* at Helgoland (North Sea). J Phycol 49:1061–1073

Blunden J, Boyer T (2024) State of the climate in 2023. Bull Am Meteror Soc 105(8):1–484

Bolton J (2010) The biogeography of kelps (Laminariales, Phaeophyceae): a global analysis with new insights from recent advances in molecular phylogenetics. Helgoland Mar Res 64:263–279

Brodie J, Williamson CJ, Smale DA, Kamenos NA, Mieszkowska N, Santos R, Cunliffe M, Steinke M, Yesson C, Anderson KM, Asnaghi V, Brownlee C, Burdett HL, Burrows MT, Collins S, Donohue PJC, Harley B, Foggo A, Noisette F, Nunes J, Ragazzola F, Raven JA, Schmidt DN, Suggett D, Teichberg M, Hall-Spencer JM (2014) The future of the Northeast Atlantic benthic flora in a high CO_2 world. Ecol Evol 4(13):2787–2798

Buchholz C, Lüning K (1999) Isolated, distal blade disc of the brown alga *Laminaria digitata* from sorus, but not discs, near to the meristematic transition zone. J Appl Phycol 16:579–584

Buck BH, Buchholz CM (2004) The offshore-ring: a new system design for the open ocean aquaculture of macroalgae. J Appl Phycol 16:355–368

Buck BH, Nevejan N, Wille M, Chambers MD, Chopin T (2017) Offshore and multi-use aquaculture with extractive species: seaweeds and bivalves. In: Buck BH, Langan R (eds) Aquaculture perspective of multi-use sites in the open ocean: the untapped potential for marine resources in the anthropocene. Springer Nature, Cham, pp 23–69

Buck BH, Troell MF, Krause G, Angel DL, Grote B, Chopin T (2018) State of the art and challenges for offshore integrated multi-trophic aquaculture (IMTA). Front Mar Sci 5:165

Cai P (1982) The summering of grown-out *Laminaria* sporophytes and culturing *Laminaria* seedlings in low temperature greenhouse in Fujian. Fujian Fis 5:1–5. (in Chinese)

Camus C, Infante J, Buschmann AH (2018) Overview of 3 year precommercial seafarming of *Macrocystis pyrifera* along the Chilean coast. Rev Aquac 10:543–559

Chen W, Chen L, Zhang X, Yang N, Guo J, Wang M, Ji S, Zhao X, Yin P, Cai L, Xu J, Zhang L, Han Y, Xiao Y, Xu G, Wang Y, Wang S, Wu S, Yang F, Jackson D, Cheng J, Chen S, Sun C, Qin F, Tian F, Fernie AR, Li J, Yan J, Yang X (2022) Convergent selection of a WD40protein that enhances grain yield in maize and rice. Science 375(6587):7985

Chung HK, Choi N, Woo H (2015) Novel ocean system for high density mass production of seaweed biomass in Korea. In: The 2015 world congress on advances in structural engineering and mechanics. Incheon, Korea, pp 4–8

Chung IK, Sondak CF, Beardall SJ (2017) The future of seaweed aquaculture in a rapidly changing world. Eur J Phycol 52(4):495–505

Clements JC, Chopin T (2017) Ocean acidification and marine aquaculture in North America: potential impacts and mitigation strategies. Rev Aquac 9:326–341

D'Agaro E, Favaro A, Matiussi S, Gibertoni PP (2021) Genomic selection in salmonids: new discoveries and future perspectives. Aquac Int 29:2259–2289

Duan D, Miao G, Wang X (2015) Aquacultural biology of *Saccharina japonica*. Science Press, Beijing. (in Chinese)

Duarte CM, Holmer M, Olsen Y, Soto D, Marbà N, Guiu J, Black K, Karakassis I (2009) Will the oceans help feed humanity? Bioscience 59:967–976

Duarte CM, Wu J, Xiao X, Bruhn A, Krause-Jensen D (2017) Can seaweed farming play a role in climate change mitigation and adaptation? Front Mar Sci 4:100

FABPRC (Fishery Administration Bureau of the Ministry of Agriculture and Rural Affairs of the People's Republic of China) (2024) China fisheries statistics yearbook. China Agriculture Press, Beijing

Fang T, Wu C, Jiang B, Li J, Ren G (1963) The breeding of a new variety of Haidai (*Laminaria japonica* Aresch.). Sci China Ser A 12(7):1011–1018

Fang T, Cui J, Ou Y, Dai J, Wang M, Liu Q, Yang Q (1983) Breeding of the new variety "Danhai No.1" of *Laminaria japonica* by using a female haploid clone of the kelp. J Shandong Coll Oceanol 13(4):63–70. (in Chinese with English abstract)

Fang T, Ou Y, Cui J (1985) Breeding of hybrid *Laminaria* "Danza No.10" -an application of the *Lamianria* haploid cell clones. J Shandong Coll Oceanol 15(1):64–72. (in Chinese with English abstract)

FAO (2014) The state of world fisheries and aquaculture 2014. Opportunities and challenges. FAO, Rome

Froehlich HE, Afflerbach JC, Frazier M, Halpern BS (2019) Blue growth potential to mitigate climate change through seaweed offsetting. Curr Biol 29:3087–3093

Fu W, Yao J, Wang X, Liu F, Fu G, Duan D (2009) Molecular cloning and expression analysis of a cytosolic Hsp70 gene from *Laminaria japonica* (Laminariaceae, Phaeophyta). Mar Biotechnol 11(6):738–747

Gao X, Kim J, Park SK, Yu OH, Kim YS, Choi HG (2019) Diverse responses of sporophytic photochemical efficiency and gametophytic growth for two edible kelps, *Saccharina japonica* and *Undaria pinnatifida*, to ocean acidification and warming. Mar Pollut Bull 142:315–320

Geng H, Kong F, Wang J, Zhang Q, Li F, Hong X, Song M, Lian Z, Cai Y, Yu R (2024) An unusual winter bloom of dinoflagellates with notable damage to kelp cultivation around Shandong peninsula, China. Mar Environ Res 201:106687

Goecke F, Klemetsdal G, Ergon Å (2020) Cultivar development of kelps for commercial cultivation- past lessons and future prospects. Front Mar Sci 8:110

Harley CDG, Anderson KM, Demes KW, Jorve JP, Kordas R, Coyle TA, Graham MH (2012) Effects of climate change on global seaweed communities. J Phycol 48:1064–1078

Hengjie T, Das SK, Zainee NFA, Yana R, Rozaimi M (2023) Ocean acidification and aquacultured seaweeds: progress and knowledge gaps. J Mar Sci Eng 11:78

Hu Z, Shan T, Zhang J, Zhang Q, Critchley AT, Choi H, Yotsukura N, Liu F, Duan D (2021) Kelp aquaculture in China: a retrospective and future prospects. Rev Aquac 13:1324–1351

Hu Z, Shan T, Zhang Q, Liu F, Jueterbock A, Wang G, Sun Z, Wang X, Chen W, Critchley AT, Ye N (2024) Kelp breeding in China: challenges and opportunities for solutions. Rev Aquac 16:855–871

Huang L, Yan X (2019) Construction of a genetic linkage map in *Pyropia yezoensis* (Bangiales, Rhodophyta) and QTL analysis of several economic traits of blades. PLoS One 14(3):e0209128

Huang M, Robbins KR, Li Y, Umanzor S, Marty-Rivera M, Bailey D, Aydlett M, Schmutz J, Grimwood J, Yarish C, Lindell S, Jannink J (2023) Genomic selection in algae with biphasic lifecycles: a *Saccharina latissima* (sugar kelp) case study. Front Mar Sci 10:1040979

Hurd CL, Harrison PJ, Bischof K, Lobban C (2014) Seaweed ecology and physiology, 2nd edn. Cambridge University Press, Cambridge

Hwang EK, Yotsukura N, Pang S, Su L, Shan T (2019) Seaweed breeding programs and progress in eastern Asian countries. Phycologia 58(5):484–495

Ichihara K, Yamazaki T, Kawano S (2022) Genome editing using a DNA-free clustered regularly interspaced short palindromic repeat-Cas9 system in green seaweed *Ulva prolifera*. Phycol Res 70:50–56

IOIMF (Section of seaweed genetics and breeding, Institute of Oceanology, Section of seaweed cultivation, Institute of Marine Fisheries) (1976) The breeding of new varieties of Haidai (*Laminaria japonica*) with high production and high iodine content. Sci Sin 3(3):224–230. (in Chinese with English abstract)

IPCC (2013) In: Stocker TF, Qin D, Plattner G-K, Tignor M, Allen SK, Boschung J, Nauels A, Xia Y, Bex V, Midgley PM (eds) Climate change 2013: the physical science basis. Contribution of working group I to the fifth assessment report of the intergovernmental panel on climate change. Cambridge University Press, Cambridge

IPCC (2021) Climate change 2021: the physical science basis. In: Masson-Delmotte V, Zhai P, Pirani A, Connors SL, Péan C, Berger S, Caud N, Chen Y, Goldfarb L, Gomis MI, Huang M, Leitzell K, Lonnoy E, Matthewa JBR, Maycock TK, Waterfield T, Yelekci O, Yu R, Zhou B (eds) Contribution of working group I to the sixth assessment report of the intergovernmental panel on climate change. Cambridge University Press, Cambridge

Ji Y, Gao K (2021) Effects of climate change factors on marine macroalgae: a review. In: Sheppard C (ed) Advances in marine biology. Academic, London, pp 91–136

Jiang T, Hong Y, Lu L, Zhu Y, Chen Z, Yang M (2022) Design and experiment of a new mode of mechanized harvesting of raft cultured kelp. Aquac Eng 99:102289

Koch M, Bowes G, Ross C, Zhang X (2013) Climate change and ocean acidification effects on seagrass and marine macroalgae. Glob Chang Biol 19:103–132

Krause-Jensen D, Duarte CM (2016) Substantial role of macroalgae in marine carbon sequestration. Nat Geosci 9:737–742

Krumhansl KA, Okamoto DK, Rassweiler A, Novak M, Bolton JJ, Cavanaugh KC, Conell SD, Johnson CR, Konar B, Ling SD, Micheli F, Norderhaug KM, Pérez-Matus A, Sousa-Pinto I, Reed DC, Salomon AK, Shears NT, Wernberg T, Anderson RJ, Barrett NS, Buschmann AH, Carr MH, Caselle JF, Derrien-Courtel S, Edgar GJ, Edwards M, Estes JA, Goodwin C, Kenner MC, Kushner DJ, Moy FE, Nunn J, Steneck RS, Vásquez J, Waston J, Witman JD, Byrnes EK (2016) Global patterns of kelp forest change over the past half-century. PNAS 113:13785–13790

Largo DB, Chung IK, Phang SM, Gerung GS, Sondak CFA (2017) Impacts of climate change on *Eucheuma-Kappaphycus* farming. In: Hurtado AQ, Critchley AT, Neish IC (eds) Tropical seaweed farming trends, problems and opportunities, Developments in applied phycology 9. Springer, Cham, pp 121–130

Leal PP, Hurd CL, Fernández PA (2017) Ocean acidification and kelp development: reduced pH has no negative effects on meiospore germination and gametophyte development of *Macrocystis pyrifera* and *Undaria pinnatifida*. J Phycol 53:557–566

Li D, Zhou Z, Liu H, Wu C (1999) A new method of *Laminaria japonica* strain selection and seedling culture by the use of vegetative gametophytes. Hydrobiologia 398(399):473–476

Li X, Cong Y, Yang G, Shi Y, Qu S, Wang G, Zhang Z, Luo S, Dai H, Xie J, Jiang G, Liu J, Wang T (2007) Trait evaluation and trial cultivation of Dongfang No.2, the hybrid of a male gametophyte clone of *Laminaria longissima* (Laminariales, Phaeophyta) and a female one of *L .japonica*. J Appl Phycol 19(2):139–151

Li X, Liu J, Cong Y, Qu S, Zhang Z, Dai H, Luo S, Han X, Huang S, Wang Q, Liang G, Sun J, Jin Y, Wang D, Yang G (2008) Breeding and trial cultivation of Dongfang No.3, a hybrid of *Laminaria* gametophyte clones with a more than intraspecific but less than interspecific relationship. Aquaculture 280(1–4):76–80

Li X, Zhang Z, Qu S, Liang G, Zhao N, Sun J, Song S, Cao Z, Li X, Pan J, Luo S, Zhang L, Cui C, Peng J, Li Y, Wu R, Zhao J, Qian P, Wang L, Sai S, Yang G (2016a) Breeding of an intraspecific kelp hybrid Dongfang No.6 (*Saccharina japonica*, Phaeophyceae, Laminariales) for suitable processing products and evaluation of its culture performance. J Appl Phycol 28(1):439–447

Li X, Zhang Z, Qu S, Liang G, Sun J, Zhao N, Cui C, Cao Z, Li Y, Pan J, Yu S, Wang Q, Li X, Luo S, Song S, Guo L, Yang G (2016b) Improving seedless kelp (*Saccharina japonica*) during its domestication by hybridizing gametophytes and seedling-raising from sporophytes. Sci Rep 6(1):21255

Li C, Wang G, Wang H, Feng D, Wei X, Xu T, Xu K (2022) Review on the development of kelp industry in Shandong province: a comparative perspective. Isr J Aquac-Bamidgeh 74:1–8

Li XD, Su L, Li XJ, Li J, Xu YJ, Chang LR, Yu RC, Yang DZ, Pang SJ (2023) Comprehensive analysis of large-scale *Saccharina japonica* damage in the principal farming area of Rongcheng in Shandong Province from 2021 to 2022. J Agric Sci Technol 25(1):206–222. (in Chinese with English abstract)

Li WJ, Wang ZY, Cui QC, Sun XH, Huang HJ (2024a) Coastal ecological disasters triggered by an extreme rainfall event thousands of kilometers inland. Commun Earth Environ 5:238

Li X, Wang Y, Xiao L, Huang E, Liu X, Shan T, Pang S (2024b) *Saccharina* sporophytes settling on kelp cultivation rafts come from wild intertidal populations and not from the farmed cultivars. Bot Mar 67(3):231–238

Lima FP, Ribeiro PA, Queiroz N, Hawkins SJ, Santos AM (2007) Do distributional shifts of northern and southern species of algal match the warming pattern? Glob Chang Biol 13:2592–2604

Liu F, Pang S (2010) Performance of growth, photochemical efficiency, and stress tolerance of young sporophytes from seven populations of *Saccharina japonica* (Phaeophyta) under short-term heat stress. J Appl Phycol 22:221–229

Liu H, Su J (2017) Vulnerability of China's nearshore ecosystems under intensive mariculture development. Enrivon Sci Pollut Res 24:8957–8966

Liu F, Wang X, Liu J, Fu W, Duan D, Yang Y (2009) Genetic mapping of the *Laminaria japonica* (Laminariales, Phaeophyta) using amplified fragment length polymorphism markers. J Phycol 45(5):1228–1233

Liu F, Shao Z, Zhang H, Liu J, Wang X, Duan D (2010) QTL mapping for frond length and width in *Laminaria japonica* Aresch (Laminariales, Phaeophyta) using AFLP and SSR markers. Mar Biotechnol 12(4):386–394

Liu F, Wang W, Sun X, Liang Z, Wang F (2014a) RNA-seq revealed complex response to heat stress on transcriptomic level in *Saccharina japonica* (Laminariales, Phaeophyta). J Appl Phycol 26(3):1585–1596

Liu F, Sun X, Wang F, Wang W, Liang Z, Lin Z, Dong Z (2014b) Breeding, economic traits evaluation, and commercial cultivation of a new *Saccharina* variety "Huangguan No.1". Aquac Int 22(5):1665–1675

Liu F, Wang W, Sun X, Liang Z, Wang F (2015) Conserved and novel heat stress-responsive microRNAs were identified by deep sequencing in *Saccharina japonica* (Laminariales, Phaeophyta). Plant Cell Environ 38(7):1357–1367

Liu F, Liang Z, Zhang P, Wang W, Sun X, Wang F, Yuan Y (2019a) Preliminary discussion on the development of *Saccharina japonica* offshore aquaculture in China. Progress Fishery Sci 40(1):161–166. (in Chinese with English abstract)

Liu T, Wang X, Wang G, Jia S, Liu G, Shan G, Chi S, Zhang J, Yu Y, Xue T, Yu J (2019b) Evolution of complex thallus alga: genome sequencing of *Saccharina japonica*. Front Genet 10:378

Liu Y, Liang Z, Yu W, Chang L, Lu L, Xiao L, Zheng Y, Liu F (2022) Effects of cultivation area of different offshore distance on the agronomic traits and breeding rope of *Saccharina japonica*. Prog Fish Sci 43(3):156–164. (in Chinese with English abstract)

Liu Y, Liang Z, Zhang P, Yuan Y, Wu Y, Zhang D, Duan M, Liu F (2023) Sorus development biology of hybrid cultivar in *Saccharina japonica*: environmental and endogenous regulation. Aquaculture 565:739165

Lovatelli A, Aguilar-Manjarrez J, Soto D (2013) Expanding mariculture farther offshore: technical, environmental, spatial and governance challenges, FAO Fisheries and Aquaculture Proceedings, FAO Technical Workshop, Orbetello

Meuwissen THE, Hayes BJ, Goddard ME (2001) Prediction of total genetic value using genome-wide dense marker map. Genetics 157:819–829

Mizuta H, Nimura K, Yamamoto H (1999a) Inducible condition for sorus formation of the sporophyte discs of *Laminaria japonica* Areschoug (Phaeophyceae). Fish Sci 65(1):104–108

Mizuta H, Nimura K, Yamamoto H (1999b) Sorus development on median and marginal parts of the sporophyte of *Laminaria japonica* Areschoug (Phaeophyceae). J Appl Phycol 11:585–591

Moscicki Z, Swift MR, Dewhurst T, MacNicoll M, Chambers M, Tsukrov I, Fredriksson DW, Lynn P, Landon ME, Zotter B, MacAdam N (2024) Design, deployment, and operation of an experimental offshore seaweed cultivation structure. Aquac Eng 105:102413

Notoya M (2010) Production of biofuel by macroalgae with preservation of marine resources and environment. In: Seckbach J, Einav R, Israel A (eds) Seaweeds and their roles in globally changing environments. Springer, Dordrecht, pp 219–228

Pang S, Liu F, Liu Q, Wang J, Sun C (2015) Breeding and genetic stability evaluation of the new *Saccharina* variety "205". China Fisheries 10:59–60. (in Chinese)

Pessarrodona A, Howard J, Pidgeon E, Wernberg T, Filbee-Dexter K (2024) Carbon removal and climate change mitigation by seaweed farming: a state of knowledge review. Sci Total Environ 918:170525

Raven JA, Caldeira K, Elderfield H, Hoegh-Guldberg O, Liss PS, Riebesell U, Shepherd J, Turley C, Watson A (2005) Ocean acidification due to increasing atmospheric carbon dioxide, Royal Society Policy Document 12/05. The Royal Society, London

Robinson N, Winberg P, Kirkendale L (2013) Genetic improvement of macroalgae: status to date and needs for the future. J Appl Phycol 25:703–716

Roe ML (1916) life histories in Laminariaceae. Botanical gazette, 62(5):422–423

Sander G, Karisson M (2023) Approaches to managing kelp forests. In: Nairobi (ed) Into the blue: securing a sustainable future for kelp forests. United Nations Environment Programme, pp 101–129

Shan T, Pang S, Li J, Li X, Su L (2015) Construction of a high-density genetic map and mapping of a sex-linked locus for the brown alga *Undaria pinnatifida* (Phaeophyceae) based on large scale marker development by specific length amplified fragment (SLAF) sequencing. BMC Genomics 16:902

Shao ZR, Wang WL, Zhang PY, Yao JT, Wang FH, Duan DL (2019) Genome-wide identification of genes involved in carbon fixation in *Saccharina japonica* and responses of putative C4-related genes to bicarbonate concentration and light intensity. Plant Physiol Bioch 137:75–83

Shen Y, Motomura T, Ichihara K, Matsuda Y, Yoshimura K, Kosugi C, Nagasato C (2023) Application of CRISPR-Cas9 genome editing by microinjection of gametophytes of *Saccharina japonica* (Laminariales, Phaeophyceae). J Appl Phycol 35:1431–1441

Simonson EJ, Scheibling RE, Metaxas A (2015) Kelp in hot water: I. Warming seawater temperature induces weakening and loss of kelp tissue. Mar Ecol Prog Ser 537:89–104

Smale DA (2020) Impacts of ocean warming on kelp forest ecosystems. New Phytol 225:1447–1454

Song X, Meng X, Guo H, Cheng Q, Jing Y, Chen M, Liu G, Wang B, Wang Y, Li J, Yu H (2022) Targeting a gene regulatory element enhances rice grain yield by decoupling panicle number and size. Nat Biotechnol 40:1403–1411

Starko S, Bailey LA, Creviston E, James KA, Warren A, Brophy MK, Danasel A, Fass MP, Townsend JA, Neufeld CJ (2019) Environmental heterogeneity mediates scale-dependent declines in kelp diversity on intertidal rocky shores. PLoS One 14(3):e0213191

Steneck RS, Johnson CR (2013) Kelp forests: dynamic patterns, processes and feedbacks. In: Bertness MD, Silliman B, Stachowitz J (eds) Marine community ecology, Sinauer, Sunderland pp 315–336

Su L, Pang SJ, Shan TF, Li X (2017) Large-scale hatchery of the kelp *Saccharina japonica*: a case study experienced at Lvshun in northern China. J Appl Phycol 29(6):3003–3103

Su L, Shan T, Li J, Gao S, Pang S, Leng X, Zhang Y (2021) Characterization of the novel hybrid cultivar E25 of *Saccharina japonica* in the northern farming region of China. J Appl Phycol 33:4011–4021

Tanaka K, Taino S, Haraguchi H, Prendergast G, Hiraoka M (2012) Warming off southwestern Japan linked to distributional shifts of subtidal canopy-forming seaweeds. Ecol Evol 2(11):2854–2865

Thomsen MS, Mondardini L, Alestra T, Gerrrity S, Tait L, South PM, Lilley SA, Schiel DR (2019) Local extinction of bull kelp (*Durvillaea* spp.) due to a marine heatwave. Front Mar Sci 6:84

Tian Z, Yuan B (1989) Experimental report of new variety "Zaohoucheng No.1" of *Laminaria japonica* Aresch. Mar Aquac 1:7–16. (in Chinese)

Troell M, Henriksson PJG, Buschmann AH, Chopin T, Quahe S (2023) Farming the ocean- seaweeds as a quick fix for the climate. Rev Fish Sci Aquac 31(3):285–295

Tseng C (1981) Commercial cultivation. In: Lobban CS, Wynne MJ (eds) The biology of seaweeds. Blackwell Scientific, Oxford, pp 680–725

Tseng C (1987) *Laminaria* mariculture in China. In: Doty MS, Caddy JF, Santelices B (eds) Case studies of seven commercial seaweed resources, FAO Fish Tech Pap 281. Food and Agriculture Organization of the United Nations, Rome, pp 239–263

Tseng C, Wu C (1962) The cultivation biology of Haidai (*Saccharina japonica*). Science Press, Beijing. (in Chinese)

Tseng C, Sun K, Wu C (1955) On the cultivation of Haitai (*Laminaria japonica* Aresch.) by summering young sporophytes at low temperature. Acta Bot Sin 4(3):255–264. (in Chinese with English abstract)

Tullberg RM, Nguyen HP, Wang CM (2022) Review of the status and developments in seaweed farming infrastructure. J Mar Sci Eng 10:1447

Van der Loos LM, Schmid M, Leal PP, McGraw CM, Britton D, Revill AT, Virtue P, Nichols P, Hurd CL (2019) Responses of macroalgae to CO_2 enrichment cannot be inferred solely from their inorganic carbon uptake strategy. Ecol Evol 9:125–140

Van der Meer JP (1983) The domestication of seaweeds. Bioscience 33:172–176

Vea J, Ask E (2011) Creating sustainable commercial harvests of *Laminaria hyperborea*. J Appl Phycol 23:489–494

Wang J, Xue H (1959) Observations on the sorus formation of mature *Laminaria* summering in low temperature culture room. Acta Bot Sin 8(4):259–261. (in Chinese with English abstract)

Wang Q, Liu H, Sui J (2018a) Mariculture: developments, present status and prospects. In: Gui JF, Tang Q, Li Z, Liu J, De Silver SS (eds) Aquaculture in China: success stories and modern trends. Wiley, Oxford, pp 38–54

Wang X, Chen Z, Li Q, Zhang J, Liu S, Duan D (2018b) High-density SNP-based QTL mapping and candidate gene screening for yield-related blade length and width in *Saccharina japonica* (Laminariales, Phaeophyta). Sci Rep 8(1):13591

Wang X, Yao J, Zhang J, Duan D (2020) Status of genetic studies and breeding of *Saccharina japonica* in China. J Oceanol Limnol 38(4):1064–1079

Wang F, Li X, Tang X, Sun X, Zhang J, Yang D, Xu L, Zhang H, Yuan H, Wang Y, Yao Y, Wang C, Guo Y, Ren Q, Li Y, Zhang R, Wang X, Zhang B, Sha Z (2023a) The seas around China in a warming climate. Nat Rev Earth Environ 4:535–551

Wang X, Yang X, Yao J, Li Q, Lu C, Duan D (2023b) Genetic linkage map construction and QTL mapping of blade length and width in *Saccharina japonica* using SSR and SNP markers. Front Mar Sci 10:1116412

Wang W, Xu L, Jiang G, Li Z, Bi Y, Zhou Z (2024a) Characterization of a novel γ-type carbonic anhydrase, SJγ-CA2, in *Saccharina japonica*: insights into carbon concentration mechanism in macroalgae. Int J Biol Macromol 263:130506

Wang H, Xie X, Gu W, Zheng Z, Zhuo J, Shao Z, Huan L, Zhang B, Niu J, Gao S, Wang X, Wang G (2024b) Gene editing of economic macroalga *Neopyropia yezoensis* (Rhodophyta) will promote its development into a model species of marine algae. New Phytol 244(5):1687–1691

Wernberg T, Smale DA, Tuya F, Thomsen MS, Langlois TJ, de Bettignies T, Bennet S, Rousseaux CS (2012) An example climatic event alters marine ecosystem structure in a global biodiversity hotspot. Nat Clim Chang 3:78–82

Wernberg T, Bennett S, Babcock RC, Bettignies T, Cure K, Depczynski M, Dufois F, Fromont J, Fulton CJ, Hovery RK, Harvey ES, Holmes TH, Kendrick GA, Radford B, Santana-Garcon J, Saunders BJ, Smales DA, Thomsen MS, Tuckett CA, Tuya F, Vanderklift MA, Wilson S (2016) Climate-driven regime shift of a temperate marine ecosystem. Science 353(6295):169–171

Wernberg T, Krumhansl K, Filbee-Dexter K, Pedersen M (2019) Status and trends for the world's kelp forests. In: Sheppard C (ed) World seas: an environmental evaluation. Elsevier, London, pp 57–78

Wernberg T, Thomsen MS, Baum JK, Bishop MJ, Bruno JF, Coleman MA, Filbee-Dexter K, Gagnon K, He Q, Murdiyarso D, Rogers K, Silliman BR, Smale DA, Starko S, Vanderklift MA (2024) Impacts of climate change on marine foundation species. Annu Rev Mar Sci 16:247–282

Wu C, Pang S, Jiang B (1994) Establishment and application of vegetative gametophytes of Laminaria japonica. In: Wang S (ed) Seaweed biotechnology. Shanghai Scientific and Technical Publishers, Shanghai, pp 104–110. (in Chinese)

Wu J, Keller DP, Oschlies A (2023) Carbon dioxide removal via macroalgae open-ocean mariculture and sinking: an earth system modeling study. Earth Syst Dynam 14:185–221

Xiao X, Agustí S, Yu Y, Huang Y, Chen W, Hu J, Li C, Li K, Wei F, Lu Y, Xu C, Chen Z, Liu S, Zeng J, Wu J, Duarte CM (2021) Seaweed farms provide refugia from ocean acidification. Sci Total Environ 776:145192

Xu B, Zhang Q, Qu S, Cong Y, Tang X (2009) Introduction of a seedling production method using vegetative gametophytes to the commercial farming of *Laminaria* in China. J Appl Phycol 21:171–178

Xu D, Wang D, Li B, Fan X, Zhang XW, Ye NH, Wang Y, Mou S, Zhuang Z (2015a) Effect of CO_2 and seawater acidification on the early stages of *Saccharina japonica* development. Environ Sci Technol 49:3548–3556

Xu Y, Huang L, Ji D, Chen C, Zheng H, Xie C (2015b) Construction of a dense genetic linkage map and mapping quantitative trait loci for economic traits of a double haploid population of *Pyropia haitanensis* (Bangiales, Rhodophyta). BMC Plant Biol 15(1):228

Xu W, Liu Y, Li M, Lu S, Chen S (2024) Advances in biotechnology and breeding innovations in China's marine aquaculture. Adv Biotechnol 2:38

Yang G, Sun Y, Shi Y, Zhang L, Guo S, Li B, Li X, Li Z, Cong Y, Zhao Y, Wang W (2009) Construction and characterization of a tentative amplified fragment length polymorphism-simple sequence repeat linkage map of *Laminaria* (Laminariales, Phaeophyta). J Phycol 45(4):873–878

Ye RX, Yu Z, Shi W, Gao H, Bi Y, Zhou Z (2014) Characterization of α-type carbonic anhydrase (CA) gene and subcellular localization of α-CA in the gametophytes of *Saccharina japonica*. J Appl Phycol 26:881–890

Ye N, Zhang X, Miao M, Fan X, Zheng X, Xu D, Wang J, Zhou L, Wang D, Gao Y, Wang Y, Shi W, Ji P, Li D, Guan Z, Shao C, Zhuang Z, Gao Z, Qi J, Zhao F (2015) *Saccharina* genomes provide novel insight into kelp biology. Nat Commun 6(1):6986

Zhang Q, Tang X, Cong Y, Qu S, Luo S, Yang G (2007) Breeding of an elite *Laminaria* variety 90-1 through inter-specific gametophyte crossing. J Appl Phycol 19:303–311

Zhang J, Liu Y, Yu D, Song H, Cui J, Liu T (2011) Study on high-temperature-resistant and high-yield *Laminaria* variety "Rongfu". J Appl Phycol 23(2):165–171

Zhang J, Fang J, Wang W, Du M, Gao Y, Zhang M (2012) Growth and loss of mariculture kelp *Saccharina japonica* in Sungo Bay, China. J Appl Phycol 24:1209–1216

Zhang J, Liu T, Bian D, Zhang L, Li X, Liu D, Liu C, Cui J, Xiao L (2016) Breeding and genetic stability evaluation of the new *Saccharina* variety "Ailunwan" with high yield. J Appl Phycol 28(6):3413–3421

Zhang Y, Chang Z, Zheng Z, Yang J (2017a) Harvesting machine for kelp culture in floating raft. Aquac Eng 78:173–179

Zhang Y, Zhang J, Liang Y, Li H, Li G, Chen X, Zhao P, Jiang Z, Zou D, Liu X, Liu J (2017b) Carbon sequestration processes and mechanisms in coastal mariculture environments in China. Sci China Earth Sci 60:2097–2107

Zhang J, Liu T, Feng R, Liu C, Jin Y, Jin Z, Song H (2018) Breeding of the new *Saccharina* variety "Sanhai" with high yield. Aquaculture 485:59–65

Zhang X, Xu D, Guan Z, Wang S, Zhang Y, Wang W, Zhang X, Fan X, Li F, Ye N (2020) Elevated CO_2 concentrations promote growth and photosynthesis of the brown alga *Saccharina japonica*. J Appl Phycol 32:1949–1959

Zhang F, Yang X, Sun Q, Yao S, Guo Q (2022) Three-dimensional structural anomalies of the western pacific subtropical high ridge and its relationship with precipitation in China during August-September 2021. Atmos 13:1089

Zhang J, Wu Q, Eléouët M, Chen R, Chen H, Zhang N, Hu Y, Sui Z (2023) CRISPR/LbCas12a-mediated targeted mutation of *Gracilariopsis lemaneiformis* (Rhodophyta). Plant Biotechnol J 21:235–237

Zhou L, Gui JF (2018) Applications of genetic breeding biotechnologies in Chinese aquaculture. In: Gui JF, Tang Q, Li Z, Liu J, De Silver SS (eds) Aquaculture in China: success stories and modern trends. Wiley, Oxford, pp 465–496

Zhu Y, Hong Y, Jiang T, Yang M, Lu L, Yu Y (2023) Design and test of an efficient automatic clip seedling system for raft aquaculture kelp. J Mar Sci Eng 11:2301

Xiuliang Wang is an Associate Professor in the Institute of Oceanology, Chinese Academy of Sciences. He has published a number of papers on *Saccharina japonica* germplasm and its quantitative traits and is interested in the evolution of brown seaweeds with climate change.

Jie Zhang is an Associate Professor in the Institute of Oceanology, Chinese Academy of Sciences. She received her Ph.D. in marine biology from the University of the Chinese Academy of Sciences. Her research interests include the evolutionary biology of kelps, conservation genetics and biodiversity of seaweeds, population genetics and biogeography, and phylogenomics and population genomics. She published ~20 papersand obtained more than five government grants.

Jianting Yao is an Associate Professor in the Institute of Oceanology, Chinese Academy of Sciences. He received his Ph.D, in marine biology from the University of the Chinese Academy of Sciences. His research interests include the evolutionary biology of kelps, seaweed cultivation and comprehensive utilization of seaweeds. He has published ~30 papers and obtained more than six governmental grants.

Delin Duan is Professor in the Institute of Oceanology, Chinese Academy of Sciences. He received his Ph.D. in marine biology from University of Chinese Academy of Sciences, and post-doc at Tokyo University. His research interests are seaweeds phylogenetics, genetics and genomics of the phylogenetics and genomics of seaweeds. He has co-published 200 SCi-cited papers and six books.

18 Sustaining Kelp Resources: The Challenges of Harvesting *Laminaria digitata*

Thomas Burel, Anaëlle Bizien, Marc Lennon, and Eric Deslandes

Abstract

The French seaweed industry, primarily driven by Brittany, plays a crucial role in local employment and alginate production. Traditionally, depending on *Laminaria digitata*, the local alginate sector now faces significant challenges. The late decline of *L. digitata* populations due to climate change and competition with other kelp species—mainly *Saccorhiza polyschides*—poses a serious threat to its harvesting sustainability. Additionally, increasing variability in the seaweed landing composition has been observed in recent years, worrying local alginate processors. In the short term, ensuring a high proportion of *L. digitata* is thus required to maintain high-quality alginate production. In response to these recent changes, this study aims to assess the quality of *L. digitata* landings and identify the factors influencing their variability. It also examines the past and present exploitation of *L. digitata* while exploring potential short- and long-term solutions to sustain the local alginate production. Results indicate that landing composition varies significantly, with *L. digitata* biomasses sometimes representing less than 50% of the total harvest. Two main factors seem to drive these variations: (1) sea conditions, with lower quality observed when wave heights exceed a certain threshold, and (2) differences among vessels, as some seaweed harvesters consistently deliver lower quality landings. This requires implementing best practices for harvesters, combined with systematic visual inspections when necessary. In the mid-term, as *L. digitata* resources will continue to decline, yet reducing reliance on this species will be necessary to support its conservation. Promoting alternative sources of alginate will be required, and a shift toward other species such as *Laminaria hyperborea*, *L. ochroleuca,* or even *S. polyschides* has to be considered.

Keywords

Alginate · Kelp harvesting · *Laminaria digitata* · *Saccorhiza polyschides* · Sustainability

18.1 Introduction

18.1.1 General Context

In Brittany, seaweeds have long been an integral part of coastal traditions, initially harvested to enrich agricultural soils through direct collection of attached or drift material, and later finding broader applications in industry and scientific advancements (Frangoudes 2011; Arzel 1994). The utilization of seaweeds in Brittany evolved alongside new discoveries about their properties and progressed in line with industrial developments as observed in neighboring countries (e.g., Ireland, Guiry and Morrison 2013) as it evolved from exploiting beach-cast seaweeds for soil fertilizers to a source of potash, via iodine production, to the current hydrocolloid extraction (Chapman 1970; Synytsya et al. 2015).

To date, the main output of brown algae exploitation is the alginate industry which currently processes over 100,000 tons of harvested or cultivated seaweeds annually across Europe (Araújo et al. 2021). Alginate is a linear heteropolysaccharide consisting of 1,4-linked β-D-mannuronic acid (M) and 1,4 α-L-guluronic acid (G). It is a primary component of the cell walls in many brown seaweeds, accounting for 17–45% of their dry weight (Black 1950; Haug et al. 1967). The key properties of alginates, such as their gelling, emulsifying, and film-forming capabilities, have made them widely utilized across various industries, including pharmaceuticals, cosmetics, textiles, and food production (Draget et al. 2006; Lee and Mooney 2012; Pereira & Cotas 2020).

T. Burel (✉) · A. Bizien · E. Deslandes
Univ Brest, CNRS, IRD, Ifremer, LEMAR, Plouzané, Brittany, France
e-mail: thomas.burel@univ-brest.fr

M. Lennon
Hytech-Imaging, Plouzané, Brittany, France

A. T. Critchley et al. (eds.), *Advances in Temperate Phyconomy: Algal Harvest and Cultivation in Globally Distributed Temperate Waters*, Developments in Applied Phycology 15, https://doi.org/10.1007/978-3-032-02955-3_18

Alginates are primarily found in brown algae belonging to the Fucales and Laminariales orders. In Europe, several species from the genus *Laminaria* (kelp) and family Fucaceae (fucoids) are extensively harvested for alginate production (Peteiro 2018). Additionally, the European sugar kelp *Saccharina latissima* is the most widely cultivated species locally, used both for direct consumption (food) and alginate production, contributing to the creation of around 200 commercially available alginate variants (Araújo et al. 2021).

The most abundant and high-quality alginates are obtained from species within the *Laminaria* genus (Vauchel et al. 2008; Bartsch et al. 2008), especially *Laminaria digitata,* called "tali" in the Breton language, along the coasts of Brittany (Arzel 2002). *Laminaria digitata* is a kelp species characteristic of the infralittoral fringe and the shallow subtidal zone (Floc'h 1970), forming a complex habitat which is a home to a wealth of biodiversity beneath its canopy (Derrien-Courtel et al. 2013a; Schaal et al. 2016).

It is rarely harvested by foot, as it can only be accessed at low tide during spring tides. From the early twentieth century, the harvesting of this species was carried out by "goémoniers" (seaweed harvesters), using flat-bottomed boats and various hand tools such as the "croc" (a large curved fork/rake) or the "guillotine" (a sickle mounted on a long handle). This type of harvesting was known to be hazardous and challenging (Arzel 1987). However, it persisted and became more professionalized with the modernization of boats in the 1950s, allowing for a more abundant harvest further from the near-shore (Colin 1964). These tools were then outpaced by the invention of the "scoubidou" in 1961, a hook specifically designed for targeting *L. digitata* by rotating around its flexible stipe, thus facilitating the hand-harvesting work (Perez 1973). The subsequent mechanization of this tool with articulated arms (the first boat with a hydraulic "scoubidou" was built in 1967) brought about a significant industrial transformation in the seaweed harvesting industry (Davoult et al. 2011). The harvesting of *Laminaria digitata* for alginate production has evolved profoundly since its inception in 1960 (Arzel 1987). This evolution has been shaped by technological advancements, shifting industrial demands, legislative changes, and the influence of extreme weather (see Fig. 18.1 and the section below, Arzel 1998; Alban et al. 2011). Over the past decade, approximately 45,000 tons have been harvested on average yearly, making it the most harvested seaweed in France (Davoult et al. 2011). Landings, however, are known to fluctuate significantly between years. For example, the winter of 2013–2014, characterized by severe storms (Smale and Vance 2015), resulted in the *Laminaria digitata* harvest dropping to 23,500 tons in 2014.

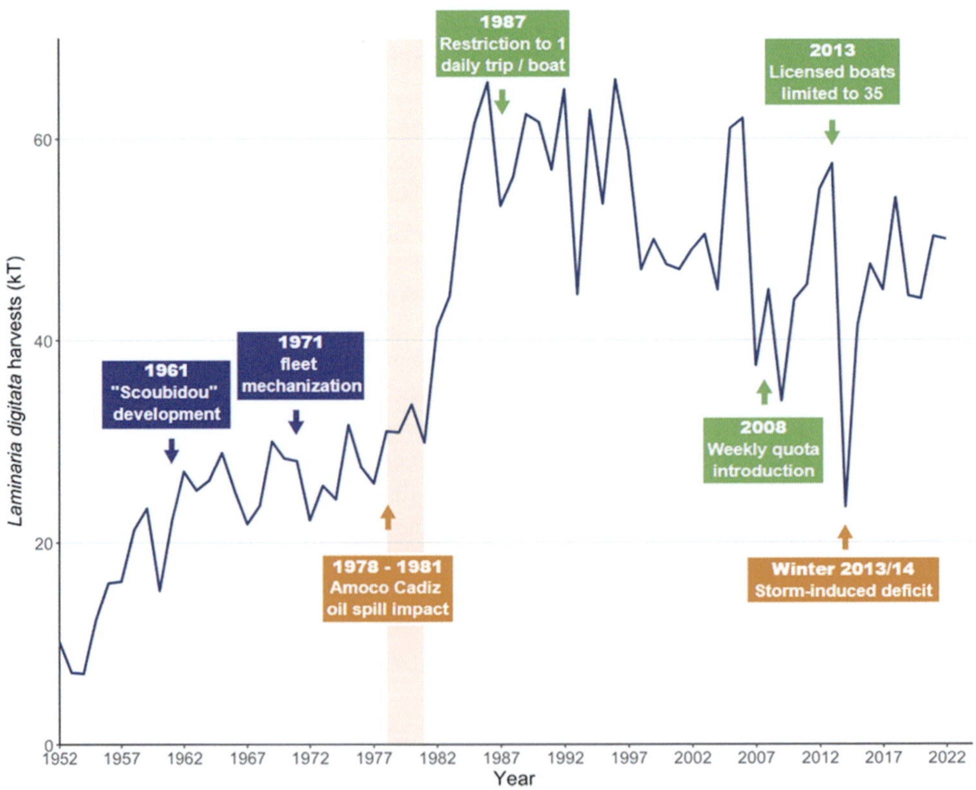

Fig. 18.1 Evolution of *Laminaria digitata* harvest volumes and key events (1952–2022). Technical advancements are shown in blue, fleet regulations are represented in green, and extreme events are marked in orange

18.1.2 *Laminaria digitata* Harvest Trends and Milestones in Brittany (1952–2022)

Until the 1960s, *Laminaria digitata* was primarily harvested for the iodine industry, despite earlier, unsuccessful attempts to industrialize and commercialize alginate production in the early twentieth century. With the closure of iodine factories and the discovery of new applications for alginates in food, cosmetics, and pharmaceuticals, the seaweed harvesting fleet naturally transitioned to alginate production. To support this shift, various harvesting tools were developed. Among them, the "scoubidou," introduced in 1961 by Yves Colin—a hook-like device attached to a handle—emerged as the most effective tool, becoming iconic in Brittany's *L. digitata* harvesting practices (Arzel 1987).

Originally relying on sailing vessels and manual tools, the fleet underwent significant modernization, starting with the gradual motorization of boats in the 1950s. This culminated in 1971 with the introduction of the first vessel equipped with a hydraulic "scoubidou," revolutionizing harvesting efficiency and scale. These advancements supported a steady increase in harvested biomass, which stabilized around 25 kT (Arzel 1998).

From the late 1970s, industrial demand for raw materials surged, spurred by a new method for processing fresh seaweed introduced in 1978. However, that same year saw a major setback with the Amoco Cadiz oil spill off the coast of Portsall, Brittany, which released over 220,000 tons of oil into the sea. This disaster directly impacted areas harvested by seaweed gatherers, significantly disrupting operations. Beyond the immediate environmental damage, the spill left a lingering negative perception that hindered recovery efforts for several years. Subsequently, a 30 kT annual quota was implemented to regulate harvests and promote resource sustainability.

Between 1982 and 1985, the harvesting of *L. digitata* experienced exceptional growth, reaching a historic peak of 62 kT in 1985. This surge was fueled by high industrial demand, favorable weather, and a modern, relatively young fleet. Alongside record harvests, the exploited area expanded, with *Saccorhiza polyschides* appearing more prominently in landings. However, the transition to fresh harvesting methods made quality control more challenging. These developments prompted the fisheries committee to introduce stricter regulations in 1987, including daily trip limits and quotas based on vessel size. For example, boats under 10 m were capped at 10 tons per day, while larger vessels (up to 12 m) were allowed 12 to 24 tons per day, depending on their configuration (Frangoudes and Garineaud 2015).

The 1980s and 1990s were marked by efforts to regulate and reorganize the seaweed harvesting sector. Locally, tensions arose from license reductions, various quota proposals, and disagreements among harvesters, industrial stakeholders, and fisheries committees. Internationally, the industry faced heightened competition due to the rapid expansion of alginate production in China during the early 1980s, the collapse of the textile industry, and the dumping of surplus alginate products by the United States in 1993. These pressures, combined with rising vessel construction and maintenance costs, led to a significant decline in the number of boats in the fleet (Alban et al. 2011). Tragic accidents resulting in the loss of lives and equipment further contributed to this decline.

In the 2000s, regulatory efforts intensified. In 2008, under the Water Framework Directive, inspections revealed elevated formaldehyde levels in wastewater from processing sites, exceeding regulatory limits due to limited treatment capacity. To address this issue, a weekly harvest quota for *L. digitata* was introduced, capping harvests at 900 tons—a significant reduction from previous levels of up to 1100 tons. These quotas had a notable impact, reducing the number of active vessels and reshaping harvesting practices.

The 2010s brought further changes, including a new quota on seaweed harvesting licenses at 35, aiming to enhance resource sustainability. The Breton coast was divided into nine distinct *L. digitata* harvesting zones to confine harvesters to specific areas, improving resource management and reducing conflicts among harvesters and other stakeholders. However, this zoning system introduced challenges related to territorial appropriation and perceived privatization of resources, fostering tensions among harvesters. Today, five of the nine zones are actively exploited, with two—the Molène Archipelago and the Iroise Sea coast—hosting approximately three-quarters of the fleet (Garineaud 2017).

Since the late 2000s, there has been an increasing emphasis on integrating fisheries with scientific research, exemplified by programs like Ecokelp, Valmer, and Slamir. In 2019, Vessel Monitoring Systems (VMS) were systematically installed on harvesting vessels, enabling detailed studies of localized spatial effects, such as how specific sites influence resource quality. This period underscored the significant impact of environmental factors on seaweed resources, as seen during the destructive winter storms of 2013–2014. It also highlighted the importance of developing environmental forecasting tools to support and sustain the industry (Henry et al. 2021).

18.1.3 Recent Changes in Kelp Populations and Their Impact on the Local Alginate Industry

Overall, *L. digitata* populations have declined at their southern limits in the northeast Atlantic (de Bettignies et al. 2021), most likely due to warming waters (Liesner et al. 2020; Smale 2020). This demise is further accelerated by the species' sensitivity to light and temperature, making it potentially more vulnerable to pathogens (Valero et al. 2006). Also undergoing a northward distribution shift, *Saccorhiza polyschides,* a more opportunistic species, is spreading in the subtidal macroalgal forests of Brittany (Engelen et al. 2011). *Saccorhiza polyschides*, locally known as "tosser," is a brown alga morphologically and ecologically close to *L. digitata* (Table 18.1) but differs in its associated biodiversity (Castric-Fey et al. 2001). Both species have large, digitate blades but differ in their stipes: cylindrical in *L. digitata* and flattened in *S. polyschides*. Another key distinction lies in the massive, bulbous holdfast of *S. polyschides*, which provides a habitat for a variety of marine organisms (Salland and Smale 2021). Despite these similarities, *S. polyschides* does not belong to Laminariales but is a Tilopteridales, a phylogenetically distant order. This distinction is by the phylogeny, but the close resemblance to other kelp species sensu stricto has led to the definition of pseudo-kelp (Sasaki et al. 2001).

Table 18.1 Characteristics of the sporophytes from the two species considered in this study, with morphology reflecting mature thalli traits. Bracketed values represent extreme observations. Depth is relative to chart datum. Ranges given in the table must be considered as approximations

	Laminaria digitata (Hudson) J. V. Lamouroux	*Saccorhiza polyschides* (Lightfoot) Batters
Morphology		
Size (m)	2 (−3)	3 (−4)
Holdfast	Round haptera, smooth	Bulbous, coriaceous
Stipe	Flexuose, smooth	Flattened, stiff, and inflexible
Blade	Digitate, smooth, and glossy	Digitate, leathery
Life history		
Sporangia location	Blade	Blade [stipe, holdfast]
Reproduction time	June/July and September/October	September/October
Life span	3–4 [−6]	1–1.5
Habitat and distribution		
Latitudinal range (°N)	77.7–47.2 (Spitsbergen–Brittany)	62.4–31.3 (South Norway–Morocco [western Mediterranean])
Thermal range (°C)	[0-] 10–15 [−22]	[4-] 13–17 [−24]
Substratum	Rocks, stones, epiphytic	Rocks, stones
Wave exposure	Exposed to sheltered	[very exposed] moderate to extremely sheltered
Depth range (m)	+1.5 to −3 [−9]	0 to −15 [−35]
References	Arzel (1988), Arzel and Castric-Fey (1997), Birkett et al. (1998), Derrien-Courtel et al. (2013a), Fletcher (2024), Lüning (1990), Norton (1977)	

Competition between *Laminaria* spp. and *Saccorhiza polyschides* has long been documented (Hawkins and Harkin 1985), with various life cycle and physiological traits explaining their competitive dynamics. Firstly, their differing lifespans play a crucial role, as *L. digitata* is a perennial species, typically living for 3 to 4 years, whereas *S. polyschides* is an annual species (Table 18.1 and references herein). Unlike *L. digitata*, which requires stable conditions to establish and persist, *S. polyschides* thrives in environments experiencing frequent disturbance; its ability to rapidly colonize available substrata enables it to outcompete *L. digitata* in areas exploited by seaweed harvesters (Castric-Fey et al. 2001). Additionally, the large, bulbous holdfasts of *S. polyschides* can persist year-round, preventing *Laminaria* sporophytes from settling and further hindering the recovery of *L. digitata* populations (Salland et al. 2024). Finally, unlike *S. polyschides,* which extends as far south as Morocco, *L. digitata* reaches its southern limit in southern Brittany. Yet, from a physiological perspective, *S. polyschides* has a higher optimal growth temperature than *L. digitata* and may gain a competitive edge in the context of rising sea temperatures (Smale et al. 2013). This thermal advantage, combined with its rapid colonization capacity, suggests that *S. polyschides* could further expand in regions where *L. digitata* is in decline, potentially reshaping the structure of kelp-dominated ecosystems in the future.

Despite its ecological success, *S. polyschides* is not currently harvested in Brittany, unlike in other nearby countries (Kaidi et al. 2022). Beyond the competitive interactions between these two species, their economic value differs significantly. *Laminaria digitata* is highly prized for its alginate content, whereas *S. polyschides* is often considered an undesirable bycatch by local seaweed harvesters. Its alginates are not only of lower commercial value, but the fibrous structure of its thallus, particularly the stipe, disrupts standard extraction processes, making it problematic for the industry. As a result, the increasing presence of *S. polyschides* instead of *L. digitata* in the harvest poses a challenge for local alginate production, emphasizing the need to monitor and assess the quality of landings at processing sites.

In response to a request from the *Chambre Syndicale des Algues & Végétaux Marins* (CSAVM—French Seaweed & Marine Plants Trade Association) and the seaweed harvesting industry, a study on the quality of kelp landings in the Iroise Sea (western Brittany) has been underway since 2021. The study aims to (1) identify the factors driving variation in landing quality and to (2) provide short-term recommendations for the harvesting practices. It also aims to (3) enhance alginate quality at processing sites by introducing innovative and efficient spectroscopy methods for analyzing and handling large quantities of harvested kelp. A first hypothesis suggests that environmental conditions are the main drivers of these fluctuations. Extreme events, such as storms (e.g., winter 2013–14), can drastically reduce population density and subsequent harvest volumes, as seen in 2014's biomass deficit. Additionally, unfavorable sea conditions during harvest seasons may force harvesters to collect from suboptimal areas, lowering yield quality. While adverse weather may partly explain these variations, a second hypothesis attributes these fluctuations to the practices and habits of the harvesters. As related in interviews, harvesters can clearly distinguish between *L. digitata* and *S. polyschides*, yet some landings are deemed "dirty" by processors upon delivery (Garineaud 2017). With a reduction in quality controls by processors, some harvesters may pay less attention to the quality of their landings. Ultimately, long-term shifts in fleet structure and composition—such as vessel size and numbers (Alban et al. 2011)—resulting from regulations or more profound changes in the harvesting profession may also play a role in yield fluctuations.

In 2021, we initiated a preliminary monitoring reporting large variations in the proportion of *L. digitata* in the landings. This monitoring was further expanded in 2022 by exploring the relationship between the amounts of *L. digitata* and *S. polyschides* and various environmental factors. In 2023, it was extended with the introduction of field spectroscopy, which could enhance the assessment of the quality of the landed biomass. The present work aims to consolidate the findings from these reports, offering a comprehensive perspective on the historical and current exploitation of *Laminaria digitata* while highlighting potential future challenges that may affect its sustainability.

18.2 Materials and Methods

18.2.1 Data on Fleet Dynamics and Evolution

A comprehensive dataset was constituted on the Breton seaweed harvester fleet, detailing the number of vessels (both traditional and mechanized fleets), their dimensions (length, width, etc.), and their port of registry, along with the dates they began and ceased seaweed harvesting activities. The dataset was compiled from various sources, including bibliographic research (primarily the work of P. Arzel, Ifremer), unpublished data (from M. Laurans, Ifremer, and from the CSAVM), online resources (notably goemoniers.e-monsite.com), and articles from the local press.

18.2.2 Methodology from Seaweed Harvest to Processing Site

The study was conducted in Brittany, France, over three consecutive harvesting seasons: from 24/06/2021 to 12/08/2021, from 31/05/2022 to 07/07/2022, and from 04/07/2023 to 09/08/2023.

Laminaria digitata is collected by specific seaweed harvesting vessels. The seaweed vessels considered here depart for an area roughly corresponding to the Molène Archipelago, typically during low tide. They harvest *L. digitata* using a hydraulic articulated crane equipped with a hook (referred to as a "scoubidou"). The hook rotates underwater, wrapping around the stipe of the *L. digitata*, which is then uprooted from its substratum and hauled back into the vessel's hold.

Harvested seaweeds are consistently unloaded at the port of Lanildut, where they are transferred onto semi-trailers. The landings are sourced from clearly identified seaweed harvesting vessels. Clearly identified tractors then transport the harvest to two processing sites (either *Algaia* in Lanilis, France, or *JRS Marine Products* in Landerneau, France). For the purposes of this study, harvesting vessels will be anonymized, and processing sites will also be anonymized.

18.2.3 Operational Methodology Performed at the Processing Site

Upon arrival at the processing site, the overall condition of the landing is visually assessed ("good," "moderate," or "poor"), along with the identification of the associated seaweed vessel. This visual assessment relies on the general observation and impression of the landing's quality, focusing primarily on the surface composition of both *L. digitata* and *S. polyschides*. The primary goal is to determine whether the quality of landings can be quickly and accurately estimated through a visual inspection upon arrival at the processing site.

Once the landing is unloaded, samples are collected using a hydraulic grab claw. Although the landings may appear relatively uniform, three distinct samples are taken: one from each end and one from the middle. This approach ensures a more accurate assessment of the harvest composition and reduces biases that could result from heterogeneity within the landing (e.g., a grab inadvertently collecting a localized cluster of *Himanthalia elongata*). Sampling methods were slightly adjusted over the 3 years to obtain the most accurate data and to perform the most appropriate statistical analyses. These adjustments are compiled in Table 18.2.

For each sample, the two primary species, *Laminaria digitata* and *Saccorhiza polyschides*, are visually identified and separated by hand into two distinct piles. Any possibly additional species present (e.g., *Laminaria hyperborea*, *L. ochroleuca*, *Saccharina latissima*, *Cystoseira* spp., *Sargassum muticum*, *Fucus serratus*, *Himanthalia elongata*, etc.) are identified and grouped into a third pile. Other factors, such as the presence of stones, are also noted. Each of

Table 18.2 Differences in sampling across years

Sampling year	2021	2022	2023
Number of landings analyzed	16	12	12
Percentages of main species	Yes	Yes	Yes
Percentages of minority species	No	No	Yes
Triplicate of samples	No	Yes	Yes
Thallus measurement	No	No	Yes
Vessel/processing site identification	No	Yes	Yes
Hydrodynamics estimates	No	Yes	Yes
Visual evaluation	No	Yes	Yes
Spectral analysis	No	No	Yes

the piles is then weighed individually to ensure accurate composition analysis.

Finally, to assess the stratification and structure of the canopy, a selection of specimens from the dominant kelp species is also measured. Specifically, between 14 and 24 individuals of *L. digitata* and between 1 and 17 individuals of *S. polyschides* were randomly measured per landing.

18.2.4 Characterization of Species Based on Field Spectroscopy at the Landing Site

In order to assess the contribution of spectroscopy to the study of harvested macroalgae, a spectral library was created on landing site. It aimed in particular to assess whether macroalgae could be differentiated based on their spectral characteristics of absorption and reflection of light but also to assess the impact of the aging of macroalgae on the landing site on these spectral characteristics. Reflectance spectra were acquired on-site in the visible to near-infrared range of the spectrum (400 to 1000 nm) using a field spectroradiometer (GER-1500) from different macroalgae species, primarily *L. digitata* and *S. polyschides*. These spectra were recorded from various parts of the thallus and at different physiological states. Occasionally, spectra from other collected brown and red macroalgae were also analyzed. The spectral library created contains multiple measurements on approximately 50 targets on site.

18.2.5 Hydrodynamics Assessment

Both the significant wave heights and wave direction values were obtained from the WaveWatch III model for the "Molène" area (48.377, −4.960) computed over the sampling period over the 3 years covered in the study. To assess the effect of hydrodynamics, the Douglas Sea Scale was used as a proxy to categorize the different sea conditions encountered during this study.

18.2.6 Data Processing and Statistical Analysis

The data were analyzed using R software (R version 4.0.3; R Core Team 2020). Data were visualized using boxplots or scatter plots to examine the distribution of values. Comparisons between landings were made using either ANOVA (parametric) or Kruskal-Wallis tests (non-parametric). If the test results were significant, a post-hoc Tukey (parametric) or Dunn (non-parametric) test was conducted to identify which landings differed significantly. Pearson's r or Kendall's τ coefficients were calculated using the "corrplot" package (Wei et Simko 2017) to assess the correlations between variables.

18.3 Results

18.3.1 Changes in the Seaweed Harvesting Fleet Over the Years

The seaweed harvesting fleet has undergone significant changes since the 1960s, especially starting in the 1970s with the mechanization of the boats, i.e., boats equipped with a hydraulic "scoubidou" (Fig. 18.2). The traditional seaweed harvesting fleet, once comprising several hundred vessels in the mid-twentieth century, gradually dwindled and completely disappeared by 1984 (data not shown). The mechanized seaweed harvesting fleet saw significant growth, expanding from 5 vessels in 1971 to 72 in 1979. It maintained this size between the 1980s and late 1990s, with an average of 70 vessels, peaking at 74 vessels in 1991. The number of vessels steadily decreased thereafter, reaching 35 in 2013, following the implementation of the licensing cap. The fleet has remained at this level ever since. These

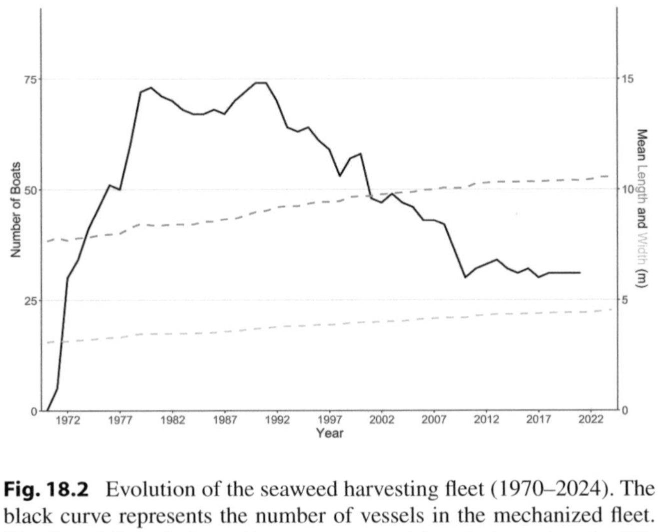

Fig. 18.2 Evolution of the seaweed harvesting fleet (1970–2024). The black curve represents the number of vessels in the mechanized fleet. The dotted lines indicate the average vessel length (green) and width (orange) for the entire fleet

changes in the number of vessels were accompanied by a steady increase in the average length and width of the boats within the fleet: the average vessel size has significantly increased, from 7.7 m in length and 3.1 m in width in 1970 to 10.6 m in length and 4.5 m in width by 2024.

18.3.2 Overview of Landing Trends: 2021–2023

Landings were dominated by *Laminaria digitata*, averaging 63 ± 20% (SD) in the analyzed samples, with percentages showing high variability (Fig. 18.3, see also Table 18.3). The maximum percentages of *L. digitata* recorded in a landing were 97% (July 1, 2021), 88% (July 21, 2023), and 83% (July 13 and 20, 2021, as well as July 7, 2022). Conversely, the minimum percentages of *L. digitata* recorded in a landing were 10% (July 13, 2023), 14% (Aug 8, 2023), and 15% (Aug 9, 2023). The second most abundant seaweed, in terms of percentage, is *Saccorhiza polyschides*. The abundances of *S. polyschides* and *L. digitata* are obviously strongly correlated, with a high abundance of one species inevitably corresponding to a low abundance of the other ($R^2 = 0.94$). Other seaweed species occasionally contributed an average of ≃ 5% of the total weight. These included other brown algae such as *Laminaria hyperborea*, *Laminaria ochroleuca*, *Saccharina latissima*, *Himanthalia elongata*, *Fucus serratus*, *Sargassum muticum*, and *Gongolaria baccata*. Smaller amounts of red algae (< 1%) were also noticed, primarily *Palmaria palmata* and *Gelidium corneum*.

Fig. 18.3 Example of sorting a landing replicate: *Laminaria digitata* is separated from *Saccorhiza polyschides* and, if present, other minority macroalgae species for subsequent weighing. Pictures illustrate the heterogeneity in landings, featuring one landing predominantly composed of *Laminaria digitata*—in the foreground—(**A**) and another primarily consisting of *Saccorhiza polyschides*—on the right side of the image—(**B**)

Table 18.3 Total biomass and percentage of *L. digitata* in the 36 samples analyzed between 2021 and 2023

Collecting date	Analyzed biomass (kg)	Percentage of *L. digitata* in the landing
24/06/2021	1580	69
28/06/2021	1900	65
28/06/2021	750	71
30/06/2021	334	61
01/07/2021	1880	97
08/07/2021	492	37
08/07/2021	380	66
13/07/2021	820	83
20/07/2021	2846	72
20/07/2021	1458	83
21/07/2021	410	68
22/07/2021	1580	76
03/08/2021	467	64
04/08/2021	1400	67
12/08/2021	232	63
12/08/2021	411	71
31/05/2022	860	76
31/05/2022	760	81
14/06/2022	860	74
16/06/2022	940	61
22/06/2022	575	62
22/06/2022	780	69
29/06/2022	760	72
29/06/2022	720	43
30/06/2022	860	52
05/07/2022	820	47
06/07/2022	680	81
06/07/2022	940	83
04/07/2023	532	72
05/07/2023	490	81
06/07/2023	275	35
10/07/2023	275	80
12/07/2023	540	61
12/07/2023	510	10
21/07/2023	760	88
24/07/2023	320	45
06/08/2023	710	14
07/08/2023	560	61
08/08/2023	715	44
09/08/2023	740	15

When analyzing the variation in *Laminaria digitata* percentages across the 3 years (Fig. 18.4) in relation to the progression of the season (typically from mid-June to late August), no correlation is detected ($\tau = -0.16$, p-value = 0.15).

18.3.3 Key Drivers Affecting Landing Quality: 2022–2023

The results of statistical tests comparing the abundance of *L. digitata* with potential key drivers influencing their variation are presented in Table 18.4.

The processing site effect did not reveal any significant differences in *L. digitata* abundance, whether for the years 2022, 2023, or for the combined data of both years. When considering both years together, one processing site had an average of 56 ± 20% *L. digitata*, while the other had 60 ± 25%. When analyzing the other key drivers, contrasting effects are observed depending on the year considered, with notable variations present.

In 2022, no significant differences were observed in the quality of landings among the identified harvesting boats. In contrast, in 2023, significant disparities were detected. Two boats reported markedly lower *L. digitata* abundances, at 21 ± 16% and 47 ± 15%, compared to the other four boats, which delivered higher quality landings with abundances ranging from 61 ± 7% to 88 ± 7%.

Laminaria digitata percentages show variations according to sea conditions (Fig. 18.5). Higher percentages of *L. digitata* are consistently associated with calm sea conditions. In contrast, lower percentages are linked to stronger wave heights, as observed on June 29 and 30, 2022, or on August 6–7, 2023. In 2022, significant differences were observed in relation to wave height. Under "moderate" sea conditions (1.25 < wave height < 2.50 m), the percentage of *L. digitata* was notably lower at 48 ± 10%. Conversely, during more favorable sea conditions, the quality of landings improved and showed greater consistency, with percentages reaching 73 ± 13% under "slight" conditions (0.50 < wave height < 1.25 m) and 69 ± 13% under "smooth" conditions (0.10 < wave height < 0.50 m). Similar trends were observed in 2023; under "moderate" sea conditions, the percentage of *L. digitata* was notably lower, averaging 29 ± 20%. In contrast, calmer sea conditions were associated with significantly higher percentages, reaching up to 75 ± 20% ("slight" conditions).

18.3.4 Estimation of Canopy Development

The size analysis of individuals revealed that *S. polyschides* are, on average, the largest kelps harvested by seaweed harvesters, with an average length of 2.6 m, while *L. digitata* averaged 2.3 m (Fig. 18.6; Table 18.5). The largest individuals recorded during this survey included an *S. polyschides* measuring 5.7 m and an *L. digitata* measuring 3.8 m.

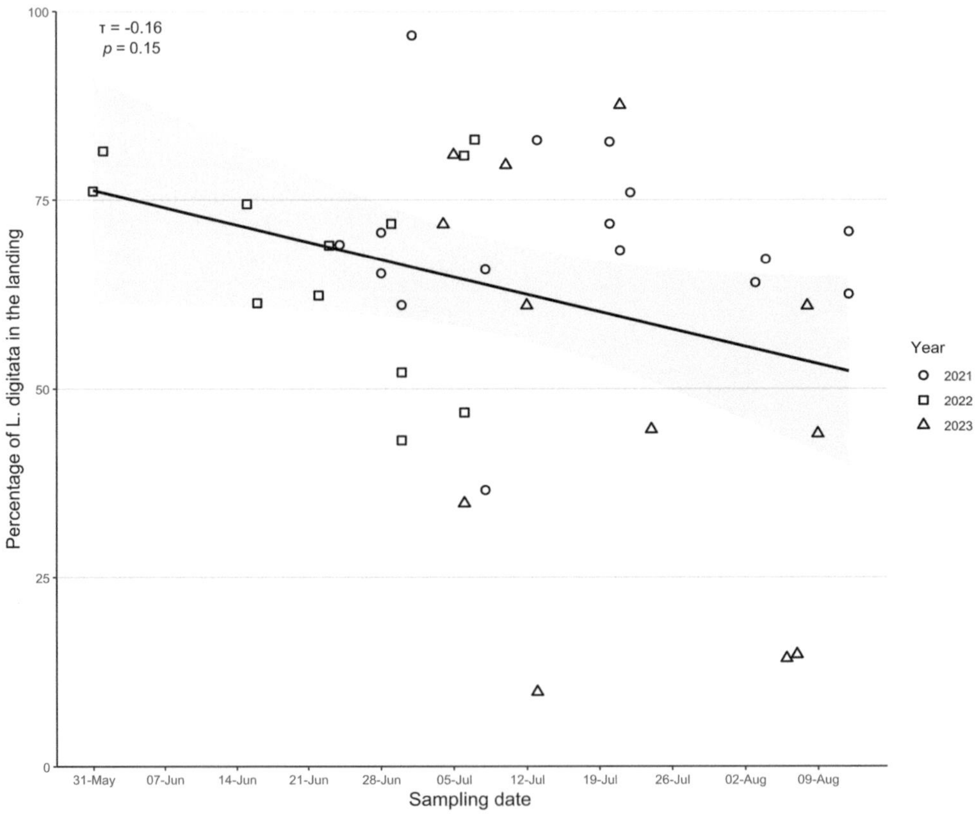

Fig. 18.4 Percentage of *Laminaria digitata* in landings based on the harvest date. The symbols correspond to the years considered: circles for 2021, squares for 2022, and triangles for 2023

Table 18.4 Statistical test results comparing *Laminaria digitata* abundances with potential key drivers influencing their values (assessed using Kruskal-Wallis or Mann-Whitney U tests with associated statistics). Statistically significant results ($p < 0.05$) are highlighted in bold

Sampling year	2022	2023	Total
Processing site	MW, $W = 156$ p-value = 0.86	MW, $W = 171$ p-value = 0.68	MW, W = 648 p-value = 1.00
Vessel	KW, $\chi^2 = 1.36$ p-value = 0.27	**KW, $\chi^2 = 29.77$ p-value < 0.01**	KW, $\chi^2 = 45.19$ p-value =< 0.01
Wave height	**KW, $\chi^2 = 7.75$ p-value = 0.02**	**KW, $\chi^2 = 6.13$ p-value = 0.04**	**KW, $\chi^2 = 14.72$ p-value < 0.01**

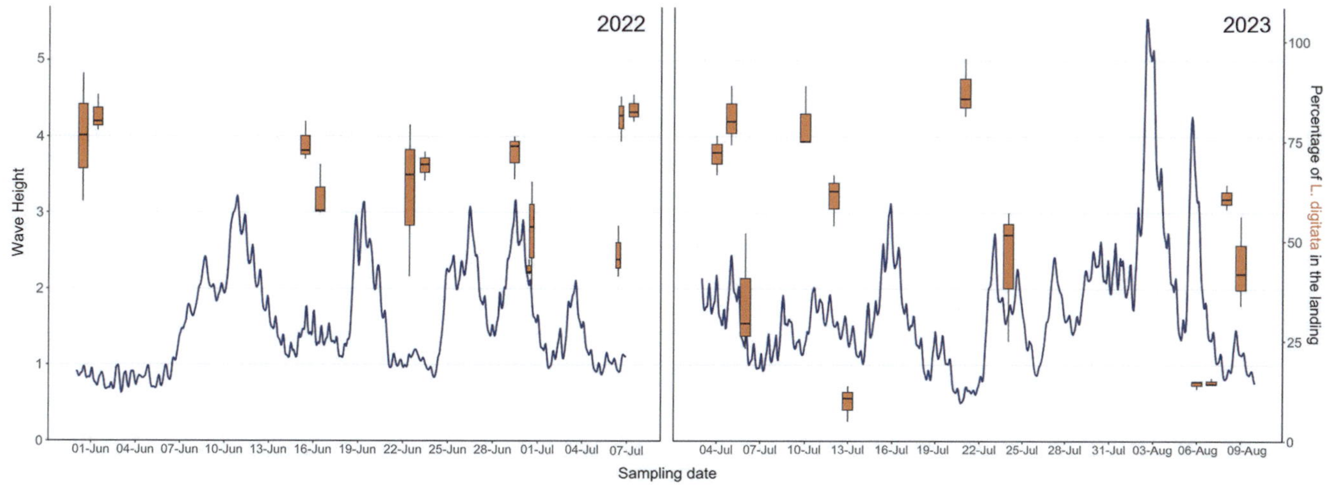

Fig. 18.5 Percentage of *Laminaria digitata* in landings and significant wave height determined by the Wave Watch 3 model for the periods from May 31 to July 7, 2022, and from July 3 to August 10, 2023

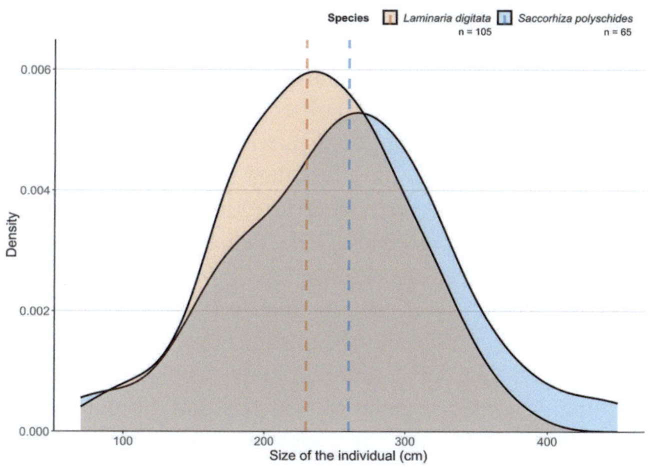

Fig. 18.6 Size distribution of two species considered in this study. The orange curve represents the density of *Laminaria digitata*, and the blue curve represents the density of *Saccorhiza polyschides*. The dashed line indicates the average size of the species

The sizes of both *L. digitata* and *S. polyschides* were globally comparable, although some anecdotal differences were noted. A general decreasing trend in size was nevertheless observed throughout the sampling period, with *L. digitata* decreasing from 2.35 ± 0.65 m in early July to 2.07 ± 0.66 m in August, and *S. polyschides* declining from 3.05 ± 0.98 m in early July to 2.36 ± 0.85 m in August. Despite this trend, our analyses did not reveal any clear relationships between individual size and other measured factors, such as vessels, processing facilities, or hydrodynamic conditions (data not shown).

18.3.5 Visual and Spectral Assessment of the Landings

Based on a priori visual assessments, landings were categorized as "poor," "moderate," or "good" quality. "Poor" qual-

Table 18.5 Average sizes (± SD) of the main kelp species found in the study

Species	*Saccorhiza polyschides* (*n* = 133)	*Laminaria digitata* (*n* = 214)	*Laminaria hyperborea* (*n* = 16)
Average size (m)	2.6 ± 0.9	2.3 ± 0.6	1.3 ± 0.1

Table 18.6 Summary of the spectra collected for each macroalg (species, thallus part, and physiological state). All species, except *L. digitata* and *S. polyschides*, were measured in their fresh state. The other brown algae species analyzed include *Saccharina latissima*, *Sargassum muticum*, *Fucus serratus*, *Laminaria hyperborea*, and *Himanthalia elongata*. The red algae species analyzed include *Palmaria palmata* and *Gelidium corneum*

Macroalgae	Physiological states	Part of the thallus	Number of samples
Laminaria digitata	Fresh	Blade	45
		Stipe	10
	Damaged	Blade	30
Saccorhiza polyschides	Fresh	Blade	20
		Stipe	10
		Holdfast	10
	Damaged	Blade	28
		Stipe	14
Other browns	Fresh	Blade	25
		Stipe	6
Other reds	Fresh	Entire	10

ity landings contained an average of 38 ± 25% *L. digitata*, while "moderate" quality landings showed an increase to 48 ± 27%, and "good" quality landings had the highest proportion, averaging 65 ± 25%. Significant differences were observed between these categories, with "good" quality landings containing significantly more *L. digitata* than "moderate" ones, which in turn contained more *L. digitata* than the "poor" category (Tukey's range test: *p*-value = 0.03, *p*-value <0.01, and *p*-value <0.01, respectively).

A total of 208 spectra have been analyzed (Table 18.6), allowing for the establishment of a spectral library (Fig. 18.7). The differences between red and brown algae are obviously distinguishable (data not shown). However, discriminating between *L. digitata* and *S. polyschides* based solely on spectral signatures remains challenging. The spectral absorption bands in the visible range (500–670 nm), which are influenced by pigment content, are highly similar for both species. However, *S. polyschides* generally exhibits steeper slopes in the near-infrared range (740–890 nm), suggesting differences in the internal structure of the thallus (Fig. 18.8).

Macroalgal degradation induces a clear and rapid shift in spectral characteristics (Fig. 18.9). The general pigment content, including chlorophyll, decreases over time, leading to changes in the absorption bands within the visible spectrum as well as in the red/infrared slope. Degradation of the internal structure of the algae results in further modifications to spectral features in the near-infrared range. Spectral analysis reveals that the degradation process is both faster and more pronounced for *S. polyschides* compared to *L. digitata*, with the stipe showing more rapid degradation than the blade.

18.4 Discussion

The French seaweed industry is largely driven by the dynamism of Brittany, where harvesting generates an estimated annual value exceeding €7 million (Metz et al. 2024), with *Laminaria digitata* representing the largest share of this market (Mesnildrey et al. 2012) supporting several hundred jobs. Along with Norway, France remains one of the few European countries that significantly exploits natural kelp stocks for locally sourced alginate production (Werner and Kraan 2004; Hurtado et al. 2022). Managing *L. digitata* exploitation requires balancing ecological sustainability with economic interests. Governance frameworks, including fishing licenses, harvest quotas, and temporal restrictions, regulate kelp harvesting while preserving ecosystem integrity (Frangoudes and Garineaud 2015). However, challenges remain in enforcing compliance and adapting regulations to environmental changes, all while ensuring the resource's long-term sustainability (Mac Monagail et al. 2017).

Since the onset of industrial harvesting of *Laminaria digitata* for alginate production in the 1960s, the sector has undergone profound transformations. Since at least the 1980s, as landed biomass increased, the industry has observed fluctuating landing composition (Arzel 1998), negatively impacting the quality of the alginate produced. The present study confirms this observation, highlighting that some landings are of poor quality, with quantities of *L. digitata* sometimes constituting less than 50% of the total. The remaining biomass is often made up of the opportunistic species *Saccorhiza polyschides*. The factors explaining these variations are of various natures.

First, the analysis of landing quality variations during the season did not confirm the observations reported by seaweed harvesters during our 3 years of study, who assess that the resource becomes scarcer toward the end of the season (Henry et al. 2021). A non-significant trend, however, suggests a decrease in the proportion of *L. digitata* as the harvesting season progresses. While this observation seems logical, the absence of a clear correlation is unexpected, as one would reasonably anticipate a reduction in the algal field's biomass after several weeks of exploitation; this underscores the need for a larger dataset to validate the trend. This also indicates that other, more influential factors are

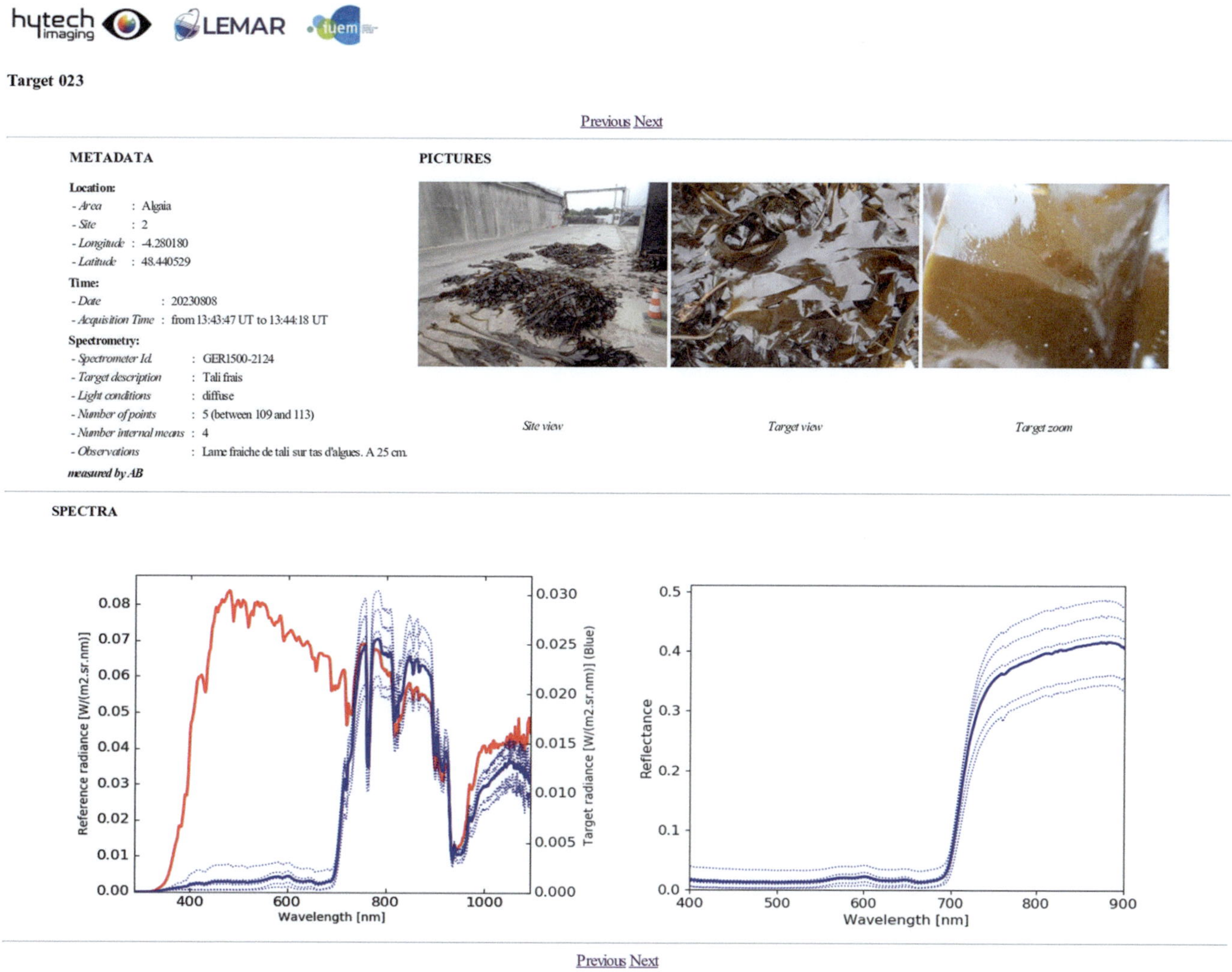

Fig. 18.7 Spectral library collected on site. Up left: metadata; up right: pictures linked to the measurement; bottom left: radiance spectra collected on algal target (blue) and reference target (red); bottom right: reflectance spectra and mean spectrum of the algal target in bold, spectral range 400–1000 nm

driving the variations in biomass quality. When analyzing wave heights, stronger hydrodynamics appear to negatively affect harvest quality. Given that the seaweed harvester fleet operates in coastal waters, the relatively small size of the vessels limits their ability to withstand rough seas (Arzel 1997). As a result, harvesters tend to seek out extremely sheltered areas, which provide more favorable conditions for the growth of *S. polyschides* over *L. digitata*. Also, in calm sea conditions, a sorting process can be carried out, where bundles of *S. polyschides* are identified and removed from the vessel. Since some vessels are exclusive to a specific processing site, one might have expected differences based on the processing site considered; this was, however, not the case here. On the other hand, when focusing on individual vessels, notable differences were observed. Some vessels consistently deliver lower quality landings. These results therefore confirm the findings reported in harvester interviews (Garineaud 2017). This highlights a dual impact: environmental conditions and the harvesting practices of seaweed harvesters. Results suggest the need for raising awareness among harvesters about better selective harvesting practices. Examples of these practices could include: (1) avoiding going to sea beyond a certain wave height threshold, (2) conducting sorting directly on the vessel at sea, (3) systematically removing *S. polyschides* directly in the field at the start of the season to promote the growth of *L. digitata*, (4) encouraging the harvesting of suitable areas, such as the currently underexploited rocky edges, and (5) establishing resting areas (fallow periods) to help strengthen the resource.

If necessary, reintroducing systematic quality controls from landing to processing could help ensure better integration into the value chain. Thus, visual inspection appears to provide satisfactory results in identifying the quality of landings. However, it presents a huge standard deviation (at least

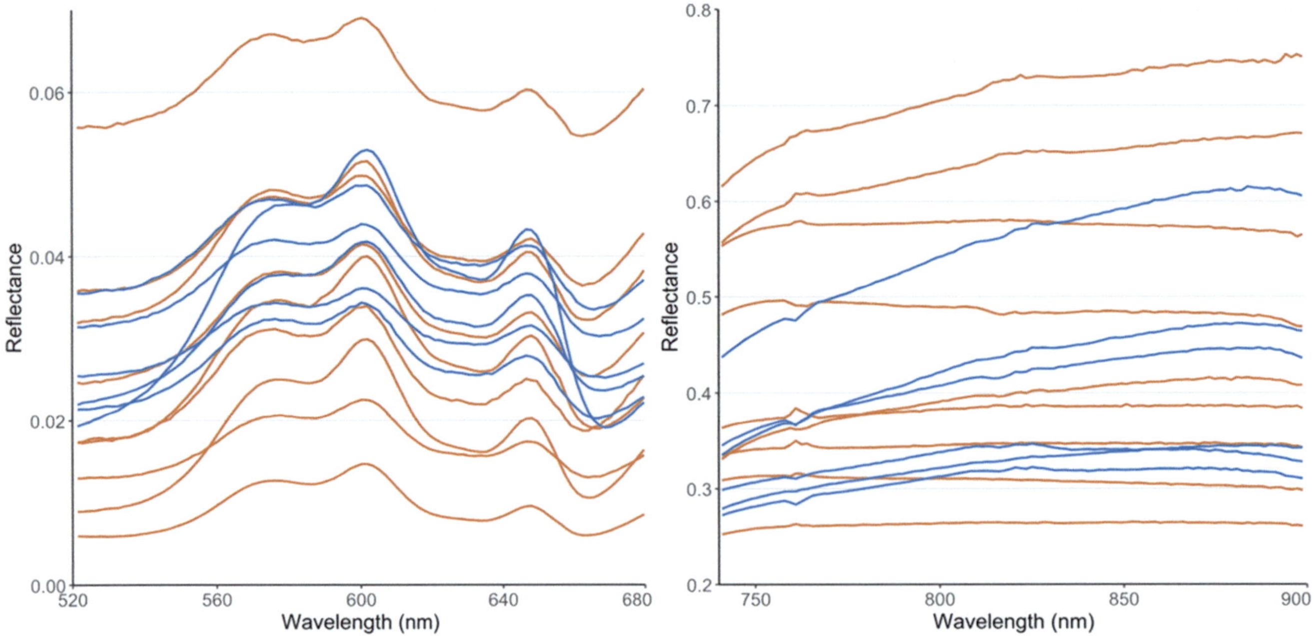

Fig. 18.8 Spectral differences between *Laminaria digitata* (orange) and *Saccorhiza polyschides* (blue). Left: absorption bands due to pigments in the visible range; right: spectral slopes due to the internal structure of algae in the near-infrared range

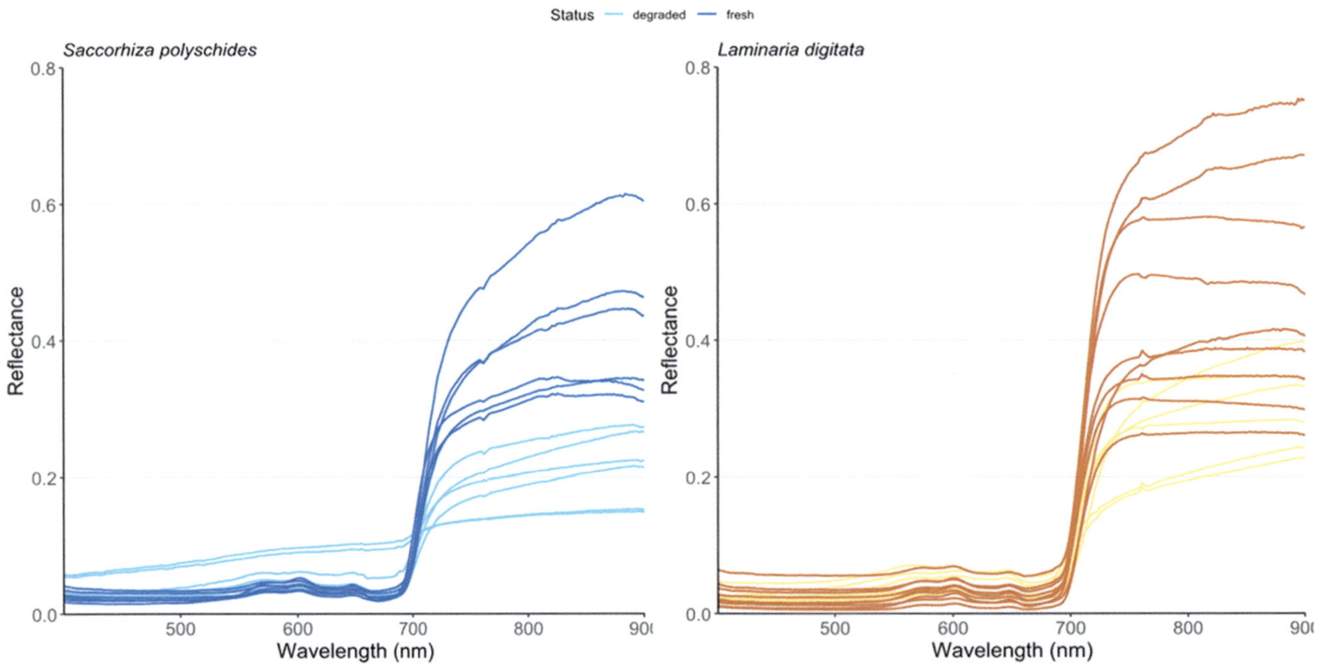

Fig. 18.9 Spectral evolution due to the aging of algae. Left in blue: *Saccorhiza polyschides*; right in orange: *Laminaria digitata*. Dark color: fresh algae; light color: degraded algae

25%, as seen previously), which is not acceptable in an industrial setting. A commonly used alternative involves optical methods for quality evaluation (Alander et al. 2013). These optical approaches, particularly those utilizing hyperspectral imaging, have already demonstrated their effectiveness in differentiating macroalgal species (Diruit et al. 2022). Preliminary results from this study, based on field spectroscopy measurements, suggest the feasibility of developing an algal degradation index using in situ spectral data. However, distinguishing between *L. digitata* and *S. polyschides* based solely on their spectral characteristics remains a challenging task; further analyses are therefore needed.

In addition to this quality control approach, modifications to the industrial alginate extraction processes could also be considered. The industrial extraction of alginate involves collecting, transporting, and storing algae, often with formaldehyde to prevent degradation (Bojorges et al. 2022). This process generates substantial waste—about 80% of the initial dry biomass—most of which is discarded (Fauziee et al. 2021). While alternative extraction methods aim to reduce chemical use or recover bioactive compounds, major changes to industrial practices remain unlikely in the short term. Meanwhile, valorizing these waste streams could support a circular bioeconomy (Mak et al. 2020; Beaumont et al. 2021; Li et al. 2021), as they may still contain valuable compounds for biotechnology and biomedical applications (Hao et al. 2019; Zargarzadeh et al. 2020). Integrating waste valorization into existing processes without disrupting alginate production is a promising avenue. Advances in spectroscopic techniques (^{1}H NMR, FT-IR) and rheological analysis (viscosity, molecular weight) provide valuable tools for optimizing the recovery of phycocolloids and low-carbohydrate fractions (Gómez-Ordóñez and Rupérez, 2011; Stiger et al. 2016).

The evolution of the seaweed harvesting fleet in Brittany has brought substantial changes to landing practices and outcomes over the years. Starting with traditional small-scale vessels, the fleet has evolved into a more mechanized structure, marked by fewer but larger and more efficient vessels, transitioning from coastal seaweed harvesters to larger island seaweed harvesters (Table 18.7). This transition has resulted in a divide between larger vessels, which are permitted to harvest both *Laminaria digitata* and *L. hyperborea* with higher quotas, and smaller boats, which are restricted to harvesting only *L. digitata* with lower quotas (Frangoudes and Garineaud 2015). Since 2013, the number of active seaweed harvesting vessels has been capped at 35 due to regulatory license limits. Concurrently, vessel dimensions have steadily increased. Between 2000 and 2008, the average vessel length grew from 9.7 m to 10.0 m (Alban et al. 2011), a trend that continued over the next 15 years. Today, the fleet has an average vessel length of 10.6 m, with broader beams that enhance harvesting capacity and efficiency.

Table 18.7 Characteristics of seaweed harvesting boats

Type of vessel	Coastal seaweed harvester	Island seaweed harvester	Large island seaweed harvester
Ship length	8 m	10 m	12 m
Deck	Backside	Frontside	Frontside
Number of cranes	1	1	2
Load capacity (t)	8	14	30

Adapted from Arzel (1987)

These structural changes in the fleet have likely influenced the proportion of the target species, *L. digitata*, relative to non-target species like *S. polyschides*, affecting both the quality and volume of landings. Several factors contribute to this shift: (1) Larger mechanized vessels, equipped with dual cranes fitted with "scoubidou," can harvest significantly greater quantities of seaweed per trip. However, this increased efficiency may come at the cost of reduced selectivity, leading to a higher proportion of non-target species in the catch. (2) The growing size of harvesting vessels presents operational challenges, particularly when maneuvering near rocky edges, where *L. digitata* is most abundant. Yet, *L. digitata* preferentially colonizes substrates between +1.5 m and −3 m relative to chart datum, whereas *S. polyschides* typically begins establishing itself below chart datum. Although *S. polyschides* can be found throughout the distribution range of *L. digitata*, including the deeper infralittoral zone, individuals settled in the infralittoral fringe tend to die off rapidly (Arzel 1997). Furthermore, the larger average size of *S. polyschides* (Fig. 18.8) provides it with a competitive advantage over *L. digitata*, as it can capture light more efficiently.

Even with a targeted approach, *L. digitata* remains a declining resource along Breton coasts, as the species reaches its southernmost range extension (Castric-Fey et al. 2001; Oppliger et al. 2014). Adapted to cooler waters, it is highly sensitive to rising temperatures, which can cause cellular damage, leading to direct mortality and impaired reproduction (Bartsch et al. 2013; Leathers et al. 2024). Beyond its commercial importance, this species provides essential ecosystem services, including coastal protection, habitat for marine biodiversity, and nursery and feeding grounds for numerous species (Pecquet et al. 2022). It might also play a crucial role in long-term carbon sequestration, significantly contributing to blue carbon exportation and climate change mitigation (Legge et al. 2020). Recognizing their ecological and cultural significance, multiple conservation and protection initiatives have been established: kelp habitats are currently listed under OSPAR (de Bettignies et al. 2021) and are included in Marine Protected Areas, such as the Iroise Marine Nature Park, where regulations help prevent overexploitation and promote the long-term sustainability of kelp forests (Couceiro et al. 2013; Vanhoutte-Brunier et al. 2016). More recently, the OBSlamir long-term observatory was created to reinforce biodiversity conservation efforts while safeguarding traditional seaweed harvesting practices. Such non-exploited areas could be an opportunity to conduct long-term scientific monitoring that would be particularly valuable in understanding the functioning of such habitats. Improving both landing quality and extraction processes, alongside *L. digitata* protection measures, could enhance alginate production yields in the near future. Yet, distribution models predict that by the end of the century, environmental conditions along France's Channel and Atlantic coasts will no longer support *L. digitata*, leading to significant population declines (Assis et al. 2017). Expanding its harvesting range to northern France (Normandy and the North Sea) is unlikely, as populations there have long been in decline or have disappeared locally (Cosson 1999; de Bettignies et al. 2021; Fouqueau 2024). Additionally, the absence of legal frameworks in these areas further restricts potential exploitation. Given this reality, solutions must be found to propose a sustainable local alginate industry.

An approach is then to diversify and optimize the harvesting of other alginophytes, with kelp species being the most promising candidates. *Laminaria hyperborea* is the most abundant kelp at the tip of Brittany (Bajjouk et al. 2015), covering approximately 11,000 hectares compared to 4500 hectares for *L. digitata*. Current yields, now averaging 20,000 tons per year (Leclerc et al. 2015), have significantly increased from the initial 1990s trials, which did not exceed 2000 tons (Arzel et al. 1996). Since its initiation in 2008, this harvest has been managed following the Norwegian model (Vea and Ask 2011), using a 3-year fallow system with zoned quotas and a standardized trawl to ensure selectivity and minimize ecological impact (5-year in Norway). While the impact of this harvesting practice on local *L. hyperborea* communities appears relatively limited (Derrien-Courtel et al. 2013b), further large-scale studies are needed to confirm these findings (Derrien-Courtel et al. 2022). Most other kelp species along the Breton coast (*Saccharina latissima, Alaria esculenta*, and *Undaria pinnatifida*) are currently either too scarce and/or in decline, preventing their sustainable exploitation for now (Lüning 1990; Castric-Fey et al. 2001; de Bettignies et al. 2021). Other seaweeds, such as *Himanthalia elongata*, have been considered for alginate harvesting using "scoubidou," with trials reaching 200 tons in 2017. However, beyond its limited biomass and its extremely limited bathymetric distribution (a very narrow band at low tide), this species has recently shown signs of decline along our coasts, with observations in 2024 indicating its replacement by the invasive *Sargassum muticum* in the Bay of Brest (Burel pers. obs.). This decline, also reported further south (Lima et al. 2007), raises concerns about the long-term viability of its potential exploitation for alginates. Sargassaceae found locally (genera *Cystoseira*, *Gongolaria*, *Ericaria,* and *Sargassum*), although considered for alginate extraction, have not yet been selected due to their overall poor alginate quality, preventing economic viability (McHugh 2003; Abka-Khajouei et al. 2022; seaweed processors pers. comm.).

Supporting alginate production through the collection of intertidal seaweed, particularly Fucales, could serve as an exceptional complementary measure but would inevitably remain limited in both scope and duration, as large-scale exploitation—such as seen in Canada (Ugarte and Sharp 2001)—is not being considered. *Ascophyllum nodosum* is currently the most extensively hand-harvested species in Brittany, with annual yields ranging from 1500 to 3000 tons over the past decade (Comité Régional des Pêches Maritimes et des Elevages Marins de Bretagne 2023). While primarily collected for soil amendments and animal feed, with only a small fraction used for alginate production, its decline over the past 20 years (Bizien et al. 2024) strongly discourages any expansion of its exploitation. Considering the cultivation of alginate-producing seaweeds for helping the alginate industry remains challenging (McHugh 2003; Buschmann et al. 2017), particularly in France, where achieving consistent and stable production has proven difficult. The primary cultivated alginate-producing species, *S.latissima*, is yet, increasingly being outcompeted by *S. polyschides* in cultivation systems (seaweed producer pers. comm.).

In light of previous findings, exploring the potential of alternative resources such as *L. ochroleuca*, which appears to be expanding its range (Smale et al. 2015; Franco et al. 2018), or *S. polyschides* for alginate production emerges as a crucial avenue of research (Silva et al. 2015; Kaidi et al. 2022). They both present several advantages that make their exploitation particularly promising: (1) they are expected to demonstrate greater resilience to rising ocean temperatures compared to other species; (2) existing harvesting tools, such as the "scoubidou" or trawl, are already well-adapted to these species, eliminating the need for new equipment. Concerning *S. polyschides,* (3) its annual life cycle allows for rapid resource renewal, although it may also introduce greater year-to-year variability. However, significant challenges remain: Industrial extraction processes must be optimized and adapted to effectively process these species, ensuring their successful integration into the alginate production industry.

The future of alginate production, and the sustainable management of kelp resources, will require a collective effort from all local stakeholders. To successfully navigate this transition, it is essential that industry representatives, resource managers, biodiversity managers, policymakers, and harvesters work together to ensure the long-term viability of this ancestral activity, which is deeply embedded in local traditions. A collaborative approach, involving key organizations such as the *Comité des Pêches*, *Cluster Algues Bretagne*, and the CSAVM, will be crucial in shaping innovative and adaptive management strategies. By fostering dialogue and cooperation, these efforts can help balance ecological preservation with economic sustainability, ensuring that kelp harvesting continues to thrive while respecting environmental constraints.

Acknowledgments This study was conducted at the request of the Chambre Syndicale des Algues & Végétaux Marins (CSAVM—French Seaweed & Marine Plants Trade Association). We sincerely thank them for their financial support, with particular appreciation to their president, Jean-Baptiste Wallaert, for his involvement and guidance. A special thank you to Alan Critchley for inviting us to contribute to this publication. We also express our deep appreciation to Yves-Marie Paulet, Jacques Grall, and Gauthier Schaal, as well as the co-editors of the book, for their time and effort in reviewing this chapter. Many thanks to Annaëlle Anquet and Mathieu Helias for their assistance. We also extend our gratitude to Arnaud Delafon from JRS Marine Products, as well as Anne-Laure Duros from Algaia. Many thanks to Erwan Bergot, François De Borggraef, and Gwen Sanquer for facilitating our work at the processing sites. Thanks to Martial Laurans for his help with *Laminaria* data. Special thanks to Sébastien Le Corre from the website goemoniers.e-monsite.com.

References

Abka-Khajouei R, Tounsi L, Shahabi N, Patel AK, Abdelkafi S, Michaud P (2022) Structures, properties and applications of alginates. Mar Drugs 20(6):364

Alander JT, Bochko V, Martinkauppi B, Saranwong S, Mantere T (2013) A review of optical nondestructive visual and near-infrared methods for food quality and safety. Int J Spectrosc 1:341402. https://doi.org/10.1155/2013/341402

Alban F, Frangoudes K, Frésard M (2011) Kelp harvesting fleet dynamics and the fleet's dependence on *Laminaria* forests in the Iroise Sea (North Finistère, France). Cah Biol Mar 52(4):507

Araújo R, Vázquez Calderón F, Sánchez López J, Azevedo IC, Bruhn A, Fluch S, Ullmann J (2021) Current status of the algae production industry in Europe: an emerging sector of the blue bioeconomy. Front Mar Sci 7:626389. https://doi.org/10.3389/fmars.2020.626389

Arzel P (1987) Les goémoniers. Chasse-Marée:305

Arzel P (1994) L'exploitation des algues en Bretagne. Études rurales 133:113–126

Arzel P (1997) La flottille goémonière et l'exploitation du champ de laminaires sur les côtes bretonnes - Essai de synthèse

Arzel P (1998) Les laminaires sur les côtes bretonnes: évolution de l'exploitation et de la flottille de pêche, état actuel et perspectives. Éditions Quae, p 139

Arzel P (2002) La laminaire digitée. Les nouvelles de l'Ifremer 33(4)

Arzel P, Castric-Fey A (1997) Les forêts de laminaires. In: Dauvin JC (ed) Les biocénoses marines et littorales françaises des côtes atlantiques, Manche et Mer du Nord. Muséum National d'Histoire Naturelle, Paris, pp 159–167

Arzel P, Mingant C, Noël P, Gourronc E (1996) Compte rendu des essais de pêche du *Laminaria hyperborea* - Campagne 1995. Direction des ressources vivantes, Ressources halieutiques, Direction de l'ingénierie, de la technologie et de l'informatique, Génie océanique Ifremer, centre de Brest, 52pp

Assis J, Araújo MB, Serrão EA (2017) Projected climate changes threaten ancient refugia of kelp forests in the North Atlantic. Glob Change Biol 24(1):e55–e66. https://doi.org/10.1111/gcb.13818

Bajjouk T, Rochette S, Laurans M, Ehrhold A, Hamdi A, Le Niliot P (2015) Multi-approach mapping to help spatial planning and management of the kelp species *L. digitata* and *L. hyperborea*: case study of the Molène archipelago. Brittany J Sea Res 100:2–21. https://doi.org/10.1016/j.seares.2015.04.004

Bartsch I, Wiencke C, Bischof K, Buchholz CM, Buck BH, Eggert A, Feuerpfeil P, Hanelt D, Jacobsen S, Karez R, Karsten U, Molis M, Roleda MY, Schubert H, Schumann R, Valentin K, Weinberger F, Wiese J (2008) The genus *Laminaria* sensu lato: recent insights and developments. Eur J Phycol 43(1):1–86. https://doi.org/10.1080/09670260701711376

Bartsch I, Vogt J, Pehlke C, Hanelt D (2013) Prevailing sea surface temperatures inhibit summer reproduction of the kelp *Laminaria digitata* at Helgoland (North Sea). J Phycol 49(6):1061–1073. https://doi.org/10.1111/jpy.12125

Beaumont M, Tran R, Vera G, Niedrist D, Rousset A, Pierre R, Shastri VP, Forget A (2021) Hydrogel-forming algal polysaccharides: from seaweed to biomedical applications. Biomacromolecules 22(3):1027–1052. https://doi.org/10.1021/acs.biomac.0c01406

de Bettignies T, Hébert C, Assis J, Bartsch I, Bekkby T, Christie H, Dahl K, Derrien-Courtel S, Edwards H, Filbee-Dexter K, Franco J, Gillham K, Harrald M, Hennicke J, Hernández S, Le Gall L, Martinez B, Mieszkowska N, Moore P, Moy F, Mueller M, Norderhaug KM, Parry M, Ramsay K, Robuchon M, Russel T, Serrão E, Smale D, Steen H, Street M, Tempera F, Valero M, Werner T, La Rivière M (2021) Case report for kelp forests habitat. OSPAR 787/2021, 39pp. ISBN 978-1-913840-16-7

Birkett DA, Maggs CA, Dring MJ, Boaden PJS, Seed R (1998) Infralittoral reef biotopes with kelp species. An overview of dynamic and sensitivity characteristics for conservation management of marine SACs. In: Scottish Association of Marine Science: UK marine SACs project, vol VII. The Scottish Association for Marine Science, Oban. 7

Bizien A, Toumi C, Boyé A, Sturbois A, Le Duff M, Grall J, Helias M, Gauthier O, Burel T (2024) Contrasted trends of intertidal macroalgal communities and sharp decline of canopy-forming species across two decades. Mar Ecol Prog Ser 742:73–86. https://doi.org/10.3354/meps14652

Black WÁ (1950) The seasonal variation in weight and chemical composition of the common British Laminariaceae. J Mar Biol Assoc UK 29(1):45–72. https://doi.org/10.1017/S0025315400056186

Bojorges H, Fabra MJ, López-Rubio A, Martínez-Abad A (2022) Alginate industrial waste streams as a promising source of value-added compounds valorization. Sci Total Environ 838:156394. https://doi.org/10.1016/j.scitotenv.2022.156394

Buschmann AH, Camus C, Infante J, Neori A, Israel Á, Hernández-González MC, Pereda SV, Gomez-Pinchetti JL, Golberg A, Critchley AT (2017) Seaweed production: overview of the global state of exploitation, farming and emerging research activity. Eur J Phycol 52(4):391–406. https://doi.org/10.1080/09670262.2017.1365175

Castric-Fey A, Girard-Descatoire A, L'Hardy-Halos MT, Derrien-Courtel S (2001) La vie sous-marine en Bretagne—Découverte des fonds rocheux. Les Cahiers Naturalistes de Bretagne n3, Conseil Régional de Bretagne, Biotope édition, 184p

Chapman VJ (1970) Seaweeds and their uses. 340p

Colin Y (1964) Avenir du métier de goémonier. Penn Ar Bed 37:178–181

Comité Régional des Pêches Maritimes et des Elevages Marins de Bretagne (2023) Evaluation de l'impact de l'exploitation d'*Ascophyllum nodosum* par les récoltants professionnels d'algues de rive et pistes pour l'adaptation de la gestion - Programme Agrid (2019–2022), Rapport final, 44 pp

Cosson J (1999) Sur la disparition progressive de *Laminaria digitata* sur les côtes du Calvados (France). Cryptogam Algol 20(1):35–34. https://doi.org/10.1016/S0181-1568(99)80005-1

Couceiro L, Robuchon M, Destombe C, Valero M (2013) Management and conservation of the kelp species *Laminaria digitata*: using genetic tools to explore the potential exporting role of the MPA "Parc naturel marin d'Iroise". Aquat Living Resour 26(2):197–205. https://doi.org/10.1051/alr/2012027

Davoult D, Engel CR, Arzel P, Knoch D, Laurans M (2011) Environmental factors and commercial harvesting: exploring possible links behind the decline of the kelp *Laminaria digitata* in Brittany, France. Cah Biol Mar 52:429

Derrien-Courtel S, Le Gal A, Grall J (2013a) Regional-scale analysis of subtidal rocky shore community. Helgoland Mar Res 67:697–712. https://doi.org/10.1007/s10152-013-0355-2

Derrien-Courtel S, Le Gal A, Catherine E (2013b) Etude d'incidence de l'utilisation du peigne à *Laminaria hyperborea* sur la biocénose à laminaires – Rapport final – Données 2012, Contrat PNMI-MNHN, mars 2013, 30p

Derrien-Courtel S, Le Gal A, Voisin A, Derrien R, Decaris FX (2022) SLAMIR: Suivi de la biodiversité associée aux champs de LAMinaires du Parc naturel marin d'IRoise. Conventions OFB-MNHN Concarneau & MNHN-UBO. Rapport final, 73 pp.

Diruit W, Le Bris A, Bajjouk T, Richier S, Helias M, Burel T, Lennon M, Guyot A, Ar Gall E (2022) Seaweed habitats on the shore: characterization through hyperspectral UAV imagery and field sampling. Remote Sens 14(13):3124. https://doi.org/10.3390/rs14133124

Draget KI, Moe ST, Skjak-Bræk G, Smidsrød O (2006) Alginates. In: Stephen AM, Phillips GO, Williams PA (eds) Food polysaccharides and their applications, 2nd edn. CRC Press, Boca Raton, pp 289–334. https://doi.org/10.1201/9781420015164.ch9

Engelen AH, Leveque L, Destombe C, Valero M (2011) Spatial and temporal patterns of recovery of low intertidal *Laminaria digitata* after experimental spring and autumn removal. Cah Biol Mar 52:441

Fauziee NAM, Chang LS, Mustapha WAW, Nor ARM, Lim SJ (2021) Functional polysaccharides of fucoidan, laminaran and alginate from Malaysian brown seaweeds (*Sargassum polycystum*, *Turbinaria ornata* and *Padina boryana*). Int J Biol Macromol 167:1135–1145. https://doi.org/10.1016/j.ijbiomac.2020.11.067

Fletcher RL (2024) Brown seaweeds (Phaeophyceae) of Britain and Ireland. Pelagic Publishing Ltd, p 816. https://doi.org/10.53061/XDPG8462

Floc'h JY (1970) Cartographie de la végétation marine dans l'archipel de Molène (Finistère). Rev Trav Inst Pêches Marit 34(1):89–120

Fouqueau L, Reynes L, Tempera F, Bajjouk T, Blanfuné A, Chevalier C, Laurans M, Mauger S, Sourisseau M, Assis J, Lévêque L, Valero M (2024) Seascape genetic study on *Laminaria digitata* underscores the critical role of sampling schemes. Mar Ecol Prog Ser 740:23–42. https://doi.org/10.3354/meps14640

Franco JN, Tuya F, Bertocci I, Rodríguez L, Martínez B, Sousa-Pinto I, Arenas F (2018) The 'golden kelp' *Laminaria ochroleuca* under global change: integrating multiple eco-physiological responses with species distribution models. J Ecol 106(1):47–58

Frangoudes K (2011) Seaweeds fisheries management in France, Japan, Chile and Norway. Cah Biol Mar 52(4):517

Frangoudes K, Garineaud C (2015) Governability of kelp forest small-scale harvesting in Iroise Sea, France. In: Jentoft S, Chuenpagdee R (eds) Interactive governance for small-scale fisheries: global reflections, pp 101–115. https://doi.org/10.1007/978-3-319-17034-3_6

Garineaud C (2017) Récolter la mer: des savoirs et des pratiques des collecteurs d'algues à la gestion durable des ressources côtières dans le Finistère (Bretagne). Doctoral dissertation, Muséum national d'histoire naturelle-MNHN Paris

Gómez-Ordóñez E, Ruperez P (2011) FTIR-ATR spectroscopy as a tool for polysaccharide identification in edible brown and red seaweeds. Food Hydrocoll 25:1514–1520. https://doi.org/10.1016/j.foodhyd.2011.02.009

Guiry MD, Morrison L (2013) The sustainable harvesting of *Ascophyllum nodosum* (Fucaceae, Phaeophyceae) in Ireland, with notes on the collection and use of some other brown algae. J Appl Phycol 25:1823–1830. https://doi.org/10.1007/s10811-013-0027-2

Hao C, Yu G, He Y, Xu C, Zhang L, Wang W (2019) Marine glycan-based antiviral agents in clinical or preclinical trials. Rev Med Virol 29(3). https://doi.org/10.1002/rmv.2043

Hawkins SJ, Harkin E (1985) Preliminary canopy removal experiments in algal dominated communities low on the shore and in the shallow subtidal on the Isle of Man. Bot Mar 28:223–230. https://doi.org/10.1515/botm.1985.28.6.223

Haug A, Myklestad S, Larsen B, Smidsrod O (1967) Studies on the sequence of uronic acid residues in alginic acid. Acta Chem. Scand 21:691–704; 768–778

Henry S, Taillens M, Laurans M, Frangoudes K (2021) Histoires d'Algues en Finistère. Cloître, Saint-Thonan 104p

Hurtado A, Aljabali AA, Mishra V, Tambuwala MM, Serrano-Aroca Á (2022) Alginate:Enhancement strategies for advanced applications. Int J Mol Sci 23(9):4486. https://doi.org/10.3390/ijms23094486

Kaidi S, Bentiss F, Jama C, Khaya K, Belattmania Z, Reani A, Sabour B (2022) Isolation and structural characterization of alginates from the kelp species *Laminaria ochroleuca* and *Saccorhiza polyschides* from the Atlantic Coast of Morocco. Colloids Interfaces 6(4):51. https://doi.org/10.3390/colloids6040051

Leathers T, King NG, Foggo A, Smale DA (2024) Marine heatwave duration and intensity interact to reduce physiological tipping points of kelp species with contrasting thermal affinities. Ann Bot-London 133(1):51–60. https://doi.org/10.1093/aob/mcad172

Leclerc JC, Riera P, Laurans M, Leroux C, Lévêque L, Davoult D (2015) Community, trophic structure and functioning in two contrasting *Laminaria hyperborea* forests. Estuar Coast Shelf Sci 152:11–22. https://doi.org/10.1016/j.ecss.2014.11.005

Lee KY, Mooney DJ (2012) Alginate: properties and biomedical applications. Prog Polym Sci 37(1):106–126. https://doi.org/10.1016/j.progpolymsci.2011.06.003

Legge O, Johnson M, Hicks N, Jickells T, Diesing M, Aldridge J, Andrews J, Artioli Y, Bakker DCE, Burrows MT, Carr N, Cripps G, Felgate SL, Fernand L, Greenwood N, Hartman S, Kröger S, Lessin G, Mahaffey C, Mayor DJ, Parker R, Queirós AM, Shutler JD, Silva T, Stahl H, Tinker J, Underwood GJC, Van Der Molen J, Wakelin S, Weston K, Williamson P (2020) Carbon on the Northwest European shelf: contemporary budget and future influences. Front Mar Sci 7:143. https://doi.org/10.3389/fmars.2020.00143

Li Y, Zheng Y, Zhang Y, Yang Y, Wang P, Imre B, Wong ACY, Hsieh YSH, Wang D (2021) Brown algae carbohydrates: structures, pharmaceutical properties, and research challenges. Mar Drugs 19:620–632. https://doi.org/10.3390/md19110620

Liesner D, Fouqueau L, Valero M, Roleda MY, Pearson GA, Bischof K, Valentin K, Bartsch I (2020) Heat stress responses and population genetics of the kelp *Laminaria digitata* (Phaeophyceae) across latitudes reveal differentiation among North Atlantic populations. Ecol Evol 10(17):9144–9177. https://doi.org/10.1002/ece3.6569

Lima FP, Ribeiro PA, Queiroz N, Hawkins SJ, Santos AM (2007) Do distributional shifts of northern and southern species of algae match the warming pattern? Glob Change Biol 13(12):2592–2604. https://doi.org/10.1111/j.1365-2486.2007.01451.x

Lüning K (1990) Seaweeds: their environment, biogeography, and ecophysiology. Wiley

Mac Monagail M, Cornish L, Morrison L, Araújo R, Critchley AT (2017) Sustainable harvesting of wild seaweed resources. Eur J Phycol 52(4):371–390. https://doi.org/10.1080/09670262.2017.1365273

Mak TM, Xiong X, Tsang DC, Iris KM, Poon CS (2020) Sustainable food waste management towards circular bioeconomy: policy review, limitations and opportunities. Bioresour Technol 297:122497. https://doi.org/10.1016/j.biortech.2019.122497

Metz S, Mommens MP, Madec A, Lenaff P (2024) Évaluation et suivi de la contribution sociale et économique de la filière des macroalgues. Rapport Pêche et Aquaculture réalisé pour FranceAgriMer.107 pp

McHugh DJ (2003) A guide to the seaweed industry. FAO Fisheries Technical Paper no. 441. FAO, Rome, p 105

Mesnildrey L, Jacob C, Frangoudes K, Reunavot M, Lesueur M (2012) Seaweed industry in France. Report. Interreg program NETALGAE. Les publications du Pôle halieutique Agrocampus Ouest, no. 9, 34 pp

Norton TA (1977) Experiments on the factors influencing the geographical distributions of *Saccorhiza polyschides* and *Saccorhiza dermatodea*. New Phytol 78(3):625–635. https://doi.org/10.1111/j.1469-8137.1977.tb02167.x

Oppliger LV, Von Dassow P, Bouchemousse S, Robuchon M, Valero M, Correa JA, Mauger S, Destombe C (2014) Alteration of sexual reproduction and genetic diversity in the kelp species *Laminaria digitata* at the southern limit of its range. PLoS One 9(7):e102518. https://doi.org/10.1371/journal.pone.0102518

Pecquet J, Mouchet M, Campagne S, Raybaud V, Baulaz Y, Gevaert F, Ben Rais Lasram F (2022) Combining ecological niche models and ecosystem services indicators to assess impacts of climate change on kelp: application to French coasts. Ecosyst People 18(1):358–377. https://doi.org/10.1080/26395916.2022.2080766

Pereira L, Cotas J (2020) Introductory chapter: alginates - a general overview. In: Pereira L, Cotas J (eds) Alginates. Recents uses of this natural polymer. InterOpen, pp 1–16. https://doi.org/10.5772/intechopen.88381

Perez R (1973) La récolte mécanisée des laminaires-Ses conséquences sur les peuplements. Science et Pêche 226:13–21

Peteiro C (2018) Alginate production from marine macroalgae, with emphasis on kelp farming. In: Rehm B, Moradali M (eds) Alginates and their biomedical applications, Springer Series in Biomaterials Science and Engineering, vol 11. Springer, Singapore, pp 29–47. https://doi.org/10.1007/978-981-10-6910-9_2

Salland N, Smale DA (2021) Spatial variation in the structure of overwintering, remnant *Saccorhiza polyschides* sporophytes and their associated assemblages. J Mar Biol Assoc UK 101(4):639–648. https://doi.org/10.1017/S0025315421000692

Salland N, Wilding C, Jensen A, Smale DA (2024) Spatiotemporal variability in population demography and morphology of the habitat-forming macroalga *Saccorhiza polyschides* in the Western English Channel. Ann Bot-London 133(1):117–130. https://doi.org/10.1093/aob/mcad181

Sasaki H, Flores-Moya A, Henry EC, Müller DG, Kawai H (2001) Molecular phylogeny of Phyllariaceae, Halosiphonaceae and Tilopteridales (Phaeophyceae). Phycologia 40(2):123–134. https://doi.org/10.2216/i0031-8884-40-2-123.1

Schaal G, Leclerc JC, Droual G, Leroux C, Riera P (2016) Biodiversity and trophic structure of invertebrate assemblages associated with understorey red algae in a *Laminaria digitata* bed. Mar Biol Res 12(5):513–523. https://doi.org/10.1080/17451000.2016.1164318

Silva M, Gomes F, Oliveira F, Morais S, Delerue-matos C (2015) Microwave-assisted alginate extraction from Portuguese *Saccorhiza polyschides*-influence of acid pretreatment. Int J Chem Nucl Mat Metallurg Eng 9:30–33

Smale DA, Vance T (2015) Climate-driven shifts in species' distributions may exacerbate the impacts of storm disturbances on North-east Atlantic kelp forests. Mar Freshw Res 67(1):65–74. https://doi.org/10.1071/MF14155

Smale DA, Burrows MT, Moore P, O'Connor N, Hawkins SJ (2013) Threats and knowledge gaps for ecosystem services provided by kelp forests: a Northeast Atlantic perspective. Ecol Evol 3(11):4016–4038. https://doi.org/10.1002/ece3.774

Smale DA, Wernberg T, Yunnie AL, Vance T (2015) The rise of *Laminaria ochroleuca* in the Western English Channel (UK) and comparisons with its competitor and assemblage dominant *Laminaria hyperborea*. Mar Ecol 36(4):1033–1044. https://doi.org/10.1111/maec.12199

Smale DA (2020) Impacts of ocean warming on kelp forest ecosystems. New Phytol, 225:1447–1454. https://doi.org/10.1111/nph.16107

Stiger V, Bourgougnon N, Deslandes E (2016) Carbohydrates from seaweeds. In: Fleurence J, Irwin I (eds) Seaweed in health and disease prevention. Academic Press Elsevier, pp 223–256. https://doi.org/10.1016/B978-0-12-802772-1.00008-7

Synytsya A, Čopíková J, Kim WJ, Park YI (2015) Cell wall polysaccharides of marine algae. In: Kim SK (ed) Springer handbook of marine biotechnology. Springer Handbooks, Springer, Berlin, Heidelberg, pp 543–582. https://doi.org/10.1007/978-3-642-53971-8_22

Ugarte RA, Sharp G (2001) A new approach to seaweed management in eastern Canada: the case of *Ascophyllum nodosum*. Cah Biol Mar 42(1/2):63–70

Valero M, Engel C, Arzel P, Creach A, Davoult D, Destombe C, Gevaert F, Leblanc C, Levavasseur G, Potin P (2006) Dynamique des champs de *Laminaria digitata*, ressource algale en Bretagne: impacts biotiques, abiotiques et anthropiques. Biodiversité et changement global - Colloque de restitution, 18–20 septembre 2006, pp 37–45

Vanhoutte-Brunier A, Laurans M, Mongruel R, Guyader O, Davoult D, Marzin A, Vaschalde D, Charles M, Niliot P (2016) Évaluation des services écosystémiques du champ de laminaires de l'archipel de Molène. Retour d'Expérience du Site du Parc Naturel Marin d'Iroise. https://doi.org/10.13140/RG.2.1.2037.1608

Vauchel P, Kaas R, Arhaliass A, Baron R, Legrand J (2008) A new process for extracting alginates from *Laminaria digitata*: reactive extrusion. Food Bioprocess Tech 1:297–300. https://doi.org/10.1007/s11947-008-0082-x

Vea J, Ask E (2011) Creating a sustainable commercial harvest of *Laminaria hyperborea*, in Norway. J Appl Phycol 23:489–494. https://doi.org/10.1007/s10811-010-9610-y

Werner A, Kraan S (2004) Review of the potential mechanisation of kelp harvesting in Ireland, Marine Environment & Health Series, vol 17. Marine Institute

Zargarzadeh M, Amaral AJ, Custódio CA, Mano JF (2020) Biomedical applications of laminarin. Carbohydr Polym 232:115774. https://doi.org/10.1080/09670262.2017.1365175

Dr. Thomas Burel is a marine biologist and lecturer at the University of Western Brittany (Université de Bretagne Occidentale), specializing in coastal ecology and macroalgae. His research focuses on the functioning of coastal habitats, monitoring the regression of macroalgal forests in the northeast Atlantic, and studying the introduction of non-indigenous species. He examines these issues through the lens of environmental and anthropogenic pressures

Anaëlle Bizien is a PhD student at the Laboratory of Environmental Marine Sciences (Plouzané, France), working on the long-term dynamics of macroalgal communities on rocky shores. She focuses on understanding their drivers, role as bioindicators, and distribution models. She has developed expertise in taxonomy, community responses, and numerical ecology while studying a wide range of coastal habitats

Marc Lennon (52 years old), CEO, co-founder of Hytech-imaging—Marc obtained a PhD in signal processing and telecommunications from the University of Rennes 1/Telecom Bretagne in 2002, dedicated to the development of hyperspectral image analysis methods for agro-environmental mapping applications. In 2003, he joined ACTIMAR, where he set up and developed the remote sensing activity and then the "Earth Observation Pole" of ACTIMAR and SUEZ Consulting from 2014. During the period from 2003 to 2013, he was an Associate Researcher within the Image and Information Processing Department of IMT Atlantique, Brest, France. From 2009 to 2016, Marc was responsible for several long-term R&D programs that led to the development of SWIM® (*S*hallow *W*ater mapp<u>I</u>ng using optical re*M*ote sensors), a suite of protocols and algorithms for airborne hyperspectral data acquisition, radiative transfer inversion in the water column, and operational production of innovative cartographic products in shallow waters. After more than 13 years within ACTIMAR, he created Hytech-imaging in April 2016. The aim of Hytech-imaging is to contribute actively to the development of spectral imagery technologies and applications for operational mapping applications by offering innovative products and services as well as innovative business models for spectral imagery exploitation

Eric Deslandes is an Emeritus Professor at the University of Brest in Brittany, France. Eric led an academic research team working on marine and coastal plants for several years at the European Institute for Marine Studies. His research topic is the ecophysiology and biochemistry of marine algae in the Brittany coasts, with a specialization in red algae and compounds from carbon metabolism. He is a referee for several international peer-reviewed journals in phycology. He has contributed to the publication of about 100 articles and several chapters of books on algae

Learning Valuable Insights from Past Experiences in Wakame Aquaculture in Japan to Advance Phyconomy

19

Yoichi Sato

Abstract

Wakame (*Undaria pinnatifida*) aquaculture has played a significant role in Japan. Traditional knowledge of culturing Wakame has been enhanced by technological advancements, leading to its widespread cultivation and commercialization. However, recent environmental changes, such as rising seawater temperatures and declining nutrient levels, have significantly impacted production, creating a demand for new approaches to sustain the industry. This chapter provides a comprehensive overview of the history of Wakame aquaculture, its current challenges, and future prospects in Japan. It highlights key innovations in sporeling production, cultivation methods, and processing technologies that have improved efficiency and product quality. Also examined is the growing interest in the potential of Wakame for a role in creating blue carbon: short-term carbon sequestration. To address declining yields and labor shortages, various new strategies are being explored, such as artificial sporeling production using free-living gametophytes, optimization of cultivation conditions, and the adoption of intensive aquaculture models. These individual-scale activities in Japan could be adapted in other countries and areas where there is interest in Wakame and other macroalgal aquaculture. By combining traditional aquaculture practices with scientific innovation, Wakame cultivation can continue to support food security, economic sustainability, and climate change mitigation. This review emphasizes the importance of leveraging historical insights to develop future-focused solutions that enhance both productivity and environmental resilience in the Wakame industry.

Keywords

Undaria pinnatifida · Cultivation · Sporelings · Processing

19.1 Introduction

Undaria pinnatifida (Harvey 1860), commonly known in Japan as Wakame, is an important aquaculture species along the coasts of Japan, Korea, and China, primarily for its value as an edible resource (Yamanaka and Akiyama 1993). However, in Europe, North and South America, Australia, and New Zealand, where this alga has expanded as an invasive species (Hay and Luckens 1987; Castric-Fey et al. 1999; Dellatorre et al. 2014; Silva et al. 2002), the focus has traditionally been on its extermination and habitat control (e.g., Wotton et al. 2004; Hesitt et al. 2005). Recently, the perception of invasive Wakame has shifted from threat to potential industrial application, leading to an increase in cultivation trials and product development (Stenton-Dozey et al. 2021). Additionally, given its valuable nutrient content and high biomass, its applications are expanding beyond edible and nutraceutical to include carbon storage, contributing to blue carbon effects globally.

Wakame is one of the many traditional food products that have long been produced by the fishing industry in Japan. From the knowledge accumulated through hundreds of years of experience, the fishing community has a deep understanding of the production methods necessary to maintain high yield, quality, and productivity. However, recent changes in the marine environment have exceeded the extent to which experience and traditional knowledge can address current fisheries-related problems, leading to declining production yields and a failure to meet domestic demand. As a contribution toward addressing the new paradigm, this chapter aims to provide a comprehensive overview of Wakame aquaculture history in Japan, along with current cultivation and processing technologies, which are supported by examples of

Y. Sato (✉)
Algal Research Innovation Centre, Riken Food Co., Ltd, Natori, Miyagi, Japan

Nishina Center for Accelerator-Based Science, RIKEN, Saitama, Japan

A. T. Critchley et al. (eds.), *Advances in Temperate Phyconomy: Algal Harvest and Cultivation in Globally Distributed Temperate Waters*, Developments in Applied Phycology 15, https://doi.org/10.1007/978-3-032-02955-3_19

recent research results and industrial applications. Also considered are answers to key questions posed by researchers and stakeholders from outside Japan (as well as those interested in Wakame *per se*), such as 'How do you produce Wakame sporelings in Japan?' 'How is Wakame processed at the primary stage?' and 'What are the routes of distribution for Wakame in Japan?' Such aspects of Wakame farming have not been disseminated widely before. With the growing international interest in macroalgal farming, this review demonstrates that the small-scale aquaculture and processing knowledge available in Japan can provide valuable insights for other countries, regions, and companies looking to enter the macroalgal farming sector.

19.2 History of Wakame Aquaculture in Japan and the Present Situation

19.2.1 The Cultivation History of Wakame

Wakame, along with other macroalgae, has been used in Japan since ancient times as a core edible resource for Japanese cuisine. Its name was even recorded as a tax item in Japan's oldest law, the Taiho Code of 701 A.D. (Miyashita 1974). Recipes using Wakame have been recorded since the middle of the Heian period (900s A.D.), including miso soup, vinegared Wakame, *tsukudani* (preserved Wakame boiled down in soy source and sugar), *ohitashi* (Wakame seasoned with blended soy source), and so on (Miyashita 1974). The natural harvest yield was first recorded at 18,653 tons in 1922, and by the early 1960s, this amount had increased to approximately 60,000 tons (Fig. 19.1). In response to the growing demand for stable harvests, the development of cultivation technology began in the early 1950s to establish a new fisheries industry. One of the pioneers in this effort, Mr. Yoshiro Ohtsuki, developed a method for producing sporelings and in-sea incubation at Konari Bay in Miyagi Prefecture. His technique was developed through observations that the mucous membrane of migratory cysts could easily be broken to obtain zoospores for sporelings production by drying mature sporophylls in shaded conditions (Funaki et al. 2021). At the same time, Mr. Tozo Komatsu of Ofunato City, Iwate Prefecture, established and disseminated technology to produce artificial sporelings. Dr. Yunosuke Saito from The University of Tokyo studied the growth and maturation of gametophytes and sporophytes during the germination stage, applying his findings to sporeling production (i.e., Saito 1958a). He made several significant discoveries that form the basis of modern Wakame aquaculture, such as the fact that if sporelings were installed too early (aiming for an earlier harvest), their mortality increased before they had grown sufficiently. Saito also emphasized that the water depth for Wakame aquaculture must be determined appropriately depending on the time of year and locality.

The contributions of these pioneering researchers facilitated the spread of Wakame aquaculture technology throughout the country. Aquaculture production was first included in national statistical data in 1965, recording 12,537 tons, since when it has grown rapidly, reaching a peak of 126,701 tons in 1976 (Fig. 19.1). As aquaculture technology has advanced, the wild harvest of Wakame has continued to decline, and since 1980, aquaculture production has accounted for over 95% of total domestic Wakame production in most years (Fig. 19.1).

19.2.2 History of Wakame Processing

These advances in aquaculture techniques have not only facilitated mass production but also led to the development

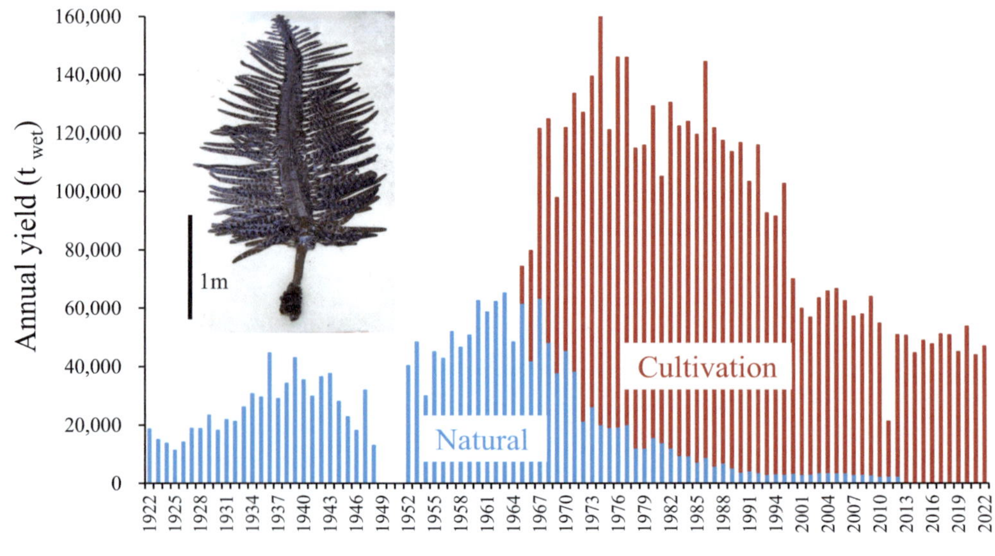

Fig. 19.1 Yearly records of natural harvest and cultivated Wakame in Japan from 1922 to 2022. (There are no data available for the years 1949 to 1951)

of new processing methods, which have significantly contributed to the popularity of Wakame. Previous historical processing methods of fresh or naturally dried Wakame had limitations regarding production volume and sales channel expansion. To address these challenges, research institutes across the country began developing innovative processing techniques. One notable advance came from the Fukushima Prefectural Fisheries Research Institute, which created 'fresh salted Wakame' in the 1960s. The method to make this product involves mixing a substantial amount of salt with freshly harvested Wakame, followed by dehydration, removal of the stipe, sorting, and packaging. This method allows the Wakame to better retain its flavor and nutrients. Excellent preservation, combined with the convenience of being able to eat it after simply rinsing off the salt in water, has led to its widespread distribution across the country.

In the 1970s, 'boiled and salted Wakame' was established. The method for this product involves

- Boiling Wakame in seawater (freshwater may be used in some places) immediately after harvesting, which changes the color of the thallus from brown to green,
- Cooling the green thalli in cold seawater,
- Sprinkling them with salt and storing them for 1 to 2 days, after which they are dehydrated (removing approximately 50–60% water content), and
- Separating the blade from the stipe.

Compared with fresh salted Wakame, this processing enables Wakame to be stored more stably and for a longer period, an innovation which triggered a further increase in demand.

In the 1980s, technology was developed to produce 'cut-dried Wakame' from boiled and salted Wakame. This process involved washing and cutting the boiled and salted Wakame before drying it, which enabled a further-extended period of storage. This cut-dried Wakame is convenient to use, as it can be rehydrated with water or hot water just before consumption. It has become popular in various culinary applications, such as instant noodles, miso soup, and seaweed salads. Furthermore, advances in technology for drying Wakame using rotary dryers have improved its handling properties after drying, enhancing ease of packaging, which is ideal for automated production. Today, cut-dried Wakame has become the standard form of Wakame products. Since the 1980s, this technology has also been adopted in Korea and China, leading to a growing global market for Wakame.

19.2.3 The Current State of Wakame Aquaculture and Industry in Japan

Wakame cultivation is practiced on both small and large scales in most sea-facing prefectures across Japan, except for the southern-most prefecture, Okinawa. The primary production area is the Sanriku coast of northeastern Honshu, which includes Iwate and Miyagi Prefectures, accounting for about 60–70% of national production (Fig. 19.2). Significant production occurs also in western Japan (the Naruto region of Tokushima Prefecture, the coastal region of Awaji Island in the Seto Inland Sea, and part of Hyogo Prefecture) and the coastal region of the Ise-Shima Peninsula in Mie Prefecture, central Japan (Fig. 19.2). The coastlines of the Sanriku region and the Ise-Shima Peninsula are rias, characterized by deep inlets that provide ideal conditions for high-quality Wakame aquaculture farms.

As mentioned previously, the demand for Wakame in Japan is rising and is driven by a growing segment of the population who are especially health conscious about food. However, in contrast to demand, production yields are declining. In the Sanriku region, where production levels can be more accurately tracked compared to other areas, the yield has fallen from approximately 50,000 t in 1980 to just 14,375 t in 2024 (Fig. 19.3a, Japan Fisheries Cooperative 2024). As a result, the price of fresh Wakame has become

Fig. 19.2 Major sites of Wakame cultivation in Japan

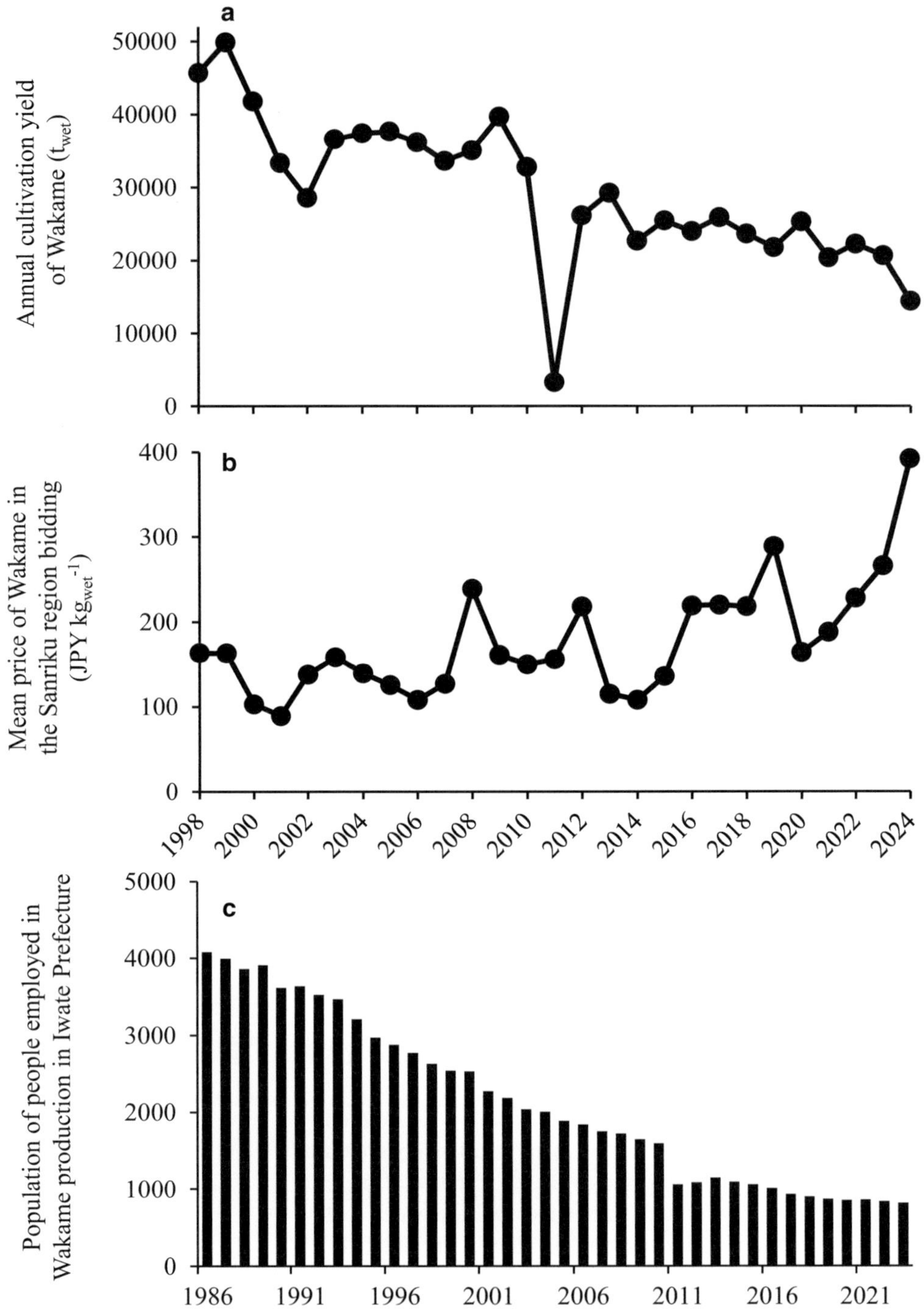

Fig. 19.3 Annual yield of Wakame in Japan, mean price of cultivated Wakame in the Sanriku region (1998–2024), and decline in the number of Wakame producers in Iwate Prefecture (1986–2023)

increasingly volatile, reaching 399 JPY per kg wet weight in 2024 (Fig. 19.3b). This price increase is largely due to environmental changes, including rising seawater temperatures, a drop in nutrient levels, and a decrease in the number of fishermen (Fig. 19.3c). These trends in declining production can be observed in other cultivation areas, too, such as in the Naruto region of western Japan. Additionally, production losses from grazing by herbivorous fish species during the early stages of Wakame aquaculture are an emerging issue.

19.3 Wakame Cultivation on the Sanriku Coast in Japan

19.3.1 The Wakame Life Cycle: Sporeling Production

Wakame possesses a heteromorphic life cycle that includes both a macroscopic sporophyte and a microscopic gametophyte stages (Fig. 19.4). In spring, sporophylls develop along

Fig. 19.4 Life cycle of Wakame. (**a**) Gametophytes, approximately 3 months after germination from spores. (**b**) Female gametophytes with freshly germinated sporophyte. (**c**) Juvenile sporophyte approximately 1 month after germination. (**d**) Mature sporophyte with sporophylls on the lower part of the stipe (red arrow)

the lower part of the stipe, resulting in the release of millions of asexual spores (Fig. 19.4d). These spores attach to substrates and germinate into dioecious gametophytes (male and female), as shown in Fig. 19.4a. Sperm from the male gametophytes fertilize the egg cells of the female gametophytes, resulting in the formation of juvenile sporophytes (Fig. 19.4b and c). The successful production of sporelings relies on achieving a large and stable number of juvenile sporophytes.

In Iwate Prefecture, most sporeling production is carried out in the natural environment (Fig. 19.5). Wild individuals forming mature sporophylls are collected between April and June. These individuals are often collected from a 'seed stock site' where individuals are collected for spore collection. Alternatively, cultivated individuals can be used for this purpose. After the sporophylls are harvested and visible, epiphytes are removed and they are placed in a shaded well-ventilated area to partially dry overnight. This drying process not only facilitates the release of migratory cells but also intended to eliminate other seaweeds and diatoms that may be adhering to the surfaces. Sporophylls are then immersed in seawater (often sterilized seawater provided by the fishery cooperative) inside a 1-tonne cultivation tank, making sure to avoid direct sunlight (Fig. 19.5a). Once the zoospores are released, knitted collectors made from palm fibers (each knitted piece is approximately 300 m long, 2 m wide, and 1.5 m deep) are soaked in the tank (Fig. 19.5b). After about 30 min to 2 h, the fishermen immediately place them in net bags and submerge them at a depth of 10–20 m (Fig. 19.5c, d). The collectors remain incubating in the natural environment until around October.

At a seawater temperature of approximately 20 °C, the collectors are removed from the mesh bags and suspended in the seawater. While the depth of initial suspension may vary among aquaculturists from person to person, it is gradually reduced over the course of about a month, starting from around 10 m. During this time, germination progresses from gametophyte to sporophyte, resulting in 'sporelings'—sporophytes that germinate at high density and can be used in aquaculture (Fig. 19.5e, f). Frequent maintenance of the collectors is essential during this period. Specifically, the collectors should be washed with seawater to eliminate any attached diatoms and other seaweeds. Additionally, the local fishermen often beat the collectors with a stick to remove other seaweeds. *Ulva* sp. (Chlorophyta) and *Polysiphonia* sp. (Rhodophyta) are particularly common contaminants. Most of these fishermen also have a good intuitive understanding of the specific

Fig. 19.5 Wakame sporeling production in the sea in the Sanriku region. (**a**) Sporophylls are immersed in sterilized seawater and left to release zoospores. (**b**) Collectors made of knitted palm fiber are soaked in the tank containing zoospores. (**c**) The collectors are then placed into a mesh bag. (**d**) The mesh bag, containing the collectors, is submerged at a depth of 10–20 m. (**e**, **f**) By regulating the depth level, the collectors germinate in the sea, resulting in a high density of juvenile sporophytes, which act as seedlings. (**g**) Zoospores are collected onto PVC frames wrapped with synthetic fiber in a tank (off the coast of Matsushima Bay in Miyagi Prefecture). (**h**) Finally, frames holding sporelings are hung from cultivation ropes in the inner bay to help acclimate them to environmental conditions at the final sites of cultivation

areas and depths where these unwanted seaweeds are less likely to attach.

Production of sporelings in Miyagi Prefecture is similar to that in Iwate Prefecture, up to the point of zoospore release. In Miyagi Prefecture, zoospores are collected in a metal or polyvinyl chloride (PVC) frame wrapped with synthetic fibers, which are immersed for 30 min to 2 h (Fig. 19.5g). They are then cultured in inner areas of the bay in natural conditions or in indoor tanks. The tanks would be placed in a well-ventilated area shaded from direct sunlight, to prevent the water temperature from rising.

In western Japan, particularly in the Naruto region, sporelings are grown in shaded inshore tanks during the summer, similar to the process in Miyagi Prefecture. By adjusting the light level provided to the tank, germination progresses from gametophytes to sporophytes. However, in recent years, the increasingly high summer temperatures are creating many challenges in producing sporelings in tanks, and establishing a new stable sporeling production technique is a critical issue. Nevertheless, in Miyagi Prefecture and the Naruto region, the germinated sporelings are hung on the cultivation rope in the inner areas of the bay to acclimate to the surrounding environment until final deployment on cultivation ropes in the ocean (Fig. 19.5h).

In every production area, as germination progresses, light intensity in the ocean or in cultivation tanks is carefully regulated. The timing for starting this process is determined by back calculating from the expected start of deployment of seeded ropes in the ocean. For example, producers in Hirota Bay, Iwate Prefecture, use a water temperature of 18 °C or lower as a reference to decide on the starting date for aquaculture. Therefore, producers determine when to start sporeling production by referring to data on seawater temperatures in the surrounding waters published by the local Prefectural government research stations and their own daily measurement data. Future improvements in water temperature forecasting systems using technology such as artificial intelligence and Internet of Things are expected to provide a powerful tool for Wakame producers.

Information regarding the state of nutrient concentrations in the ocean is also essential during the nursery period. In Iwate Prefecture, approximately 2 μM has been set as the threshold at which healthy sporeling growth is possible, and nutrient concentration forecasting is being put to practical use, utilizing records from the past (Kaga et al. 2022).

Periodically, prepared sporelings need to be transferred to other cultivation sites. In recent years, long-distance transport of seeded ropes has become necessary to supply aquaculture sites that are unable to prepare sufficient quantities of sporelings. For such transfers, producers empirically wrap sporelings in advance with old newspaper or old rush mats moistened with seawater. Sporelings are not immersed directly in seawater because fisherman have noted that juvenile sporophytes transported in seawater showed poor growth. This phenomenon is supported by the photosynthetic activity of the Wakame juvenile sporophyte stage as measured by the effective quantum yields of photosystem II ($\Delta F/Fm'$, Sato et al. 2022b): the value at 50% humidity decreased markedly 20 min after emersion and was not restored even after subsequent 10-day immersion, but the value at 99% humidity remained at the initial level even after 72 h. To summarize, the thalli were not overly dehydrated under saturated humidity, so photosynthetic activity was maintained for several days, even without immersion in seawater. Furthermore, cultivation experiments showed that sporelings wrapped in a paper towel moistened with seawater exhibited higher growth than those immersed in seawater without aeration, and the difference was more marked at higher temperatures (above 12 °C, Sato et al. 2022b). These results indicate that maintaining moisture levels for this alga at low temperatures is probably essential for successful sporeling transport.

19.3.2 Cultivation

19.3.2.1 Selection of the Aquaculture Site and Aquaculture Substrates

Typical cultivation sites in the Sanriku rias are located near the mouth of the bay and beyond, where the water depth is approximately 30–60 m. Sites with higher current velocities tend to have higher productivity and increased blade thickness, which is one indicator of Wakame quality (Nanba et al. 2011). Therefore, an essential factor in selecting a cultivation site is balancing water current conditions with suitability for installing cultivation facilities.

The deployment synthetic fiber ropes (long lines) of 10 to 18 mm in diameter is the most ubiquitous form of cultivation. This method involves placing cultivation ropes near the sea surface and mooring them at both ends with anchors of several tones made of concrete blocks and other material. The length of the cultivation ropes varies from place to place, with chosen lengths from 50 to 200 m, and around 100 m the most common. Generally, a diameter of 10–18 mm is chosen.

Although, a single cultivation rope is fixed by one set of anchor and mooring ropes (Fig. 19.6) is the typical deployment method, double cultivation ropes are not unusual. With double cultivation ropes, a pair of long lines are placed parallel to each other and fixed by one set of mooring ropes and anchors (Fig. 19.7). Experience has shown that single ropes are stronger against wave action, and Wakame growth is better. However, the double method tends to increase the yield per unit area, so it is often used by groups adopting large-scale aquaculture. In contrast, raft-type facilities are used in relatively calm water areas, such as in the Naruto region, where square sections are set up with ropes of about 100 m, between which cultivation ropes are placed at intervals of about 10 m (Fig. 19.8).

19.3.2.2 Setting of Sporelings

The palm strings of sporelings germinated in the sea are directly wound around the cultivation ropes (Fig. 19.9a). In contrast, the synthetic fiber string of indoor-grown sporel-

Fig. 19.6 Single-line cultivation in the Sanriku region. Rope lengths supporting sporelings have been inserted into the main rope at regular intervals. (**a**) Young sporophytes in the early stages of growth in Matsushima Bay, Miyagi Prefecture, on 19 November 2017. (**b**) Mature sporophytes at the harvesting stage in Hirota Bay, Iwate Prefecture, on 21 April 2021

ings is fixed to the rope by cutting the string into pieces of 3–5 cm and pinching them between twists in the lay of the cultivation ropes or attaching them with staples from a construction industrial gun tacker (Fig. 19.9b,c). Alternatively, a soft rope (such as polystyrene of diameter about 5 mm) is used to insert cut lengths of sporeling string, which is wound around the cultivation rope. In either method, it is important to ensure that the sporeling threads are firmly fixed to the cultivation rope to prevent displacement due to movement or expansion of the rope caused by waves. The method of automatic coiling used in Europe (Mooney-McAuley et al. 2016) has been implemented on a small scale in Akita Prefecture, northwestern Honshu. In South Korea, the process of inserting cut sporelings has been mechanized (https://www.youtube.com/watch?v=c1-9hSYQuVg). Such labor-saving initiatives are also under consideration in Japan.

At the sporeling setting stage of Wakame aquaculture in Japan, the growth of juvenile sporophytes was found to require a particular seawater depth (i.e., 1.5–2 m), with growth decreasing near the seawater surface (0–0.5 m; Saito 1958b). This practical knowledge was supported by experimental findings concerning differences in photosynthetic performance between the different growth stages. During recovery from photoinhibition in low and high Photosynthetically Active Radiation treatments, juvenile sporophytes at low light levels show much better growth than mature sporophytes (Sato et al., 2021b). Given that the cultivation depth is a controllable environmental factor during the aquaculture, it is crucial that 'aquaculture methods must pay attention to water depth in particular' (Saito 1958b). This underscores the significant role of aquaculture research in Wakame cultivation.

19.3.2.3 Maintenance During Aquaculture

After deployment of cultivation ropes, Wakame fishermen periodically survey the ropes and provide maintenance to ensure good yields. Two of the most common maintenance operations are regulating the cultivation density of Wakame and regulating the depth of Wakame growing on the ropes.

Regulation of Cultivation Density Wakame morphogenesis stabilizes when the thallus is around 70 cm in length (Ishikawa 1993). In the Sanriku region, thinning is typically conducted by late January to ensure that the thalli are of uniform size (Sato et al. 2023). Fishermen must decide whether to thin out large, overgrown individuals or small, slowly growing ones. Many of the thinned Wakame are purchased by fishing cooperatives and traders and are often sold in the market as 'sprout Wakame'.

Regulation of Water Level for Cultivation As Wakame grows on the cultivation ropes, the increased biomass tends to cause the cultivation ropes to sink, increasing the cultivation depth. This can lead to reduced light availability and hence reduced yields. Therefore, to maintain the appropriate depth, besides thinning excess Wakame, the number of floats and the length of the rope connecting the floats to the cultivation are adjusted as necessary.

19.3.2.4 Harvest

In Japan, harvests are manual, using one of two main methods. For single long-line cultivation, a ship is positioned parallel to the cultivation rope, and a winch is used to lift the rope up and over the ship, where the fisherman can then harvest the Wakame from the ropes (see Fig. 19.10a). For double long-line cultivation, ships are positioned perpendicular to the two ropes. The ropes are allowed to pass over the ship, and as the ropes travel over the ship, the thalli are harvested (Fig. 19.10b and c). Although these methods can be costly due to the number of personnel required and the use of relatively large vessels, it achieves high productivity. In some

Fig. 19.7 Double-line cultivation in the Sanriku region. (**a**) Maintenance being performed on land, showing the steel pipe positioned between two ropes to maintain a consistent separation. (**b**) In situ in the sea at Ryori Bay, Iwate Prefecture. (**c**) Mature sporophytes in Ryori Bay just before harvesting, on 3 April 2023

areas of the Naruto region, the cultivation ropes are transported to the wharf for harvesting on land, an approach which provides the advantage of safer working conditions, as there is no need for the laborers to board a boat. Trials are being conducted in Iwate Prefecture with the aim of promoting better efficiency in the future.

19.3.3 Boiled and Salted Wakame: Primary Processing

Typical postharvest processing for Wakame is as follows (Fig. 19.11). While nearly all producers carry out this processing independently, large-scale facilities have been estab-

Fig. 19.8 Raft-type cultivation facilities in the Naruto region. (**a**) During the growing period off the coast of Naruto City, on 21 February 2016, showing 10 m aquaculture ropes installed at 2 m intervals. Parts of the rope devoid of Wakame thallus in the foreground have been denuded due to fish-feeding damage. (**b**) Wakame thalli growing well in other parts of the facility at the same time and location. (**c**) Harvesting the Wakame by pulling the rope up over the boat and dropping the thalli on board

lished in certain areas. The principles and basic methods used are consistent, regardless of production scale.

1. *Treatment after harvest*: The thalli are separated into blades with midribs and sporophylls, either on the boat or after landing at the fishing port. The sporophylls are removed from the core and kept frozen.
2. *Heat treatment* (Fig. 19.11a): This step occurs immediately after harvesting, and the Wakame is typically boiled in seawater. The heat treatment temperature is usually above 90 °C, and the duration of boiling is adjusted depending on the size and thickness of the thallus. Overboiling can soften the texture of the blades, while underboiling leaves them partly raw. As the thickness of the blade and midrib can vary significantly, it is crucial to find a balance ensuring that both components are adequately treated while retaining their desired texture. Additionally, continuously using the same seawater to boil harvested Wakame leads to a decrease in the pH of seawater used to boil the Wakame: boiling in low-pH seawater can dull the green color of the boiled Wakame; hence, it is necessary to periodically change the seawater. When changing the water, to prevent large fluctuations in the water temperature used to boil harvested Wakame, fresh seawater is either introduced gradually or is brought up to temperature by preheating in a separate tank.
3. *Cooling* (Fig. 19.11b): After boiling, the Wakame thalli are cooled immediately by pumping large quantities of seawater over the thalli.

Fig. 19.9 Setting Wakame sporelings into the cultivation rope. (**a**) Wound directly around the cultivation rope. (**b**, **c**) Inserting pieces of seedlings into the cultivation rope by twisting open the rope lay at intervals

4. *Salting(Ordinarily)*: After cooling and draining the water, approximately 40% of the weight of the Wakame in salt is sprinkled onto the thalli. A batch concrete mixer or a cylindrical rotating drum is typically used to mix in the salt. The thalli are then placed in tanks for 1 to 2 days, with a significant amount of salt adhering to the thallus surface, which extracts water from the thalli. The excess salt on the surface of the thalli is then transferred to a saturated salt medium, and the Wakame is subsequently moved on to the next stage of processing.

   Salting (high-speed salting method; Fig. 19.11c, d): The ordinary method, which takes two days, requires significant labor when producers move the Wakame in and out of the tank while washing it with the saturated salt medium. To reduce labor during this process, a 'High-Speed Salting Method' was developed in Iwate Prefecture (Onodera et al. 2008), which involves preparing saturated saltwater in a tank with an agitator paddle. Portions of about 10 kg of Wakame are added after boiling and cooling, and the mixture is agitated for about 1 h to salt the Wakame. Producers can then immediately move the treated Wakame to the following dehydrating process. The weight of Wakame per batch and the mesh size of the portion bag are specified by Iwate Prefecture guidelines to ensure that the quality of salted Wakame matches that produced by the ordinary method. Today, this new method enjoys support from various producers, ranging from individuals to companies.
5. *Compressing and draining*: Water is removed from the salted Wakame using a weight or a mechanical press. Boiled and salted Wakame have approximately 50% of their original water content removed by treatment for one or two days. Then, the Wakame is stored in a freezer at −18 °C. Generally, fish and seafood freezers are set to −25 °C in Japan, but it is known empirically that a temperature as low as this leads to lower quality for Wakame, characterized by a poorer texture and brownish color.
6. *Separating blades and stipes*: Each thallus of boiled and salted Wakame is separated manually into blades and stipes (Fig. 19.11e). This process is still carried out manually, even in South Korea and China, whereas other processes are increasingly automated. A few companies in Japan are attempting to develop automated processing equipment. However, due to extensive variation in mor-

Fig. 19.10 Harvesting Wakame in the Sanriku region. (**a**) For the single-rope cultivation method, producers harvest Wakame from the rope hoisted with a port-mounted winch. (**b**) For the double-rope cultivation method, producers harvest Wakame from rope being passed over the boat. (**c**) Producers dropping thalli from the rope into containers on the boat

phology and the thickness of the raw material, even within the same production area, such machines have yet to be used on a wide scale.

7. *Packing*: Blades and stipes are packed separately. In the Sanriku region, the amount per box is 15 kg for cooperative bidding.

19.3.4 Distribution of Boiled and Salted Wakame

1. *Bidding:* In the Sanriku region, boiled and salted Wakame (both the blades and stipes) is typically sold to processors (or buyers) through auctions organized by the fisheries cooperatives in Miyagi and Iwate Prefectures. Wakame producers deliver their boiled and salted product to regional fisheries cooperatives, where inspectors evaluate the quality for each producer. These evaluations are based on criteria such as color tone, hydration level (i.e., water content), and the efficiency of stipe removal. Once assessed, the products are collected at one location in Miyagi Prefecture and two locations in Iwate Prefecture, where approximately ten bidding rounds take place from the end of February to the end of April during the aquaculture season. Buyers, who receive prior approval to bid, inspect the extracted samples and submit their bids.

   This auction method offers the advantage of saving producers time in finding buyers for their products, and it is considered a fair approach due to the quality assessments made by inspectors. However, it also places a significant burden on aquaculture producers, who must manage the entire process from harvesting to separating of blades and stipes before each bidding round. To alleviate this burden, recent years have seen changes in pricing practices. For instance, fisheries cooperatives may allocate quantities to representative buyers at the raw material stage—prior to boiling and salting—with prices determined at the end of the season's bidding rounds. Alternatively, buyers can purchase the raw material before the separation of blades and stipes, allowing producers to carry out the separation process without the pressure of a delivery deadline for the bidding rounds. Regardless of these changes, the number of producers is declining, and the lack of mechanization in the separation process remains a significant challenge. Thus, improvements through such distribution innovations are essential for the future development of the industry.

2. *Direct trade:* In the Naruto region and elsewhere, producers and buyers trade directly without auction procedures.

19.4 Optimum Conditions for Sporeling Production and Maintenance of the Supply System

Recent changes to the marine environment have markedly influenced the production of Wakame sporelings in the sea; and the lack of sporelings is becoming a problem in the Sanriku region. In this situation, the sporeling production technique using free-living gametophytes in Japan (reviewed by Dan et al. 2015) has attracted attention as a stable production method because it can be controlled indoors. The following section presents examples of how to produce industrial-scale sporelings from these free-living gametophytes by optimizing their growth and maturation conditions.

Fig. 19.11 Steps in the primary postharvest process for small-scale Wakame production: (**a**) Boiling the Wakame in seawater at a temperature exceeding 90 °C, using a tank with approximately 1 ton of water. (**b**) Cooling the Wakame by mixing it with seawater pumped from the shore. (**c**) High-speed salting and packing the cooled thallus into mesh bags. (**d**) Churning the mesh bags containing Wakame with saturated seawater. (**e**) Manually separating the blades and midribs before shipping the product to auction

19.4.1 Optimum Conditions of Growth and Maturation for Wakame Gametophytes

In the free-living gametophyte method, gametophytes can act in the same way as 'seeds' for terrestrial plants. Unlike land plants, gametophytes undergo repeated cell division when conditions are favorable. Therefore, increasing the number of cells in the gametophyte, particularly in females, ensures a higher production of egg cells for sporophytes. If both male and female gametophytes can mature with increased cell numbers, a larger quantity of sporophytes can be produced and used as sporelings. For mass production, it is not necessary to optimize individual environmental conditions, such as water temperature and light intensity; instead, it is essential to consider the interactions among these factors.

To identify the optimal conditions for growth and maturation of male and female Wakame gametophytes, Sato et al. (2020) utilized Design of Experiments (DoE) and Response Surface Methodology (RSM). The findings revealed the following:

1. There is an interaction between water temperature and light intensity.
2. The optimum water temperature varies among regional strains, with southern strains generally requiring higher seawater temperatures, which reflects their latitudinal distribution.
3. The response to light hue was similar among regional strains and sexes: green light promoted growth without maturation, red light impeded both growth and maturation, and blue light accelerated both growth and maturation (Fig. 19.12).

Based on these results, stable sporeling production can be achieved by incubating gametophytes under green light at the optimal water temperature and light intensity, which helps increase cell numbers (or 'increase seeds'). Juvenile sporophytes (or 'growth sporelings') can then be germinated under blue light, in conditions set at the optimal seawater temperature and light intensity for maturation, with both male and female gametophytes present. Additionally, male gametophytes exhibit a higher optimum growth temperature than female gametophytes, indicating sexual dimorphism in Wakame gametophytes. This dimorphism may reduce the risk of lost fertilization opportunities for this species, as sporangia germinate in the wild during declining seawater temperatures. By allowing males to grow and mature first to produce sperm, followed by females forming egg cells, the likelihood of successful fertilization is enhanced.

19.4.2 The Maintenance of a Sporeling Production System at an Industrial Scale

The basic design of the sporeling production system using free-living Wakame gametophytes follows that of the sporeling production system for *Saccharina japonica* (Kombu) in Hokkaido and Iwate Prefecture, as follows:

1. A large tank is filled with water, which is circulated by a heat/cool pump chiller system to maintain temperature.
2. Smaller tanks of seawater are placed inside a larger tank, which maintain a constant seawater temperature.
3. A frame with synthetic fiber substrate is placed in each small tank.
4. Seawater with zoospores is poured into each small tank.
5. Sporeling development is induced and controlled by adjusting seawater temperature and light intensity, which germinates gametophytes into sporophytes.

Wakame sporeling production facilities function effectively by integrating a fundamental design with optimal conditions established through DoE and RSM (Sato et al. 2020). The seawater used in this process is filtered and sterilized. Furthermore, a commercially available LED light source

Fig. 19.12 Growth and maturation of Wakame female gametophytes after being raised under (**a**) red, (**b**) green, or (**c**) blue light

with a relatively intense blue spectrum (to accelerate both growth and maturation) was chosen for incubation, which helps to reduce both labor requirements and initial investment costs. Using this system and equipment, an aquaculture company can produce viable sporelings—from the setting of gametophytes to the fiber stage—within 38 days. This process enables an annual supply of approximately 500 t of adequately germinated sporelings as raw material (see Fig. 19.13). The widespread adoption of this system to facilitate stable sporeling production will be crucial for maintaining reliable aquaculture output, especially given the challenges of poor growth rates observed in sporeling production in natural waters due to rising seawater temperatures during summer and autumn.

19.5 From Tacit Knowledge to Formal Knowledge: Initiatives to Improve Productivity

19.5.1 Cultivation Density

Wakame exhibits significant variations in its morphological features based on cultivation density (Sato et al. 2023). At low densities, individuals tend to be wider and larger, featuring shorter stipes and larger sporophylls. Conversely, at higher cultivation densities, the individuals become narrower, with longer stipes and smaller sporophylls (Fig. 19.14). Producers in the Sanriku region have developed an intuitive understanding of the cultivation densities that yield the best

Fig. 19.13 Experimental Wakame sporeling production system operated by optimizing conditions during the gametophyte stage. Key components of this system include: (**a**) Large tanks filled with seawater, equipped with a heat-pump chiller to maintain the optimal temperature. Inside the large tank in the foreground can be seen smaller tanks (100 liters each) for different experiments at the same temperature. LED lights are commercially available units that emit relatively high-intensity blue and low-intensity red wavelengths. (**b**) Sporeling strings wound around PVC frames. (**c**) Close-up of (b). (**d**) Sporelings undergoing acclimation at the cultivation site: upper row, after five days acclimation; middle, after 10 days; bottom, after 15 days, from which they are ready for further development

Fig. 19.14 Photographs of representative Wakame sporophytes cultivated at different densities per 1 m of cultivation rope: (**a**) 10; (**b**) 30; (**c**) 50; (**d**) 80; (**e**) 100; (**f**) 120; (**g**) 200. Scale bar 100 cm

quality and quantity at their specific sites. To quantify this rule of thumb, the profitability of Wakame in Okirai Bay (Ofunato City, Iwate Prefecture) was assessed by adjusting the Wakame cultivation density during the early growth stages and analyzing the relationship between density and profitability from selling boiled salted products.

The results revealed that without density adjustments, the maximum number of individuals per meter (a total yield of over 200 individuals per meter of cultivation rope, as shown in Fig. 19.14g) was recorded in the test area. However, the highest profitability was observed at a density of approximately 80 to 120 individuals per meter (Fig. 19.14d–f). This discrepancy arises from the labor-intensive process of separating blades from their stipe. As this separation is performed manually, the time required for each individual remains consistent regardless of size. Smaller thalli can be more difficult to handle and are more prone to tearing, leading to increased time and effort in processing. Thus, the notion that greater quantity translates into higher profitability does not hold true. Furthermore, during density adjustments, many fishermen work to remove excessively large or small individuals, resulting in a more uniform morphology. This homogeneity reduces inconsistencies during postharvest boiling and salting, which enhances the efficiency of the separating process. Properly adjusting the density proves to be a highly beneficial practice, effectively achieving multiple advantages.

19.5.2 Expanding Harvesting Period Through the Use of Different Cultivation Strains

To maximize productivity with limited labor, it is essential to find ways to minimize the workload during cultivation processes such as setting sporelings and harvesting. Many producers in the Sanriku region address this challenge by utilizing multiple strains with different growth and maturation characteristics. For instance, producers in Utatsu Bay, Miyagi Prefecture, harvest early maturing strains from southern regions, such as those sourced from Naruto, in February, while late-maturing strains sourced from Iwate Prefecture are harvested in April (Kusaka et al. 2007; Gao et al. 2013). Additionally, an intermediate strain, which matures between these two periods, is also selected to enable continuous harvesting from February to April (Fig. 19.15). As water temperatures continue to rise with the trend of global warming, the window of conditions suitable for aquaculture in each region is becoming shorter. This situation highlights an increasing need for further breeding of strains that grow relatively quickly or exhibit distinct characteristics, such as either early or late maturation. To facilitate this breeding, researchers have developed a novel tank system, which can simulate the current velocity found at aquaculture sites in the Sanriku region (Fig. 19.16, Sato et al. 2017a, b, 2021a). This system enables 'common garden experiments' to determine whether genetic or environmental factors influence these traits.

19.5.3 Initiatives for Intensive Large-Scale Aquaculture (Fig. 19.17)

Wakame aquaculture in South Korea and China is largely conducted on an industrial scale, focusing on mass production. In contrast, Wakame aquaculture in Japan is conducted on a much smaller scale, is typically privately managed and (due to a decline in the number of fisherman) its productivity is comparatively low. Therefore, a paradigm shift in Wakame aquaculture in Japan is needed and a more efficient and productive system must be implemented. One example of increasing the productivity of Wakame production is the collaboration among a fishing cooperative, a fishing company, and a food company in Ryori Bay, Ofunato City, Iwate Prefecture.

Ryori Bay has historically provided suitable environmental conditions for cultivating high-quality Wakame. The intuition and expertise in carefully and skillfully boiling and salting Wakame to maintain a high-quality product have been passed down through generations of local fishermen. However, the number of active fishermen has decreased significantly. As of 2024, out of the 2100 approved cultivation units (each consisting of 100 m of rope), approximately 500 units remain vacant. In Japan, fishery cooperatives lease cultivation sites to fishermen, earning income from commissions levied on the Wakame produced. The growing number of idle units is also placing financial pressure on the management of these cooperatives. Therefore, there is a desire to minimize the number of idle units.

One initiative that is underway at Ofunato City involves the Marukatsu Fisheries Corporation and Riken Food Co., Ltd. Marukatsu Fisheries Cooperation, based at Ryori Port, has traditionally focused on fishing for salmon and saury. However, in recent years, their catch has declined significantly due to rising seawater temperatures and shifts in fish populations. Consequently, they are considering transitioning from a catch-only operation to aquaculture. Riken Food Co., Ltd., a key buyer of Wakame in the Sanriku region, has been providing processing technology for boiled and salted Wakame to producers and fishing cooperatives since the inception of Wakame aquaculture in Sanriku. This support also includes lending extensive storage facilities for boiled and salted Wakame to several fishing cooperatives in the region, aimed at enhancing production capacity.

The collaborative efforts of these two companies and Ryori Fisheries Cooperative began in 2020, driven by their shared objectives to implement large-scale aquaculture. Initially, Marukatsu Fisheries Co. adopted technology developed in Miyagi Prefecture to implement a double-line aquaculture facility tailored for large-scale operations (Figs. 19.7 and 19.10b,c). They also constructed a boat specifically

Fig. 19.15 The extended harvesting period achieved by utilizing various cultivation strains. The early maturing, middle-maturing, and late-maturing strains, selected from regional varieties, are harvested at different times: the early strains at the end of February to the beginning of March, the middle strains from mid-March to the end of March, and the late strains in April. The yield figures displayed below the photographs represent the harvesting yield of each strain from Utatsu Bay in the Sanriku region for the year 2021

Fig. 19.16 Wakame cultivation using a novel tank system: the circulation and floating culture system (CFCS) for breeding experiments

designed for this method. This was facilitated by members of the Ryori Fisheries Cooperative, who arranged the aquaculture anchoring points to maximize the installation of double facilities, creating the first instance of this type of system in the area. Additionally, Riken Food Co., Ltd. conducted trials to cultivate several strains of Wakame with different growth characteristics, which when used in combination can ensure an extended harvesting period. They also provided loans for boiling and salting production facilities and offered technical guidance by quality control operators. As a result of these initiatives, the harvests increased substantially, yielding 40 tons in 2022 and 130 tons in 2023. Despite the overall reduction of Wakame production in the Sanriku region due to unusually high seawater temperatures in the 2023–2024 period, the Ryori Bay collaborative group managed to harvest 130 tons, maintaining the production levels from the previous year. This initiative is therefore poised to serve as a successful model for large-scale aquaculture production across the country.

Although there are considerable challenges in changing traditional practices, the push for intensive production is vital for advancing and sustaining aquaculture into the future and advancing the Wakame industry in Japan. It may also offer valuable insights for similar efforts in other nations.

19.6 Efforts to Harmonize Food Production with CO_2 Fixation Through theAquaculture of Macroalgae

In recent years, there has been a growing global interest in the CO_2-fixing capacity of seaweed (Duarte and Cebrián 1996; Froehlich et al. 2019). Various initiatives have been proposed to enhance CO_2 fixation by exporting seaweed sourced from natural beds or through large-scale seaweed farming to the deep ocean (Filbee-Dexter et al. 2023). However, it is important to carefully examine the ecological, economic, and sociological risks associated with the rapid

Fig. 19.17 Facilities for large-scale intensive aquaculture in Ryori Bay, in the Sanriku region. (**a**) Customized fishing boat designed for the double-rope type cultivation method, featuring two pulleys (blue-arrows) that accommodate the spacing of the double ropes. The boat has a broad expanse of deck space to hold containers during harvesting. (**b**) Conveyor belts transporting raw Wakame up and into the boiling pot. (**c**) Boiling pot (red-arrow) alongside two cooling tanks (blue-arrows), which are connected by a conveyor. (**d**) Tanks equipped with paddles designed for high-speed salting

and large-scale expansion of seaweed beds and farming (Gallagher et al. 2022; Fujita et al. 2023).

Comparisons of the net ecological productivity of natural seaweed beds versus cultivated seaweed ecosystems have shown significant variability among species, regions, and time of year (Sato et al. 2022a). In Japan, the J-Blue Credit system was launched much earlier than in many other countries. This is a system that allows for the assessment and trading of carbon credits based on the CO_2 fixation capacity of seagrass and seaweed beds (Kuwae et al. 2022); an initiative that has generated considerable interest from fisheries producers and local authorities. To achieve carbon neutrality by 2050, there is an urgent need for efforts to reduce CO_2 emissions, alongside initiatives and technological innovations that utilize biological resources for CO_2 absorption. Given the burgeoning interest in the blue carbon potential of macroalgae, it is essential to deepen our understanding by conducting more research to strengthen the scientific foundation of the blue carbon effects of macroalgae (Hurd et al. 2022).

Traditionally, seaweed has been utilized primarily as food in Japan, making it challenging to embrace the concept of exporting large quantities of otherwise edible seaweed into the deep ocean for CO_2 fixation. The optimal approach would be to ensure that seaweed farming remains as a source of food, while simultaneously achieving blue carbon effects during seaweed production. Degradation resistant polysaccharides produced by many seaweeds are expected to play a crucial role in this process (Salmeán et al. 2022). Large brown algae (kelp) such as Wakame and Kombu (*Saccharina*) are rich in degradation resistant polysaccharides, such as fucoidan. Research conducted on kelp beds in the Baltic Sea suggests that fucoidan from kelps may remain embedded in the seafloor sediment for over a century (Salmeán et al. 2022). Consider that kelp exhibits an intergrowth pattern, where cells divide at the meristem and the apical part erodes from the early growth stages (Mizuta et al. 1992, 1994; Li et al. 2007, 2009; Sato and Agatsuma 2016). Therefore, by quantifying the amount of degradation resistant polysaccharides released into the seawater by kelp during growth, it is possible to improve the quantitative assessment of CO_2 fixation capacity of seaweed farms. Perhaps it is possible that seaweed farming can provide both food and a place to capture and sequester CO_2 simultaneously. However, very little is known about the viability of this approach, especially with regards to the types of degradation resistant polysaccharides, the quantity of these compounds, and the time-scale of sequestration and degradation. Such knowledge can enhance our understanding of the dual roles that seaweed farming and seaweed beds can play in contributing to environmental sustainability.

19.7 Conclusion

Almost all Wakame producers in Japan view the entire process of production—from cultivating sporelings to shipping boiled and salted products—as a single farming operation. They have accumulated extensive knowledge and intuition at every stage of this process because Wakame has been a staple food and a source of livelihood since ancient times. However, there is a pressing need to enhance efficiency and scale-up production and include more support from corporations. Wakame aquaculture currently faces significant challenges due to environmental changes and a decline in the number of producers. In the future, it will be essential to adopt some of the mass production systems utilized by countries such as South Korea and China. This can be extended to other seaweed cultivation systems in Japan, such as those for Nori (*Pyropia*) and Kombu.

The title of this chapter was inspired by the teachings of Confucius, conveying the idea that learning from the wisdom of our predecessors can provide insights into current issues and pave the way for the future. The wealth of knowledge held by Wakame producers will serve as the foundation for the future development of the seaweed aquaculture industry. However, this traditional wisdom must be complemented by advances based on scientific experiments and technology. Thus, researchers must move forward and focus on the issues outlined in this chapter to ensure the success of this industry. The concept of 'Phyconomy', which was proposed by Dr. Alan Critchley, Dr. Anicia Hurtado, and Dr. Iain Neish and other contributors in this book, represents an academic field where traditional and region-specific knowledge of the seaweed industry can be gathered and developed into a new industry. Creating a sustainable seaweed industry that benefits people, various plant and animal species, and the global environment is the vital role that 'Phyconomy' should fulfill around the world.

Acknowledgments I would like to express my sincere gratitude to Dr. Yukio Agatsuma, Dr. Gregory N. Nishihara, and Dr. Ian G. Gleadall for their many helpful suggestions and comments in preparing this manuscript. I would also like to express my gratitude to the following authorities for their collective support in this collation of the history of and information about Wakame aquaculture and processing: Hironobu Watanabe, Naotaka Noda, Ryoichi Yamanaka, Junichi Sato, Katsunori Ono, Masakazu Honda, and Takeshi Suzuki of Riken Food Co., Ltd.; all staff at the Algal Research Innovation Centre of Riken Food Co. Ltd.; Yutaka Ishikawa, Tadashi Takahashi, Takayuki Saido, and Takayuki Fujiwara of the Iwate Prefectural government office; and Miho Honjo and Michiko Nagaki of Miyagi Prefectural government office. In addition, I thank the Fisheries Cooperatives and fishermen who have always given us much guidance on Wakame aquaculture: Miyagi Fisheries Association, Hirota-Bay Fisheries Association, Ryori Fisheries Association, Fudai Fisheries Association, Hiroshi Akama (Seafoods Akama Co., Ltd.), Fumihiro Oikawa, Hiromi Oikawa (Miyagi Fisheries Association), Osamu Hatakeyama (Hirota-Bay Fisheries Association), and Akio Sasaki (Marukatsu Fisheries Cooperation).

Funding Support The writing of this chapter was supported in part by grants F-REI (JPFR23-03-01-02 and JPFR 24-03-01-02) and JST-CREST (JPMJCR24J5).

References

Castric-Fey A, Beaupoil C, Bouchain J, Pradier E, L'Hardy-Halos MT (1999) The introduced alga *Undaria pinnatifida* (Laminariales, Alariaceae) in the rocky shore ecosystem of the St. Malo area: growth rates and longevity of the sporophyte. Bot Mar 42:83–96

Dan A, Ohno M, Matsuoka M (2015) Changes of the research and development on the resources of *Undaria* and *Laminaria* in the culture ground of Tokushima coasts. Bull Tokushima Pref Fish Res Ins 10:25–48. (in Japanese)

Dellatorre FG, Amoroso RO, Saravia J, Orensanz JM (2014) Rapid expansion and potential range of the invasive kelp *Undaria pinnatifida* in the southwestern Atlantic. Aquat Invasions 9:467–478

Duarte CM, Cebrián J (1996) The fate of marine autotrophic production. Limnol Oceanogr 41:1758–1766. https://doi.org/10.4319/lo.1996.41.8.1758

Filbee-Dexter K, Pessarrodona A, Duarte CM, Krause-Jensen D, Hancke K, Smale D, Wernberg T (2023) Seaweed forests are carbon sinks that may help mitigate CO_2 emissions: a comment on Gallagher et al. (2022). ICES J Mar Sci 0:1–6. https://doi.org/10.1093/icesjms/fsad107

Froehlich HE, Afflerbach JC, Frazier M, Halpern BS (2019) Blue growth potential to mitigate climate change through seaweed offsetting. Curr Biol 29:3087–3093.e3. https://doi.org/10.1016/j.cub.2019.07.041

Fujita R, Brittingham P, Cao L, Froehlich H, Thompson M, Voorhees T (2023) Toward an environmentally responsible offshore aquaculture industry in the United States: ecological risks, remedies, and knowledge gaps. Mar Pol 147:105351

Funaki M, Otsuki K, Ohno M (2021) The early cultivation of *Saccharina japonica* var. *japonica* and *Undaria pinnatifida* in China and Japan by Mr. Yoshiro Otsuki and its evaluation. Algal Res 14:29–47

Gallagher JB, Shelamoff V, Layton C (2022) Seaweed ecosystems may not mitigate CO_2 emissions. ICES J Mar Sci 0:1–8

Gao X, Endo H, Taniguchi K, Agatsuma Y (2013) Genetic differentiation of high-temperature tolerance in the kelp *Undaria pinnatifida* sporophytes from geographically separated populations along the Pacific coast of Japan. J Appl Phycol 25:567–574

Hay CH, Luckens PA (1987) The Asian kelp *Undaria pinnatifida* (Phaeophyta: Laminariales) found in a New Zealand harbour. NZ J Bot 25:364–366

Hesitt CL, Campbell MLC, McEnnulty F, Moore KM, Murfet NB, Robertson B, Schaffelke B (2005) Efficacy of physical removal of a marine pest: the introduced kelp *Undaria pinnatifida* in a Tasmanian marine reserve. Biol Invas 7:251–263

Hurd CL, Law CS, Bach LT, Britton D, Hovenden M, Paine ER, Raven JA, Tasmitt V, Boyd PW (2022) Forensic carbon accounting: assessing the role of seaweeds for carbon sequestration. J Phycol 58:347–363

Ishikawa Y (1993) A simple method for growth estimation of blades in *Undaria pinnatifida*. Nippon Suisan Gakkaishi 59:1331–1336. (in Japanese with English abstract)

Japan Fisheries Cooperative (2024) Annual report of Wakame in the Sanriku region. URL: https:// www. zengy oren. or. jp/

Kaga S, Kakehi S, Naiki K, Kodama T, Wagawa T, Segawa S, Watanabe S, Musashi T, Kuroda H, Ito S (2022) Seasonal variations in nutrient concentrations in Sanriku coastal waters, Japan: effects on *Undaria pinnatifida* (Laminariales; Phaeophyta) seaweed farms. Reg Stud Mar Sci 54:102484

Kusaka K, Sasaki R, Tsukada T (2007) Growth and morphological characteristics of native Wakame (*Undaria pinnatifida*) seed collected from eight locations around Japan coast. Miyagi Pref Rep Fish Sci 7:17–28. (in Japanese)

Kuwae T, Watanabe A, Yoshihara S, Suehiro F, Sugimura Y (2022) Implementation of blue carbon offset crediting for seagrass meadows, macroalgal beds, and macroalgae farming in Japan. Mar Poll 138:104996

Li J-Y, Murauchi Y, Ichinomiya M, Agatsuma Y, Taniguchi K (2007) Seasonal changes in photosynthesis and nutrient uptake in *Laminaria japonica* (Laminariaceae; Phaeophyta). Aquacult Sci 55:587–597

Li J-Y, Agatsuma Y, Nagai T, Sato Y, Taniguchi K (2009) Differences in resource storage pattern between *Laminaria longissima* and *Laminaria diabolica* (Laminariaceae; Phaeophyta) reflecting their morphological characteristics. J Appl Phycol 21:215–224

Miyashita A (1974) Kaisou. Hosei University Press

Mizuta H, Maita Y, Yanada M (1992) Seasonal changes of nitrogen metabolism in the sporophyte of *Laminaria japonica* (Phaeophyceae). Nippon Suisan Gakkaishi 58:2345–2350

Mizuta H, Maita Y, Kuwada K (1994) Nitrogen recycling mechanism within the thallus of *Laminaria japonica* (Phaeophyceae) under the nitrogen limitation. Fis Sci 60:763–767

Mooney-McAuley KM, Edwards MD, Champenois J, Gorman E (2016) Best practice guidelines for seaweed cultivation and analysis, public output report of the EnAlgae project, Swansea, June 2016, 36pp, www.enalgae.eu

Nanba N, Fujiwara T, Kuwano K, Ishikawa Y, Ogawa H, Kado R (2011) Effect of water flow velocity on growth and morphology of cultured Undaria pinnatifida sporophytes (Laminariales, Phaeophyceae) in Okirai Bay on the Sanriku coast, Northeast Japan. J Appl Phycol 23:1023–1030

Onodera M, Yoshie-Stark Y, Suzuki T (2008) Development of high-speed salting method for boiled and salted *Undaria pinnatifida* and *Laminaria religiosa*. Nippon Suisan Gakkaishi 74:671–677

Saito Y (1958a) An ecological study of *Undaria pinnatifida* Sur. -III. On the effects of light-intensity and water-temperature upon the rate of photosynthesis-1. Nippon Suisan Gakkaishi 24:484–486. (in Japanese)

Saito Y (1958b) Cultivation of *Undaria pinnatifida*. Aquacult Sci 5:16–20

Salmeán AA, Willats WGT, Ribeiro S, Anderson TJ, Ellegaard M (2022) Over 100-year preservation and temporal fluctuations of cell wall polysaccharides in marine sediments. Front Mar Sci 13. https://doi.org/10.3389/fpls.2022.785902

Sato Y, Agatsuma Y (2016) Resource accumulation of the kelp *Saccharina ochotensis* based on photosynthetic rate and specific nutrient uptake kinetics. J Appl Phycol 28:499–509

Sato Y, Hirano T, Ichida H et al (2017a) Morphological and physiological characterization of *Undaria pinnatifida* originating from five cultivation sites in Japan through common garden experiments using a tank culture system. J Appl Phycol 29:2287–2295

Sato Y, Yamaguchi M, Hirano T, Fukunishi N, Abe T, Kawano S (2017b) Effect of water velocity on *Undaria pinnatifida* and *Saccharina japonica* growth in a novel tank system designed for macroalgae cultivation. J Appl Phycol 29:1429–1436

Sato Y, Endo H, Oikawa H, Kanematsu K, Naka H, Mogamiya M, Kawano S, Kazama Y (2020) Sexual difference in the optimum environmental conditions for growth and maturation of the brown alga *Undaria pinnatifida* in the gametophyte stage. Genes 11:944

Sato Y, Hirano T, Ichida H, Fukunishi N, Abe T, Kawano S (2021a) Extending the cultivation period of *Undaria pinnatifida* by using regional strains with phenotypic differentiation along the Sanriku Coast in Northern Japan. Phycology 1:129–142

Sato Y, Nagoe H, Nishihara GN, Terada R (2021b) The photosynthetic response of cultivated juvenile and mature *Undaria pinnatifida* (Laminariales) sporophytes to light and temperature. J Appl Phycol 33:3437–3448

Sato Y, Nishihara GN, Tanaka A, Belleza DFC, Kawate A, Inoue Y, Hinode K, Matsuda Y, Tanimae S, Tozaki K, Terada R, Endo H (2022a) Variability in the net ecosystem productivity (NEP) of seaweed farms. Front Mar Sci 9:861932

Sato Y, Saito D, Nishihara GN, Terada R (2022b) Effects of different stressors on the PSII photochemical efficiency and application to sporeling transportation in cultured young sporophytes of *Undaria pinnatifida*. J Appl Phycol 34:551–563

Sato Y, Fujiwara T, Endo H (2023) Density regulation of aquaculture production and its effects on commercial profit and quality as food in the cosmopolitan edible seaweed *Undaria pinnatifida*. Front Mar Sci 10. https://doi.org/10.3389/fmars.2023.10850

Silva PC, Woodfield RA, Cohen AN, Harris LH, Goddard JHR (2002) First report of the Asian kelp *Undaria pinnatifida* in the northeastern Pacific Ocean. Bio Invasions 4:333–338

Stenton-Dozey JME, Heath P, Ren JS, Zamora LN (2021) New Zealand aquaculture industry: research, opportunities and constraints for integrative multitrophic farming. N Z J Mar Freshw Res 552:265–285. https://doi.org/10.1080/00288330.2020.1752266

Wotton DM, O'Brien C, Stuart MD, Fergus DJ (2004) Eradication success down under: heat treatment of a sunken trawler to kill the invasive seaweed *Undaria pinnatifida*. Mar Poll Bull 49:844–849

Yamanaka R, Akiyama K (1993) Cultivation and utilization of *Undaria pinnatifida* (wakame) as food. J Appl Phycol 5:249–253

Yoichi Sato has been the Director of Riken Food Co., Ltd., Miyagi, Japan, since 2024, and manager of the research institute Algal Research and Innovation Centre (ARIC), since 2021. He has worked at Riken Food since 2005. Before 2005, he was an officer in the Ministry of Agriculture, Forestry and Fisheries (2001–2002) and an assistant professor at the Graduate School of Agricultural Science Faculty of Agriculture, Tohoku University (2002–2004). His main interests are the physiology and ecology of macroalgae, predominantly brown seaweeds (Wakame, Kombu, Mozuku). His efforts have elucidated the plasticity and adaptability of seaweeds to environmental changes and developed and improved cultivation technology. Additional fields of interest include developing the optimum processing methods of seaweeds suitable for markets and customer demands. He obtained Ph.D. (Life Science) at the University of Tokyo in 2016.

20 Planning for an Expanding Industry: Opportunities for a Successful Kelp Economy in British Columbia

Sarah B. Gutzmann
and Andrés M. Cisneros-Montemayor

Abstract

Ocean industries must be grounded within environmental and social boundaries to meet the goals of a Blue Economy, which requires economic and social equity frameworks to transform the status quo. Kelp economies have been increasingly emerging worldwide, particularly with respect to aquaculture, harvest, and restoration operations, and provide an opportunity to use regenerative development as a framework for transformation. As a foundational species with numerous market applications, kelps can provide social-ecological benefits with the economic potential to meet several sustainable development goals. However, social, ecological, and economic objectives may conflict within siloed regulatory processes. Using British Columbia—where the kelp sector is rapidly expanding—as an example, we use regenerative development principles to ground the kelp sector socially and ecologically in place. This involves taking inspiration from Indigenous teachings, systems thinking, and the concept and critiques of planetary boundaries to illustrate a boundary-enclosed spectrum of kelp–human interactions that make up the kelp sector. This spectrum considers limits to growth both environmentally for kelp itself and as defined by human values and regulation, which is important in equitable governance decision-making. Finally, we provide recommendations for future research to progress towards a regenerative, blue kelp economy. Overall, this work provides an alternative cognitive framework to inspire integrated, regenerative management and governance for the kelp sector in BC and comparable social-ecological systems.

Keywords

Blue economy · Regenerative development · Planetary boundaries · Values-based governance · Systems thinking

20.1 Introduction

Although current economic systems that were built in search of unceasing growth have brought many benefits to humanity, they are ultimately poorly equipped to handle, and have indeed led to, pivotal social-ecological crises such as biodiversity loss, wealth inequalities, and climate change (Huntjens 2021; Shannon et al. 2022). Moreover, while the window of opportunity for climate resilient development shrinks in step with the severity of anthropogenic climate change (Singh et al. 2019, Calvin et al. 2023), societies must reconcile equity and economic development needs with diverse worldviews, governance, and ways of knowing across cultural landscapes. As such, it is imperative that emerging resource sectors and associated resource management systems are situated in this intersection of climate (environmental) and societal factors and promote a more sustainable and equitable economy. This is particularly true in the ocean given its high interconnectedness and need for policy cooperation (High Level Panel for a Sustainable Ocean Economy 2022).

Seaweed is an emerging aquaculture sector that has the potential to restore social-ecological systems (The Nature Conservancy 2021). Seaweed (macroalgae) farming is the fastest growing aquaculture industry in the world. In 2022, global seaweed aquaculture produced ~36.5 million t of product, equal to ~27.8% of global aquaculture production (~130.9 million t), of which the vast majority was produced in Asia (97–99%) (FAO 2024; Mouga 2024). Globally, seaweed aquaculture has an estimated 48 million km^2 of suitable area (Froehlich et al. 2019) and at present is primarily farmed by nearshore operations (<3 nm from shore) (Bermejo et al. 2022; Fricke et al. 2024), although interest in offshore

The original version of the chapter has been revised. A correction to this chapter can be found at https://doi.org/10.1007/978-3-032-02955-3_29

S. B. Gutzmann (✉) · A. M. Cisneros-Montemayor
Simon Fraser University, School of Resource and Environmental Management, 8888 University Drive, Burnaby, BC, Canada
e-mail: sarah_gutzmann@sfu.ca

A. T. Critchley et al. (eds.), *Advances in Temperate Phyconomy: Algal Harvest and Cultivation in Globally Distributed Temperate Waters*, Developments in Applied Phycology 15, https://doi.org/10.1007/978-3-032-02955-3_20

aquaculture is increasing (Visch et al. 2023; Fricke et al. 2024).

Kelps are brown algae in the order Laminariales and considered some of the most promising seaweeds for regenerative industries (further discussed below). Kelp forests are found from polar through temperate regions, providing unique and ecologically invaluable habitat for fish, marine mammals, and invertebrates (United Nations Environment Programme 2023). Moreover, species of kelp have been stewarded by people for food, materials, and as a cultural relation for millennia (Turner and Bell 1973; Turner 2000; Erlandson et al. 2007; Dillehay et al. 2008; Hale Hendlin et al. 2024). Presently, kelps have diverse market applications, such as medicine and cosmetics, fertilizers, and biotechnology applications (e.g., plastic alternatives), as well as being eaten directly by humans and livestock (Sultana et al. 2023).

More recently, kelps have attracted interest because they sequester carbon in their biomass. Although it is unlikely that much kelp-carbon is sequestered at a climate relevant timescale (e.g., in deep ocean sediments) (Krause-Jensen and Duarte 2016; Filbee-Dexter and Wernberg 2020; Fisheries and Oceans Canada 2022; Kelly et al. 2023; Filbee-Dexter et al. 2024), downstream applications for carbon sequestration and/or emissions avoidance have been gaining attention, creating synergies between kelp industires and other industrial sectors. For example, kelp can potentially be used for carbon sequestration through the direct sinking of biomass (Ross et al. 2022; Fujita et al. 2023), or sequestration in building materials (Schikan and Gwóźdź 2022), and for emission avoidance as biofuel (Gao et al. 2020), a synthetic fertilizer replacement (Duarte et al. 2017), and in applications for shipping vessel carbon capture (Syrgkani 2024). Thus, kelp aquaculture is heralded as being environmentally low impact (e.g., it requires no land-base or fertilizer input), an enabler of other sectors, and able to enhance habitat, fisheries productivity, carbon capture, and the livelihoods of coastal communities worldwide.

Despite these potential benefits, the combined challenges of simultaneously managing for climate change, environmental health, and social equity are clearly seen in ocean-based industries such as kelp aquaculture. Therefore, we suggest that the economic frameworks of a Blue Economy and "regenerative development and design" are suited for developing kelp aquaculture and navigating multiple social-ecological crises. Unlike other ocean development frameworks, a Blue Economy explicitly calls for social equity goals in addition to environmental sustainability (Cisneros-Montemayor et al. 2022). Likewise, regenerative development and design paradigms focus on system transformation to restore and renew human and ecological health and relationships, emphasizing social justice and a fair distribution of benefits (Mang et al. 2016; Shannon et al. 2022). These types of approaches were developed due to evidence of poor ecological and social outcomes from a narrow focus on economic growth (Ota et al. 2022), but existing businesses and economies have struggled to adopt or translate blue/regenerative economic theory into practice.

Kelp industries have the potential to be an example of a Blue/regenerative economic sector, in part because they can meet multidisciplinary development goals (Khan et al. 2024) which may be a bridge to regenerative development (Huntjens 2021). For example, kelp industries fit within 7 of 10 focus areas for the United Nations Decade of Ocean Science for Sustainable Development 2021–2030 (the Ocean Decade, which is a targeted, global effort at progressing towards SDG14 through these 10 focus areas known as challenges; Table 20.1). To explore this regenerative potential, this chapter provides an overview of the kelp economic system developing in British Columbia, Canada, and discusses the potential opportunities and challenges for this sector within an equitable and sustainable Blue Economy.

Table 20.1 The seaweed/kelp sector addresses 7/10 of the Ocean Decade's challenges

Decade challenge	Potential role of seaweed sector
Challenge 1: Understand and beat marine pollution	Nutrient absorption (e.g., Khan et al. 2024)
Challenge 2: Protect and restore ecosystems and biodiversity	Restoration (Eger et al. 2020, 2022; Filbee-Dexter et al. 2022)
Challenge 3: Sustainably nourish the global population	Agricultural biostimulants, livestock feed/supplements, direct consumption (e.g., Ali et al. 2021; Kelly et al. 2023; Singh et al. 2018; Yakhin et al. 2017)
Challenge 4: Develop a sustainable, resilient and equitable ocean economy	Discussed herein
Challenge 5: Unlock ocean-based solutions to climate change	Carbon cycling, hypoxia and ocean acidification at a local scale (Evans et al. 2023; Kelly et al. 2023; Sultana et al. 2023), shipping carbon capture (Syrgkani 2024)
Challenge 6: Increase community resilience to ocean and coastal risks	Sea-level rise adaptation (e.g., Sultana et al. 2023)
Challenge 10: Restore society's relationship with the ocean	E.g., tourism, education, cultural knowledge and practise, inspiration, etc. (Pecquet et al. 2022; Gutzmann 2023; Cotas et al. 2023; Fricke et al. 2024)

20.2 British Columbia Context

British Columbia (BC) has a strong coastal economy and a budding kelp sector. Three-quarters of BC's population live in coastal areas, and the ocean economy is worth more than $21 billion ($CAD 2021), making the development of guidelines in support of a Blue Economy a priority (Government of BC 2024). Although kelps have been stewarded by Indigenous People for at least 14,000 years (Turner 2020), the commercialized kelp sector in BC is quite young with some of the largest kelp cultivation operations having started between 2017 and 2019. However, BC's kelp sector is expanding; for instance, harvests from BC's largest commercial producer alone were expected to grow from roughly ~50,000 kg in 2022 to an estimated ~200,000 kg in 2023 (Jones 2023). The sector also includes both small- and large-scale hand-harvest (i.e., kelp is cut by hand) of wild kelp, where the largest harvest averages at ~7000 kg per working day from May to October (Allchurch 2023). Finally, several organizations are focused on kelp restoration and/or kelp propagation.

The potential role of a kelp sector in BC sits within the broader context for kelp and Blue Economy planning in Canada, which is in its initial stages. Canada has made international commitments to the 2030 Agenda for Sustainable Development, Paris Agreement on Climate Change, Kunming-Montreal Global Biodiversity Framework, and High-Level Panel for a Sustainable Ocean Economy. Alongside these goals, Canada is in the process of developing a blue economic strategy and through consultations has seen an expressed interest in kelp and seaweed industries (Fisheries and Oceans Canada 2022). During Blue Economy engagement, Canadians expressed the importance of ocean sectors which prioritize healthy environments, are aligned with climate policy, are fully valued (i.e., more than via economic/market values), are collaborative, equitable, and have adequate, accessible funding (Fisheries and Oceans Canada 2022). Yet, key challenges identified in current ocean sectors included difficulties in accessing capital, poor government coordination, and outdated policies, as well as the social-ecological impacts of climate change and unsustainable practices (Fisheries and Oceans Canada 2022).

20.2.1 Environmental Context

British Columbia is environmentally well suited to support a kelp sector. The BC coast spans more than 26,000 km within the Cold Temperate Northeast Pacific Ocean (US Geological Survey 2016; Government of BC 2024). Overall, the abundance of brown algae in the order Laminariales in BC is predicted to fair relatively well under Shared Socioeconomic Pathway (SSP) scenario SSP3–7 (moderate-high climate change); however, species composition turnover is likely with the percentage of species richness decreasing in some areas (Assis et al. 2024; Fig. 20.1a–c). For kelp (Laminariales) specifically, BC supports an impressive diversity with at least 30 species (Bates 2004; Druehl and Clarkston 2016), including the iconic canopy-forming bull kelp (*Nereocystis luetkeana*) and giant kelp (*Macrocystis pyrifera).*

Yet, as interest and the number of entities in the kelp sector expand, there is a greater potential to perpetuate and/or exacerbate injustices like "ocean grabbing" which occurs when ocean-based actions, policies, or initiatives dispossess or further disadvantage marginalized groups, such as Indigenous Peoples (Bennett et al. 2015, 2021). Moreover, kelp cultivation may alter local ecosystems via changes in water quality, genetic material, biological communities (Gruenthal and Habicht 2022), pathogens, and/or invasive species (Behera et al. 2022; Sugumaran et al. 2022; Visch et al. 2023). Further, the development of kelp cultivation and harvest infrastructure may introduce macro- and microplastic pollution (Lusher et al. 2017) and additional vessel traffic (e.g., noise), as well as impacts from the land-based development required to support the sector (e.g., processing; Campbell et al. 2019; Grebe et al. 2019; Visch et al. 2020; Bhuyan 2023; Waqas et al. 2024). Although additional monitoring and study is needed to determine the scale at which kelp industries may contribute to these stressors, potential impacts highlight the need for appropriate policy and legislation within the sector, particularly as it scales up.

20.2.2 Regulatory Context

In BC, kelp harvest and cultivation activities are regulated by the aquaculture team of the BC Ministry of Water, Land, and Resource Stewardship (WLRS) under the Fish and Seafood Act, where the possession of >100 kg wet weight or (any amount) for human consumption requires a Wild Aquatic Plant Harvester license. Culture licenses are required when growing for human consumption. Culture licenses may last up to 10 years, whereas harvest licenses are typically renewed annually. Cultivation operations which install infrastructure (i.e., kelp farms) are additionally licensed through a harmonized review process under Transport Canada's Navigation Protection Program, as well as tenured under the BC Land Act if in waters over Crown Land. Moreover, Fisheries and Oceans Canada is drawn in if the Fisheries Act is triggered, for example, if operations are expected to impact wild habitat or species. Kelp propagation requires an upland aquatic plant culture license, and kelp restoration as well as scientific work may require a permission depending on proposed activities (Outhwaite and Pawluk 2024).

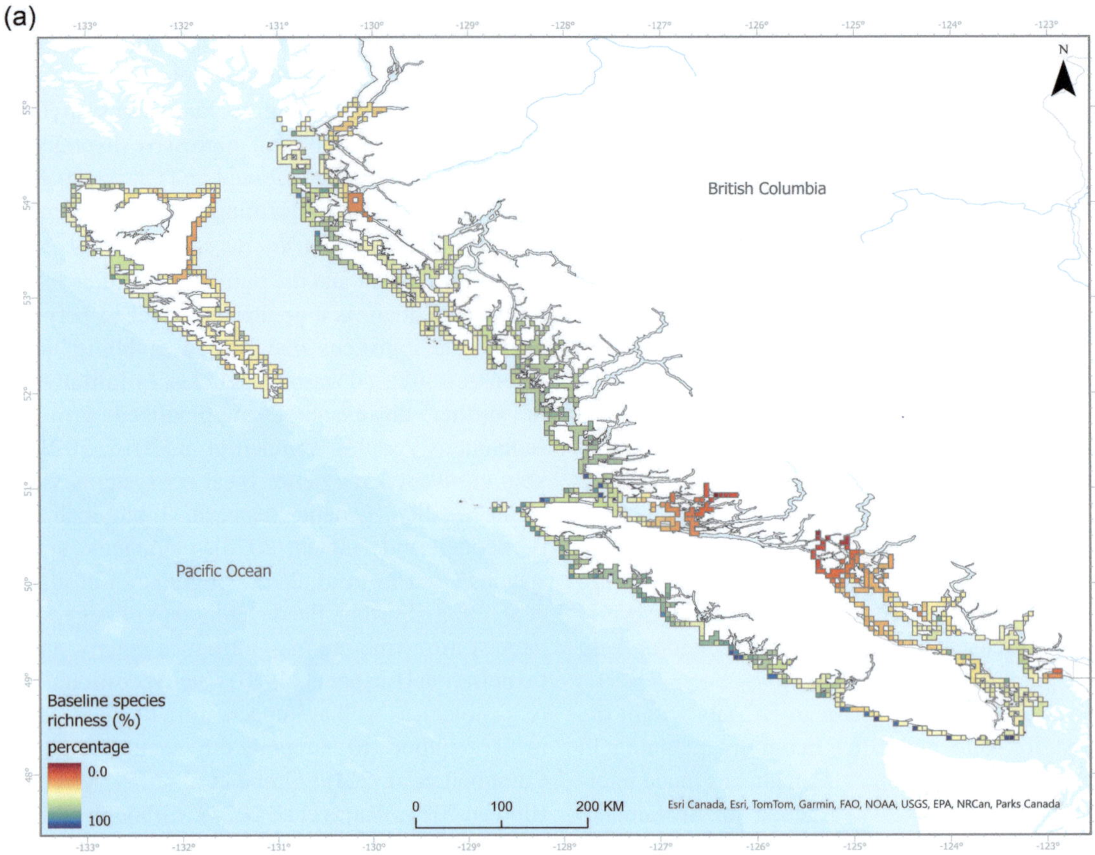

Fig. 20.1 Projected kelp species richness in BC. Figures created using ArcGIS using data modeled by Assis et al. (2024). Data was exported using R (version 4.4.2; R Core Team 2024) after being manually subset to species in Laminariales. Species richness is presented as a percentage out of the 44 species modeled to keep the scale consistent between figures. (**a**) indicates baseline (years 2010–2020) species richness, (**b**) indicates species richness by the end of the century (2090–2100) under projected SSP1-1.9 and (**c**) indicates species richness by the end of the century (2090–2100) under projected SSP3–7

Although kelp management in BC is overall under provincial jurisdiction, there are mechanisms in place for collaborative decision making to take place within the sector. For instance, the Province of BC partners with 17 member First Nations through the Marine Plan Partnership for the North Pacific Coast (MaPP) to run a Regional Kelp Monitoring program and develop marine spatial plans (Tamburello and Semmens 2023; Marine Plan Partnership for the North Pacific Coast 2025), where kelp habitats are a priority for monitoring within the province (Government of BC 2024). In addition, BC's Declaration on the Rights of Indigenous People's Act (DRIPA) and associated Declaration Act Action Plan (2022) require that the province "take all measures necessary to ensure the laws of BC are consistent with the UN Declaration [on the Rights of Indigenous People]" (Government of BC and Ministry of Indigenous Relations and Reconciliation 2022).

An ongoing challenge for resource management and governance is fitting together conflicting social, economic, and environmental objectives (Potts 2020), which can limit decision-making from a systems perspective. As outlined above, the current policy landscape for BC's kelp sector is segmented into separate activities, although application processes and kelp-management staff acknowledge the intersection of these activities (Outhwaite and Pawluk 2024). However, integrated management is needed, for instance, to balance kelp cultivation with wild kelp populations (Grebe et al. 2019); further, a link between cultivation and restoration is needed to scale up kelp restoration initiatives (Filbee-Dexter et al. 2022). Here we provide an alternative conceptual framework that may help inspire integrated management and governance for BC's kelp sector, where we use "governance" to refer to both the formal and informal societal rules, norms, processes, and institutions that determine how decisions are made and for whom (Bennett and Satterfield 2018).

20.3 Social-Ecological Kelp System

A transformative kelp sector is ecologically regenerative and socially reconciliatory. Uniting these notions, transformation is not fully possible without humanity's reconciliation with

(b)

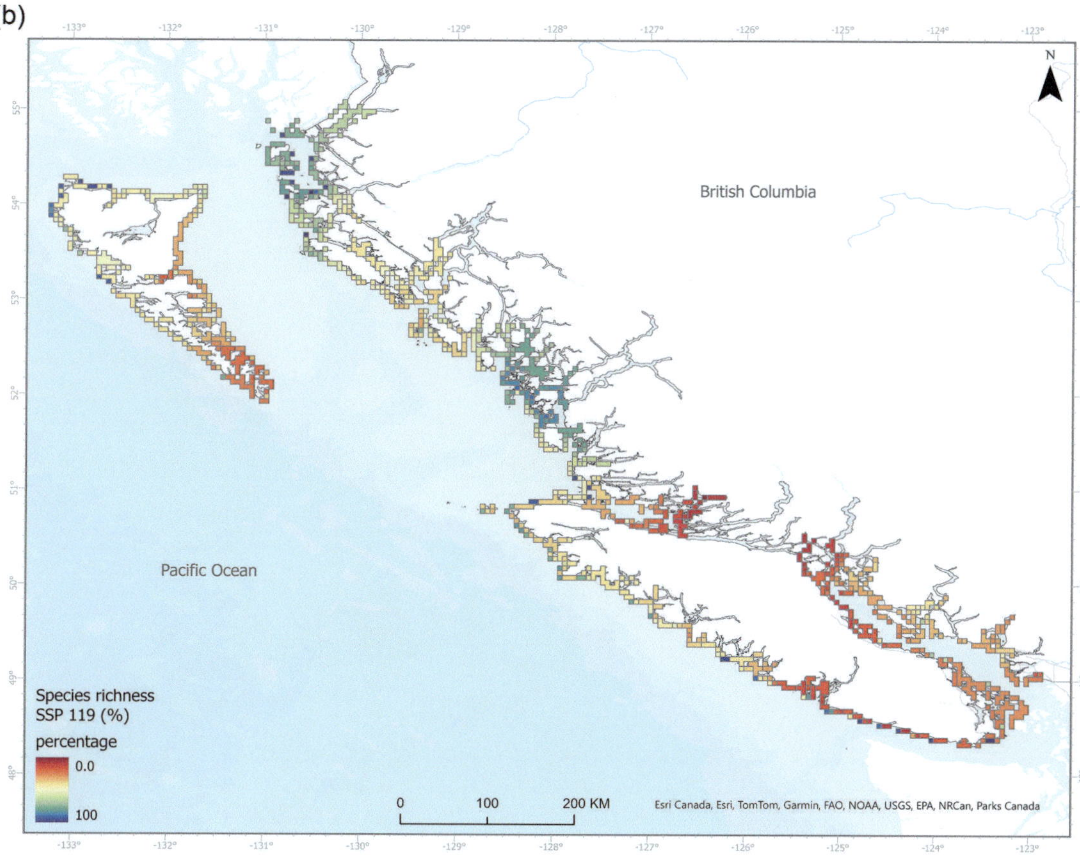

Fig. 20.1 (continued)

nature (Truth Reconciliation Commission of Canada 2015; McGregor 2018; Murdock 2018), including rooting the well-being of nature within governance and justice (McGregor 2016; McGregor et al. 2020). Thus, the kelp sector may draw on the teachings from many Indigenous Peoples, for example, from Indigenous food systems in the Pacific Northwest, which blurred the boundaries between "wild" and cultivated with various plant and animal species intentionally stewarded in forest gardens (e.g., Armstrong et al. 2022, 2023) and sea gardens (Jackley et al. 2016; Levy 2023).

Taking inspiration from Coastal Indigenous teachings (e.g., W̱SÁNEĆ Calendar seasonal wheel (Tsawout First Nation 2019), Kwakwaka'wakw seasonal wheel (Snively and Williams 2016), systems thinking and complex adaptive systems, as well as the concept and critiques of planetary boundaries, we herein consider kelp–human interactions to be a continuum (Fig. 20.2). This spectrum is illustrated through a spiral of increasing interaction between humans and kelp. We note that this definition of interaction strength is not necessarily bound by time, personal value, or feelings of connection, but rather the degree to which humans directly/intentionally alter the kelp or its environment. We call this a "stewardship" spectrum following the BC Coastal Marine Strategy's (2024) definition, where stewardship is

> the responsible use and protection of the environment… the recognition of our responsibility to care for and maintain the quality and natural function of ecosystems. Environmental stewardship includes restoration and protection, research and monitoring, community and civic action, and everyday choices (p.10)

We expand upon this stewardship definition to more explicitly include mariculture (i.e., propagation and cultivation).

20.3.1 Stewardship Spectrum

The stage with the least direct interaction is "wild" kelp; however, we acknowledge that wild kelp is still and has been influenced by anthropogenic stressors and may have a baseline legacy of Indigenous stewardship. The "monitor" stage consists of interactions between humans and kelp in which kelp is observed, studied, and/or measured (etc.) but not intentionally altered (unintentional impacts to kelp may occur, such as damage from boat propellers). The "tend" stage refers to ancestral, revitalized, and proposed practices of active kelp management where the primary objective is not to retain the cut kelp, such as thinning kelp beds (e.g., to control for bryozoans; Denley et al. 2022). Ancestral tending of kelp also included transplanting kelp to promote roe on kelp fisheries, as well as clearing kelp to corral salmon (Stewart 1977).

(c)

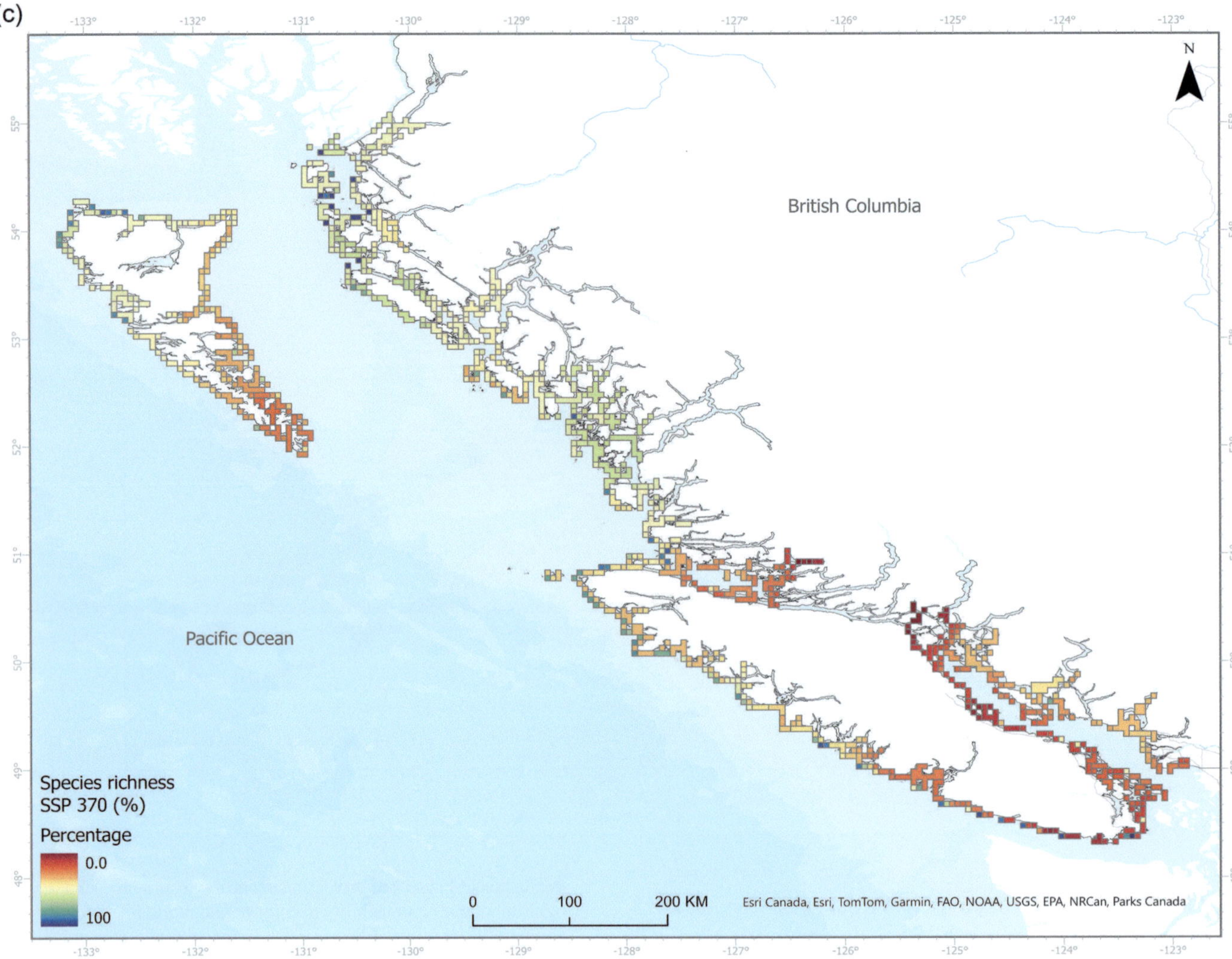

Fig. 20.1 (continued)

The "harvest" stage refers to more prominent human-kelp interaction where the objective is to obtain kelp for various purposes, including for food, medicine, processing, ceremony, etc. The "restore" stage and "cultivate" stages are often beyond a threshold of "propagation," where humans have collected reproductive material from kelp and grown it out to a kelp plant. Restoration includes kelp being planted (e.g., on rocks, settling plates, or gravel) with the intent to revive or supplement wild kelp populations in a given area. Finally, the most intensive kelp-human interaction is what we consider "cultivation," where more extensive human-built structures are used to grow out kelp plants in a farm-like setting. Each of these 6 stages has varying levels of intensity within it (e.g., a small vs. large harvest, restoration, or cultivation project), and actions may fall between or incorporate multiple of the defined stages. For instance, a Haida-values-centered kelp restoration project in Gwaii Haanas National Park Reserve, National Marine Conservation Area Reserve, and Haida Heritage Site restored kelp forests via sea urchin removal (Lee et al. 2021), which can be considered both tending and restoration; likewise, ancestral and continuing Indigenous wild kelp harvest can include tending practices (Turner 2020; Kobluk et al. 2021). Finally, we point to downstream kelp product applications (e.g., food, chemical applications, bioplastics, etc.) at the stages where kelp is intentionally collected (i.e., the harvest and cultivate stages) for these purposes. We note that while, to our knowledge, bryozoan-encrusted kelp cut during tending processes in BC is not currently used in downstream applications, the potential exists.

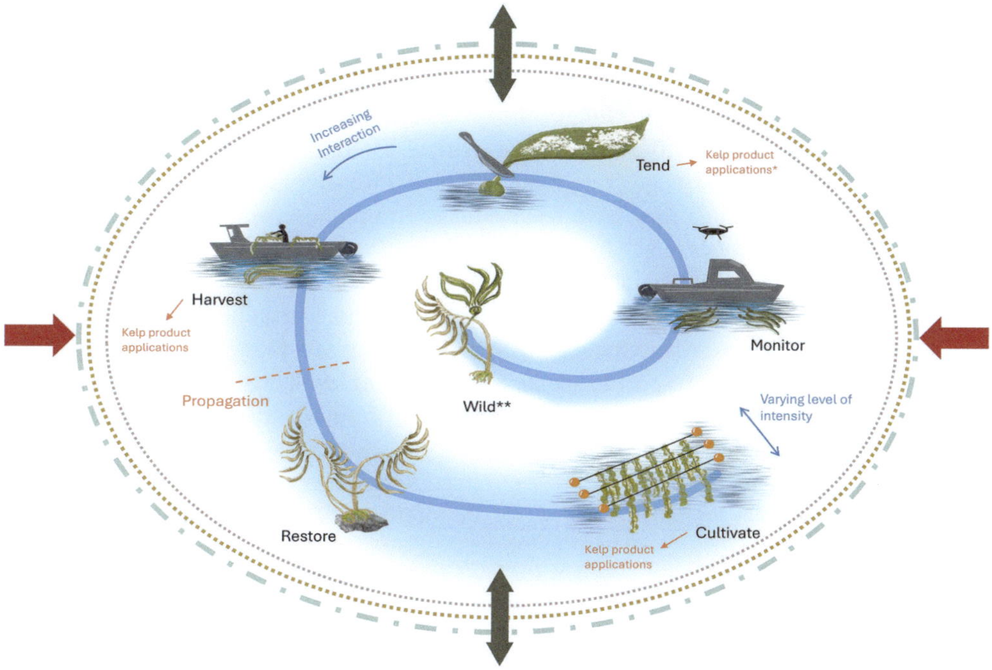

Fig. 20.2 Kelp stewardship spectrum and system boundaries. The blue spiral illustrates an increasing spectrum of interaction between humans and kelp (see text for details). The outermost dashed line represents the unknown "true" environmental boundary, the middle-dashed line is the value-derived boundary, and the innermost dashed line is the regulatory boundary (e.g., set by the BC government). Red arrows (left and right) represent constraining factors on the environmental boundary such as climate change, conventional economic practices (e.g., overexploitation, habitat loss from development, and pollution) which can reduce the ability of the environment to support kelp. Regenerative practices (dark green arrows, top and bottom) are reciprocal with the environment and could benefit/expand the environmental boundary through rewilding (e.g., see rewilding in du Toit and Pettorelli 2019). * applications possible, ** wild kelp is influenced by anthropogenetic stressors and may have a baseline legacy of indigenous stewardship even if it is no longer actively/directly interacted with

20.3.2 True Ecological Boundary

The "true" ecological boundary for kelp in BC is unknown. Yet, climate change is expected to negatively impact kelps overall as they have a narrow tolerance for ocean temperature (Valdez et al. 2003; Krumhansl et al. 2016; Cavanaugh et al. 2019; Smale 2020; Kobluk et al. 2021; Tamburello and Semmens 2023; Hill et al. 2025). However, as kelp abundance is modulated by other environmental factors such as flow, substrate type, depth, salinity, and herbivory, future kelp populations are expected to be site-specific. For instance, Krumhansl et al. (2016) found an instantaneous loss of 1.8% globally with climate change and other stressors, but it was highly variable. Likewise, kelp population changes in BC are expected to be site-specific, with declines being observed in some areas (Hollarsmith et al. 2022; Starko et al. 2022, 2024; Tamburello and Semmens 2023; Mora-Soto et al. 2024).

20.3.3 Value-Derived Boundaries

Biodiversity-related planetary boundaries are implicitly defined by the biodiversity needed to support human welfare (Rockström et al. 2009; Mace et al. 2014), and do not often directly address the assumptions made in determining whose values best describe "welfare" and the policies used to maintain it (Kim and Kotzé 2021). Human values also determine what matters enough to be accounted for and regulated at all (Heikkurinen et al. 2021). Thus, in essence, the value-derived boundary describes how much and what type(s) of kelp humans want. However, values differ depending on the underlying ontologies of the people making the decision, and some values may be marginalized. For example, indigenous values have been and continue to be deprioritized in marine management and governance processes compared to Western value systems (Turner et al. 2013; Parsons et al. 2021; Dick et al. 2022; Salomon et al. 2023).

Kelp–human relationships (and therefore values) are fluid, having evolved and will continue to evolve through time and with climate change. And for many, human-kelp relationships in the present have narrowed to be dominated by a perceived dichotomy of economic opportunity or conservation. While conservation and economic ventures are important human-kelp interactions, for BC's kelp sector to be successful and equitable, the spectrum of kelp values/uses/relations needs to be considered in its governance. For example, kelps have significant cultural importance and associated practices for many Coastal Indigenous Nations in BC; for instance, the Kwakiutl traditionally used bull kelp to carry eulachon grease, for making rope, fishing line, harpoon lines, and in ceremony (e.g., Gutzmann 2023; Gutzmann et al. in prep). Cultural values are not currently represented in economic valuations of kelp to our knowledge (e.g., Eger et al. 2023; Cevallos et al. 2024), although they have been represented using other valuation techniques (e.g., Gutzmann 2023; Gutzmann et al. in prep). Yet, underlying values influence the level of interaction intensity and at what stage on the stewardship spectrum industry occurs, where the regulatory boundary sits (e.g., precautionary with respect to social-ecological boundaries or not), as well as the transition between conventional extraction and regenerative practices.

20.4 Recommendations

As discussed, the current ocean economy and its governance require transformation if we are to develop the BC kelp sector into a regenerative one that meets goals of a Blue Economy. Therefore, research must look at all three pillars of the Blue Economy, namely social, economic, and ecological factors but prioritize the social factors that the approach emphasizes. Moreover, applied, interdisciplinary research is needed to enact transformation and decrease gaps between academia and practitioners because as Evans et al. (2024a p. 1244) state, "achieving transformation though governance is inseparable from transforming ocean governance." Following the principles of regenerative design as laid out by Mang et al. (2016), one must start by situating themselves in place. We have begun that process here by conceptualizing a stewardship spectrum, thus creating a cognitive connection between parts of the BC kelp system which are currently being regulated to the broader range of kelp–human interactions. We have also provided an overview of the social and ecological environments a kelp economy is a part of.

The developing BC kelp system presents a unique opportunity to define what Blue/regenerative success looks like from the start because the kelp sector is less socially and politically ingrained than other ocean industries. Comparatively, finfish and shellfish aquaculture are well established in BC, and future research may learn from comparing these industries to regenerative development and design frameworks. For instance, considering the social-ecological failures and successes of finfish and shellfish aquaculture, as well as seaweed aquaculture internationally, may inform how to bolster the regulatory boundary that aligns with, and protects, the environmental and human value-derived boundaries for kelp (Kim and Kotzé 2021). That said, future research will need to investigate where the value-derived boundary for kelp in BC sits. Hence, additional research into the holistic valuation of kelp (i.e., covering a spectrum of value types; see Tadaki et al. 2017), particularly with respect to trade-offs between stages in the stewardship spectrum, is needed.

Finally, when planning for a successful kelp sector, it is critical to consider how it will perpetuate into the future. Socially, we suggest that future research address the sector's long-term requirements for financial and human capital, for instance, with research into various funding mechanisms (e.g., payments for ecosystem services and carbon and biodiversity credits). In addition, given the uncertainty of climate change's impacts to the BC seascape, research should assist in wrestling with the ethical and environmental considerations of resource management strategies that direct change (e.g., Clifford et al. 2022) such as the assisted migration (e.g., Vranken et al. 2021), active rewilding (e.g., du Toit and Pettorelli 2019), and/or afforestation of kelp genotypes or species with greater thermal tolerance (e.g., Hill et al. 2025).

Developing the kelp Blue Economy in BC in accordance with principles of regenerative development and design will require directed research—technological, economic, ecological, and regulatory—to support the scaling up of the industry (Mang et al. 2016). However, we believe that the most important priority is determining the sector's collective vocation, the guiding vision and values which will underpin decisions—and therefore governance—of the system. To further define this vocation, case studies looking for collaborations and shared values between Indigenous, private, public, academic, and non-profit organizations would be useful. Indeed, values have been identified as a leverage point in ocean governance transformation (Evans et al. 2024b), and there is a unique opportunity in BC to center regenerative practices and climate resilience in kelp governance as the sector begins to scale.

Acknowledgments We thank Mohammad Nasir Tighsazzadeh for the creation of Fig. 20.1a–c in ArcGIS, and Dr. Kylee Pawluck for reviewing the text on BC's regulatory context. Funding for this work is provided to SBG through Social Sciences and Humanities Research Council (SSHRC) Doctoral Fellowship and Mitacs Accelerate Internship programs.

References

Ali O, Ramsubhag A, Jayaraman J (2021) Biostimulant properties of seaweed extracts in plants: implications towards sustainable crop production. Plants (Basel) 10:531. https://doi.org/10.3390/plants10030531

Allchurch A (2023) Investigating the resilience of kelp forests to harvest amid a rapidly changing ocean. https://summit.sfu.ca/item/37971. Accessed 4 Jul 2024

Armstrong CG, Earnshaw J, McAlvay AC (2022) Coupled archaeological and ecological analyses reveal ancient cultivation and land use in Nuchatlaht (Nuu-chah-nulth) territories, Pacific Northwest. J Archaeol Sci 143:105611. https://doi.org/10.1016/j.jas.2022.105611

Armstrong CG, Lyons N, McAlvay AC et al (2023) Historical ecology of forest garden management in Laxyuubm Ts'msyen and beyond. Ecosyst People 19:2160823. https://doi.org/10.1080/26395916.2022.2160823

Assis J, Fragkopoulou E, Gouvêa L et al (2024) Kelp forest diversity under projected end-of-century climate change. Divers Distrib 30:e13837. https://doi.org/10.1111/ddi.13837

Bates C (2004) An Introduction to the Algae of British Columbia. In: Klinkenberg B (ed) E-Flora BC: Atlas of the plants of British Columbia [www.eflora.bc.ca]. Lab for Advanced Spatial Analysis, Department of Geography, University of British Columbia, Vancouver

Behera DP, Ingle KN, Mathew DE et al (2022) Epiphytism, diseases and grazing in seaweed aquaculture: a comprehensive review. Rev Aquac 14:1345–1370. https://doi.org/10.1111/raq.12653

Bennett NJ, Satterfield T (2018) Environmental governance: A practical framework to guide design, evaluation, and analysis. Conserv Lett 11. https://doi.org/10.1111/CONL.12600

Bennett NJ, Govan H, Satterfield T (2015) Ocean grabbing. Mar Policy 57:61–68. https://doi.org/10.1016/j.marpol.2015.03.026

Bennett NJ, Blythe J, White CS, Campero C (2021) Blue growth and blue justice: Ten risks and solutions for the ocean economy. Mar Policy 125. https://doi.org/10.1016/j.marpol.2020.104387

Bermejo R, Buschmann A, Capuzzo E et al (2022) State of knowledge regarding the potential of macroalgae cultivation in providing climate-related and other ecosystem services. In: EPIC3Eklipse. https://eklipse.eu/wp-content/uploads/website_db/Request/Macro-Algae/EKLIPSE_DG-Mare-Report-PrintVersion_final.pdf. Accessed 12 Aug 2024

Bhuyan MS (2023) Ecological risks associated with seaweed cultivation and identifying risk minimization approaches. Algal Res 69:102967. https://doi.org/10.1016/j.algal.2022.102967

Calvin K, Dasgupta D, Krinner G et al (2023) IPCC, 2023: Climate Change 2023: Synthesis Report. Contribution of Working Groups I, II and III to the Sixth Assessment Report of the Intergovernmental Panel on Climate Change [Core Writing Team, H. Lee and J. Romero (eds.)]. IPCC, Geneva. Intergovernmental Panel on Climate Change (IPCC)

Campbell I, Macleod A, Sahlmann C et al (2019) The environmental risks associated with the development of seaweed farming in Europe-prioritizing key knowledge gaps. Front Mar Sci 6:107. https://doi.org/10.3389/fmars.2019.00107

Cavanaugh KC, Reed DC, Bell TW et al (2019) Spatial variability in the resistance and resilience of giant kelp in southern and Baja California to a multiyear heatwave. Front Mar Sci 6. https://doi.org/10.3389/fmars.2019.00413

Cevallos B, Aller-Rojas O, Aponte H, Moreno B (2024) Carbon capture by stipitate kelp forests in Peru: insights from population assessment of Lessonia trabeculata at 15° S. J Appl Phycol 36(5):3057–3076

Cisneros-Montemayor AM, Croft F, Issifu I et al (2022) A primer on the "blue economy": Promise, pitfalls, and pathways. One Earth 5:982–986. https://doi.org/10.1016/j.oneear.2022.08.011

Clifford KR, Cravens AE, Knapp CN (2022) Special Section on the Resist-Accept-Direct Framework. Responding to ecological transformation: mental models, external constraints, and manager decision-making. BioScience 72. https://doi.org/10.1093/biosci/biab086

Cotas J, Gomes L, Pacheco D, Pereira L (2023) Ecosystem services provided by seaweeds. Hydrobiology 2:75–96. https://doi.org/10.3390/hydrobiology2010006

Denley D, Frid A, Salomon A (2022) Hidden impacts of climate change on Canada's undersea forests: Chapter 1: Warm sea water and high kelp density magnify Bryozoan abundance. Central Coast Indigenous Resource Alliance

Dick CA, Sewid-Smith D, Recalma-Clutesi K et al (2022) "From the beginning of time": The colonial reconfiguration of native habitats and Indigenous resource practices on the British Columbia Coast. FACETS 7:543–570. https://doi.org/10.1139/facets-2021-0092

Dillehay TD, Ramírez C, Pino M et al (2008) Monte Verde: Seaweed, food, medicine, and the peopling of South America. Science 320:784–786. https://doi.org/10.1126/SCIENCE.1156533/SUPPL_FILE/DILLEHAY.SOM.PDF

Druehl L, Clarkston B (2016) Pacific Seaweeds: Updated and Expanded Edition, 2nd edn. Harbour Publishing. https://harbourpublishing.com/products/9781550177374?srsltid=AfmBOoohbgrTvBb6zSDAGDF1rRhE5gJAqT8OfK2iINeJZbAhlKac9VgD

du Toit JT, Pettorelli N (2019) The differences between rewilding and restoring an ecologically degraded landscape. J Appl Ecol 56:2467–2471. https://doi.org/10.1111/1365-2664.13487

Duarte CM, Wu J, Xiao X et al (2017) Can seaweed farming play a role in climate change mitigation and adaptation? Front Mar Sci 4. https://doi.org/10.3389/fmars.2017.00100

Eger AM, Vergés A, Choi CG et al (2020) Financial and institutional support are important for large-scale kelp forest restoration. Front Mar Sci 7

Eger AM, Marzinelli EM, Christie H et al (2022) Global kelp forest restoration: past lessons, present status, and future directions. Biol Rev. https://doi.org/10.1111/BRV.12850

Eger AM, Marzinelli EM, Beas-Luna R et al (2023) The value of ecosystem services in global marine kelp forests. Nat Commun 14:1894. https://doi.org/10.1038/s41467-023-37385-0

Erlandson JM, Graham MH, Bourque BJ et al (2007) The kelp highway hypothesis: marine ecology, the coastal migration theory, and the peopling of the Americas. J Island Coast Archaeol 2:161–174. https://doi.org/10.1080/15564890701628612

Evans W, Giménez I, Christian J et al (2023) Scientific assessment for the British Columbia Ocean Acidification and Hypoxia Action Plan. https://www2.gov.bc.ca/assets/gov/environment/climate-change/adaptation/climate-risks/bc-ocean-acidification-hypoxia-action--plan.pdf

Evans T, Fletcher S, Failler P et al (2024a) Journeys of change towards the blue economy: evaluating process in transformational change. Reg Environ Change 24:170. https://doi.org/10.1007/s10113-024-02336-y

Evans T, Fletcher S, Failler P et al (2024b) Radical and incremental, a multi-leverage point approach to transformation in ocean governance. Sustain Sci 19:1243–1258. https://doi.org/10.1007/s11625-024-01507-4

FAO (2024) The state of world fisheries and aquaculture 2024. https://openknowledge.fao.org/items/06690fd0-d133-424c-9673-1849e414543d

Filbee-Dexter K, Wernberg T (2020) Substantial blue carbon in overlooked Australian kelp forests. 10:12341. https://doi.org/10.1038/s41598-020-69258-7

Filbee-Dexter K, Wernberg T, Barreiro R et al (2022) Leveraging the blue economy to transform marine forest restoration. J Phycol 58:198–207. https://doi.org/10.1111/jpy.13239

Filbee-Dexter K, Pessarrodona A, Pedersen MF et al (2024) Carbon export from seaweed forests to deep ocean sinks. Nat Geosci. https://doi.org/10.1038/s41561-024-01449-7

Fisheries and Oceans Canada (2022) Engaging on Canada's blue economy strategy: what we heard. Fisheries and Oceans Canada, Ottawa. https://waves-vagues.dfo-mpo.gc.ca/library-bibliotheque/41030503.pdf

Fricke A, Capuzzo E, Bermejo R et al (2024) Ecosystem services provided by seaweed cultivation: state of the art, knowledge gaps, constraints and future needs for achieving maximum potential in Europe. Rev Fish Sci Aquac:1–19. https://doi.org/10.1080/23308249.2024.2399355

Froehlich HE, Afflerbach JC, Frazier M, Halpern BS (2019) Blue growth potential to mitigate climate change through seaweed off-

setting. Curr Biol 29:3087–3093.e3. https://doi.org/10.1016/j.cub.2019.07.041
Fujita R, Augyte S, Bender J et al (2023) Seaweed blue carbon: Ready? or not? Mar Policy 155:105747. https://doi.org/10.1016/j.marpol.2023.105747
Gao G, Wu M, Wang S, Gao K (2020) Using macroalgae as biofuel: current opportunities and challenges. Bot Mar 63. https://doi.org/10.1515/bot-2019-0065
Government of BC (2024) B.C. Coastal Marine Strategy. https://www2.gov.bc.ca/assets/gov/environment/air-land-water/water/coastal-marine-strategy/coastal_marine_strategy.pdf
Government of BC, Ministry of Indigenous Relations and Reconciliation (2022) Declaration on the Rights of Indigenous Peoples Act Action Plan: 2022-2027. https://www2.gov.bc.ca/assets/gov/government/ministries-organizations/ministries/indigenous-relations-reconciliation/declaration_act_action_plan.pdf
Grebe GS, Byron CJ, Gelais AS et al (2019) An ecosystem approach to kelp aquaculture in the Americas and Europe. Aquac Reports 15. https://doi.org/10.1016/J.AQREP.2019.100215
Gruenthal KM, Habicht C (2022) Literature review for implementation of the 50-50 rule for cultivation of seaweeds and other aquatic plants in Alaska. Alaska Department of Fish and Game, Division of Commercial Fisheries. https://www.adfg.alaska.gov/FedAidPDFs/RIR.2A.2022.01.pdf
Gutzmann SB (2023) Centering Indigenous governance principles, values, and objectives to build a just blue economy decision space. MRM, Simon Fraser University
Hale Hendlin Y, Weggelaar J, Derossi N, Mugnai S (2024) Being algae: transformations in water, plants. BRILL. https://brill.com/display/title/68608
Heikkurinen P, Clegg S, Pinnington AH et al (2021) Managing the anthropocene: relational agency and power to respect planetary boundaries. Organ Environ 34:267–286. https://doi.org/10.1177/1086026619881145
High Level Panel for a Sustainable Ocean Economy (2022) Tracking blue: from ambition to action for a sustainable ocean economy. https://oceanpanel.org/publication/tracking-blue/
Hill G, Gauci C, Assis J, Jueterbock A (2025) Turning the tide: a 2°C increase in heat tolerance can halve climate change induced losses in four cold-adapted kelp species. Ecol Evol 15, e71271. https://doi.org/10.1002/ece3.71271
Hollarsmith JA, Andrews K, Naar N et al (2022) Toward a conceptual framework for managing and conserving marine habitats: A case study of kelp forests in the Salish Sea. Ecol Evol 12. https://doi.org/10.1002/ECE3.8510
Huntjens P (2021) Towards a natural social contract: transformative social-ecological innovation for a sustainable, healthy and just society. Springer Nature Link. DOI: 10.1007/978-3-030-67130-3
Jackley J, Gardner L, Djunaedi AF, Salomon AK (2016) Ancient clam gardens, traditional management portfolios, and the resilience of coupled human-ocean systems. E&S 21:art20. https://doi.org/10.5751/ES-08747-210420
Jones N (2023) The promises and perils of the seaweed farming boom. In: Hakai Magazine – republished in The Tyee, 24 March. https://thetyee.ca/News/2023/03/24/Promise-Peril-Seaweed-Farming-Boom/. Accessed 12 Aug 2024
Kelly B, Currie J, Fish M et al (2023) Coastal blue carbon in Canada: state of knowledge. https://wwf.ca/wp-content/uploads/2023/06/BlueCarbon_StateofKnowledge_Report.pdf
Khan N, Sudhakar K, Mamat R (2024) Macroalgae farming for sustainable future: navigating opportunities and driving innovation. Heliyon 10:e28208. https://doi.org/10.1016/j.heliyon.2024.e28208
Kim RE, Kotzé LJ (2021) Planetary boundaries at the intersection of Earth system law, science and governance: a state-of-the-art review. Rev Eur Compar Int Environ Law 30:3–15. https://doi.org/10.1111/reel.12383
Kobluk HM, Gladstone K, Reid M et al (2021) Indigenous knowledge of key ecological processes confers resilience to a small-scale kelp fishery. People Nat 3:723–739. https://doi.org/10.1002/pan3.10211
Krause-Jensen D, Duarte CM (2016) Substantial role of macroalgae in marine carbon sequestration. Nat Geosci 9:737–742. https://doi.org/10.1038/ngeo2790
Krumhansl KA, Okamoto DK, Rassweiler A et al (2016) Global patterns of kelp forest change over the past half-century. Proc Natl Acad Sci U S A 113:13785–13790. https://doi.org/10.1073/PNAS.1606102113/-/DCSUPPLEMENTAL/PNAS.1606102113.SAPP.PDF
Lee LC, Daniel McNeill G, Ridings P et al (2021) Chiixuu Tll iinasdll: indigenous ethics and values lead to ecological restoration for people and place in Gwaii Haanas. Ecol Rest 39:45–51. https://doi.org/10.3368/er.39.1-2.45
Levy J (2023) Diversity and Resilience of Sea Gardens Across the Pacific Ocean. MRM Thesis. Simon Fraser University. https://summit.sfu.ca/_flysystem/fedora/2023-06/etd22428.pdf
Lusher A, Hollman PCH, Mendoza-Hill J (2017) Microplastics in fisheries and aquaculture: status of knowledge on their occurrence and implications for aquatic organisms and food safety. Food and Agriculture Organization of the United Nations, Rome
Mace GM, Reyers B, Alkemade R et al (2014) Approaches to defining a planetary boundary for biodiversity. Glob Environ Change 28:289–297. https://doi.org/10.1016/j.gloenvcha.2014.07.009
Mang P, Haggard B, Regenesis Group (2016) Regenerative development and design: a frameworks for evolving sustainability. Wiley, New Jersey
Marine Plan Partnership for the North Pacific Coast (2025) Marine Plan Partnership for the North Pacific Coast. In: MaPP. https://mappocean.org/. Accessed 28 Jan 2025
McGregor D (2016) Living well with the Earth: Indigenous rights and the environment. In: Handbook of indigenous peoples' rights. Routledge
McGregor D (2018) Reconciliation and environmental justice. J Glob Ethics 14:222–231. https://doi.org/10.1080/17449626.2018.1507005
McGregor D, Whitaker S, Sritharan M (2020) Indigenous environmental justice and sustainability. Curr Opin Environ Sustain 43:35–40. https://doi.org/10.1016/j.cosust.2020.01.007
Mora-Soto A, Schroeder S, Gendall L et al (2024) Kelp dynamics and environmental drivers in the southern Salish Sea, British Columbia, Canada. Front Mar Sci 11. https://doi.org/10.3389/fmars.2024.1323448
Mouga T (2024) Seaweed harvesting and aquaculture: an overview of the past 70 years. In: Galvão JRDCS, Brito P, Neves FDS et al (eds) Proceedings of the 3rd International Conference on Water Energy Food and Sustainability (ICoWEFS 2023). Springer Nature Switzerland, Cham, pp 365–375
Murdock EG (2018) Unsettling reconciliation: decolonial methods for transforming social-ecological systems. Environ Values 27:513–533. https://doi.org/10.3197/096327118X15321668325948
Ota Y, Singh GG, Clark T et al (2022) Finding logic models for sustainable marine development that deliver on social equity. PLOS Biol 20:e3001841. https://doi.org/10.1371/journal.pbio.3001841
Outhwaite J, Pawluk K (2024) Q&A with the Province of BC about kelp aquaculture and harvest. CARTI Seaweed Series. North Island College. Virtual. https://www.youtube.com/watch?time_continue=101&v=C0n_FyK4Y3c&embeds_referring_euri=https%3A%2F%2Fwww.nic.bc.ca%2F&source_ve_path=MTM5MTE3LDM2ODQyLDEzOTExNywyODY2Ng
Parsons M, Taylor L, Crease R (2021) Indigenous environmental justice within marine ecosystems: A systematic review of the literature on indigenous peoples' involvement in marine governance and management. Sustainability. https://doi.org/10.3390/su13084217

Pecquet J, Mouchet M, Campagne S et al (2022) Combining Ecological Niche Models and ecosystem services indicators to assess impacts of climate change on kelp: application to French coasts. Ecosyst People 18:358–377. https://doi.org/10.1080/26395916.2022.2080766

Potts R (2020) Disconnected dots?: A systematic review of governance challenges for natural resource management. J Environ Plan Manage 63:1356–1374. https://doi.org/10.1080/09640568.2019.1663723

R Core Team (2024) R: A language and environment for statistical computing. R Foundation for Statistical Computing. Vienna, Austria. https://www.R-project.org/

Rockström J, Steffen W, Noone K et al (2009) Planetary boundaries: exploring the safe operating space for humanity. Ecol Soc 14(2). https://www.jstor.org/stable/26268316?seq=1

Ross F, Tarbuck P, Macreadie PI (2022) Seaweed afforestation at large-scales exclusively for carbon sequestration: Critical assessment of risks, viability and the state of knowledge. Front Mar Sci 9. https://doi.org/10.3389/fmars.2022.1015612

Salomon AK, Okamoto DK, Wilson Ḵii'iljuus Barbara J et al (2023) Disrupting and diversifying the values, voices and governance principles that shape biodiversity science and management. Philos Trans R Soc B 378:20220196. https://doi.org/10.1098/RSTB.2022.0196

Schikan J, Gwóźdź B (2022) The seaweed archives: A material study of seaweed as a building material and its implementation on two buildings on North Koster, Sweden. Masters in Architecture, Chalmers University of Technology

Shannon G, Issa R, Wood C, Kelman I (2022) Regenerative economics for planetary health: A scoping review. Int Health Trends Persp 2:81–105. https://doi.org/10.32920/ihtp.v2i3.1704

Singh I, Anand KGV, Solomon S et al (2018) Can we not mitigate climate change using seaweed based biostimulant: A case study with sugarcane cultivation in India. J Clean Prod 204:992–1003. https://doi.org/10.1016/j.jclepro.2018.09.070

Singh GG, Hilmi N, Bernhardt JR, et al (2019) Climate impacts on the ocean are making the Sustainable Development Goals a moving target travelling away from us.People and Nature 1:317–330. https://doi.org/10.1002/pan3.26

Smale DA (2020) Impacts of ocean warming on kelp forest ecosystems. New Phytologist 225:1447–1454. https://doi.org/10.1111/nph.16107

Snively G, Williams L (eds) (2016) Knowing Home: Braiding Indigenous Science with Western Science, Book 1. University of Victoria. Victoria, BC.https://pressbooks.bccampus.ca/knowinghome/

Starko S, Neufeld CJ, Gendall L et al (2022) Microclimate predicts kelp forest extinction in the face of direct and indirect marine heatwave effects. Ecol Appl. 32(7):1-18. https://doi.org/10.1002/EAP.2673

Starko S, Timmer B, Reshitnyk L et al (2024) Local and regional variation in kelp loss and stability across coastal British Columbia. Mar Ecol Prog Ser 733:1–26. https://doi.org/10.3354/meps14548

Stewart H (1977) Indian fishing: early methods on the Northwest Coast. Douglas & McIntyre

Sugumaran R, Padam BS, Yong WTL et al (2022) A retrospective review of global commercial seaweed production—current challenges, biosecurity and mitigation measures and prospects. Int J Environ Res Public Health 19:7087. https://doi.org/10.3390/ijerph19127087

Sultana F, Wahab MA, Nahiduzzaman M et al (2023) Seaweed farming for food and nutritional security, climate change mitigation and adaptation, and women empowerment: A review. Aquac Fish 8:2468–2550. https://doi.org/10.1016/j.aaf.2022.09.001

Syrgkani M (2024) Assessment and optimization of marine algae-based carbon capture systems for vessel emissions reduction. MSc, ALBA Graduate Business School

Tadaki M, Sinner J, Chan KMA (2017) Making sense of environmental values: a typology of concepts. Ecol Soc 22(1):7. https://doi.org/10.5751/ES-08999-220107

Tamburello N, Semmens C (2023) MaPP kelp monitoring conceptual model and roadmap. ESSA Technologies for Marine Plan Partnership. https://mappocean.org/wp-content/uploads/2024/06/MaPP_KelpMonitoringModelRoadmap_May2023_Public.pdf

The Nature Conservancy (2021) Global principles of restorative aquaculture. https://www.nature.org/content/dam/tnc/nature/en/documents/TNC_PrinciplesofRestorativeAquaculture.pdf

Truth Reconciliation Commission of Canada (2015) Honouring the truth, reconciling for the future: Final report of the Truth and Reconciliation Commission of Canada. https://ehprnh2mwo3.exactdn.com/wp-content/uploads/2021/01/Principles_English_Web.pdf

Tsawout First Nation (2019) Tsawout First Nation Marine Use Study. https://iaac-aeic.gc.ca/050/evaluations/document/130284?culture=en-CA

Turner NJ (2000) Coastal Peoples and Marine Plants on the Northwest Coast. In: Tides of Technology: Proceedings of the 26th Annual Conference of the International Association of Aquatic and Marine Science Libraries and Information Centers (IAMSLIC). pp 69–76

Turner NJ (2020) Stories from The Front Lines From "taking" to "tending": learning about Indigenous land and resource management on the Pacific Northwest Coast of North America. ICES J Mar Sci 77:2472–2482. https://doi.org/10.1093/icesjms/fsaa095

Turner N, Bell A (1973) The Ethnobotany of the Southern Kwakiutl Indians of British Columbia. Econ Bot 27:257–310

Turner NJ, Berkes F, Stephenson J, Dick J (2013) Blundering intruders: extraneous impacts on two indigenous food systems. Hum Ecol 41:563–574. https://doi.org/10.1007/s10745-013-9591-y

United Nations Environment Programme (2023) Into the blue: Securing a sustainable future for kelp forests. https://wedocs.unep.org/handle/20.500.11822/42255

US Geological Survey (2016) USGS OFR 2010-1251: Moderate-resolution sea surface temperature data for the nearshore North Pacific, Fig. 3. https://pubs.usgs.gov/of/2010/1251/figure3.html. Accessed 8 Aug 2024

Valdez MC, Zaragoza ES, Belda DL et al (2003) Effect of climatic change on the harvest of the kelp *Macrocystis pyrifera* on the Mexican Pacific coast. Bull Mar Sci 73:546–556. https://www-ingentaconnect-com.proxy.lib.sfu.ca/content/umrsmas/bullmar/2003/00000073/00000003/art00002

Visch W, Kononets M, Hall POJ et al (2020) Environmental impact of kelp (*Saccharina latissima*) aquaculture. Mar Pollut Bull 155:110962. https://doi.org/10.1016/j.marpolbul.2020.110962

Visch W, Layton C, Hurd CL et al (2023) A strategic review and research roadmap for offshore seaweed aquaculture—a case study from southern Australia. Rev Aquac 15:1467–1479. https://doi.org/10.1111/raq.12788

Vranken S, Wernberg T, Scheben A et al (2021) Genotype–Environment mismatch of kelp forests under climate change. Mol Ecol 30:3730–3746. https://doi.org/10.1111/mec.15993

Waqas MA, Hashemi F, Mogensen L, Knudsen MT (2024) Environmental performance of seaweed cultivation and use in different industries: a systematic review. Sustain Prod Consump 48:123–142. https://doi.org/10.1016/j.spc.2024.05.001

Yakhin OI, Lubyanov AA, Yakhin IA, Brown PH (2017) Biostimulants in plant science: a global perspective. Front Plant Sci 7:2049. https://doi.org/10.3389/FPLS.2016.02049/BIBTEX

Sarah B. Gutzmann (she/her) is a social-ecological systems scientist with an interdisciplinary background spanning coastal ecology, conservation biology, social science, toxicology, public education, and communication. She holds a BSc (Hons) in Environmental Science as well as a Master of Resource Management (MRM). Currently, Sarah is a PhD candidate in Simon Fraser University's Ocean Equity Lab and is a Student Fellow with the Nippon Foundation Ocean Nexus. Sarah's doctoral research focuses on fostering regenerative kelp stewardship by collaborating with communities, regulators, and industries. Sarah's research is supported by awards including the SSHRC Doctoral Fellowship and the Mitacs Accelerate Internship, which is conducted in partnership with Ocean Regenerative Aquaculture.

Dr. Andrés M. Cisneros-Montemayor specializes in applied resource economics and is engaged in research and policy on human development and the Blue Economy, including artisanal and Indigenous fisheries, ecotourism, and emerging ocean sectors in developing and developed regions all over the world. Dr. Cisneros-Montemayor has published over 100 peer-reviewed studies and book chapters and is an active participant and advisor in international and national ocean policy.

21 Starting from Scratch: Evaluating Research Investment to Underpin Seaweed Aquaculture in Aotearoa New Zealand (A-NZ)

Cecilia Biancacci, Daniel Crossett, Nicholas Scott, and Paul M. South

Abstract

Seaweed aquaculture is one of the fastest-growing sectors globally, contributing to the production of food, feed, biostimulants, biomaterials and ecosystem services. The sector has tripled its value over the past two decades, with most seaweed production concentrated in Asia. In other regions seaweed aquaculture production is limited or in early development. This is particularly true for Aotearoa New Zealand (A-NZ), where there is only a small seaweed sector based on wild-harvested and beach-cast resources. However, the A-NZ Government has signalled significant intent to develop a seaweed sector and has been funding research to that end. This chapter explores the status of the seaweed aquaculture research in A-NZ, focusing on identifying current areas of research, sources of investment, ongoing efforts and barriers. We emphasize the critical role of research investment, as it serves to bridge academia and industry, promoting innovation and supporting the development of a sustainable and economically viable seaweed sector in A-NZ. While some initial investments have been made, scaling up the sector effectively will require greater and more diversified funding sources. These will be essential to advance fundamental and applied research, upscale current initiatives and position seaweed as a central pillar of A-NZ's aquaculture sector.

Keywords

Phyconomy · Macroalgae · Cultivation · Applications · Investment · Aquaculture development

C. Biancacci (✉) · D. Crossett · N. Scott · P. M. South
Cawthron Institute, Nelson, New Zealand
e-mail: cecilia.biancacci@cawthron.org.nz; daniel.crossett@cawthron.org.nz; nicholas.scott@cawthron.org.nz; paul.south@cawthron.org.nz

21.1 Introduction

Global seaweed production has more than tripled in the past 20 years reaching an annual production of more than 32 t wwt in 2018 and is dominated (86%) by countries in Asia (FAO 2022). Nonetheless, the seaweed aquaculture sector is emerging globally providing seaweed biomass for food and feed production, extraction of valuable compounds, fertilisers, bioremediation, biomaterials and ecosystem services as part of a circular economy (Buschmann et al. 2017; Zhang et al. 2022; Fricke et al. 2024). There is an increasing demand for seaweeds and seaweed products worldwide, which requires significant investment in research, development and infrastructure to support the cultivation of seaweed resources at scale. Aotearoa New Zealand (A-NZ) has one of the largest Exclusive Economic Zones (EEZ—around 4.2 million km^2) worldwide, an extensive coastline (around 15,000 to 18,000 km), and high seaweed biodiversity of more than 1,000 species, of which around 40% are endemic (Nelson 2013). Together, these features position the country for a sustainable seaweed sector, yet there is only a very small seaweed-based industry in A-NZ that is predominantly based on the wild-harvest of beach-cast seaweed, or the harvesting of an invasive species from existing (mussel) aquaculture structures (Fig. 21.1). To date, there are 21 seaweed focused industries in A-NZ with most producing fertilisers and biostimulants (52.4% of total, Fig. 21.1). Currently, seaweed aquaculture in A-NZ is even less well developed than its wild-harvest industries, for example, there are no commercial seaweed nurseries, and the limited sea-based aquaculture relies entirely on seaweed recruitment from the wild. However, there is significant government and commercial interest in developing a sustainable seaweed aquaculture industry in A-NZ (ANZSA 2022; Stenton-Dozey et al. 2021) and as with any new industry, the journey from concept to commercial scalability and viability is complicated and starts with a crucial step: research.

A. T. Critchley et al. (eds.), *Advances in Temperate Phyconomy: Algal Harvest and Cultivation in Globally Distributed Temperate Waters*, Developments in Applied Phycology 15, https://doi.org/10.1007/978-3-032-02955-3_21

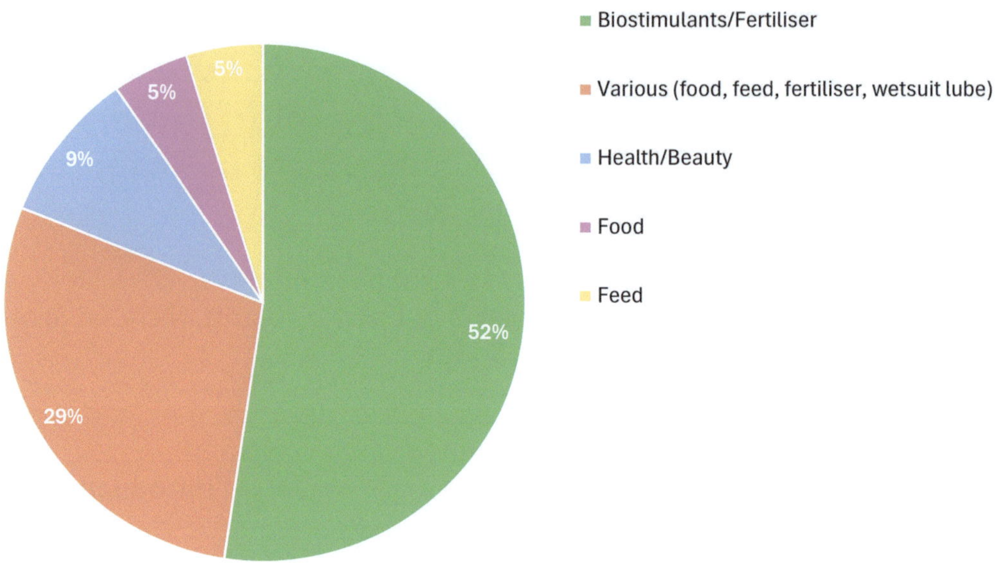

Fig. 21.1 Commercial seaweed sectors in A-NZ. (Updated from Bradly et al. 2021). Contribution of each sector is presented as percentage of the total

In 2019, the A-NZ government released a strategy for accelerating growth of its aquaculture sector from NZ$1.5 billion to NZ$3 billion in annual sales by 2035 (NZ government Aquaculture Strategy, MPI 2019; Accelerate the Aquaculture Strategy, MPI, New Zealand Government 2021). One of the key steps toward achieving this ambitious goal is crop and product diversification aimed at building resilience across export markets. To this end, seaweed aquaculture has been highlighted as a strong candidate for future development and is envisioned as a low-impact, highly sustainable addition to the aquaculture in A-NZ, which is currently focused on three species: green-lipped mussels (*Perna canaliculus*), Pacific oysters (*Crassostrea gigas*) and Chinook salmon (*Oncorhynchus tshawytscha*).

This chapter provides a case study of the seaweed aquaculture research and funding landscape in A-NZ that is currently supporting the development of a seaweed sector. Following a roadmap (Fig. 21.2) that illustrates how scientific knowledge can support and boost a thriving seaweed-based economy in A-NZ, we will identify and explore key areas for current and future research investment and analyse potential pathways to commercialize seaweed-based products that leverage A-NZ's unique environment and resources. Additionally, we highlighted the critical role of collaboration among the different parties that are fundamental players, such as research institutions, government agencies, indigenous communities and private enterprises in overcoming industry challenges and fostering sustainable growth.

21.2 Overview of Current Seaweed Aquaculture Research in New Zealand

21.2.1 Current State of Research

There is a rich history of seaweed research in A-NZ and the well-established research sector tackles a diverse and growing range of topics (Hurd et al. 2004) including aspects of seaweed biology (Bonin et al. 1987; Nelson and Knight 1995), physiology (Phillips and Hurd 2003; Tait and Schiel 2010; Hepburn et al. 2011), ecology (Schiel 1988; Thomsen et al. 2019; Cornwall et al. 2023), biogeography (Parsons 1985; Nelson et al. 2021), biochemistry (Hurd 2000; Battershill 2024; Hrstich-Manning and Aguirre 2024) and management (Nelson et al. 2015; South et al. 2017). By contrast, fewer studies have researched aspects of seaweed cultivation in A-NZ although the number of aquaculture-focused studies is growing (Table 21.1). Overall, published aquaculture-related work in A-NZ has addressed three species of kelp (*Macrocystis pyrifera*, *Ecklonia radiata* and *Lessonia variegata*) (Phaeophyceae), two species of *Gracilaria* (Rhodophyta), one fucoid (*Hormosira banksii*) (Phaeophyceae), *Asparagopsis armata* (Rhodophyta) and *Ulva* spp. (Chlorophyta). Interest in seaweed aquaculture appears to have begun in the early 1990s and was initially focused on Gracilariales and Gigartinales and to a lesser extent on the kelp, *L. variegata*, although there was a paucity of publications in A-NZ between 2005 and 2017. The great-

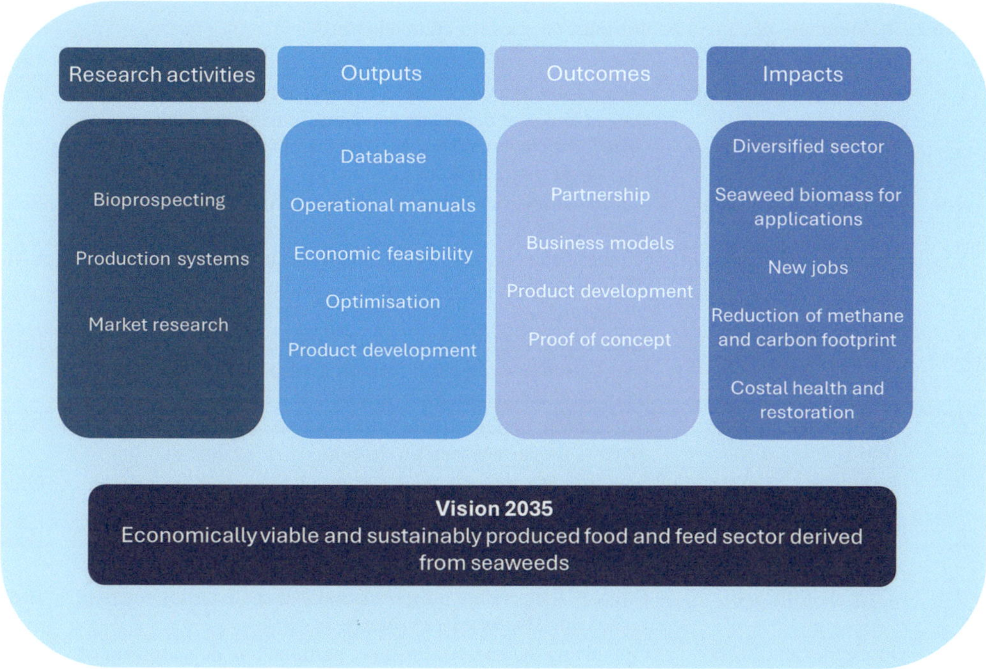

Fig. 21.2 Roadmap for seaweed aquaculture in Aotearoa New Zealand (A-NZ)

est numbers of papers about seaweed aquaculture in A-NZ were published in 2022 (3) and 2024 (4). Some of the key findings have delineated aspects of the cultivation of *Asparagopsis armata*, a red seaweed with anti-methanogenic potential when fed to ruminants (Mihaila et al. 2023a, b, 2024; Hall et al. 2025), laid some of the groundwork for fucoids (Loy and Jeffs 2025; Crossett et al. 2025) and kelp aquaculture in A-NZ from the nursery to the sea (Nelson et al. 2005; Le et al. 2022a, b; Praeger et al. 2022; Nepper-Davidsen et al. 2023b; Biancacci et al. 2024; Lawton and Magnusson 2024; Neill et al. 2024; Lawton and Praeger 2025) and explored the aquaculture of novel carrageenan and agar producing species (Pickering et al. 1993, 1995; McNeill et al. 2003, 2017; Ross et al. 2024). However, the overall small number of studies Lawton and Magnusson (2025) indicated that seaweed cultivation and aquaculture have not been a strong focus in A-NZ and clearly more fundamental and applied research will be necessary to kick-start a seaweed aquaculture industry.

There has been a strong history of exploration of the biochemical properties of A-NZ seaweeds, with some having been shown to be valuable resources for compounds such as proteins, carbohydrates, fibres, minerals and pigments (Hurd et al. 2023). Polysaccharides are some of the most researched compounds in the A-NZ seaweed flora due to their wide use as stabilisers and gelling agents (i.e. agar, alginates and carrageenans). Research has also focused on bioactive compounds, particularly fucoidans extracted from *Undaria pinnatifida* (Phaeophyceae) for applications in the medical, nutraceutical, biomaterial and food sector (Wang et al. 2014; Lu et al. 2018; Zhao et al. 2018), bromoform from *Asparagopsis armata* as feed supplement to reduce methane emissions in cattle and sheep (Glasson et al. 2022; Mihaila et al. 2022; Fennessy et al. 2025) and mycosporine-like amino acids as UV sunscreen compounds from *Pyropia plicata* (Rhodophyta) (Diehl et al. 2019).

21.2.2 Key Stakeholders in Seaweed Aquaculture Research

There are a wide range of stakeholders in the A-NZ seaweed sector who contribute a diverse set of perspectives and resources to seaweed research. Developing a successful seaweed aquaculture industry in A-NZ requires a collaborative effort among the stakeholders, which include research institutes, government agencies, iwi and Māori enterprises and the private sector.

The principal stakeholders in seaweed research and activities in New Zealand currently include

- Government agencies (e.g., New Zealand Ministry of Primary Industries, New Zealand Ministry of Business, Innovation and Employment)
- Crown research institutes (CRIs, the National Institute of Water and Atmospheric Research (NIWA), Plant and Food Research, SCION)
- Universities (University of Auckland, University of Waikato, University of Otago, Victoria University of Wellington, Auckland University of Technology)
- Independent research organisations (IROs, e.g. Cawthron Institute)

Table 21.1 Summary of current publications on seaweed cultivation in A-NZ

#	References	Title	Group	Species Studied
1	Pickering et al. (1993)	Comparison of growth rate in culture, dry matter content, agar content and agar quality of two New Zealand red seaweeds, *Gracilaria chilensis* Bird, McLachlan et Oliveira and *Gracilaria truncata* Kraft	Red	*Gracilaria chilensis, G. truncata*
2	Pickering et al. (1995)	A preliminary trial of a spray culture technique for growing the agarophyte *Gracilaria chilensis* (Gracilariales, Rhodophyta)	Red	*Gracilaria chilensis*
3	McNeill et al. (2003)	Field trials to optimise timing and frequency of pruning for cultivation of a New Zealand carrageenophyte, *Gigartina atropurpurea*.	Red	*Sarcothalia atropurpurea*
4	Nelson (2005)	Life history and growth in culture of the endemic New Zealand kelp *Lessonia variegata* J. Agardh in response to differing regimes of temperature, photoperiod and light	Brown	*Lessonia variegata*
6	McNeill and Falshaw (2017)	Aquaculture of *Sarcothalia atropurpurea* tetraspores in New Zealand for carrageenan production	Red	*Sarcothalia atropurpurea*
7	Lawton et al. (2021)	Selection of temperate *Ulva* species and cultivars for land-based cultivation and biomass applications	Green	*Ulva* spp.
8	Le et al. (2022a, b)	Reproduction, hatchery and culture applications for the giant kelp *(Macrocystis pyrifera)*: a methodological appraisal	Brown	*Macrocystis pyrifera*
9	Le et al. (2022a, b).	Effect of temperature on sporulation and spore development of giant kelp (*Macrocystis pyrifera*)	Brown	*M. pyrifera*
10	Praeger et al. (2022)	Optimising the zoospore release, germination, development of gametophytes and formation of sporophytes of *Ecklonia radiata*	Brown	*Ecklonia radiata*
11	Mihaila et al. (2023a, b)	Early hatchery protocols for tetrasporogenesis of the antimethanogenic seaweed *Asparagopsis armata*	Red	*Asparagopsis armata*
12	Nepper-Davidsen et al. (2023a, b)	Line configuration and farming depth markedly affect survival and growth in the kelp *Ecklonia radiata*	Brown	*Ecklonia radiata*
13	Biancacci et al. (2024)	Intrapopulation variation in reproduction of *Lessonia variegata*	Brown	*Lessonia variegata*
14	Lawton and Magnusson (2024)	Effects of seeding twine type and seeding density on hatchery performance and initial at-sea cultivation performance of the kelp *Ecklonia radiata*	Brown	*Ecklonia radiata*
15	Mihaila et al. (2024)	Moderate temperature and water flow increase growth during the nursery phase of *Asparagopsis armata*.	Red	*Asparagopsis armata*
16	Neill et al. (2024)	Notes on the experimental laboratory and field culture of *Lessonia variegata* (Laminariales)	Brown	*Lessonia variegata*
17	Ross et al. (2024)	Seasonal biomass productivity and bioremediation performance of the novel aquaculture target species *Gracilaria transtasmanica* in nutrient-enriched and turbid conditions	Red	*Gracilaria transtasmanica*
18	Hall et al. (2025)	The effects of water motion on fragmentation of *Asparagopsis armata*	Red	*Asparagopsis armata*
19	Lawton and Praeger (2025)	Effects of nutrient concentration, germanium dioxide and seeding density on sporophyte production in the kelp *Ecklonia radiata*	Brown	*Ecklonia radiata*
20	Loy and Jeffs (2025)	Assessment of the potential of the fucoid seaweed, *Hormosira banksii*, as a source of fucoidan and coculture on a Pacific oyster farm in New Zealand	Brown	*Hormosira banksii*
21	Crossett et al. (2025)	Experimental assessment of growing substrates for the early life stages of five large brown seaweeds	Brown	*Ecklonia radiata, Macrocystis pyrifera, Cystophora scalaris, Cystophora retroflexa, Landsburgia quercifolia*

- Iwi (Māori 'tribes', e.g. Te Whānau-ā-Apanui, Ngai Tahu)
- Private organisations and companies (e.g., Agrisea NZ Seaweed Ltd, Premium Seas Ltd)

Each of these stakeholders has fundamental roles in the development of the seaweed sector in A-NZ. The government agencies, for example, legislate and regulate aquaculture activities and fund research and development. Funded research is typically done by the CRIs, universities and IROs, often in conjunction with iwi, and private organisations, with the direction of the research being driven by wider stakeholder interests that range from 'blue-sky' science to targeted, commercially focused activities such as hatchery development.

21.3 Current Funding Landscape in New Zealand

The funding landscape for seaweed aquaculture research and industry development in New Zealand reflects a mix of public and private investment with the A-NZ government playing a pivotal role via three key ministries. Overall, the A-NZ Government has administered around NZ$46 million dollars in funding to support seaweed aquaculture and industries between 2018 and 2024.

21.3.1 Current Public Sector Funding

21.3.1.1 Government Agencies

Key funding bodies such as A-NZ Ministry of Business, Innovation and Employment (MBIE) and the A-NZ Ministry for Primary Industries (MPI) offer competitive access to funds such as MBIE Endeavour Fund Smart Ideas (https://www.mbie.govt.nz/science-and-technology/science-and-innovation/funding-information-and-opportunities/investment-funds/endeavour-fund/smart-ideas-call-for-proposals-2025-investment-round-endeavour-fund), MBIE Endeavour Fund Research Programmes (https://www.mbie.govt.nz/science-and-technology/science-and-innovation/funding-information-and-opportunities/investment-funds/endeavour-fund/smart-ideas-call-for-proposals-2025-investment-round-endeavour-fund) and the Sustainable Food and Fibre Futures (https://www.mpi.govt.nz/funding-rural-support/sustainable-food-fibre-futures/about-sustainable-food-and-fibre-futures/) (SFFF). For example, the MBIE Endeavour Fund Smart Ideas and Research Programmes had a total investment of NZ$72 million in 2024–2025, of which NZ$17 million was assigned to smart ideas and NZ$55 million to research programmes. Together, these are the largest competitive research funding mechanism in A-NZ. In 2024–2025 10% of the MBIE Endeavour Fund was allocated to algal research (including macro and microalgae) related projects, accounting for a total of around NZ$20.3 million, although only one of these programmes was focused on phyconomy (seaweed biochemistry and production valued at around NZ$11 million), whereas the other was focused on the coastal ecology of phytoplankton (MBIE 2025). For smart ideas, 3.8% of the funding was allocated to algae-related research (novel microalgal feed for aquaculture species and on carbon removal and effects of carbonate on phytoplankton) for a total investment of around NZ$2 million (MBIE 2025).

The SFFF funding scheme aims to support problem-solving and innovation in A-NZ's food and fibre sector by co-investing in initiatives that make a positive and lasting difference. Up to NZ$75 million is available each year and the scheme is based on a hybrid model of co-funding by government and industrial entities. In 2024–2025 no SFFF funding was allocated to algal research according to a database search, with the most recent investment of NZ$2 million in 2021 for a three-year project to develop a pilot programme on seaweed cultivation in the Hauraki Gulf and Bay of Plenty, in collaboration with University of Waikato, Premium Seas Limited, AgriSea and EnviroStrat Ltd (Lawton and Magnusson 2024). Other government agencies providing funding that is accessible to seaweed research includes the Callaghan Innovation (https://www.callaghaninnovation.govt.nz/) and the Royal Society of New Zealand. Some funds, such as the Royal Society of New Zealand's Catalyst Leader and Strategic and Seeding (https://www.royalsociety.org.nz/what-we-do/funds-and-opportunities/catalyst-fund/) funds, are specifically designed to foster international collaboration. Around NZ$4 million in 2024 were available within the catalyst seeding funding, with only NZ$75,500 allocated to research relevant to algae.

The 10-year long Science for Technological Innovation National Science Challenge (SfTI) investment programme funded by MBIE has supported five seaweed-related research projects. One key project sought to establish a Seaweed Sector Framework with a compendium of 'stocktake and characterisation of A-NZ's seaweed sector' documents that addressed environmental considerations for seaweed aquaculture (Clark et al. 2021; Howarth et al. 2023), focusing on the market and regulatory environments (Bradly et al. 2021), and species characteristics and consideration of the role and worldview of Māori with respect to seaweed aquaculture (Wheeler et al. 2021).

Additionally, the A-NZ government provides strategic loans to aid industry development in the regions (typically rural areas). In 2022, for example, the Kānoa Regional Strategic Partnership Fund (RSPF) supported a commercial seaweed-based nanocellulose manufacturing plant with an investment loan of NZ$750,000. In 2019–2020, RSPF administered government funding that invested NZ$500,000 to conduct a scale assessment of cultivating and harvesting *Asparagopsis armata*, and NZ$6 million in early 2020 to enable the Cawthron Institute to develop a National Algae Research Centre with research and production facilities, for macroalgae and microalgae (Nissanka and Stevenson 2023).

Information on the investment focus, funding recipients and organization from government organisations in the period from 2018 to 2024, were compiled using source data from MPI (Fig. 21.3). Overall, there were 29 funded programmes although three were government-funded loans for research infrastructure development. Of these 29 programmes, funding values ranged from NZ$5,000 to NZ$11,432,930 and had a mean value of NZ$1,592,643 ± NZ$2,510,441 SD. The most successful funding recipients were based at universities, followed by independent research institutes (IRIs), commercial entities and lastly crown research institutes (CRIs). The main body allocating funding for seaweed aquaculture research was MBIE (Fig. 21.3b). It was also evident that funding was heavily weighted towards products (e.g. bioactive compounds), followed by infrastructure investments (largely loans), cultivation and lastly economics (Fig. 21.3c). An interesting trend in terms of the total investment per year from 2018 to 2024 showed that this decreased almost linearly from 2018 when the total investment was NZ$11 million, to the lowest values in 2022 and 2023, when investment was below NZ$2 million followed by a sharp increase in 2024 when investment in seaweed research leapt to its greatest value in this period at just below NZ$12.5 million (Fig. 21.3d, values are expressed in NZD, 2025).

21.3.2 Current Private and Commercial Investment

Private funding for seaweed aquaculture is lacking and limited compared to investment in the more established aquaculture sectors in A-NZ. Little information is available publicly on private investment in seaweed aquaculture. A related example is the NZ$7 million in private investment secured by Supreme Biotechnologies for production of the microalga *Haematococcus lacustris* (formerly *H. pluvialis*) (Chlorophyta) for extraction of astaxanthin, which can be incorporated in dietary supplements (Idealog 2015). Often, private investment occurs under the co-investment model of MPI's SFFF, although the relative contributions of the gov-

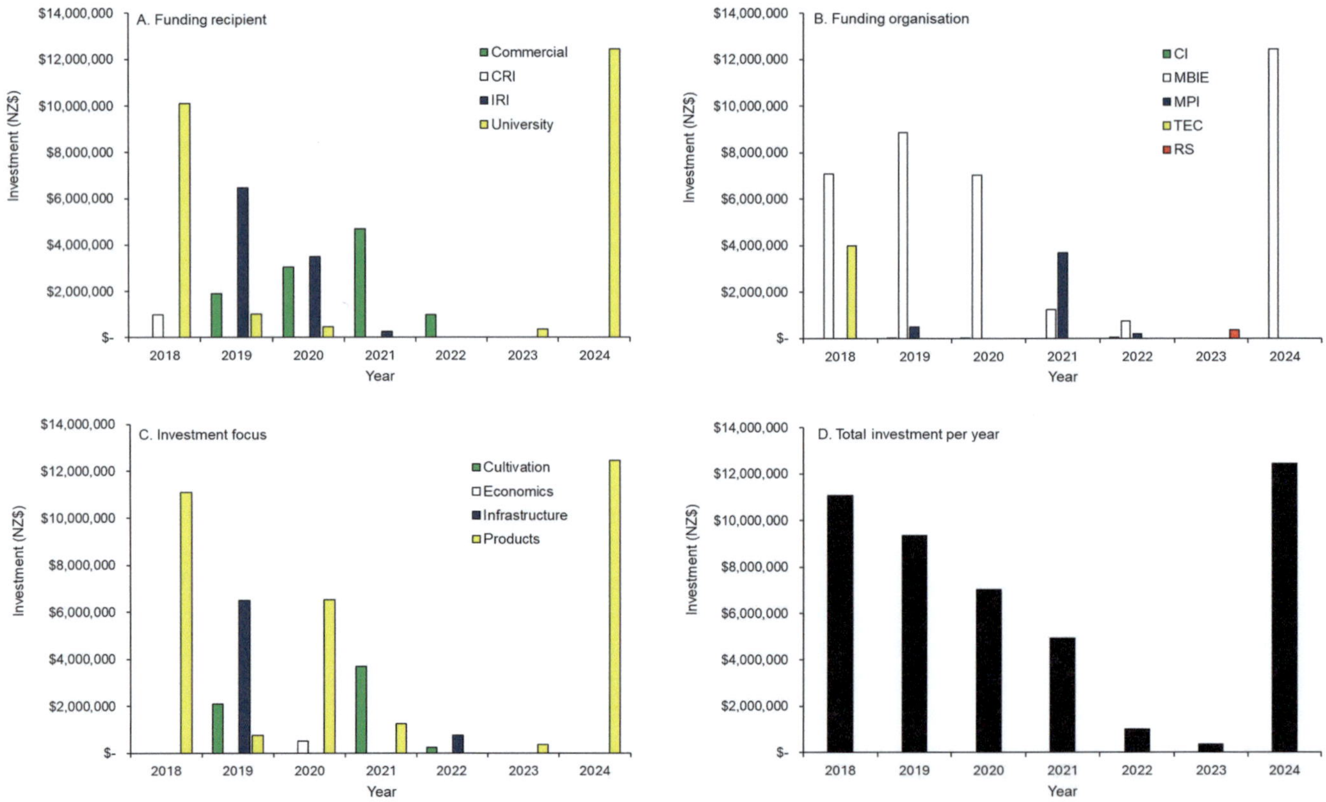

Fig. 21.3 Public sector (government) investment to develop the New Zealand phyconomy between 2018 and 2024. (**a**) shows funding investment sectors (CRI Crown Research Institutes, IRI Independent Research Institutes). (**b**) shows the government organisations providing the funding (CI Callaghan Innovation, MBIE the NZ ministry of Business, Innovation and Employment, MPI the NZ Ministry of Primary Industries, TEC the Tertiary Education Commission, RS the Royal Society of New Zealand). (**c**) shows the primary focus of the awarded funding. (**d**) shows the total investment in seaweed aquaculture research by the government from 2018 to 2024

ernment and private funder vary among grants. In the period between 2018 and 2024 there were seven seaweed-related SFFF grants with values of between NZ$75,200 and NZ$1,993,920 that have funded projects including an investigation into harvesting of *Undaria pinnatifida* (Phaeophyceae), seaweed aquaculture for bioremediation and the establishment of regenerative aquaculture following the US-based GreenWave initiative (GreenWave 2024). It is possible that there is considerable uncertainty around the prospects for profitability that is preventing significant private and commercial investment in A-NZ.

21.3.3 Current International Collaborations and Funding Opportunities

Opportunities for funding and knowledge-sharing exist through partnerships with global research networks, such as through the Royal Society's Catalyst funds and more recently with access to schemes such as Horizon Europe—Research and innovation funding. Since February 2024, New Zealand is associated with Pillar II of the Horizon Europe funding scheme, which has a total budget of €53.5 billion (NZ$95 billion) and that could be an opportunity to secure funding for seaweed research, leveraging collaboration with international partners. However, to date no projects on seaweeds have been reported among the successful bids. Other schemes through which A-NZ can collaborate with international partners include the Cooperative Research Centre (CRC) funding scheme in Australia. An example is the Blue Economy CRC with a total investment of almost AUD$12.5 million and a focus on sustainable aquaculture practices including seaweed, marine renewable energy, maritime engineering, environmental assessments, policy and regulation that includes partners from nine countries including A-NZ (Blue Economy CRC 2022).

21.4 Research Activities

Developing a novel seaweed aquaculture sector in A-NZ requires significant foundational research to establish a strong scientific and technical groundwork necessary for long-term success. By addressing key knowledge gaps and developing innovative methodologies and technologies, A-NZ can develop a sustainable seaweed industry that contributes to global food security, climate resilience and marine conservation. Collaboration will play a pivotal role in success of the seaweed sector and is recognised and embedded in the core investment streams. Partnerships with iwi and Māori enterprises, international research institutions, government and private investors can support the sector by sharing costs, diversifying the investment strategies and accelerating progress. This section examines and identifies the critical research activities, including (1) Bioprospecting, (2) Production systems and (3) Market research (Fig. 21.1).

21.4.1 Bioprospecting

21.4.1.1 Current Species and Databases

Bioprospecting could be a pivotal step in enabling seaweed research and subsequent investment in A-NZ. Although the country boasts a rich diversity of seaweeds (Fig. 21.4), with more than 1,000 reported species and around 40% endemism, very little is known about the biology and biochemistry of many of these seaweeds. Similarly, cultivation has only been explored for six species (Table 21.1) and there is often only limited data for species that have been previously identified as candidates for commercial aquaculture (Table 21.2). It follows that significant research is needed to unravel the aquaculture potential of A-NZ seaweeds.

A key goal of bioprospecting is to develop a database of the biochemical components of a suite of seaweed species, with a view to developing products and industries around species with promising characteristics. Live culture collections can be a key tool for bioprospecting exercises as they enable continuous assessment of the cultures and pave the way for selective breeding and scale up of promising species. The Cawthron Institute Culture Collection of Microalgae (CCCM; Cawthron Culture Collection of Microalgae—Cawthron Institute—https://www.cawthron.org.nz/ciccm/#:~:text=The%20Cawthron%20Institute%20Culture%20Collection%20of%20Microalgae%20%28CICCM%29,to%20ensure%20the%20safety%20of%20New%20Zealand%20seafood) is A-NZ's largest live-culture collection of microalgae and is a repository of over 700 species of living and cryopreserved microalgae (Rhodes et al. 2016). However, there are currently no significant biochemistry databases or live-culture collections for seaweeds in A-NZ, likely due to the high costs and uncertainty around success associated with such exercises. Despite this, a model similar to that of the CCCM could be envisioned for seaweeds, with species being classified based on key attributes including their biology, ecology, current threats, distribution, biochemical composition and potential applications (Table 21.2). While there are some examples of worldwide seaweed databases (e.g. AlgaeBase (Guiry and Guiry, 2025); Seaweed.ie (Guiry, 2025); Seaweed Metabolite database, Davis and Vasanthi 2011) with some related to specific regions (e.g. Paul et al. 2020; Bursic et al. 2022), a comprehensive database of key characteristics of A-NZ seaweeds is likely essential to support the development of its seaweed industry.

Fig. 21.4 Pictures of common seaweeds in Aotearoa New Zealand waters (**a**) *Lessonia variegata*, (**b**) juveniles of *Macrocystis pyrifera* and *Durvillaea* sp., (**c**) *Ecklonia radiata*, (**d**) reproductive blades at the base of an adult *Macrocystis pyrifera*, (**e**) assemblage of seaweeds including *Caulerpa* sp. and fucoids, (**f**) *Asparagopsis armata* gametophytes, (**g**) *Ulva* sp., (**h**) *Pyropia plicata*. (Photo credits Dan Crossett, Paul South, Cecilia Biancacci)

Table 21.2 Summary of the main species of seaweed identified as potential candidates for further cultivation and commercial applications in A-NZ (White and White 2020) and relative publications on cultivation and biochemistry in A-NZ and for A-NZ species

Species	Control of life cycle	Biochemistry	Cultivation	References
Macrocystis pyrifera	Green	Green	Green	McKee et al. (1992); Lamare and Wing (2001); Smith et al. (2010); Le et al. (2022)a, b; Battershill (2024); Bunting et al. (2024); DM et al. (2024)
Ecklonia radiata	Green	Green	Green	Novaczek (1984); Smith et al. (2010); White and White (2020); Praeger et al. (2022); Nepper-Davidsen et al. (2023)b, a; Lawton and Magnusson (2025) to Ecklonia radiata
Lessonia variegata	Green	Green	Green	Nelson et al. (2005); White and White (2020); Biancacci et al. (2024); Neill et al. (2024)
Durvillaea spp.	Red	Green	Red	Smith et al. (2010); White and White (2020)
Hormosira banksii	Red	Green	Red	Cooke et al. (2004); Smith et al. (2010); Edmunds (2023); Loy and Jeffs (2025)
Undaria pinnatifida	Red	Green	Red	Hemmingson et al. (2006); Smith et al. (2010); Billakanti et al. (2013); Mak et al. (2013 2014); Morelissen et al. (2013); Boulom et al. (2014); Wang et al. (2014); Balbas et al. (2015); Zhou et al. (2015); Lu et al. (2018); Koh et al. (2019); White and White (2020); Gan and Baroutian (2022); Liu et al. (2022); Battershill (2024)
Ulva spp.	Green	Green	Green	Lamare and Wing (2001); Smith et al. (2010); Magnusson et al. (2016 2019); White and White (2020); Kidgell et al. (2021); Lawton et al. (2021); Battershill (2024); Hrstich-Manning and Aguirre (2024)
Codium spp.	Red	Green	Red	(Lamare and Wing (2001); Hrstich-Manning and Aguirre (2024)
Caulerpa sp.	Red	Green	Red	Hurd et al. (2004)
Asparagopsis armata	Green	Green	Green	White and White (2020); Mihaila et al. (2022 2023b a2024); Hrstich-Manning and Aguirre (2024); Fennessy et al. (2025); Hall et al. (2025)
Porphyra and *Pyropia* spp.	Red	Green	Red	Nelson and Conroy (1989); Friedlander et al. (1989); Brown et al. (1990); Smith et al. (2010); Diehl et al. (2019); White and White (2020); Gambichler et al. (2021); Battershill (2024)
Gigartina spp.	Red	Green	Red	Pickmere et al. (1973); Parsons et al. (1977); Falshaw et al. (2003); Falshaw and Furneaux (2009); Neill et al. (2016 2018); Mcneill and Falshaw (2017); White and White (2020)
Coralline algae	Red	Green	Red	Hrstich-Manning and Aguirre (2024)

Red data deficient, *green* data available

21.4.1.2 Consents and Permits

An important first step in all research activities in A-NZ is obtaining appropriate licenses, consents and permits. Research is typically conducted under the Special Permitting systems outlined in the Fisheries Act 1996, which is managed by the New Zealand Ministry of Primary Industries (MPI). This special permit provides multiple schedules of use, based on factors such as regions and species of interest, and whether the research is for commercial gain. However, projects may be amended to increase their scope and research focus. For specific projects, further consultation may be necessary, such as with regional councils, mana whenua (indigenous peoples of New Zealand) and iwi (tribes) and other end users associated with the research. A necessary next step is to gather appropriate resource consents, which may be required if the research involves deploying material into the marine environment, for example to a pilot-scale seaweed farm. Generally, researchers should expect the permit and consent process to take 3–12 months. A summary of some generalised steps that might be required when submitting an application for a permit is listed below.

In A-NZ, applications may need to be made to multiple key government agencies for sample collection and marine research permits. Points to consider when submitting an application for a permit are listed below:

- The Ministry of Primary Industries (MPI)[1] regulates biosecurity permits and any research involved in sampling, moving and studying marine resources. As part of a permit application, it may be required to demonstrate that local iwi have been consulted and support the project.
- Fisheries New Zealand (FNZ; part of MPI)[2] regulates special permits for sampling and translocation of marine organisms.
- The Department of Conservation (DOC)[3] regulates permits for work within conservation areas and marine reserves, or on species protected by the Wildlife Act 1953.
- A resource consent may be required for the project activity under the Resource Management Act 1991 (RMA).[4] A resource consent is permission from the local council for an activity that might affect the environment and that is not allowed 'as of right' in the district or regional plan (Ministry for the Environment 2021a, b). Information on permitted and restricted activities in the marine environment can be found within each region's regional coastal plan.

Identifying and forming appropriate partnerships early in the development of a project can increase the likelihood of successfully obtaining the relevant permits. For example, identifying and engaging with iwi in regions of interest is not only a necessary part of the process but will also highlight key mātauranga (cultural knowledge) around species of interest. An important consideration in A-NZ is the Wai 262 treaty-settlement claim. The Waitangi Tribunal released a report in 2011 that brought about the Wai 262 claim, which addresses Māori cultural intellectual property and the use of Māori knowledge pertaining to indigenous flora and fauna (Wai 262 2025; Potter and Māngai 2022). The management of this claim is still evolving within the A-NZ government (Wheeler et al. 2021); however, there will likely be implications for the propagation and transplantation of seaweed and the commercial applications of some species. For example, many species hold cultural significance, often due to their extensive historical use by Māori (Zemke-White et al. 2005). For example, *Durvillaea* spp. (Bull kelp or Rimurapa) (Phaeophyceae) have been used for generations by mana whenua for food and pōhā, or bull kelp bags, for preservation of food such as mutton birds (Te Runanga o Kaikoura 2007; Reid and Rout 2019).

21.4.2 Production Systems

21.4.2.1 Biological and Ecological Studies

Identifying and mapping the distributions of native seaweeds, their habitats and population dynamics in A-NZ waters represents fundamental groundwork for meaningful and applicable seaweed aquaculture research. Such data are important for site and species selection and predicting future distributions of various seaweeds based on the habitat preferences of the potential crop species. As such, determining the thermal tolerances of candidate species for seaweed aquaculture, either via correlating current distribution data to seawater temperature data or via physiological studies in the laboratory, is a high priority.

Significant progress has been made toward understanding seaweed growth, reproduction and cultivation methods, particularly for key species such as kelps, *Ulva* spp., *Gracilaria* and *Gigartina* spp. and *Asparagopsis armata*. Similarly, characterization of species based on their biochemical composition, especially their potential for agar, carrageenans and fucoidans production, has progressed in A-NZ (Tables 21.1 and 21.2). Despite this progress, the A-NZ seaweed industry is still some distance from being a viable economic concern.

[1] MPI permit information can be found at https://animalplantimportpermit.mpi.govt.nz/

[2] FNZ permit information can be found at https://www.mpi.govt.nz/fishing-aquaculture/fisheries-management/how-to-apply-fisheries-special-permit/

[3] DOC permit information can be found at https://www.doc.govt.nz/get-involved/apply-for-permits/marine-reserve-research/

[4] RMA consent information can be found at https://environment.govt.nz/publications/applying-for-a-resource-consent/

Further research is needed to validate initial findings, optimize large-scale farming, understand and enhance climate resilience and ecosystem services and unlock new biotechnological applications that could expand the ecological and commercial potential of seaweed aquaculture in A-NZ.

Understanding seaweed life cycles and reproductive strategies is critical knowledge for seaweed aquaculture. Seaweeds have complex life cycles that vary across phyla and families, and which must be controlled, if closed life cycle aquaculture is a goal. Understanding the influence of how environmental parameters and culture conditions (e.g., water temperature, photoperiod, nutrient resources) affect the reproduction and the development and growth of new generations is required to establish long-term (parental) cultures, methods to induce reproduction in the laboratory (or nursery) and scale the process towards commercial production. Once achieved, control over the reproductive and early life stages of seaweeds can enable a wealth of applications including commercial aquaculture development and ecological restoration initiatives. Further studies and research are needed to investigate the life cycles and early life stages of A-NZ seaweeds in order to add to the few studies that have been conducted to date (Tables 21.1 and 21.2). One barrier, as is often the case, is the inherent cost and substantial initial investment associated with early life stage studies, which are often relatively long in duration and require repeating through time, to develop a full understanding of the effects of seasonality and other environmental parameters on the performance of the seaweed of interest.

Strain selection for specific characteristics (e.g. heat-resilient strains) will likely be an important next research step and has been the focus of significant research and implementation in other parts of the world (Hwang et al. 2019). However, strain selection has not been explored in the published A-NZ seaweed literature despite being commonplace for other commercial-aquaculture species, such as shellfish and finfish (Camara and Symonds 2014; Symonds et al. 2019; Chi et al. 2022). Strain selection could allow growers to identify, culture and selectively breed strains with desirable traits such as rapid growth, resilience or concentration of high-value compounds. While costly and requiring skilled personnel and dedicated facilities, this approach has the potential to significantly boost the seaweed aquaculture sector and production in A-NZ. One barrier to a strain selection and selective breeding approach is potentially the resource consenting process in A-NZ which has the capacity to dictate where broodstock or farmed strains are to be sourced.

An important line of research for developing a seaweed aquaculture sector is the assessment of environmental impact which will help to determine and contribute to the management of impacts from seaweed farming on the surrounding environment. Very few studies outside of Asia have assessed the effects of seaweed farms on aspects of the environment such as the biodiversity of benthic organisms, marine mammal entanglement or the spread of invasive species, as they have done for the finfish and shellfish aquaculture sectors (Suplicy 2020; Braña et al. 2021). In A-NZ, it is currently impossible to directly assess the impacts of seaweed aquaculture because there are no pilot seaweed farms operating at sufficient scale to impact the environment. It seems likely that assessments of environmental impact will develop alongside the seaweed aquaculture industry.

It is also expected that seaweed farming will play a significant role in the development of integrated multitrophic aquaculture (IMTA) systems, which could contribute to lower impact production systems through the uptake of effluent nutrients from higher trophic-level crop species such as mussels and fish (Chopin et al. 2001; Stenton-Dozey et al. 2021; Pereira et al. 2024). Yet in A-NZ, application and implementation of IMTA is far from been realised and only one published study has assessed the co-culture of selected seaweeds with a higher trophic level in A-NZ (Loy and Jeffs 2025). Similarly, it has been proposed that seaweed aquaculture could be used for nutrient remediation and a range of other ecosystem services in coastal waters (Campbell et al. 2019; Howarth et al. 2023), developing the promising field of bioremediation by seaweed as part of land-based IMTA systems (Lawton et al. 2013).

Finally, an area that is gaining significant interest is understanding the role seaweeds play in carbon sequestration and climate change mitigation by absorbing and storing carbon dioxide through photosynthesis. To that end, seaweed cultivation potentially offers a sustainable pathway toward reducing greenhouse gas emissions and there is currently significant global interest in establishing carbon and nitrogen credits to incentivise farmers into low-trophic aquaculture of species such as seaweeds (Crowe 2021; Hughes et al. 2025). However, the sequestration potential of seaweed biomass is dependent on the final use of the biomass and is currently a contentious issue among seaweed scientists. For example, most of the carbon stored in seaweed stock is likely to be released back into the atmosphere when it is processed or consumed (Park et al. 2021). However, carbon and nitrogen credits for seaweed crops have been adopted at a national scale in some countries, such as Japan (Kuwae et al. 2022). Significant research is still needed to assess seaweed aquaculture-based carbon sequestration and its potential inclusion in carbon credit markets (Ross et al. 2022; Hurd et al. 2023) and while seaweed farming has clear potential as a carbon sink, accurately assessing its long-term impact remains a scientific and logistical challenge in A-NZ and around the world (Duarte et al. 2025).

21.4.2.2 Nursery and Cultivation Systems

Seaweed aquaculture (e.g. kelp, Fucales, nori) often requires an interactive land- and sea-based production system to ensure a continuous supply chain. The establishment of such a system, which involves land-based hatchery and nursery capability and infrastructure and sea-based marine farms for growing crops to a marketable size, ensures consistent and continuous access to biomass without relying on wild harvesting to obtain new broodstock. Complete life cycle control can lead to the identification of interesting strains and populations with sought-after characteristics which can be tailored to produce biomass specifically targeted to the needs of consumers, markets and potentially coastal managers.

Research in A-NZ has focused on investigating methods for nursery-based propagation, applying existing protocols from Asia and Europe or developing new ones for species such as *Asparagopsis armata* (Mihaila et al. 2023a, b; Hall et al. 2025). In A-NZ, research has focused on understanding life cycle and production methods for *A. armata*, *Ulva* spp., *Macrocystis pyrifera*, *Lessonia variegata*, *Ecklonia radiata* and a few species of fucoids (Tables 21.1 and 21.2). At the Cawthron Institute for example, there has been a significant focus on early life stage cultivation, hatchery development and grow-out trials for the three A-NZ kelps (Case Study 1, Figs. 21.5, 21.6 and 21.7). Further research including seasonal deployments onto marine farms and scalability of the protocols is now required to underpin implementation seaweed aquaculture at scale. One significant envisaged expansion of scale includes open ocean aquaculture (OOA). Currently, there is one programme underway in A-NZ that aims to develop aquaculture structures for exposed coast and offshore environments for a range of species including seaweed (https://openocean.cawthron.org.nz/). Expanding aquaculture into A-NZ's extensive EEZ is a significant opportunity for the development of seaweed aquaculture in the country.

Case Study 1 Cawthron Institute Kelp Nursery Development

Kelps have heteromorphic life histories that alternate between microscopic (gametophytes) and macroscopic life stages (sporophyte) (North 1994). Reproductive blades of kelps release spores seasonally (usually during the autumn-winter, with variation between species and locations) that once germinated develop into male and female gametophytes. Subsequently, through fertilisation the oogonia on the female gametophytes develop into a sporophyte (Fig. 21.5).

At the Cawthron Institute National Algae Research Centre near Nelson we hold a collection of gametophyte cultures of three native NZ species, *Lessonia variegata*, *Macrocystis pyrifera* and *Ecklonia radiata* and the non-native *Undaria pinnatifida* (Fig. 21.6) and we have investigated sporophyte production in the nursery for the three native species (Fig. 21.7).

In the culture collection, gametophyte cultures are kept in continuous aerated flasks under red light and as separate cultures of male and female gametophytes and are then used for sporophyte production accordingly. For the sporophyte production, gametophytes developing into sporophytes are sprayed onto seeding polyester twine wrapped around tubular PVC pipes (also referred to as 'spools'). Both production of sporophytes from direct seeding of spores and gametophyte-spraying have been trialled for *L. variegata*, with a focus instead on production only from sprayed gametophytes for the other two kelps (Fig. 21.6). The methods trialled were adapted and developed from existing manuals (e.g. Visch et al. 2024).

Once ready, the spools were deployed at sea to understand potential for cultivation of the three species in different sites in the top of A-NZ's South Island. These preliminary trials were promising and underline the necessity to invest in similar research efforts and the potential for further development of seaweed aquaculture in A-NZ.

21.4.2.3 Technology and System Development

Another area of major investment will be the development of technology and systems that are cost-effective, scalable, adaptable and that can support the product chain from cultivation, through to deployment, harvesting, processing and packaging. Two main approaches are likely: (1) the development of new infrastructure, entirely designed and developed for the purpose of seaweed growing, harvesting and processing (Case Study 2) and (2) the modification of existing structures, processes and facilities used for other species to reduce the initial investment capital and fast-track the nascent industry. Both have pros and cons that will be dependent on the scale of the activity and the final product. At sea, for example, the double backbone longline systems used for mussel farming could readily be adapted to kelp farming. Similarly, existing processing factories (e.g., milled grain processing, milk drying, winemaking) could be used as starting point to process seaweed biomass. However, it will be important to consider that the approaches to farming and processing will need to be sustainable, cost-effective and scalable, if they are to support the development of the seaweed aquaculture industry in A-NZ.

Case Study 2 Development of a Modular Seaweed Drying Units for Seaweed Biomass in Australia

Example of an interdisciplinary collaboration between scientists and an agricultural drying company in Australia as part of a project that was funded by the commercial and government sectors. Titled, 'Seaweed Solutions for Sustainable Aquaculture CRC-P', this project developed cultivation protocols for three Australian kelps, *Lessonia corrugata*, *Macrocystis pyrifera* and *Ecklonia radiata* and a modular, movable, sustainable and scalable system for seaweed drying (Fig. 21.8), which consisted of a partially ventilated drying system built into containers that used partially recycled air and control of moisture levels for fast drying of food and feed grade seaweed biomass with low energy consumption. In this case, existing technologies have been adapted and tailored to the emerging industry, offering a replicable and cost-effective unit that allowed for processing of up to 250–300 kg of wet seaweed within 24–48 h.

Similar systems could be developed and adapted to the A-NZ context.

21.4.2.4 Quantifying Research Costs for Future Development

The costs associated with research in seaweed aquaculture and subsequent development of the seaweed sector in A-NZ will be influenced by the complexity, range and scale of the studies, as well as the need for interdisciplinary collaboration. Broad categories of investment are identified and discussed in Case Study 3 and include

Surveys and Mapping: of habitats and seaweed populations, biodiversity and environmental impacts. In this category the major costs will include permits, equipment and skilled personnel.

Fundamental and Applied Biology Studies: including species selection (e.g. bioprospecting), life cycle and reproduction research, genetic analysis, and nursery and biobank development. Major costs in this category will include specialized personnel and facilities, equipment and long-term investment.

Technology Development: for designing, developing, testing and applying cultivation systems, protocols and infrastructures on land and at sea (coastal and offshore) and processing equipment (harvesting, processing, including washing, drying, packaging, etc., and analyses) (Fig. 21.9). Major costs in this category will include infrastructure, skilled personnel, technological expertise, equipment, scale-up procedures, access to space (on land and at sea), biosecurity systems and continuous investment.

Case Study 3 Seaweed Aquaculture as Part of the Open Ocean Aquaculture (OOA) project, Ngā Punga o te Moana: Anchoring Our Open Ocean Future

The 'Ngā Punga o te Moana—Anchoring Our Open Ocean Future' project was funded by Ministry of Business, Employment and Innovation (MBIE). This research project includes a component of growing various seaweeds and developing seaweed farming practices by establishing a pilot system relevant and directly applicable to commercial aquaculture infrastructure for seaweed production offshore (with a focus on Kelps) in A-NZ. Based on the estimates provided by the project and considering the spending categories identified above, the summary of the overall research project budget is presented below.

Consenting and Permitting processing times were 6 to 12 months. The costs for consenting and permitting were around NZ$50,000, including consultancy (biosecurity advisors, industry governing bodies-Aquaculture New Zealand (AQNZ), contractors-Aquaculture Direct Limited (ADL), site consent holders-MacLab Ltd), environmental assessment of possible impacts (e.g., shading of infrastructures on benthic ecosystems, marine mammal entanglement, nutrient profiles), mapping and species selection (e.g., the addition of target aquaculture species, in this case *L. variegata*, to the resource consent). While the project was able to leverage space on an existing mussel farm to kick-start activities, the cost of developing a new consented marine farm without established infrastructures was calculated as part of the total presented. The cost of seaweed farming could be reduced if activities are performed on existing consented farms.

Nurseries and Personnel: the set-up of an operational research nursery, running costs and personnel for production of seeded twine for deployment at sea for research purposes can account to NZ$250,000-300,000, annually. In addition, maintenance of the structures per year can account to NZ$50,000, for a total of NZ$300,000–350,000 yearly.

Boating and Personnel: The cost to hire a commercial aquaculture vessel and divers was NZ$10,000 a day. For example, a monthly site visit and stock assessment would cost NZ$120,000 a year. However, the commercial vessel without the diver team was NZ$4,500 per day, and smaller research and operational vessels can cost NZ$750–1,000 per day and can be used for checking crops. Moreover,

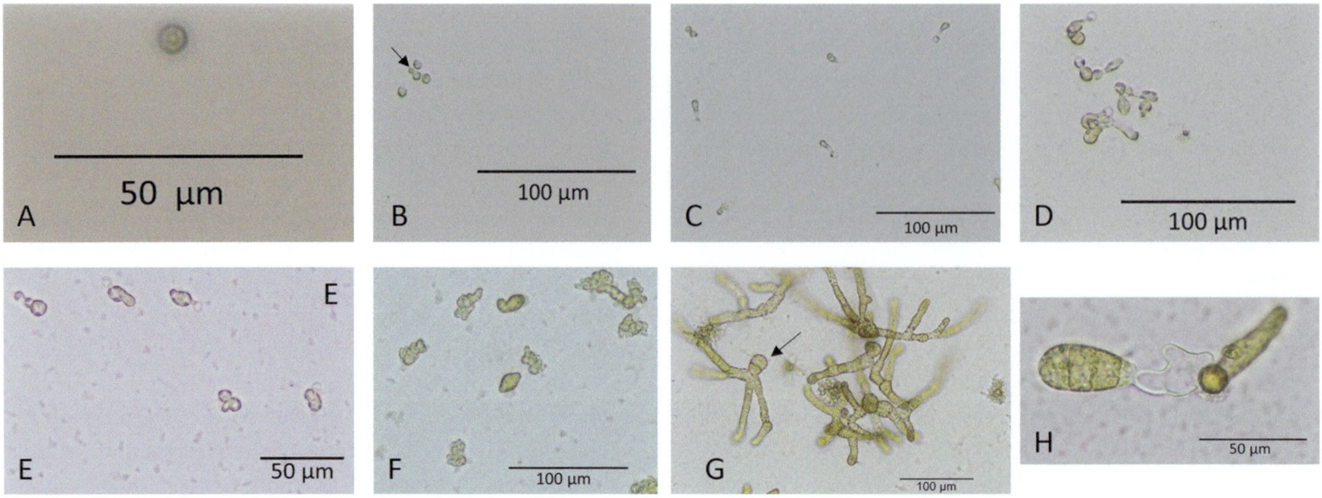

Fig. 21.5 Example of *Lessonia variegata* life cycle, from spores (**a–c**) to gametophytes (**d–g**) and finally sporophyte (**h**). (Pictures from Biancacci et al. (2024))

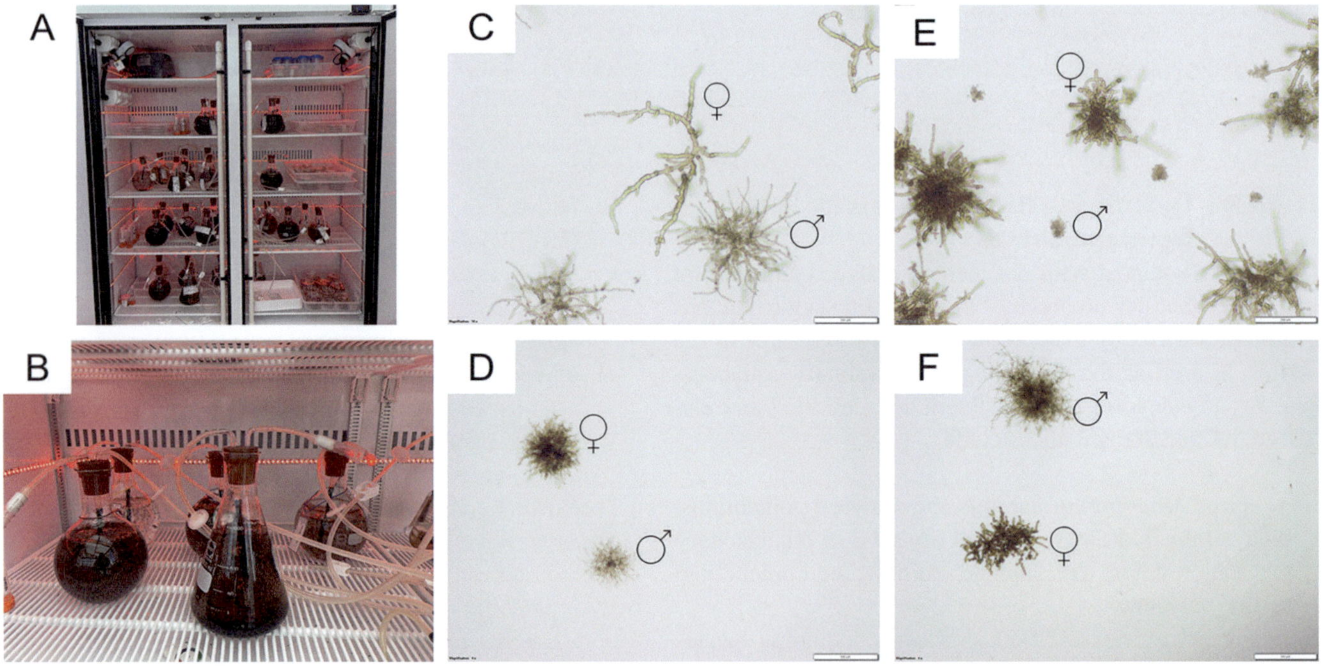

Fig. 21.6 Gametophytes cultures held at the Cawthron Institute National Algae Research Centre (NARC) near Nelson, A-NZ. (**a**) Culture cabinet with gametophytes cultures of different species and different provenance; (**b**) Continuously bubbled flasks containing gametophytes; Examples of male and female gametophytes of (**c**) *Ecklonia radiata*, (**d**) *Macrocystis pyrifera*, (**e**) *Lessonia variegata*, (**f**) *Undaria pinnatifida*

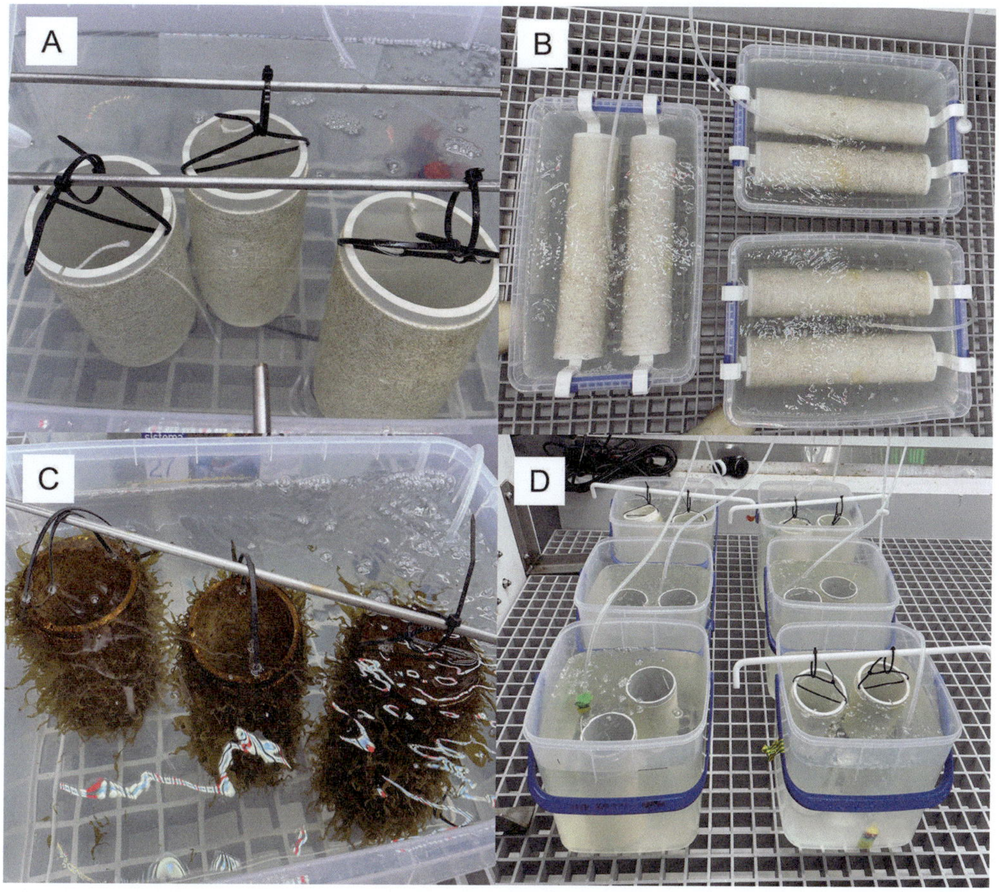

Fig. 21.7 Examples of a nursery system for kelp production at the Cawthron Institute National Algae Research Centre near Nelson, A-NZ. The kelps are grown on polyester (PE) twine wrapped around polyvinyl chloride (PVC) pipes (commonly referred to as 'spools') after a solution of gametophytes or gametophytes and sporophytes has been sprayed onto the twine. They are kept under controlled light, temperature and nutrient conditions until they are ready for deployment. (**a**) Spools with developing gametophytes of *Macrocystis pyrifera*; spools of *Lessonia variegata* (**b**) incubated horizontally, (**c** and **d**) vertically, after two months and two weeks respectively in the nursery. (Photo credits: Cecilia Biancacci)

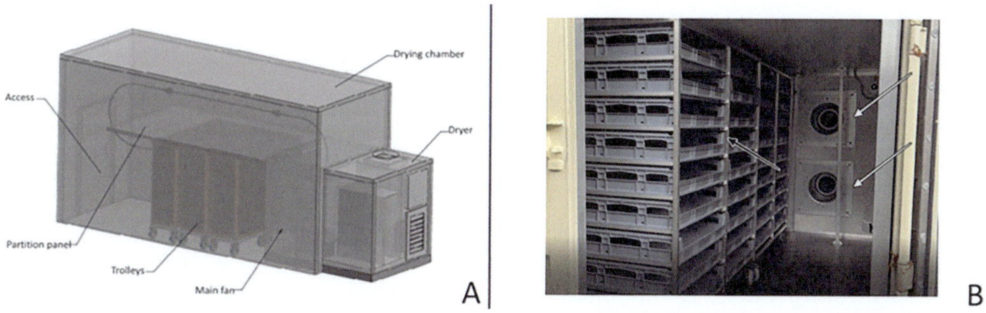

Fig. 21.8 (**a**) Schematic of the dryer unit and (**b**) dryer unit inside a 20-foot container at the processing facility at the Institute for Marine and Antarctic Studies (IMAS)-Taroona, Tasmania. White arrows indicate the fans, and the black arrow highlights the trolleys and trays custom-made for the unit on which the seaweed is placed. (Figure credit: Technical report: 'Aquaculture Production of Australian Laminarian Kelps: A manual and research recommendations for *Ecklonia radiata*, *Lessonia corrugata*, and *Macrocystis pyrifera*', (Visch et al. 2024))

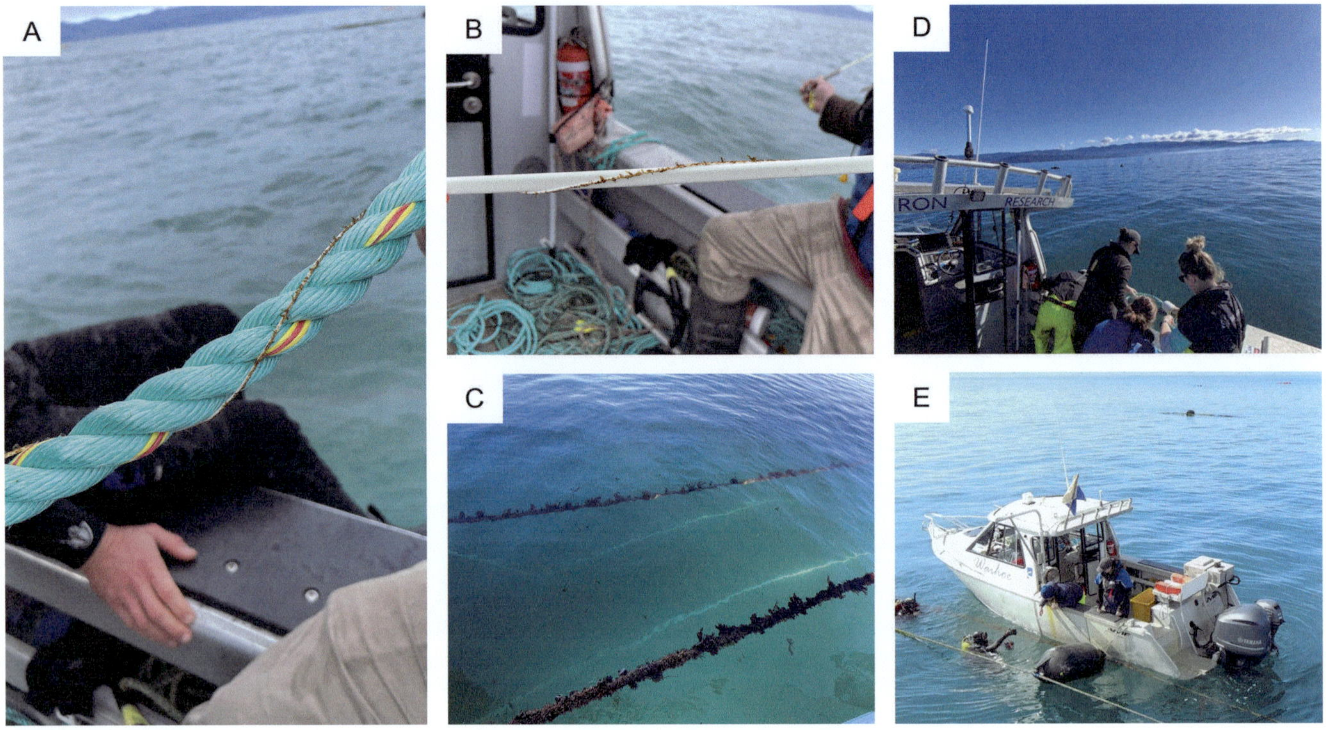

Fig. 21.9 Fieldwork activities as part of the OOA project in Tasman Bay, Nelson, New Zealand. (**a**) seeding of a deployment rope for *Lessonia variegata* sporophytes; (**b**) testing of alternative substrate to rope for seeding of *Lessonia variegata*; (**c**) example of horizontal long line for seaweed deployment; (**d** and **e**) the field work team during sampling and monitoring operations. (Photo credits: Cecilia Biancacci and Nicholas Scott)

a team of four staff for research project costs between NZ$5,000 and NZ$10,000 per day, which was between NZ$60,000 and NZ$120,000, yearly, given monthly visits.

Infrastructure: the deployment and set up of the infrastructures at sea cost between NZ$10,000 (for a 150 m^2 farm) and NZ$50,000 (for a 750 m^2 farm). This can be reduced if activities are set up on existing or repurposed infrastructures such as mussel farms. Annual maintenance of the research structures costs NZ$50,000–NZ$150,000, depending on their size and complexity.

Environmental Monitoring and Technological Support: water sampling and water monitoring as part of the research project together with acquisition, deployment and maintenance of sensors for light irradiance, nutrients and temperature were NZ$15,000 a year.

In summary, and based on this case study, a year of small to medium-scale seaweed aquaculture research activities costs between NZ$600,000 and NZ$900,000.

21.4.3 Market Research

Seaweed aquaculture worldwide is experiencing rapid growth due to its potential applications that include food and feed, biomaterials, pharmaceuticals, cosmetics, bioremediation, ecosystem services, fertilisers and biostimulants and biofuels (Buschmann et al. 2017; Khan et al. 2024). The global seaweed market was valued at approximately NZ$17 billion in 2022 (UN Trade and Development 2024) with a growth rate of 7% per year on average over the last decade (Sustainable Sea Challenge 2022). Demand worldwide is being driven by increasing industry and consumer interest in sustainable and nutritious food and feed sources, as algae are rich in vitamins, minerals and protein (Case Study 4). Additionally, the role of seaweeds in carbon sequestration and ecosystem restoration is attracting investment from governments, communities and environmental organizations (Clark et al. 2021; Qu et al. 2023).

Challenges in scaling New Zealand's seaweed aquaculture industry include regulatory approval, supply chain development, access to investment and identification of suitable markets. While there is strong international demand for premium seaweed products, particularly in health food markets, A-NZ processing capabilities will need improvement and scalability to compete globally (Bradly et al. 2021). Hence, market

Case Study 4 *Asparagopsis armata* Production for Bromoform Extraction in New Zealand

Land-based cultivation of *Asparagopsis armata* is the only commercial scale seaweed production in A-NZ. This species has attracted significant interest and investment in the last decade due to its potential to mitigate methane emissions from cattle, contributing to a more sustainable farming sector. More than NZ$387 m has been invested globally to facilitate the development of land-based *Asparagopsis* spp. cultivation to produce methane-inhibiting feed supplements for cattle, yet to make this commercially viable and economically feasible additional investments and research are needed (Fig. 21.10). For example in A-NZ, based on conservative bromoform content (4 mg/g dry matter) and conservative biomass productivity (2.5 t dry matter/ha), NZ$100—400 m would be needed to develop and maintain an *Asparagopsis* supply chain to reduce biogenic methane by 10% for all beef and dairy cattle in A-NZ (based on the estimates reported by Agrifutures and the Commonwealth Bank in their scoping study for commercial *Asparagopsis* farming in Australia, Ball et al. 2022). This investment could lead to thousands of jobs and could strengthen the aquaculture sector in A-NZ by >NZ$100 m per annum.

research that identifies customer product demand, analyses competition and market trends and identifies priority products is a necessary step in A-NZ to support the development of a thriving and diversified seaweed sector.

A 2021 study that addressed market research for the A-NZ seaweed sector identified that the main source of biomass into the domestic seaweed market was a mix of beach-cast seaweeds (Bradly et al. 2021) (Fig. 21.1). The report identified a rising demand for products for which markets and applications already exist in the country, such as fertiliser and biostimulants (Fig. 21.1, Table 21.2). However, due to the limited local market and the potential for A-NZ to contribute to the global production of these relatively low-value commodities, it appears likely that higher high-value products (e.g. food, feed and supplements) will be required to drive investment in aquaculture development.

21.5 Outputs

There is a key set of strategic outputs that seaweed aquaculture research could provide to support the nascent seaweed industry in A-NZ (Fig. 21.2).

21.5.1 Database and Operational Manuals

Transforming foundational research into a viable seaweed aquaculture industry requires bridging the gap between scientific discovery and commercial application (Fig. 21.11). The research promoted by invested capital will contribute to the production of databases (e.g. detailing distribution, bioactive compounds, proximate components), technical reports and seaweed aquaculture manuals that will support the establishment and development of the seaweed sector in A-NZ. While databases of seaweed species exist in other parts of the world, these are absent beyond simple species lists in A-NZ. There are many highly useful and applicable seaweed aquaculture manuals worldwide (e.g. Juanich 1988; Chen and Xu 2009; Redmond et al. 2014; Rolin et al. 2014; Callo and Darag 2015; Flavin 2021; Visch et al. 2024). We expect that similar A-NZ-focused publications will be pivotal in implementing seaweed aquaculture in the country.

21.5.2 Economic Feasibility, Optimisation and Product Development

While some seaweed products are currently produced and commercialised in A-NZ (Fig. 21.1), the overall quantity in tonnes produced and the economic value of the local supply is not reported in publicly accessible resources. The main products used and produced in the country include food, feed, fertiliser and biostimulants, and health and beauty products (Fig. 21.1). Nonetheless, it is likely that A-NZ will need to identify and invest in a diverse range of products to align its seaweed production with global needs and demand. A-NZ currently imports most of the seaweed products that are available in the country, with these coming predominantly from the People's Republic of China, the Republic of Korea and Japan. In 2021, the combined volume of edible seaweed and seaweed products imported into A-NZ was a little over 1,000 t, with 56% being nori sheets (from dried *Porphyra* and *Pyropia* spp.), 19% from other dried species (e.g. *Sargassum fusiforme*), 14.1% as frozen seaweed salad (e.g. *Undaria pinnatifida*) and 10.2% as undefined frozen and dried seaweed (Cressey et al. 2023). These data show the potential for A-NZ to shift from import to production, consuming and potentially exporting its own seaweed products, leveraging its already established internationally reputation for sustainably produced seafood and a relatively pristine ocean environment, while developing a sustainable and economically viable seaweed sector.

Tapping into existing high-value global markets will be an essential first step for the development of commercial sectors in A-NZ (Bradly et al. 2021). These are listed in value proposition and priority order below:

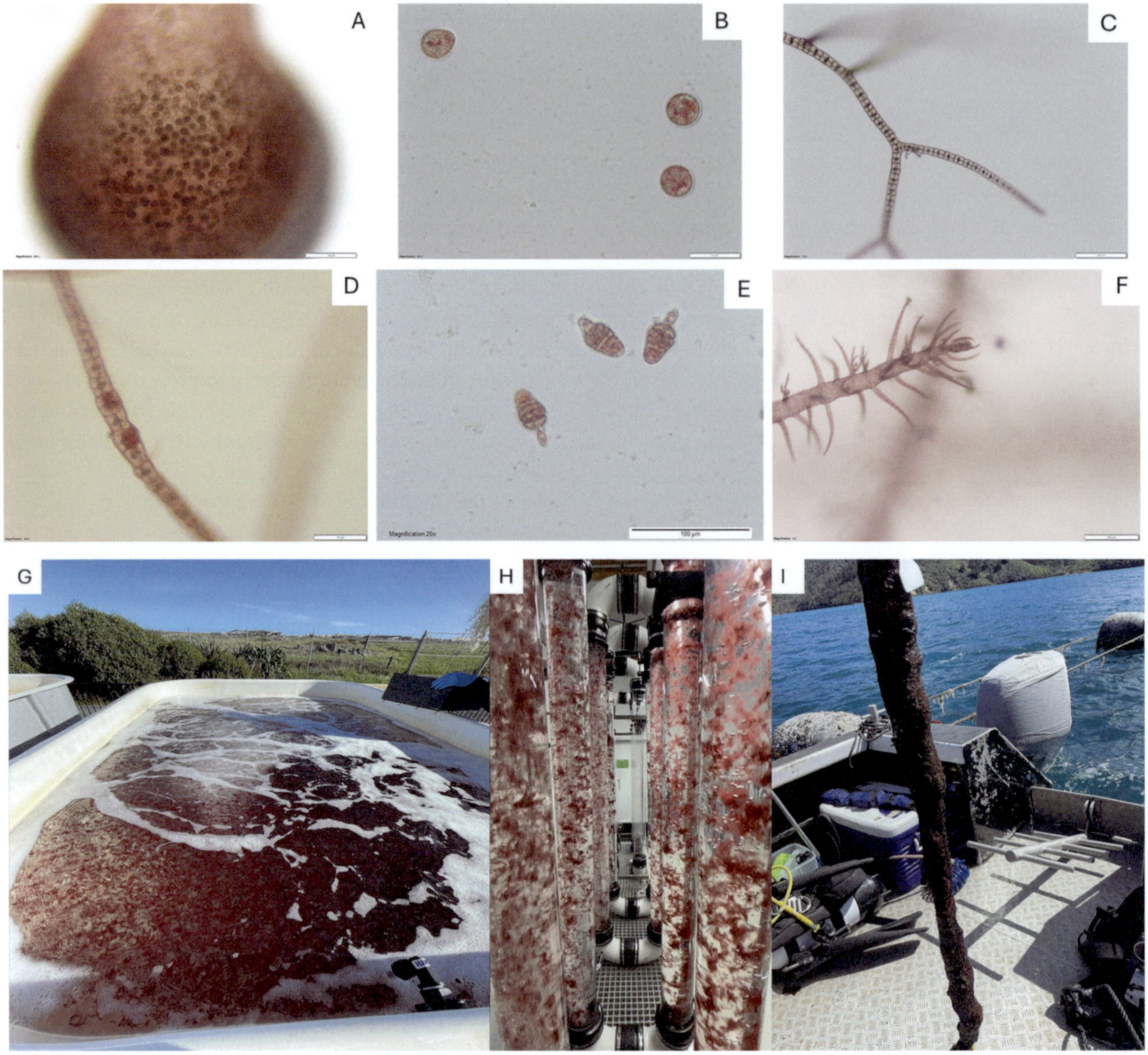

Fig. 21.10 Examples of *Asparagopsis armata* cultivation at Cawthron Institute, Nelson, New Zealand. Through sexual reproduction (**a**–**f**) and vegetative propagation (**g**–**i**): (**a**) mature wild-collected carposporophyte with carpospores on a female gametophyte; (**b**) released carpospores; (**c**) developing tetrasporophytes; (**d**) mature tetrasporophytes with visible tetrasporangia; (**e**) tetraspores developing into early stage gametophytes; (**f**) later stage developing gametophytes; (**g**) production of tetrasporophytes via fragmentation in an outdoor tumbling system; (**h**) production of tetrasporophytes via fragmentation in an indoor photobioreactor system; (**i**) cultivation of wild-harvested and fragmented gametophytes at sea. (Photo credits: Cecilia Biancacci)

- *Health and Beauty*: seaweed extracts for use in beauty, nutraceutical or pharmaceutical products (e.g. antioxidants, anti-inflammatory compounds, skincare and haircare products, wound healing and dietary supplements, etc.). The seaweed cosmetic ingredients global market was valued at approximately USNZ$500 million in 2023 (Transparency Market Research 2025).
- *Human Food*: while consumption of seaweed in A-NZ is still low and mainly centred towards imported products from low-cost Asian suppliers of snack and sushi wraps, there is growing interest in more sustainable and highly nutritious food, rich in vitamins, minerals and proteins.
- *Animal Feed Supplements*: to date, New Zealand seaweed-based feed supplements exist for cattle, horses, pigs, goats, poultry and bees, with an increasing interest and investment in methane-reducing feed additives (such as bromoform from *Asparagopsis armata*).
- *Biostimulants and Fertilisers*: this sector is the most developed in A-NZ (Fig. 21.2), with most of the existing companies processing wild-harvested or beach-cast sea-

Fig. 21.11 Cawthron Institute field-based seaweed aquaculture research to develop kelp cultivation in the Marlborough Sounds, New Zealand. (**a**) A mussel farm used for pilot kelp farming experiments; (**b**) Example of a vertical dropper system for cultivation of native kelps; (**c**) Close-up of seeded lines of *Lessonia variegata*; (d) seeded spool with *L. variegata* sporophytes during the deployment, (**e**–**h**) sampling the seaweed biomass after 3 months of at-sea cultivation, (**e**) *Macrocystis pyrifera* blades, (**f** and **g**) *Lessonia variegata* blades, (**h**) *Macrocystis pyrifera* bladders, (**i** and **j**) seaweed harvesting, (**k**) assessment of yield and quality of harvested biomass on land. (Photo credits: Petra Reid, Courtney Moir, Cecilia Biancacci)

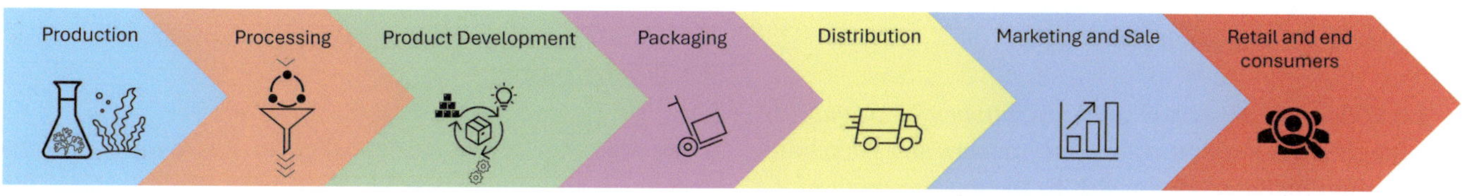

Fig. 21.12 Schematic of an example of a seaweed product value chain, from production to the end consumers

weed biomass for the extraction of biostimulants. While the development of these products could expand quickly into the well-established A-NZ agriculture sector, growth of the seaweed-based biostimulant industry is currently constrained by limited availability of seaweed biomass. Efficiently establishing cultivation of seaweed biomass will support the growth of the sector.
- *Ecosystem Services and Restorative Economies*: while ecosystem services such as nutrient removal, coastal protection, carbon sequestration of seaweed aquaculture are well studied elsewhere in the world (e.g. Kim et al. 2017; Cotas et al. 2023), there are significant barriers around translating research findings into effective and measurable benefits that, for example, in the case of carbon and nitrogen, could be monetised as national and international credits.

The global seaweed industry is predicted to grow 12% per annum reaching USD37.8 billion by 2028 (GrandviewResearch 2021), with food expected to be the largest and fastest-growing market in 2025 (MarketsandMarkets 2020). Globally, biostimulants, animal feed additives and pet food are projected to reach US$ 4.4 billion, nutraceuticals US$ 6 billion and pharmaceuticals US$ 1.4 billion by 2030 (The World Bank 2023). Undoubtedly, seaweed applications and products offer a multifaceted source of significant revenue. Achieving economic viability, however, likely requires a tailored approach for A-NZ and a focus on identifying viable high-value products, sectors and markets, while optimizing production costs and efficiently scaling seaweed aquaculture operations.

21.6 Outcomes

If research investment is successful, outcomes such as strong partnerships and proof of concept for seaweed aquaculture industries (Fig. 21.2) will help to overcome some of the barriers that are currently hindering the successful establishment of the A-NZ seaweed sector. Presently these barriers include a lack of investment in research and development, gaps in practical and theoretical knowledge, a not fit-for-purpose regulatory system and mismatch between production goals and market needs. We envisage that highly collaborative (i.e., iwi, industry, stakeholder and government agencies) funded research could target

- *Regulatory Compliance and Environmental Standards*: through collaboration with policymakers and stakeholders, a significant next step will be the reassessment of the regulatory landscape for seaweed aquaculture in A-NZ (e.g., resource consent processes, permitting, marine farm licensing, etc.) to ensure a supportive, collaborative and clear framework. This research will benefit from baseline data obtained from environmental, commercial and social assessments to delineate sustainability guidelines and facilitate regulatory compliance. Ultimately, clear and flexible regulations and alignment with international standards will facilitate import and export of seaweed biomass and products.
- *Business Model and Proof of Concept*: through the knowledge and information gathered via research, collaboration and market research it will be possible to create flexible business models that can adapt to new opportunities (species, products) and capture the whole seaweed aquaculture value chain (Fig. 21.12). This could start from a low-risk high probability species (based on existing knowledge, methodologies, biomass availability and existing market demand) that can be used to provide a model and proof of concept, which can then be applied and adjusted to commercial products made from other species. A proof of concept business model will likely have a full-value chain focus that includes *production* including cultivation (in nurseries on land and at sea) and wild harvesting (of beach-cast biomass); *processing* such as cleaning, drying and processing seaweed into various forms (e.g. powder, flakes, whole blades or extracts); *product development* focusing on moving from the raw biomass to a final product (e.g. food, cosmetics, nutraceuticals, agricultural products); *packaging* and *distribution* which refer to the transportation and storage of the finished products; *marketing and sales* to promote awareness of seaweed products via different channels until the products are sold to the *end consumers* (Fig. 21.12).

- *Capacity Building and Partnership*: one of the areas that will require investment and that will directly enable the development of the seaweed sector is the establishment of educational and training programs, workshops and school and community events to build and foster local expertise. To calculate the required investment, which goes beyond the scope of this chapter, it would be necessary to estimate the economic contribution of seaweed aquaculture, the workforce required to support the industry, the cost of training programs and infrastructure investment.

21.7 Impacts

If the research activities described in this chapter can be resourced and implemented, they will likely enable the development of a prosperous seaweed aquaculture sector in A-NZ.

- *Diversification of the Sector*: while the three main pillars of the A-NZ aquaculture sector, salmon, mussels and oyster are profitable and commercially established in the country, the A-NZ government has identified ambitious goals to almost double the current revenue, and this will require diversification of the aquaculture sector. Seaweeds can contribute significantly to this diversification and the development of novel production systems such as co-culture or IMTA (Chopin et al. 2001; Ridler et al. 2007; Ben-Ari et al. 2014; Fossberg et al. 2018). Yet in A-NZ application and implementation of this practice is far away from been realised (Stenton-Dozey et al. 2021). De-risking the seaweed industry via investment in research and development will open the path to diversified production and markets, while providing environmental benefits such as nutrient bioremediation, increased biodiversity and carbon sequestration.
- *Biomass Production and Algal Products*: one of the barriers to the further growth of existing seaweed companies, and the development of new ones, is the access to consistent and continuous supply of seaweed biomass. While some of the companies currently operating in A-NZ entirely or mainly rely on wild-harvested and beach-cast biomass, this approach is not sustainable as it can impact coastal ecosystems and is highly variable in terms of biomass availability through time, making it difficult to establish a consistently profitable business. Seaweed aquaculture should remove uncertainties related to natural variability of biomass provision. If comprehensive cataloguing and database generation is done as we recommend here, the current number and value of A-NZ seaweed products will increase, and new products made from novel species could broaden demand for A-NZ grown seaweed.
- *Job Creation*: the current aquaculture sector in A-NZ provided more than 3,000 jobs in 2024 (New Zealand government 2024a, b). As for the development of new business and production areas, the establishment of the seaweed sector in A-NZ will add to this and will result in a tangible impact of creating more jobs and benefits particularly for coastal communities. As defined in Fig. 21.12, the value chain has multiple levels, each of which will require a trained workforce and will support the social and economic growth of A-NZ, particularly in remote, rural and coastal areas.
- *Contribution to Costal Ecosystems, Restoration and Reduction of Methane and Carbon Emissions*: seaweed aquaculture is widely considered to be a low impact or even ecologically beneficial industry due to its provision of ecosystem services, which have often been lost in the wild due to the pressures of overfishing, pollution and the warming seas (Case Study 5). A new and important line of research will be required to determine whether commercial seaweed aquaculture can be used to redress some of the damage humans have done to coastal environments in A-NZ and around the world (Hamilton et al. 2022; Cornwall et al. 2023; Wear et al. 2023, Fig. 21.13). Two key types of regenerative seaweed aquaculture, restorative aquaculture and conservation aquaculture, appear to be broadly applicable in A-NZ with restorative aquaculture seeking to leverage ecosystem services from commercial aquaculture operations and conservation

Case Study 5 Advancing Seaweed Restoration with Conservation Aquaculture

Rocky reefs in temperate zones are dominated by large brown algae (LBA) that act as crucial foundation species and ecosystem engineers (Bennett et al. 2016; Blamey and Bolton 2018; Zheng et al. 2019). There has been widespread decline of LBA due to climate-related temperature increases (e.g. Thomsen et al. 2019; Thoral et al. 2022), terrigenous runoff (e.g. Tait et al. 2021), natural hazards such as earthquakes (Crossett et al. 2023) and herbivore proliferation due to overfishing of their predators (e.g. Miller et al. 2022, Fig. 21.13a and b). In response to the declines in LBA around the world, there is a growing interest in using seaweed aquaculture techniques in restoration efforts.

Most recent research has focused on developing techniques to rear in nurseries juvenile kelps and fucoids that can be deployed in *masse* onto rocky reefs. Identifying appropriate substrates that enable seeding at scale has been a key goal in recent research (Wood et al. 2024; Crossett et al. 2025, Fig. 21.13c and d). One promising restoration technique is Green Gravel, which involves seeding small rocks with kelp and later deploying them into the ocean (Wood et al. 2024). Unlike traditional restoration approaches, Green Gravel does not require scuba diving, specialized field-workers or engineered structures, making it an effective strategy for large-scale application (Green Gravel 2022).

Such initiatives are achieving great strides in understanding how to grow and deploy a wide range of seaweeds into natural environments, their success will ultimately depend on whether the causes of the losses of wild populations can be managed (Coleman et al. 2008). If stressors such as overfishing or warmer waters cannot be managed, it might be essential to turn to selective breeding to develop less palatable or warm-water resilient strains of selected seaweeds, if the function of coastal selected seaweed forests is to be restored (Coleman et al. 2020).

Fig. 21.13 Examples of (**a**) sea urchins (kina) overgrazing and (**b**) denuded kelp forests at D'Urville Island, A-NZ; (**c**) photographs of four substrates used in experiments to trial alterative substrata for cultivation of kelps and fucoids for aquaculture and restoration purposes, Crossett et al. 2025. These include (1) a Greenshell™ mussel valve, (2) pieces of recycled concrete (3) fibreglass and (4) polyester twine. (**d**) Cement-fibre tile with juveniles of *Lessonia variegata*. (Photo credits Dan Crossett)

aquaculture focused on the provision of seedlings for use in restoration programmes (Overton et al. 2024), although there are currently no such programmes receiving significant government support in A-NZ (Table 21.1).

21.8 Vision

While there is still a long way to go before seaweed aquaculture in A-NZ is realised, the seaweed sector can learn from experiences in other countries, adapting and applying overseas knowledge and know-how to the A-NZ context. Given the rich biodiversity and endemism of seaweeds and the extensive water space in A-NZ, there is significant opportunity for innovation and growth. The global seaweed sector is characterised by success stories providing inspirational models toward a vision of a thriving and sustainably seaweed aquaculture sector in A-NZ (Case Studies 6 and 7).

21.8.1 Comparisons with International Realities: Lessons to Learn

While some investment has been allocated in the last ecade to develop the A-NZ phyconomy, this has been fluctuating and not equally distributed among research areas. Determining whether the current investment has been successful requires several key economic metrics that go beyond the scope of this chapter. However, non-financial metrics highlight that the current investment scenario has already contributed to the development of the A-NZ seaweed sector:

Environmental Impacts: environmental benefits are better understood, particularly the potential for methane reduction (Eason and Fennessy 2023).

Knowledge Transfer: the successful dissemination of research findings to the broader scientific, community, government agencies and industry stakeholders is a measure of success of the A-NZ seaweed research. More recently, a special issue on 'Seaweed aquaculture in A-NZ: Current advances, benefits, challenges, and perspectives' (New Zealand Journal of Marine and Freshwater Research 2025) has collated the largest compendium of seaweed-focused research in A-NZ to date.

Community Engagement: in 2021, The Aotearoa New Zealand Seaweed Association was established to bring together research and industry stakeholders to build a sector that 'delivers healthy marine ecosystems, thriving communities and is a significant contributor to the New Zealand economy' (ANZSA 2022). Community engagement has grown with prominent examples being around conservation aquaculture for kelp-forest restoration. For example, conservation aquaculture techniques are being co-devel-

Case Study 6 Seaweed Aquaculture in Norway: An Example of Investment Towards Growth

Seaweed aquaculture in Europe has gained significant attention due to its economic, environmental and social benefits. Countries like Norway, France and the United Kingdom have been at the forefront of developing sustainable seaweed farming. Norway in particular has invested in large-scale, mechanized seaweed farming to cultivate kelp species such as *Saccharina latissima* (sugar kelp) and *Alaria esculenta* (Phaeophyceae) in coastal and offshore regions (Broch et al. 2019). In 2017, 309 permits were allocated for seaweed cultivation in Norway, of which roughly half were granted for kelp cultivation (*S. latissima*, *L. digitata*, *A. esculenta*) (Broch et al. 2019), with *S. latissima* as the commercially most important species. It is forecasted that by 2050, the Norwegian aquaculture industry could produce 20 million t FW of kelp with a turnover of 4 billion Euros annually (Olafsen et al. 2012).

Companies in Norway specialize in sustainable production and industrial-scale offshore seaweed cultivation to supply industries, including food, cosmetics and biofuel. Moreover, the Norwegian government has been actively supporting the seaweed industry through a wide range of initiatives including research funding and public–private partnerships. An example was the Norwegian Seaweed Biorefinery Platform (SBP-N) funded by the Research Council of Norway in 2019. This was a national consortium created to coordinate the efforts of research institutions toward an increased and sustainable Norwegian seaweed-based industry.

Additionally, the Norwegian Seaweed Technology Centre has been recently established and is committed to addressing the global demand for food, feed, materials and energy via seaweed aquaculture. The centre provides dedicated research infrastructure to generate new insights in producing (land-based and at-sea), harvesting and processing of kelps (https://norwegianseaweedcentre.com). These initiatives serve as models that could be applied in A-NZ and highlight the importance of government's recognition of seaweed potential in the blue economy and its commitment to fostering sustainable growth in this sector.

Case Study 7 Seaweed Aquaculture in Indonesia: A Story for Success

Seaweed farming in Indonesia is established throughout much of the archipelago and is mainly undertaken by smallholder farmers. And yet, Indonesia accounts for 25% of global seaweed production, with 9.6 million t of seaweed (live weight) per annum (MMAF) and is the largest global producer of the red seaweeds *Kappaphycus* and *Eucheuma* for production of different carrageenans, and a major producer of *Gracilaria* for agar production (FAO 2022).

The main products besides raw biomass are food products including salads (e.g., made from *Caulerpa* spp.) and snacks (made from dried seaweeds) for local and export markets (Perryman et al. 2017; Larson et al. 2021).

Seaweed farming in Indonesia provides economic alternatives in remote areas where coastal communities can have a limited and variable regular income. Here seaweed production using relatively simple culture techniques provides financial resilience.

Two key features are the pillar of seaweed production in Indonesia: its scale and cooperative approach. The majority of coastal areas in the archipelago are occupied by seaweed farming activities involving large numbers of smallholder farmers, with an estimated 267,000 households being supported in some way by seaweed production (Langford et al. 2021) from relatively small (0.1 ha to 1.0 ha) seaweed farms (Neish 2013).

Production costs are low, and the production cycles are kept relatively simple and short, making the sectors profitable. Overall, the sector contributes positively to the society, the environment and the economy of the country and offers an example of how systems can be implemented with relatively simple approaches (Rimmer et al. 2021).

oped by the Massey University and Te Runga o Te Whānau (a federated sub-tribe (hapu) group) and Te Wairua O Te Moananui-Ocean Spirit Charitable Trust (Wood et al. 2024; Green Gravel 2022). Education programmes have also initiated to highlight the importance of seaweed for coastal ecology and their potential for aquaculture and product development (Love RimuRimu 2023).

Regulatory and Policy Impact: due to the growing interest in seaweed research activities, and probably the increased investment, the current policies and regulations that control the licenses and permits for seaweed collection as stock and subsequently deployment at sea are currently under review with the hope to make them more flexible and responsive to the developing seaweed sector (MPI 2025).

While current efforts and investments have made valuable contributions, the relatively smaller funding opportunities that are common in the current funding landscape have not enabled focused development, rather this funding model has produced a piecemeal understanding of the potential of a few species of seaweeds. It is clear that significant additional investment in research will be paramount in the near and long-term future to truly support and promote the development and growth of the seaweed sector in A-NZ. For example, between 2018 and 2029, approximately NZ$15 million will have been invested in *Ulva* spp. cultivation research in A-NZ (MBIE), bringing cultivation capability to a reliable pilot or small commercial scale and supporting the development of high-value products. This level of investment is likely essential for advancing fundamental and applied research around all target crop species. For example, we envisage that a similar amount of research investment will be needed to refine kelp production to support dependable crops to produce high-quality and high-value products. Similar or greater investments will also be required for research activities that focus solely on fucoids, IMTA or any novel or emerging species or aquaculture system in A-NZ. Hence, continued investment, innovation and policy support will be crucial to up-scaling current initiatives, overcoming emerging challenges and ensuring the sustained success of the A-NZ seaweed sector in the future.

With the necessary investment, we envision that by 2035 seaweeds could become an established and accepted part of the NZ diet and consumption of algae has made a significant impact on human health (Fig. 21.14a). We expect that A-NZ will produce thousands of tons of various seaweeds (Fig. 21.14b) contributing to 1,000 new jobs (Fig. 21.14c). This diversified economy will be more resilient to global stressors (Fig. 21.14 d and e) and seaweed aquaculture will significantly contribute to coastal ecosystems (Fig. 21.14f) and a low-carbon economy (Fig. 21.14g). Our vision is that seaweed aquaculture provides at least 10% of NZ aquaculture production by value (Fig. 21.14h) and, while ambitious, we believe that by implementing, supporting and investing in critical areas, the country can move towards this vision.

A-NZ waters can provide ideal conditions for high-quality seaweed cultivation, and the increasing global demand for sustainably sourced seaweed products aligns well with NZ reputation for environmental stewardship and quality. A-NZ has the potential to leverage existing aquaculture activities, such as mussel, salmon and oyster farming, for integrating seaweed aquaculture creating sustainable production, synergies and shared costs and infrastructures. Moreover, leveraging the Māori and iwi knowledge systems and ensuring

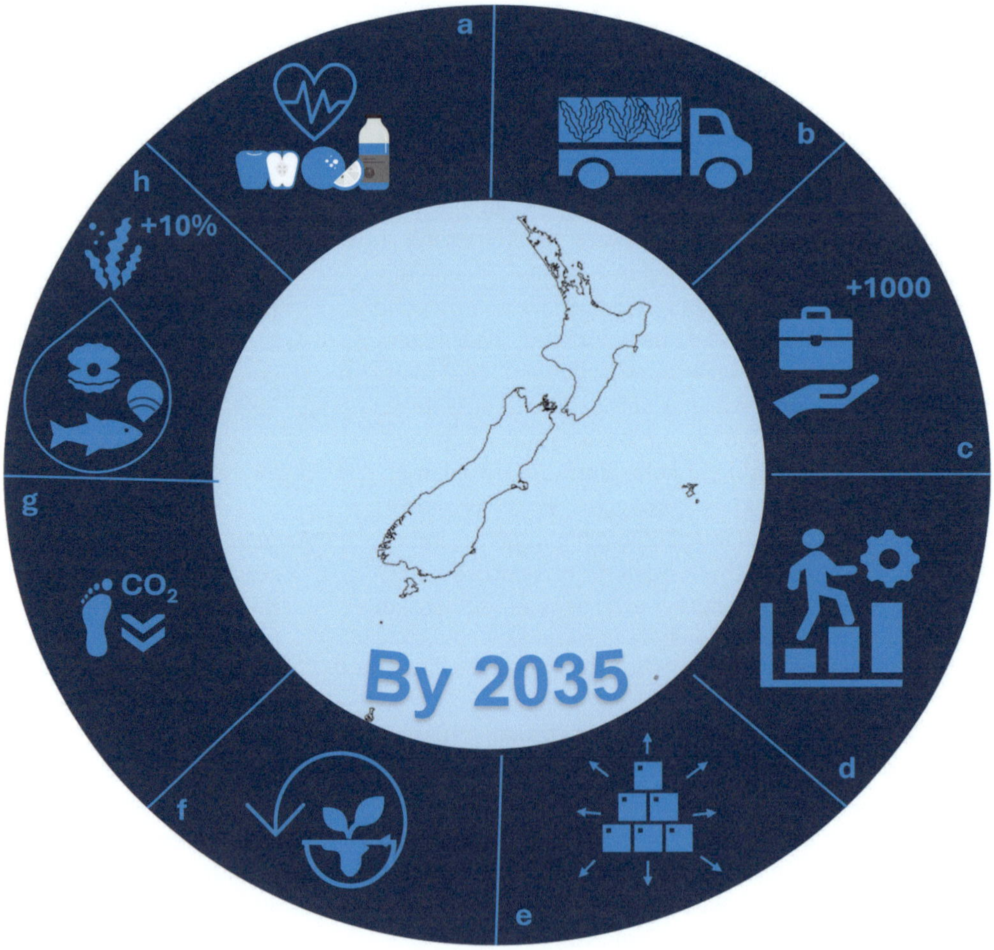

Fig. 21.14 Diagram of 2035 vision for a sustainable, economic valuable and socially accepted seaweed sector in A-NZ

equitable benefits for indigenous communities, the country can unlock innovative approaches to cultivation, processing and commercialization. By focusing on a circular economy and sustainable innovations such as biomaterial, bioremediation, high-end products, reduction of methane emissions, restoration, carbon sequestration and investing in biotechnology and monitoring systems for increased efficiency and scalability, A-NZ can aspire to become a key player in the evolving global seaweed sector.

21.9 Conclusions

This chapter has underscored the critical role of research investments in laying the groundwork for the development of a sustainable seaweed sector. From ecological and biological studies to applied research, the research has led to significant outputs although it still requires additional investments to achieve an established, profitable and sustainable A-NZ seaweed sector. Pathways to commercialization highlight the importance of applying the current research efforts to achieve economic viability, scalability, strategic partnerships and supportive policies. By addressing barriers such as high initial costs and regulatory hurdles, A-NZ can unlock significant opportunities for innovation in global markets and future prospectives. By integrating cutting-edge research, innovative technologies, a collaborative approach and indigenous knowledge, A-NZ can be a pivotal part of the global seaweed sector. The vision goes beyond the establishment of an industry, envisioning a collaborative approach that aligns environmental stewardship and cultural values with economic viability, creating long-term benefits.

Acknowledgement We thank Hsiao-Heng Pan, Senior Aquaculture Analyst at Ministry for Primary Industries (MPI), Fisheries New Zealand for providing the data on current public sector investment to develop the New Zealand phyconomy. We acknowledge the financial support of Cawthron Institute and the technical support of the fieldwork and technical teams at Cawthron Institute.

References

Aotearoa New Zealand Seaweed Association (ANZSA) (2022). https://anzsa.org.nz/. Accessed 20 Feb 2025

Balbas J, Hamid N, Liu T, Kantono K, Robertson J, White WL, Ma Q, Lu J (2015) Comparison of physicochemical characteristics, sensory properties and volatile composition between commercial and New Zealand made wakame from *Undaria pinnatifida*. Food Chem 186:168–175. https://doi.org/10.1016/j.foodchem.2015.03.079

Ball A, Williams S, Pattinson R (2022) Scoping study of the capital requirements for commercial production of *Asparagopsis* for methane reduction in cattle. AgriFutures and Commonwealth Bank, pp 1–23

Battershill ZV (2024) Nutritional profiling of five New Zealand seaweeds—a preliminary assessment. Front Mar Sci 11:1410005. https://doi.org/10.3389/fmars.2024.1410005

Ben-Ari T, Neori A, Ben-Ezra D, Shauli L, Odintsov V, Shpigel M (2014) Management of *Ulva lactuca* as a biofilter of mariculture effluents in IMTA system. Aquaculture 434:493–498. https://doi.org/10.1016/j.aquaculture.2014.08.034

Bennett S, Wernberg T, Connell SD, Hobday AJ, Johnson GR, Poloczanska ES (2016) The "Great Southern Reef": social, ecological and economic value of Australia's neglected kelp forests. Mar Freshw Res 67:47–56. https://doi.org/10.1071/MF15232

Biancacci C, South PM, Scott N, Crossett D, Heasman K (2024) Intrapopulation variation in reproduction of *Lessonia variegata* J. Agardh. NZJMFR:1–16. https://doi.org/10.1080/00288330.2024.2381754

Billakanti JM, Catchpole OJ, Fenton TA, Mitchell KA, MacKenzie AD (2013) Enzyme-assisted extraction of fucoxanthin and lipids containing polyunsaturated fatty acids from *Undaria pinnatifida* using dimethyl ether and ethanol. Process Biochem 48:1999–2008. https://doi.org/10.1016/j.procbio.2013.09.015

Blamey LK, Bolton JJ (2018) The economic value of South African kelp forests and temperate reefs: past, present and future. J Mar Syst 188:172–181. https://doi.org/10.1016/j.jmarsys.2017.06.003

Blue Economy Cooperative research Centre (2022) Underpinning the growth of our blue economy. https://blueeconomycrc.com.au/. Accessed 20 Feb 2025

Bonin DR, Hawkes MW, Hawkes MW (1987) Systematics and life histories of New Zealand Bonnemaisoniaceae (Bonnemaisoniales, Rhodophyta): I. The genus *Asparagopsis*. NZJ Bot 25:577–590. https://doi.org/10.1080/0028825X.1987.10410088

Boulom S, Robertson J, Hamid N, Ma Q, Lu J (2014) Seasonal changes in lipid, fatty acid, α-tocopherol and phytosterol contents of seaweed, *Undaria pinnatifida*, in the Marlborough Sounds, New Zealand. Food Chem 161:261–269. https://doi.org/10.1016/j.foodchem.2014.04.007

Bradly N, Syddall V, Ingram C, Clarkson R, Elliot A, Major R, Adams S (2021) Stocktake and characterisation of Aotearoa New Zealand's seaweed sector: market and regulatory focus. Report for Sustainable Seas National Science Challenge project. National Science Challenge, pp 1–58

Braña CBC, Cerbule K, Senff P, Stolz IK (2021) Towards environmental sustainability in marine finfish aquaculture. Front Mar Sci 8:1–24. https://doi.org/10.3389/fmars.2021.666662

Broch OJ, Alver MO, Bekkby T, Gundersen H, Forbord S, Handå A, Skjermo J, Hancke K (2019) The kelp cultivation potential in coastal and offshore regions of Norway. Front Mar Sci 5:1–15. https://doi.org/10.3389/fmars.2018.00529

Brown MT, Frazer AWJ, Brasch DJ, Melton LD (1990) Growth and reproduction of *Porphyra columbina* Mont. (Bangiales, Rhodophyceae) from southern New Zealand. J Appl Phycol 2:35–44. https://doi.org/10.1007/BF02179767

Bunting I, Kok YY, Krieger EC, Bury SJ, D'Archino R, Cornwall CE (2024) Marine heatwave intensity and duration negatively affect growth in young sporophytes of the giant kelp *Macrocystis pyrifera*. Front Mar Sci 11:1423595. https://doi.org/10.3389/fmars.2024.1423595

Bursic J, Fulton S, Bellgrove A, Biancacci C, Francis P, Macreadie P, Carnell P (2022) Assessing the potential for seaweed aquaculture in Gippsland. Guidance in selecting species and locations. Summary Report, Deakin University, Geelong, pp 1–23

Buschmann AH, Camus C, Infante J, Neori A, Israel A, Hernández-González M, Pereda SV, Gomez-Pinchetti JL, Golberg A, Tadmor-Shalev N, Critchley AT (2017) Seaweed production: overview of the global state of exploitation, farming and emerging research activity. Eur J Phycol 52:391–406. https://doi.org/10.1080/09670262.2017.1365175

Callo DP, Darag AN (2015) Manual of seaweed production discovery-based exercises for farmer field schools. Peace and Equity Foundation, Philippines and Social Action Center Northern Quezon, Philippines pp 1–261

Camara MD, Symonds JE (2014) Genetic improvement of New Zealand aquaculture species: programmes, progress and prospects. NZJMFR 48:466–491. https://doi.org/10.1080/00288330.2014.932291

Campbell I, Macleod A, Sahlmann C, Neves L, Funderud J, Øverland M, Hughes AD, Stanley M (2019) The environmental risks associated with the development of seaweed farming in Europe - prioritizing key knowledge gaps. Front Mar Sci 6:107. https://doi.org/10.3389/fmars.2019.00107

Chen J, Xu P (2009) *Porphyra* spp. In: Cultured aquatic species fact sheets. FAO, Rome, pp 1–12

Chi Y, Jiang G, Liang Y, Xu C, Li Q (2022) Selective breeding for summer survival in Pacific oyster (*Crassostrea gigas*): genetic parameters and response to selection. Aquaculture 556:738271. https://doi.org/10.1016/j.aquaculture.2022.738271

Chopin T, Buschmann AH, Halling C, Troell M, Kautsky N, Neori A, Kraemer GP, Zertuche-González JA, Yarish C, Neefus C (2001) Integrating seaweeds into marine aquaculture systems: a key toward sustainability. J Phycol 37:975–986. https://doi.org/10.1046/j.1529-8817.2001.01137.x

Clark D, Newcombe E, Clement D, Magnusson M, Lawton RJ, Glasson CRK, Major R, Adams S (2021) Stocktake and characterisation of Aotearoa New Zealand' s seaweed sector: environmental effects of seaweed wild-harvest and aquaculture. Report for Sustainable Seas National Science Challenge project. National Science Challenge, pp 1–74

Coleman M, Kelaher B, Steinberg P, Millar A (2008) Absence of a large brown macroalga on urbanized rocky reefs around Sydney, Australia, and evidence for historical decline. J Phycol 44:897–901. https://doi.org/10.1111/j.1529-8817.2008.00541.x

Coleman MA, Wood G, Filbee-Dexter K, Minne AJP, Goold HD, Vergés A, Marzinelli EM, Steinberg PD, Wernberg T (2020) Restore or redefine: future trajectories for restoration. Front Mar Sci 7:1–12. https://doi.org/10.3389/fmars.2020.00237

Cooke RRM, Hurd CL, Lord JM, Peake BM, Raven JA, Rees TAV (2004) Iron and zinc content of *Hormosira banksii* in New Zealand. NZJMFR 38:73–85. https://doi.org/10.1080/00288330.2004.9517219

Cornwall CE, Nelson WA, Aguirre JD, Blain CO, Coyle L, D'Archino R, Desmond MJ, Hepburn CD, Liggins L, Shears NT, Thomsen MS (2023) Predicting the impacts of climate change on New Zealand's seaweed-based ecosystems. NZJ Bot 63:1–27. https://doi.org/10.1080/0028825X.2023.2245786

Cotas J, Gomes L, Pacheco D, Pereira L (2023) Ecosystem services provided by seaweeds. Hydrobiol 2(1):75–96. https://doi.org/10.3390/hydrobiology2010006

Cressey P, Wright J, Gautam A, Horn B, Wheeler T, Hampton H, Harwood T, Lymburn J, Nicolas J, Thomas K, Esguerra C (2023) Evaluation of food safety risks associated with seaweed and seaweed products. New Zealand Food Safety Tech Paper No: 2023/01, pp 1–180

Crossett D, Dunmore R, Schiel D (2023) Reef recovery and resilience: interpreting abiotic effects on critical early life stages of large brown algae. Mar Ecol Prog Ser 710:27–42. https://doi.org/10.3354/meps14293

Crossett D, Biancacci C, Clark DE, South PM (2025) Experimental assessment of growing substrates for the early life stages of five large brown seaweeds. NZJMFR. Accepted for publication 18-03-2025

Crowe L (2021) A case for including deep-water marine permaculture in agricultural emissions pricing negotiations, pp 1–11. https://gen.org.nz/wp/wp-content/uploads/2021/03/Gen2021_Essay_Lennox_Crowe_Agricultural_Emissions.pdf

Davis GDJ, Vasanthi AHR (2011) Seaweed metabolite database (SWMD): a database of natural compounds from marine algae. Bioinfo 5:361–364. https://doi.org/10.6026/97320630005361

Diehl N, Michalik D, Zuccarello GC, Karsten U (2019) Stress metabolite pattern in the eulittoral red alga *Pyropia plicata* (Bangiales) in New Zealand-mycosporine-like amino acids and heterosides. J Exp Mar Biol Ecol 510:23–30. https://doi.org/10.1016/j.jembe.2018.10.002

Duarte CM, Delgado-Huertas A, Marti E, Gasser B, San Martin I, Cousteau A, Neumeyer F, Reily-Cayten M, Boyce J, Kuwae T, Hori M, Miyajima T, Price NN, Arnold S, Ricart AM, Davis S, Surugau N, Abdul AJ, Wu J, Xiao X, Chung IK, Choi CG, Sondak CFA, Albasri H, Krause-Jensen D, Bruhn A, Boderkov T, Hancke K, Funderud J, Borrero-Santiago AR, Pascal F, Joanne P, Ranivoarivelo L, Collins WT, Clark J, Gutierrez JF, Riquelme R, Avila M, Macreadie P, Masque P (2025) Carbon burial in sediments below seaweed farms matches that of Blue Carbon habitats. Nat Clim Chang 15:180–187. https://doi.org/10.1038/s41558-024-02238-1

Eason CT, Fennessy P (2023) Methane reduction, health and regulatory considerations regarding *Asparagopsis* and bromoform for ruminants. NZJ Agric Res 68:1–30. https://doi.org/10.1080/00288233.2023.2248948

Edmunds R (2023) A comprehensive review of publications about kelp and fucoids from New Zealand and subsequent experiments on the early life stages of *Hormosira banksii*. MSc Thesis. University of Canterbury, Christchurch, p 133

Falshaw R, Furneaux RH (2009) Chemotaxonomy of New Zealand red algae in the family Gigartinaceae (Rhodophyta) based on galactan structures from the tetrasporophyte life-stage. Carbohydr Res 344:210–216. https://doi.org/10.1016/j.carres.2008.10.020

Falshaw R, Bixler HJ, Johndro K (2003) Structure and performance of commercial κ-2 carrageenan extracts. Part III. Structure analysis and performance in two dairy applications of extracts from the New Zealand red seaweed, *Gigartina atropurpurea*. Food Hydrocoll 17:129–139. https://doi.org/10.1016/S0268-005X(02)00045-0

FAO (2022) The State of World Fisheries and Aquaculture 2022. FAO, Rome, pp 1–232

Fennessy PF, Proctor LE, Muetzel S (2025) Methane inhibition in sheep fed *Asparagopsis armata* and Mootral™. NZJ Agric Res. https://doi.org/10.1080/00288233.2024.2445811

Flavin K, Flavin N, Flahive B (2021) Kelp Farming Manual a guide to the processes, techniques, and equipment for farming kelp in new England waters. Ocean approved, Portland, pp 1–123. https://maineaqua.org/wp-content/uploads/2020/06/OceanApproved_KelpManualLowRez.pdf

Fossberg J, Forbord S, Broch OJ, Malzahn AM, Jansen H, Handå A, Førde H, Bergvik M, Fleddum A, Skjermo J, Olsen Y (2018) The Potential for upscaling kelp (*Saccharina latissima*) cultivation in salmon-driven Integrated Multi-Trophic Aquaculture (IMTA). Front Mar Sci 5:418. https://doi.org/10.3389/fmars.2018.00418

Fricke A, Capuzzo E, Bermejo R, Hoffmann LC, Hernández I, Pereira R, Van den Burg SWK, Pereira T, Buschmann AH, Cottier-Cook EJ (2024) Ecosystem services provided by seaweed cultivation: state of the art, knowledge gaps, constraints and future needs for achieving maximum potential in Europe. Rev Fish Sci Aqua. https://doi.org/10.1080/23308249.2024.2399355

Friedlander S, Melton L, Brown M (1989) Ascorbic acid in the New Zealand seaweeds *Porphyra columbina* Mont. & *Porphyra subtumens* J. Ag. (Rhodophyceae). Jpn J Phycol 37:295–301

Gambichler V, Zuccarello GC, Karsten U (2021) Seasonal changes in stress metabolites of native and introduced red algae in New Zealand. J Appl Phycol 33:1157–1170. https://doi.org/10.1007/s10811-020-02365-0

Gan A, Baroutian S (2022) Subcritical water extraction for recovery of phenolics and fucoidan from New Zealand Wakame (*Undaria pinnatifida*) seaweed. J Supercrit Fluids 190:105732. https://doi.org/10.1016/j.supflu.2022.105732

Glasson CRK, Kinley RD, de Nys R, King N, Adams SL, Packer MA, Svenson J, Eason CT, Magnusson M (2022) Benefits and risks of including the bromoform containing seaweed *Asparagopsis* in feed for the reduction of methane production from ruminants. Algal Res 64:102673. https://doi.org/10.1016/j.algal.2022.102673

Grand View research (2021) Commercial Seaweed Market Size, share & trends analysis report by product (brown seaweeds, red seaweeds, green seaweeds), by form, by application, by region, and segment forecasts, 2021–2028. https://www.grandviewresearch.com/industry-analysis/commercial-seaweed-market. Accessed 20 Feb 2025

Green Gravel (2022). https://www.greengravel.org. Accessed 20 Feb 2025

Green Wave (2024). https://www.greenwave.nz. Accessed 20 Feb 2025

Guiry MD (2025) Seaweed Site. World-wide electronic publication, University of Galway, Galway. https://www.seaweed.ie/algae/algae.html. Accessed 20 Feb 2025

Guiry MD, Guiry GM (2025) AlgaeBase. World-wide electronic publication, National University of Ireland, Galway. http://www.algaebase.org. Accessed 20 Feb 2025

Hall AM, South PM, Elvy JE, Jeffs AG (2025) The effects of water motion on fragmentation of *Asparagopsis armata*. NZJMFR. https://doi.org/10.1080/00288330.2025.2458508

Hamilton SL, Gleason MG, Godoy N, Eddy N, Grorud-Colvert K (2022) Ecosystem-based management for kelp forest ecosystems. Mar Policy 136:104919. https://doi.org/10.1016/j.marpol.2021.104919

Hemmingson JA, Falshaw R, Furneaux RH, Thompson K (2006) Structure and antiviral activity of the galactofucan sulfates extracted from *Undaria Pinnatifida* (Phaeophyta). J Appl Phycol 18:185–193. https://doi.org/10.1007/s10811-006-9096-9

Hepburn C, Pritchard D, Cornwall C, McLeod RJ, Beardall J, Raven JA, Hurd CL (2011) Diversity of carbon use strategies in a kelp forest community: implications for a high CO_2 ocean. Glob Chang Biol 17:2488–2497. https://doi.org/10.1111/j.1365-2486.2011.02411.x

HowarthL,MajorR,(2023)Keyenvironmentalconsiderationsforseaweed aquaculture in Aotearoa New Zealand. Report for Sustainable Seas National Science Challenge. National Science Challenge, pp 1–15. https://www.sustainableseaschallenge.co.nz/tools-and-resources/key-environmental-considerations-for-seaweed-aquaculture/

Hrstich-Manning G, Aguirre JD (2024) Nutritional composition of common, coastal seaweeds from northeastern New Zealand nutritional composition of common, coastal seaweeds from northeastern New Zealand. NZJMFR. https://doi.org/10.1080/00288330.2024.2353211

Hughes AD, Fendenko J, O'Dell A, Swanlund H (2025) Seaweed aquaculture for carbon farming: an assessment under the EU's QU.A.L.ITY (Quantification, Additionality, Long-Term Storage, Sustainability) Framework. Corp Soc Responsib Environ Manag:1–10. https://doi.org/10.1002/csr.3098

Hurd C (2000) Water motion, marine macroalgal physiology, and production. J Phycol 36:453–472. https://doi.org/10.1046/j.1529-8817.2000.99139.x

Hurd C, Nelson W, Falshaw R, Neill K (2004) History, current status and future of marine macroalgal research in New Zealand: taxonomy, ecology, physiology and human uses. Phycol Res 52:80–106. https://doi.org/10.1111/j.1440-183.2004.00331.x

Hurd CL, Gattuso JP, Boyd PW (2023) Air-sea carbon dioxide equilibrium: will it be possible to use seaweeds for carbon removal offsets? J Phycol 60:4–14. https://doi.org/10.1111/jpy.13405

Hwang EK, Yotsukura N, Pang SJ, Su L, Shan TF (2019) Seaweed breeding programs and progress in eastern Asian countries. Phycol 58:484–495. https://doi.org/10.1080/00318884.2019.1639436

Idealog (2015) How a company turning algae into supplements attracted over NZ$7 million from investors. https://idealog.co.nz/tech/2015/01/algae-boom. Accessed on 20 Feb 2025

Juanich GL (1988) Manual on seaweed farming 1. *Eucheuma* spp. FAO, Rome, pp 1–30. https://www.fao.org/4/ac417e/AC417E00.htm

Khan N, Sudhakar K, Mamat R (2024) Macroalgae farming for sustainable future: navigating opportunities and driving innovation. Heliyon 10(7):e28208. https://doi.org/10.1016/j.heliyon.2024.e28208

Kidgell JT, Carnachan SM, Magnusson M, Lawton RL, Sims IM, Hinkley SFR, de Nys R, Glasson CRK (2021) Are all ulvans equal? A comparative assessment of the chemical and gelling properties of ulvan from blade and filamentous *Ulva*. Carbohydr Polym 264:118010. https://doi.org/10.1016/j.carbpol.2021.118010

Kim JK, Yarish C, Hwang EK, Park M, Kim Y (2017) Seaweed aquaculture: cultivation technologies, challenges and its ecosystem services. Algae 32(1):1–13. https://doi.org/10.4490/algae.2017.32.3.3

Koh HSA, Lu J, Zhou W (2019) Structure characterization and antioxidant activity of fucoidan isolated from *Undaria pinnatifida* grown in New Zealand. Carbohydr Polym 212:178–185. https://doi.org/10.1016/j.carbpol.2019.02.040

Kuwae T, Watanabe A, Yoshihara S, Suehiro F (2022) Implementation of blue carbon offset crediting for seagrass meadows, macroalgal beds, and macroalgae farming in Japan. Mar Policy 138:104996. https://doi.org/10.1016/j.marpol.2022.104996

Lamare MD, Wing SR (2001) Calorific content of New Zealand marine macrophytes NZJMFR. 35:335–341. https://doi.org/10.1080/00288330.2001.9517004

Langford A, Waldron S, Sulfahri SH (2021) Monitoring the COVID-19-affected Indonesian seaweed industry using remote sensing data. Mar Policy 127:104431. https://doi.org/10.1016/j.marpol.2021.104431

Larson S, Stoeckl N, Fachry ME, Mustafa MD, Lapong I, Purnomo AH, Rimmer MA, Paul NA (2021) Women's well-being and household benefits from seaweed farming in Indonesia. Aquaculture 530:735711. https://doi.org/10.1016/j.aquaculture.2020.735711

Lawton RJ, Magnusson M (2024) Effects of seeding twine type and seeding density on hatchery performance and initial at-sea cultivation performance of the kelp *Ecklonia radiata*. Algal Res 84:103777. https://doi.org/10.1016/j.algal.2024.103777

Lawton RJ, Praeger C (2025) Effects of nutrient concentration, germanium dioxide and seeding density on sporophyte production in the kelp *Ecklonia radiata*. J Appl Phycol. https://doi.org/10.1007/s10811-025-03451-x

Lawton RJ, Magnusson M (2025) Optimisation of gametophyte culturing and seeding methods forcultivation of the kelp Ecklonia radiata; Group: Brown, Species: *Ecklonia radiata*. https://doi.org/10.1016/j.algal.2025.104240

Lawton RJ, Mata L, de Nys R, Paul NA (2013) Algal bioremediation of waste waters from land-based aquaculture using *Ulva*: selecting target species and strains. PLoS One 8. https://doi.org/10.1371/journal.pone.0077344

Lawton RJ, Sutherland JE, Glasson CRK, Magnusson ME (2021) Selection of temperate Ulva species and cultivars for land-based cultivation and biomass applications. Algal Res 56:102320. https://doi.org/10.1016/j.algal.2021.102320

Le DM, Desmond MJ, Buschmann AH, Pritchard D (2022a) Reproduction, hatchery and culture applications for the giant kelp (*Macrocystis pyrifera*): a methodological appraisal. Appl Phycol 3(1):368–382. https://doi.org/10.1080/26388081.2022.2086823

Le DM, Desmond MJ, Pritchard DW, Hepburn CD (2022b) Effect of temperature on sporulation and spore development of giant kelp (*Macrocystis pyrifera*). PLoS One 17:e0278268

Liu L, Yang X, Yuan P, Cai S, Bao J, Zhao Y, Aimaier A, Aipire A, Lu J, Li J (2022) In Vitro and in vivo dendritic cell immune stimulation effect of low molecular weight fucoidan from New Zealand. Mar Drugs 20:197. https://doi.org/10.3390/md20030197

Love RimuRimu (2023). https://www.loverimurimu.org/. Accessed 20 Feb 2025

Loy C, Jeffs A (2025) Assessment of the potential of the fucoid seaweed, *Hormosira banksii*, as a source of fucoidan and coculture on a Pacific oyster farm in New Zealand. NZJMFR. https://doi.org/10.1080/00288330.2025.2472807

Lu J, Shi KK, Chen S, Wang J, Hassouna A, White LN, Merien F, Xie M, Kong Q, Li J, Ying T, White WL, Nie S (2018) Fucoidan extracted from the New Zealand *Undaria pinnatifida*-physicochemical comparison against five other fucoidans: unique low molecular weight fraction bioactivity in breast cancer cell lines. Mar Drugs 16:461. https://doi.org/10.3390/md16120461

Magnusson M, Carl C, Mata L, de Nys R, Paul NA (2016) Seaweed salt from *Ulva*: a novel first step in a cascading biorefinery model. Algal Res 16:308–316. https://doi.org/10.1016/j.algal.2016.03.018

Magnusson M, Glasson CRK, Vucko MJ, Angell A, Neoh TL, de Nys R (2019) Enrichment processes for the production of high-protein feed from the green seaweed *Ulva ohnoi*. Algal Res 41:101555. https://doi.org/10.1016/j.algal.2019.101555

Mak W, Hamid N, Liu T, Lu J, White WL (2013) Fucoidan from New Zealand *Undaria pinnatifida*: monthly variations and determination of antioxidant activities. Carbohydr Polym 95:606–614. https://doi.org/10.1016/j.carbpol.2013.02.047

Mak W, Wang SK, Liu T, Hamid N, Li Y, Lu J, White WL (2014) Anti-proliferation potential and content of fucoidan extracted from sporophyll of New Zealand *Undaria pinnatifida*. Front Nutr 10(1):9. https://doi.org/10.3389/fnut.2014.00009

Markets and Markets (2020) Market Research Report, FB 6702. accessed: 12 September 2025. https://www.marketsandmarkets.com/Market-Reports/commercial-seaweed-market-152763701.html

McKee JWA, Kavalieris L, Brasch DJ, Brown MT, Melton LD (1992) Alginate content and composition of *Macrocystis pyrifera* from New Zealand. J Appl Phycol 4:357–369. https://doi.org/10.1007/BF02185794

McNeill SE, Falshaw R (2017) Aquaculture of *Sarcothalia atropurpurea* tetraspores in New Zealand for carrageenan production. NZJMFR 51(4):528–542. https://doi.org/10.1080/00288330.2017.1294085

McNeill SE, Page M, Falshaw R (2003) Field trials to optimise timing and frequency of pruning for cultivation of a New Zealand carrageenophyte, *Gigartina atropurpurea*. J Appl Phycol 15:381–405. https://doi.org/10.1023/A:1026033317850

Mihaila AA, Glasson CRK, Lawton R, Muetzel S, Molano G, Magnusson M (2022) New temperate seaweed targets for mitigation of ruminant methane emissions: an in vitro assessment. Appl Phycol 3:274–284. https://doi.org/10.1080/26388081.2022.2059700

Mihaila AA, Lawton RJ, Glasson CRK, Magnusson M (2023a) Early hatchery protocols for tetrasporogenesis of the antimethanogenic seaweed *Asparagopsis armata*. J Appl Phycol 35:2323–2335. https://doi.org/10.1007/s10811-023-03029-5

Mihaila AA, Magnusson M, Glasson CRK, Lawton RJ (2023b) The reproductive phenology of *Asparagopsis armata* in New Zealand-Potential shifts 35 years later. Algal Res 76:103318. https://doi.org/10.1016/j.algal.2023.103318

Mihaila AA, Lawton RJ, Glasson CRK, Magnusson M (2024) Moderate temperature and water flow increase growth during the nursery phase of *Asparagopsis armata*. Algal Res 78:103380. https://doi.org/10.1016/j.algal.2023.103380

Miller KI, Blain CO, Shears NT (2022) Sea urchin removal as a tool for macroalgal restoration: a review on removing "the Spiny Enemies.". Front Mar Sci 9:1–18. https://doi.org/10.3389/fmars.2022.831001

Ministry for Primary Industries (MPI) (2025). https://www.mpi.govt.nz/. Accessed 20 Feb 2025

Ministry for the Environment (2021a) Everyday guide to the Resource Management Act. 3.1. New Zealand Government, pp 1–19

Ministry for the Environment (2021b) Applying for a resource consent. Resource Management Act 2.1. New Zealand Government, pp 1–18

Ministry of Business, Innovation and Employment (MBIE) (2025). https://www.mbie.govt.nz/. Accessed 20 Feb 2025

Morelissen B, Dudley BD, Geange SW, Phillips NE (2013) Gametophyte reproduction and development of *Undaria pinnatifida* under varied nutrient and irradiance conditions. J Exp Mar Biol Ecol 448:197–206. https://doi.org/10.1016/j.jembe.2013.07.009

Neill KF, Nelson WA, Falshaw R, Hurd CL (2016) Clump structure, population structure and non-destructive biomass estimation of the New Zealand carrageenophyte *Sarcothalia lanceata* (Gigartinaceae, Rhodophyta). Bot Mar 59(5):373–385. https://doi.org/10.1515/bot-2016-0059

Neill K, Nelson W, Hurd C, Falshaw R (2018) Growth and carrageenan composition of two populations of the New Zealand carrageenophyte *Sarcothalia lanceata* (Gigartinaceae, Rhodophyta). J Appl Phycol 30:1–13. https://doi.org/10.1007/s10811-018-1416-3

Neill K, Miller S, Nelson W (2024) Notes on the experimental laboratory and field culture of *Lessonia variegata* (Laminariales). NZJMFR. https://doi.org/10.1080/00288330.2024.2353916

Neish I (2013) Social and economic dimensions of carrageenan seaweed farming in Indonesia. FAO Technical Paper 580, Rome, pp 1–204

Nelson WA (2013) New Zealand seaweeds: an illustrated guide. Te Papa Press, Wellington, p 352

Nelson WA, Conroy AM (1989) Effect of harvest method and timing on yield and regeneration of Karengo (*Porphyra* spp.) (Bangiales, Rhodophyta) in New Zealand. J Appl Phycol 1:277–283. https://doi.org/10.1007/BF00003654

Nelson WA, Knight GA (1995) Endosporangia-a new form of reproduction in the *genus* Porphyra (Bangiales, Rhodophyta). Bot Mar 38(1–6):17–20. https://doi.org/10.1515/botm.1995.38.1-6.17

Nelson WA, Islands C, Snares LCHH (2005) Life history and growth in culture of the endemic New Zealand kelp *Lessonia variegata* J. Agardh in response to differing regimes of temperature, photoperiod and light. J Appl Phycol 17:23–28. https://doi.org/10.1007/s10811-005-5521-8

Nelson WA, Neill KF, D'Archino R (2015) When seaweeds go bad: an overview of outbreaks of nuisance quantities of marine macroalgae in New Zealand. NZJMFR 49(4):472–491. https://doi.org/10.1080/00288330.2015.1064975

Nelson WA, D'Archino R, Neill KF, Robinson NM (2021) Introduced marine macroalgae: new perspectives on species recognition and distribution in New Zealand. Bot Mar 64(5):379–393. https://doi.org/10.1515/bot-2021-0042

Nepper-Davidsen J, Glasson CRK, Lawton RJ, Magnusson M (2023a) High spatial and temporal variation in biomass composition of the novel aquaculture target *Ecklonia radiata*. J Appl Phycol 35:1755–1768. https://doi.org/10.1007/s10811-023-02969-2

Nepper-Davidsen J, Magnusson M, Glasson CRK, Lawton RJ (2023b) Line configuration and farming depth markedly affect survival and growth in the kelp *Ecklonia radiata*. NZJMFR. https://doi.org/10.1080/00288330.2023.2256685

New Zealand Government (2021) Accelerate the aquaculture strategy: investment roadmap. New Zealand Government, pp 1–20

New Zealand Government (2024a) Aquaculture strategy. Ministry of Fisheries, pp 1–20

New Zealand Government (2024b) New Zealand aquaculture development plan. New Zealand Government, pp 1–16

New Zealand Journal of Marine and Freshwater Research (2025) Special issue on Seaweed aquaculture in New Zealand: current advances, benefits, challenges, and perspectives. https://www.tandfonline.com/journals/tnzm20. Accessed 20 Feb 2025

Nissanka M, Stevenson B (2023) The contribution of Kānoa- administered government funding to Aotearoa New Zealand's Aquaculture Industry. Allen+Clarke, Kānoa Impact study-Aquaculture, vol 6, pp 1–56

North WJ (1994) Review of *Macrocystis* biology. In: Akatsuka I (ed) Biology of economic algae. Academic Publishing, Hague, pp 447–527

Novaczek I (1984) Development and phenology of *Ecklonia radiata* at two depths in Goat Island Bay, New Zealand. Mar Biol 81:189–197. https://doi.org/10.1007/BF00393117

Olafsen T, Winther U, Olsen Y, Skjermo J (2012) Value created from productive oceans in 2050. DKNVS and NTVA, pp 1–83

Overton K, Dempster T, Swearer SE, Morris RL, Barrett LT (2024) Achieving conservation and restoration outcomes through ecologically beneficial aquaculture. Conserv Biol 38:e14065. https://doi.org/10.1111/cobi.14065

Park JS, Shin SK, Wu H, Yarish C, Yoo HI, Kim JK (2021) Evaluation of nutrient bioextraction by seaweed and shellfish aquaculture in Korea. J World Aquac Soc 52:1118–1134. https://doi.org/10.1111/jwas.12786

Parsons MJ (1985) New Zealand seaweed flora and its relationships. NZJMFR 19:131–138. https://doi.org/10.1080/00288330.1985.9516079

Parsons MJ, Pickmere SE, Bailey RW (1977) Carrageenan composition in New Zealand species of *Gigartina* (Rhodophyta): geographic variation and interspecific differences. NZJ Bot 15(3):589–595. https://doi.org/10.1080/0028825X.1977.10429632

Paul NA, Wegner A, Tuart I, Tamuera K, Teata T, Tioti T, Temari T, Morris C, Bala S, Tiitii U, Tanielu E, Aiafi J, Ravitu N, Vuakaca J, Batisbasag A (2020) Biochemical database and product sheets for seaweeds from Fiji, Samoa and Kiribati. University of the Sunshine Coast, Sunshine Coast, pp 1–11. https://doi.org/10.25907/5e13b12523e43

Pereira R, Yarish C, Critchley AT (2024) Chapter 6 - Seaweed aquaculture for human foods in land based and IMTA systems. In: Hefft DI, Adetunji CO (eds) Applications of seaweeds in food and nutrition. Elsevier, Amsterdam, pp 77–99

Perryman S, Lapong I, Mustafa A, Sabang R, Rimmer MA (2017) Potential of metal contamination to affect the food safety of seaweed (*Caulerpa* spp.) cultured in coastal ponds in Sulawesi, Indonesia. Aquac Rep 5:27–33. https://doi.org/10.1016/j.aqrep.2016.12.002

Phillips J, Hurd C (2003) Nitrogen ecophysiology of intertidal seaweeds from New Zealand: N uptake, storage and utilisation in relation to shore position and season. Mar Ecol Prog Ser 264:31–48. https://doi.org/10.3354/meps264031

Pickering TD, Sladden VH, Furneaux, Hemmingson JA, Redfearn P (1993) Comparison of growth rate in culture, dry matter content, agar content and agar quality of two New Zealand red seaweeds, *Gracilaria chilensis* Bird, McLachlan et Oliveira and *Gracilaria truncata* Kraft. J Appl Phycol 5:85–91

Pickering TD, Gordon ME, Tong LJ (1995) A preliminary trial of a spray culture technique for growing the agarophyte *Gracilaria chilensis* (Gracilariales, Rhodophyta). Aquaculture 130:43–49. https://doi.org/10.1016/0044-8486(94)00111-Z

Pickmere SE, Parsons MJ, Bailey RW (1973) Composition of *Gigartina* carrageenan in relation to sporophyte and gametophyte stages of the life cycle. Phytochem 12:2441–2444. https://doi.org/10.1016/0031-9422(73)80452-2

Potter H, Mangai R (2022) A Wai 262 best practice guide for science partnerships with kaitiaki for research involving taonga. Lessons from Māori voices in the New Zealand science sector. Rauika Māngai, Dunedin, pp 1–40

Praeger C, Magnusson M, Lawton R (2022) Optimising the zoospore release, germination, development of gametophytes and formation of sporophytes of *Ecklonia radiata*. J Appl Phycol 34:2535–2549. https://doi.org/10.1007/s10811-022-02806-y

Qu Z, Thrush S, Blain C, Lewis N (2023) Assessing the carbon storage value of kelp forest restoration in the Hauraki Gulf Marine Park, New Zealand: lessons from no-take marine protected areas. Mar Policy 154:105682. https://doi.org/10.1016/j.marpol.2023.105682

Redmond S, Green L, Yarish C, Kim J, Neefus C (2014) New England seaweed culture handbook. Seaweed cultivation. Paper 1. University of Connecticut Sea Grant, pp 1–92

Reid J, Rout M (2019) Tribal Economies—Ngāi Tahu: an examination of the historic and current tītī and pounamu institutional frameworks. Originally published: NTRC, pp 1–143

Rhodes L, Smith KF, MacKenzie L, Sa W, Ponikla K, Harwood DT, Packer M, Munday R (2016) The Cawthron Institute Culture Collection of Micro-algae: a significant national collection. NZJMFR 50:291–316. https://doi.org/10.1080/00288330.2015.1116450

Ridler N, Wowchuk M, Robinson B, Chopin T (2007) Integrated multi-trophic aquaculture (IMTA): a potential strategic choice for farmers. Aquacult Econ Manag 11:99–110. https://doi.org/10.1080/13657300701202767

Rimmer MA, Larson S, Lapong I, Purnomo AH, Pong-Masak PR, Swanepoel L, Paul NA (2021) Seaweed aquaculture in Indonesia contributes to social and economic aspects of livelihoods and community wellbeing. Sustainability 13:1–22. https://doi.org/10.3390/su131910946

Rolin C, Inkster R, Laing J, Hedges J, McEvoy L (2014) Seaweed cultivation manual. Shetland Seaweed Growers Project 2014–16 39. University of the Highlands and Islands, pp 1–38

Ross F, Tarbuck P, Macreadie PI (2022) Seaweed afforestation at large-scales exclusively for carbon sequestration: critical assessment of risks, viability and the state of knowledge. Front Mar Sci 9:1–15. https://doi.org/10.3389/fmars.2022.1015612

Ross BG, Magnusson ME, Glasson CRK, Lawton RJ (2024) Seasonal biomass productivity and bioremediation performance of the novel aquaculture target species *Gracilaria transtasmanica* in nutrient-enriched and turbid conditions. Algal Res 79:103488. https://doi.org/10.1016/j.algal.2024.103488

Schiel DR (1988) Algal interactions on shallow subtidal reefs in northern New Zealand: a review. NZJMFR 22:481–489. https://doi.org/10.1080/00288330.1988.9516317

Smith JL, Summers G, Wong R (2010) Nutrient and heavy metal content of edible seaweeds in New Zealand. NZJ Crop Hortic Sci 38(1):19–28. https://doi.org/10.1080/01140671003619290

South PM, Floerl O, Forrest BM, Thomsen MS (2017) A review of three decades of research on the invasive kelp *Undaria pinnatifida* in Australasia: an assessment of its success, impacts and status as one of the world's worst invaders. Mar Environ Res 131:243–257. https://doi.org/10.1016/j.marenvres.2017.09.015

Stenton-Dozey JME, Heath P, Ren JS, Leonardo N (2021) New Zealand aquaculture industry: research, opportunities and constraints for integrative multitrophic farming. NZJMFR 55(2):265–285. https://doi.org/10.1080/00288330.2020.1752266

Suplicy FM (2020) A review of the multiple benefits of mussel farming. Rev Aquac 12:204–223. https://doi.org/10.1111/raq.12313

Sustainable Sea Challenge (2022) Aotearoa New Zealand's seaweed sector framework. National Science Challenge, New Zealand, pp 1–20

Symonds JE, Clarke SM, King N, Walker SP, Blanchard B, Sutherland D, Roberts R, Preece MA, Tate M, Buxton P, Dodds KG (2019) Developing successful breeding programs for New Zealand aquaculture: a perspective on progress and future genomic opportunities. Front Genet 10:1–7. https://doi.org/10.3389/fgene.2019.00027

Tait L, Schiel D (2010) Primary productivity of intertidal macroalgal assemblages: comparison of laboratory and in situ photorespirometry. Mar Ecol Prog Ser 416:115–125. https://doi.org/10.3354/meps08781

Tait LW, Thoral F, Pinkerton MH et al (2021) Loss of giant kelp, *Macrocystis pyrifera*, driven by marine heatwaves and exacerbated by poor water clarity in New Zealand. Front Mar Sci 8:1–13. https://doi.org/10.3389/fmars.2021.721087

Te Runanga o Kaikoura (2007) Te Rūnanga o Kaikōura Environmental Management Plan. Te Rūnanga o Kaikōura, pp 1–334

The World Bank (2023) Global Seaweed: new and emerging markets report. https://doi.org/10.1596/40187
Thomsen MS, Mondardini L, Alestra T, Gerrity S, Tait L, South PM, Lilley SA, Schiel DR (2019) Local extinction of bull kelp (*Durvillaea* spp.) due to a marine heatwave. Front Mar Sci 6:1–10. https://doi.org/10.3389/fmars.2019.00084
Thoral F, Montie S, Thomsen MS, Tait LW, Pinkerton MH, Schiel DR (2022) Unravelling seasonal trends in coastal marine heatwave metrics across global biogeographical realms. Sci Rep 12:1–13. https://doi.org/10.1038/s41598-022-11908-z
Transparency Market Research (2025) Seaweed Cosmetic Ingredients Market, Global Industry analyses, size, share, growth, trend and forecast, 2024–2034. https://www.transparencymarketresearch.com/seaweed-cosmetic-ingredients-market.html. Accessed 20 Feb 2025
UN Trade and development (UNCTDA) (2024) Seaweed holds huge potential to bring economic, climate and gender benefits. https://unctad.org/news/seaweed-holds-huge-potential-bring-economic-climate-and-gender-benefits. Accessed 20 Feb 2025
Visch W, Biancacci C, Sanderson JC, et al (2024) Aquaculture production of australian laminarian kelps: a manual and research recommendations for *Ecklonia radiata*, *Lessonia corrugata*, and *Macrocystis pyrifera*. IMAS/UTAS, Tasmania, Technical report, pp 1–72
Wai 262 (2025). https://www.wai262.nz/ko-aotearoa-tenei. Accessed 20 Feb 2025
Wang SK, Li Y, White WL, Lu J (2014) Extracts from New Zealand *Undaria pinnatifida* containing fucoxanthin as potential functional biomaterials against cancer in vitro. J Funct Biomater 5:29–42. https://doi.org/10.3390/jfb5020029
Wear B, O'Connor NE, Schmid MJ, Jackson MC (2023) What does the future look like for kelp when facing multiple stressors? Ecol Evol 13:1–9. https://doi.org/10.1002/ece3.10203
Wheeler T, Romanazzi D, Adams S (2021) Stocktake and characterisation of New Zealand's seaweed sector: species characteristics and Te Tiriti o Waitangi considerations New Zealand. National Science Challenge, pp 1–51
White L, White WL (2020) Seaweed utilisation in New Zealand. Bot Mar 63(4):303–313. https://doi.org/10.1515/bot-2019-0089
Wood GV, Filbee-Dexter K, Coleman MA, Valckenaere J, Aguirre JD, Bentley PM, Carnell P, Dawkin PD, Dykman LN, Earp HS, Ennis LB, Francis P, Franco JN, Hayford H, Lamb JB, Ling SD, Layton C, Lis E, Masters B, Miller N, Moore OJ, Neufeld C, Pocklington JB, Smale D, Stahl F, Starko S, Steel SC, Verbeek J, Vergés A, Wilding CM, Wernberg T (2024) Upscaling marine forest restoration: challenges, solutions and recommendations from the Green Gravel Action Group. Front Mar Sci 11:1–15. https://doi.org/10.3389/fmars.2024.1364263
Zemke-White WL, Speed SR, McClary DJ (2005) Beach-cast seaweed: a review. New Zealand Fisheries Assessment Report 2005144. Ministry of Fisheries, pp 1–47
Zhang L, Liao W, Huang Y, Wen Y, Chu Y, Zhao C (2022) Global seaweed farming and processing in the past 20 years. Food Prod Proc Nut 4:23. https://doi.org/10.1186/s43014-022-00103-2
Zhao Y, Zheng Y, Wang J, Ma S, Yu Y, White WL, Yang S, Yang F, Lu J (2018) Fucoidan extracted from *Undaria pinnatifida*: Source for nutraceuticals/functional foods. Mar Drugs 16:321. https://doi.org/10.3390/md16090321
Zheng Y, Jin R, Zhang X, Wang Q, Wu J (2019) The considerable environmental benefits of seaweed aquaculture in China. Stoch Environ Res Risk Assess 33:1203–1221. https://doi.org/10.1007/s00477-019-01685-z
Zhou AY, Robertson J, Hamid N, Ma Q, Lu J (2015) Changes in total nitrogen and amino acid composition of New Zealand *Undaria pinnatifida* with growth, location and plant parts. Food Chem 186:319–325. https://doi.org/10.1016/j.foodchem.2014.06.016

Dr Cecilia Biancacci is a seaweed aquaculture scientist at Cawthron Institute in New Zealand, working on sustainable aquaculture of seaweeds with valuable applications. She has a Ph.D. in Marine Science focused on red seaweed aquaculture from the University of Aberdeen and SAMS. Before moving to NZ, she worked as post-doctoral research at Deakin University in Australia on the 'Seaweed Solutions for Sustainable Aquaculture CRC-P project'. She is experienced in macroalgae, microalgae and invertebrate cultivation. Her research primarily explores seaweed cultivation across nursery, laboratory and open-sea environments for large-scale production. She investigates how cultivation conditions influence the biochemical composition of macroalgae and examines the potential of seaweeds in Integrated Multi-Trophic Aquaculture (IMTA), as well as their ecology and physiology.

Dr. Daniel Crossett is a marine ecologist in Cawthron's Restoration Ecology Team. He focuses on state of the environment monitoring and bottom-up ecological processes that may enhance recruitment of foundation species. Dan is interested in how multiple stressors impact coastal foundation species, like seaweed, and the biodiversity they support. Dan's research focus includes the effects of natural disturbances, such as earthquakes and severe weather events, and anthropogenic impacts, such as sedimentation, warming seas and overfishing, on coastal ecosystems. The core goal of Dan's research is to identify and implement strategies for restorative measures in areas affected by multiple, interacting stressors, such as through seaweed-seeding programmes for rocky reef restoration. He also advocates for community outreach, public awareness, and educational initiatives to invoke stewardship among community groups, future scientists and managers.

Nicholas Scott is the Operations Coordinator for the *Ngā Punga o te Moana* Open Ocean Aquaculture research program at Cawthron Institute in Nelson, New Zealand. He holds a Master's in Marine Conservation and has over 8 years of experience in offshore aquaculture development. His work focuses on advancing structural, technological, and biological research to support sustainable aquaculture. Specializing in shellfish and macroalgae farming, Nicholas plays a key role in developing large-scale cultivation methods and infrastructure. As Operations Coordinator, he manages project logistics and research operations, working closely with Cawthron scientists and industry partners to ensure the success of applied aquaculture initiatives. Additionally, Nicholas contributes to Cawthron's *Blue Technologies* team, supporting the development and deployment of monitoring buoys and other ocean technology solutions to enhance aquaculture operations.

Dr. Paul M. South is a marine ecologist and the Algae Team Leader working in the Aquaculture Group at Cawthron Institute. Paul's research ranges from fundamental questions about the ecology of coastal ecosystems to more applied research in the mussel and seaweed aquaculture sectors. His current research seeks to support the development of a sustainable seaweed aquaculture industry in Aotearoa New Zealand by developing the capability to grow a wide range of species for a wide range of aquaculture and coastal restoration initiatives.

22 Integrating Breath-Hold Diving Techniques into a Low-Cost Community-Based Methodology for Kelp Restoration in Te Whanganui-a-Tara, Aotearoa New Zealand

Rachel L. Parry, Journey Luond, Hannah Ryall, Jorge Jiménez Senén, Joseph Warmington, and Zoe Studd

Abstract

Kelp forests provide essential habitat for temperate flora and fauna; however, they are experiencing global declines. Restoration efforts are typically costly, short-lived, and predominantly led by academics, highlighting the need for more cost-effective community-based approaches that foster long-term engagement. In this study, we present a novel low-cost methodology for kelp restoration through the integration of breath-hold diving techniques. This methodology was used for the first time to out-plant *Macrocystis pyrifera* sporophytes (green gravel) in Te Whanganui-a-Tara, Aotearoa New Zealand. We trained community members in subtidal scientific monitoring techniques, which were performed using breath-hold diving. Although integrating breath-hold techniques introduced some logistical challenges, it proved more cost-effective than traditional scuba-based subtidal field methodologies and removed some of the existing academic barriers to participation in science and restoration efforts. Community members were able to collect robust scientific data while maintaining high standards of health and safety, validating the utility of our novel methodology for kelp restoration. Trained community members (citizen scientists) assisted in completing four plant outs, successfully out-planting 387 green gravel cobbles. Although we observed high mortality (60–90%) within the first 4–6 weeks after planting, surviving sporophytes demonstrated success, with many reaching reproductive maturity and some growing over 5 m long. This study demonstrates the potential of integrating breath-hold diving into subtidal scientific monitoring for kelp restoration, offering an accessible, cost-effective alternative to traditional approaches while enhancing community engagement and minimizing restoration costs.

Keywords

Seaweed · Restoration · *Macrocystis pyrifera* · Citizen science · Aotearoa New Zealand · Breath-hold diving · Community monitoring · Kelp · Green gravel

22.1 Introduction

22.1.1 Kelp Forests and Their Ecological Importance

Kelp forests create structurally complex, three-dimensional habitats. These habitats offer shelter, nurseries, and food sources for numerous taxa across benthic and pelagic communities (Layton et al. 2020; Cotas et al. 2023; Grime et al. 2023). Seaweed forests, together with phytoplankton, represent the most productive vegetative ecosystems in the ocean in terms of per-unit-area productivity (Filbee-Dexter et al. 2024). Upon the transportation of seaweed detritus to the deep ocean, it can function as a long-term carbon sink (van der Mheen et al. 2024), capable of sequestering carbon for

R. L. Parry (✉)
Love Rimurimu, Mountains to Sea Wellington Trust, Wellington, Aotearoa, New Zealand

Greater Wellington Regional Council, Wellington, Aotearoa, New Zealand

J. Luond
Love Rimurimu, Mountains to Sea Wellington Trust, Wellington, Aotearoa, New Zealand

Te Herenga Waka–Victoria University of Wellington, Wellington, Aotearoa, New Zealand

H. Ryall · J. J. Senén · J. Warmington · Z. Studd
Love Rimurimu, Mountains to Sea Wellington Trust, Wellington, Aotearoa, New Zealand

A. T. Critchley et al. (eds.), *Advances in Temperate Phyconomy: Algal Harvest and Cultivation in Globally Distributed Temperate Waters*, Developments in Applied Phycology 15, https://doi.org/10.1007/978-3-032-02955-3_22

over 100 years (Filbee-Dexter et al. 2024). It is estimated that seaweed-derived organic material comprises approximately 2% of the global carbon sink (Filbee-Dexter et al. 2024). Beyond carbon sequestration, seaweed offers critical ecosystem services such as nutrient cycling (Bayley et al. 2021), coastal defence (Morris et al. 2019), and pharmaceutical applications (Vásquez et al. 2024). Ongoing research continues to uncover wide-ranging applications and elucidate the ecological importance of seaweed. For example, Pfister et al. (2019) observed that *Macrocystis pyrifera* kelp beds promote unique microbial communities, influencing seawater chemistry, phytoplankton production, and microbial assemblages which leads to higher taxonomic and phylogenetic diversity.

As a foundation species, *M. pyrifera* plays a crucial role in forming kelp forests (Wernberg et al. 2019). These forests provide vital habitat for numerous species, including shelter, food, and recruitment grounds for larval stages (Jackson and Winant 1983; Morton and Anderson 2013). In Te Whanganui-a-Tara (Wellington Harbour), *M. pyrifera* forms small kelp forests, historically extending several hundred square metres, and is an important species in the local marine ecosystem (Hay 1990). *M. pyrifera* is highly sensitive to environmental stressors, making it an effective ecological indicator for monitoring the health and resilience of kelp ecosystems (D'Archino and Piazzi 2021). This region marks the northernmost limit of the species, which is confined to coastal areas with rocky substrate and within specific temperature ranges (Hay 1990; Tait et al. 2021).

Globally, kelp forests are facing declines, with temperature being a key determinant of their distribution (Wilson et al. 2015; Krumhansl et al. 2016; Assis et al. 2017). Under both current and projected future ocean temperatures, kelp forests are projected to experience a range of adverse ecological impacts (Smale et al. 2013). These habitats are particularly vulnerable to the rise in sea surface temperatures, marine heatwaves, and the expansion of herbivorous taxa, such as kina (sea urchins), which exacerbate stressors (Vergés et al. 2014; Vergés et al. 2016; Wernberg et al. 2019). At a local level, factors such as nutrient availability, light for photosynthesis, and types of substrata further influence the distribution of *M. pyrifera* (Steneck et al. 2002; Schiel and Foster 2015). Additionally, in Te Whanganui-a-Tara, pollution and sediment loading, two significant inputs, may contribute to the distribution of *M. pyrifera*. Te Awa Kairangi (Hutt River) has been identified as the primary source of sediment in Te Whanganui-a-Tara (van der Linden 1967). The Kaiwharawhara River is another source of sedimentation in the harbour and additionally presents concerns due to elevated levels of nutrients and toxic metals (particularly zinc and copper) resulting from stormwater run-off (Milne and Watts 2008). Elevated zinc and copper concentrations from this river can lead to chronic exposure toxicity in aquatic life, affecting not only *M. pyrifera* but also other marine organisms, such as invertebrates and fish species (Milne and Watts 2008). These elevated concentrations are particularly concerning as toxic metals can substantially diminish the quantity and size of *M. pyrifera* sporophyte holdfasts and blades, potentially leading to mortality (Jara-Yáñez et al. 2021). Furthermore, both increased sediment loading and nutrient enrichment can induce indirect negative effects on *M. pyrifera* through the facilitation of nutrient-driven algal blooms and coastal darkening (Fleming-Lehtinen and Laamanen 2012).

Coastal darkening is an anthropogenic phenomenon characterized by increased turbidity resulting from terrigenous matter, often linked to urban development and agricultural practices. This increase in turbidity poses significant threats to coastal ecosystems (Kavanaugh et al. 2009; Harley et al. 2012; Desmond et al. 2015; Deregibus et al. 2020). Furthermore, these threats do not operate in isolation; climate change contributes to heightened rainfall, increased storm activity, and rising sea levels, which are expected to exacerbate coastal darkening (Klein et al. 2012; Fabricius et al. 2014; Seers and Shears 2015). The loss of light due to darkening is critically important, as light is the primary driver of photosynthesis, which limits growth and distribution (Nielsen et al. 2002; Cebrian et al. 2013). While *M. pyrifera* can thrive at depths of up to 30 m (Steneck et al. 2002), its habitat within the harbour is predominantly restricted to shallower depths, typically between 2 and 15 m due to varying turbidity (Hay 1990). With light penetration reduced in these conditions, photosynthesis is similarly limited, potentially restricting the spatial distribution of *M. pyrifera* to the shallower areas of the harbour. Both global and local factors can contribute independently or interactively to the decline of kelp, as noted by Rogers-Bennett and Catton (2019), and these influences may vary across different spatial and temporal scales. These variances highlight both the need for further research on what these factors are and how they influence kelp growth and the importance of kelp restoration and conservation.

22.1.2 Global and Local Restoration Efforts

Kelp restoration initiatives are increasing worldwide (Layton et al. 2020). While active restoration methods such as transplanting or out-planting laboratory-cultured kelp tend to be more effective, they are also costly. Integrating citizen science into kelp restoration and monitoring can help reduce these expenses and expand restoration efforts (Layton et al. 2020). Many kelp restoration and monitoring programmes are conducted by scuba (Sayer 2007; Eger et al. 2022). In Tasmania, Australia, restoration efforts for *M. pyrifera* have utilized scuba techniques for over 20 years (Sanderson et al.

2003). Scuba diving is an effective method for conducting underwater work. However, outsourcing paid scientific scuba divers can be exceptionally expensive for daily hires (Fredriksen et al. 2020). Additionally, the cost of acquiring and maintaining diving equipment, including tank refills, can also be substantial. These costs associated with equipment and training can exclude participants and, along with strict health and safety regulations, can be prohibitively expensive and burdensome for smaller organizations involved in restoration. In areas where scuba-based tourism is popular, community members are likely to have their own equipment to facilitate their participation in scuba-based citizen science. In regions such as Te Whanganui-a-Tara, New Zealand, however, scuba tourism is less popular, and winter water temperatures average around 12 °C, presenting additional challenges (Chappell 2014). The integration of breath-hold techniques for subtidal fieldwork allows for a more cost-effective approach than traditional subtidal field methodologies (using scuba) and removes some of the existing barriers to science and restoration, engaging a wider range of participants in underwater research. Breath-hold diving requires minimal equipment—often just a wetsuit, mask, snorkel, and fins—avoiding many of the high costs associated with scuba diving.

Restoration efforts for kelp forests are typically expensive, with the average cost of restoring one hectare often reaching hundreds of thousands of dollars (Eger et al. 2022). When compared to other marine ecosystem restoration initiatives, such as seagrass meadows, coral reefs, and mangroves, Eger et al. (2022) found that kelp restoration typically incurs the highest costs. Methods such as green gravel deployment have the potential to offer significant cost savings in marine restoration projects (Fredriksen et al. 2020). The green gravel technique reduces the need for extensive in-water labour by allowing most of the work to be conducted on land. In this approach, laboratory-cultured *M. pyrifera* gametophytes are grown and attached to gravel in controlled environments before being deployed into the ocean (Fredriksen et al. 2020). Traditionally, out-planting involved dropping the green gravel with attached kelp from a boat, often with a diver in the water to mark the location (Fredriksen et al. 2020). More recently, an alternative method has emerged in which divers deploy the green gravel directly on hard substrate. While both surface drops and direct deployment of green gravel to the substrate demonstrate similar growth rates (Fredriksen et al. 2020), anecdotal evidence (Green Gravel Action Group, personal communication) suggests there may be differences in survival rates between these methods, with hand placement by divers yielding more successful outcomes. Green gravel deployment offers a promising alternative for decreasing both the cost and complexity of restoration efforts.

The *Love Rimurimu* project, launched in 2020, demonstrates how community engagement can not only raise awareness of local kelp ecosystems but also foster long-term ecological restoration through collective action. This model offers a blueprint for kelp restoration efforts in other coastal regions. In 2023, with community involvement in mind, the *Love Rimurimu* began a collaboration with science institutions, and mana whenua (for this context: Indigenous Māori) to initiate a community kelp restoration project. The community remains at the centre of the project, and they have been active participants in restoration and education activities since the project's inception. The development of a subtidal monitoring programme involving the wider community sought to democratize science and remove the barriers that often limit public access to science and conservation efforts (Bonney et al. 2015; de Sherbinin et al. 2021). Additionally, it addressed two wider purposes: the wider community wanted opportunities to contribute, and *Love Rimurimu* needed monitoring information with limited resources, time, and capacity for infield work. We trained community members in scientific fieldwork and subtidal monitoring techniques including sorus collection, sporophyte out-planting, and restoration monitoring which were performed using breath-hold diving.

In this chapter, we examine the application of breath-hold techniques in subtidal scientific monitoring and kelp restoration to enhance community engagement and reduce historically high restoration costs. Furthermore, we describe a novel low-cost methodology for kelp restoration through the integration of breath-hold diving techniques. This approach was employed to out-plant *M. pyrifera*—locally called rimu kakauroa—sporophytes (green gravel) at a restoration site in Te Whanganui-a-Tara, Aotearoa New Zealand, for the first time. The kaupapa (for this context: set of values, principles, and ideas) of the restoration project was to restore rimu kakauroa by involving the local community and kaitiaki (for this context: guardians or custodians) of the area. Our methodology was designed to be simple, inclusive, cost-effective, and scientifically robust while minimizing environmental impact by using natural materials wherever possible.

22.2 Materials and Methods

22.2.1 Site Description

Te Whanganui-a-Tara (41.2924° S, 174.7787° E), in the southernmost end of the North Island of Aotearoa New Zealand (Fig. 22.1.), represents the uppermost thermal limit for *Macrocystis pyrifera* (Hay 1990). Within Te Whanganui-a-Tara area, shallow (no deeper than 5 m), subtidal restoration sites were initially scoped in consultation with Taranaki

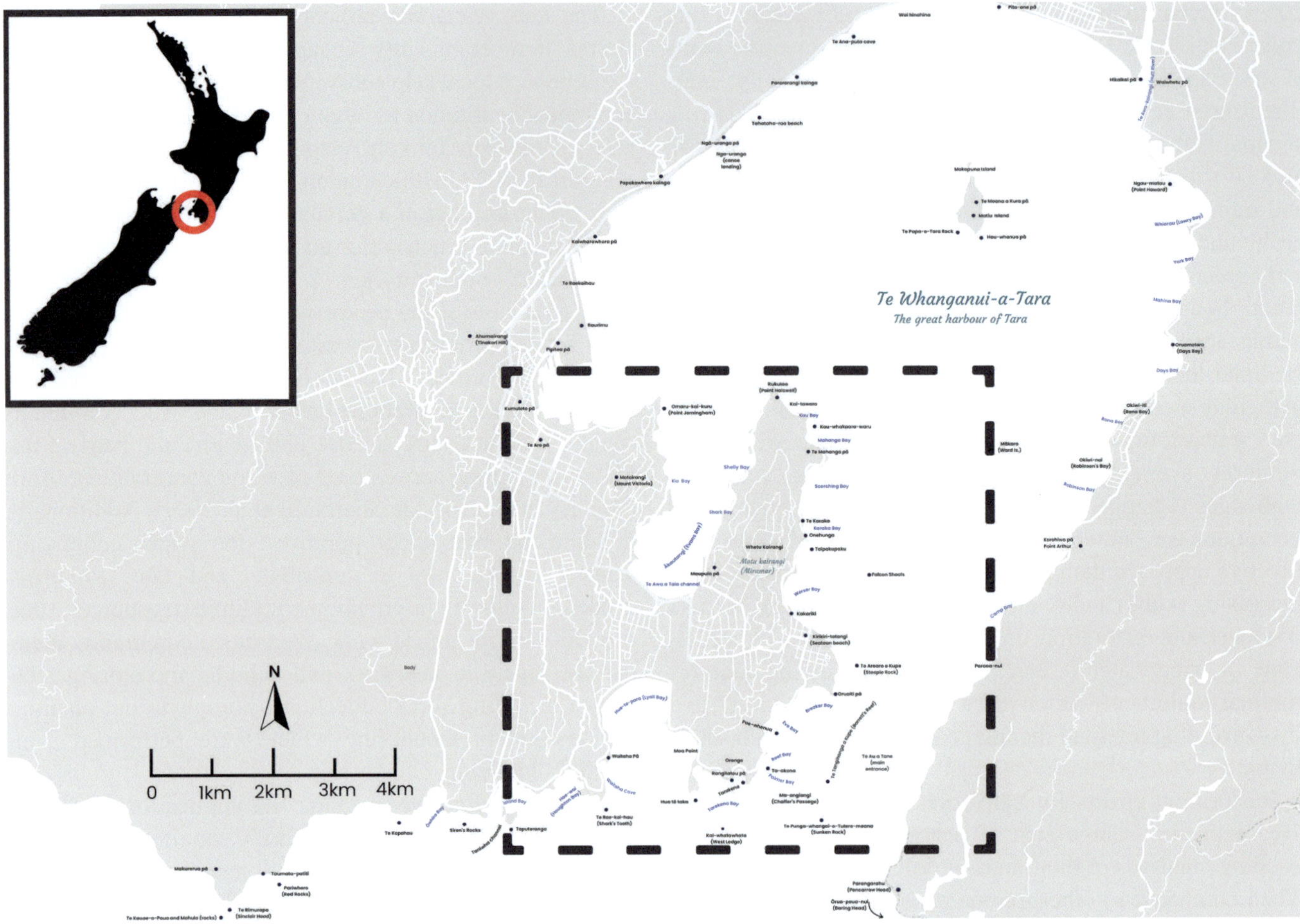

Fig. 22.1 Broadscale map of the project location of Te Motu Kairangi in Te Whanganui-a-Tara. The inset depicts a map of Aotearoa New Zealand with the red circle indicating the location of Te Whanganui-a-Tara

Whānui (mana whenua—for this context: Indigenous people of the area) and other stakeholders. In 2023, Kākāriki Hutia (Worser Bay) was selected as the first restoration site for this project due to factors which were deemed favourable to successful kelp restoration, such as the historical (Hay 1990) and current presence of *M. pyrifera*, ease of access for community members, substrate suitability, and the presence of local kaitiaki. Kākāriki Hutia is situated on the southeast coast of Te Motu Kairangi (Miramar peninsula). The average sea surface temperature in this area ranges from 10.3 °C in the winter to 17.7 °C in the summer (Chiswell and Grant 2018). The local community uses the site for recreational sailing, fishing, and spearfishing. Within this site, shallow subtidal areas (between 1.5 m and 4 m relative to MLWS) with rock, boulder, and cobble substrates, and proximity to existing *M. pyrifera*, were chosen as restoration sites.

22.2.2 Breath-Hold Techniques in Scientific Diving

Breath-hold techniques for scientific monitoring and kelp restoration were utilized for all fieldwork and out-planting. A total of ten community members were recruited for training through advertisements on websites, newsletters, and social media campaigns. The criteria for selection were a basic freediving certification, ability to commit to training and monitoring activities, and a desire to be involved in marine conservation. The group of community members (citizen scientists) was called 'Kelpers' (Fig. 22.2). Although all Kelpers were required to be certified freedivers, they were additionally required to undergo further water safety and competency training, as well as training in breath-hold diving and water rescue techniques with *Love Rimurimu* staff.

Fig. 22.2 Kelpers, Kura Kelpers, and *Love Rimurimu* interns. (**a**) First intake of Kelpers out on a monitoring dive, (**b**) breath-hold training in the pool, (c) *Love Rimurimu* staff and Kelpers, (**d**) *Love Rimurimu* interns, and (**e**) the Kura Kelpers from Te Kura Kaupapa Māori o Ngā Mokopuna

To minimize observer error and observer bias, rigorous training was provided to all data collectors. On-land training in marine conservation and ecology concepts and seaweed identification was provided. They were required to undergo extensive training on species identification, data collection, and monitoring methodology both on land and in the water. All training was practised on land before completing the in-water training.

Kelpers were initially trained to perform baseline environmental surveys of four sites (including Kākāriki hutia) for the *Love Rimurimu* baseline project over a period of 4–8 weeks. Upon completion of the Kelper Monitoring Programme, Kelpers were able to demonstrate proficiency in the following monitoring skills (with a buddy, at depths of up to six metres, and using only breath-hold techniques):

- Species identification of local seaweed and benthic organisms.
- Collection of environmental data (on land and in water).
- Completion of video transects and photos of 1-m^2 quadrats.
- Point intercept quadrat sampling of species presence and abundance.
- Laying a 20-m transect line underwater.
- Completing a 20-m belt transect survey to count and measure kelp underwater.
- Recording of structured observational data in water.

Data collectors were allowed to monitor and contribute to the *Love Rimurimu* baseline project dataset when they demonstrated proficiency in the data collection and had passed quality assurance tests. Once Kelpers had been trained in baseline monitoring, they could progress to monitoring the success of the restoration. To achieve this, they were required to demonstrate proficiency in the following skills:

- Identification of *Macrocystis pyrifera* recruits
- Ability to maintain steady buoyancy underwater and move sporophytes on rocks onto suitable substrate
- Ability to maintain steady buoyancy underwater and measure sporophytes from 2 cm to 2 m in length using slates and measuring rope
- Site surveys to locate and monitor the presence and abundance of sporophytes

All data were quality-checked during training (to calibrate and standardize the data collection) and post-training, ensuring both accuracy and precision. In addition to recruiting a group of adult Kelpers, we also trained a group of eight rangatahi (for this context: young people) from a local kura (for this context: school), the 'Kura Kelpers' (Fig. 22.2). This group of rangatahi is considered kaitiaki of the area and was trained in many of the same scientific monitoring skills as the Kelpers. The sporophytes in plant out 1 were entirely out-planted by Kura Kelpers who continue to play a significant role in the *Love Rimurimu* project. Over time, a more structured and dedicated *Love Rimurimu* internship programme was developed, and two early-career scientists were recruited from the Kelpers. These interns received further training and were instrumental in the collection of data for the *Love Rimurimu* baseline project and for the restoration project.

22.2.3 Sorus Collection

Sori (fertile materials) were collected from existing healthy *M. pyrifera* forests around Te Motu Kairangi, including the plant out site Kākāriki Hutia (−41.313045, 174.83045), and the South coast (Fig. 22.3). Sites for sorus collection were chosen based on ease of access and density of *M. pyrifera*. Most fertile material was found on *M. pyrifera* between 1 and 5 m.

Collection was conducted under government permit using breath-hold diving. Divers took individual blades from the base of the individual algae (up to a maximum of five blades). The sori were collected in small mesh bags and taken directly to the National Institute for Water and Atmospheric Research (NIWA) laboratory in buckets for processing. At the NIWA laboratory, fertile material was extracted and grown to a suitable size before out-planting.

22.2.4 Laboratory Methods

In the NIWA laboratory, fertile material was first sterilized (placed into a solution of 2 L 14-degree fresh water with 5 drops of betadine or similar wound disinfectant for 1 min), the sorum was allowed to drip-dry in a colander, and the extraneous material adjacent to the sori was removed when the sorum was no longer dripping wet. The sorum was then placed between paper towels and into a sealed container into a refrigerator overnight (creating dark and cold stress). The next day, the sori were cut into small segments and extracted into nutrient-enriched seawater creating a spore slurry. The

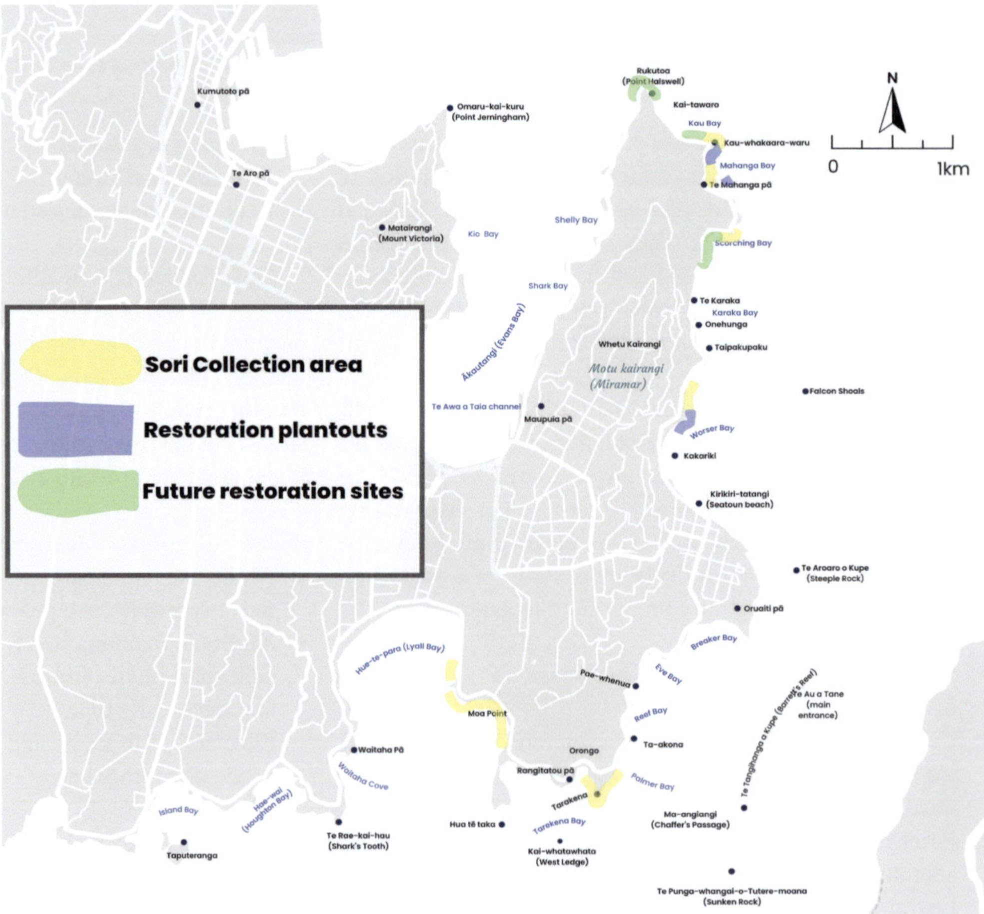

Fig. 22.3 Te Motu Kairangi: Sorus collection areas are depicted in yellow; restoration plant out areas are depicted in lavender; and future restoration areas are depicted in green

spore slurry was then cultured over several weeks, with the spores themselves developing into their gametophyte stage within a flask or a container. This gametophyte culture could then be fragmented and added to a seeding tank, along with cobble substrate. The cobbles used were all collected by Kura Kelpers, Kelpers, and community members from local beaches. Once the gametophytes settled, the cobble was moved to the grow tanks where they were held until the new sporophytes were at least 2 cm in length. For plant out 1, many of the seeded cobbles were then planted out as they were. For subsequent plant outs (2, 3, and 4), the small sporophytes were detached and reattached to separate rocks using the 'string theory' method (below).

22.2.5 'String Theory' Methods

When rocks are seeded, the density of sporophytes on any individual rock is initially high, but as the plants grow, the density becomes unsustainable, with sporophytes competing for limited resources. Monitoring green gravel after out-planting reveals that this competition leads to a decline in survival, likely a result of self-thinning (Fredriksen et al. 2020; Earp et al. 2024). Larger individuals appear to outcompete smaller ones, which contributes to this decline. Thinning the seeded rocks by removing some of the sporophytes could reduce competition and allow for greater growth, as the remaining plants would not need to allocate energy towards competing for resources. The use of the 'string theory' facilitated the out-planting of a larger quantity of sporophytes. However, the process of carefully detaching the holdfast from seed rocks, winding it onto a cotton string, and wrapping this around a cobble can be laborious and time-consuming (see Fig. 22.4 for the 'string theory' process).

Twenty-five fully seeded rocks were out-planted at the site during plant out 1, and 25 rocks that used 'string theory' were also out-planted. The 'string theory' method was subsequently used for plant outs 2, 3, and 4. Most of these rocks had only one or two sporophytes individually attached using 100% biodegradable cotton string, with a small quantity having up to five sporophytes attached on the same string.

22.2.6 Plant Outs

22.2.6.1 Plant Out Site

The first plant out series for the *Love Rimurimu* project included four plant outs at the Kākāriki Hutia site using sporophytes grown in the NIWA laboratory. We out-planted a total of 387 rocks. These rocks included some densely seeded larger cobbles and pebbles and cobbles with between one and five sporophytes attached with a string. Plant outs were performed over a series of months from September 2023 to January 2024 and grouped into a total of 16 whānau clusters (between 0.25 m^2 and 1 m^2 in size). The rocks used ranged in size from approximately 0.0016 m^2 to 0.005 m^2, and these were placed in a stratified random pattern within these clusters. The first sporophytes were out-planted in September 2023 (seventh, eighth, and 22nd) and totalled 50 rocks across five clusters. On 30 November 2023, we completed our second plant and out-planted 120 rocks into four clusters (including 10 larger outlier sporophytes). On 13 December (2 weeks after plant out 2), 173 rocks were out-planted into five clusters (including 10 larger and 43 smaller outlier sporophytes). On 25 January 2024, the remaining sporophytes (44 rocks) that had not been used for other experiments were out-planted into two clusters (see Fig. 22.5).

22.2.6.2 Whānau (For This Context: Family) Clusters

M. pyrifera is not typically a solitary kelp and tends to grow in large forests which extend outwards (Gaylord et al. 2007). Kelp restoration efforts are more likely to be successful in areas with existing kelp forests (Eger et al. 2022). Consequently, our restoration strategy employed a whānau-based approach, where sporophytes were out-planted in small family groups near mature *M. pyrifera*, referred to as 'whānau clusters'. Cluster locations were predetermined and marked by divers on the day of out-planting before transporting the sporophytes to the site. Surface marker buoys, attached to small weights, were used to mark these areas from the surface, while labelled rocks and quadrats were used to delineate the areas underwater. To facilitate subsequent monitoring, all sporophytes were out-planted into defined 1-m^2 clusters. A 1-m^2 PVC quadrat, marked with different coloured tapes at 50-cm intervals, was used to ensure even placement of sporophytes within each 50 cm x 50 cm section. A label rock was placed in the centre of the quadrat to mark individual clusters. The PVC quadrats and all other equipment were removed from the water after out-planting; however, the label rock remained as a visual marker for subsequent monitoring.

22.2.6.3 Out-Planting Using Breath-Hold Diving

Out-planting of the sporophytes was achieved using breath-hold diving. Accordingly, the developed methodology was precise and practised ensuring the safety of the divers and the consistency and quality of the scientific process. A standardized health and safety briefing, including emergency procedures, evacuation routes, meeting points, common hazards, environmental considerations, breath-hold safety, and other relevant health and safety considerations, was provided by staff members before entering the water. In addition, every diver was asked to perform a wairua (for this context: spirit) and tinana (for this context: body) evaluation, which served

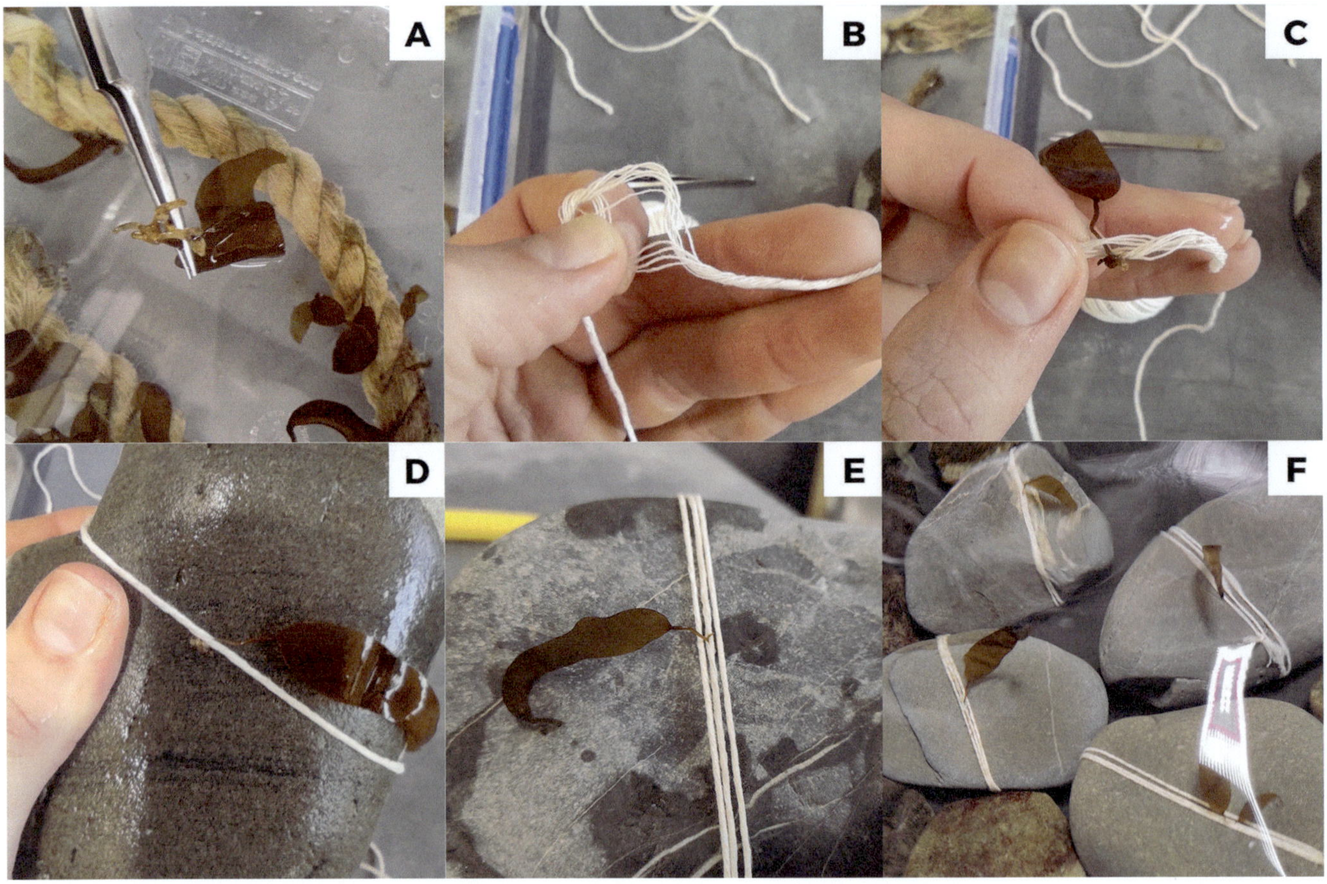

Fig. 22.4 'String theory' detachment and reattachment method for creating green gravel rocks for out-planting. (**a**) Sporophyte is detached from initial seeded substrate, (**b**) cotton string is split, (**c**) the stipe of the seaweed is fed through the hole in the string, (**d**) string is wrapped around new rock, (**e**) string is tied tightly with the seaweed holdfast pressed tight against the rock, and (**f**) rocks are returned to the nursery tanks to allow holdfast to re-establish

as a physical and mental health check and grounding exercise before entering the water. Karakia (for this context: blessing or prayer) was performed before entering the water. Divers adhered to the 'one up, one down method', taking turns to duck dive down and complete the out-planting or monitoring tasks. Diver safety was paramount, and protocols outlined in the relevant health and safety documents (see discussion) were strictly followed, including gold standard freediving surface safety protocols and breath-up techniques. Due to the shallow depths of the plant out areas (between 2 and 5 metres), and the colder water (between 10° and 18°) necessitating wetsuits with thicker neoprene, divers experienced buoyancy issues and found it difficult to stay underwater to complete tasks. Adding extra weights to divers posed a health and safety risk; therefore, we used a kettlebell or extra weight belt tied in a loop as an unattached prop to hold on to underwater. These were placed at the edge of the quadrat or near the work site allowing the divers to stabilize themselves underwater and perform tasks without floating to the surface. A safety buoy with a diver flag was deployed when divers were in the water serving as a floatation device for attaching equipment and for divers to rest and relax before duck diving.

22.2.6.4 Out-Planting Sporophytes for Restoration Using Breath-Hold Diving

For all plant outs, tikanga (for this context: customary Māori protocol) was followed and was led by mana whenua. Karakia was recited as the sporophytes were lowered into the sea and by divers before entering the ocean. The first plant out was led by Kura Kelper citizen scientists from the local kura, and additional karakia and waiata (for this context: songs) were used to mark the significance of the occasion. Sporophytes were moved from the transportation container to a plastic gridded tray (a repurposed supermarket bread tray) or a kono (for this context: a basket woven out of harakeke flax) and lowered into the ocean. Divers received the tray below the surface, which facilitated underwater transportation. Divers swam the tray to the corresponding marked

Fig. 22.5 Map of the Kākāriki Hutia restoration plant out site indicating the locations of each cluster from the four plant outs: plant out 1: September 2023, 50 rocks out-planted in five clusters; Plant out 2: 30 November 2023, 120 rocks out-planted in four clusters; Plant out 3: 13 December (2 weeks after plant out 2), the remaining 173 rocks from the laboratory out-planted in five clusters; and plant out 4: 25 January 2024, 44 rocks out-planted in two clusters. Bottom right depicts the plant out arrangement within the 1-m^2 quadrat

cluster. At the cluster area, the basket was carefully lowered to the substrate next to the PVC quadrat and weighted down to prevent it from floating up as the sporophytes were removed and out-planted into the 1-m^2 cluster area (Fig. 22.6).

Divers transferred the sporophytes (measuring between ~2 cm and ~10 cm) from the basket to the PVC quadrat, positioning them to minimize potential movement from surge. Where the substrata permitted, sporophytes were wedged into uneven, rocky areas (such as small holes, valleys, or overhangs) to stabilize them while allowing light penetration. Sporophytes were arranged in a stratified random pattern within the quadrat. For instance, at cluster 6a, 40 sporophytes were out-planted, with ten sporophytes placed in each quarter of the quadrat. This stratified random distribution facilitated the subsequent monitoring of the whānau clusters. Figure 22.7 illustrates an example of sporophyte placement, with five sporophytes in each coloured quarter.

On the day of plant outs, all site preparation and diver briefings were completed prior to the arrival of the sporophytes at the site to streamline the transfer process and prevent prolonged exposure of the sporophytes to sunlight or heat. The sporophytes used for plant outs 2 and 3 were grown in the NIWA laboratory from the same gametophyte culture. These sporophytes were divided into two groups of similar sizes and planted out 2 weeks apart to compare sporophyte growth in the ocean with varying durations of laboratory growth. Although efforts were made to standardize variables between these two plantings, standardization was difficult to achieve completely.

At each whānau cluster, details of the location, depth, the number of sporophytes planted out, observations, and environmental features were recorded on a slate along with the cluster's identifying name (i.e., 'cluster 1'). GPS coordinates for each whānau cluster were taken and mapped. Photos of the cluster and any identifying natural features above and below water were taken. Additionally, photos were taken of the sporophytes to document their general health and appear-

Fig. 22.6 Out-planting of green gravel. (**a**) Lowering the basket of *M. pyrifera* to a waiting diver, (**b**) a fully seeded cobble covered in sporophytes and a singular 'string theory' *M. pyrifera* sporophyte, both in a kono being swum out to the restoration site, (**c**) a diver placing the quadrat to delineate the plant out cluster area, (**d**) the weighted basket containing sporophytes next to the plant out quadrat. A kettlebell sits in the middle of the quadrat to help divers, and (**e**) a fully seeded cobble and 'string theory' pebble after out-planting

ance at the time of out-planting. To ensure data integrity during review, labelling, and filing, a photo of the whānau cluster name (written on a slate or as a photo of the label rock) was taken first, followed by photos of the cluster and location. Videos were also used to document this information, with divers recording oral descriptions of the cluster location and additional observations of the sporophytes upon out-planting. Using videos enabled a safer, more streamlined process, as divers did not need to spend extended time documenting data on slates. Thorough documentation of the locations of sporophytes and whānau clusters was crucial for the success of future monitoring efforts.

22.2.6.5 Label Rocks

The challenges in identifying and monitoring different rocks were somewhat mitigated through thorough mapping and documentation of the plant outs at each cluster, as well as the use of label rocks. Labelled rocks were placed in the centre of each whānau cluster as a visual marker to facilitate monitoring by indicating the underwater locations. In the first plant out, we used small rocks that were distinctly shaped (i.e. triangular, oval) as label rocks, deploying them at each cluster to mark them. For subsequent plant outs, 'label rocks' were trialled. A permanent marker was used to write directly onto larger dry cobbles which were then placed at cluster sites as 'label rocks' to mark them. We trialled marking small boulders (collected from local coastal areas) with a black permanent marker pen (while dry) and deployed 'label tests' at our restoration site (see Fig. 22.12 in results).

A study by Frias-Torres et al. 2023 used reflective tiles to identify coral reef restoration sites instead of numbered tags due to the expected amount of biofouling growing on the tag surface. We decided to trial this methodology in a temperate setting to facilitate the presence of long-term cluster markers underwater. We collected large (roughly 10 cm^2) greywacke rocks from local coastal areas. Large, stable rocks

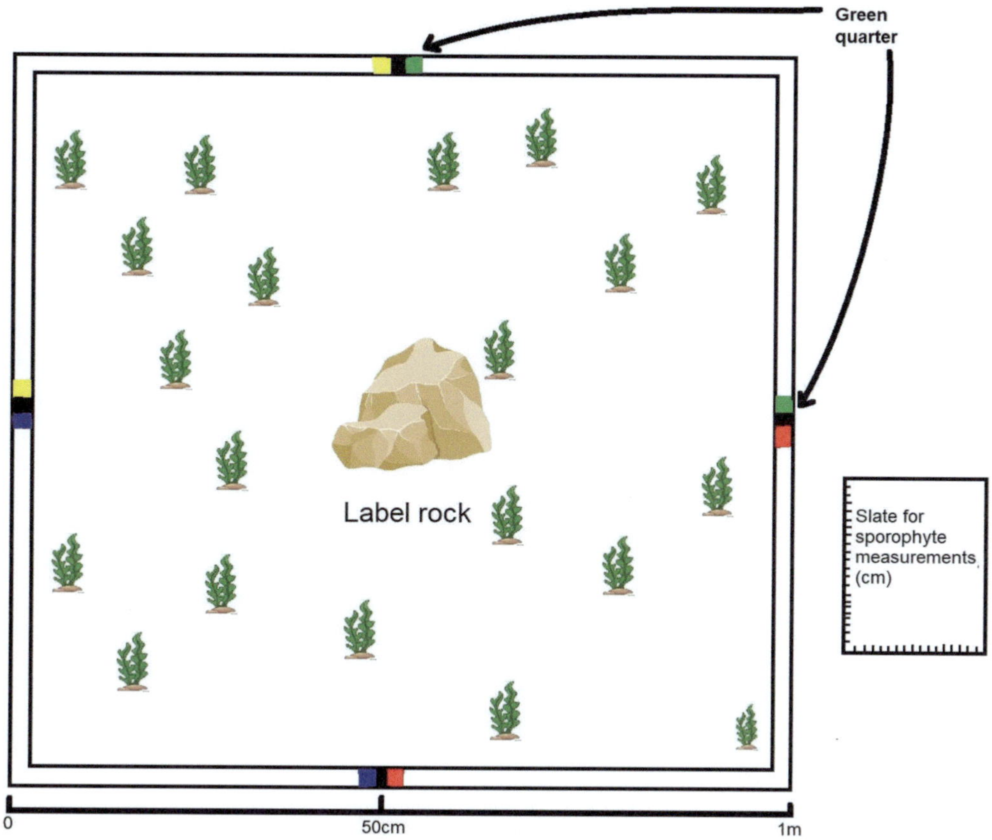

Fig. 22.7 Quadrat demonstrating the out-planting methods. Note the label rock in the centre, coloured markings to denote the quadrat quarters, and an example of the stratified random pattern placement of sporophytes within the quadrat (here, five sporophytes per quarter). The arrows highlight how each quarter is colour-coded. Right, a slate with ruler measurements on the *x*- and *y*-axis for monitoring sporophyte growth

with a relatively flat base and flat surface areas for the best application of tiles were chosen. These were cleaned and dried. We used the marine eco-crete 'Rockbond' to affix the tiles to the rocks. We mixed approximately 50 g (¼ cup) of Rockbond cement powder with 11 ml of water for approximately 10 s to form cement with a toothpaste-like consistency. This was used to cement the white, reflective tiles to the rocks. We used a combination of reflective tiles broken into assorted sizes and shapes (Frias-Torres et al. (2023) used 8.2 cm × 8.2 cm tiles). The tiles were broken into small irregular pieces and arranged in distinct patterns for each rock (Fig. 22.8). Before the Rockbond application, the rock and tile were dampened with water. This mixture was applied on the bottom surface of the tile. Pressure was applied to the rocks for 1 min immediately after application. These rocks were placed in buckets of freshwater and left to set for 48 h.

22.2.6.6 Temperature and Light Data Collection

To track temperature and light fluctuations, we used HOBO temperature and light loggers (Pendant MX Temperature/light Data Logger—IC-MX2202). Due to some technical difficulties, the first HOBO logger was not deployed until the beginning of summer (after plant out 3). We affixed it to a large boulder with cable ties and deployed it at Kākāriki Hutia on 14 December 2023 (Fig. 22.9). Approximately 1 month after deployment, the logger disappeared and was presumed stolen.

We observed that in publicly accessible areas (such as Kākāriki Hutia), loggers may attract attention and be removed by curious snorkellers. Discussions with local academic institutions and researchers revealed that the removal and disappearance of these loggers were a relatively common occurrence. Accordingly, we initiated a small social media public awareness campaign (see Fig. 22.10). An informal missing poster was created and displayed in various public areas within the community and dive shops, and at the restoration site.

Our intentions were twofold:

1. To locate our HOBO logger by offering a no-questions-asked policy and reminding the public that the device itself has no resale value
2. To educate the public on the purpose of these devices to deter future removals

Despite our efforts, the missing HOBO logger was not recovered, and another logger was deployed on 21 January 2024 at the same site. This logger has remained in place for over 10 months which we attribute to the success of the public awareness campaign and the education of the local community regarding its purpose.

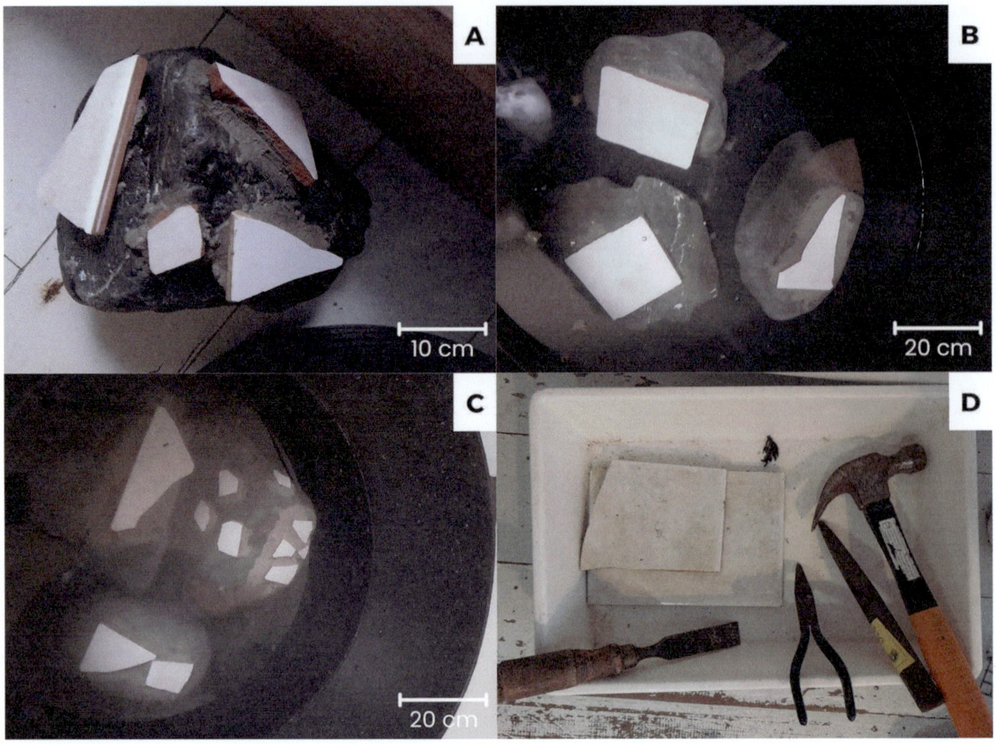

Fig. 22.8 Photos of the new tile marker rocks. (**a**–**c**) Depicting different configurations and shapes for ease of identification and (**d**) the tools used in making the tiled marker rocks

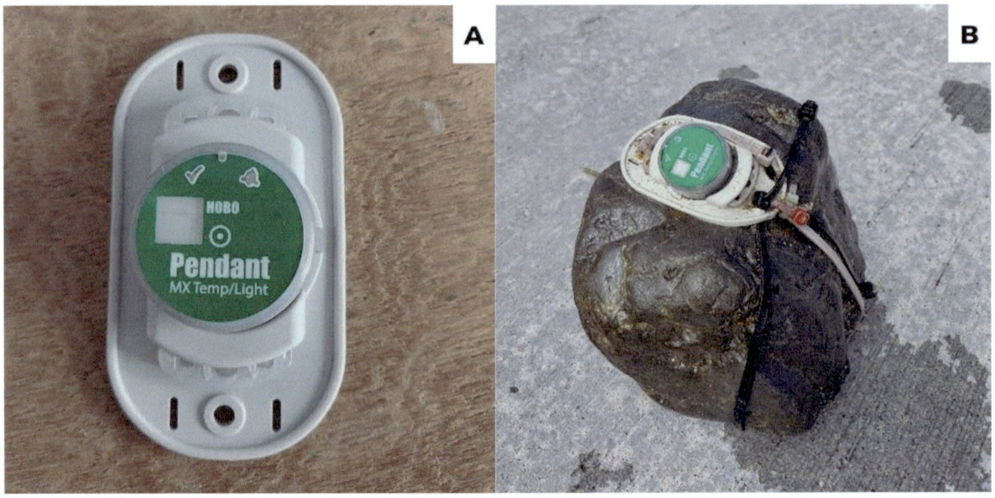

Fig. 22.9 (**a**) New and unused HOBO logger and (**b**) the HOBO logger affixed to a boulder for deployment

22.2.7 Scientific Monitoring Using Breath-Hold Diving

Monitoring of sporophyte survival and growth (described below) was initially conducted weekly, then every 2–3 weeks. Sporophyte survival was consistently monitored, and the monitoring of individual sporophyte growth began after 4 weeks in situ. Monitoring sporophyte growth using breath-hold diving presents several challenges. Considerable care and fine motor skills are needed to carefully measure very small sporophytes while holding your breath upside down, often in areas with dense algal coverage, swell, surge, and cold water. Divers were trained to maintain slow and gentle movements and to minimize excessive handling of the sporophytes. Minimal in-water equipment was used for monitoring and recording growth and survival rates to ensure safe and accurate monitoring. Excessive equipment could overburden divers, as one hand is needed for equalizing, and a certain degree of streamlining is required to descend using less oxygen, allowing for safer, more comfortable, and longer monitoring periods. Accordingly, A4 slates with marked rulers were used to measure the growth of sporophytes in the initial phase after out-planting when individual sporophytes measured less than 25 cm. These slates had maps drawn on one side which detailed the positions of the clusters relative to

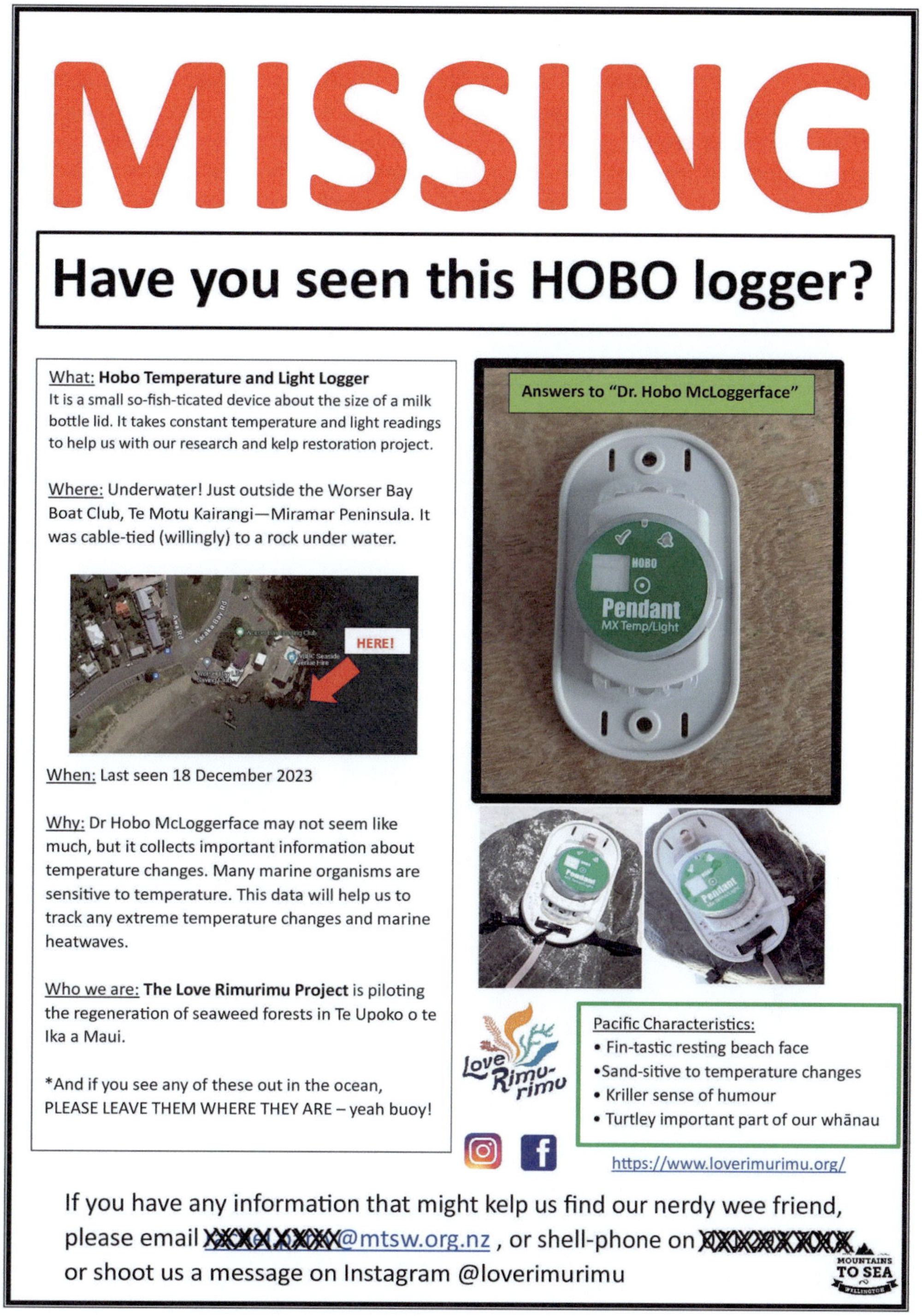

Fig. 22.10 Poster used in the 'Missing HOBO Logger' public awareness campaign

the site and the position of surviving sporophytes (according to the previous monitor) relative to the clusters. The other side of the slate was used for recording the measurements and the survival (or mortality) of the sporophytes. Rulers were marked on both sides of the slate both on the X- and Y-axis to enable quick measuring and less manoeuvring of the slate underwater. Kelpers could often measure up to five small sporophytes in a single breath-hold dive. This number decreased as the sporophytes grew and measuring became more complex. Once sporophytes exceeded 25 cm in length, flexible measuring strings were used to measure growth.

Once in the water, the divers navigated to the first whānau cluster using the slate map, label rocks, and natural navigation to locate the correct area. The clusters were hand-drawn on the one side of the monitoring slate to indicate the expected sporophyte numbers and placement. In addition, the PVC quadrat was placed on the approximate position of the whānau cluster and a kettlebell, or weight belt was placed underwater to enable divers to hold on with one hand if required. Divers took turns diving down to monitor growth and survival. For survival monitoring, mortality was determined by the absence of a blade. Typically, one diver would survey the quadrat for mortality, while the other diver began taking measurements. To avoid over- or undercounting and measuring, each diver focused on one quarter of the quadrat at a time. Sporophytes were measured from the base of the holdfast to the tip of the longest blade. Measurements were to the nearest centimetre to ensure efficient and effective monitoring and to avoid handling the blades (see Fig. 22.11 for the monitoring process).

Once the sporophytes grew to larger sizes of over ~30 cm, a flexible measuring tape with a 250-gram weight attached to one end was used for measurements. Divers alternated diving down to place the weighted end at the holdfast of the sporophyte, then slid the tape upwards from the base of the holdfast along the stipe to the tip of the longest blade. They marked the tip position with their fingers and surfaced safely. At the surface, they noted the length, which their buddy recorded on the data slate. In addition to measuring sporophyte growth and survival, divers made additional observations, including changes in blade colour, sporophyte health, evidence of grazing, and general observations. Environmental variables, such as water temperature, air temperature, wind speed and direction, weather conditions, and water visibility, were also recorded at each monitor. The sporophytes from these plant outs were monitored closely with sporophyte measurements regularly taken of each surviving individual after 1 month. For plant outs 2 and 3, the first in situ sporophyte monitoring occurred the day after planting out. Due to the number of sporophytes out-planted at each cluster, initial measurements were taken of 25% of the sporophytes. When the total number of surviving sporophytes was less than 10 or after 1 month (whichever came first), all the surviving sporophytes were measured. Due to grazing and water movement, survival numbers for plant outs 2, 3, and 4 were small enough to monitor all the sporophytes at each cluster after the second monitor for each plant out. At each monitor, if out-planted rocks were found to be barren of sporophytes, they were removed unless they already formed part of the substrate (i.e. had other organisms growing on them). Additionally, any unattached cotton string was removed.

Fig. 22.11 Monitoring of out-planted kelp. (**a**) Diver places the flexible measuring tape at the holdfast of a kelp, (**b**) measuring a smaller sporophyte with the slate ruler, (**c**) recording additional observation, and (**d**) running out the flexible measuring tape along the length of the sporophyte

22.2.8 Kina (Sea Urchin) Density Surveys

In addition to plant out monitoring, Kelpers participated in executing baseline monitoring of the benthic assemblages at various sites including the restoration site (Kākāriki hutia). This chapter will only detail the part of this baseline monitoring as it relates to the restoration project. Kina density at the site was surveyed quarterly using point intercept quadrat surveys along a subtidal transect line. Five haphazardly placed 1-m^2 quadrats were monitored along a 20-m transect line. This transect line ran parallel to shore at a depth of 2–5 m. This was repeated twice at the site giving three replicates and a total of 15 quadrats monitored. Before entering the water, five numbers were generated between 0 and 19 (inclusive). The quadrats were placed so that the bottom edge of the quadrat touched the generated number, with the transect line running through the middle, eliminating any unconscious bias in quadrat placement.

To mitigate any entanglement with *M. pyrifera* and to facilitate easy placement of quadrats for the breath-hold divers, quadrats were separated into two halves, which could be easily assembled underwater or on the surface. Quadrats had holes drilled at 10-cm intervals along each edge to facilitate underwater mobility and provide a size reference for later photo analysis. Kina within the quadrat were counted using a methodical sweeping method, starting at the quadrat side closest to the start of the transect and scanning the quadrat from left to right and right to left until the whole quadrat was searched. The quadrat with the quarters marked with different coloured tapes was used. Where the benthic assemblage and substrate were complex or there was abundant kina and the diver was unable to scan a whole quadrat in one breath, they completed only half (using the colour markings to measure this).

22.3 Results

22.3.1 Label Rocks

The label rocks deployed demonstrated variance in the degradation of the ink. Figure 22.12 depicts two different label rocks, label test 3 at 2 and 4 weeks in situ and label rock 7a at 3 weeks in situ. The ink on label rock 7 (sandstone) appears to be more faded than the ink on label test 3 (greywacke).

Fig. 22.12 (**a**) Label test 3 rock on 15/11/2023, (**b**) label test 3 rock on 30/11/2023, (**c**) label test 3 rock on 10/12/2023; depicting minimal fading of original markings, (**d**) cluster 7a label rock on 23/11/2023 newly placed in situ, and (**e**) cluster 7a again on 13/12/2023 3 weeks later; severe degradation of markings can be seen

We postulate that the variance in ink quality over time could be attributed to the type of rock used or the thickness of ink applied. It should be noted that the label rocks that demonstrated minimal fading were still eventually subject to biofouling of organisms such as crustose coralline algae (CCA). After 2 months of being fully submerged in the ocean, these rocks often exhibited significant biofouling making it difficult to discern the ink markings and, consequently, to identify the rocks and locate our plant out areas with precision. Greater success (i.e. less biofouling) was observed using the tile marker rocks following the methodology outlined by Frias-Torres et al. (2023).

22.3.2 Sporophyte Survival

As of 28 March 2024, of 387 rocks out-planted, 34 retained healthy sporophytes, resulting in an overall survival rate of 8.78 per cent (see Table 22.1). For all plant outs, the number of surviving sporophytes tended to decline dramatically within the first four to 5 weeks post-out-planting, after which survival rates stabilized. Figure 22.13 (below) illustrates the percentage survivorship of the sporophytes from each plant out for the first 16 weeks post-out-planting. Sporophytes that could not be located were excluded from the survival counts, although it is possible that some survived but were displaced from the site by sea surge.

Grazing was identified as the primary threat to sporophyte survivorship during the initial 4–6 weeks. Grazers such as gastropods, sea stars, and fish were observed on and around sporophytes (see Fig. 22.14). Although many kelp restoration projects experience significant grazing pressure from kina, this issue was not encountered in our study. No kina was observed near any of the clusters from plant out 1 in the first 3 months, and very few were observed at the site. The results from the kina density surveys, which were conducted quarterly from the time of plant outs for 1 year, indicate a kina density of less than 1 per square metre (see Fig. 22.15).

Table 22.1 Number of rocks out-planted (with sporophytes attached) at each plant out and their respective survival rates as of 28 March 2024

Plant out	Number of rocks out-planted	Number of rocks with surviving sporophytes (as of 28 March 2024)	Survival rate
1 (September 2023)	50	2	4%
2 (30 November 2023)	120	8	6.66%
3 (13 December 2023)	173	18	10.4%
4 (25 January 2024)	44	6	13.63%
Total	387	34	8.78%

22.3.3 Sporophyte Growth

Our methodology proved effective for monitoring the growth of individual sporophytes as demonstrated by the collection of rigorous data. Divers were able to take precise measurements of sporophytes ranging from under 5 cm to over 2 metres long while using breath-hold techniques. The chart below (Fig. 22.16) illustrates the growth progression of individual sporophytes in clusters 2 and 4 from plan out 1, from 10 November 2023 until 28 March 2024.

Plant outs 2 and 3 were conducted 2 weeks apart to assess the effects of extended laboratory growth on subsequent growth and survival in natural conditions. It was hypothesized that an additional 2 weeks in the laboratory would enhance holdfast attachment to substrate and promote blade growth, providing a competitive advantage in the natural environment. However, no significant differences in growth were observed between the two plant outs. Observations from plant out 2 revealed signs of blade degradation and discoloration post-transplantation. Specifically, many blade tips exhibited a brighter green hue, indicative of bleaching, and evidence of erosion and deterioration was noted within a day post-transplantation (Fig. 22.17). One week post-transplantation, further monitoring indicated extensive blade damage, with some blades more than 50% eroded. This damage was attributed to blade deterioration rather than herbivory, as evidenced by the irregular, eroded blade edges, which contrasted with the clean cuts typical of grazing observed in plant out 1. Conversely, on the day after out-planting, plant out 3 sporophytes appeared healthy, with dark pigmentation and no noticeable deterioration.

Clusters 8a and 8b were observed to be the most successful clusters with 4 and 13 surviving sporophytes, respectively, on 28 March 2024. A plant out monitor was conducted on 28 March 2024, with the largest sporophytes from the surviving clusters (clusters 2, 7a, 7b, 8a, 8b, 8c, 8d, 9a, and 10) measuring 203 cm, 88 cm, 184 cm, 114 cm, 157 cm, 145 cm, and 84 cm, respectively. A further site check on 24 May 2024 revealed that many of these sporophytes had grown to lengths of over 3 m and were creating canopy cover (see Fig. 22.18).

Once a sporophyte reached over 1 m in length, most were observed to have fertile sorus tissue (see Fig. 22.19c). We saw many of our surviving sporophytes become reproductively mature. Furthermore, we were able to collect fertile sori from the sporophyte C2S4 (from plant out 1 located at cluster 2) and, after processing it in the laboratory, harvest the sori to seed new rocks.

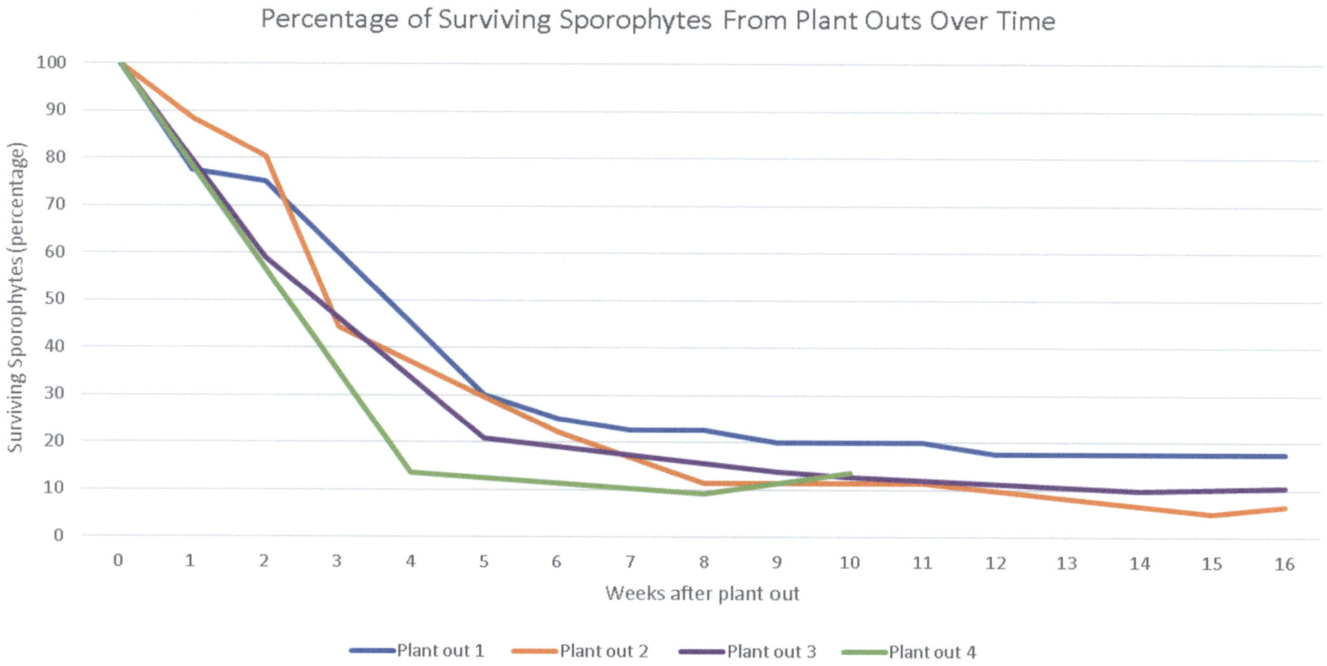

Fig. 22.13 Percentage of surviving sporophytes from plant outs 1–4 over time. The *y*-axis depicts the survival percentage, and the *x*-axis depicts the number of weeks after plant out

Fig. 22.14 Grazers encountered on and around out-planted sporophytes. (**a**, **b**) *Patiriella regularis* or common cushion star, (**c**) gastropod covered by coralline algae (Corallinales), and (**d**) *Forsterygion* or triple fin seen under the rock edge

We observed that some sporophytes demonstrated increased growth at the base near the holdfast and became bushier, without demonstrating significant vertical growth. Although blade erosion was a factor for most of the sporophytes at some stage, only some sporophytes tended to grow *bushy* instead of *tall*. For example, on 28 March 2024, we monitored the sporophytes C2S4 and C4B2 and found that although they had been out-planted on the same day (both measuring ~3 cm), C2S4 measured 206 cm and C4B2 measured 90 cm and was observed to be more 'bushy'.

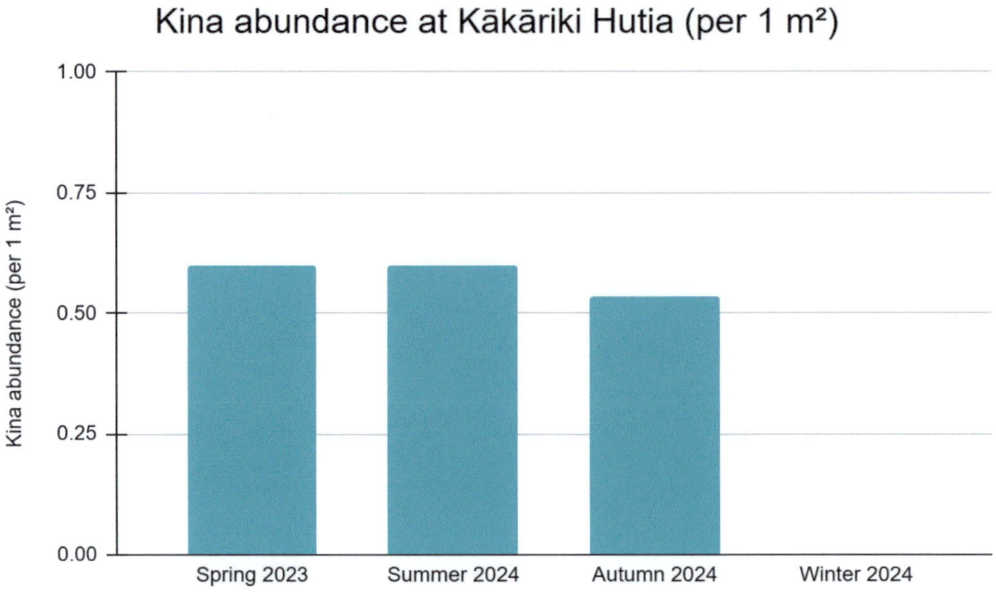

Fig. 22.15 Kina abundance at Kākāriki Hutia (per 1 m^2). The x-axis depicts the quarters over a year where spring 2023 was September, October, and November 2023; summer was December 2023, January 2024, and February 2024; autumn was March, April, and May 2024; and winter was June, July, and August 2024

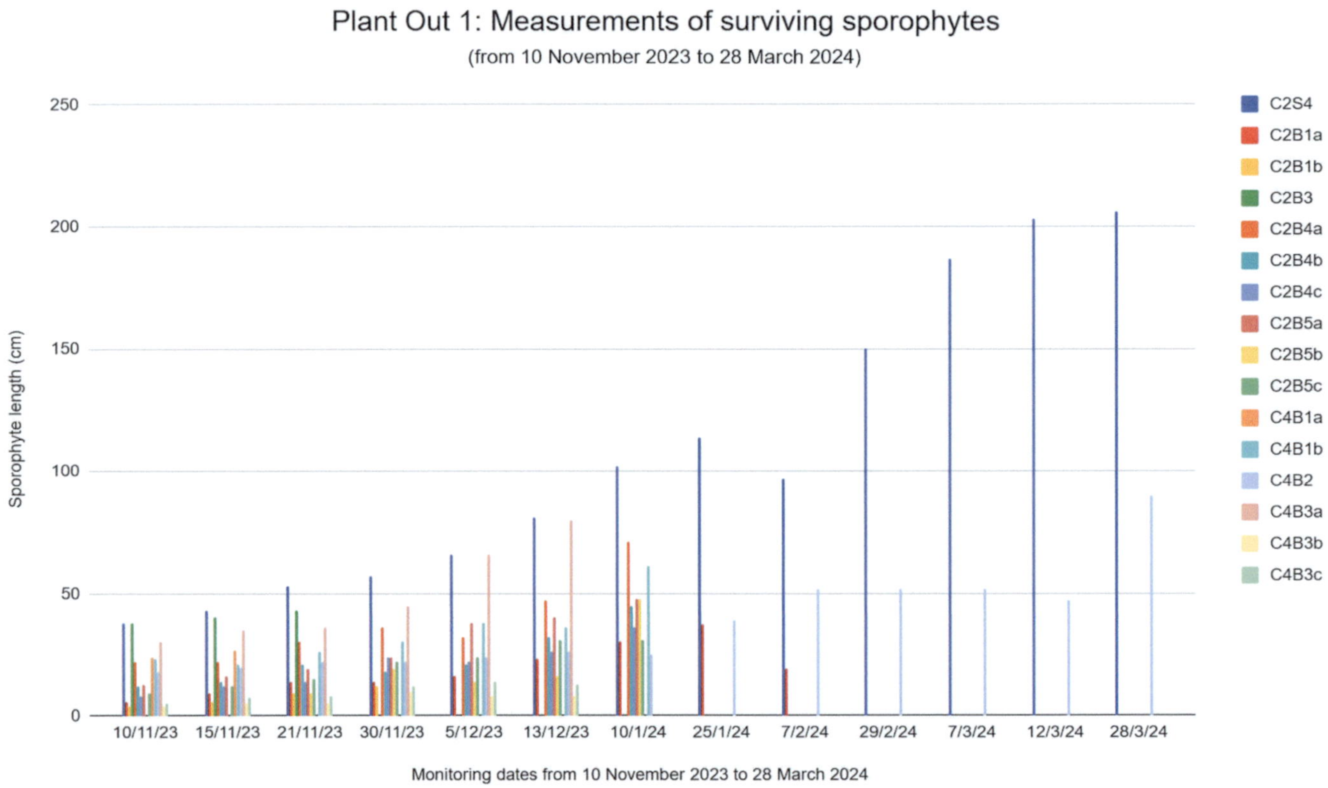

Fig. 22.16 Sporophyte measurements (cm) from clusters 2 and 4 in plant out 1. The different colours represent the names of different sporophytes

22.4 Discussion

22.4.1 Integrating Breath-Hold Diving into Community-Based, Subtidal Scientific Monitoring

Published research through citizen science initiatives has exponentially increased since 2003, with multiple case studies demonstrating success by motivating community members to learn, raise awareness, enhance social connections, and find solutions to significant issues while participating in the research process and strengthening connections between people and their local environments (Bonney et al. 2015; Morris et al. 2020; Walker et al. 2020; Larkins 2024). Similarly, this project experienced success in these areas. Our experience in training community members and inte-

Fig. 22.17 (**a**–**c**) Examples of bleaching and blade erosion observed on the blades of sporophytes from plant out 2. The degradation can be seen working its way from the tip of the blades

Fig. 22.18 Photos of out-planted sporophytes creating canopy cover at clusters 8a and 8b

grating breath-hold diving into a kelp restoration project and scientific monitoring programme has proven valuable for numerous reasons, extending beyond scientific benefits. Incorporating breath-hold diving into this initiative provided an effective and cost-efficient way to expand the workforce for both restoration and monitoring. This allowed for more extensive and frequent monitoring which can lead to a bigger dataset, improving robustness and reliability of scientific findings. Involving community members brings valuable local knowledge that can provide context and depth to scientific data while leading to more open sharing of information and greater public engagement with scientific research.

Eger et al. (2022) discuss the need to explore more cost-effective community-based methods in kelp restoration to increase community engagement and awareness and to help ease the historically high costs of restoration. Involving the public in dialogue and decision-making around environmental issues reinforces the importance of autonomy and behaviour within conservation efforts (Jordan et al. 2011). Our novel method allowed for cost-effective restoration and inclusive community participation, which is crucial for the long-term success and sustainability of restoration projects. The involvement of community members, including the Kelpers and Kura Kelpers, not only provided valuable sup-

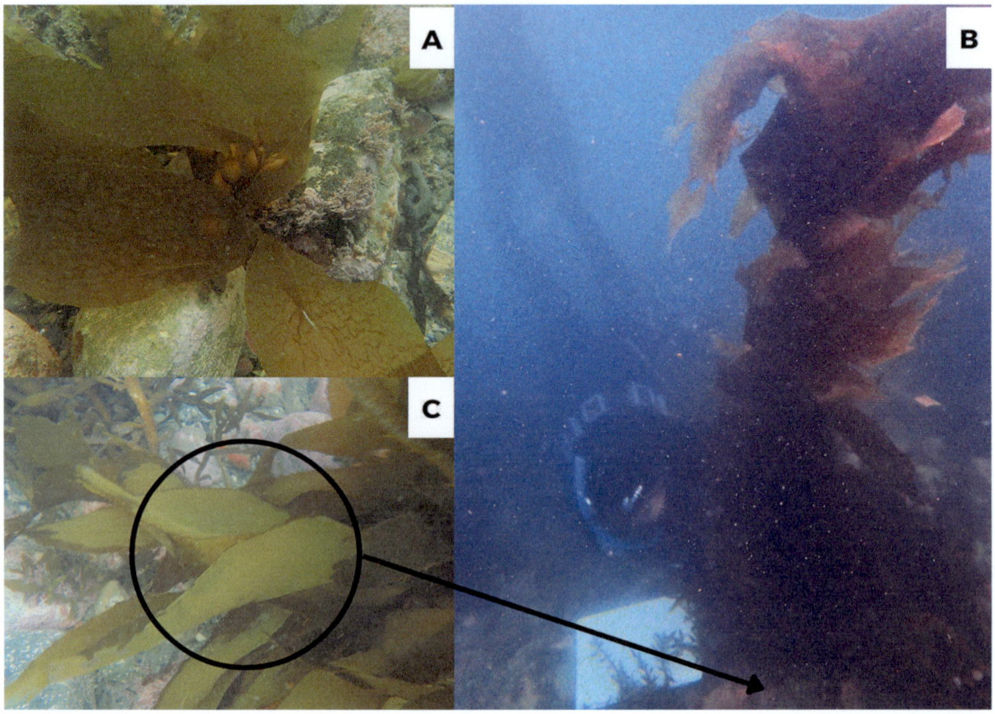

Fig. 22.19 Sporophyte C2S4 (**a**) on 30 November 2023 five weeks after out-planting and (**b**) on 12 March 2024 over five months after out-planting with a diver for size reference, and (**c**) fertile sori observed on C2S4, which were harvested, and the spores were extracted

port but also fostered a sense of ownership and stewardship among local residents.

Breath-hold techniques were effectively employed for all aspects of the project, with no elements conducted using scuba. This ensured that schools and community members could participate in out-planting efforts under existing health and safety protocols. While the training process can be time-intensive and rigorous to ensure data quality, the use of breath-hold diving techniques proved effective for both the out-planting and monitoring of sporophytes. Breath-hold diving reduces the financial barriers to participation, allowing a broader range of community members to engage in marine conservation efforts. Through a progressive and comprehensive training programme, we successfully trained a team of scientific breath-hold divers. These citizen scientists assisted in out-planting sporophytes at our restoration site and collected robust data while passing frequent quality control and assurance checks. While initial participation was high, we found that maintaining a high level of community involvement for frequent monitoring was challenging long term. By developing an internship programme, *Love Rimurimu* was able to mitigate these challenges, demonstrating the importance of structured training and dedicated personnel in achieving successful long-term monitoring and restoration outcomes. Interns were recruited from the Kelper group, and this represented a natural progression for the project. The internship programme provided a consistent and reliable workforce, which was crucial for maintaining the continuity and quality of the monitoring activities. This model highlights the potential benefits and opportunities of integrating internship programmes into community-based conservation projects, ensuring both scientific rigor and community engagement.

Kosmala et al. (2016) assert that with strong project design and participant training, citizen scientists have proven their ability to collect robust and reliable data (Kosmala et al. 2016). We demonstrated that breath-hold diving can be effectively used for precise scientific tasks, such as point intercept quadrat surveys, measuring sporophyte growth, and monitoring survival rates. Divers were able to perform detailed observations and data collection while holding their breath, highlighting the potential for this technique to be adapted for underwater research activities. The success of breath-hold diving in this project suggests that it could be a viable alternative to scuba diving in other marine restoration and monitoring initiatives, particularly in shallow water environments.

22.4.2 The Challenges of Subtidal Scientific Monitoring on Breath-Hold

We conducted point intercept quadrat surveys along a subtidal belt transect to survey Kina densities. As these subtidal surveys are typically performed on scuba, adaptations were required when designing a methodology to survey on breath-hold. The primary barriers were limited underwater time and the physical demands of maintaining precision while breath-holding. To overcome these limitations, we used pegs to help pace our belt transect and prevent losing position along the belt. Additionally, adding colours to the quadrats helped minimize potential miscounts and improve accuracy. Integrating breath-hold diving into subtidal scientific monitoring required ensuring that every diver was confident with the methodical steps of the monitoring process before entering the water. This allowed divers to channel their mental focus into scientific tasks. Winter dives presented additional challenges, as colder temperatures impacted breath-hold duration, reducing the time divers could spend underwater. Furthermore, weather conditions, such as strong wind, variable turbidity, and swell, presented additional challenges and could restrict breath-hold techniques and, therefore, data collection. Slow, methodical steps and extensive surface communication were crucial, and this was emphasized during training. Ultimately, the greatest challenge of breath-hold diving lies in managing human limitations and respecting the body, mind, and lungs as our primary 'equipment'. In addition to a consistent and thorough health and safety briefing, we found that taking the time to connect with our wairua and tinana prior to entering the water served as a grounding exercise, promoting a mindful and steady approach to monitoring which improved data quality. Furthermore, consistently performing karakia at the water's edge prior to monitoring served to strengthen our connection to the moana (for this context: ocean) and unite us as a team leading to safer and more efficient data collection. Proper training, a healthy body, and a focused mind became essential tools for our research, enabling us to collect robust data despite the inherent limitations of breath-hold diving.

22.4.3 Sporophyte Survival and the Impact of Kina

The results of this study highlight several key findings regarding the restoration of *Macrocystis pyrifera* in Te Whanganui-a-Tara. The monitoring data revealed that sporophytes could grow to significant lengths (over 3 m) creating essential canopy cover and considerable complex habitat and reach reproductive maturity within a relatively short period. This rapid growth of sporophytes indicates that restored kelp forests can quickly contribute to the local ecosystem's health and resilience. The overall survival rate of 8.78% for out-planted sporophytes indicates the challenges associated with kelp restoration in this region. The initial decline in sporophyte survival within the first 4–6 weeks post-out-planting, followed by stabilization, suggests that early-stage mortality is a significant challenge. Grazing by gastropods, sea stars, and fish was identified as the primary threat during this period, although kina was not a major issue at the restoration site.

Kina (*Evechinus chloroticus*), a key herbivore of New Zealand coastlines (Miller et al. 2024), did not pose a significant threat at our restoration site. However, rising sea temperatures and the potential range extension of kina could threaten future restoration efforts (Vergés et al. 2014; Vergés et al. 2016). While the primary cause of kina overpopulation in New Zealand is the reduction of their natural predators, other environmental factors, such as coastal darkening, increased sedimentation, and rising sea temperatures, also contribute to the development of barrens and the decline of kelp forests (Bulmer et al. 2024). Kina barrens form when overgrazing by excessive kina populations strip rocky reefs of their kelp forests, leaving expanses of bare rock. This transformation disrupts natural habitats needed by a wide range of marine species, including the kina themselves. While there is not yet a universally standardized kina density that defines a kina barren, once a regime shift has occurred, they need to be reduced to a critical density of <1 per m^2 to reverse the barren and recover kelp-dominated habitats (Shears and Babcock 2003; Bulmer et al. 2024). Our surveys indicated that kina densities at the restoration site were low (<0.75/m^2), which suggests the site has not yet reached a 'kina barren' state. This is in contrast to many other restoration sites where overgrazing by kina has been a major factor in kelp decline (Steneck et al. 2002).

22.4.4 Plant Out Variability

The initial strategy was to out-plant all sporophytes during the winter season. However, due to various factors, including delays in laboratory procedures and weather variability, the out-planting schedule was delayed. Consequently, plant outs 2, 3, and 4 were conducted in the spring and summer. This introduced numerous variables to consider between each out-planting event, and more studies should be executed in the future to understand the effects of these variables on sporophyte growth. Heat and light exposure varied significantly between plant outs, particularly between plant outs 2 and 3. The differences in weather conditions and transit times between plant outs appeared to contribute to variability in sporophyte survival. Plant out 2 faced higher heat and UV stress, leading to blade deterioration, while plant out 3 benefited from cooler, overcast conditions. As the process of transporting sporophytes from the laboratory to the ocean baskets became more streamlined, the duration of transport in tubs, which may have experienced varying temperatures, was minimized for plant out 3 compared to plant out 2, theoretically reducing heat stress on the sporophytes. Additionally, due to the observed blade deterioration in plant out 2, the cluster locations for plant out 3 were adjusted to less exposed areas, providing more opportunities for intermittent shade from natural features such as rocky ledges or other algae, such as *Carpophyllum* sp.

22.4.5 The 'String Theory' Method

We identified several advantages and disadvantages associated with the 'string theory' method. The biggest disadvantage of the 'string theory' method was that it was more time-intensive and laborious to detach and reattach sporophytes to cobbles with a string compared to using seeded rocks. The initial use of smaller pebbles (3–6 cm) resulted in significant loss, leading to no long-term survival success. In contrast, the use of larger cobbles (10–15 cm) minimized rock movement and increased the survival rate of sporophytes, with all surviving sporophytes out-planted using this method. The string exhibited a higher propensity to slip off when not adequately wound around the rock or very round or small pebbles were used. We adjusted our methods and quality checking of 'string theory' accordingly; however, initially this led to sporophyte loss in some instances and posed a marine pollution risk if the cotton string became entangled with other seaweed. Instances were observed where the cotton string detached from the rock and attached to other seaweed, notably *Carpophyllum* sp. In two cases, the *M. pyrifera* sporophyte remained attached to the string and began growing on the *Carpophyllum* sp. (Fig. 22.20). In these instances, the string with the sporophyte was left in situ and monitored. In all other cases where the string was found unattached to rocks, it was removed from the environment.

We observed significant success with the 'string theory' method, with all surviving sporophytes originating from this group. An additional advantage of using the 'string theory' method was the facilitated identification and monitoring of these sporophytes due to the presence of the string. In contrast, seeded rocks were more challenging to distinguish from naturally occurring recruits, particularly after biofouling occurred. Laboratory challenges with sporophyte growth limited the production of seeded rocks, highlighting the utility of the 'string theory' method in increasing the number of rocks out-planted into restoration sites.

22.4.6 Challenges in Data Collection

Although we were able to collect substantial data on the survival and growth of the out-planted sporophytes, we experienced a few limitations. Logistical challenges resulted in being unable to collect temperature and light data during critical stages of the plant outs and over the summer months of December and January, leaving gaps in the environmental data that could have provided insights into sporophyte survival. Complications with growing the sporophytes in the laboratory led to a reduced number of sporophytes out-planted, as well as plant outs occurring over a larger temporal scale and in the spring and summer months instead of winter. Due to the small sample size of surviving sporophytes at each plant out after the first 4 weeks, statistically significant conclusions could not be drawn and more research should be performed to investigate different variables involved in sporophyte growth and survival. The 'hole punch method' is widely utilized as a method for measuring individual blade growth due to its consideration of blade erosion. A study by Fieler et al. (2021) demonstrated that the hole punch method is effective for assessing growth in erosion-prone seaweeds such as kelp. We decided against this method for several reasons: the absence of research on potential disease and damage risks, the complexity of underwater application using breath-hold diving, and the focus on investigating the habitat and canopy cover generated by the sporophytes, which made sporophyte length a more suitable metric for our study. This approach did not allow us to account for blade erosion or physical degradation of the kelp. As sea temperatures rise and ocean acidification intensifies due to global climate change, blade erosion may increase, and further studies should be performed to investigate the effect that these may have on kelp growth.

Fig. 22.20 (**a**, **b**) *M. pyrifera* sporophyte detached from cobble and reattaching and subsequently growing on *Carpophyllum* sp.

22.4.7 Health and Safety Considerations

In Aotearoa New Zealand, all in-water activities must adhere to stringent health and safety protocols established by the conducting organization and approved by WorkSafe NZ. The *Love Rimurimu* project, managed by the Mountains to Sea Wellington Trust (MTSWT) under the national Mountains to Sea Conservation Trust (MTSCT), follows these regulations. A comprehensive health and safety framework was developed for monitoring and out-planting activities using breath-hold diving. A comprehensive health and safety framework, building on existing protocols from the EMR programme, was developed for breath-hold diving at depths of 2–6 m. While not classified as freediving, participants were trained in freediving techniques to ensure safety.

The EMR Advanced Snorkelling standard operating procedures (SOP), approved by WorkSafe NZ, include health and safety protocols for these depths. A robust risk assessment and management system (RAMS) was implemented to anticipate hazards and ensure informed decision-making for in-water activities (Fig. 22.21). Our novel methodology demonstrated the ability of citizen scientists to collect robust datasets and advance restoration projects while complying with and maintaining high standards of health and safety.

22.5 Conclusion

This novel, low-cost methodology utilizing breath-hold diving techniques for subtidal scientific monitoring and community involvement has proven effective for kelp restoration. Breath-hold techniques were employed for all aspects of the project, enabling participation by schools and community members under existing health and safety protocols. The project's outreach fostered collaboration and broadened participation in marine conservation, creating opportunities previously limited to those with scuba diving or scientific backgrounds while facilitating the restoration of rimu kakauroa (*Macrocystis pyrifera*) through the out-planting of green gravel.

The findings indicate that while early-stage mortality and grazing present significant challenges, successful growth and reproductive maturity of out-planted sporophytes were observed, demonstrating the potential for restored kelp forests to contribute to and enhance local ecosystem health. Additionally, this study demonstrates that the use of breath-hold diving offers a cost-effective and accessible method for scientific monitoring and restoration. Its minimal equipment requirements and lower training barriers make it an attractive option for low-cost, community-based projects. The success of this technique in the *Love Rimurimu* project suggests its potential for broader application as a scientific monitoring method, particularly in shallow coastal environments.

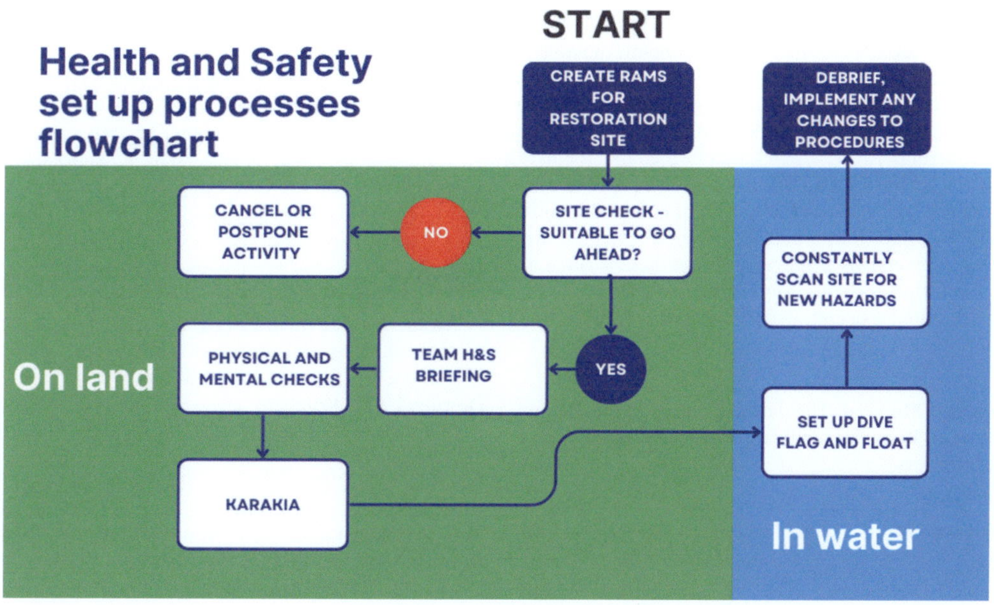

Fig. 22.21 Summary of health and safety procedures for in-water activities. (1) RAMS is developed for the restoration site. (2) Site check is conducted on the day, and a pre-site assessment is performed to determine the suitability of diving based on weather, hazards, water quality, etc. (3) Health and safety briefing is given. (4) Team completes physical and mental checks. (5) Karakia is performed prior to entering the water. (6) Once in the water, a dive flag and float are deployed. ((7) All equipment is cleared from the site upon completion. (8) A debrief is conducted with the team to discuss any incidents or near-misses

This study highlights the potential of breath-hold diving and community involvement to enhance the feasibility and sustainability of restoration efforts. The *Love Rimurimu* project serves as a model for community-based restoration, emphasizing local involvement, cost-effective methodologies, and the integration of traditional knowledge within scientific research. This approach fosters a more inclusive, informed, and engaged society, contributing to effective and sustainable conservation outcomes.

The restoration of rimu kakauroa in Te Whanganui-a-Tara is ecologically imperative and empowers local communities through engagement and collaboration. By integrating cost-effective methods such as green gravel deployment and breath-hold diving, we established a sustainable approach to kelp forest regeneration. Ongoing monitoring and adaptive management will be key to assessing the long-term success of these efforts and refining the methods for broader applications in marine conservation.

Acknowledgements The authors would like to give our thanks to the mana whenua of this region, particularly Taranaki Whānui ki te Upoko o te Ika, who have generously shared their knowledge and supported our research. We acknowledge the traditional custodians of the land and waters where this study was conducted, and we pay our respects to their elders past, present, and emerging. In particular, we would like to mihi the young kaitiaki, the Kura Kelpers, from Te Kura Kaupapa Māori o Ngā Mokopuna, who have been the drivers of this project and who were responsible for planting out the sporophytes at the first plant out of rimu kakauroa in Te Whanganui-a-Tara: Hawea Sullivan, Kapua Strickland, Hemaima-te-Wai MacDougall, Te Ao Rangi Tawhara, Ngawaierua Campbell, Tutahi Devonshire, Mihirangi Kohatu, Rakaipo Tapiata, with special thanks to their kaiako, teachers, Muri Sullivan, Darcy Sullivan, and Renee Campbell. We are grateful to all the community members, Kelpers, and Kura Kelpers who helped in the field, particularly Reimi Konda and Kat Greager, and to the freediving instructors Sacha Williamson and Kristine Zipfel. Thanks to Siddarth Ravishankar, Yun Yi Kok, Neill Barr, Roberta D'Archino, and Laura Bornemann for all of their mahi in the NIWA laboratory, and Lee Rauhina-August, Māia-te-oho Holman-Wharehoka, and our collaborating partners at Taranaki Whānui ki te Upoko o te Ika, NIWA, and Te Herenga Waka Victoria University Wellington. The authors acknowledge that there is often no matching English translation for kupu Māori (Te Reo Māori words). Many kupu are open-textured concepts, defined by their context as they are in this text.

Availability of Data and Material The datasets generated and/or analysed during the current study are available from the corresponding author on reasonable request.

Authors' Contributions RP: conceptualization, methodology, fieldwork, formal analysis, investigation, visualization, and writing—original draft, review, and editing. JL: fieldwork, investigation, and writing—original draft, review, and editing. HR: fieldwork, investigation, and writing—original draft, review, and editing. JJS: conceptualization, fieldwork, visualization, and writing—content contribution. JW: fieldwork, visualization, and writing—content contribution. ZS: conceptualization, fieldwork, funding acquisition, and writing—content contribution. All authors read and approved the final manuscript.

Funding The *Love Rimurimu* project has received funding from the Clare Foundation and Eurofins Foundation. The authors declare no conflicts of interest. The authors declare that the research was conducted in the absence of any commercial or financial relationships that could be construed as a potential conflict of interest.

References

Assis J, Araújo MB, Serrão EA (2017) Projected climate changes threaten ancient refugia of kelp forests in the North Atlantic. Glob Chang Biol 24(1):e55–e66

Bayley D, Brickle P, Brewin P, Golding N, Pelembe T (2021) Valuation of kelp forest ecosystem services in The Falkland Islands: a case study integrating blue carbon sequestration potential. One Ecosyst 6:e62811

Bonney R, Phillips TB, Ballard HL, Enck JW (2015) Can citizen science enhance public understanding of science? Public Underst Sci 25(1):2–16

Bulmer R, Pilditch C, Stephenson F, Shears N, Flowers G (2024) Assessing potential recovery solutions to shift from kina barrens to kelp forests. Sustainable Seas. https://www.sustainableseaschallenge.co.nz/assets/dms/Reports/Kina-barrens/Kina-barrens-to-kelp-forests.pdf. Accessed 18 Nov 2024

Cebrian, Corcoran D, Lartigue J (2013) Eutrophication-driven shifts in primary producers in shallow coastal systems: implications for system functional change. Ecol Coasts 37(S1):180–197

Chappell P (2014) The climate and weather of Wellington 2nd edition. https://niwa.co.nz/sites/default/files/Wellington_Climate_WEB.pdf

Chiswell S, Grant B (2018) New Zealand coastal sea surface temperature prepared for NZ Ministry for the Environment. NIWA. https://environment.govt.nz/assets/Publications/Files/nz-coastal-sea-surface-temperature.PDF

Cotas J, Gomes L, Pacheco D, Pereira L (2023) Ecosystem services provided by seaweeds. Hydrobiology 2(1):75–96

D'Archino R, Piazzi L (2021) Macroalgal assemblages as indicators of the ecological status of marine coastal systems: a review. Ecol Indic 129:107835

de Sherbinin A, Bowser A, Chuang T-R, Cooper C, Danielsen F, Edmunds R, Elias P, Faustman E, Hultquist C, Mondardini R, Popescu I, Shonowo A, Sivakumar K (2021) The critical importance of citizen science data. Front Clim 3. https://doi.org/10.3389/fclim.2021.650760

Deregibus D, Zacher K, Bartsch I, Campana GL, Momo FR, Wiencke C (2020) Carbon balance under a changing light environment. In: Gómez I, Huovinen P (eds) Antarctic seaweeds: diversity, adaptation and ecosystem services. Springer, Cham, pp 173–191. https://doi.org/10.1007/978-3-030-39448-6_9

Desmond MJ, Pritchard DW, Hepburn CD (2015) Light limitation within southern New Zealand kelp forest communities. PLoS One 10(4):0123676

Earp HS, Smale DA, Catherall HJ, Moore PJ (2024) An assessment of the utility of green gravel as a kelp restoration tool in wave-exposed intertidal habitats. J Mar Biol Assoc UK 104:e28

Eger AM, Marzinelli EM, Christie H, Fagerli CW, Fujita D, Gonzalez A, Hong SW, Kim JH, Lee LC, McHugh TA, Nishihara GN, Tatsumi M, Steinberg PD, Vergés A (2022) Global kelp forest restoration: past lessons, present status, and future directions. Biol Rev 97(4):1449

Fabricius KE, Logan M, Weeks S, Brodie J (2014) The effects of river run-off on water clarity across the central Great Barrier Reef. Mar Poll Bull 84(1–2):191–200

Fieler R, Greenacre M, Matsson S, Neves L, Forbord S, Hancke K (2021) Erosion dynamics of cultivated kelp, *Saccharina latissima*, and implications for environmental management and carbon sequestration. Front Mar Sci 8(8):632725. https://doi.org/10.3389/fmars.2021.632725

Filbee-Dexter K, Pessarrodona A, Pedersen MF, Wernberg T, Duarte CM, Assis J, Bekkby T, Burrows MT, Carlson DF, Gattuso J-P, Gundersen H, Hancke K, Krumhansl KA, Kuwae T, Middelburg JJ, Moore PJ, Queirós AM, Smale DA, Sousa-Pinto I, Suzuki N (2024) Carbon export from seaweed forests to deep ocean sinks. Nat Geosci 17:552–559. https://doi.org/10.1038/s41561-024-01449-7

Fleming-Lehtinen V, Laamanen M (2012) Long-term changes in Secchi depth and the role of phytoplankton in explaining light attenuation in the Baltic Sea. Estuar Coast Shelf Sci 102–103:1–10

Fredriksen S, Filbee-Dexter K, Norderhaug KM, Steen H, Bodvin T, Coleman MA, Moy F, Wernberg T (2020) Green gravel: a novel restoration tool to combat kelp forest decline. Sci Rep 10(1):3983

Frias-Torres S, Reveret C, Henri K, Shah N, Maya PHM (2023) A low-tech method for monitoring survival and growth of coral transplants at a boutique restoration site. PeerJ 11:15062

Gaylord B, Rosman JH, Reed DC, Koseff JR, Fram J, MacIntyre S, Arkema K, McDonald C, Brzezinski MA, Largier JL, Monismith SG, Raimondi PT, Mardian B (2007) Spatial patterns of flow and their modification within and around a giant kelp forest. Limnol Oceanogr 52(5):1838–1852

Grime BC, Sanders R, Ford T, Burdick H, Claisse JT (2023) Urchin gonad response to kelp forest restoration on the Palos Verdes Peninsula, California. Bull South Calif Acad Sci 122(1):1–18

Harley CDG, Anderson KM, Demes KW, Jorve JP, Kordas RL, Coyle TA, Graham MH (2012) Effects of climate change on global seaweed communities. J Phycol 48(5):1064–1078

Hay CH (1990) The distribution of *Macrocystis* (Phaeophyta: Laminariales) as a biological indicator of cool sea surface temperature, with special reference to New Zealand waters. J R Soc NZ 20(4):313–336

Jackson GA, Winant CD (1983) Effect of a kelp forest on coastal currents. Cont Shelf Res 2(1):75–80

Jara-Yáñez R, Meynard A, Acosta G, Latorre-Padilla N, Oyarzo-Miranda C, Castañeda F, Piña F, Rivas J, Bulboa C, Contreras-Porcia L (2021) Negative consequences on the growth, morphometry, and community structure of the kelp *Macrocystis pyrifera* (Phaeophyceae, Ochrophyta) by a short pollution pulse of heavy metals and PAHs. Toxics 9(8):190–190

Jordan RC, Gray SA, Howe DV, Brooks WR, Ehrenfled JG (2011) Knowledge gain and behavioural change in citizen-science programs. Conserv Biol 25(6):1148–1154

Kavanaugh MT, Nielsen KJ, Chan FT, Menge BA, Letelier RM, Goodrich LM (2009) Experimental assessment of the effects of shade on an intertidal kelp: do phytoplankton blooms inhibit growth of open coast macroalgae? Limnol Oceanog 54(1):276–288

Klein RD, Lewis J, Buffleben MS (2012) Logging and turbidity in the coastal watersheds of northern California. Geomorphology 139-140:136–144

Kosmala M, Wiggins A, Swanson A, Simmons B (2016) Assessing data quality in citizen science. Front Ecol Environ 14(10):551–560

Krumhansl KA, Okamoto DK, Rassweiler A, Novak M, Bolton JJ, Cavanaugh KC, Connell SD, Johnson CR, Konar B, Ling SD, Micheli F, Norderhaug KM, Pérez-Matus A, Sousa-Pinto I, Reed DC, Salomon AK, Shears NT, Wernberg T, Anderson RJ, Barrett NS (2016) Global patterns of kelp forest change over the past half-century. Proc Natl Acad Sci 113(48):13785–13790

Larkins D (2024) Engagement beyond inclusion: reframing citizen science in Aotearoa New Zealand. MSc Dissertation. ResearchGate Auckland. https://doi.org/10.13140/RG.2.2.28314.66240

Layton C, Coleman MA, Marzinelli EM, Steinberg PD, Swearer SE, Vergés A, Wernberg T, Johnson CR (2020) Kelp forest restoration in Australia. Front Mar Sci 7:74. https://doi.org/10.3389/fmars.2020.00074

Miller KI, Balemi CA, Blain CO, Spyksma AJ, Shears NT (2024) Sea urchin roe quality within urchin barrens and improvement through kelp restoration. Ecosphere 15(6):e4911

Milne JR, Watts L (2008) Stormwater contaminants in urban streams in the Wellington region. Greater Wellington Regional Council, Environment

Morris RL, Graham TDJ, Kelvin J, Ghisalberti M, Swearer SE (2019) Kelp beds as coastal protection: wave attenuation of *Ecklonia radiata* in a shallow coastal bay. Ann Bot 125(2):235–246

Morris RL, Hale R, Strain EMA, Reeves SE, Vergés A, Marzinelli EM, Layton C, Shelamoff V, Graham TDJ, Chevalier M, Swearer SE (2020) Key principles for managing recovery of kelp forests through restoration. Bioscience 70(8):688–698

Morton DN, Anderson TW (2013) Spatial patterns of invertebrate settlement in giant kelp forests. Mar Ecol Prog Ser 485:75–89

Nielsen SL, Sand-Jensen K, Borum J, Geertz-Hansen O (2002) Depth colonization of eelgrass (*Zostera marina*) and macroalgae as determined by water transparency in Danish coastal waters. Estuaries 25(5):1025–1032

Pfister CA, Altabet MA, Weigel BL (2019) Kelp beds and their local effects on seawater chemistry, productivity, and microbial communities. Ecology 100(10). https://doi.org/10.1002/ecy.2798

Rogers-Bennett L, Catton CA (2019) Marine heat wave and multiple stressors tip bull kelp forest to sea urchin barrens. Sci Rep 9:15050. https://doi.org/10.1038/s41598-019-51114-y

Sanderson JC, Ibbott S, Foster J (2003) Restoration of String Kelp (*Macrocystis pyrifera*) habitat on Tasmania's east and south coasts. Final report to Natural Heritage Trust for Seacare. Seacare

Sayer M (2007) Scientific diving: a bibliographic analysis of underwater research supported by SCUBA diving, 1995-2006. Underw Technol 27(3):75–94

Schiel DR, Foster MS (2015) The biology and ecology of giant kelp forests. University of California Press, Oakland, California

Seers BM, Shears NT (2015) Spatio-temporal patterns in coastal turbidity – long-term trends and drivers of variation across an estuarine-open coast gradient. Est Coast Shelf Sci 154:137–151

Shears N, Babcock R (2003) Continuing trophic cascade effects after 25 years of no-take marine reserve protection. Mar Ecol Prog Ser 246:1–16

Smale DA, Burrows MT, Moore P, O'Connor N, Hawkins SJ (2013) Threats and knowledge gaps for ecosystem services provided by kelp forests: a Northeast Atlantic perspective. Ecol Evol 3(11):4016–4038

Steneck RS, Graham MH, Bourque BJ, Corbett D, Erlandson JM, Estes JA, Tegner MJ (2002) Kelp forest ecosystems: biodiversity, stability, resilience and future. Environ Conserv 29(4):436–459

Tait LW, Thoral F, Pinkerton MH, Thomsen MS, Schiel DR (2021) Loss of Giant Kelp, *Macrocystis pyrifera*, driven by marine heatwaves and exacerbated by poor water clarity in New Zealand. Front Mar Sci 8:721087. https://doi.org/10.3389/fmars.2021.721087

van der Linden WJM (1967) A textural analysis of Wellington Harbour sediments. NZ J Mar Freshw Res 1(1):26–37

van der Mheen M, Wernberg T, Pattiaratchi C, Pessarrodona A, Janekovic I, Simpkins T, Hovey R, Filbee-Dexter K (2024) Substantial kelp detritus exported beyond the continental shelf by dense shelf water transport. Sci Rep 14(1):839

Vásquez V, Martínez R, Rodríguez-Núñez K, Bernal C (2024) Angiotensin-I converting enzyme (ACE) inhibitory bioaccessible peptides obtained from protein extracts of *Macrocystis pyrifera* and *Chondracanthus chamissoi* seaweeds. Algal Res 81:103575–103575

Vergés A, Steinberg PD, Hay ME, Poore AGB, Campbell AH, Ballesteros E, Heck KL, Booth DJ, Coleman MA, Feary DA, Figueira W, Langlois T, Marzinelli EM, Mizerek T, Mumby PJ, Nakamura Y, Roughan M, van Sebille E, Gupta AS, Smale DA

(2014) The tropicalization of temperate marine ecosystems: climate-mediated changes in herbivory and community phase shifts. Proc R Soc B: Biol Sci 281(1789):20140846

Vergés A, Doropoulos C, Malcolm HA, Skye M, Garcia-Pizá M, Marzinelli EM, Campbell AH, Ballesteros E, Hoey AS, Vila-Concejo A, Bozec Y-M, Steinberg PD (2016) Long-term empirical evidence of ocean warming leading to tropicalization of fish communities, increased herbivory, and loss of kelp. Proc Natl Acad Sci USA 113(48):13791–13796

Walker DW, Smigaj M, Tani M (2020) The benefits and negative impacts of citizen science applications to water as experienced by participants and communities. WIREs Water 8(1):e1488. https://doi.org/10.1002/wat2.1488

Wernberg T, Krumhansl K, Filbee-Dexter K, Pedersen MF (2019) Status and trends for the world's kelp forests. In: Sheppard C (ed) World seas: an environmental evaluation, vol 3. Elsevier, pp 57–78. https://www.sciencedirect.com/book/9780128050521/world-seas-an-environmental-evaluation#book-description

Wilson KL, Kay LM, Schmidt AL, Lotze HK (2015) Effects of increasing water temperatures on survival and growth of ecologically and economically important seaweeds in Atlantic Canada: implications for climate change. Mar Biol 162(12):2431–2444

Rachel L. Parry has spent over a decade weaving her passion for marine conservation into impactful projects across the globe. From leading invasive species management in Mexico, to directing whale shark and marine research in Honduras, and spearheading coral restoration efforts in Thailand, her career reflects her commitment to the protection and restoration of coastal and marine ecosystems.

Now back in Aotearoa, Rachel fell in love with the kaupapa of the *Love Rimurimu* project with Mountains to Sea Wellington and worked as the Research and Community Monitoring Lead for the first series of plant outs and is currently working as Coastal Scientist for Greater Wellington Regional Council.

Rachel holds a Master of Marine Conservation from Te Herenga Waka—Victoria University of Wellington. Her published research on microplastics in coastal sponges emphasizes the interconnectedness of terrestrial, freshwater, and marine conservation. She specializes in collaborative marine restoration and monitoring of coastal environments, and designing and refining monitoring methodologies. She is PADI Master Scuba Instructor, SSI Advanced Scuba Instructor, and Master Freediver. She believes that when robust science incorporates local and indigenous knowledge and empowers community, it fosters holistic and inclusive environmental stewardship and has the power to drive transformative conservation outcomes.

Although her mahi is marine-focused, she believes in a whole-systems approach to the sustainable management of water: from the mountains to the sea, *ki uta ki tai*. Outside of her work, Rachel enjoys surfing, diving, beach cleans, and exploring new ways to protect and connect with the moana.

Journey Luond began her involvement in this mahi as a volunteer Kelper with Mountains to Sea Wellington and later became an intern with the *Love Rimurimu* project. Passionate about the marine world and conservation, she completed her undergraduate degree in Marine Biology, Ecology & Biodiversity at Te Herenga Waka—Victoria University of Wellington. Through her studies, she gained extensive experience in technical laboratory work, over 100 h of fieldwork, research methodologies, report writing, and data analysis for scientific insight as well as completing an exchange programme with the University of Hawaii, Hilo.

Growing up surrounded by the ocean, Journey learned to swim before she could walk or talk. She was raised in the waters of Kaipara Harbour and hopes to give back to this community through marine conservation efforts later in her career. Journey has worked as a Surveillance Technician for the Ministry of Primary Industries, collecting data from farms and bovine animals suspected of carrying Mycobacterium bovis disease. This role provided her with comprehensive biosecurity training and valuable experience working on a large-scale conservation project.

Journey is grateful for her time as a *Love Rimurimu* intern, an experience that fuelled her passion for seaweed research and deepened her understanding of its ecological importance.

As an early-career scientist, she is eager to pursue a thesis-based master's degree, with hopes of contributing meaningfully to marine ecosystem research—ideally focused on seaweed! Journey holds an AIDA 1 freediving certification, and her passions include surfing, freediving, swimming, and all things moana-related.

Hannah Ryall is Marine Conservation Enthusiast with experience in marine education, field research, and conservation projects. She worked at *Mountains to Sea Wellington*, as Education Coordinator and supported fieldwork for the *Love Rimurimu* project, later going on to work as the Research and Community lead. Hannah holds a Bachelor of Science, majoring in Marine Biology and Environmental Studies, and a Master's degree in Marine Conservation, both from Te Herenga Waka—Victoria University of Wellington, Aotearoa New Zealand. As part of her Master's research, she conducted an independent study on the spatial and temporal abundance of the broadnose sevengill shark in Fiordland, New Zealand.

Beyond her academic and professional work, Hannah has been involved in several voluntary conservation efforts both locally in Wellington and internationally. Notably, she volunteered for 4 months at the Madagascar Research and Conservation Institute, where she contributed to coral and benthic invertebrate surveys. In addition, Hannah served as an Executive for the Society for Conservation Biology Wellington Chapter, where she led the *Places for Penguins* project under Forest and Bird in Island Bay, Wellington. In this role, she was responsible for recruiting and training volunteers to monitor and scan microchips of Little Blue Penguins and their chicks, as well as assessing egg survivability. Hannah is a trained Marine Mammal Medic with Project Jonah, specializing in whale strandings, and holds certifications as a PADI Open Water Diver, PADI Advanced Open Water Diver, and AIDA 1 Introduction to Freediving.

Jorge Jiménez Senén grew up in Madrid, Spain. Although far from the ocean, he was always interested in the marine environment and spent every summer by the beach in the south of Spain. Growing up, he always said he would be an oceanographer or ocean explorer one day. After school, he completed his Bachelor of Science in Biotechnology at the University of Nottingham, UK. After graduating, Jorge took a job in Denmark working at a pharmaceutical laboratory, but his passion and curiosity for the ocean were still there. In 2014, he travelled to South-East Asia to volunteer at a marine conservation project. This experience made him realize that taking action for and educating people about the marine environment were what he was really meant to do.

In 2015, he moved to Wellington, Aotearoa New Zealand, and completed a Master of Science in Marine Conservation at Te Herenga Waka—Victoria University of Wellington. While studying, he volunteered for the marine education programme Experiencing Marine Reserves, and in 2017, when Mountains to Sea Wellington Trust was established, he was hired as a programme coordinator. Since then, he has been involved in this mahi, leading programmes that help educate people about marine conservation and enabling communities to take action for their environment. He has led projects in collaboration with partners such as NIWA and Sustainable Coastlines regarding litter pollution in freshwater and marine environments and has been part of the *Love Rimurimu* project since it launched in 2020. Jorge is currently the Marine Lead for Mountains to Sea Wellington Trust and the Community Engagement Lead for the *Love Rimurimu* project.

Joseph Warmington Originally from South Wales in the UK, Joe is a soggy boy who has worked and played on the coasts of Te Whanganui-a-Tara (Wellington Harbour) for the past 8 years. A true water baby, his place of peace, is drifting through the kelp forests that dot the Wellington coastline. His work consists of teaching school kids and community groups about the marine environment and sharing his passion through hands-on immersive education. He is also the aquaculture technician with the *Love Rimurimu* kelp restoration project, growing *Macrocystis pyrifera* for restoration purposes and helping lead plant outs around Te Motu Kairangi. 'I never expected to be living in Aotearoa New Zealand, but I'm enamored with this place, its nature and its people, and I even ended up falling in love with a real special human. I'm so joyful that I can combine two of the things that excite me the most… growing stuff and spending time in the ocean. I like to think of myself as an underwater gardener, helping to revitalise the kelp forests that bring life to the coastal waters of the region. I'm just here trying to lend a "kelping" hand'.

Zoe Studd is the Co-Founder and Director of the Mountains to Sea Wellington Trust and Project Lead for Love Rimurimu, www.loverimurimu.org. Zoe's love for the sea started early, snorkelling around the Wellington region from a young age. She has a background as Dive Instructor, working both in New Zealand and in Samoa. She has obtained degrees in Environmental Science (AppSci), Marine, and a Diploma in Education and Learning. Her work experience has also seen her in roles in environmental policy at the Ministry for the Environment in New Zealand, and Department of Sustainability and Environment in Victoria, Australia.

For the past 16 years, she has worked in marine and freshwater education and restoration bringing together her experience and expertise in connecting people with the environment, and finding ways for community engagement in addressing pressing environmental challenges.

In 2017, Zoe co-founded the Mountains to Sea Wellington Trust, and went on to initiate the *Love Rimurimu* restoration project. Love Rimurimu began in 2020 as an education programme and is now a community-led restoration effort with many partners. Its collaborative approach brings people together to restore the well-being and mauri of the Wellington Harbour.

23 Seeding Technologies for Seaweed Species: New Insights

Job Schipper and Alexander Ebbing

Abstract

The seaweed farming industry is steadily growing in the Western world, fueled by rising interest in advanced farming techniques and premium applications. Similar to agriculture, a significant breakthrough in large-scale farming can be anticipated once the farming process becomes mechanized. However, considerable manual labor is still required in today's seaweed farming, which limits opportunities for scaling. In this chapter, we assess the methods of propagating, seeding, and deploying kelp while evaluating how the process has become mechanized.

We discuss traditional methods of using twine and other vectors to aid the attachment of juvenile seaweeds to the cultivation substrate. Additionally, we evaluate the alternative approach of applying MultiAnnual Delayed (MAD) gametophytes as a "seed" utilizing the potential of the full generative cycle of kelp. Applying MAD gametophytes for propagation enables cost-effective and relatively rapid genetic improvements by leveraging stable gametophyte parental strains to produce F1 hybrids. Induced gametophytes and juvenile sporophytes can be generated through low-maintenance operations and are employed in precision direct seeding. The genetic and physiological advancements (e.g., priming) achievable through the propagation of gametophytes instead of meiospores are substantial, leading to a significant increase in the predictability of sporophyte density.

Over the past decade, significant progress has been made in alternative propagation and seeding technologies for kelp species. However, the implementation of these technologies remains relatively limited, particularly in Asia, where they are seldom utilized in production. The use of MAD gametophytes presents an opportunity to scale seaweed farming for the future economically, turning potential impacts on the environment, climate, and food security into a reality. Reforestation projects will also benefit from knowledge about direct seeding, as this method is applicable to natural substrates like rock. Technological advancements are happening rapidly and are currently being tested and implemented.

Keywords

Gametophytes · Meiospores · Seeding strategies · Direct seeding · Kelp · Seaweed cultivation · Mechanization · Reforestation

23.1 Introduction

The fundamentals of propagation and seeding methods for cultivating kelp and other species have been established in Japan and Korea for the past 400 years. In China, the cultivation of kelp gained popularity in the second half of the last century. Recently, these methods have advanced, especially concerning the scale of their hatcheries. The growing population and the tradition of consuming seaweed have fostered the development of aquaculture as an alternative to wild harvesting, which has been practiced for many centuries (Cai et al. 2021).

Understanding the life cycle of kelp species (*Saccharina japonica* and *Undaria pinnatifida*) (Phaeophyceae) and *Neopyropia/Porphyra* spp. (Nori) (Rhodophyta) strongly bolstered commercial farming in the temperate seas, leading to impressive industry growth in Asia during the second half of the twentieth century (Chopin and Tacon 2021). While breeding for high-yielding varieties has significantly improved output, farming technology still requires considerable manual labor (Hwang et al. 2020). This is partly due to the methods used for propagating, seeding, and deploying

J. Schipper (✉)
Seawiser BV, Burgerbrug, The Netherlands
e-mail: job.schipper@seawiser.nl

A. Ebbing
Ebbing Tides Vof, Dalfsen, The Netherlands

A. T. Critchley et al. (eds.), *Advances in Temperate Phyconomy: Algal Harvest and Cultivation in Globally Distributed Temperate Waters*, Developments in Applied Phycology 15, https://doi.org/10.1007/978-3-032-02955-3_23

kelp. In this chapter, we will discuss the currently employed methods for propagation and deployment at sea and the innovations made in recent years to enhance the efficiency and cost-effectiveness of these farming practices.

23.2 History of Hatching and Deployment Methods

Kelp species have a well-manageable reproductive cycle compared to, for instance, several species of red seaweed. This characteristic allows for the use of different stages in the propagation method of kelps. The most traditional approach involves using the meiospores released from sori on the blade (Kim et al. 2017). A sorus develops from the blade during specific times of the year or can be artificially induced by keeping blade pieces under short day lengths (<8 h of light per day) in tanks with low light intensity (<100 μmol photons m^{-2} s^{-1}) and temperatures below 10 °C (Forbord et al. 2012).

A common practice in China and South Korea, however, is to harvest the blades of selected specimens of the varieties that have continued to grow from the previous year. The typical meiospore extraction process includes successive cleaning and disinfection steps of the sori, followed by a slow superficial dehydration step. Next, the sori are placed back in seawater, creating such an osmotic shock that the capsules holding the meiospores burst open, releasing the meiospores (Lüning and Dring 1975). Although this method is robust, it requires significant labor to keep the mother specimen alive throughout the year, along with sampling the sori, cutting, cleaning, and inducing spore release. This process is repeated every year.

After the spore release, the spores are sprayed over thin twine (2–3 mm) or coconut fiber rope (5 mm) coiled on frames, which are then placed in tanks where the spores germinate and form small juvenile sporophytes. In China and South Korea, these frames are incubated inside greenhouse-like facilities (Fig. 23.1; see https://seaweedinsights.com/seeding/). This method of propagation and hatchery process is labor-intensive and requires ample space. The method is primarily utilized by Asian kelp farms, where large, sometimes cooperative, hatcheries have been developed (Table 23.1). In China and Korea, large (greenhouse) facilities have been constructed to accommodate this seasonal activity.

Twines used in South Korea and Japan measure only 2–3 mm in thickness and are made from kuralon (polyvinyl alcohol) (Edwards et al. 2016). The fabric should provide good adhesion properties for the meiospores and should not be overly fluffy, as this can obstruct adherence to the core of the twine. Research has shown that twisted twine outperforms braided varieties (Kerrison et al. 2019b; Lawton and Magnusson 2024). Just before deployment at sea, a short piece of twine is manually or mechanically inserted into the final cultivation rope. In Europe and the Americas, the most commonly utilized method features spools filled with twine

Fig. 23.1 Kelp hatchery in China showing the maintenance of racks coiled with twine keeping numerous kelp juveniles. (Figure courtesy of Hatch/Innovation Norway)

Table 23.1 Seeding Information for *Saccharina japonica*; based on 14 interviews conducted across four major producing regions in three countries. Data by Hatch Seaweed Insight

	South Korea	Japan	China
Seed supply	Seed companies	Production by fishery cooperatives	Own hatcheries (Shandong) or external seed companies Fujian)
Format of seed material	Seed frames (uniform dimensions 45 × 55 cm) with seeded fibers (150–200 m)	Seed frames (different dimensions) with seeded fibers (15–80 m)	Seed frames/ curtains made from palm fiber rope in Shandong (70 m) or Vinylon fiber rope in Fujian (40 m)
Nursery period	30 days	N/A	45–60 days
Seeding method	Inserting 2–3 cm long seeded fibers into the cultivation rope	Inserting 2–3 cm long seeded fibers into the cultivation rope	Inserting 20 cm juvenile plants into the twist of the culture rope
Density (intervals on rope)	20–30 cm	20 cm	20 cm
Machine used for seeding	Yes	Manual (with stapler)	Manual
Deploying at sea	Manually (150 m per h)	Manually (100 m per h)	Manually (100 m per h)

that are handled on a relatively small scale in specialized hatcheries or by the seaweed farmers themselves. These coils are deployed by either pulling the cultivation rope through them or twisting the coil around the rope (Fig. 23.4).

Due to its simplicity, this method remains popular (Kim et al. 2019) Additionally, the presence of a visible sporophyte at the time of deployment makes it a relatively reliable and yielding approach (Kerrison et al. 2020). However, one disadvantage is that spore suspensions can easily harbor diseases and other (micro)algal species, which pose a risk to the success of the entire hatchery. More importantly, maintaining genetically uniform varieties, like those developed in China and Korea over the last decades, is quite challenging when following the meiospore route. The variety may gradually diverge from its original genetic composition—due to limited numbers of sporophytes—leading to species mixing, cross-fertilization, and inbred depletion, as observed in practice (Li et al. 2015). Consequently, each spore (male and female) carries a slightly different genome.

Other seaweed species are also often propagated directly using spores. *Neopyropia* spp., *Pyropia* spp., and *Porphyra* spp. are interesting examples, as the sporophyte life stage grows inside an oyster shell. The conchospores formed during this stage are collected and seeded on twine nets. Impressive large spools are utilized to hold the nets and turn them through the spore suspension (see https://seaweedinsights.com/seeding/ and Figs. 23.2 and 23.3).

23.3 Innovations at Deployment of Twine Seeded Kelps

In Asia, machines that mechanically insert (just before deployment) pieces of twine into a rope have been developed and are in use in Korea; however, in China, this remains a manual process. In Europe, both simple and complex machines have been designed to wind twine around the rope (Sintef). Most European farmers pull the cultivation rope through the coil, depending on the twine's wrap being short enough to remain close to the cultivation rope (see Fig. 23.4). In practice, this can be quite problematic and necessitates additional binding. Other methods of using twine include gluing it to the cultivation rope (Seawiser) or, similar to the Asian method, using smaller pieces of twine inserted into the cultivation rope. Principally, the idea is that every method is the same, using a "vector" to bring and attach the juveniles to

Fig. 23.2 Oyster shells keeping conchocelis phase of *Neopyropia* spp., hanging in tanks. (Figures courtesy Hatch/Innovation Norway)

Fig. 23.3 Large 4 m diameter wheels coiled with conchospores seeded netting. (Figures courtesy Hatch/Innovation Norway)

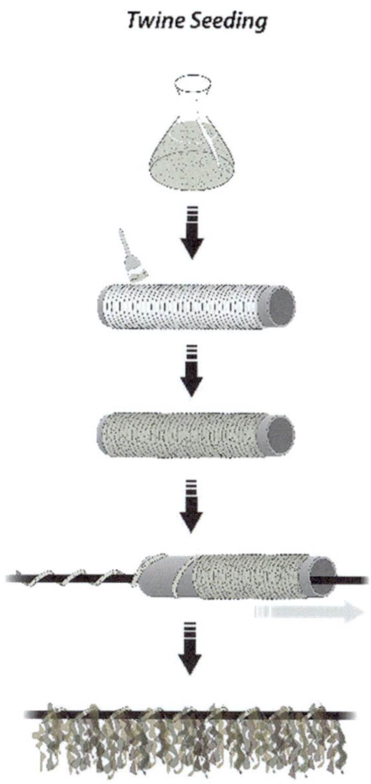

Fig. 23.4 Twine seeding and deployment process

the final substrate (rope). These methods are named "vector seeding" (Skjermo et al. 2019).

23.4 Utilizing the Full Generative Cycle of Kelps

To date, two fundamentally different propagation methods are employed in kelp aquaculture (Valero et al. 2017; Forbord et al. 2018; Goecke et al. 2020): the uncontrolled reproductive method (traditional kelp farming) and the controlled reproductive method. In traditional kelp farming (Shan et al. 2013), large numbers of meiospores are released into large tanks. These meiospores first attach to strings and develop into gametophytes, which then form gametes. The female gametophytes are fertilized, subsequently forming juvenile sporophytes that are attached to these strings (Su et al. 2017). We refer to this as the uncontrolled reproductive method because the farmer does not have direct control over the sexual reproduction occurring on the ropes (Ebbing et al. 2021).

In contrast, when the meiospores of kelps are cultivated in flasks in vitro with the appropriate media, light, and aeration, they develop into the next life stage, forming a gametophyte (Ratcliff et al. 2017; Forbord et al. 2020). Each spore, whether male or female, initially produces a single-celled spherical gametophyte, which later grows into multicellular filamentous gametophytes over time (Avila-Petroche et al. 2022). These gametophytes can also be seeded on twine or other substrates and successively exposed to controlled conditions that are favorable for the induction of fertility (Ebbing et al. 2020).

Recent studies have provided critical insights into the pace at which gametophytes progress toward fertilization. When optimal conditions—such as adequate light intensity (photon flux), nutrient availability, and proper culture density—are met, gametophytes can advance directly from their initial spherical phase to the formation of oogonia (female) and antheridia (male), resulting in immediate fertilization and sporophyte development (Ebbing et al. 2020). Conversely, when these criteria are not met, gametophytes tend to remain in a vegetative state and continue growing toward multicellular filamentous algae (Motomura and Sakai 1984; Lüning 1980; Ratcliff et al. 2017; Edwards 2022). After one year of vegetative growth, this stage is currently defined as a MultiAnnual Delayed, or MAD, gametophyte (Liang et al. 2023; Ebbing et al. 2025).

The long-lived MAD gametophytes exhibit unique characteristics: they can easily fragment and propagate vegetatively, growing continuously under favorable conditions (at rates of 1–6% biomass increase per day, depending on strain, density, and culture settings, as noted from my observations). However, keeping them in a vegetative state requires strict control of environmental parameters, as gametophytes will quickly enter the fertile phase when exposed to a photon flux above 20 μmol m^{-2} s^{-1}, particularly at low culture densities (Ebbing et al. 2020).

Effective and commercially sustainable techniques for large-scale production of these MAD gametophytes in bioreactors with volumes of 18 l or more have been developed (Fig. 23.5-Seawiser). These bioreactors can produce gametophytes continuously throughout the year at an ideal propagation density for optimal growth (Ebbing et al. 2022; Yun et al. 2019). The entire system is automated and requires very little maintenance during production. Typically, a culture begins with about 10 g/l ww gametophytes, which doubles in weight in just 3 weeks; however, higher yields may be expected depending on the type of bioreactor and loading. The gametophytes are harvested and kept in specialized containers until they are needed. Storage, following a low-maintenance protocol, can last for a year or longer.

The vegetatively propagated MAD gametophytes represent a life stage that can be used at the seaweed crop start, allowing for seeding on twine, rope, or other substrates. The gametophyte may either be left to enter the fertilization process after seeding, or it can be induced for fertilization prior to seeding:

Fig. 23.5 Tubular bioreactor filled with gametophytes (Seawiser)

(a) Direct seeding of MAD gametophytes: Directly deploying gametophytes onto a substrate allows for in situ fertilization, depending on environmental triggers to initiate the formation of sporophytes.
(b) Direct seeding of young sporophytes: In vitro reproduction of MAD gametophytes under controlled conditions results in sporophytes before seeding, allowing for precision dosing.

In both scenarios (Fig. 23.6), the gametophyte will be induced to produce generative male and female cells (antheridia and oogonia), and subsequently, the fertilization process will take place (Ebbing et al. 2020). The second approach ensures synchronized fertilization, with developing sporophytes reaching 20–100 μm within 2 weeks under laboratory conditions. These juvenile sporophytes can then be immediately seeded or stored for several days until offshore deployment conditions are ideal (Ebbing et al. 2020). These large juvenile sporophytes clearly possess a competitive advantage over naturally arriving fouling organisms.

23.5 Precision Seeding and Deployment

Utilizing MAD gametophytes provides the opportunity to precisely control the timing, size, and quantity of sporophytes that will be seeded or deployed per meter of cultivation rope (Xu et al. 2009). This enables the mechanization of the seeding process and promotes further optimization of the crop for higher yields and improved economic outcomes. We distinguish between:

1. Vector seeding, which includes a nursery or hatchery period
2. Direct seeding with precision dosing, using an adhesive, deployed immediately at sea

23.5.1 Vector Seeding Using Gametophytes

The traditional way of twine seeding can be performed using gametophytes (Xu et al. 2009), see Fig. 23.6, lower route. Since gametophytes need time to adhere, the induction process is scheduled in situ on the twine.

Alternative methods involve using a small propagule for seeding, which can be attached or plugged into a rope once established in the propagule (Seawiser). The advantage is the precise positioning/spacing of a cluster of seaweed (e.g., for *Macrocystis*) on the cultivation rope. In our lab, we developed a plug that can be inserted into the rope. It worked well, but technically, it remained too complicated and needed further improvement (Fig. 23.7).

23.5.2 Direct Seeding Using Gametophytes or Sporophytes

Gametophytes and juvenile sporophytes offer a viable approach for seeding kelp species by applying them directly to the final substrate, which can include rope, netting, ribbon, or even stones (Kerrison et al. 2018). To facilitate this process, it is advisable to use an appropriate adhesive to support the gametophytes and juvenile sporophytes, as natural bonding may take a considerable amount of time, too long to keep the sporophyte attached in open sea conditions (Kerrison et al. 2019a).

The initial application of the direct seeding method involved immersing a rope in a mixture of seeds and adhesive. However, this approach resulted in seed losses exceeding 90%. To address this issue, the method has evolved with the development of a seeding machine that employs a two-step process: seeds are first precisely dosed onto the rope (Fig. 23.8), followed by the application of adhesive (*SEASEEDER*, by Seawiser). The direct seeding method also works to seed hard substrates like pieces of rock (Fig. 23.9), which is of interest for reforestation projects.

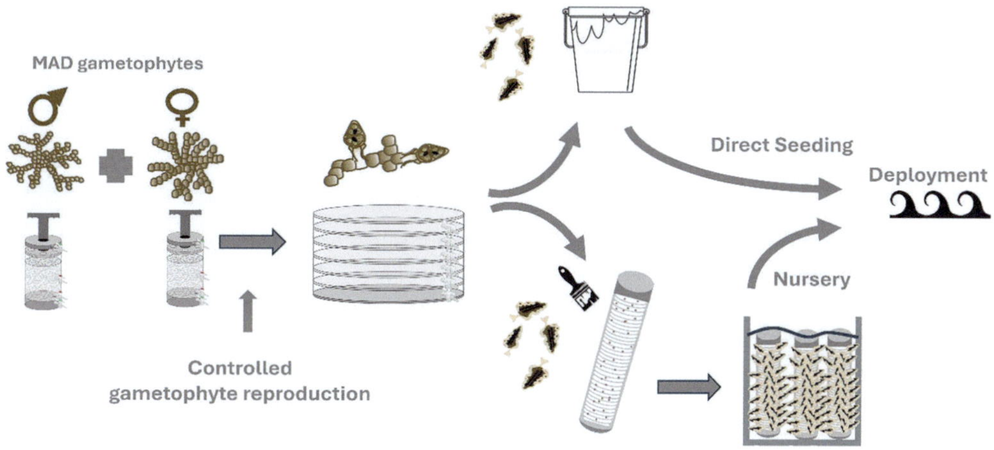

Fig. 23.6 Illustration of the bulk production of male and female gametophytes in bioreactors, the induction process, and two main routes for seeding: direct seeding or vector seeding requiring a nursery stage

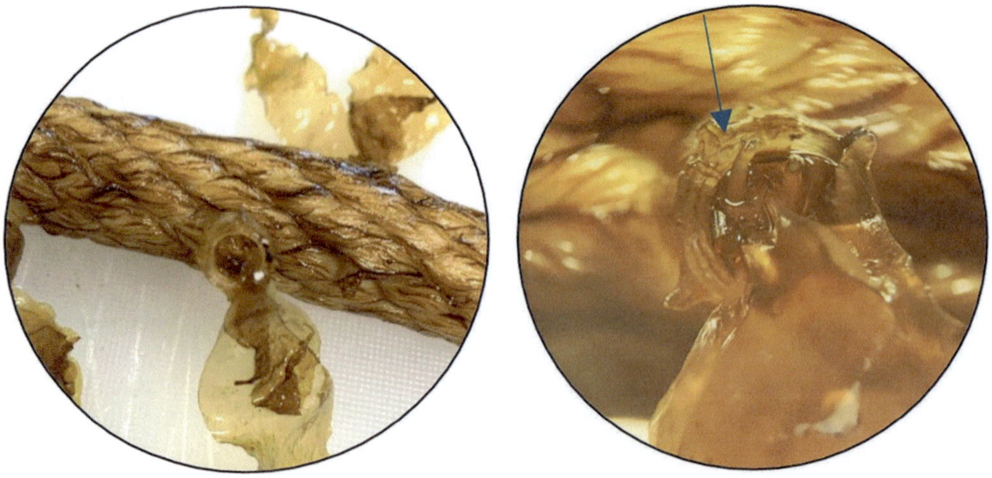

Fig. 23.7 Seawiser plugs containing juvenile *Macrocystis pyrifera* (Phaeophyceae) inserted in a rope (left) and holdfast commencing to grow outside the plug (right)

Fig. 23.8 Direct-seeded ropes hatched in a tank for 28 days; different seed densities (Seawiser)

Fig. 23.9 Direct seeded and hatched on gravel (Seawiser)

Both the two-step process and the use of the seeding machine significantly enhance reproducibility and minimize seed losses during deployment. This refinement ensures that seeds are embedded more deeply within the rope, as demonstrated in the accompanying image of the *SEASEEDER* and the microscopic cavities within the rope (Figs. 23.10 and 23.11).

Fig. 23.10 Piece of rope after 14 days of outgrowth in a tank seeded with juvenile sporophytes using the SEASEEDER (Seawiser)

Fig. 23.11 SEASEEDER (Seawiser)

Several strategies have been suggested to improve the low recovery rates of seeded gametophytes and sporophytes. It is crucial to utilize small gametophyte fragments, around 50–60 μm in size. Each fragment can produce one to three sporophytes. By opting for smaller fragments rather than larger clusters, their distribution along the rope will improve. In contrast, if the fragments are too large, there is a significant risk of losing up to 80–90% of them, resulting in a very scattered distribution along the rope. In our lab, just before starting the induction process, the gametophytes will be ground into smaller pieces. This method enhances their effectiveness compared to leaving them intact (Seawiser).

23.5.3 Adhesive

Multiple research teams have extensively studied the recommended adhesive (Kerrison et al. 2018; Umanzor et al. 2020), yet none of the tested compounds have proven entirely satisfactory. Since the invention of direct seeding (Schipper pers comm, Hortimare), Algae Binder, supplied by AtSeaNova, has been identified as the first and currently most effective compound. It has since been compared to various alternatives, including alginate- and protein- or starch-based adhesives (Umanzor et al. 2020, Seawiser).

In field applications at sea, Algae Binder has consistently outperformed other substances like alginates and starch-based adhesives. Its mechanism of action is believed to involve the formation of a protective layer around the rope and seed, reducing washout while offering additional protection during handling and deployment.

However, the adhesive properties of Algae Binder remain quite limited, as users report. During the initial days after deployment, approximately 90% of the seeded units are typically lost, as illustrated in Fig. 23.12, in a test that simulates at-sea conditions.

In addition to utilizing adhesives, enhancing the surface properties of the rope or substrate, along with the proper adhesive that forms covalent bonds, can significantly improve adherence. For this test, specific polymer primers were used to enhance the molecular adherence points of the rope polymers, which had a considerable impact on polyamide (PA) and polyethylene terephthalate glycol (PETG), as depicted in Fig. 23.13. In the test, we used polymer slides made from the same material as the ropes and assessed those in a flume channel with a current of 0.5 m s^{-1}. After 2 weeks, the remaining sporophytes or clusters were evaluated (Figs. 23.14 and 23.15; Seawiser pers comm).

Furthermore, Kerrison et al. (2020) demonstrated improved surface activity of the rope through corona treatment. The choice of the appropriate polymer, combined with the type of rope braiding and pre-treatment, emerges as a critical factor for the success of direct seeding.

These experiments highlight the need for research on improved compounds that provide both better adhesion and enhanced protection for gametophytes and sporophytes. The search for more effective alternatives remains ongoing.

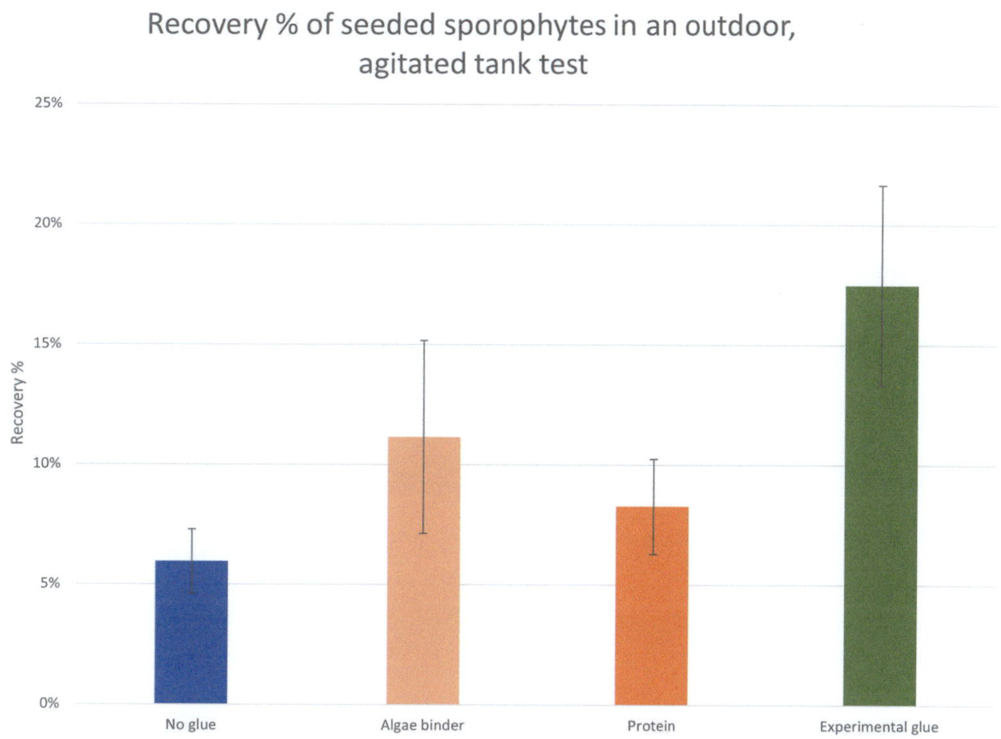

Fig. 23.12 Comparison of various glues used in a direct seeding system on 12 mm ropes, tested in a heavily stirred outdoor tank. All ropes were seeded in the same manner by the SEASEEDER using either Algae Binder, a protein-based glue, or an experimental glue. The differences in the established sporophytes are still too small to be of practical significance. The standard deviation is indicated by error bars

Fig. 23.13 Comparison of various substrate polymers (PA, PE, PETG) tested solely as solid slides. Prior to seeding, they were coated with or without a primer, showing a significant improvement in the adherence of gametophyte fragments. The standard deviation is represented by error bars

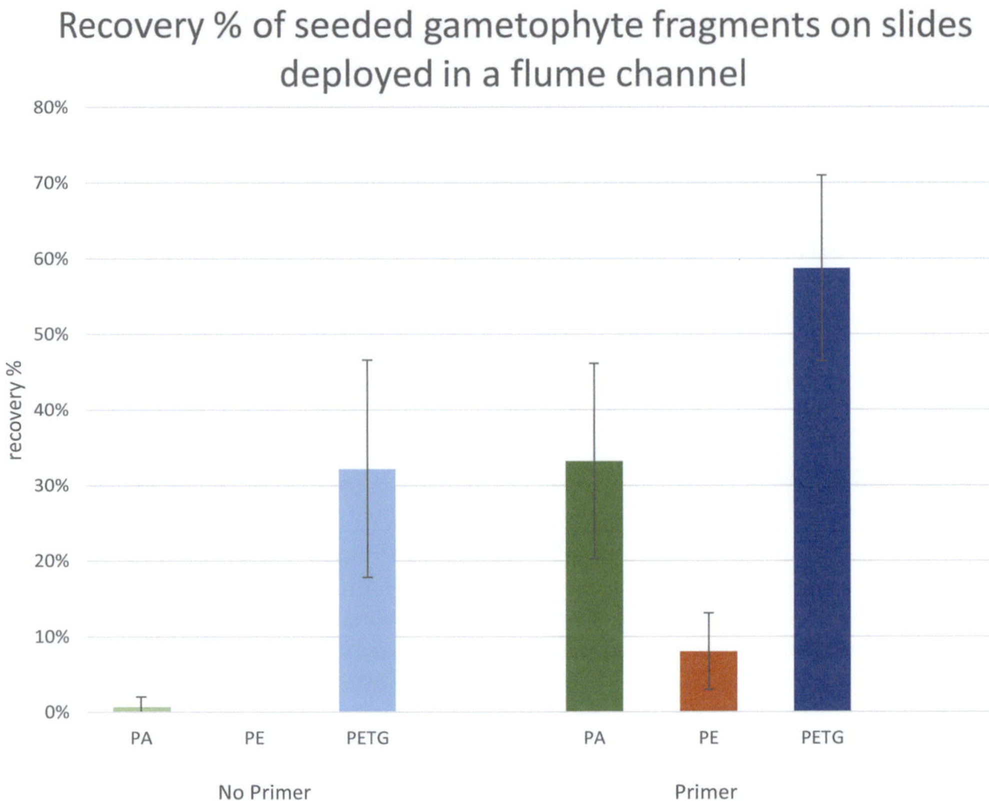

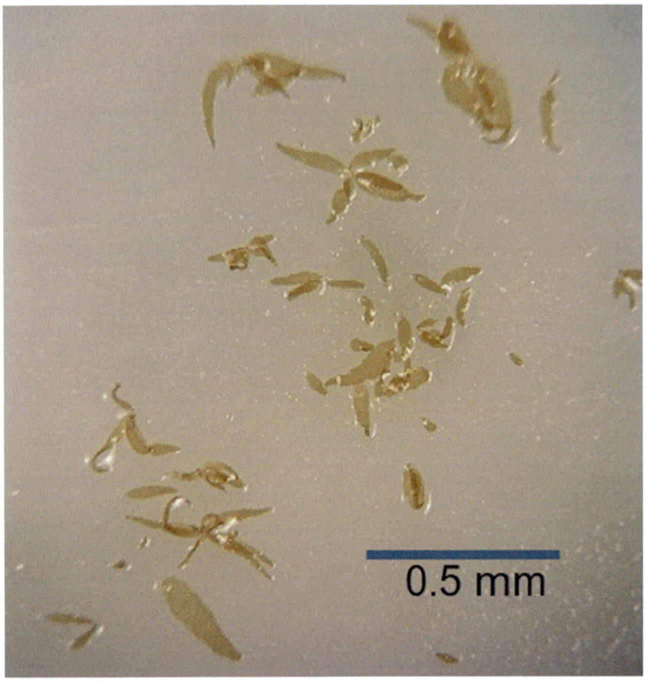

Fig. 23.14 Microscopic sporophytes of *Saccharina latissima* (Phaeophyceae) glued on a slide (Picture Seawiser)

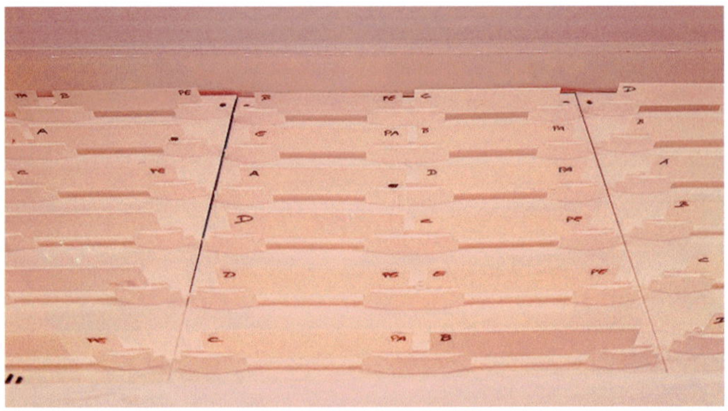

Fig. 23.15 Flume channel at a current speed of 0.5 m s^{-1} (Pictures Seawiser). Flume channel with slides: a current runs parallel to the slides in an almost laminar, non-turbulent flow. It is a compact but very reproducible testing system for adhesion tests

23.6 Discussion

In this paper, we review several approaches for deploying kelp on ropes. The traditional method of seeding twine or thin rope has been well-developed and is now partially mechanized with cutting devices that produce small pieces and mechanically insert them into the rope. However, this technique is considered too slow (a few hundred meters per hour), labor-intensive, and requires large facilities, especially during the hatching stage. While these methods can be effective for relatively small farms and high-value products, they are not seen as the most promising solution for scaling seaweed farming to an industrial level, which necessitates cultivating offshore with 1000 km of ropes per farm. The industry's off-takers require a nearly constant, year-round supply of seaweed biomass or extracts to minimize costs. We suggest shifting away from meiospore seeding and toward utilizing gametophytes as the starting point for the crop. In the following sections, we discuss the implications and benefits.

23.6.1 Enabling Stable F1 Varieties

The MAD gametophyte propagation and the subsequent use of a direct seeding technique provide an effective method for scaling and industrializing seaweed farming. This approach offers the opportunity to mechanize the reproduction and seeding process, thereby reducing costs per unit of farmed seaweed (Xu et al. 2009). Additionally, given the effective large-scale production of seaweed biomass using selected varieties, it is more sensible to focus on gametophyte propagation rather than meiospore seeding (Li et al. 2015). Breeding kelp species requires following the complete generative cycle and cultivating male and female gametophyte clone cultures separately to produce unique specimens with identical genetic makeup (Goecke et al. 2020). The MAD gametophyte method is particularly promising now that unique varieties are being developed and introduced to the market as distinct, uniform, and stable (DUS). Another advantage is that the unique genetics obtained can be protected against duplication, similar to practices in agriculture with F1 hybrids, by maintaining the parental lines within the breeding company to prevent copying.

The propagation method enables optimal maintenance of selected male and female gametophyte strains in relatively small laboratory cultures or through cryopreservation (Visch et al. 2019), allowing for expansion to larger volumes when necessary for annual seeding. This process has shown the capability to be automated with minimal maintenance, resulting in an annual production of over 10 kg of gametophytes that can fit into a 20-foot container. Surprisingly, this method has not yet been adopted in seaweed propagation, as hatcheries and farmers often continue to rely on meiospore propagation and do not fully utilize the potential of the selected varieties (Li et al. 2015). Therefore, there is significant potential for improvement in China, Korea, and Japan if propagation and seeding methods for kelps are modified to accommodate the direct seeding of gametophytes or to employ gametophytes instead of meiospores during twine seeding (Xu et al. 2009).

23.6.2 New Opportunities during Gametophyte Induction

Every kelp species—and even local strains—has its own optimal conditions. For example, strains adapted to warmer or colder water can thrive and induce fertility at temperatures 2 to 4 degrees higher or lower than the average strain. This enables the hatchery to optimize propagation for each strain and run the facility more efficiently. This relates to another intriguing opportunity presented by Gauci et al. (2024), who primed/weaned the gametophytes at supra-optimal temperatures. It seems that the gametophytes developed a tolerance to higher temperatures, which was then inherited by the sporophytes that developed subsequently from them.

To date, we have a clearer understanding of the MAD gametophyte induction process (Ebbing et al. 2020), but induction remains the most cost-intensive part of the gametophyte-to-sporophyte transition. Development of automation and mechanization for this aspect of the reproduction process is underway to be scaled to commercially feasible systems.

23.6.3 Direct Seeding

Direct seeding primarily aims to be the most effective method for significantly reducing the costs of hatching, propagation, and planting while enabling large-scale farming. It offers considerable flexibility in the seeding process and operations. Since the direct seeding process can theoretically occur just seconds before deploying the rope into the sea, operations at sea can be shorter, permitting more rope to be deployed within a limited window of opportunity. Currently, the top speed of the available seeding machine exceeds 3 km/h, requiring only two operators.

The weakest point remains the adhesive used, which leads to excessive seed loss and reduces predictability. Presently, improved glue and new approaches are being developed. Promising results indicate a recovery of up to 60% when tested on a small scale in a flow channel (flume) at approximately 0.5 m/s. This glue is now being prepared for testing under practical conditions at sea.

We discussed the direct seeding technique here mainly for kelp species, but it is likely that it will also be effective for other species such as *Sargassum* (Phaeophyceae), *Porphyra*, and *Asparagopsis* (Rhodophyta), and several tropical species, as long as the ability to form a holdfast of the seaweed propagule is retained. At this stage, we expect that the seaweed juvenile will ultimately adhere its own holdfast to the substrate, so the artificial adhesion needs to last for a few weeks to allow the seaweed to establish itself.

23.6.4 Seeding Density and Strategies

In the context of cultivating, for example, *Saccharina latissima*, particularly concerning seeding, it is advisable to rely on the count of sporophytes seeded per unit length of rope. Current guidelines recommend aiming for a final count of no fewer than 200 healthy sporophytes per meter of rope to optimize yield over a 6-month growing period. Other species and conditions have different recommendations. If establishing a permanent crop is feasible (specific site selection needed), targeting a lower count may be more appropriate. Larger kelps, such as *Macrocystis*, can be seeded in spots instead of continuously along the rope, with just a few spots per meter giving them a better chance to establish. Direct seeding with a precision dosing machine allows the farmer to optimize the densities for their unique conditions.

Each kelp species varies and requires experience for optimization. A common question is the size of the sporophyte to be seeded; the larger it is, the more quickly it establishes and competes with other fouling algae that naturally settle on the rope. However, seeding larger sporophytes also makes them more susceptible to being rinsed off. One strategy to mitigate the risk of losing the initially induced sporophytes, if at-sea conditions appear too hazardous right after deployment, is to seed complementary small fragments of non-induced male and female gametophytes together (backdrop). These fragments can hide in small cavities on the rope and secure a crop if the sporophytes initially fail, although their development will be slower. If all seeded sporophytes emerge, they will remain small as an under-canopy, filling the gaps.

23.7 Conclusion

Over the past decade, significant progress has been made in alternative propagation and seeding technologies for kelp species. However, the implementation of these technologies remains relatively limited, particularly in Asia, where they are seldom utilized in production.

Utilizing MAD gametophytes in propagation enables cost-effective and relatively quick genetic improvements by using stable gametophyte parental strains to produce F1 hybrids. Induced gametophytes and juvenile sporophytes can be produced with low-maintenance operations and are employed in precision direct seeding. The genetic and physiological improvements achievable through the propagation of gametophytes instead of meiospores are substantial, leading to a notable increase in the predictability of sporophyte density.

The use of MAD gametophytes presents an opportunity to scale seaweed farming economically for the future, making the promised impacts on the environment, climate, and food security a reality.

Reforestation projects will also be supported by knowledge about direct seeding, as this method is applicable to natural substrates like rock. Technological advancements are rapidly occurring and are currently being tested and implemented.

References

Avila-Peltroche J, Won WY, Cho TO (2022) An improved protocol for protoplast production, culture, and whole plant regeneration of the commercial brown seaweed *Undaria pinnatifida*. Algal Res 67: 102851

Cai J, Lovatelli A, Aguilar-Manjarrez J, Cornish L et al (2021) Seaweeds and microalgae: an overview for unlocking their potential in global aquaculture development. FAO Fisheries and Aquaculture Circular, Rome

Chopin T, Tacon A (2021) Importance of seaweeds and extractive species in global aquaculture production. Rev Fish Sci Aquacult 29:139–148

Ebbing A, Pierik R, Bouma T, Kromkamp JC, Timmermans K (2020) How light and biomass density influence the reproduction of delayed *Saccharina latissima* gametophytes (Phaeophyceae). J Phycol 56(3):709–718. https://doi.org/10.1111/jpy.12976

Ebbing A, Fivash G, Pierik R, Bouma T, Kromkamp JC, Timmermans K (2021) The SeaCoRe system for large scale kelp aquaculture: a plug-and-play, compatible, open-source system for the propagation and transport of clonal gametophyte cultures. J Appl Phycol 34:517–527

Ebbing A, Fivash GS, Pierik R et al (2022) The SeaCoRe system for large scale kelp aquaculture: a plug-and-play, compatible, open-source system for the propagation and transport of clonal gametophyte cultures. J Appl Phycol 34:517–527. https://doi.org/10.1007/s10811-021-02638-2

Ebbing A, Lindell S, Holm H, Sato Y, Timmermans K (2025) Unravelling the secret life of multi annual delayed gametophytes in the order of the *Laminariales*. J Exp Mar Biol Ecol 583:152081

Edwards MS (2022) It's the little things: the role of microscopic life stages in maintaining kelp populations. Front Mar Sci 9:871204. https://doi.org/10.3389/fmars.2022.871204

Edwards M, Mooney K, Healey GE, Champenois J (2016) Standard operating protocol manual for seaweed biomass cultivation and analysis. EnAlgae project. https://pure.qub.ac.uk/en/publications/standard-operating-protocol-manual-for-seaweed-biomass-cultivatio

Forbord S, Steinhovden KB, Solvang T, Handå A, Skjermo J (2019) Effect of seeding methods and hatchery periods on sea cultivation of *Saccharina latissima* (Phaeophyceae): a Norwegian case study. J Appl Phycol 32:2201–2212

Forbord S, Skjermo J, Arff J, Handå A, Reitan KI, Bjerregaard R, Lüning K (2012) Development of *Saccharina latissima* (Phaeophyceae) kelp hatcheries with year-round production of zoospores and juvenile sporophytes on culture ropes for kelp aquaculture. J Appl Phycol 24(3):393–399

Gauci C, Jueterbock A, Khatei A, Hoarau G, Bartsch I (2024) Thermal priming ofS accharina latissima: A promising strategy to improve seaweed production and restoration in future climates. Mar Ecol Prog Ser 745:59–71. https://doi.org/10.3354/meps14683

Goecke F, Klemetsdal G, Ergon Å (2020) Cultivar development of kelps for commercial cultivation—past lessons and future prospects. Front Mar Sci 8:110. https://doi.org/10.3389/fmars.2020.00110

Hwang EK, Choi HG, Kim JK (2020) Seaweed resources of Korea. Bot Mar 63:395–405. https://doi.org/10.1515/bot-2020-0007

Kerrison PD, Stanley MS, Hughes AD (2018) Textile substrate seeding of *Saccharina latissima* sporophytes using a binder: an effective method for the aquaculture of kelp. Algal Res 33:352–357

Kerrison PD, Stanley MS, De Smet D, Buyle G, Hughes AD (2019a) Holding (not so) fast: surface chemistry constrains kelp bioadhesion. Eur J Phycol 54:291–299

Kerrison PD, Twigg G, Stanley M (2019b) Twine selection is essential for successful hatchery cultivation of *Saccharina latissima*, seeded with either meiospores or juvenile sporophytes. J Appl Phycol 31:3051–3060. https://doi.org/10.1007/s10811-019-01793-x

Kerrison PD, Innes M, Macleod A (2020) Comparing the effectiveness of twine- and binder-seeding in the Laminariales species *Alaria esculenta* and *Saccharina latissima*. J Appl Phycol 32:2173–2181. https://doi.org/10.1007/s10811-020-02069-5

Kim JK, Yarish C, Hwang EK, Park M, Kim Y, Kim JK (2017) Seaweed aquaculture: cultivation technologies, challenges and its ecosystem services. Algae 32(1):1–3. https://doi.org/10.4490/algae.2017.32.3.3

Kim JK, Stekoll M, Yarish C (2019) Opportunities, challenges and future directions of open-water seaweed aquaculture in the United States. Phycologia 58:446–461

Lawton RJ, Magnusson M (2024) Effects of seeding twine type and seeding density on hatchery performance and initial at-sea cultivation performance of the kelp Ecklonia radiata. Algal Res 84:103777. https://doi.org/10.1016/j.algal.2024.103777

Li X, Zhang Z, Qu S, Liang G, Zhao N, Sun J, Song S, Cao Z, Li X, Pan J, Luo S, Zhang L, Cui C, Peng J, Li Y, Wu R, Zhao J, Qian R, Wang L, Yang G (2015) Breeding of an intraspecific kelp hybrid Dongfang no. 6 (*Saccharina japonica*, Phaeophyceae, Laminariales) for suitable processing products and evaluation of its culture performance. J Appl Phycol 28:439–447. https://doi.org/10.1007/s10811-015-0562-0

Liang Z, Wang X, Zhang P, Liu W, Wang W, Liu F (2023) Chronological development of the morphological, physiological, biochemical, and transcriptomic changes provides insights into the mechanisms of gametogenesis in Saccharina japonica. J Appl Phycol 35(2):785–802. https://doi.org/10.1007/s10811-023-02921-4

Lüning K (1980) Critical levels of light and temperature regulating the gametogenesis of three *Laminaria* species (Phaeophyceae). J Phycol 16:1–15

Lüning K, Dring MJ (1975) Reproduction, growth and photosynthesis of gametophytes of *Laminaria saccharina* grown in blue and red light. Mar Biol 29:195–200

Motomura T, Sakai Y (1984) Ultrastructural studies of gametogenesis in *Laminaria angustata* (Laminariales, Phaeophyta) regulated by iron concentration in the medium. Phycologia 23(3):331–343. https://doi.org/10.2216/i0031-8884-23-3-331.1

Ratcliff JJ, Soler-Vila A, Hanniffy D, Johnson MP, Edwards MD (2017) Optimisation of kelp (*Laminaria digitata*) gametophyte growth and gametogenesis: effects of photoperiod and culture media. J Appl Phycol 29:1957–1966. https://doi.org/10.1007/s10811-017-1070-1

Shan TF, Pang SJ, Gao SQ (2013) Novel means for variety breeding and sporeling production in the brown seaweed Undaria pinnatifida (Phaeophyceae): Crossing female gametophytes from partheno sporophytes with male gametophyte clones. Phycol Res 61:154–161

Skjermo J, Forbord S, Steinhovden K, Solvang T, Handå A (2019) Effect of vector and direct seeding on growth of sugar kelp *Saccharina latissima* at sea. Presented at the International Seaweed Symposium, Jeju

Su l, Pang SJ, Shan TF, Li X (2017) Large-scale hatchery of the kelp Saccharina japonica: A case study experience at Lvshun in northern China. J Appl Phycol 29:3003–3013. https://doi.org/10.1007/s10811-017-1154-y

Umanzor S, Li Y, Yarish C (2020) Effect of direct "seeding" binders and embryonic sporophyte sizes on the development of the sugar kelp, *Saccharina latissima*. J Appl Phycol 32:4137–4143. https://doi.org/10.1007/s10811-020-02277-z

Valero M, Guillemin ML, Destombe C, Jacquemin B, Gachon CMM, Badis Y, Buschmann AH, Camus C, Faugeron S (2017) Perspectives on domestication research for sustainable seaweed aquaculture. Persp Phycol 4:33–46

Visch W, Rad-Menendez C, Nylund GM, Pavia H, Ryan MJ, Day J (2019) Underpinning the development of seaweed biotechnology: cryopreservation of brown algae (*Saccharina latissima*) gametophytes. Biopreserv Biobank 17(5):378–386. https://doi.org/10.1089/bio.2018.0147

Xu B, Zhang QS, Qu SC et al (2009) Introduction of a seedling production method using vegetative gametophytes to the commercial farming of *Laminaria* in China. J Appl Phycol 21:171–178. https://doi.org/10.1007/s10811-008-9347-z

Yun Z, Yingdi Z, Mengying L, Li L, Xianming T, Jiangtao G (2019) Homogenization significantly enhances growth of macroalga *Saccharina japonica* female gametophytes. Algal Res 43:101640. https://doi.org/10.1016/j.algal.2019.101640

Job Schipper Since 2008, Job has actively participated in the seaweed business. With the vision that seaweed has significant potential to become a major industry in Europe as well, he founded the seaweed company Hortimare BV in the Netherlands and Norway. With his background in horticultural science and business, he had the ideal expertise to launch a seaweed 'seed' company. Job and his team learned to farm, propagate, and breed seaweeds at Hortimare's 2-hectare seaweed farm in Solund, Norway, an Integrated Multi-Trophic Aquaculture (IMTA) farm that combines salmon and seaweed farming. He invented the direct seeding method for kelps and tested various cultivation methods with his team. The first candidate Saccharina latissima varieties were developed, and they explored ways to harvest and preprocess seaweed to establish a standard in the market. The Dutch-based seaweed hatchery, Hortimare, became well-known as a supplier of starting materials to the seaweed industry. After transferring the company to the next generation of management, Job served as an advisor/project manager in seaweed farming, bringing his hands-on practical expertise. With the start of the consultancy company Seawiser, he focused on advising new seaweed farmers and continues to improve the direct seeding technology in his lab. This led to the development of the precision seeding machine, SEASEEDER, which has been introduced to the market. Scaling gametophyte and juvenile production is another important research topic in his lab.

Job offers innovative solutions for addressing complex strategic and technological challenges. His mission in the coming years is to scale up seaweed cultivation and ensure a year-round supply of seaweed biomass from farms to buyers, making the industry profitable. Holding an MSc in Horticultural Science from Wageningen University in the Netherlands, Job has served as a researcher and project manager at Syngenta Seed, sales and marketing manager for De Groot & Slot Allium seed company, and head of laboratories at Naktuinbouw. Agriculture and horticulture are his primary frameworks in the approach to developing the seaweed farming industry, but he stresses the importance of adopting sustainable practices from day one!

Alexander Ebbing is a researcher and consultant specializing in seaweed aquaculture, with over 13 years of experience studying the microscopic life stages of kelp. His work focuses on the controlled cultivation of gametophytes and sporophytes, aiming to make seed production more reliable and scalable for the industry. His research contributes to understanding the resilience and behavior of these early life stages under varying environmental conditions, providing insights crucial for both commercial aquaculture and marine rewilding efforts.

Alexander's work on multiannual delayed (MAD) gametophytes explores how long-term cultivated gametophytes respond to stress, nutrient availability, and light conditions. By studying their reproductive timing and viability over extended periods, he has contributed to the growing body of knowledge on their potential role in improving hatchery efficiency and adaptation strategies in a changing climate. His research has also examined the impact of light spectra, temperature, culture density, and genetic variance on gametophyte growth and fertility, demonstrating how controlled environmental conditions can optimize seedstock production.

Beyond his research, Alexander collaborates closely with seaweed companies and industry partners to translate scientific findings into practical applications. As the founder of Ebbing Tides, he advises aquaculture businesses on optimizing hatchery techniques, scaling seed production, and integrating scientific advancements into commercial operations. His broader vision is to help establish a resilient and sustainable seaweed sector that moves beyond wild-stock reliance toward controlled, scalable production methods. By bridging research and practice, Alexander aims to contribute to a deeper understanding of kelp's life cycle and its potential for food security, marine restoration, and climate resilience.

In addition to his work, Alexander enjoys sharing his perspectives on seaweed's role in the future economy through his writings on www.ebbing-tides.com.

Conserving Marine Biodiversity in the North American Pacific: Biobanking Macroalgae and Their Associated Microbiome

24

Siobhan Schenk, S Clay Steell, Jasmin M. Schuster, Lingjie Zhou, Liam Coleman, Jennifer Smith, Laura W. Parfrey, and Emily G. Aguirre

Abstract

Here, we examine the current state and future directions of biobanking initiatives for preserving macroalgal biodiversity along the North American Pacific coast. We summarize the efforts of existing biobanking programs across non-profit organizations, academic institutions, and commercial entities, highlighting their diverse approaches to preserving brown and red macroalgae through both traditional culture-based methods and cryopreservation techniques. While several organizations are successfully maintaining gametophyte cultures, current biobanking efforts cover only a fraction of the region's genetic diversity. We also highlight how understanding population genetics and local adaptation are important in the maintenance of regionally distinct germplasm collections. This chapter also explores the potential functional roles of microbial communities in macroalgal health and development, emphasizing the importance of preserving both host and microbiome for successful conservation, restoration, and aquaculture outcomes. Some key challenges we identified include limited geographical coverage, lack of standardization in preservation methods, and the need for greater coordination among biobanking facilities. We conclude by outlining key research priorities and advocating for standardized practices, while encouraging the expansion of biobanking initiatives along the Pacific coast.

Keywords

Seedbank macroalgae · Biobanking · Restoration · Conservation · Mariculture · Microbial symbiosis

24.1 Background

Biobanks are key infrastructure that safeguards biological knowledge and variation, especially in marine environments. The marine environment is vast, covering approximately 70% of Earth's surface and harboring a myriad of diverse life forms, many of which remain poorly known, or undescribed. Considering rapid ecological changes and associated biodiversity loss, access to marine samples through biobanks is key for more effective research, conservation, and restoration of biodiversity. For example, as ocean temperatures continue to rise and extreme temperature events become more frequent, preserving and researching variation in thermal tolerance across local populations is essential for their long-term survival.

Coastal habitats, though occupying a relatively small portion of the ocean, play an outsized role in supporting marine biodiversity and ecosystem functions. Much of the habitable space in coastal waters is created by habitat formers, such as kelp forests, seagrass beds, mangroves, or coral reefs. Biogenic, coastal habitats then create biodiversity hotspots by providing food and shelter, increasing productivity, and

S. Schenk, Ph.D. · L. W. Parfrey, Ph.D.
The University of British Columbia, Vancouver, BC, Canada
e-mail: sschenk@student.ubc.ca; lwparfrey@botany.ubc.ca

S. C. Steell, M.S. · J. M. Schuster, Ph.D.
The Kelp Rescue Initiative, Bamfield Marine Sciences Centre, Bamfield, BC, Canada
e-mail: steellclay@gmail.com; jasmin.schuster@kelprescue.org

L. Zhou, Ph.D. · J. Smith, Ph.D.
Scripps Institution of Oceanography, University of California, San Diego, CA, USA
e-mail: liz047@ucsd.edu; jes013@ucsd.edu

L. Coleman, Ph.D.
Simon Fraser University, Pacific Salmon Foundation, Vancouver, BC, Canada
e-mail: lcoleman@psf.ca

E. G. Aguirre, Ph.D. (✉)
Kelp Ark, Los Angeles, CA, USA
e-mail: emily@kelpark.org

A. T. Critchley et al. (eds.), *Advances in Temperate Phyconomy: Algal Harvest and Cultivation in Globally Distributed Temperate Waters*, Developments in Applied Phycology 15, https://doi.org/10.1007/978-3-032-02955-3_24

supporting ecosystem services and functioning (Jones et al. 1994; Hastings et al. 2007; Smale et al. 2013). These habitats not only sustain marine life but also contribute to global carbon storage through their "blue carbon" potential, as vegetated coastal ecosystems are highly effective at sequestering carbon (Mcleod et al. 2011; Duarte et al. 2013; Alongi 2014; Filbee-Dexter and Wernberg 2020). Their role in climate mitigation highlights the urgency of preserving and understanding these sensitive ecosystems.

Among these coastal habitats, kelp forests stand out as particularly important in temperate and polar waters (Thomson et al. 2015; Wernberg et al. 2016). Underwater macroalgae forests, such as kelp forests, cover ~2 million km^2 globally and are found along one third of temperate and polar coastlines (Krumhansl et al. 2016; Starko et al. 2021; Duarte et al. 2022). Kelp forests, formed by various species of brown algae (Order Laminariales), are critically important coastal ecosystems that provide important ecosystem services: as kelps are prolific primary producers, cycle nutrients and carbon, and support biodiversity and fisheries production. Additionally, kelp forests provide coastal protection, clean water, and habitat for countless marine species (Paddack and Estes 2000; Cebrian 2002). Globally, kelp ecosystem services are estimated to generate ~$500 billion on average per year, primarily through fisheries production, nitrogen, and carbon removal (Eger et al. 2022).

24.1.1 Seedbanks and Biobanking as Insurance Against Species Loss

When kelp forests and other macroalgae decline, they are often replaced by unproductive sea urchin barrens or turf covered reefs (Scheibling et al. 1999; Filbee-Dexter and Scheibling 2014; Filbee-Dexter and Wernberg 2018). Sea urchins can persist in barrens for years or decades on minimal food by lowering their metabolic rates (Schuster et al. 2022a; Brandl et al. 2023), leading to a sudden and long-lasting regime shift that is difficult to reverse and has significant consequences for ecosystem structure and functioning (Scheibling et al. 1999; Johnson et al. 2013). Once kelp forests disappear, so do many of the animals that live in them. In addition to their inherent value as diverse life forms, many kelp-associated species hold significant cultural and economic value (e.g., salmon, herring, and abalone) that are at risk of loss. Paths to kelp recovery are often limited or require human intervention through urchin removal and/or kelp restoration efforts. With kelp restoration still in its infancy, this requires significant financial and human resources (Eger et al. 2022).

The rapid disappearance of kelp forests from large swaths of coastline around the world has prompted kelp conservation and restoration efforts in many places. Organizations across the world are embarking on critical missions to develop innovative approaches to kelp restoration. While relatively new, the field of kelp restoration has developed rapidly in recent years, and there is now a global movement to restore four million hectares of kelp forest by 2040 (the Kelp Forest Challenge, https://kelpforestalliance.com/kelp-forest-challenge). Kelp restoration involves collecting wild kelp tissue, which is then either directly used (e.g., transplants or spore bags) or cultured in the lab. The latter requires inducing spore release in a nursery, cultivating juvenile kelp on a restoration substrate (e.g., gravel, rocks, tiles) under controlled conditions, and then out-planting them in the wild. Sourcing kelp tissue is a key step in either of these approaches, so it's important to consider genetic diversity of the material used. Higher genetic diversity can increase resilience to stressful conditions and adaptive potential and help preserve different phenotypes. As donor populations are shrinking or disappearing, it is important to preserve this genetic diversity in biobanks. The distance between a source population and the area where kelp is outplanted on restoration materials is also a consideration, important for biosecurity reasons.

Long-term storage of viable genetic and reproductive tissue (i.e., biobanking) thus constitutes a major part of the conservation and preservation of species. Biobanks preserve evolutionary significant units and can be used to store genetic diversity while long-term conservation and restoration plans are developed for species in decline. Even in cases where restoration efforts are well underway, restoration alone cannot halt biodiversity loss because, as a species' numbers dwindle, populations become too fragmented, and low genetic diversity threatens long-term species viability. In such cases, bio-banked material and assisted reproductive technology (ART) can be used to reinstate lost genetics back into a population (Howard et al. 2016; Bolton et al. 2022). While this technology is not yet available for all taxa, biobanks can fill the gap between species loss in the wild today and future technological developments (Bolton et al. 2022). Furthermore, bio-banked material can be used in aquaculture or other industries, thereby reducing dependency on often lethal sampling from natural stocks.

24.2 Current State of Kelp Seedbanks

Kelp biobanks, commonly referred to as seedbanks, store germplasm by preserving the microscopic gametophyte stage of kelp for use in restoration and aquaculture efforts. Just like terrestrial seedbanks, kelp seedbanks also serve as a safeguard against habitat destruction and environmental changes by maintaining genetic diversity. However, whereas terrestrial seedbanks primarily store desiccated seeds that can remain viable for years or even decades, kelp seedbanks usually rely on maintaining living gametophytes in a semi-suspended state, requiring monitoring and specific temperature and light conditions. Kelp undergoes a complex life cycle involving an alternation of generations, where the large, familiar kelp fronds represent the diploid sporophyte stage, and the microscopic haploid gametophytes serve as the reproductive stage (Fig. 24.1). By preserving gametophytes in controlled conditions (<11 °C, red light) and cultivating them in nutrient-rich media designed for photosynthetic marine organisms (PES, L1, or F/2) (Guillard 1975; Bold and Wynne 1985; Guillard and Hargraves 1993; Hallegraeff et al. 2004) within laboratory settings, phycologists can sustain genetic diversity, ensuring a reliable supply of viable kelp for restoration initiatives or commercial cultivation (Fig. 24.2). This method allows for the strategic out-planting of kelp sporophytes when environmental conditions are most favorable, aiding in the recovery of declining kelp forests.

24.2.1 Storage Methods: Ex-Situ Cultivation and Cryopreservation

Traditional culture-based methods have long been utilized and are a generally effective approach to kelp biobanking. Storage of live gametophyte cultures is reliable, methodologically simple, and easy and inexpensive to initially set up, making it suitable for many biobanking operations (Wade et al. 2020). However, it also comes with a few disadvantages that may make it less feasible in specific situations. One potential limitation of culture-based kelp biobanking methods is the maintenance required. Even under low light and temperature conditions, gametophyte cultures still metabolize and grow, which creates the need for regular media changes and other upkeep. While this is not overly burdensome when culture collections are relatively small, performing this sort of maintenance with a very large culture

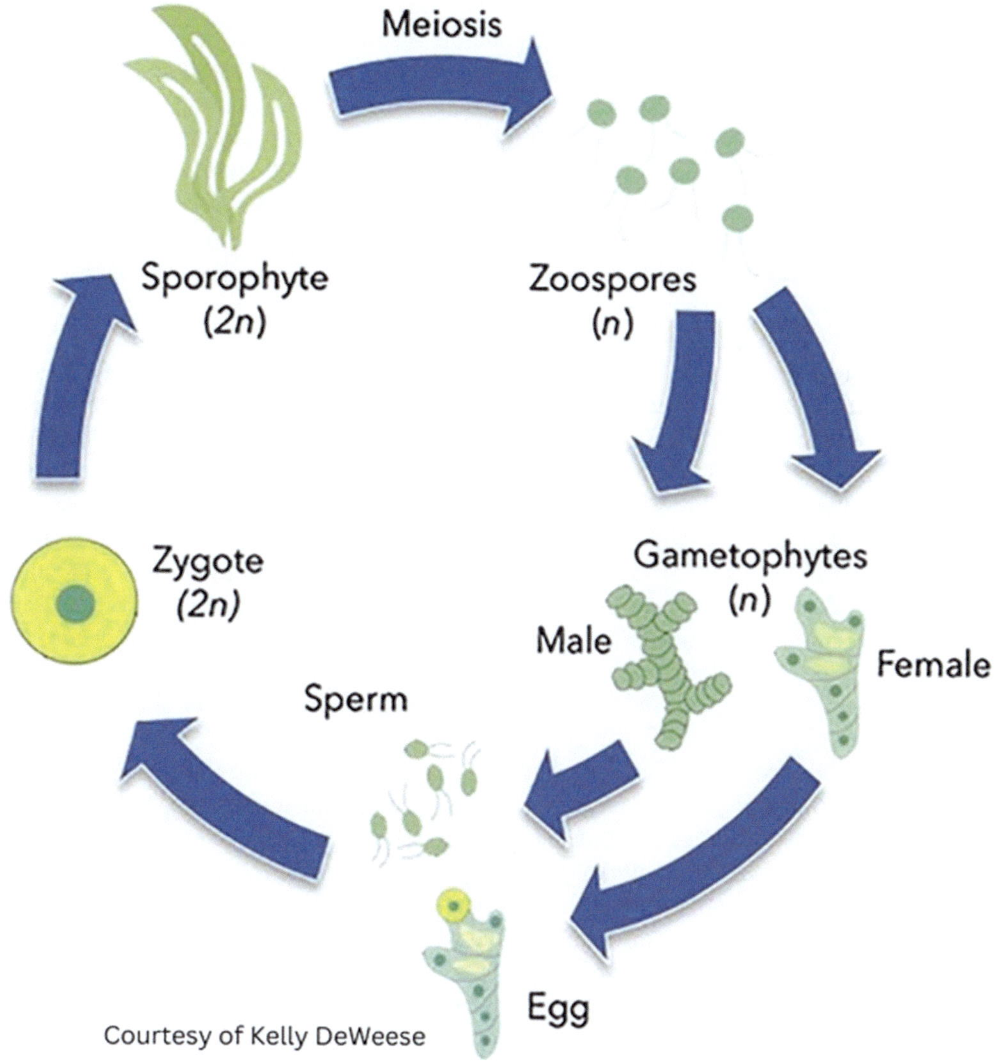

Fig. 24.1 The reproductive cycle of giant kelp: Adult sporophytes (2*n*) undergo meiosis to produce haploid zoospores (*n*), which settle and develop into male and female gametophytes. Fertilization occurs when sperm from a male gametophyte reaches the egg of a female gametophyte, forming a zygote that grows into a young sporophyte, completing the cycle. (Image by Dr. Kelly DeWeese and adapted from the original publication by Molano et al. 2022)

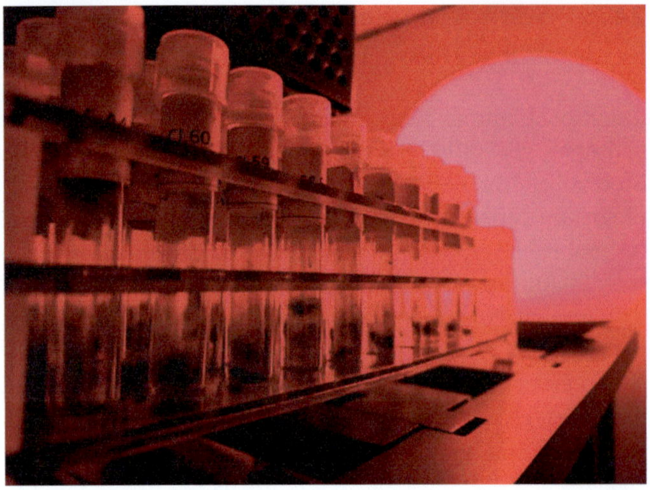

Fig. 24.2 Kelp germplasm, "gametophytes", kept under red light and cold storage in test tubes at the nonprofit organization, Kelp Ark. Location: AltaSea at the Port of Los Angeles, San Pedro CA USA. (Image by Michael Marty-Rivera)

collection can become prohibitively time consuming and may require more personnel, which has potential to significantly increase costs (Yang et al. 2021). Furthermore, maintaining live cultures comes with constant risk of contamination by pest organisms or culture loss due to failure of environmental control systems (Yang et al. 2021). Additionally, when any live, growing culture is kept for long periods of time, it has potential to accumulate mutations as a result of normal cell division processes (Lakeman et al. 2009), which may be problematic for the preservation of biodiversity.

An alternative approach to kelp biobanking that has potential to remedy some of the limitations inherent in culture-based methodologies is cryopreservation. This refers to the practice of storing live biological material at ultracold temperatures (Day 2008). Under these conditions, metabolic activity is almost completely halted, which permits organisms to be held in near-unchanging state for long periods of time (Jaiswal and Vagga 2022). Cryopreservation has found favor in many fields of biology and is now routinely and reliably used in a variety of settings, including medicine. It has also been used successfully in several species of red, green, and brown algae, including multiple kelp species (Day et al. 2008).

Cryopreserved biological material is most commonly stored in liquid nitrogen at a temperature of −196 °C (Day et al. 2008; Jaiswal and Vagga 2022). Under normal circumstances, the introduction of any living cell into these conditions would be lethal, as formation of ice crystals and changes in concentrations of intra- and extracellular solutes that occur during the freezing process can rupture plasma membranes (McGann et al. 1988). For this reason, methods have had to be developed to protect cells from cryoinjury during cooling. The two main approaches for safely cryopreserving cells are (1) slow cooling and (2) vitrification (Day et al. 2008; Jaiswal and Vagga 2022). In slow cooling, cells are exposed to chemicals called cryoprotective agents (CPAs) that mitigate the damaging effects of the freezing process (Day et al. 2008; Jaiswal and Vagga 2022). The temperature of the cells is then slowly reduced, usually at a rate of approximately -1 °C min^{-1}, before the cells are ultimately immersed in liquid nitrogen. In a vitrification approach to cryopreservation, the biological sample is dehydrated using relatively high concentrations of CPAs before being quickly plunged into liquid nitrogen (Day et al. 2008; Jaiswal and Vagga 2022). If the sample is sufficiently dehydrated, the liquid phase forms a glass instead of crystals when rapidly cooled, thereby mitigating cryoinjury. In both slow cooling and vitrification, if the freezing protocol works as intended, cryopreserved cells are not fatally damaged and are in a stable state when in liquid nitrogen storage. The samples can then be carefully removed from cryogenic conditions when needed and returned to a culture environment.

There is a growing body of literature on cryopreservation of kelps. Kelp species with published cryopreservation protocols include *Saccharina japonica* (Sakanishi and Saga 1994; Kuwano et al. 2004; Zhang et al. 2007a, b, 2008), *Saccharina latissima* (Visch et al. 2019), *Saccharina longissima* (Kuwano et al. 2004), *Macrocystis pyrifera* (Piel et al. 2015), *Undaria pinnatifida* (Kuwano et al. 2004; Nanba et al. 2009; Wang et al. 2011), *Eisenia bicyclis* (Kono et al. 1998), *Ecklonia cava* (Kuwano et al. 2004), and *Kjellmaniella crassifolia* (Kuwano et al. 2004). Almost all published studies cryopreserve gametophytes of the target species (e.g., Kuwano et al. 2004; Zhang et al. 2007b; Visch et al. 2019), but one demonstrates successful cryopreservation of spores (Zhang et al. 2007a). Most of these studies utilized a slow-cooling approach to cryopreservation, but vitrification-based protocols have also proven successful (Wang et al. 2011).

Cryopreservation presents several potential advantages for kelp biobanking when compared with traditional culture-based methods. One significant advantage is that cryopreserved samples require much less maintenance than live cultures. Once introduced into liquid nitrogen storage, samples do not generally need to be accessed again until they are required to be for their intended use. This could substantially reduce labor and maintenance costs for biobanking efforts, especially when working with very large collections, and may also limit opportunities for valuable samples to be contaminated by pest organisms. Additionally, as liquid nitrogen storage does not require electrically powered environmental control systems, cryopreserved kelp collections may be more resilient to loss due to power outages or other catastrophic system failures compared to live culture collections. Furthermore, since live cells have almost no metabolic activity under cryogenic conditions, cryopreserved kelp cultures may be at less risk of unwanted mutations or other genetic changes compared to live cultures.

Despite the noteworthy advantages presented by cryopreservation, it has not been widely adopted for use in kelp biobanking efforts at time of writing, and most existing operations continue to exclusively utilize traditional live culture methods. There are several possible reasons for this. A likely explanation is that cryopreservation ultimately remains understudied and insufficiently optimized for kelps compared to live culture protocols. While numerous cryopreservation protocols exist for kelp (e.g., Kuwano et al. 2004; Zhang et al. 2007b; Day et al. 2008; Visch et al. 2019), there are many knowledge gaps that may limit the trust that practitioners are willing to place in the process. For instance, there is no research documenting how kelps fare when cryopreserved for timelines measured in years. The longest published period of time a kelp has been cryopreserved successfully to our knowledge is 5 days (Sakanishi and Saga 1994), and most studies only report cryopreserving kelps for approximately 24 h (e.g., Kono et al. 1998; Wang et al. 2011; Visch et al. 2019). It will be important to ensure that kelps can be viably removed from cryogenic conditions after years of storage before practitioners are likely to be willing to adopt the process for biobanking. Additionally, while virtually all papers documenting kelp cryopreservation report success, rates of survivorship following removal from cryogenic conditions are highly variable. Many papers report 50% survivorship at best for female gametophytes (Kono et al. 1998; Zhang et al. 2008; Wang et al. 2011; Visch et al. 2019). Better optimizing kelp cryopreservation protocols to yield stronger survivorship would likely make the approach more useful for biobanking. This, however, calls attention to a challenge inherent to cryopreservation in general, which is that protocols are difficult to develop because there are numerous variables, such as choice of CPAs and cooling rates, which need to be experimentally optimized for individual species and cell types (Taylor and Fletcher 1998). Furthermore, the equipment and facilities required to biobank kelp utilizing cryopreservation are less accessible than those needed to culture kelp. It may not be feasible for all would-be kelp biobanking operations to have consistent access to liquid nitrogen storage, for instance. Finally, no research has been conducted examining whether there may be any unintended side effects that cryopreservation might have on kelp that could be problematic for biobanking. In other settings, it has been found that cryopreservation can cause genetic and epigenetic changes in organisms (Riesco and Robles 2013).

Overall, while we argue that cryopreservation has a great deal to offer kelp biobanking practitioners, it currently faces several obstacles that are likely limiting its adoption over live culture techniques. We recommend that more research be conducted to optimize cryopreservation protocols for multiple kelp species and demonstrate the effectiveness of these protocols for storing kelp for long periods of time. Work will also need to be done to examine whether there could be unintended consequences of incorporating cryopreservation into biobanking programs. Smaller biobanking operations with limited resources are likely best off using traditional culture techniques, as these are more accessible than cryopreservation methods. However, cryopreservation may be advantageous for more developed operations with larger collections.

24.2.2 Key Programs and Organizations Spearheading Kelp Biobanking

24.2.2.1 The State of Kelp Biobanks in North America

The northeastern Pacific Ocean is a hotspot of biodiversity for macroalgae, particularly kelps (Druehl and Clarkston 2016; Fragkopoulou et al. 2022). Along the shores of the many First Nations territories that make up the area thought of today as British Columbia, Canada, dozens of kelp species occur in a wide diversity of forms and habitats. These include the canopy-forming species bull kelp (*Nereocystis luetkeana*) and giant kelp (*M. pyrifera sensu lato*), which create highly productive and ecologically critical habitats (Rennick et al. 2022; Schuster et al. 2022b; Eger et al. 2023). These species and others, such as sugar kelp (*Saccharina latissima*) and winged kelp (*Alaria marginata*), are also increasingly cultivated in ocean-based farms, supporting a nascent mariculture industry.

In the last 20 years, and particularly following the 2014–2016 marine heatwave known as "The Blob," declines in kelps and other seaweed species have been recorded throughout the Pacific region (Starko et al. 2019). The canopy-forming *N. luetkeana* and *M. pyrifera*, in particular, have declined widely, prompting concern from conservationists, coastal communities, and resource managers over the loss of habitat and ecosystem services this represents (Starko et al. 2022; Mora-Soto et al. 2024). In response, several organizations in the Pacific Canada region and the United States involved in phycological research and mariculture have started culturing gametophytes from various kelp species to preserve living samples of genetic diversity from populations at risk of local extinction. In doing so, they have created their own respective "seed banks" or "biobanks" for archiving specific strains or culture lines that can be subsequently withdrawn and amplified for applications in restoration, mariculture, and research (Barrento et al. 2016; Wade et al. 2020). Recent findings (Bemmels et al. 2025) confirmed that in at least two species, *N. luetkeana* and *M. pyrifera*, population genetic structure in this region is localized at scales of tens to hundreds of kilometers, indicating the importance of preserving at-risk genotypes with potential resilience to climate warming (Wernberg et al. 2016; Starko et al. 2022).

While these organizations are diverse and varied, many of their reasons for biobanking kelp gametophytes overlap, and they can be broadly categorized into either (1) non-profit organizations, (2) university labs and educational institutes, or (3) mariculture companies. Disclaimer: This is not a comprehensive list but a selection of key agencies due to space constraints.

24.2.3 Non-profit Organizations

24.2.3.1 Pacific Canada

Two non-profit organizations operate biobanks that are among the most extensive in the region for the number of separately cultured genotypes they hold. While overlapping in the species and regions they cover, each has developed its biobank and uses it for specific ends in research and restoration, along with conserving genotypes from populations of interest.

The Kelp Rescue Initiative (KRI), based at the non-profit Bamfield Marine Sciences Centre in traditional Huu-ay-aht territory, formed in response to declines of canopy kelps following "The Blob" marine heatwave. Founded in 2021 by Dr. Chris Neufeld with start-up funding from the Ngan-Page Foundation, KRI is supported today primarily by a grant from Fisheries and Oceans Canada. They began by culturing gametophyte cultures of *N. luetkeana* and *M. pyrifera* locally in the Barkley Sound region of western Vancouver Island, then expanded to include cultures of *N. luetkeana* collected across the northern and central Salish Sea as part of an experiment on local adaptation and population-specific performance (Dykman et al. n.d.). At its outset, little was understood of these species' population structure in the region or the role that local genotypes may play in resilience to climate-related stressors and future persistence, so KRI maintained biobanked gametophyte lines derived from each of the wild sporophytes it collected. Most of these were held in small tubes with many mixed gametophytes, and in a subset of individuals, three to five gametophytes of each sex were isolated and cultured separately, should any single-spore clonal culture be needed for future applications. More recently, Bemmels et al. (2025) found that some of the populations from which these cultures were derived do indeed differ in their genetic structure and indices of genetic health. While experiments testing the phenotypic relevance of these genotypes would be necessary to understand their significance to conservation and restoration, these findings highlight the potential importance of preserving genotypes that are likely to experience near-term temperatures higher than they can survive (Fales et al. 2023). As of 2025, KRI had bio-banked cultures derived from over 200 individual wild bull kelp and over 60 individual wild giant kelp, including from populations that had experienced recession and then recovery from "The Blob" and populations that formed new range edges following die-offs in other areas during the marine heatwave.

Another non-profit organization biobanking kelps in the region is the Hakai Institute. Their mesocosm laboratory on Quadra Island, on Wei Wai Kai and Wei Wai Kum territory, opened in 2018 and has maintained a growing number of gametophyte cultures. They research a breadth of topics related to kelp ecology and restoration and have fully genotyped their germplasm biobank to facilitate experiments linking genotype to performance metrics and for assessing impacts on genetic diversity.

24.2.3.2 Pacific United States

Puget Sound Restoration Fund (PSRF) (Coast Salish Territory, Washington, USA) and Kelp Ark (traditional Tongva Territory, California, USA) are two active non-profit organizations that focus on restoration and conservation, respectively. PSRF is in Puget Sound and mostly focused on local bull kelp recovery. They have a conservation hatchery, located at NOAA's Manchester Research Station, where they house a bull kelp seed bank in collaboration with the Alberto lab at University of Wisconsin-Milwaukee (UWM). The seedbank contains 29 bull kelp collections from the Southern Salish Sea. They regularly collaborate with other research institutions and government agencies to implement and carry out conservation efforts for not only macroalgae but also other marine species as well.

Kelp Ark is located at AltaSea in the Port of Los Angeles and is an initiative dedicated to the conservation of kelp canopy cover along the US Pacific coastline. They currently hold over 2600 germplasm from 12 different brown macroalgal species, similarly in collaboration with Dr. Alberto at UWM, as well as the Nuzhdin Lab at the University of Southern California, Dan Reed at the University of California, Santa Barbara, and the Lewis Lab at Wheaton College. In addition to biobanking, Kelp Ark is also interested in supporting mariculture and is involved in research to better understand kelp biology, resilience, and microbiome influences on gametophyte viability.

24.2.4 Universities and Educational Institutes

24.2.4.1 Pacific Canada

As hubs of scientific research, several universities in the region contain groups that focus on biobanking kelps along with their research. One of these is led by Dr. Sherryl Bisgrove at Simon Fraser University, on the territories of several Coast Salish peoples. This lab has developed a cryopreservation technique for storage of gametophytes and is focused on biobanking bull kelp populations.

Nearby the University of British Columbia, also on the territories of several Coast Salish peoples, the research lab led by Dr. Patrick Martone maintains gametophyte cultures from collections they have performed for laboratory and field experiments. These primarily consist of *N. luetkeana* from populations across southern British Columbia, as well as localized populations of *A. marginata*, three-ribbed kelp (*Cymathaere triplicata*), Gardner's kelp (*Pleurophycus gardneri*), sea cabbage (*Hedophyllum sessile*), and five-ribbed kelp (*Costaria costata*). The research from which these collections began is expansive and varied and has included experiments in gametophyte and sporophyte physiology, novel culturing techniques, and applied restoration in the field (Supratya and Martone 2024).

Along with the university groups, one college in the region is biobanking kelps as well. With four campuses and a learning center spread across a combined 35 First Nations territories, North Island College's applied research department, CARTI, provides training and research support to a variety of mariculture initiatives on BC's coast. They are banking kelp gametophyte cultures and are researching how to scale biobanked cultures to industrial volumes with seaweed bioreactors. They are also supporting the development of a regional decentralized seaweed biobanking network in BC.

24.2.4.2 Pacific United States

Research on kelp in the U.S. is advancing through various academic labs dedicated to understanding genetic diversity, sustainable aquaculture, and the ecological roles of microorganisms in host health. We will highlight some of the research labs that conduct macroalgae research and biobank macroalgae. The Alberto lab at the UWM has isolated distinct gametophyte stocks and collaborated with Kelp Ark as well as PSRF. The lab's research focuses on understanding the genetic diversity of macroalgae, including exploring life history strategies and dispersal mechanisms. Another academic research lab exploring genetic diversity and genotype–phenotype interactions in macroalgae is the Nuzhdin lab located at the University of Southern California (traditional Tongva land), where they are actively breeding improved non-GMO kelp varieties and establishing cultivation practices with diverse gametophyte stocks to advance commercial production and growth of brown macroalgae. Similarly, the Lewis Lab at Wheaton College also focuses on kelp growth and physiology while maintaining stocks of select brown macroalgae collected from various coastal regions throughout the USA. Lastly, at the University of California, San Diego, Scripps Institution of Oceanography (on traditional Kumeyaay land), the Smith lab is working to characterize the genetic diversity and thermal tolerance of giant kelp and other understory kelp species in collaboration with the Braybrook lab at UCLA and the Michael lab at Salk Institute. The Smith lab currently maintains biobanks of kelp as well as other red seaweed taxa.

24.2.5 Mariculture Companies

24.2.5.1 Pacific Canada

Likely the first and longest-standing mariculture company to undertake biobanking in the region was Canadian Kelp Resources (CKR), a small business based in traditional Huu-ay-aht territory in the town of Bamfield, in western Vancouver Island. Another small business based in Bamfield is West Coast Kelp, which opened in 2022 and produces kelp seed, along with operating an ocean-based kelp farm. Both companies have collected and stored gametophyte cultures representing populations of several kelp species from throughout Pacific Canada. Vital Kelp Inc. is another small company that works with conservation groups, indigenous and government partners to provide kelp cultivation and gametophyte biobanking services for research, aquaculture, and restoration activities on the Sunshine Coast, BC. Vital Kelp primarily maintains *N. luetkeana* cultures, as well as some *A. marginata*, from which they seed spools for ocean-based kelp farms and bottom-attached substrates for restoration projects. The largest mariculture company in the region biobanking kelps is Cascadia Seaweed, established in 2019 and based in Sidney in southeast Vancouver Island, on the traditional territories of the W̱SÁNEĆ First Nations. As of 2025, they had bio-banked cultures of sugar kelp (*S. latissima*), winged kelp (*A. marginata*), giant kelp (*M. pyrifera*), and bull kelp (*N. luetkeana*) from across Vancouver Island, the Prince Rupert area in northern British Columbia, and the indigenous territories therein. Cascadia Seaweed operates several ocean-based farms in the region, growing these species for commercial production and to trial the use of biobanks in direct gametophyte seeding and restoration.

These companies are biobanking gametophytes of several kelp species not only to conserve genotypes from wild populations at risk of declining with climate warming but also to use them in their farming and business. They are among the growing number of commercial kelp growers worldwide to seed spools for ocean-based farming with cultured gametophytes, rather than with spores released from reproductive sporophytes (Ebbing et al. 2021; Sæther et al. 2024). Such an approach offers greater flexibility when growing species with highly seasonal reproduction, potentially lowering overhead costs compared to collecting sporophytes from the wild or ocean-based farms, and greater control over which genotypes are grown (Ebbing et al. 2021; Supratya and Martone 2024).

24.2.5.2 Pacific United States

Although several mariculture companies on the Eastern Atlantic coast of the United States are involved in biobanking kelp germplasm (e.g., Greenwave, Springtide Seaweed, and Atlantic Sea Farms) along with their daily operations, there are no similar initiatives on the Pacific side, which is the central focus of our interest here.

24.2.5.3 Current Opportunities, Limitations, and Future Directions

Thanks to the efforts of these organizations, many genotypes from populations of kelp species at risk of local extinction are now being kept alive as gametophyte cultures. While this offers the promise of re-establishing such genotypes following any future extirpations through habitat restoration, scalable restoration has proven difficult for these species and in this region, and many of these groups are actively researching how to conduct restoration successfully. Similarly, some of these genotypes may show promise for "future-proofing" restored habitats, and further testing for phenotypic traits and performance may identify important cultures to maintain (Alsuwaiyan et al. 2021; Wood et al. 2024). A relatively small part of the Pacific Canadian region has had kelp populations sampled for gametophyte cultures, primarily around Vancouver Island and the southern mainland region. Due to its extensive inlets, fjords, and islands, British Columbia has a total coastline area longer than that of the United States' Pacific coast by an order of magnitude, with many enclosed waterways where reproductive isolation and genotypic diversity may be high (Howes 2001; Starko et al. 2019). As such, the genotypic diversity currently bio-banked for the region is likely a small fraction of that found in the wild, and expanding it to a wider geographical area may ensure potentially valuable genotypes are not lost (Bemmels et al. 2025). Broadening this approach would strengthen conservation efforts while also establishing a more robust genetic foundation for future ecological restoration and adaptation strategies along the Pacific coast, including U.S. coastlines, where biobanking local populations has so far been limited as well. Some international cooperation is in place to share and distribute kelp germplasm (Wade et al. 2020), but, similar to the organizations discussed in this article, most biobanks currently operate independently, maintaining their stocks with minimal redundancy or backup infrastructure. This lack of coordination increases vulnerability to equipment failures or culture losses. As a result, some biobanks could benefit from exploring collaborative opportunities to distribute duplicate cultures across multiple facilities, enhancing preservation efforts and reducing the risk of irreversible loss.

24.3 Biobanking Red Macroalgae

24.3.1 Ecological and Economic Importance of Red Macroalgae

Like brown macroalgae, red macroalgae (Phylum Rhodophyta) are key habitat formers in marine environments (Cotas et al. 2023). Some calcified forms even help construct reefs and facilitate settlement in a number of important invertebrate taxa. They are also economically significant, serving as food sources and supporting diverse industrial applications (Rawiwan et al. 2022; Carpena et al. 2022; Polat et al. 2023). However, with increasing pressures such as climate change, habitat loss, and pollution, red macroalgae biobanking has become more and more essential to maintain and preserve marine biodiversity and ensure the continued availability of their economic benefits (Brakel et al. 2021).

Red macroalgae significantly contribute to carbon sequestration and may help to mitigate climate change by absorbing atmospheric CO_2 (Ould and Caldwell 2022; Pessarrodona et al. 2022). For instance, *Gracilaria* species exhibit high carbon absorption capacity due to their rapid growth rates and efficient photosynthetic pathways, making them vital for potential blue carbon strategies (Yang et al. 2015). Additionally, many species such as *Gracilaria*, *Corallina*, and *Gelidium* enhance ecosystem stability by providing substrates and habitats for marine organisms, recycling nutrients critical to coastal ecosystem health, and acting as biofilters to improve water quality (Akioka et al. 1999; Johnston and Lipcius 2012; Indraswari and Adharini 2022). Historically, red macroalgae have been an important food source in human societies across the globe. Coastal indigenous groups have traditionally harvested them to provide an invaluable source of vitamins and minerals to coastal communities for centuries. Further, species in the genus *Porphyra* are among the most historically significant red macroalgae, cultivated for centuries in Asia and consumed globally as nori. Its high nutritional value, rich in protein, vitamins, and minerals, makes it a cornerstone of sustainable food production (Blouin et al. 2011). Beyond food, red macroalgae are economically valuable, as they support many commercial industries due to their production of the hydrocolloids agar and carrageenan. *Kappaphycus alvarezii*, *Chondrus crispus*, and *Eucheuma denticulatum* are key sources of carrageenan, used as emulsifiers and thickening agents in food, cosmetics, and pharmaceuticals, while *Betaphycus gelatinum* provides high-quality beta carrageenan for use in biomedicine, microbiology, and biotechnology (Naseri et al. 2020; Bahari et al. 2021; Rupert et al. 2022). The pharmaceutical potential of sulfated polysaccharides derived from *Gracilaria* species is evident through demonstrated antiviral and anticoagulant properties (Souza et al. 2012). Additionally, *Asparagopsis taxiformis* plays a critical role in methane reduction by producing halogenated compounds, including bromoform, which inhibits methane production in livestock digestion (Hargrave et al. 2024). These diverse applications emphasize the dual ecological and economic importance of red macroalgae and their potential for innovative uses in sustainable aquaculture in the future.

24.3.2 Red Macroalgae Biobanking, Cryopreservation, and *Ex-Situ* Cultivation

Given the ecological and economic significance of red macroalgae, biobanking is vital for preserving species and genetic diversity as well as ensuring they can be used in perpetuity for a variety of purposes. Live tissue preservation, particularly through liquid cultures, is a widely adopted method for keeping tissue alive for years. For instance, the seedbank of *A. taxiformis* keeps the live tissue of distinct strains in autoclaved seawater with nutrients, enabling the selection and preservation of high-performing strains (Fig. 24.3) (Dishon et al. 2023a; Hargrave et al. 2024; Resetarits et al. 2024). In the Pacific United States, a notable seedbank of *Asparagopsis* exists in the Smith laboratory at University of California, San Diego, Scripps Institution of Oceanography, where they use an ex-situ approach.

In red macroalgae, the innovation of cryopreservation technology has been an alternative to expanding biobanking capabilities. Cryopreservation of gametophytes and sporophytes has been optimized for *Porphyra* species, focusing on enhancing resilience to diseases and environmental stressors (Yang et al. 2021). It's also been used for species such as *Gracilaria* and *Chondrus*, minimizing ice crystal formation and preserving cellular integrity (Yang et al. 2021). Through cryopreservation methods, *Gracilaria tikvahiae* is embedded in alginate beads, enabling it to enter a dormant state with minimized metabolic activities. Upon rehydration, the seaweed can return to an active state, supporting long-term preservation and safeguarding both tissues and associated microbiomes for future use (van der Meer and Simpson 1984).

Beyond preserving phenotypic traits, maintaining genetic diversity is also important for red macroalgae biobanking. Advances in genomic technologies have enabled comprehensive genetic profiling, revealing genetically distinct strains and critical traits within species. Through genetic profiling, various strains of *A. taxiformis* have been distinguished globally, enabling targeted aquaculture efforts that select strains with optimal physiological traits for a particular region (Nahor et al. 2022; Patwary 2023; Dishon et al. 2023b; Hargrave et al. 2024; Resetarits et al. 2024). Studies on *Gracilariopsis lemaneiformis* (formerly *Gracilaria lemaneiformis*) have identified genetic factors for temperature tolerance, while *Chondrus crispus* genomic research has elucidated pathways for carrageenan synthesis (Fu et al. 2014; Lipinska et al. 2020). Similarly, *Pyropia yezoensis* has been found to harbor stress tolerance genes beneficial for cultivation under variable conditions (Sun et al. 2015). These insights inform breeding programs and the preservation of wild strains as genetic reservoirs for future applications.

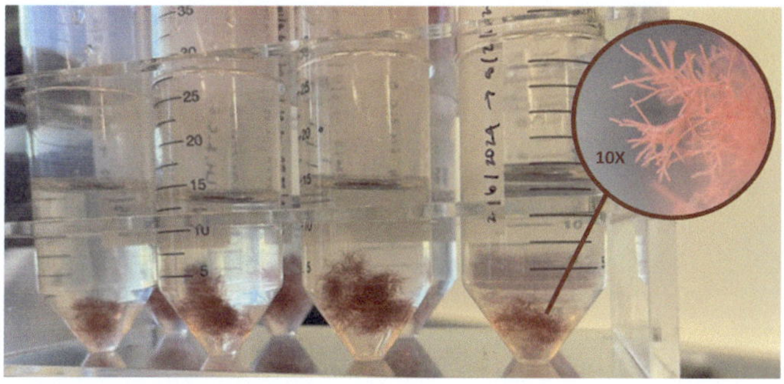

Fig. 24.3 Ex-situ seedbank of *Asparagopsis taxiformis*. *Asparagopsis* micrograph (right) at 10X magnification. (Courtesy of Isabella Wu)

Collaborative frameworks between research institutions and aquaculture industries further strengthen these efforts by developing innovative preservation protocols and sharing resources for large-scale restoration and sustainable use. Programs like the SeaStrains initiative (https://www.awi.de/en/science/special-groups/aquaculture/marine-aquaculture/research/translate-to-english-seastrains.html) and biobanks such as Culture Collection of Algae and Protozoa (CCAP) (http://www.ccap.ac.uk) exemplify progress in this field.

24.3.3 Biobanking Challenges and Opportunities for Red Macroalgae

Preserving red macroalgae presents unique challenges due to their complex life cycles and tissue fragility. Most red algae exhibit a triphasic life cycle. In biphasic life cycles, alternation occurs between haploid gametophytes and diploid sporophytes. The triphasic life cycle, unique to red algae, includes an additional carposporophyte stage, which is often surrounded by the female gametophyte and plays a crucial role in completing the reproductive process (Fig. 24.4). Tailored strategies are required to ensure the viability of all life stages, particularly given their reliance on non-motile gametes and dependence on water currents for fertilization (Mayanglambam and Sahoo 2015). Maintaining viable microbiomes also adds complexity to preservation efforts.

Meanwhile, biobanking efforts face legislative hurdles and permitting challenges in some locations, necessitating coordinated advocacy and policy to streamline these processes. These challenges often include navigating complex legal frameworks and obtaining necessary permits for collecting and preserving seaweed resources (Nakhate and van der Meer 2024). Additionally, biobanking raises questions about the intellectual property/ownership and equitable use of genetic resources, particularly for species collected in international waters. Therefore, ensuring compliance with the Nagoya Protocol on access and benefit-sharing is imperative at an international level. This framework promotes fair distribution of resources and benefits, fostering collaboration

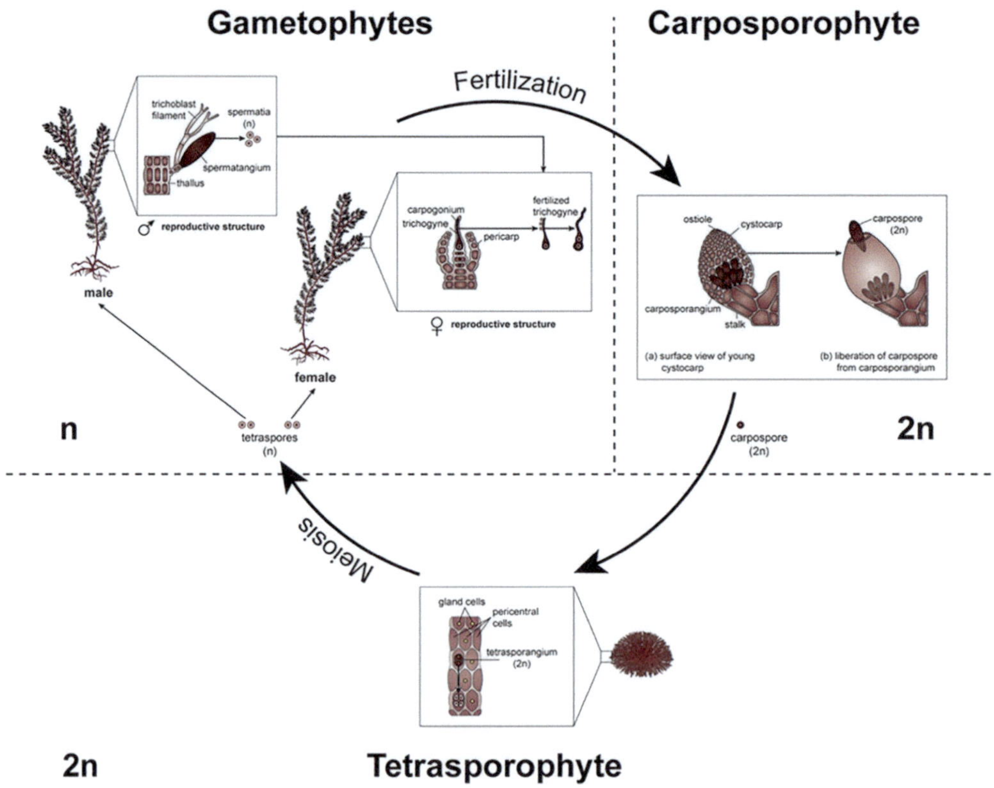

Fig. 24.4 *Asparagopsis taxiformis* life cycle. (Image by Dr. Adi Khen)

among nations and stakeholders while safeguarding the rights of indigenous communities and protecting biodiversity (Brakel et al. 2021).

Red macroalgae have immense potential for future commercialization across many industries and are an inherently important component of marine biodiversity and global sustainability. Future research should focus on developing advanced biobanking techniques, understanding red macroalgae-microbiome interactions, and leveraging biobanked material for restoration and synthetic biology. International collaboration will be essential for the establishment of robust biobanking systems, ensuring the conservation and sustainable use of red macroalgae resources in the future. By addressing current challenges and capitalizing on cutting-edge research, red macroalgae biobanking can deliver long-term benefits for both marine ecosystems and human societies.

24.4 Biobanking Microbes Associated with Macroalgae

24.4.1 Microbial Contributions to Macroalgae Health and Growth

Like plants and animals, macroalgae (brown, green, and red) rely on their associated microbiota for a variety of essential ecological roles (Egan et al. 2013; Apprill 2017; Sánchez-Cañizares et al. 2017; Saha et al. 2024), such as facilitating the uptake of key nutrients like iron and nitrogen (Kaur et al. 2023), producing antimicrobial and antifouling compounds (Busetti et al. 2017), synthesizing antifungals (Valverde et al. 2022), and contributing to the exchange of vitamins, micronutrients, heavy metal sequestration, and phytohormone precursor molecules (Wang et al. 2022; Weigel et al. 2022; Aguirre and Schwartzman 2024). Reciprocally, microorganisms benefit from a stable and consistent carbon source, which they obtain from the complex sugars secreted by their macroalgal host (Egan et al. 2013). Additionally, mutualisms may also be based on the exchange of other bioactive molecules. A classic example comes from (Provasoli and Pintner 1980) where different bacterial isolates influenced the morphology of *Ulva lactuca* into typical long and narrow thallus (foliose morphology) or clump (ball), an atypical morphology. Years later, the causative agent for these morphological differences was identified as "thallusin," produced by some bacterial strains, and is responsible for the development of the foliose morphology of *Gayralia oxysperma* (formerly *Monostroma oxyspermum*), *Ulva lactuca* (formerly *U pertusa*), and *U intestinalis* (formerly *Enteromorpha intestinalis*) (green algae, commonly known as sea lettuce) (Matsuo et al. 2005).

Nitrogen availability in these systems has long been suspected of playing a critical role in regulating kelp growth, yet research on the microbial processes involved in nitrogen cycling in algae remains limited. There is genetic evidence of ammonification genes in the kelps *M. pyrifera* and *N. luetkana* (Miranda et al. 2022; Hochroth and Pfister 2024). However, microbial interactions related to nitrogen transformation could involve more than just ammonium production. For example, iron-scavenging bacteria, which interact with nitrogen-cycling microbes, may influence both nitrogen and iron availability for the kelp host (Kolody et al. 2022). Because iron is a necessary cofactor for nitrogenase enzymes involved in nitrogen fixation, these bacteria could facilitate nitrogen fixation by increasing the iron accessible to nitrogen-cycling microbes, leading to a more abundant nitrogen supply that supports kelp growth and potentially enhances the production of Indole-3-acetic acid (IAA), an important hormone for plant development (Padhy and Pattnaik 1975). However, disruptions in these host–microbe interactions can intensify the effects of environmental stress, potentially leading to the decline or collapse of both farmed and wild kelp populations (Minich et al. 2018; Mancuso et al. 2023).

Although a disruption of the microbiome can have detrimental impacts on the host, this is not always the case, and it is important to investigate this in algae. In anemones (*Nematostella vectensis*), adding microbes from a warmer temperature increased the survival of anemones that were not acclimated to the warmer temperature, after controlling for genetics (Baldassarre et al. 2022). This effect shows that microbes play a significant role in host acclimation and survival at different temperatures. For algae, one of the main threats is consistent heat waves, brought on by climate change, so the possibility that microbes could increase algal tolerance of higher temperatures is an exciting avenue of future research.

Likewise, it is also noteworthy to acknowledge that in some seaweeds, a change in the microbiome may be less impactful on host health. Work by Schenk et al. (2023) found that although low-salinity stress led to changes in the bacterial community of *Saccharina latissima*, the stable (core) section of the microbiome persisted, suggesting that the changes observed in low-salinity stress are due to shifts in relative abundance and rare microbial taxa, rather than a complete disruption of the microbiome.

Considering this information, we believe that exploring how the native microbiota can be manipulated to enhance kelp growth, improve stress tolerance, and aid growth is a promising area of research. While recent studies have suggested that microbes can influence kelp health, growth, and transitions between life stages, the specific interactions between seaweed and their microbiota remain poorly understood. Biobanking strategies must therefore address the preservation of these symbiotic relationships.

24.4.2 Importance of Microbial Biobanking

Microbial biobanking advances our understanding of microbial diversity, pathogenesis, and the development of innovative therapies (De Paoli 2005). Microbial biobanks provide researchers with access to diverse microbial strains, facilitating the study of microbial behaviors and self or host interactions in a controlled environment (De Paoli 2008; Smith and Bussas 2021). This is important for multiple research fields, including the discovery of new microorganisms, metabolite production, and resistance mechanisms, all of which are becoming increasingly relevant amid global health challenges from shelf-stable food microbiomes to antimicrobial resistance (AMR) (De Vero et al. 2019; Nikoloudaki et al. 2024). Additionally, biobanks provide a vital resource for the development of diagnostic tools and therapeutic interventions for hosts, both in the marine and terrestrial environments (Betsou et al. 2011; Barzegari et al. 2014; Lee et al. 2016). By maintaining well-characterized strains and non-model microbes, researchers can identify biomarkers, track disease outbreaks, and monitor the evolution of pathogens (Day and Stacey 2008). Access to a comprehensive collection of bacterial strains can help elucidate genomic variations that may contribute to an organism's virulence or resistance to treatment (De Paoli 2008).

For marine biologists like us, microbial biobanks serve as valuable resources in environmental microbiology, aiding in the study of microbial ecology. By preserving microbial samples from various ecosystems, biobanks facilitate research into microbial functions within different aquatic habitats, such as lakes, estuaries, oceans, and animal/host microbiomes. This can inform sustainability efforts, including the development of bioremediation strategies or the identification of microbes with potential biotechnological applications. Access to microbe biobanks advances fundamental research while also contributing to public health and environmental sustainability, making it an invaluable tool in modern microbiology.

24.4.3 Current State of Macroalgae-Associated Microbial Biobanking

Microbial biobanking for seaweed-associated microbes remains underdeveloped compared to terrestrial biobanks (Chandel et al. 2021), despite increasing awareness of the ecological and biotechnological significance of marine microorganisms. To the best of our knowledge, beyond academic research labs at universities (e.g., the Parfrey Lab at the University of British Columbia, the Braybrook Lab at the

University of California, Los Angeles, the Weigel Lab at Oceans Stanford University Hopkins Marine Station, and the Nuzhdin Lab at the University of Southern California), the only organization outside of academia dedicated solely to biobanking macroalgae and its associated microbiota in Pacific North America is the non-profit kelp seedbank, Kelp Ark (kelpark.org), which has cultivated 120+ bacterial strains and fungi from kelp in partnership with the McCauley Lab (California State University, Dominguez Hills).

Researchers have previously focused on isolating and characterizing microbial communities associated with kelp (Lin et al. 2018; Weigel and Pfister 2021; Lemay et al. 2021; Park et al. 2023). However, the development of biobanks for kelp-associated microbial communities is still in its infancy, and standardization in collection methods, preservation techniques, and taxonomic characterization has not been widely implemented.

Various techniques have been employed to isolate and preserve marine microbes, each with its advantages and caveats, but the most common cultivation approach that balances throughput with the challenges of culturing is the dilution to extinction method (Zhang et al. 2021b). The defining factor in this culturing process lies in the cultivability of marine microorganisms, as many species are non-culturable under standard laboratory conditions, which limits the scope of microbial collections (Zhang et al. 2021b; Thrash 2019). The potential for incomplete or biased representation of the kelp microbiome is then present, and cultivation methods may not capture all microbial diversity, leading to an underrepresentation of certain beneficial microbes. Efforts to systematically biobank microbes could provide vital incentive and resources to isolate and cultivate these microbes. This could lead to the development of new microbe cultivation techniques, which would likely be employed by microbiologists interested in non-algal hosts (corals, sponges, anemones, etc.). In this sense, biobanking kelp-associated microbes reaches beyond kelp-focused interests and have potential to benefit broader research, conservation, and restoration efforts.

In response to this limitation, cryopreservation (low % glycerol or DMSO/glycerol mix) and lyophilization can be used, where the microbial fraction of a wild kelp can be preserved and later analyzed using molecular techniques, like polymerase chain reaction (PCR)-based methods. Microbial species can then be identified through their 16S rRNA gene profiles, providing insights into the diversity of non-culturable microbes, estimated to be ~80% (Rodrigues and de Carvalho 2022; Schultz et al. 2023). Preserving microbial communities in their natural habitat allows for the complexity and diversity of marine microbial ecosystems to be better maintained, without the need for culturing. It is important, however, to acknowledge that utilizing this approach can result in issues with replicability for future applications like restoration or farming.

24.4.4 Applications of Microbial Biobanking

Because marine microorganisms have developed specialized mechanisms to cope with salinity, temperature, and pressure extremes, conditions that are rarely encountered by terrestrial microbes, they have the ability to produce biochemicals that have unique structural features, making them excellent candidates for novel drug and compound discovery (Persson et al. 2011; Singh et al. 2015; Busetti et al. 2017). Kelp-associated microbes have shown promise in producing antimicrobial, anti-inflammatory, and antioxidative compounds, which could lead to the development of next-generation pharmaceuticals (Valliappan et al. 2014; Menaa et al. 2020; Tan 2023). Additionally, enzymes produced by these marine microbes have shown superior stability and functionality under harsh conditions, offering potential applications in industrial processes that require extreme temperature and chemistry, like bioconversion of complex sugars into usable ethanol as a biofuel (Zhang et al. 2019, 2021a). In fact, kelp-associated bacteria are known to target various polysaccharides in kelp, including alginate, cellulose, and fucoidan (Michel et al. 2006). Among these polysaccharides, alginate is the most abundant in *M. pyrifera* (McKee et al. 1992) and is identified as the primary target for degradation during the initial stages of the brown kelp (*Saccharina japonica*) decomposition by bacteria containing alginate lyases. Bioconversion of alginate (bioenergy) by alginate lyases of bacteria is an emerging field but has been hindered by a lack of (1) a standardized method of bioconversion, (2) functional studies in alginate degraders, and (3) available communities of alginate-degrading bacteria.

The novelty of marine microbes also extends to their potential applications in environmental sustainability and climate change mitigation (Hutchins et al. 2019; Dumka et al. 2020; Shi et al. 2022). For example, certain bacteria associated with kelp have been identified as capable of breaking down complex organic compounds, such as hydrocarbons, offering potential solutions for marine pollution, including petroleum and byproducts (Atlas 1981; Brooijmans et al. 2009; Xue et al. 2015). In addition, other seaweed-associated microbes have been metagenomically linked to heavy metal cycling (Aguirre and Schwartzman 2024; Mathew et al. 2025). These microbes could be essential for developing more efficient and sustainable bioremediation methods, a significant aspect of environmental protection amidst increasing anthropogenic pollution. Therefore, preserving them alongside kelps is vital.

24.4.5 Harnessing Symbiotic Relationships for Kelp Restoration

To utilize kelp-associated microbes to improve restoration and aquaculture outcomes, it is imperative to disentangle how bacterial inoculants will influence kelp development. Although microbial communities are diverse and dynamic, a bank of well-tested strains is a necessary element that will enable us to fully assess the feasibility of microbial approaches for improving industry and restoration outcomes. In this context, biobanking microbes is less focused on preventing extinction of specific strains (as seen in macroalgae biobanking) and leans more toward enabling the study of locally available microorganisms.

Due to the putatively influential roles of bacteria on seaweed hosts, we surmise that banked bacterial isolates can potentially be exploited to overcome common challenges in seaweed cultivation, including disease mitigation and growth rates. Recent work by Park et al. (2023) shows that adding select bacterial isolates to *Saccharina latissima* (sugar kelp) can shorten the time required for sporophytes to develop in culture (Fig. 24.5). In red macroalgae, the addition of cultured bacterial isolates can protect against bleaching disease (Li et al. 2022) and red rot disease (Weng et al. 2024) in red macroalgae. These studies highlight the influence of bacteria, augmenting the classical work in *Ulva* in morphological development (Provasoli and Pintner 1980) and in *Ectocarpus* on salinity tolerance (Dittami et al. 2016).

Critically, the ability of microbes to influence macroalgal health and growth may be functionally redundant since many cultured isolates induce similar phenotypes (Provasoli and Pintner 1980; Park et al. 2023). Simply put, a broad bank of bacterial strains collected from wild and cultivated seaweeds is highly likely to contain an isolate that impacts a key seaweed function. This may also be the case even when microbial isolates were not specifically collected to target specific phenotypes. This redundancy in functional traits suggests that suitable isolates can likely thrive across a range of abiotic/biotic conditions, from those in nurseries to both current and future ocean conditions, indicating higher adaptability when outplanted to the open ocean for cultivation or restoration.

The studies featured above highlight the roles bacteria have on algal development, which could also be beneficial to aquaculture and restoration efforts. Kelps grown in cultivation for restoration or farming likely have a distinct microbial community from wild-grown kelps. Preliminary explorations into this question have provided us insight into these differences between kelp populations (Park et al. 2023, Fig. 24.5). Although wild kelp populations have slightly different microbial communities, preprocessing steps conducted in cultivated kelps (cleaning the sorus patch and keeping hosts in media made from filtered seawater) reduce microbial load prior to releasing spores. This likely modifies the kelp-associated microbial community. Although reducing microbial load in cultivation is important, as a high microbial load can inhibit kelp gametophyte development (Morris et al. 2016), it likely alters the pool of available microbes to colonize the kelp during development while in cultivation.

Beyond potentially differing microbiomes of cultivated kelps (Fig. 24.5), the land-based nursery is a dynamically varying environment from the open ocean where wild kelps naturally develop. It is important to note that some microbes may be better suited to the nursery environment or to the open ocean. Ultimately, biobanking microbes allows kelp farmers and restoration practitioners access to a well-characterized pool of microbes, which enables optimal strain selection to use for their targeted environment (land or open ocean farms), application use, and kelp population type. Additionally, we recommend that biobanking efforts target both wild and cultivated kelp because cultivation seems to alter the microbial community of seaweeds. Collecting and preserving samples from diverse source populations increases the likelihood of capturing microbes suited to the different environmental conditions present in both nursery and wild settings.

24.5 Future of Biobanking Macroalgae and Microbes

As kelp biobanking grows, a more centralized database, similar to NCBI, GenBank, ENA, and ATCC, could make it easier for researchers to access and exchange strains. This system could not only streamline the process of obtaining strains for study but also create a repository that facilitates easier comparisons across studies, improving reproducibility and consistency in kelp research. However, a key challenge is the development and application of effective storage techniques, like cryopreservation, that can maintain genetic integrity over extended periods. Improving methods such as freezing and thawing techniques, evaluating the effects of cryoprotectants or media, and enhancing long-term storage conditions will be crucial in determining the success of biobanks that store high-quality, reproducible strains long-term. These improvements would not only make it easier to store and access kelp strains but also help safeguard genetic diversity while controlling for somatic mutations, crucial for future research.

In addition to storage and preservation advancements, there is a growing need for establishing community-driven guidelines and best practices. Currently, kelp biobanking lacks widely accepted best practices, which may lead to variability in experimental results between organizations and ineffective attempts at strain isolation. Research aimed at

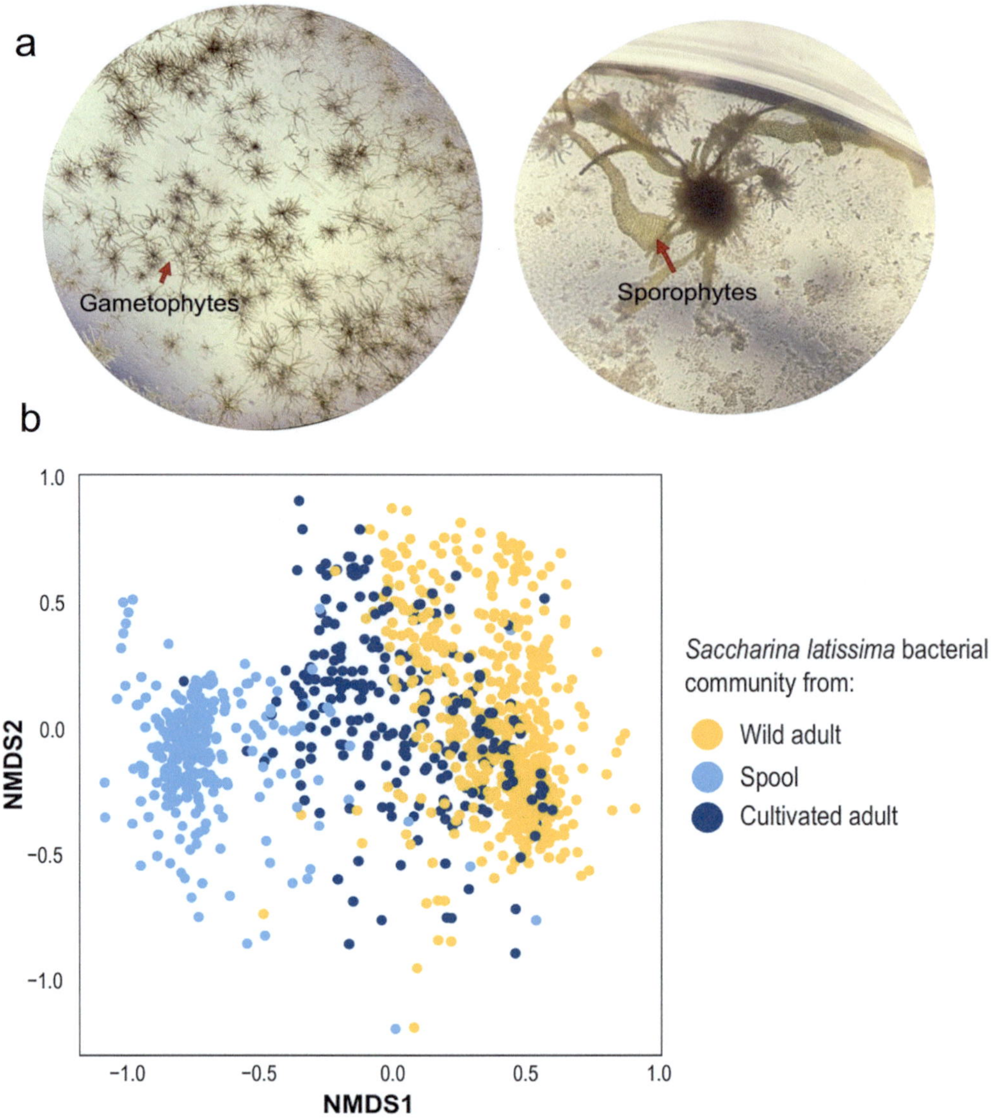

Fig. 24.5 (**a**) Kelp gametophytes with no bacterial inoculants (control) and with a bacterial inoculum (*Maribacter* sp.) at the same age. *Maribacter* sp. Decreases the time to sporophyte formation in lab culture. (**b**) NMDS plot (Bray-Curtis ordination) showing the bacterial community of *Saccharina latissima*. Yellow points are from wild-grown adult individuals from multiple studies (Schenk et al. 2023; Park et al. 2023). Blue points are from spools (juveniles in a kelp nursery) or adult individuals outplanted on an open-ocean farm (Schenk et al. 2023; Davis et al. 2023). PERMANOVA $p = 0.001$. All kelp types are significantly different ($p = 0.006$). This plot shows separation of the bacterial community by the environment where the kelps are reared across multiple studies, indicating that this is a consistent effect

developing universal methods for strain cultivation, sequencing, and processing will be vital for standardizing this field. Establishing community-driven guidelines will provide a framework for researchers around the globe to adopt common methodologies. Clear priorities for standardization should focus on optimizing growth/viability conditions, streamlining genetic analysis, and refining storage conditions (e.g., cryopreservation techniques, media considerations). Standardizing these processes will not only lead to facilitating collaborative efforts between researchers but also ensure that biobanks contain reliable, reproducible samples for supporting advancements in macroalgae environmental research, conservation, and biotechnology applications.

Glossary

Assisted Reproductive Technology (ART) Techniques used to achieve insemination by artificial or partially artificial means, applied here to marine species for conservation purposes.

Biobank Infrastructure that preserves biological materials (such as tissues, cells, or genetic material) for future research, conservation, and restoration purposes.

Blue Carbon Carbon captured and stored by coastal and marine ecosystems, particularly in vegetation like kelp forests, seagrass beds, and mangroves.

Cryoinjury Damage to cells caused by the formation of ice crystals during the freezing process in cryopreservation.

Cryopreservation The process of preserving biological material at ultra-low temperatures (typically −196 °C in liquid nitrogen) to maintain viability for long periods.

Cryoprotective Agents (CPAs) Chemicals used to protect cells from damage during the freezing process in cryopreservation.

Gametophyte In kelps and other algae, the sexual reproductive phase of the life cycle that produces gametes; typically microscopic in kelps.

Germplasm Living genetic resources such as seeds, tissues, or cells that can be used for breeding, conservation, or research.

Hydrocolloids Substances (like agars, alginates, and carrageenans) derived from algae that form gels or thickeners when combined with water, used in various industrial applications.

Mariculture The cultivation of marine organisms for food and other products in seawater, either in the open ocean, coastal waters, or in tanks.

Microbiome The community of microorganisms (bacteria, fungi, etc.) living in association with a host organism.

Phenotype The observable characteristics of an organism resulting from the interaction of its genetic makeup with the environment.

Regime Shift A sudden, persistent change in the structure and function of an ecosystem, such as the transformation of kelp forests to urchin barrens.

Seedbank In the context of kelp, a (biobank) facility that preserves gametophyte cultures for future use in restoration or research.

Sporophyte The asexual reproductive phase in the kelp life cycle; the large, visible kelp fronds commonly seen in kelp forests.

Sorus A reproductive structure on kelp fronds that produces and releases spores.

Strain A genetic variant or subtype of a particular species, often with distinct characteristics.

Vitrification A cryopreservation technique where biological samples are transformed into a glass-like state rather than forming ice crystals during freezing.

Appendix

Macroalgae and Microbial Biobanks Mentioned in This Chapter

Pacific Canada

Non-Profit Organizations:

- Kelp Rescue Initiative (https://kelprescue.org/)
- Hakai Institute (https://hakai.org/)

Universities and Educational Institutes:

- Simon Fraser University- Bisgrove lab (https://www.sfu.ca/biology/people/faculty/sbisgrov.html)
- University of British Columbia—Laura W. Parfrey lab (https://www.zoology.ubc.ca/~parfrey/parfrey_lab/)
- University of British Columbia—Patrick Martone lab (https://www3.botany.ubc.ca/martone/)
- North Island College- Resource Trade Programs (https://www.nic.bc.ca/programs/trades-apprenticeship-and-technical/resource-trades/)

Mariculture Companies:

- Canadian Kelp Resources (https://canadiankelp.com/)
- West Coast Kelp (https://westcoastkelp.ca/)
- Cascadia Seaweed (https://www.cascadiaseaweed.com/)
- Vital Kelp (contact Lee-Ann Ennis, vitalkelp@gmail.com)

Pacific United States

Non-profit Organizations:

- Puget Sound Restoration Fund (https://restorationfund.org/)
- Kelp Ark, funded by the Grantham Foundation (https://kelpark.org/)
- Salk Institute for Biological Sciences—Michael Lab

Universities:

- University of Wisconsin-Milwaukee—Alberto lab (https://alberto-lab.blogspot.com/)
- California State University Dominguez Hills—McCauley Lab (http://www.mccauleylab.com/)
- University of Southern California- Nuzhdin Lab (https://dornsife.usc.edu/nlab/)
- University of California, Santa Barbara—Dan Reed (https://msi.ucsb.edu/people/principal-investigators/dan-reed)
- University of California, Los Angeles—Braybrook Lab (https://sites.lifesci.ucla.edu/mcdb-braybrook)
- Stanford University Hopkins Marine Station—Weigel Lab (https://weigel-lab.stanford.edu/)
- Wheaton College—Lewis Lab (https://www.wheaton.edu/academics/faculty/raymond-lewis/)
- Scripps Institution of Oceanography—Smith Lab (https://seaweedecologylab.ucsd.edu/)

Eastern U.S. Mariculture Companies:

- Greenwave (https://www.greenwave.org/)
- Springtide Seaweed (https://www.springtideseaweed.com/)
- Atlantic Sea Farms (https://atlanticseafarms.com/)

International Collections

- Culture Collection of Algae and Protozoa (https://www.ccap.ac.uk/)
- SeaStrains Initiative (https://www.awi.de/en/science/special-groups/aquaculture/marine-aquaculture/research/translate-to-english-seastrains.html)

Acknowledgements We would like to acknowledge Dr. Kelly DeWeese, Dr. Adi Khen, Isabella Wu, and Michael Marty-Rivera for their contributions to the figures in this chapter.

Author Contributions We used the international standard CRediT taxonomy (where applicable) to determine author contributions as follows: Conceptualization—EGA. Data Curation—SCS, JMS, LJMC, SS, LZ, and EGA. Project Administration—EGA. Visualization—SS, LZ, and EGA. Writing, original draft—SCS, JMS, LJMC, SS, LZ, and EGA. Writing, review, and editing—SCS, JMS, LJMC, SS, LZ, LWP, JS, and EGA. Authors SS, SCS, JMS, LZ, LJMC, and EGA contributed equally to the development of this chapter.

Competing Interests The authors are affiliated with various kelp seedbanks, both in the non-profit and academic sectors: SCS for KRI, JMS for KRI, LJMC for PSF/SFU, SS for UBC Parfrey Lab, LWP for UBC Parfrey Lab, LZ for SIO Smith Lab, JS for Smith Lab, and EGA for KA. However, the authors declare no competing interests, as this chapter was assembled independently, and no specific organization influenced the review and assessments presented in this manuscript.

Funding No funding was received to assist with the preparation of this manuscript.

References

Aguirre EG, Schwartzman JA (2024) Metagenome-assembled genomes of -associated bacteria. Microbiol Resour Announce 13:e0071524

Akioka H, Baba M, Masaki T, William Johansen H (1999) Rocky shore turfs dominated by *Corallina* (Corallinales, Rhodophyta) in northern Japan. Phycol Res 47:199–206

Alongi DM (2014) Carbon cycling and storage in mangrove forests. Annu Rev Mar Sci 6:195–219

Alsuwaiyan NA, Vranken S, Filbee-Dexter K et al. (2021) Genotypic variation in response to extreme events may facilitate kelp adaptation under future climates. Mar Ecol Prog Ser. https://doi.org/10.3354/meps13802

Apprill A (2017) Marine animal microbiomes: toward understanding host–microbiome interactions in a changing ocean. Front Mar Sci 4. https://doi.org/10.3389/fmars.2017.00222

Atlas RM (1981) Microbial degradation of petroleum hydrocarbons: an environmental perspective. Microbiol Rev 45:180–209

Bahari A, Moelants K, Wallecan J et al. (2021) Understanding the effect of time, temperature and salts on carrageenan extraction from *Chondrus crispus*. Algal Res 58:102371

Baldassarre L, Ying H, Reitzel AM et al. (2022) Microbiota mediated plasticity promotes thermal adaptation in the sea anemone *Nematostella vectensis*. Nat Commun 13:3804

Barrento S, Camus C, Sousa-Pinto I, Buschmann AH (2016) Germplasm banking of the giant kelp: our biological insurance in a changing environment. Algal Res 13:134–140

Barzegari A, Saeedi N, Saei AA (2014) Shrinkage of the human core microbiome and a proposal for launching microbiome biobanks. Future Microbiol 9:639–656

Bemmels JB, Starko S, Weigel BL et al. (2025) Population genomics reveals strong impacts of genetic drift without purging and guides conservation of bull and giant kelp. Curr Biol 35:688–698.e8

Betsou F, Parida SK, Guillerm M (2011) Infectious diseases biobanking as a catalyst towards personalized medicine: *Mycobacterium tuberculosis* paradigm. Tuberculosis (Edinb) 91:524–532

Blouin NA, Brodie JA, Grossman AC et al. (2011) *Porphyra*: a marine crop shaped by stress. Trends Plant Sci 16:29–37

Bold HC, Wynne MJ (1985) Introduction to the algae: structure and reproduction. Prentice Hall

Bolton RL, Mooney A, Pettit MT et al. (2022) Resurrecting biodiversity: advanced assisted reproductive technologies and biobanking. Reprod Fertil 3:R121–R146

Brakel J, Sibonga RC, Dumilag RV et al. (2021) Exploring, harnessing and conserving marine genetic resources towards a sustainable seaweed aquaculture. Plants People Planet 3:337–349

Brandl SJ, Lefcheck JS, Bates AE et al. (2023) Can metabolic traits explain animal community assembly and functioning? Biol Rev Camb Philos Soc 98:1–18

Brooijmans RJW, Pastink MI, Siezen RJ (2009) Hydrocarbon-degrading bacteria: the oil-spill clean-up crew. Microb Biotechnol 2:587–594

Busetti A, Maggs CA, Gilmore BF (2017) Marine macroalgae and their associated microbiomes as a source of antimicrobial chemical diversity. Eur J Phycol 52:452–465

Carpena M, Garcia-Perez P, Garcia-Oliveira P et al. (2022) Biological properties and potential of compounds extracted from red seaweeds. Phytochem Rev:1–32

Cebrian J (2002) Variability and control of carbon consumption, export, and accumulation in marine communities. Limnol Oceanogr 47:11–22

Chandel A, Mann R, Kaur J et al. (2021) Implications of seed vault storage seed vault storage strategies for conservation of seed bacterial seed bacterial microbiomes. Front Microbiol 12:784796

Cotas J, Gomes L, Pacheco D, Pereira L (2023) Ecosystem services provided by seaweeds. Hydrobiology 2:75–96

Day JG, Stacey GN (2008) Biobanking. Mol Biotechnol 40:202–213

Day JG, Harding KC, Nadarajan J, Benson EE (2008) Cryopreservation. In: Springer protocols handbooks. Humana Press, Totowa, pp 917–947

Davis KM, Zeinert L, Byrne A, Davis J, Roemer C, Wright M, Parfrey LW (2023) Successional dynamics of the cultivated kelp microbiome. J Phycol 59(3):538–551

De Paoli P (2005) Bio-banking in microbiology: from sample collection to epidemiology, diagnosis and research. FEMS Microbiol Rev 29:897–910

De Paoli P (2008) Future of biobanking in microbiology for medical research. Future Microbiol 3:79–86

De Vero L, Boniotti MB, Budroni M et al. (2019) Preservation, characterization and exploitation of microbial biodiversity: the perspective of the Italian Network of Culture Collections. Microorganisms 7. https://doi.org/10.3390/microorganisms7120685

Dishon G, Resetarits HM, Tsai B et al. (2023a) Image-based analysis and quantification of biofouling in cultures of the red alga *Asparagopsis taxiformis*. J Appl Phycol 35:209–218

Dishon G, Resetarits HM, Tsai B et al. (2023b) The effect of light intensity, spectrum, and photoperiod on the physiological performance of *Asparagopsis taxiformis* tetrasporophytes. Algal Res 76:103304

Dittami SM, Duboscq-Bidot L, Perennou M et al. (2016) Host-microbe interactions as a driver of acclimation to salinity gradients in brown algal cultures. ISME J 10:51–63

Druehl L, Clarkston B (2016) Pacific seaweeds, Updated and Expanded edn. Harbour Publishing

Duarte CM, Losada IJ, Hendriks IE et al. (2013) The role of coastal plant communities for climate change mitigation and adaptation. Nat Clim Chang 3:961–968

Duarte CM, Gattuso J-P, Hancke K et al. (2022) Global estimates of the extent and production of macroalgal forests. Glob Ecol Biogeogr 31:1422–1439

Dumka S, Gupta C, Kamboj A (2020) Utilizing bacteria to mitigate global climate change: a review. Int J Curr Microbiol Appl Sci 9:1385–1394

Dykman, Lauren N, Steell Clay S, Zielinski Amanda, Zielinski Rob, Wiebe Alexandra, Csordas Matthew, Baum Julia K, Neufeld Christopher J. (n.d.) "Evidence of local adaptation to extreme warming in bull kelp (*Nereocystis luetkeana*) with applications to restoration." In Review

Ebbing APJ, Fivash GS, Pierik R et al. (2021) The SeaCoRe system for large scale kelp aquaculture: a plug-and-play, compatible, open-source system for the propagation and transport of clonal gametophyte cultures. J Appl Phycol. https://doi.org/10.1007/s10811-021-02638-2

Egan S, Harder T, Burke C et al. (2013) The seaweed holobiont: understanding seaweed-bacteria interactions. FEMS Microbiol Rev 37:462–476

Eger AM, Marzinelli EM, Christie H et al. (2022) Global kelp forest restoration: past lessons, present status, and future directions. Biol Rev Camb Philos Soc 97:1449–1475

Eger AM, Marzinelli EM, Beas-Luna R et al. (2023) The value of ecosystem services in global marine kelp forests. Nat Commun 14:1894

Fales RJ, Weigel BL, Carrington E et al. (2023) Interactive effects of temperature and nitrogen on the physiology of kelps (*Nereocystis luetkeana* and *Saccharina latissima*). Front Mar Sci 10. https://doi.org/10.3389/fmars.2023.1281104

Filbee-Dexter K, Scheibling RE (2014) Sea urchin barrens as alternative stable states of collapsed kelp ecosystems. Mar Ecol Prog Ser 495:1–25

Filbee-Dexter K, Wernberg T (2018) Rise of turfs: a new battlefront for globally declining kelp forests. Bioscience 68:64–76

Filbee-Dexter K, Wernberg T (2020) Substantial blue carbon in overlooked Australian kelp forests. Sci Rep 10:12341

Fragkopoulou E, Serrão EA, De Clerck O et al. (2022) Global biodiversity patterns of marine forests of brown macroalgae. Glob Ecol Biogeogr 31:636–648

Fu F, Sui Z-H, Zhou W et al. (2014) UV-irradiation mutation of tetraspores of *Gracilariopsis lemaneiformis* and screening of thermotolerant strains. J Appl Phycol 26:647–656

Guillard RRL (1975) Culture of phytoplankton for feeding marine invertebrates. In: Culture of marine invertebrate animals. Springer US, Boston, pp 29–60

Guillard RRL, Hargraves PE (1993) *Stichochrysis immobilis* is a diatom, not a chrysophyte. Phycologia 32:234–236

Hallegraeff GM, Anderson DM, Cembella AD (2004) Manual on harmful marine microalgae. In: IOC Manuals and Guides, 33

Hargrave MS, Islam S, Agarwal V, Smith JE (2024) The influence of light stress on bromoform synthesis and concentration in the red seaweed *Asparagopsis taxiformis*. J Appl Phycol 36:321–329

Hastings A, Byers JE, Crooks JA et al. (2007) Ecosystem engineering in space and time. Ecol Lett 10:153–164

Hochroth A, Pfister CA (2024) Ammonification by kelp associated microbes increases ammonium availability. PLoS One 19:e0296622

Howard JG, Lynch C, Santymire RM et al. (2016) Recovery of gene diversity using long-term cryopreserved spermatozoa and artificial insemination in the endangered black-footed ferret. Anim Conserv 19:102–111

Howes D (2001) BC biophysical shore-zone mapping system—A systematic approach to characterize coastal habitats in the Pacific Northwest. In: Proceedings of the 2001 Puget Sound Research Conference, Seattle, Washington, 11p

Hutchins DA, Jansson JK, Remais JV et al. (2019) Climate change microbiology - problems and perspectives. Nat Rev Microbiol 17:391–396

Indraswari N, Adharini R (2022) Reducing nitrate and phosphate concentrations in shrimp wastewater aquaculture using *Gelidium corneum*. Ecol Eng Environ Technol 23:137–144

Jaiswal AN, Vagga A (2022) Cryopreservation: a review article. Cureus 14:e31564

Johnson TR, Wilson JA, Cleaver C et al. (2013) Modeling fine scale urchin and kelp dynamics: implications for management of the Maine Sea urchin fishery. Fish Res 141:107–117

Johnston CA, Lipcius RN (2012) Exotic macroalga *Gracilaria vermiculophylla* provides superior nursery habitat for native blue crab in Chesapeake Bay. Mar Ecol Prog Ser 467:137–146

Jones CG, Lawton JH, Shachak M (1994) Organisms as ecosystem engineers. Oikos 69:373

Kaur M, Saini KC, Mallick A, Bast F (2023) Seaweed-associated epiphytic bacteria: diversity, ecological and economic implications. Aquat Bot 189:103698

Kolody BC, Smith SR, Zeigler Allen L et al. (2022) Nitrogen and Iron availability drive metabolic iron availability drive metabolic remodeling and natural natural selection of diverse diverse phytoplankton during experimental upwelling. mSystems 7:e0072922

Kono S, Kuwano K, Saga N (1998) Cryopreservation of *Eisenia bicylis*, (Laminariales, Phaeophyta) in liquid nitrogen. J Mar Biotechnol 6:220–223

Krumhansl KA, Okamoto DK, Rassweiler A et al. (2016) Global patterns of kelp forest change over the past half-century. Proc Natl Acad Sci USA 113:13785–13790

Kuwano K, Kono S, Jo Y-H et al (2004) Cryopreservation of the gametophytic cells of Laminariales (Phaeophyta) in liquid nitrogen. J Phycol 40:606–610

Lakeman MB, von Dassow P, Cattolico RA (2009) The strain concept in phytoplankton ecology. Harmful Algae 8:746–758

Lee S, Jung PE, Lee Y (2016) Publicly-funded biobanks and networks in East Asia. Springer Plus 5:1080

Lemay MA, Davis KM, Martone PT, Parfrey LW (2021) Kelp-associated microbiota are structured by host anatomy. J Phycol 57:1119–1130

Li J, Majzoub ME, Marzinelli EM et al. (2022) Bacterial controlled mitigation of dysbiosis in a seaweed disease. ISME J 16:378–387

Lin JD, Lemay MA, Parfrey LW (2018) Diverse bacteria utilize alginate within the microbiome of the Giant Kelp. Front Microbiol 9:1914

Lipinska AP, Collén J, Krueger-Hadfield SA et al. (2020) To gel or not to gel: differential expression of carrageenan-related genes between the gametophyte and tetasporophyte life cycle stages of the red alga *Chondrus crispus*. Sci Rep 10:11498

Mancuso FP, Morrissey KL, De Clerck O, Airoldi L (2023) Warming and nutrient enrichment can trigger seaweed loss by dysregulation of the microbiome structure and predicted function. Sci Total Environ 879:162919

Mathew DE, Soni A, Dhimmar A et al. (2025) Characterization, bioprospection, and comparative metagenomics of bacterial communities revealing the predictive functionalities in wild and cultured samples of industrially important red seaweed *Gracilaria dura*. Curr Microbiol 82:85

Matsuo Y, Imagawa H, Nishizawa M, Shizuri Y (2005) Isolation of an algal morphogenesis inducer from a marine bacterium. Science 307:1598

Mayanglambam A, Sahoo D (2015) Red algae. In: The algae world. Springer, Dordrecht, pp 205–234

McGann LE, Yang HY, Walterson M (1988) Manifestations of cell damage after freezing and thawing. Cryobiology 25:178–185

McKee JWA, Kavalieris L, Brasch DJ et al. (1992) Alginate content and composition of *Macrocystis pyrifera* from New Zealand. J Appl Phycol 4:357–369

Mcleod E, Chmura GL, Bouillon S et al. (2011) A blueprint for blue carbon: toward an improved understanding of the role of vegetated coastal habitats in sequestering CO_2. Front Ecol Environ 9:552–560

Menaa F, Wijesinghe PAUI, Thiripuranathar G et al. (2020) Ecological and industrial implications of dynamic seaweed-associated microbiota interactions. Mar Drugs 18. https://doi.org/10.3390/md18120641

Michel G, Nyval-Collen P, Barbeyron T et al. (2006) Bioconversion of red seaweed galactans: a focus on bacterial agarases and carrageenases. Appl Microbiol Biotechnol 71:23–33

Minich JJ, Morris MM, Brown M et al. (2018) Elevated temperature drives kelp microbiome dysbiosis, while elevated carbon dioxide induces water microbiome disruption. PLoS One 13:e0192772

Miranda K, Weigel BL, Fogarty EC et al. (2022) The diversity and functional capacity of Microbes microbes associated with coastal macrophytes. mSystems 7:e0059222

Mora-Soto A, Schroeder S, Gendall L et al. (2024) Back to the past: long-term persistence of bull kelp forests in the strait of Georgia, Salish Sea, Canada. Front Mar Sci 11. https://doi.org/10.3389/fmars.2024.1446380

Morris MM, Haggerty JM, Papudeshi BN et al. (2016) Nearshore pelagic microbial community abundance affects recruitment success of giant kelp. Front Microbiol 7:1800

Nahor O, Luzzatto-Knaan T, Israel Á (2022) A new genetic lineage of *Asparagopsis taxiformis* (Rhodophyta) in the Mediterranean Sea: as the DNA barcoding indicates a recent Lessepsian introduction. Front Mar Sci 9. https://doi.org/10.3389/fmars.2022.873817

Nakhate P, van der Meer Y (2024) Global initiatives, future challenges, and trends for the wider use of seaweed. In: Seaweeds and seaweed-derived compounds. Springer International Publishing, Cham, pp 521–544

Nanba N, Fujiwara T, Kuwano K et al. (2009) Effect of pre-incubation irradiance on survival of cryopreserved gametophytes of *Undaria pinnatifida* (Phaeophyta) and morphology of sporophytes formed from the gametophytes. Aquat Bot 90:101–104

Naseri A, Jacobsen C, Sejberg JJP et al. (2020) Multi-extraction and quality of protein and carrageenan from commercial Spinosum. Foods 9. https://doi.org/10.3390/foods9081072

Nikoloudaki O, Aheto F, Di Cagno R, Gobbetti M (2024) Synthetic microbial communities: a gateway to understanding resistance, resilience, and functionality in spontaneously fermented food microbiomes. Food Res Int 192:114780

Ould E, Caldwell GS (2022, 2022) The potential of seaweed for carbon capture. CAB Rev Perspect Agric Vet Sci Nutr Nat Resour. https://doi.org/10.1079/cabireviews202217009

Paddack MJ, Estes JA (2000) Kelp forest fish populations in marine reserves and adjacent exploited areas of Central California. Ecol Appl 10:855–870

Padhy SN, Pattnaik H (1975) Effect of IAA on growth and nitrogen fixation of *Westiellopsis prolifica* Janet. Z Pflanzenphysiol 75:460–463

Park J, Schenk S, Davis K et al. (2023) Exploring the impact of microbial manipulation on the early development of kelp (*Saccharina latissima*) using an ecological core microbiome framework. bioRxiv

Patwary ZP (2023) Multi-omics investigation across the life-history stages of the red seaweed, *Asparagopsis taxiformis*. https://www.researchgate.net/publication/368145336_A_Proteomic_Analysis_for_the_Red_Seaweed_Asparagopsis_taxiformis

Persson F, Svensson R, Nylund GM et al. (2011) Ecological role of a seaweed secondary metabolite for a colonizing bacterial community. Biofouling 27:579–588

Pessarrodona A, Filbee-Dexter K, Krumhansl KA et al (2022) A global dataset of seaweed net primary productivity. Sci Data 9:484

Piel MI, Avila M, Alcapán A (2015) Criopreservación de estadios iniciales de gametofitos de *Macrocystis pyrifera* (Laminariales, Ochrophyta) en condiciones controladas de laboratorio. Rev Biol Mar Oceanogr 50:157–162

Polat S, Trif M, Rusu A et al. (2023) Recent advances in industrial applications of seaweeds. Crit Rev Food Sci Nutr 63:4979–5008

Provasoli L, Pintner IJ (1980) Bacteria induced polymorphism in an axenic laboratory strain of *Ulva lactuca* (Chlorophycease). J Phycol 16:196–201

Rawiwan P, Peng Y, Paramayuda IGPB, Quek SY (2022) Red seaweed: a promising alternative protein source for global food sustainability. Trends Food Sci Technol 123:37–56

Rennick M, DiFiore BP, Curtis J et al. (2022) Detrital supply suppresses deforestation to maintain healthy kelp forest ecosystems. Ecology 103:e3673

Resetarits HM, Dishon G, Agarwal V, Smith JE (2024) The effects of temperature and CO enrichment on the red seaweed *Asparagopsis taxiformis* from Southern California with implications for aquaculture. J Phycol 60:1567–1584

Riesco MF, Robles V (2013) Cryopreservation causes genetic and epigenetic changes in Zebrafish Genital Ridges. PLoS One 8:e67614

Rodrigues CJC, de Carvalho CCCR (2022) Cultivating marine bacteria under laboratory conditions: overcoming the "unculturable" dogma. Front Bioeng Biotechnol 10:964589

Rupert R, Rodrigues KF, Thien VY, Yong WTL (2022) Carrageenan from (Rhodophyta, Solieriaceae): metabolism, structure, production, and application. Front Plant Sci 13:859635

Sæther M, Diehl N, Monteiro C et al. (2024) The sugar kelp *Saccharina latissima* II: recent advances in farming and applications. J Appl Phycol 36:1953–1985

Saha M, Dittami SM, Chan CX et al (2024) Progress and future directions for seaweed holobiont research. New Phytol 244:364–376

Sakanishi Y, Saga N (1994) Survival of female gametophytic cells of *Laminaria diabolica* Miyabe (Phaeophyta) in liquid nitrogen. Fish Sci 60:623–624

Sánchez-Cañizares C, Jorrín B, Poole PS, Tkacz A (2017) Understanding the holobiont: the interdependence of plants and their microbiome. Curr Opin Microbiol 38:188–196

Scheibling RE, Hennigar AW, Balch T (1999) Destructive grazing, epiphytism, and disease: the dynamics of sea urchin - kelp interactions in Nova Scotia. Can J Fish Aquat Sci 56:2300–2314

Schenk S, Wardrop CG, Parfrey L (2023) Low salinity destabilizes the bacterial community of sugar kelp, *Saccharina latissima* (Phaeophyceae). bioRxiv

Schultz J, Modolon F, Peixoto RS, Rosado AS (2023) Shedding light on the composition of extreme microbial dark matter: alternative approaches for culturing extremophiles. Front Microbiol 14:1167718

Schuster JM, Kurt Gamperl A, Gagnon P, Bates AE (2022a) Distinct realized physiologies in green sea urchin (*Strongylocentrotus droebachiensis*) populations from barren and kelp habitats. Facets (Ott) 7:822–842

Schuster JM, Stuart-Smith RD, Edgar GJ et al. (2022b) Tropicalization of temperate reef fish communities facilitated by urchin grazing and diversity of thermal affinities. Glob Ecol Biogeogr 31:995–1005

Shi P, Wang H, Feng M et al. (2022) The coupling response between different bacterial metabolic functions in water and sediment improve the ability to mitigate climate change. Water (Basel) 14:1203

Singh RP, Kumari P, Reddy CRK (2015) Antimicrobial compounds from seaweeds-associated bacteria and fungi. Appl Microbiol Biotechnol 99:1571–1586

Smale DA, Burrows MT, Moore P et al. (2013) Threats and knowledge gaps for ecosystem services provided by kelp forests: a Northeast Atlantic perspective. Ecol Evol 3:4016–4038

Smith D, Bussas V (2021) Preserving the reference strains. In: Trends in the systematics of bacteria and fungi. CABI, pp 55–68
Souza BWS, Cerqueira MA, Bourbon AI et al. (2012) Chemical characterization and antioxidant activity of sulfated polysaccharide from the red seaweed *Gracilaria birdiae*. Food Hydrocoll 27:287–292
Starko S, Bailey LA, Creviston E et al. (2019) Environmental heterogeneity mediates scale-dependent declines in kelp diversity on intertidal rocky shores. PLoS One 14:e0213191
Starko S, Wilkinson DP, Bringloe TT (2021) Recent global model underestimates the true extent of Arctic kelp habitat. Biol Conserv 257:109082
Starko S, Neufeld CJ, Gendall L et al. (2022) Microclimate predicts kelp forest extinction in the face of direct and indirect marine heatwave effects. Ecol Appl 32:e2673
Sun P, Mao Y, Li G et al. (2015) Comparative transcriptome profiling of *Pyropia yezoensis* (Ueda) M.S. Hwang & H.G. Choi in response to temperature stresses. BMC Genomics v:463
Supratya VP, Martone PT (2024) Kelps on demand: closed-system protocols for culturing large bull kelp sporophytes for research and restoration. J Phycol 60:73–82
Tan LT (2023) Impact of marine chemical ecology research on the discovery and development of new pharmaceuticals. Mar Drugs 21:174
Taylor R, Fletcher RL (1998) Cryopreservation of eukaryotic algae – a review of methodologies. J Appl Phycol 10:481–501
Thomson JA, Burkholder DA, Heithaus MR et al. (2015) Extreme temperatures, foundation species, and abrupt ecosystem change: an example from an iconic seagrass ecosystem. Glob Chang Biol 21:1463–1474
Thrash JC (2019) Culturing the uncultured: risk versus reward. mSystems 4. https://doi.org/10.1128/mSystems.00130-19
Valliappan K, Sun W, Li Z (2014) Marine actinobacteria associated with marine organisms and their potentials in producing pharmaceutical natural products. Appl Microbiol Biotechnol 98:7365–7377
Valverde S, Williams PL, Mayans B et al. (2022) Comparative study of the chemical composition and antifungal activity of commercial brown seaweed extracts. Front Plant Sci 13:1017925
van der Meer JP, Simpson FJ (1984) Cryopreservation of *Gracilaria tikvahiae* (Rhodophyta) and other macrophytic marine algae. Phycologia 23:195–202
Visch W, Rad-Menéndez C, Nylund GM et al. (2019) Underpinning the development of Seaweed biotechnology: cryopreservation of brown algae (gametophytes). Biopreserv Biobank 17:378–386
Wade R, Augyte S, Harden M et al. (2020) Macroalgal germplasm banking for conservation, food security, and industry. PLoS Biol 18:e3000641
Wang B, Zhang E, Gu Y et al. (2011) Cryopreservation of brown algae gametophytes of *Undaria pinnatifida* by encapsulation–vitrification. Aquaculture 317:89–93
Wang J, Tang X, Mo Z, Mao Y (2022) Metagenome-assembled genomes from microbiome provide insights into the potential metabolic potential to the seaweed. Front Microbiol 13:857901
Weigel BL, Pfister CA (2021) Oxygen metabolism shapes microbial settlement on photosynthetic kelp blades compared to artificial kelp substrates. Environ Microbiol Rep 13:176–184
Weigel BL, Miranda KK, Fogarty EC et al. (2022) Functional insights into the kelp microbiome from metagenome-assembled genomes. mSystems 7:e0142221
Weng P, Yang H, Mo Z et al. (2024) Application and evaluation of probiotics against red rot disease in *Pyropia*. Aquaculture 578:740050
Wernberg T, Bennett S, Babcock RC et al. (2016) Climate-driven regime shift of a temperate marine ecosystem. Science 353:169–172
Wood G, Steinberg PD, Campbell AH et al. (2022) Host genetics, phenotype and geography structure the microbiome of a foundational seaweed. Mol Ecol 31:2189–2206
Wood GV, Filbee-Dexter K, Coleman MA et al. (2024) Upscaling marine forest restoration: challenges, solutions and recommendations from the Green Gravel Action Group. Front Mar Sci 11. https://doi.org/10.3389/fmars.2024.1364263
Xue J, Yu Y, Bai Y et al. (2015) Marine Oil-Degrading oil-degrading microorganisms and biodegradation process of petroleum hydrocarbon in marine environments: a review. Curr Microbiol 71:220–228
Yang Y, Chai Z, Wang Q et al. (2015) Cultivation of seaweed *Gracilaria* in Chinese coastal waters and its contribution to environmental improvements. Algal Res 9:236–244
Yang H, Huo Y, Yee JC, Yarish C (2021) Germplasm cryopreservation of macroalgae for aquaculture breeding and natural resource conservation: a review. Aquaculture 544:737037
Zhang QS, Cong YZ, Qu SC et al. (2007a) A simple and highly efficient method for the cryopreservation of *Laminaria japonica* (Phaeophyceae) germplasm. Eur J Phycol 42:209–213
Zhang QS, Cong YZ, Qu SC et al. (2007b) Cryopreservation of gametophytes of *Laminaria japonica* (Phaeophyta) with two-step cooling: interactions between variables related to post-thaw survival. Cryo Letters 28:217–224
Zhang Q, Cong Y, Qu S et al. (2008) Cryopreservation of gametophytes of *Laminaria japonica* (Phaeophyta) using encapsulation-dehydration with two-step cooling method. J Ocean Univ China 7:65–71
Zhang F, Zhang W, Qian D-K et al. (2019) Synergetic alginate conversion by a microbial consortium of hydrolytic bacteria and methanogens. Water Res 163:114892
Zhang L, Li X, Zhang X et al. (2021a) Bacterial alginate metabolism: an important pathway for bioconversion of brown algae. Biotechnol Biofuels 14:158
Zhang X-H, Ahmad W, Zhu X-Y et al. (2021b) Viable but nonculturable bacteria and their resuscitation: implications for cultivating uncultured marine microorganisms. Mar Life Sci Technol 3:189–203

Siobhan Schenk is a recent Ph.D. graduate from the a University of British Columbia. She studies the bacterial community of kelps in the context of climate change, disease, and kelp cultivation. She also studies the macroalgal biodiversity in Vancouver to increase the amount of long-term biodiversity data available locally. She holds an M.Sc. in Cell Biology from McGill, where she studied microtubule dynamics to answer basic science questions about how human diseases (neurodegeneration and cancer) affect cell cycle dynamics.

S. Clay Steell is the founder of Blue Futures Kelp, a restoration initiative, based in British Columbia. He formerly served as Restoration and Culturing Technician for the Kelp Rescue Initiative (KRI) based at the Bamfield Marine Sciences Centre, a non-profit organization located in the traditional territory of the Huu-ay-aht First Nations in western Vancouver Island. He holds an M.Sc. from Carleton University, where he studied fish and crustacean ecophysiology, and a B.Sc. from Mount Allison University, where he studied the reproductive ecology of intertidal seaweeds.

Dr. Jasmin M. Schuster is a marine ecologist interested in temperate reef ecosystems and how marine communities therein respond to habitat and environmental change, with a particular focus on kelp and urchin dynamics. In her current work, Jasmin leads kelp research and restoration projects across Vancouver Island. She is currently the program manager for the Kelp Rescue Initiative (KRI), a non-profit organization based at the Bamfield Marine Sciences Centre, located in the traditional territory of the Huu-ay-aht First Nations in western Vancouver Island. She holds a Ph.D. in marine biology from Memorial University of Newfoundland and Labrador and an M.Sc. from the University of Southampton.

Dr. Lingjie Zhou is a phycologist specializing in both micro- and macroalgae. Her research focuses on using molecular biology techniques to investigate the physiological traits, biogeochemical roles, and genetic diversity of algae. Currently, she is a postdoctoral researcher in Dr. Jennifer Smith's lab at the Scripps Institution of Oceanography, where she studies the genetic diversity, genomics, and associated microbiome of the red alga *Asparagopsis taxiformis*. She holds a Ph.D. in Oceanography, an M.S. in marine science, and a B.S. in Bioengineering.

Dr. Liam Coleman is an organismal biologist interested in various aspects of kelp biology, especially growth and morphological development. He is currently a staff biologist at the Pacific Salmon Foundation, where he is working on projects to support salmon recovery through kelp forest conservation. He holds a B.Sc. in biology from the University of Victoria and a Ph.D. in Botany from the University of British Columbia.

 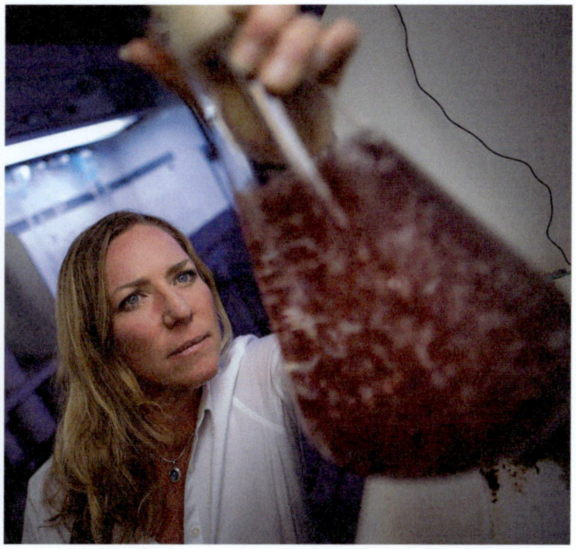

Dr. Jennifer Smith is a Professor in marine ecology and conservation in the Center for Marine Biodiversity and Conservation at Scripps Institution of Oceanography, UC San Diego. Dr. Smith's lab studies human impacts on marine ecosystems from coral reefs to kelp forests. Her research often goes beyond basic ecology by integrating conservation, restoration, management, and sustainability. While much of the Smith Lab is focused on addressing pressing issues in marine ecology, lab members are increasingly interested in seafood sustainability, coastal aquaculture, and habitat restoration. Dr. Smith and her team are largely focused on the study of seaweeds because of their immense potential to improve global sustainability and contribute to a carbon neutral future.

Research Interests

- Phycology, the Study of Seaweeds
- Coral Reef, Intertidal, and Kelp Forest Ecology
- Conservation Biology
- Aquaculture
- Human Impacts on Marine Communities
- Global Change
- Ocean Solutions, Sustainability
- Blue Carbon

Dr. Laura W. Parfrey is a biologist studying microbial ecology and evolution. Her lab at the University of British Columbia works to understand how symbiosis with kelp and other macroalgae shapes the ecology and evolution of microbes and how these relationships change in changing environmental conditions and aims to use this knowledge to develop microbiome-based strategies to enhance the productivity and sustainability of macroalgal aquaculture. She holds a Ph.D. in organismal and evolutionary biology from the University of Massachusetts and B.S. in Biology from SUNY at Albany.

Dr. Emily G. Aguirre is an interdisciplinary scientist interested in developing tractable systems to study symbiotic associations between hosts and their microbes. Now, she is passionate about engineering symbiotic interactions between bacteria and hosts to enhance health and fitness in kelp ecosystems and aid conservation and aquaculture efforts. She is currently serves as the Research Lead at Kelp Ark and adjunct faculty at Santa Monica College, where she teaches aquaculture microbiology. She holds a Ph.D. in Biology, a B.S. in. Microbiology, and an A.A.S. in Chemical Technology.

Opportunities, Challenges and Future Directions for Restoring Kelp Forests Using Green Gravel

25

Hannah S. Earp, Georgina V. Wood, Karen Filbee-Dexter, and Thomas Wernberg

Abstract

Extending along almost one-third of the world's coastlines, kelp forests rank among the most diverse and productive ecosystems on Earth, supporting a wealth of ecosystem goods and services. Despite their importance, kelp forests are declining in many regions, fuelling interest in restoration as a tool to mitigate losses. Green gravel approaches—broadly defined as techniques involving ex situ seeding/rearing of juvenile kelp on substrates that can be out-planted at restoration sites—offer an efficient and scalable means of restoring kelp forests in some areas and have potential applications for other marine ecosystems facing similar pressures. This chapter synthesizes recent research on the application and efficacy of green gravel approaches, highlighting opportunities and challenges for their implementation. Key considerations include characterizing and addressing environmental conditions at restoration sites, identifying genetically appropriate and resilient source material, establishing robust metrics for evaluating restoration success, working within regulatory frameworks, and expanding the scope, scale and funding of restoration efforts. The importance of multidisciplinary collaboration and technological advances is also emphasized. Sustained financial support is critical to restoration efforts going forward, with promising opportunities emerging from sectors within the expanding blue economy alongside novel funding schemes. Most importantly, despite its potential, restoration alone cannot reinstate kelp forests, as many regions of kelp loss are facing prolonged and intensifying stressors. Instead, a multifaceted approach that combines local restoration with proactive management/conservation, while also addressing broader climate change and other stressors is essential. Only in this way can we halt kelp forest declines and secure their associated services into the future.

Keywords

Macroalgae · Seaweed · Transplant · Seeding · Afforestation · Recovery · Resilience · Ecology · Future-proof · Blue economy

25.1 Kelp Forests in the Anthropocene

25.1.1 What Is Kelp and Where Is it Found?

'Kelp' is a non-taxonomic term for large, brown, habitat-forming macroalgae, often used specifically in reference to species within the order Laminariales (Fraser 2012; Bolton 2016; Wernberg and Filbee-Dexter 2019). Laminarian kelp exhibits a diversity of growth forms, with some possessing a single stipe and blade (e.g. some Laminariaceae) while others have multiple stipes (e.g. Lessoniaceae), wing-like blades along their stipe (e.g. Alariaceae), or gas-filled pneumatocysts (e.g. some Laminariaceae). Most kelp species are perennial, but some are annual; however, all exhibit a heteromorphic diplohaplontic lifecycle with seasonal reproduction (Fig. 25.1) that often corresponds to shifts in environmental conditions (e.g. temperature changes and light availability) (Kain 1975; Mohring et al. 2013; United Nations Environment Programme 2023). Reproduction begins with the formation of reproductive tissues (sporangia) that release zoospores which disperse and settle on the bottom before developing

H. S. Earp (✉)
Institute of Marine Research, His, Norway
e-mail: hannah.earp@hi.no

G. V. Wood
UWA Ocean Institute and School of Biological Sciences, University of Western Australia, Crawley, WA, Australia

Flinders University, Adelaide, SA, Australia

K. Filbee-Dexter · T. Wernberg
Institute of Marine Research, His, Norway

UWA Ocean Institute and School of Biological Sciences, University of Western Australia, Crawley, WA, Australia

A. T. Critchley et al. (eds.), *Advances in Temperate Phyconomy: Algal Harvest and Cultivation in Globally Distributed Temperate Waters*, Developments in Applied Phycology 15, https://doi.org/10.1007/978-3-032-02955-3_25

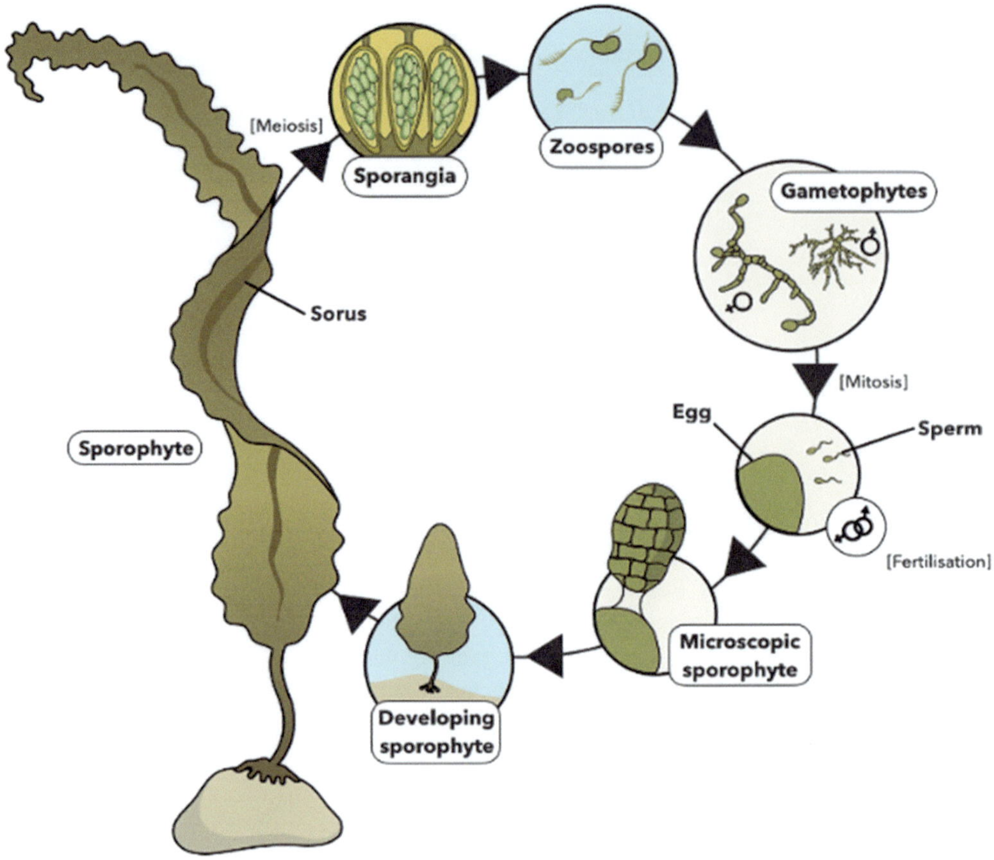

Fig. 25.1 The heteromorphic diplohaplontic lifecycle of Laminarian kelp. (Image credit: Antoine Minne)

into male or female gametophytes (Fig. 25.1). The gametophytes reproduce sexually, releasing gametes that bind together and develop into a juvenile sporophyte that matures and completes the cycle (Fig. 25.1). Only the grown sporophyte is macroscopic.

Kelp are found along approximately one-third of the world's coastlines, dominating rocky reefs in temperate and sub-polar regions where environmental conditions including temperature, wave exposure, light and water quality favour their growth and reproduction (Fig. 25.2) (Wernberg et al. 2019; Jayathilakea and Costello 2021; Fragkopoulou et al. 2022). In these areas, many kelp species form forests that represent one of the most diverse and productive habitats on Earth and the largest marine biome, covering an area five times greater than coral reefs (Steneck et al. 2002; Wernberg and Filbee-Dexter 2019; United Nations Environment Programme 2023).

25.1.2 Why Is Kelp Important?

Kelp forests provide a wealth of ecologically, socially and economically valuable ecosystem goods and services (Smale et al. 2013; Bennett et al. 2016; United Nations Environment Programme 2023) (Fig. 25.3). As foundation species, kelp form a complex three-dimensional biogenic habitat that modifies local environmental conditions including light availability, nutrient retention, flow dynamics, sediment accumulation, physical disturbance and water chemistry (Eckman et al. 1989; Wernberg et al. 2005; Irving and Connell 2006; Kim et al. 2015; Krause-Jensen et al. 2016). As such, diverse and abundant assemblages of macroflora and macrofauna, including species of commercial importance, are often found associated with kelp holdfasts, stipes and blades, as well as within epilithic understorey macroalgal assemblages (Christie et al. 2003; Anderson et al. 2005, 2006; Blight and Thompson 2008; Tuya et al. 2011; Leclerc et al. 2015; Arnold et al. 2016; Schaal et al. 2016; Teagle et al. 2017, 2018; Bué et al. 2020; Smale et al. 2020; King et al. 2021; Smale et al. 2022; Jackson-Bué et al. 2023; Earp et al. 2024a, b). For example, kelp forests are known to be crucial habitats for the European and American lobsters (*Homarus gammarus* and *H. americanus*) that collectively support fisheries valued at US$ 690 million per annum (Davies and Wootton 2018; Waller et al. 2021).

The primary productivity of kelp forests rivals that of other marine ecosystems (Mann 1979; Gao and McKinley 1994; Pessarrodona et al. 2022), but only a small proportion of this material is consumed directly, with the majority lost as detritus (Krumhansl and Scheibling 2012). Kelp detritus is a key component of coastal food webs, but in some cases, it is transported to deep ocean regions or sequestered into

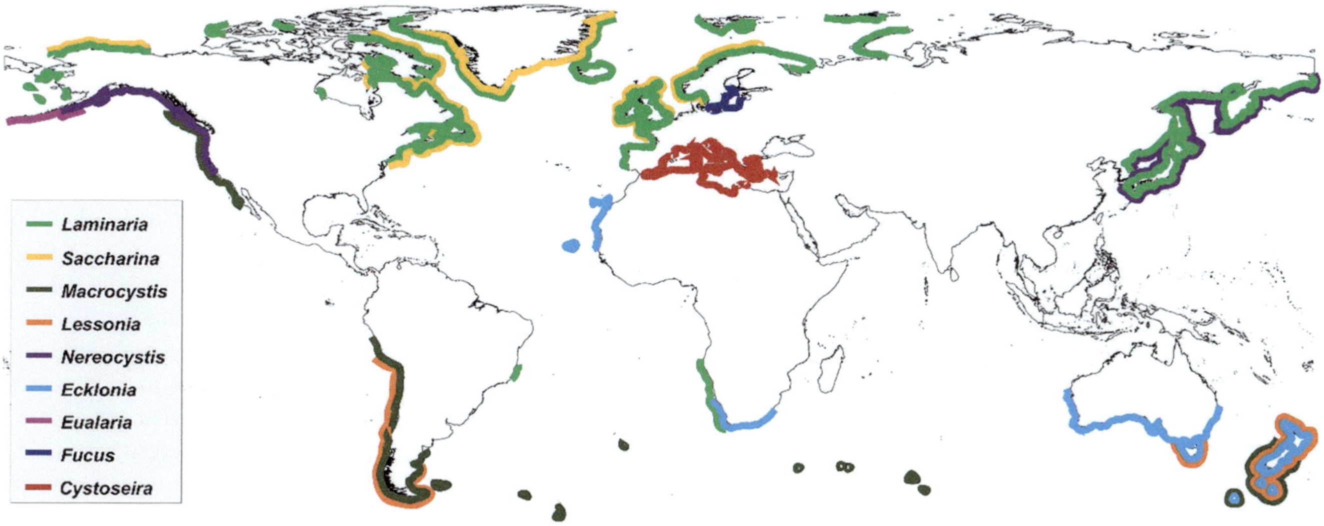

Fig. 25.2 Global distribution of marine forests and the dominant genera in each region. Note: Laminarian genera *Alaria*, *Agarum*, *Eisenia*, and *Undaria* are not shown (see UNEP 2023 for further information); *Fucus* and *Cystoseira* are forest-forming Fucalean genera. (Adapted from Wernberg et al. 2019)

Fig. 25.3 Ecosystem services provided by kelp forests. (Adapted from Filbee-Dexter et al. 2022)

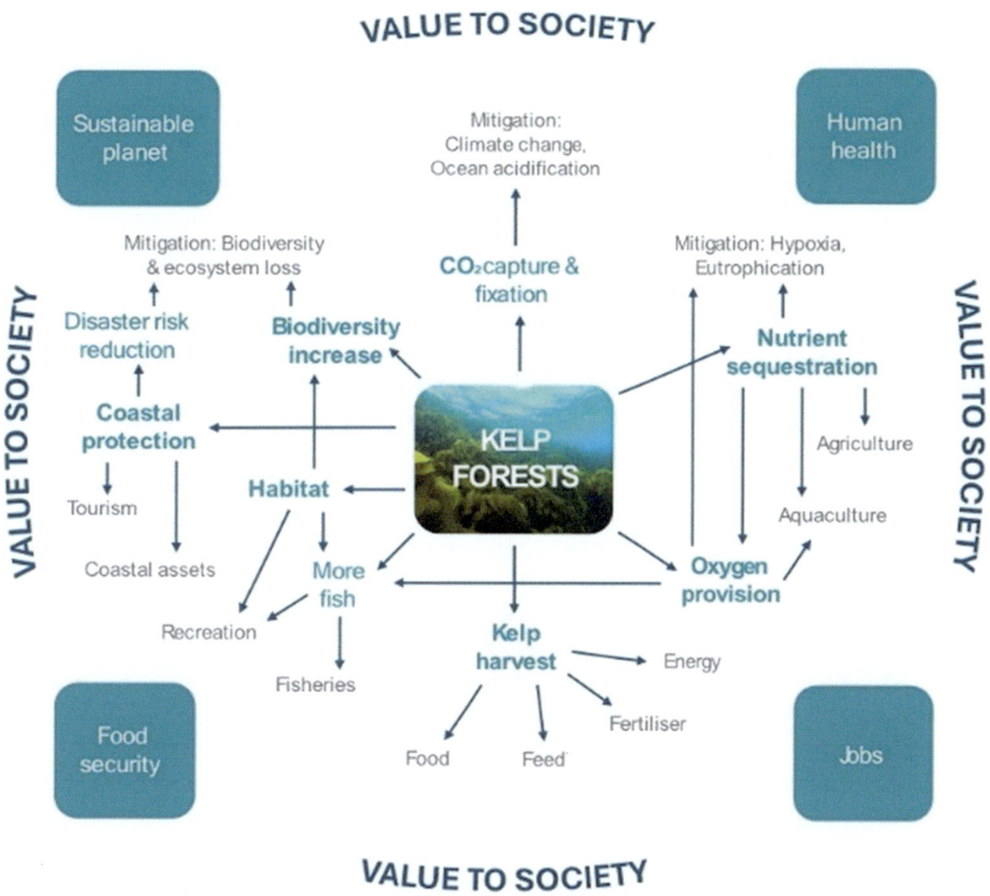

ocean sediment, forming a carbon sink (Krause-Jensen and Duarte 2016; Filbee-Dexter et al. 2024a). The role of kelp in cycling carbon and other nutrients such as nitrogen and phosphorous is also of significant economic value, particularly in areas with excess coastal run-off (Bennett et al. 2016; Bayley et al. 2021). Owing to its high levels of primary productivity, kelp is also harvested, both commercially and artisanally, for purposes including human and animal consumption, as a fertilizer, or for extracting compounds for applications such as pharmaceuticals. In Chile alone, over

300,000 dry tonnes of wild kelp, which represents ~40% of the global brown seaweed harvest, are landed annually and are valued at over US$ 70 million per annum (Vásquez et al. 2014; Lotze et al. 2019; Gouraguine et al. 2021).

Kelp forests also have the potential to dampen or attenuate waves, shielding coastlines from erosion, although the degree of protection provided is believed to be less than other coastal ecosystems and depends on an array of factors including the physical structure and health of the kelp forest, as well as abiotic factors such as depth and topography (Narayan et al. 2016). As such, some kelp forests have a negligible effect on wave attenuation, while others may reduce hydrodynamic forces by up to 60% (Mork 1996; Morris et al. 2020a).

Kelp-dominated areas support an array of tourism and recreational activities particularly fishing, SCUBA-diving and wildlife watching. Recreational fishing along Australia's Great Southern Reef has an estimated value of ~US$ 700 million per year, while ecotourism associated with South Africa's kelp forests may be worth over US$ 90 million per year (Bennett et al. 2016; Blamey and Bolton 2018; Eger et al. 2022a). Beyond their economic benefits, kelp forests hold intrinsic cultural and scientific value, providing inspiration for art, a source of information for science, and enhancing human wellbeing (Wernberg et al. 2019; United Nations Environment Programme 2023).

25.1.3 Kelp Forest Trends and Threats

Over the past 50 years, the rate of kelp forest loss has accelerated, with declines occurring in up to 60% of ecoregions for which long-term data are available, and at present, over 10,000 square kilometres of kelp forests are considered to be in a degraded state (Krumhansl et al. 2016; Wernberg et al. 2019; Wood et al. 2024a). This degradation and/or loss has been attributed to the combined effects of global, regional and local scale stressors (Wernberg et al. 2019, 2024). Global climate change, particularly ocean warming and marine heatwaves, can elevate temperatures beyond the thermal tolerance threshold of kelp (Smale and Wernberg 2013; Wernberg et al. 2013; Arafeh-Dalmau et al. 2019; Thomsen et al. 2019; Filbee-Dexter et al. 2020; Smith et al. 2024), alongside facilitating the range-expansion tropical herbivores (Vergés et al. 2016; Wernberg et al. 2016), non-native and invasive species (Saier and Chapman 2004; Scheibling and Gagnon 2009; Smale et al. 2015; Arnold et al. 2016; O'Brien and Scheibling 2018) and potentially diseases (Dhir and Graham 2022). For example, increases in the frequency and intensity of marine heatwaves have driven declines in *Saccharina latissima* in southern Norway and along the east coast of the USA (Filbee-Dexter et al. 2020). While in Western Australia, the 2010–11 marine heatwave that followed four decades of ocean warming caused extensive kelp forest decline, with losses persisting due to the tropicalization of fish communities that graze on kelp recruits and inhibit recovery (Wernberg et al. 2013, 2016; Bennett et al. 2015). Increases in the frequency and duration of storm events can also cause shifts in kelp forest structure and functioning (Byrnes et al. 2011; Filbee-Dexter and Scheibling 2012; Earp et al. 2024b).

At local scales, pollution, sediment-laden run-off and industrial effluents can have negative impacts on seawater quality, resulting in phenomena such as coastal darkening and eutrophication that can be detrimental to kelp growth and productivity (Coleman et al. 2008; Diez et al. 2013; Blain et al. 2021), with kelp recruitment often suppressed by turf algae species that can thrive in these altered environmental conditions (Filbee-Dexter and Wernberg 2018). For example, in Chile, industrial pollution has been shown to negatively impact the abundance, morphology and development of *Lessonia spicata* and *Macrocystis pyrifera* (Oyarzo-Miranda et al. 2020; Jara-Yáñez et al. 2021). Overharvesting and overfishing also pose significant risks both directly through dislodgement of kelp sporophytes, and indirectly through the removal of key predators (Christie et al. 1998; Estes et al. 1998; Tegner and Dayton 2000; Ling et al. 2009; Gouraguine et al. 2021; Pérez-Matus et al. 2025).

Historically many kelp forests have exhibited a degree of resilience to changes in environmental conditions and stressors (Friedlander et al. 2020), but in recent years their capacity to withstand and recover appears to have been eroded (Wernberg et al. 2010; Ling et al. 2015). In some areas, this has resulted in shifts from rich and complex kelp forest ecosystems to alternative, less complex systems dominated by algal turfs and/or sea urchin barrens, with reinforcing feedback mechanisms favouring the persistence of the less desirable systems (Filbee-Dexter and Scheibling 2014; Filbee-Dexter and Wernberg 2018). Future predictions suggest significant range-contractions and local extinctions of kelp forests (Martínez et al. 2018), although in some areas, these ecosystems are considered stable or have expanded (Krause-Jensen and Duarte 2014; Krumhansl et al. 2016), with future extensions into some Arctic regions likely (Filbee-Dexter et al. 2019; Krause-Jensen et al. 2020; Assis et al. 2022).

25.1.4 Traditional Kelp Restoration

In response to the degradation and decline of habitats around the world, interest in restoration has increased (Society for Ecological Restoration 2020). Restoration can be defined as 'the process of initiating or accelerating the recovery of an ecosystem that has been degraded, damaged or destroyed' (Society for Ecological Restoration 2020). The restoration of terrestrial and freshwater systems has generally outpaced that of coastal marine ecosystems, where restoration is often perceived as expensive and prone to failure (Saunders et al. 2020; Earp et al. 2022a; Eger et al. 2022d). Despite this, innovative strategies are essential for rebuilding our oceans (Saunders et al. 2020). In the marine realm, restoration has primarily been undertaken in tropical ecosystems such as mangrove forests (Kamali and Hashim 2011; Lovelock et al. 2022), seagrass meadows (Bull et al. 2004; Marion and Orth 2010; van Katwijk et al. 2016; Unsworth et al. 2019; Tan et al. 2020) and coral reefs (Rinkevich 2005; Young et al. 2012; Omori 2019; Boström-Einarsson et al. 2020), with key temperate systems including oyster reefs and kelp forests somewhat overlooked (Morris et al. 2020b; Saunders et al. 2020; Filbee-Dexter et al. 2022).

The restoration of kelp forests, however, has a long history, extending back to eighteenth-century Japan before spreading to numerous countries and ecoregions and encompassing an array of kelp genera, restoration techniques and sectors of society (Earp et al. 2022a; Eger et al. 2022d; UNEP 2023; Wood et al. 2024a). Compared to mangroves, saltmarshes and seagrass, kelp forests have been restored over much smaller spatial scales (Saunders et al. 2020), although this number is increasing (see Kelp Forest Alliance 2025). To date, the majority of kelp restoration activities have focused on restoring subtidal areas, with *Macrocystis* and *Ecklonia* the primary target genera (Morris et al. 2020c; Earp et al. 2022a; Eger et al. 2022d; Kelp Forest Alliance 2025).

Kelp restoration can be divided into two categories, namely 'passive' and 'active'. Passive restoration involves ameliorating environmental conditions through activities such as pollution mitigation or rebuilding predators through marine protected areas. These techniques rely on a natural source of kelp propagules reaching restoration sites once conditions have improved, and alone they may not be sufficient for natural kelp forest recovery. In areas where the natural recovery of kelp forests has been limited, active restoration involving the manipulation of kelp and/or their consumers through seeding, transplantation and/or culling may be beneficial. In Korea, a combination of active restoration techniques and afforestation (i.e. creating new areas of kelp forest habitat) has successfully reinstated ~200 km^2 of kelp forests (Lee 2019; UNEP 2023), while in Norway, a 0.7 km^2 of kelp forest was successfully restored by using quicklime to clear a long-term urchin barren (Christie et al. 2024).

Restoring kelp forests can be costly and challenging to scale up to match the scale of degradation/decline (Earp et al. 2022a; Eger et al. 2022c, d; Wood et al. 2024a). It has been estimated that to match the current rate of loss of 1.8% per year (Krumhansl et al. 2016), over 30,000 km^2 kilometres of kelp forest would need to be restored annually, and the cost of this could exceed US$ 17 billion per annum (Filbee-Dexter et al. 2022). Moreover, efforts are often restricted by the dynamic and often inaccessible nature of coastal environments and limited by legislation and policy that remains somewhat underdeveloped regarding restoration (Eger et al. 2022c, d; Earp et al. 2024c; Wood et al. 2024a). As such, the majority of kelp restoration efforts have been academically motivated and limited in scale and/or scope (Eger et al. 2020b, 2022d; Earp et al. 2022a; Wood et al. 2024a). For example, there has been limited comparative research across species, regions, timepoints and/or environmental contexts (but see Westermeier et al. 2016), with a lack of standardized monitoring and reporting guidelines further challenging comparisons across studies (Earp et al. 2022a). In the face of changing environmental conditions, deciding when, where and to what baseline to restore to maximize the long-term persistence of kelp forests and their associated services can represent a hurdle for restoration efforts (Filbee-Dexter and Smajdor 2019; Wood et al. 2019; Coleman et al. 2020; UNEP 2023).

25.1.5 Kelp Restoration Targets

In recent years, awareness of the loss and the importance of marine ecosystems has increased and so too has interest in restoration. This is demonstrated by a number of international commitments such as the 30 × 30 target of the Kunming-Montreal Global Biodiversity Framework that aims to effectively conserve 30% of the world's habitats by 2030 (Dinerstein et al. 2019; CBD 2022), alongside the United Nations Decade on Ecosystem Restoration (2021–2030) (United Nations General Assembly 2019). In terms of kelp forests, there has also been rising interest in restoration, with more projects in the 10 years between 2009 and 2019 than ever before (UNEP 2023). Collectively, these activities led to the development of the Kelp Forest Challenge that aims to align kelp restoration efforts to protect 30,000 km^2 and restore 10,000 km^2 of kelp forests by 2040 (Eger et al. 2024a). There have also been several recent syntheses of kelp restoration efforts to date (Morris et al. 2020b; Earp et al. 2022a; Eger et al. 2022d), alongside the generation of kelp restoration and monitoring guidebooks (Eger et al. 2022c, 2024b), and the formation of kelp restoration working groups such as the Green Gravel Action Group (Wood et al. 2024a).

To meet restoration targets and restore kelp forests at ecologically meaningful scales, significant long-term financial investments to cover personnel, infrastructure and resources are required. For example, kelp restoration efforts in Korea, which have restored 40% of their 500 km^2 by 2030 target, have an annual budget of almost US$ 30 million (Eger et al. 2022d; UNEP 2023). While other nationally motivated targets are expected in regions such as Norway and Canada in the future, for the most part, there remains a mismatch between the value and duration of restoration funding (often 1–3 years) and the realistic timeframe required to plan, implement and monitor restoration efforts beyond experimental scales (Wood et al. 2024a). In terms of kelp restoration, from the time of seeding it may take ~1.5 years or more before individuals become reproductive, and significantly longer for associated assemblages to re-establish, meaning projects are often unable to determine key attributes of successful restoration including self-sustainability and biodiversity (Christie et al. 1998; Ruiz-Jaen and Aide 2005; Marzinelli et al. 2016; Earp et al. 2022a; Wood et al. 2024a). Collaborations between scientists, industry and community groups have emerged as an effective means of overcoming some of the challenges associated with limited restoration funding and project durations—for example, local volunteer and citizen science groups can support implementation and monitoring over greater spatio-temporal scales (Watanuki et al. 2010; Vergés et al. 2020; Eger et al. 2022c; Filbee-Dexter et al. 2022). More recently, there has been an interest in 'crediting' methodologies, allowing companies to purchase credits that help fund kelp restoration activities (Reklev 2023; Urchinomics 2023). However, to maximize the benefits of these collaborations and funding mechanisms, there is a need to develop cost-effective, scalable kelp restoration techniques that are suitable for a range of kelp species, applicable for a range of environmental contexts, that have minimal impact on donor sites, are feasible for restoration practitioners with varying skills and access to facilities, and with the possibility to incorporate resilient genotypes.

25.2 The Development of 'Green Gravel'

25.2.1 Original Trials

The 'green gravel' kelp restoration technique was developed to overcome some of the challenges facing traditional active kelp restoration techniques, particularly costs, scalability and the need for SCUBA-diving (Fredriksen et al. 2020). Simply, the technique involves collecting reproductive material from the field, inducing zoospore release in the laboratory, seeding small rocks with the spore solution, rearing them in aquaria, and then outplanting them at restoration sites (Fig. 25.4). The first trials of the technique involved the sugar kelp *Saccharina latissima*, with the gravel outplanted by divers or dropped from a boat in a semi-protected area on the southern coast of Norway (Fredriksen et al. 2020). The gravel was well retained at the sites and *S. latissima* increased in length and in some cases overgrew the gravel and attached directly to the underlying natural substrate (Fredriksen et al. 2020), suggesting that they could lead to the formation of a self-sustaining forest. In this way, the technique allows large volumes of kelp recruits to be introduced at a restoration site, without the need for extensive culturing facilities or the use of costly and laborious SCUBA-diving. It also has additional benefits including limited impacts on 'donor' sites as only reproductive material needs to be sampled (not whole individuals), providing a means of 'future proofing' individuals/forests through the installation of genotypes that are resilient to future environmental conditions, and is applicable to marine forests comprised of Laminarian and Fucalean species (Fredriksen et al. 2020).

25.2.2 Methodological Developments

Following the initial trials, green gravel has expanded as a concept to encompass multiple restoration approaches whereby substrates are seeded and reared in aquarium facilities before being outplanted. Often multiple green gravel approaches are tested in the same system to determine which methods are best suited to different environmental contexts and kelp species. Since its inauguration in 2020, the Green Gravel Action Group (www.greengravel.org/action-group; GGAG herein) has spearheaded global efforts to refine and share green gravel protocols and scale-up restoration activities across multiple kelp species, regions and habitats. The network involves researchers, industry and local community groups and encompasses at least 25 projects spanning 9 countries and 17 ecoregions (Spalding et al. 2007) (Wood et al. 2024a). While the majority of these initiatives are ongoing and have yet to publish findings (see www.greengravel.org for further information), publications from aquarium and field trials in Australia, England, Norway and Portugal (Table 25.1), as well as reports from New Zealand, Canada and the USA have yielded valuable insights and highlighted key considerations for future efforts.

While a large body of literature concerning kelp cultivation exists (Edwards and Watson 2011; Forbord et al. 2012; Flavin et al. 2013; Redmond et al. 2014; Rolin et al. 2014; Kerrison et al. 2016; Visch et al. 2023, 2024), this has often focused on optimizing processes for aquaculture applications rather than ecological restoration (but see Suebsanguan et al. 2021; Chemello et al. 2023; Supratya and Martone 2024). In the case of green gravel, the cultivation of several species ($n = 6$) has been successful (Table 25.1; K. Filbee-Dexter, personal communication), with the influence of seeding density on recruit survival and growth a common focus of research efforts (Table 25.1). For *S. latissima* and *Ecklonia radiata*, seeding density had little impact on sporophyte growth, or the reproductive success and detachment rate of gametophytes (Fredriksen et al. 2020; Alsuwaiyan et al. 2022). However, *Laminaria ochroleuca* exhibited a different pattern, with high seeding densities resulting in fewer recruits that were greater in length compared to those seeded

Fig. 25.4 Process of producing green gravel: Fertile plants are identified (**a**) and reproductive material is excised for zoospore release (**b**). Gravel is inoculated with the spore solution (**c**) and reared in the aquarium (**d**). After a few months, the green gravel is deployed at the restoration site where they continue to grow (**e**). (Photo credit: Hannah Earp)

Table 25.1 Summary of key factors in published green gravel restoration efforts

Location	Kelp species	Variables	Seeding density	Rock type	Outplanting environment	Cost	Reference
Norway (S)	*S. latissima*	Seeding density Outplanting technique (hand placed, surface drop) Outplanting depth (3 m, 7 m) Encased versus open plots	14,814, 1235 and 148 individuals cm^{-2}	Granite	Sheltered, subtidal at depths of 3–7 m, cleared (bare rock) and turf covered plots	US$ 6.75 m^{-2} (exclusive of bench fees, vessel hire and monitoring)	Fredriksen et al. (2020)
England (NE)	*S. latissima*	Substrate size (gravel, cobble)	500,000 spores mL^{-1} (diluted for seeding)	Mixed (from local beach)	Exposed, intertidal	US$ 1135 m^{-2} (excluding monitoring)	Earp et al. (2024c)
England (SW)	*S. latissima*	Substrate type	Not stated	Not stated	Subtidal	US$ 1857 m^{-2} (excluding monitoring)	Wilding et al. (2023)
Australia (SW)	*E. radiata*	Seeding density Substrate type/texture Substrate size (small, large)	230 and 115 gametophyte fragments ml^{-1}	Basalt, laterite, limestone	NA—Aquarium only	Not stated	Alsuwaiyan et al. (2022)
Portugal (NW)	*L. ochroleuca*	Light intensity (130 µmol m^{-2} s^{-1}, 40 µmol m^{-2} s^{-1})	Not stated	Granite	NA—Aquarium only	Not stated	Chemello et al. (2023)
Portugal (NW)	*L. ochroleuca*	Seeding density	120 and 90 gametophyte fragments ml^{-1}	Granite	NA—Aquarium only	Not stated	Chemello et al. (2024)
Portugal (NW)	*L. ochroleuca*	Substrate type	Not stated	Granite, limestone, quartz, schist	NA—Aquarium only	Not stated	Pinheiro (2023)

at lower densities (Chemello et al. 2024). While most trials have observed some degree of self-thinning of recruits over time, initial findings suggest that seeding at lower densities may not hinder restoration outcomes, while also reducing costs associated with spore production and limiting the impact on donor populations by requiring less reproductive material (Fredriksen et al. 2020; Alsuwaiyan et al. 2022; Earp et al. 2024c). In terms of cost reduction, another promising approach involves cultivating under lower light intensities until recruits are visible to the naked eye. This reduces the risk of contamination/overgrowth (and subsequent loss) by opportunistic algal species while ensuring kelp growth, albeit at a lower rate compared to those cultivated under higher light levels (Chemello et al. 2023).

While initial green gravel efforts involved cultivation on small (~ 3 × 3 × 3 cm) granite rocks, trials involving different rock sizes and types more representative of nearby reefs have become commonplace (Table 25.1). For example, Alsuwaiyan et al. (2022) seeded *E. radiata* onto ballast, laterite and quarried limestone, while Earp et al. (2024c) used a haphazard assortment of rocks collected from the outplanting sites. The latter may prove useful in cases where there are concerns about the impact of introducing large quantities of foreign material/rocks to a restoration site. Irrespective of substrata type, it is considered best practice to wash and/or soak the material prior to seeding to prevent contamination, and a range of techniques have been used (i.e. boiling water, deionized water, bleach; Alsuwaiyan et al. 2022; Chemello et al. 2023; Earp et al. 2024c). Recruits have successfully adhered to all green gravel substrata tested thus far, but growth and health have varied across substrata, potentially due to the composition and consolidation of the substrata although further research is required to support this (Alsuwaiyan et al. 2022; Pinheiro 2022). Disparities in findings across green gravel substrata demonstrate the importance of pilot experiments before investing in large-scale restoration activities. In a bid to reduce substrata costs, trials are underway in the UK to test the effectiveness of scallop shells—a naturally occurring seafloor substrata, but a by-product of the seafood industry—as an alternative substratum for green gravel style restoration (Wilding and Earp 2024). Moreover, research into the influence of substrata texture, colour and chemistry on the survival and growth of kelp recruits is required.

Substrate size represents a trade-off between restoration success and the scale and cost of the restoration effort. Smaller substrates can be seeded in greater quantities within a given aquarium space, with recruit growth generally comparable to that of individuals seeded on larger substrates (Alsuwaiyan et al. 2022). However, field trials found that *S. latissima* seeded on larger rocks reach greater sizes than those on smaller rocks (Earp et al. 2024c), although the likelihood of reattachment to the underlying substrate may be reduced for individuals on larger rocks. This issue may not impede restoration success should the individuals persist long enough to mature and seed the underlying substrate. One solution for increasing substrata size without using more aquarium space is to seed kelp on biodegradable lines that are then wrapped around larger rocks—this has been effective in New Zealand (www.loverimurimu.org) and Australia. The retention of seeded substrata at restoration sites is critical to restoration success, and while small substrata may readily settle in cracks and crevices, in areas with elevated hydrodynamic activity they are more susceptible to displacement compared to larger substrata.

To date, published results from field out-planting of green gravel has been limited (Table 25.1), although a range of deployment techniques have been trialled (Fredriksen et al. 2020; Earp et al. 2024c). In intertidal zones, deployment was done by hand at low tide and required heavy materials to be transported across hazardous terrain, meaning it would be challenging and costly to scale up for the restoration of a larger area (Earp et al. 2024c). In subtidal environments, both hand placement by divers and dropping from the surface have been tested, with comparable kelp growth rates occurring for each technique, but retention rates were not quantified (Fredriksen et al. 2020). Hand placement by divers offers the advantage of strategic positioning with higher retention potential, but this comes at an increased financial cost to fund a technical dive team.

Field out-planting has occurred across an array of environmental contexts including sheltered and wave-exposed shores, intertidal and subtidal areas, under the canopy of adult conspecifics or other canopy-forming species, and in turf-dominated environments (Table 25.1) (Fredriksen et al. 2020; Earp et al. 2024c; G. Wood, personal communication). In turf-covered areas, gravel retention and kelp growth rates were greater, but recruit density and attachment to the underlying substrate did not differ between habitats after four months (Fredriksen et al. 2020). In wave-exposed intertidal areas, larger cobbles were better retained than smaller gravel (Earp et al. 2024c), suggesting that modifications to the green gravel methodology that may increase costs could also enhance restoration success in some environmental contexts. For example, work is underway in Chile to assess the benefit of attaching green gravel to the substrate in wave-exposed areas (A. Pérez-Matus, personal communication), while in Australia, seeded plugs that can be attached to novel restoration units are being trialled (Wood, unpublished; J. Valckenaere, personal communication).

25.3 Challenges and Opportunities in the Application of Green Gravel

25.3.1 Environmental Conditions

A foremost challenge facing kelp restoration initiatives is deciding whether restoration is feasible, and if so where and when to undertake restoration actions (Johnson et al. 2017). Critical to this is selecting appropriate restoration sites, ideally where the original driver of kelp decline/loss has been mitigated and where environmental conditions favour kelp recruitment, growth and survival. Without these conditions restoration will likely fail. In areas where the driver of decline has been mitigated but kelp forests have not re-established, active restoration techniques such as green gravel could be beneficial, as a lack of recovery may reflect factors such as propagule limitation. This was likely the case in Sydney (Australia) where improvements in water quality did not facilitate 'crayweed' (*Phyllospora comosa,* Order Fucales) recovery following declines in the 1970s, meaning active restoration techniques were required (Coleman et al. 2008; Campbell et al. 2014).

In some areas, it may not be possible to mitigate the driver of decline/loss, meaning defining what baseline to restore to (i.e. whether to focus on historical conditions, current ecological potential, or future projections) is challenging. What constituted a healthy kelp forest in the past may no longer be feasible or relevant in the context of existing/new environmental conditions, particularly in areas subject to stressors such as ocean warming. In these areas, restoration efforts should focus on restoring ecosystem functions and promoting resilient kelp forests that can withstand future environmental conditions, rather than recreating historical baselines (Wood et al. 2019, 2021). Trial out-plantings of local water-tolerant kelp strains are underway in Tasmania and Western Australia, and while the Tasmanian trials involve twine and plastic plate substrates (Layton and Johnson 2021; Wood et al. 2024a), green gravel approaches could represent an effective means of deploying such strains in the future, although further research is required to support this. The use of resilient kelp strains within restoration efforts, however, remains largely unexplored, with a lack of scientific understanding, ethical concerns, and policy regulations among the reasons for this (Filbee-Dexter and Smajdor 2019; Wood et al. 2024a).

The success of green gravel approaches to date has been site and context-dependent, likely due to environmental conditions at restoration sites (Wood et al. 2024a). However, the monitoring and/or reporting of environmental conditions at restoration sites both prior to and following the initiation of restoration interventions has been limited, with the majority of studies not reporting on key attributes such as temperature, light and nutrient availability, and exposure to wave action (Earp et al. 2022a). This makes comparisons across studies or environmental contexts challenging, but more importantly, it makes it difficult for restoration practitioners to determine the most appropriate restoration techniques for their focus area. The GGAG is working to overcome some of these barriers by testing the applicability of green gravel across an array of environmental contexts and kelp species to identify drivers of success and points of failure—for example the technique that works best with fast-growing species such as *S. latissima* and *M. pyrifera* may be less successful in areas exposed to wave action or with dense herbivore populations. Interactions between and among species are likely to influence restoration outcomes, for example the presence of adult kelp is indicative of suitable environmental parameters for restoration, and these adult individuals can modify understorey environmental conditions and facilitate the growth and survival of recruits in the subcanopy, meaning restoration may be more successful should it occur in proximity to a healthy kelp forest (Wernberg et al. 2010; Layton et al. 2019). Others have suggested that undertaking restoration in areas with intact predator populations, or co-restoration of both kelp and predatory species could enhance success (Eger et al. 2020a), although a more comprehensive understanding of the role of inter- and intra-specific interactions on restoration outcomes is required.

Seasonal variation in environmental conditions can also influence restoration outcomes. In temperate regions, kelp exhibit seasonal growth and reproduction, typically thriving during cooler months (Jupp and Drew 1974; Brown et al. 1997; Schiel and Foster 2015), implying active restoration success may be greater in spring when temperatures are cool, but light intensity and wave action are more favourable than in winter months. Timing should also align with other seasonal factors, for example, turf algae abundance, which if high, may hinder kelp recruitment and adherence to the substrate (Kennelly 1987; Diaz-Pulido and Garzón-Ferreira 2002; Burek et al. 2018). Despite this, limited comparative research investigating the influence of restoration timing on success has been undertaken. Unpredictable stochastic events, including storms or heatwaves, which can be detrimental to restoration efforts, are however, likely to remain as significant challenges to restoration interventions (De La Fuente et al. 2019; Earp et al. 2024c).

25.3.2 Identifying Source Material

One of the key benefits of the green gravel approach is that it is less likely to induce long-term impacts on natural kelp forests from repeated harvesting of fertile kelp material. Once gametophyte cultures are established, the cycle of boosting biomass under hatchery conditions and seeding onto substrates for deployment may theoretically continue indefinitely. In practice, it is likely that ongoing collections of reproductive sorus tissue will need to be made on an ongoing basis to ensure sufficient genetic diversity and avoid bottleneck issues commonly found in captive breeding programmes (e.g. Price and Hadfield 2014; Stuart et al. 2024). Choosing appropriate source stock for each restoration site is also crucial to ensure restoration success and long-term resilience (Wood et al. 2020). Ideally, source stocks should be genetically diverse and locally adapted to environmental conditions at the restoration site. Genetic diversity underpins the capacity of restored populations to cope with dynamic conditions, such as changing water temperatures, nutrient availability or disease, while local adaptation maximizes suitability to site-specific conditions like salinity, temperature and light availability, improving recruitment and survival rates. Conversely, maladaptation may cause poor recruitment or survival at the outset, while improper genetic management practices in any of the steps of the green gravel process could cause genetic bottlenecks in subsequent generations, both of which have the potential to impair the short- and long-term success of restoration projects.

Sourcing local stock in an era of rapid environmental change also, however, presents challenges. In areas where kelp populations are heavily degraded and genetically impoverished, or where the environmental baseline conditions have shifted, the introduction of genetically diverse individuals from neighbouring areas or climate-adapted genotypes through assisted gene flow may enhance resilience. Such approaches must be carefully managed to avoid negative impacts, such as outbreeding depressions that occur when the introduction of non-local genes disrupts locally adapted traits and co-adapted gene complexes. To mitigate these risks, the development of tools that provide dynamic sourcing guidelines is essential (e.g. www.reef-adapt.org; Wood et al. 2024b).

A recent survey of green gravel practitioners and researchers highlighted significant gaps in current restoration practices, with many projects neglecting genetic diversity and provenance considerations (Wood et al. 2024a), despite evidence that they can affect even the early stages of restoration projects (Wood et al. 2020). Innovative digital tools such as the new reef adapt webtool (www.reefadapt.org; Wood et al. 2024b) can enable marine managers and restoration practitioners to select genetically appropriate kelp stock that aligns with environmental conditions and future climate scenarios. Trials that compare the success of restoration when using source stocks of different provenance (source locations) are also increasing, particularly in areas that have already experienced negative impacts of a changing climate (e.g. various parts of Australia and Canada; Wood et al. 2024a).

25.3.3 Quantifying Success

Restoration success can be quantified on ecological, social and economic scales, but for the most part, activities have been small in scale and have primarily focused on ecological benefits, particularly kelp abundance and morphology (Morris et al. 2020b; Earp et al. 2022a; Eger et al. 2022d). However, these factors may not be fully indicative of restoration success. Self-sustainability and diversity relative to a reference system are key attributes of successfully restored ecosystems (Ruiz-Jaen and Aide 2005), yet these factors are often not assessed within restoration monitoring. Research from Sydney (Australia) revealed the importance of monitoring reproduction/recruitment as the growth and survival of *P. comosa* transplants (attached to plastic mesh using zip ties) was variable, but in some areas, they were able to persist long enough to reproduce and shelter young recruits, meaning the restoration action could be considered a success (Campbell et al. 2014). Monitoring biodiversity within a restoration area is more complex and can take considerably longer to re-establish, meaning it may be beyond the timeframe or funding capacity of projects (Marzinelli et al. 2016; Earp et al. 2022a).

In terms of green gravel approaches, efforts are considered successful if either seeded individuals overgrow the substrate and attach to the underlying reef or if they survive long enough to reproduce. Although at present, monitoring field deployments is somewhat challenging as seeded substrates are not fixed to the seafloor and may be moved by hydrodynamic forces, meaning additional field time to locate deployments may be required and this can be costly, particularly where diving is involved. Recent efforts have tried to overcome this by using markers or different colours of rocks to ease identification of experimental deployments in the field, with limited success (Earp et al. 2024c, K. Filbee-Dexter, personal communication). As such experiments are underway to assess the feasibility and benefit of attaching deployments to the seafloor (e.g. glues, plugs; A. Pérez-Matus and J. Valckenaere, personal communication). Should this challenge be overcome, future efforts could incorporate additional metrics of success including restoration extent, genetic diversity, biodiversity benefits and ecosystem service provision.

Going forward, to better understand and compare outcomes across green gravel endeavours and restoration efforts more broadly, the use of standardized monitoring and reporting guidelines such as those outlined in the *Kelp Restoration Guidebook* and the *Monitoring Kelp Forest Ecosystems Guidebook* would be beneficial (Eger et al. 2022b, c, 2024b). Moreover, the reporting of restoration outcomes (both successes and failures) on accessible platforms and repositories would allow projects to learn from each other, prevent the duplication of effort and encourage cross-project collaborations. In the case of green gravel, protocols and lessons learned are shared across the project website and annual meetings, while information on more diverse kelp restoration efforts can be found on the Kelp Forest Alliance website (https://kelpforestalliance.com/). Reporting of outcomes on such portals can, however, be time-consuming and is often not considered a priority for restoration efforts, meaning there is often a lag between data collection, publication and sharing.

25.3.4 Financing and Regulatory Processes

Despite increases in funding for kelp restoration in recent years, many projects have been academically motivated and characterized by short timescales (usually 1–3 years) with economic limitations that can restrict their access to key resources such as personnel and equipment/technology (Eger et al. 2022d; Wood et al. 2024a). As such, restoration efforts are often limited in scope, scale and/or duration, and our understanding of the ecosystem-wide benefits of restoration remains somewhat limited. To address some of these factors, it is suggested that a minimum funding duration of 5 years is necessary, with canvassing across a broad selection of funding strategies (e.g. research grants, government investments, philanthropic donations and environmental crediting schemes) more likely to result in long-term funding (Wood et al. 2024a). Multiple GGAG members are funding their efforts though a mixture of governmental, NGO and philanthropic grants (Wood et al. 2024a).

There is growing recognition that partnerships between conservation practitioners and the private sector can deliver valuable conservation gains (Seidl et al. 2024). The potential use of carbon credits has been suggested as a possible mechanism to fund kelp restoration, as kelp forests are emerging blue carbon ecosystems (Pessarrodona et al. 2022; Howard et al. 2023; Filbee-Dexter et al. 2024a) and good candidates for eventual inclusion carbon credit schemes. However, most of these schemes follow the IPCC's definition of blue carbon that does not include kelp forests (Vanderklift et al. 2022) and the research effort needed to effectively document kelp carbon pathways (Hurd et al. 2022) would be prohibitively expensive to finance restoration efforts. Therefore, beyond working to amend the IPCC guidelines, the most rapid method for carbon credit for kelp restoration is through the voluntary market (e.g. Verra) or through other environmental credits. The Western Australia Kelp Restoration project innovated a credit financing mechanism to support their green gravel deployments, through the development of a synthetic kelp credit that supported the documented restoration action and could be bundled with a high integrity carbon credit (www.canopyblue.org; K. Filbee-Dexter, personal communication). Similar funding could come through Biodiversity credits or certificates through the new Nature Positive markets (Whitton and Waterford 2023), as kelp forests are high biodiversity habitats. Globally, numerous national and private biodiversity credit initiatives are emerging (e.g. Open Earth, CarbonZ, South Pole, CreditNature).

Permitting and regulatory processes vary across regions, but in general, restoration policy remains critically underdeveloped, with restoration efforts often intersecting or competing with current legislation (Wood et al. 2024a). This lack of restoration-specific legislation can be a significant barrier to restoration efforts (Alsuwaiyan et al. 2022), particularly as projects scale up from research-focused activities. It can also be a lengthy process, taking anywhere from weeks to years (Eger et al. 2022c). Early engagement and consultation with local authorities and policy makers to co-develop permits and guidelines in advance of restoration efforts is advisable (Wood et al. 2024a). Moreover, projects should also ensure that any relevant permissions from local Indigenous rights holders are obtained (Lee et al. 2021).

25.3.5 Stronger Together

To date, the costs associated with green gravel efforts have been variable, ranging from US$ 6.75 to US$ 1856.25 per m^2 excluding long-term monitoring (Fredriksen et al. 2020; Wilding et al. 2023; Earp et al. 2024c) (Table 25.1). Despite this, the approach offers a high potential for upscaling through collaborations, with meaningful dialogues between partners facilitating access to resources, promoting knowledge sharing, leveraging community capacity and encouraging environmental stewardship. For example, collaboration with commercial aquaculture and seed banks can offer valuable resources such as culturing protocols, spore solutions and access to aquarium facilities, alongside support for innovation, all of which can have a positive influence on restoration and efficacy and scalability (Filbee-Dexter et al. 2022). Similarly, members of the GGAG have developed mobile culturing facilities to support the restoration of kelp forests in remote regions using green gravel approaches (Wood et al. 2024a).

Community-based engagement is also a promising means of increasing the scope of restoration efforts, with almost three-quarters of green gravel approaches incorporating some form of outreach (Wood et al. 2024a). Community involvement can range from educational events that build support for restoration activities (Vergés et al. 2020) to hands-on roles for volunteers in deploying and/or monitoring restoration activities (Watanuki et al. 2010; Hesley et al. 2017; Eger et al. 2022c). Engaging volunteers can be a cost-effective means of scaling-up restoration activities, but it is essential that volunteers are provided with training and protocols necessary to safeguard data quality, and that appropriate health and safety factors are considered/mitigated (Earp et al. 2022b; Wood et al. 2024a).

Given that kelp forests are culturally important to many indigenous communities, empowering these stakeholders and ensuring restoration actions incorporate their local knowledge and respect customary practices, although not always a formal requirement, can positively influence restoration outcomes (Lee et al. 2021; Eger et al. 2022c; United Nations Environment Programme 2023; Wood et al. 2024a). To date, green gravel approaches have involved Indigenous communities and First Nations from several countries including New Zealand and Canada.

25.3.6 Technological Advances

The potential of technological developments to enhance the effectiveness of green gravel approaches from the cultivation to the monitoring level is considerable. Emerging molecular techniques including genetic engineering and selective breeding for environmental resilience have been proposed as tools to develop kelp strains that can withstand current/predicted environmental conditions at restoration sites, potentially bolstering resistance to future stressors (Coleman et al. 2020). While promising, these methods are not yet widely incorporated into restoration efforts (except for work in Tasmania; Layton and Johnson 2021; Wood et al. 2024a) as the introduction of genetically modified individuals into natural ecosystems must be carefully managed to prevent detrimental impacts on native kelp populations (Filbee-Dexter and Smajdor 2019). Innovations in substrate design also offer potential benefits for green gravel approaches, for example 3D printing technologies enable the creation of novel substrates including those that replicate habitat features known to promote kelp attachment and survival (Evans et al. 2021). Advances in biodegradable materials could allow substrates that biodegrade naturally over time, reducing the footprint of restoration efforts.

In addition, advances in remote sensing and drone technology are enhancing the resolution of satellite and underwater imagery, while machine learning and predictive modelling can now process vast datasets—harnessing these technologies could allow the identification of optimal restoration sites and timings (Bennion et al. 2019; Reshitnyk et al. 2024). Autonomous underwater vehicles (AUVs) could streamline the deployment of restoration interventions such as green gravel by identifying areas with greater substrata retention potential and increasing the spatial coverage of activities, thereby reducing the dependency on costly dive activities. Such approaches could also be applied to the monitoring of restoration efforts, while coupling with machine learning approaches could enhance the efficacy of data processing (Campero Juardo et al. in review).

25.4 Summary

To meet international restoration commitments and to preserve kelp forests and the vital ecosystem services they support, there is a need to scale up kelp restoration efforts beyond experimental scales using time- and cost-effective methods that have been scientifically tested and are more likely to result in successful outcomes. Green gravel approaches represent a promising means of restoring kelp forests with a significant body of collaborative research underway as part of the GGAG to investigate its efficacy across a range of environmental contexts, kelp species and substrates, and to develop robust protocols for its application and monitoring. The concept also has the potential to transcend ecosystems to become a globally relevant restoration approach (Fredriksen et al. 2020). Successful application of green gravel approaches, however, requires the establishment of appropriate baselines upon which to measure success, alongside careful consideration of factors such as site selection, seasonality, and collaborations across sectors likely to boost future efforts. In particular, connecting sectors within the blue economy and novel funding strategies (e.g. restoration credits) that facilitate sustained support for restoration efforts are likely to be important going forward (Filbee-Dexter et al. 2022; Wood et al. 2024a).

Green gravel approaches can therefore be considered a valuable addition to the marine restoration toolkit; however, restoration itself is not panacea and alone is unlikely to resolve declines in kelp forests. Instead, a multifaceted approach that combines local restoration actions with proactive management/conservation strategies, while also addressing broader climate change stressors, is imperative (Filbee-Dexter et al. 2024b). Collectively, such actions can promote functional and resilient kelp forests, support their associated biodiversity, and secure the array of critical ecosystem services they support for future generations.

References

Alsuwaiyan NA, Filbee-Dexter K, Vranken S, Burkholz C, Cambridge M, Coleman MA, Wernberg T (2022) Green gravel as a vector of dispersal for kelp restoration. Front Mar Sci 9:1–11. https://doi.org/10.3389/fmars.2022.910417

Anderson MJ, Diebel CE, Blom WM, Landers TJ (2005) Consistency and variation in kelp holdfast assemblages: spatial patterns of biodiversity for the major phyla at different taxonomic resolutions. J Exp Mar Biol Ecol 320(1):35–56. https://doi.org/10.1016/j.jembe.2004.12.023

Anderson RJ, Rothman MD, Share A, Drummond H (2006) Harvesting of the kelp *Ecklonia maxima* in South Africa affects its three obligate, red algal epiphytes. J Appl Phycol 18(3–5):343–349. https://doi.org/10.1007/s10811-006-9037-7

Arafeh-Dalmau N, Montaño-Moctezuma G, Martinez JA, Beas-Luna R, Schoeman DS, Torres-Moye G (2019) Extreme marine heatwaves alter kelp forest community near its equatorward distribution limit. Front Mar Sci 6:1–18. https://doi.org/10.3389/fmars.2019.00499

Arnold M, Teagle H, Brown MP, Smale DA (2016) The structure of biogenic habitat and epibiotic assemblages associated with the global invasive kelp *Undaria pinnatifida* in comparison to native macroalgae. Biol Invasions 18(3):661–676. https://doi.org/10.1007/s10530-015-1037-6

Assis J, Serrão EA, Duarte CM, Fragkopoulou E, Krause-Jensen D (2022) Major expansion of marine forests in a warmer Arctic. Front Mar Sci 9:1–10. https://doi.org/10.3389/fmars.2022.850368

Bayley DTI, Brickle P, Brewin PE, Golding N, Pelembe T (2021) Valuation of kelp forest ecosystem services in The Falkland Islands: a case study integrating blue carbon sequestration potential. One Ecosystem 6(e62811). https://doi.org/10.3897/oneeco.6.e62811

Bennett S, Wernberg T, Harvey ES, Santana-Garcon J, Saunders BJ (2015) Tropical herbivores provide resilience to a climate-mediated phase shift on temperate reefs. Ecol Lett 18(7):714–723. https://doi.org/10.1111/ele.12450

Bennett S, Wernberg T, Connell SD, Hobday AJ, Johnson CR, Poloczanska ES (2016) The 'great southern reef': social, ecological and economic value of Australia's neglected kelp forests. Mar Freshw Res 67(1):47–56. https://doi.org/10.1071/MF15232

Bennion M, Fisher J, Yesson C, Brodie J (2019) Remote sensing of kelp (Laminariales, Ochrophyta): monitoring tools and implications for wild harvesting. Rev Fish Sci Aquac 27(2):127–141. https://doi.org/10.1080/23308249.2018.1509056

Blain CO, Hansen SC, Shears NT (2021) Coastal darkening substantially limits the contribution of kelp to coastal carbon cycles. Glob Chang Biol 27(21):5547–5563. https://doi.org/10.1111/gcb.15837

Blamey LK, Bolton JJ (2018) The economic value of South African kelp forests and temperate reefs: past, present and future. J Mar Syst 188:172–181. https://doi.org/10.1016/j.jmarsys.2017.06.003

Blight AJ, Thompson RC (2008) Epibiont species richness varies between holdfasts of a northern and a southerly distributed kelp species. J Mar Biol Ass UK 88(3):469–475. https://doi.org/10.1017/S0025315408000994

Bolton JJ (2016) What is aquatic botany?- And why algae are plants: the importance of non-taxonomic terms for groups of organisms. Aquat Bot 132:1–4. https://doi.org/10.1016/j.aquabot.2016.02.006

Boström-Einarsson L, Babcock RC, Bayraktarov E, Ceccarelli D, Cook N, Ferse SCA, Hancock B, Harrison P, Hein M, Shaver E, Smith A, Suggett D, Stewart-Sinclair PJ, Vardi T, McLeod IM (2020) Coral restoration – a systematic review of current methods, successes, failures and future directions. PLoS One 15(1):e0226631. https://doi.org/10.1371/journal.pone.0226631

Brown MT, Nyman MA, Keogh JA, Chin NKM (1997) Seasonal growth of the giant kelp *Macrocystis pyrifera* in New Zealand. Mar Biol 129(3):417–424. https://doi.org/10.1007/s002270050182

Bué M, Smale DA, Natanni G, Marshall H, Moore PJ (2020) Multiple-scale interactions structure macroinvertebrate assemblages associated with kelp understory algae. Divers Distrib 26(11):1551–1565. https://doi.org/10.1111/ddi.13140

Bull JS, Reed DC, Holbrook SJ (2004) An experimental evaluation of different methods of restoring *Phyllospadix torreyi* (surfgrass). Restor Ecol 12(1):70–79. https://doi.org/10.1111/j.1061-2971.2004.00258.x

Burek KE, O'Brien JM, Scheibling RE (2018) Wasted effort: recruitment and persistence of kelp on algal turf. Mar Ecol Prog Ser 600:3–19. https://doi.org/10.3354/meps12677

Byrnes JE, Reed DC, Cardinale BJ, Cavanaugh KC, Holbrook SJ, Schmitt RJ (2011) Climate-driven increases in storm frequency simplify kelp forest food webs. Glob Chang Biol 17(8):2513–2524. https://doi.org/10.1111/j.1365-2486.2011.02409.x

Campbell AH, Marzinelli EM, Vergés A, Coleman MA, Steinberg PD (2014) Towards restoration of missing underwater forests. PLoS One 9(1):e84106. https://doi.org/10.1371/journal.pone.0084106

CBD (2022) Decision adopted by the conference on the parties to the convention on biological diversity. 15/4. Kunming-Montreal Global Biodiversity Framework, Montreal, pp 1–23

Chemello S, Pinto IS, Pereira TR (2023) Optimising kelp cultivation to scale up habitat restoration efforts: effect of light intensity on 'Green gravel' production. Hydrobiology 2(2):347–353. https://doi.org/10.3390/hydrobiology2020022

Chemello S, Santos D, Sousa-pinto I (2024) Unlocking the potential of Green gravel production for efficient kelp restoration: how seeding density affects the development of the Golden Kelp *Laminaria ochroleuca*. Phycology 4:443–449. https://doi.org/10.3390/phycology4030025

Christie H, Fredriksen S, Rinde E (1998) Regrowth of kelp and colonization of epiphyte and fauna community after kelp trawling at the coast of Norway. Hydrobiologica 375(376):49–58. https://doi.org/10.1023/A:1017021325189

Christie H, Jørgensen NM, Norderhaug KM, Waage-Nielsen E (2003) Species distribution and habitat exploitation of fauna associated with kelp (*Laminaria hyperborea*) along the Norwegian coast. J Mar Biol Ass UK 83(4):687–699. https://doi.org/10.1017/S0025315403007653h

Christie H, Moy FE, Fagerli CW, Rinde E, Strand M, Tveiten LA, Strand HK (2024) Successful large-scale and long-term kelp forest restoration by culling sea urchins with quicklime; and supported by crab predation. Mar Biol 171:211. https://doi.org/10.1007/s00227-024-04540-0

Coleman MA, Kelaher BP, Steinberg PD, Millar AJK (2008) Absence of a large brown macroalga on urbanized rocky reefs around Sydney, Australia, and evidence for historical decline. J Phycol 44(4):897–901. https://doi.org/10.1111/j.1529-8817.2008.00541.x

Coleman MA, Wood G, Filbee-Dexter K, Minne AJP, HughD G, Vergés A, Marzinelli EM, Steinberg PD, Wernberg T (2020) Restore or redefine: future trajectories for restoration. Front Mar Sci 7:237. https://doi.org/10.3389/fmars.2020.00237

Davies CE, Wootton EC (2018) Current and emerging diseases of the European lobster (*Homarus gammarus*): a review. Bull Mar Sci 94(3):959–978. https://doi.org/10.5343/bms.2017.1142

De La Fuente G, Chiantore M, Asnaghi V, Kaleb S, Falace A (2019) First ex situ outplanting of the habitat-forming seaweed *Cystoseira amentacea* var. Stricta from a restoration perspective. PeerJ 7:e7290. https://doi.org/10.7717/peerj.7290

Dhir SSJ, Graham J (2022) Death of a star: the sea star wasting disease epidemic and its impact on kelp and climate change. J Student Res 11(3). https://doi.org/10.47611/jsrhs.v11i3.2908

Diaz-Pulido G, Garzón-Ferreira J (2002) Seasonality in algal assemblages on upwelling-influenced coral reefs in the Colombian Caribbean. Bot Mar 45:284–292. https://doi.org/10.1515/BOT.2002.028

Diez I, Santolaria A, Muguerza N, Gorostiaga JM, Muguerza JM, Gorostiaga JM (2013) Measuring restoration in intertidal macrophyte assemblages following sewage treatment upgrade. Mar Environ Res 84:31–42. https://doi.org/10.1016/j.marenvres.2012.11.006

Dinerstein E, Vynne C, Sala E, Joshi AR, Fernando S, Lovejoy TE, Mayorga J, Olson D, Asner GP, Baillie JEM, Burgess ND, Burkart K, Noss RF, Zhang YP, Baccini A, Birch T, Hahn N, Joppa LN, Wikramanayake E (2019) A global deal for nature: guiding principles, milestones, and targets. Sci Adv 5(4):1–18. https://doi.org/10.1126/sciadv.aaw2869

Earp HS, Smale DA, Pérez-Matus A, Gouraguine A, Shaw PW, Moore PJ (2022a) A quantitative synthesis of approaches, biases, successes, and failures in marine forest restoration, with considerations for future work. Aquat Conserv 32:1717–1731. https://doi.org/10.1002/aqc.3880

Earp HS, Vye S, Bohn K, Burrows M, Chenery J, Dickens S, Foster C, Grist H, Lamont P, Long S, Morrall Z, Pocklington J, Scott A, Watson G, West V, Jenkins S, Delany J, Sugden H (2022b) Do you see what I see? Quantifying inter-observer variability in an intertidal marine citizen science experiment. Citiz Sci 7(1):1–13. https://doi.org/10.5334/cstp.483

Earp HS, Smale D, King N, Bestwick J, Fazel B, Moore P (2024a) Spatio temporal variation in the diversity and structure of understorey macrofaunal assemblages within *Laminaria hyperborea* forests in the Northeast Atlantic. J Exp Mar Biol Ecol 578:152034. https://doi.org/10.1016/j.jembe.2024.152034

Earp HS, Smale DA, Almond PM, Catherall HJN, Gouraguine A, Wilding C, Moore PJ (2024b) Temporal variation in the structure, abundance, and composition of *Laminaria hyperborea* forests and their associated understorey assemblages over an intense storm season. Mar Environ Res 200(July):106652. https://doi.org/10.1016/j.marenvres.2024.106652

Earp HS, Smale DA, Catherall HJN, Moore PJ (2024c) An assessment of the utility of green gravel as a kelp restoration tool in wave-exposed intertidal habitats. J Mar Biol Ass UK 104(e28):1–11. https://doi.org/10.1017/S0025315424000225

Eckman JE, Duggins DO, Sewell AT (1989) Ecology of under story kelp environments. I. Effects of kelps on flow and particle transport near the bottom. J Exp Mar Biol Ecol 129(2):173–187. https://doi.org/10.1016/0022-0981(89)90055-5

Edwards M, Watson L (2011) Aquaculture explained. Aquaculture 1(26) https://bim.ie/wp-content/uploads/2021/02/BIM, Aquaculture, Explained, Issue, 26,-, Cultivating,Laminaria,digitata.pdf

Eger AM, Marzinelli E, Gribben P, Johnson CR, Layton C, Steinberg PD, Wood G, Silliman BR, Vergés A (2020a) Playing to the positives: using synergies to enhance kelp Forest restoration. Front Mar Sci 7:544. https://doi.org/10.3389/fmars.2020.00544

Eger AM, Vergés A, Choi CG, Christie H, Coleman MA, Fagerli CW, Fujita D, Hasegawa M, Kim JH, Mayer-Pinto M, Reed DC, Steinberg PD, Marzinelli EM (2020b) Financial and institutional support are important for large-scale kelp Forest restoration. Front Mar Sci 7(September):1–7. https://doi.org/10.3389/fmars.2020.535277

Eger AM, Bennett S, Zimmerhackel J, Rogers A, Burton M, Filbee-Dexter K, Wernberg T, Gacutan J, Milligan B, Vergés A (2022a) Quantifying the ecosystem services of the Great Southern Reef. University of New South Wales, Sydney

Eger AM, Earp HS, Friedman K, Gatt Y, Hagger V, Hancock B, Kaewsrikhaw R, Mcleod E, Moore AM, Niner HJ, Razafinaivo F, Sousa AI, Stankovic M, Worthington TA, Bayraktarov E, Saunders M, Vergés A, Reeves S (2022b) The need, opportunities, and challenges for creating a standardized framework for marine restoration monitoring and reporting. Biol Conserv 266. https://doi.org/10.1016/j.biocon.2021.109429

Eger AM, Layton C, McHugh TA, Gleason M, Eddy N (2022c) Kelp restoration guidebook: lessons learned from kelp projects around the world. The Nature Conservancy, Arlington

Eger AM, Marzinelli EM, Christie H, Fagerli CW, Fujita D, Gonzalez AP, Hong SW, Kim JH, Lee LC, McHugh TA, Nishihara GN, Tatsumi M, Steinberg PD, Vergés A (2022d) Global kelp forest restoration: past lessons, present status, and future directions. Biol Rev 97(4):1449–1475. https://doi.org/10.1111/brv.12850

Eger A, Aguirre JD, Altamirano M, Arafeh-Dalmau N, Arroyo NL, Bauer-Civiello AM, Beas-Luna R, Bekkby T, Bellgrove A, Bennett S, Bernal B, Blain CO, Boada J, Branigan S, Bursic J, Cevallos B, Choi CG, Connell SD, Cornwall CE, Earp HS, Eddy N, Ennis LA, Falace A, Ferreira AM, Filbee-Dexter K, Forbes H, Francis P, Franco JN, Geisler KG, Giraldo-Ospina A, Gonzalez AV, Hingorani S, Hohman R, Ivеša L, Kaleb S, Keane JP, Koch SJI, Krumhansl K, Ladah L, Lafont DJ, Layton C, Le DM, Lee LC, Ling SD, Lonhart SI, Malpica-Cruz L, Mangialajo L, McConnell A, McHugh TA, Micheli F, Miller KI, Monserrat M, Montes-Herrera J, Moreno B, Neufeld CJ, Orchard S, Peabody B, Peleg O, Pessarrodona A, Pocklington JB, Reeves SE, Ricart AM, Ross F, Schanz FR, Schreider M, Sedarat M, Smith SM, Starko S, Strain EMA, Tamburello L, Timmer B, Toft JE, Uribe RA, van den Burg SWK, Vásquez JA, Veenhof RJ, Wernberg T, Wood G, Zepeda-Domínguez JA, Vergès A (2024a) The Kelp Forest challenge: a collaborative global movement to protect and restore 4 million hectares of kelp forests. J Appl Phycol 36(2):951–964. https://doi.org/10.1007/s10811-023-03103-y

Eger AM, Bauer-Civeillo A, Bernal B, Bohachyk S, Janke D, Arroyo NL, Schreider M, Earp HS (2024b) Monitoring Kelp Forest ecosystems: a guidebook to quantifying biodiversity, ecosystem health, and ecosystem benefits. Sydney

Estes JA, Tinker MT, Williams TM, Doak DF (1998) Killer whale predation on sea otters linking oceanic and nearshore ecosystems. Science 282(5388):473–476. https://doi.org/10.1126/science.282.5388.473

Evans AJ, Lawrence PJ, Natanzi AS, Moore PJ, Davies AJ, Crowe TP, McNally C, Thompson B, Dozier AE, Brooks PR (2021) Replicating natural topography on marine artificial structures – a novel approach to eco-engineering. Ecol Eng 160:106144. https://doi.org/10.1016/j.ecoleng.2020.106144

Filbee-Dexter K, Scheibling RE (2012) Hurricane-mediated defoliation of kelp beds and pulsed delivery of kelp detritus to offshore sedimentary habitats. Mar Ecol Prog Ser 455:51–64. https://doi.org/10.3354/meps09667

Filbee-Dexter K, Scheibling RE (2014) Sea urchin barrens as alternative stable states of collapsed kelp ecosystems. Mar Ecol Prog Ser 495:1–25. https://doi.org/10.3354/meps10573

Filbee-Dexter K, Smajdor A (2019) Ethics of assisted evolution in marine conservation. Front Mar Sci 6:20. https://doi.org/10.3389/fmars.2019.00020

Filbee-Dexter K, Wernberg T (2018) Rise of turfs: a new battlefront for globally declining kelp forests. Bioscience 68(2):64–76. https://doi.org/10.1093/biosci/bix147

Filbee-Dexter K, Wernberg T, Fredriksen S, Norderhaug KM, Pedersen MF (2019) Arctic kelp forests: diversity, resilience and future. Glob Planet Change 172:1–14. https://doi.org/10.1016/j.gloplacha.2018.09.005

Filbee-Dexter K, Wernberg T, Grace SP, Thormar J, Fredriksen S, Narvaez CN, Feehan CJ, Norderhaug KM (2020) Marine heatwaves and the collapse of marginal North Atlantic kelp forests. Sci Rep 10(1):1–11. https://doi.org/10.1038/s41598-020-70273-x

Filbee-Dexter K, Wernberg T, Barreiro R, Coleman MA, de Bettignies T, Feehan CJ, Franco JN, Hasler B, Louro I, Norderhaug KM, Staehr PAU, Tuya F, Verbeek J (2022) Leveraging the blue economy to transform marine forest restoration. J Phycol 58(2):198–207. https://doi.org/10.1111/jpy.13239

Filbee-Dexter K, Pessarrodona A, Pedersen MF, Wernberg T, Duarte CM, Assis J, Bekkby T, Burrows MT, Carlson DF, Gattuso JP, Gundersen H, Hancke K, Krumhansl KA, Kuwae T, Middelburg JJ, Moore PJ, Queirós AM, Smale DA, Sousa-Pinto I, Suzuki N, Krause-Jensen D (2024a) Carbon export from seaweed forests to deep ocean sinks. Nat Geosci 17(6):552–559. https://doi.org/10.1038/s41561-024-01449-7

Filbee-Dexter K, Starko S, Pessarrodona A, Wood G, Norderhaug KM, Piñeiro-Corbeira C, Wernberg T (2024b) Marine protected areas can be useful but are not a silver bullet for kelp conservation. J Phycol 60(2):203–213. https://doi.org/10.1111/jpy.13446

Flavin K, Flavin N, Flahive B (2013) Kelp Farming Manual. https://maineaqua.org/wp-content/uploads/2020/06/OceanApproved_KelpManualLowRez.pdf

Forbord S, Skjermo J, Arff J, Handå A, Reitan KI, Bjerregaard R, Lüning K (2012) Development of Saccharina latissima (Phaeophyceae) kelp hatcheries with year-round production of zoospores and juvenile sporophytes on culture ropes for kelp aquaculture. J Appl Phycol 24(3):393–399. https://doi.org/10.1007/s10811-011-9784-y

Fragkopoulou E, Serrão EA, De Clerck O, Costello MJ, Araújo MB, Duarte CM, Krause-Jensen D, Assis J (2022) Global biodiversity patterns of marine forests of brown macroalgae. Glob Ecol Biogeogr 31(4):636–648. https://doi.org/10.1111/geb.13450

Fraser CI (2012) Short review: is bull-kelp kelp? The role of common names in science. N Z J Mar Freshw Res 46(2):279–284. https://doi.org/10.1080/00288330.2011.621130

Fredriksen S, Filbee-Dexter K, Norderhaug KM, Steen H, Bodvin T, Coleman MA, Moy F, Wernberg T (2020) Green gravel: a novel restoration tool to combat kelp forest decline. Sci Rep 10(1):1–7. https://doi.org/10.1038/s41598-020-60553-x

Friedlander AM, Ballesteros E, Bell TW, Caselle JE, Campagna C, Goodell W, Hüne M, Muñoz A, Salinas-DeLeón P, Sala E, Dayton PK (2020) Kelp forests at the end of the earth: 45 years later. PLoS One 15(3):e0229259. https://doi.org/10.1371/journal.pone.0229259

Gao K, McKinley KR (1994) Use of macroalgae for marine biomass production and CO_2 remediation: a review. J Appl Phycol 6(1):45–60. https://doi.org/10.1007/BF02185904

Gouraguine A, Moore P, Burrows MT, Velasco E, Ariz L, Figueroa L, Muñoz R, Italo C, Cisternas F, Smale D, Pérez A (2021) The intensity of kelp harvesting shapes the population structure of the foundation species *Lessonia trabeculata* along the Chilean coastline. Mar Biol 168:1–9. https://doi.org/10.1007/s00227-021-03870-7

Hesley D, Burdeno D, Drury C, Schopmeyer S, Lirman D (2017) Citizen science benefits coral reef restoration activities. J Nat Conserv 40:94–99. https://doi.org/10.1016/j.jnc.2017.09.001

Howard J, Sutton-Grier AE, Smart LS, Lopes CC, Hamilton J, Kleypas J, Simpson S, McGowan J, Pessarrodona A, Alleway HK, Landis E

(2023) Blue carbon pathways for climate mitigation: known, emerging and unlikely. Mar Policy 156:105788. https://doi.org/10.1016/j.marpol.2023.105788

Hurd CL, Law CS, Bach LT, Britton D, Hovenden M, Paine ER, Raven JA, Tamsitt V, Boyd PW (2022) Forensic carbon accounting: assessing the role of seaweeds for carbon sequestration. J Phycol 58(3):347–363. https://doi.org/10.1111/jpy.13249

Irving AD, Connell SD (2006) Physical disturbance by kelp abrades erect algae from the understorey. Mar Ecol Prog Ser 324:127–137. https://doi.org/10.3354/meps324127

Jackson-Bué M, Smale DA, King NG, Rushton AG, Moore PJ (2023) Spatial variability in the structure of fish assemblages associated with *Laminaria hyperborea* forests in the NE Atlantic. J Exp Mar Biol Ecol 564:151899. https://doi.org/10.1016/j.jembe.2023.151899

Jara-Yáñez R, Meynard A, Acosta G, Latorre-Padilla N, Oyarzo-Miranda C, Castañeda F, Piña F, Rivas J, Bulboa C, Contreras-Porcia L (2021) Negative consequences on the growth, morphometry, and community structure of the kelp *Macrocystis pyrifera* (Phaeophyceae, Ochrophyta) by a short pollution pulse of heavy metals and PAHs. Toxics 9(8). https://doi.org/10.3390/toxics9080190

Jayathilakea DRM, Costello MJ (2021) Version 2 of the world map of laminarian kelp benefits from more Arctic data and makes it the largest marine biome. Biol Conserv 257:109099. https://doi.org/10.1016/j.biocon.2021.109099

Johnson CR, Chabot RH, Marzloff MP, Wotherspoon S (2017) Knowing when (not) to attempt ecological restoration. Restor Ecol 25(1):140–147. https://doi.org/10.1111/rec.12413

Jupp BP, Drew EA (1974) Studies on the growth of *Laminaria hyperborea* (Gunn.) Fosl. I. Biomass and productivity. J Exp Mar Biol Ecol 15(2):185–196. https://doi.org/10.1016/0022-0981(74)90044-6

Kain JM (1975) The biology of *Laminaria hyperborea* VII. Reproduction of the sporophyte. J Mar Biol Assoc U K 55(3):567–582. https://doi.org/10.1017/S0025315400017264

Kamali B, Hashim R (2011) Mangrove restoration without planting. Ecol Eng 37(2):387–391. https://doi.org/10.1016/j.ecoleng.2010.11.025

Kelp Forest Alliance (2025) Kelp Forest Challenge. https://kelpforestalliance.com/kelp-forest-challenge. Accessed 12 Jan 2025

Kennelly SJ (1987) Inhibition of kelp recruitment by turfing algae and consequences for an Australian kelp community. J Exp Mar Biol Ecol 112(1):49–60. https://doi.org/10.1016/S0022-0981(87)80014-X

Kerrison PD, Stanley MS, Kelly M, MacLeod A, Black KD, Hughes AD (2016) Optimising the settlement and hatchery culture of *Saccharina latissima* (Phaeophyta) by manipulation of growth medium and substrate surface condition. J Appl Phycol 28(2):1181–1191. https://doi.org/10.1007/s10811-015-0621-6

Kim JK, Kraemer GP, Yarish C (2015) Use of sugar kelp aquaculture in Long Island sound and the Bronx River Estuary for nutrient extraction. Mar Ecol Prog Ser 531:155–166. https://doi.org/10.3354/meps11331

King N, Moore P, Wilding C, Jenkins H, Smale D (2021) Multiscale spatial variability in epibiont assemblage structure associated with stipes of kelp *Laminaria hyperborea* in the Northeast Atlantic. Mar Ecol Prog Ser 672:33–44. https://doi.org/10.3354/meps13794

Krause-Jensen D, Duarte CM (2014) Expansion of vegetated coastal ecosystems in the future Arctic. Front Mar Sci 1:1–10. https://doi.org/10.3389/fmars.2014.00077

Krause-Jensen D, Duarte CM (2016) Substantial role of macroalgae in marine carbon sequestration. Nat Geosci 9(10):737–742. https://doi.org/10.1038/ngeo2790

Krause-Jensen D, Marbà N, Sanz-Martin M, Hendriks IE, Thyrring J, Carstensen J, Sejr MK, Duarte CM (2016) Long photoperiods sustain high pH in arctic kelp forests. Sci Adv 2(12):1–9. https://doi.org/10.1126/sciadv.1501938

Krause-Jensen D, Archambault P, Assis J, Bartsch I, Bischof K, Filbee-Dexter K, Dunton KH, Maximova O, Ragnarsdóttir SB, Sejr MK, Simakova U, Spiridonov V, Wegeberg S, Winding MHS, Duarte CM (2020) Imprint of climate change on Pan-Arctic marine vegetation. Front Mar Sci 7:1–28. https://doi.org/10.3389/fmars.2020.617324

Krumhansl KA, Scheibling RE (2012) Production and fate of kelp detritus. Mar Ecol Prog Ser 467:281–302. https://doi.org/10.3354/meps09940

Krumhansl KA, Okamoto DK, Rassweiler A, Novak M, Bolton JJ, Cavanaugh KC, Connell SD, Johnson CR, Konar B, Ling SD, Micheli F, Norderhaug KM, Pérez-Matus A, Sousa-Pinto I, Reed DC, Salomon AK, Shears NT, Wernberg T, Anderson RJ, Barrett NS, Buschmann AH, Carr MH, Caselle JE, Derrien-Courtel S, Edgar GJ, Edwards M, Estes JA, Goodwin C, Kenner MC, Kushner DJ, Moy FE, Nunn J, Steneck RS, Vásquez JA, Watson J, Witman JD, Byrnes JEK (2016) Global patterns of kelp forest change over the past half-century. Proc Natl Acad Sci USA 113(48):13785–13790. https://doi.org/10.1073/pnas.1606102113

Layton C, Johnson CR (2021) Assessing the feasibility of restoring giant kelp forests in Tasmania. Institute for Marine and Antarctic Studies, University of Tasmania. https://www.nespmarine.edu.au/system/files/Layton%20et%20al_E7_M5_Assessing%20the%20feasibility%20of%20restoring%20giant%20kelp%20forests%20in%20Tas.pdf

Layton C, Shelamoff V, Cameron MJ, Tatsumi M, Wright JT, Johnson CR (2019) Resilience and stability of kelp forests: the importance of patch dynamics and environment-engineer feedbacks. PLoS One 14(1):e0210220. https://doi.org/10.1371/journal.pone.0210220

Leclerc JC, Riera P, Laurans M, Leroux C, Lévêque L, Davoult D (2015) Community, trophic structure and functioning in two contrasting *Laminaria hyperborea* forests. Estuar Coast Shelf Sci 152:11–22. https://doi.org/10.1016/j.ecss.2014.11.005

Lee SG (2019) Marine stock enhancement, restocking, and sea ranching in Korea. In: Kideghesho JR, Rija AA (eds) Wildlife management: failures, successes and prospects. IntechOpen, London

Lee LC, McNeill GD, Ridings P, Featherstone M, Okamoto DK, Spindel NB, Galloway AWE, Saunders GW, Adamczyk EM, Reshitnyk L, Pontier O, Post M, Irvine R, Wilson N (2021) Chiixuu tll iinasdll: indigenous ethics and values lead to ecological restoration for people and place in gwaii haanas. Ecol Restor 39(1–2):45–51. https://doi.org/10.3368/er.39.01-02.36

Ling SD, Johnson CR, Frusher SD, Ridgway KR (2009) Overfishing reduces resilience of kelp beds to climate-driven catastrophic phase shift. Proc Natl Acad Sci USA 106(52):22341–22345. https://doi.org/10.1073/pnas.0907529106

Ling SD, Scheibling RE, Rassweiler A, Johnson CR, Shears N, Connell SD, Salomon AK, Norderhaug KM, Pérez-Matus A, Hernández JC, Clemente S, Blamey LK, Hereu B, Ballesteros E, Sala E, Garrabou J, Cebrian E, Zabala M, Fujita D, Johnson LE (2015) Global regime shift dynamics of catastrophic sea urchin overgrazing. Philos Trans R Soc B Biol Sci 370(1659):1–10. https://doi.org/10.1098/rstb.2013.0269

Lotze HK, Milewski I, Fast J, Kay L, Worm B (2019) Ecosystem-based management of seaweed harvesting. Botanica Marina 62:1–16. https://doi.org/10.1515/bot-2019-0027

Lovelock CE, Barbier E, Duarte CM (2022) Tackling the mangrove restoration challenge. PLoS Biol 20(10):1–15. https://doi.org/10.1371/journal.pbio.3001836

Mann KH (1979) (1973) seaweeds: their productivity and strength for growth. Science 182(4116):975–981

Marion SR, Orth RJ (2010) Innovative techniques for large-scale seagrass restoration using *Zostera marina* (eelgrass) seeds. Restor Ecol 18(4):514–526. https://doi.org/10.1111/j.1526-100X.2010.00692.x

Martínez B, Radford B, Thomsen MS, Connell SD, Carreño F, Bradshaw CJA, Fordham DA, Russell BD, Gurgel CFD, Wernberg T (2018) Distribution models predict large contractions of habitat-forming seaweeds in response to ocean warming. Divers Distrib 24(10):1350–1366. https://doi.org/10.1111/ddi.12767

Marzinelli EM, Leong MR, Campbell AH, Steinberg PD, Verges A (2016) Does restoration of a habitat-forming seaweed restore

associated faunal diversity? Restor Ecol 24(1):81–90. https://doi.org/10.1111/rec.12292
Mohring MB, Wernberg T, Kendrick GA, Rule MJ (2013) Reproductive synchrony in a habitat-forming kelp and its relationship with environmental conditions. Mar Biol 160(1):119–126. https://doi.org/10.1007/s00227-012-2068-5
Mork M (1996) The effect of kelp in wave damping. Sarsia 80(4):323–327. https://doi.org/10.1080/00364827.1996.10413607
Morris RL, Graham TDJ, Jaya K, Marco G, Swearer SE (2020a) Kelp beds as coastal protection: wave attenuation of *Ecklonia radiata* in a shallow coastal bay. Ann Bot 125(2):235–246. https://doi.org/10.1093/aob/mcz127
Morris RL, Hale R, Strain EMA, Reeves SE, Vergés A, Marzinelli EM, Layton C, Shelamoff V, Graham TDJ, Chevalier M, Swearer SE (2020b) Key principles for managing recovery of kelp forests through restoration. Bioscience 70(8):688–698. https://doi.org/10.1093/biosci/biaa058
Morris RL, Robin H, Strain EMA, Reeves SE, Adriana V, Marzinelli EM, Cayne L, Victor S, Graham TDJ, Mathilde C, Swearer SE (2020c) Key principles for managing recovery of kelp forests through restoration. Bioscience 70(8):688–698. https://doi.org/10.1093/biosci/biaa058
Narayan S, Beck MW, Reguero BG, Losada IJ, van Wesenbeeck B, Pontee N, Sanchirico JN, Ingram JC, Lange G-M, Burks-Copes KA (2016) The effectiveness, costs and coastal protection benefits of natural and nature-based defences. PLoS One 11(5). https://doi.org/10.1371/journal.pone.0154735
O'Brien JM, Scheibling RE (2018) Low recruitment, high tissue loss, and juvenile mortality limit recovery of kelp following large-scale defoliation. Mar Biol 165(10):1–19. https://doi.org/10.1007/s00227-018-3423-y
Omori M (2019) Coral restoration research and technical developments: what we have learned so far. Mar Biol Res 15(7):377–409. https://doi.org/10.1080/17451000.2019.1662050
Oyarzo-Miranda C, Latorre N, Meynard A, Rivas J, Bulboa C, Contreras-Porcia L (2020) Coastal pollution from the industrial park Quintero bay of Central Chile: effects on abundance, morphology, and development of the kelp *Lessonia spicata* (Phaeophyceae). PLoS One 15:1–24. https://doi.org/10.1371/journal.pone.0240581
Pérez-Matus A, Muñoz-Cordovez R, Ruz CS, Gonzalez-Alonso-de-Linaje C, Siñuela JP, Earp HS, Gouraguine A,Hinostroza D, Gammara A, Uribe RA, Smale DA, Burrows MT, Gelcich S, Faugeron S, Valdivia N, Moore PJ(2025). Management regime influences the recovery of subtidal kelp forests following harvesting. J App Eco 00:1–12. https://doi.org/10.1111/1365-2664.70134
Pessarrodona A, Assis J, Filbee-Dexter K, Burrows MT, Gattuso JP, Duarte CM, Krause-Jensen D, Moore PJ, Smale DA, Wernberg T (2022) Global seaweed productivity. Sci Adv 8(37). https://doi.org/10.1126/sciadv.abn2465
Pinheiro TF (2022) Developing kelp reforestation methods. MSc Thesis, Universidade do Porto
Price MR, Hadfield MG (2014) Population genetics and the effects of a severe bottleneck in an *ex situ* population of critically endangered Hawaiian tree snails. PLoS One 9(12):1–16. https://doi.org/10.1371/journal.pone.0114377
Redmond S, Green L, Yarish C, Kim J, Neefus C (2014) New England seaweed culture handbook. Seaweed Cultivation (Paper 1):1–92
Reklev S (2023) Start-up sells first kelp reforestation credits, eyes registration under leading carbon standards. In: Carbon Pulse. https://carbon-pulse.com/196301/. Accessed 13 Sept 2024
Reshitnyk LY, Denouden TA, Holmes KR (2024) The Kelp-O-Matic: a novel artificial intelligence convolutional neural network fully automates the detection of canopy-forming Kelp species (*Macrocystis Pyrifera* and *Nereocystis Luetkeana*) in high resolution Rgb Imagery. https://doi.org/10.2139/ssrn.4838692
Rinkevich B (2005) Conservation of coral reefs through active restoration measures: recent approaches and last decade progress. Environ Sci Technol 39(12):4333–4342. https://doi.org/10.1021/es0482583
Rolin C, Inkster R, Laing J, Hedges J, Mcevoy L (2014) Seaweed cultivation manual. Shetland Seaweed Growers Project 2014–16. Shetland Seaweed Growers Project 2014–16:39
Ruiz-Jaen MC, Aide TM (2005) Restoration success: how is it being measured? Restor Ecol 13(3):569–577. https://doi.org/10.1111/j.1526-100X.2005.00072.x
Saier B, Chapman AS (2004) Crusts of the alien bryozoan *Membranipora membranacea* can negatively impact spore output from native kelps (*Laminaria longicruris*). Bot Mar 47(4):265–271. https://doi.org/10.1515/BOT.2004.031
Saunders MI, Doropoulos C, Bayraktarov E, Babcock RC, Gorman D, Eger AM, Vozzo ML, Gillies CL, Vanderklift MA, Steven ADL, Bustamante RH, Silliman BR (2020) Bright spots in coastal marine ecosystem restoration. Curr Biol 30(24):R1500–R1510. https://doi.org/10.1016/j.cub.2020.10.056
Schaal G, Leclerc JC, Droual G, Leroux C, Riera P (2016) Biodiversity and trophic structure of invertebrate assemblages associated with understorey red algae in a Laminaria digitata bed. Mar Biol Res 12(5):513–523. https://doi.org/10.1080/17451000.2016.1164318
Scheibling RE, Gagnon P (2009) Temperature-mediated outbreak dynamics of the invasive bryozoan *Membranipora membranacea* in Nova Scotian kelp beds. Mar Ecol Prog Ser 390:1–13. https://doi.org/10.3354/meps08207
Schiel DR, Foster MS (2015) The biology and ecology of Giant kelp forests. California University Press, Oakland
Seidl A, Cumming T, Arlaud M, Crossett C, van den Heuvel O (2024) Investing in the wealth of nature through biodiversity and ecosystem service finance solutions. Ecosyst Serv 66(July 2023):101601. https://doi.org/10.1016/j.ecoser.2024.101601
Smale DA, Wernberg T (2013) Extreme climatic event drives range contraction of a habitat-forming species. Proc R Soc B Biol Sci 280(1754). https://doi.org/10.1098/rspb.2012.2829
Smale DA, Burrows MT, Moore P, O'Connor N, Hawkins SJ (2013) Threats and knowledge gaps for ecosystem services provided by kelp forests: a Northeast Atlantic perspective. Ecol Evol 3(11):4016–4038. https://doi.org/10.1002/ece3.774
Smale DA, Wernberg T, Yunnie ALE, Vance T (2015) The rise of *Laminaria ochroleuca* in the Western English Channel (UK) and comparisons with its competitor and assemblage dominant Laminaria hyperborea. Mar Ecol 36(4):1033–1044. https://doi.org/10.1111/maec.12199
Smale DA, Epstein G, Hughes E, Mogg AOM, Moore PJ (2020) Patterns and drivers of understory macroalgal assemblage structure within subtidal kelp forests. Biodivers Conserv 29(14):4173–4192. https://doi.org/10.1007/s10531-020-02070-x
Smale DA, King NG, Jackson-Bué M, Moore PJ (2022) Quantifying use of kelp forest habitat by commercially important crustaceans in the United Kingdom. J Mar Biol Assoc U K 102(8):627–634. https://doi.org/10.1017/s0025315422001023
Smith KE, Aubin M, Burrows MT, Filbee-Dexter K, Hobday AJ, Holbrook NJ, King NG, Moore PJ, Sen GA, Thomsen M, Wernberg T, Wilson E, Smale DA (2024) Global impacts of marine heatwaves on coastal foundation species. Nat Commun 1–14. https://doi.org/10.1038/s41467-024-49307-9
Society for Ecological Restoration (2020) What is ecological restoration? www.ser-rrc.org/what-is-ecological-restoration/. Accessed 2 Jun 2020
Spalding MD, Fox HE, Allen GR, Davidson N, Ferdaña ZA, Finlayson M, Halpern BS, Jorge MA, Lombana A, Lourie SA, Martin KD, McManus E, Molnar J, Recchia CA, Robertson J (2007) Marine ecoregions of the world: a bioregionalization of coastal and shelf areas. Bioscience 57(7):573–583. https://doi.org/10.1641/B570707

Steneck RS, Graham MH, Bourque BJ, Corbett D, Erlandson JM, Estes JA, Tegner MJ (2002) Kelp forest ecosystems: biodiversity, stability, resilience and future. Environ Conserv 29(4):436–459. https://doi.org/10.1017/S0376892902000322

Stuart OP, Cleave R, Pearce K, Magrath MJL, Mikheyev AS (2024) Gene flow stimulates recovery of reproductive fitness in a captive bred insect. Insect Conserv Divers:1–14. https://doi.org/10.1111/icad.12790

Suebsanguan S, Strain EMA, Morris RL, Swearer SE (2021) Optimizing the initial cultivation stages of kelp *Ecklonia radiata* for restoration. Restor Ecol 29(5):0–1. https://doi.org/10.1111/rec.13388

Supratya VP, Martone PT (2024) Kelps on demand: closed-system protocols for culturing large bull kelp sporophytes for research and restoration. J Phycol 60(1):73–82. https://doi.org/10.1111/jpy.13413

Tan YM, Dalby O, Kendrick GA, Statton J, Sinclair EA, Fraser MW, Macreadie PI, Gillies CL, Coleman RA, Waycott M, van Dijk KJ, Vergés A, Ross JD, Campbell ML, Matheson FE, Jackson EL, Irving AD, Govers LL, Connolly RM, McLeod IM, Rasheed MA, Kirkman H, Flindt MR, Lange T, Miller AD, Sherman CDH (2020) Seagrass restoration is possible: insights and lessons from Australia and New Zealand. Front Mar Sci 7. https://doi.org/10.3389/fmars.2020.00617

Teagle H, Hawkins SJ, Moore PJ, Smale DA (2017) The role of kelp species as biogenic habitat formers in coastal marine ecosystems. J Exp Mar Biol Ecol 492:81–98. https://doi.org/10.1016/j.jembe.2017.01.017

Teagle H, Moore PJ, Jenkins H, Smale DA (2018) Spatial variability in the diversity and structure of faunal assemblages associated with kelp holdfasts (*Laminaria hyperborea*) in the Northeast Atlantic. PLoS One 13(7):1–25. https://doi.org/10.1371/journal.pone.0200411

Tegner MJ, Dayton PK (2000) Ecosystem effects of fishing in kelp forest communities. ICES J Mar Sci 57(3):579–589. https://doi.org/10.1006/jmsc.2000.0715

Thomsen MS, Mondardini L, Alestra T, Gerrity S, Tait L, South PM, Lilley SA, Schiel DR (2019) Local extinction of bull kelp (*Durvillaea* spp.) due to a marine heatwave. Front Mar Sci 6:1–10. https://doi.org/10.3389/fmars.2019.00084

Tuya F, Larsen K, Platt V (2011) Patterns of abundance and assemblage structure of epifauna inhabiting two morphologically different kelp holdfasts. Hydrobiologia 658(1):373–382. https://doi.org/10.1007/s10750-010-0527-x

United Nations Environment Programme (2023) Into the blue: securing a sustainable future for kelp forests. Nairobi

United Nations General Assembly (2019) United Nations Decade on Ecosystem Restoration (2021–2030). Resolution adopted by the General Assembly on 1 March 2019. In: Resolution 73/284: united nations decade on ecosystem restoration –2030). https://undocs.org/A/RES/73/284. Accessed 13 Sept 2024

Unsworth RKF, Bertelli CM, Cullen-Unsworth LC, Esteban N, Jones BL, Lilley R, Lowe C, Nuuttila HK, Rees SC (2019) Sowing the seeds of seagrass recovery using hessian bags. Front Ecol Evol 7:1–7. https://doi.org/10.3389/fevo.2019.00311

Urchinomics (2023) The impact of kelp restoration benefits many globally. In: Kelp Restoration. https://www.urchinomics.com/kelp-restoration. Accessed 13 Sept 2024

van Katwijk MM, Thorhaug A, Marbà N, Orth RJ, Duarte CM, Kendrick GA, Althuizen IHJ, Balestri E, Bernard G, Cambridge ML, Cunha A, Durance C, Giesen W, Han Q, Hosokawa S, Kiswara W, Komatsu T, Lardicci C, Lee KS, Meinesz A, Nakaoka M, O'Brien KR, Paling EI, Pickerell C, Ransijn AMA, Verduin JJ (2016) Global analysis of seagrass restoration: the importance of large-scale planting. J Appl Ecol 53(2):567–578. https://doi.org/10.1111/1365-2664.12562

Vanderklift MA, Herr D, Lovelock CE, Murdiyarso D, Raw JL, Steven ADL (2022) A guide to international climate mitigation policy and finance frameworks relevant to the protection and restoration of blue carbon ecosystems. Front Mar Sci 9(July):1–14. https://doi.org/10.3389/fmars.2022.872064

Vásquez JA, Zuñiga S, Tala F, Piaget N, Rodríguez DC, Vega JMA (2014) Economic valuation of kelp forests in northern Chile: values of goods and services of the ecosystem. J Appl Phycol 26(2):1081–1088. https://doi.org/10.1007/s10811-013-0173-6

Vergés A, Doropoulos C, Malcolm HA, Skye M, Garcia-Pizá M, Marzinelli EM, Campbell AH, Ballesteros E, Hoey AS, Vila-Concejo A, Bozec YM, Steinberg PD (2016) Long-term empirical evidence of ocean warming leading to tropicalization of fish communities, increased herbivory, and loss of kelp. Proc Natl Acad Sci USA 113(48):13791–13796. https://doi.org/10.1073/pnas.1610725113

Vergés A, Campbell AH, Wood G, Kajlich L, Eger AM, Cruz D, Langley M, Bolton D, Coleman MA, Turpin J, Crawford M, Coombes N, Camilleri A, Steinberg PD, Marzinelli EM (2020) Operation Crayweed: ecological and sociocultural aspects of restoring Sydney's underwater forests. Ecol Manag Restor 21(2):74–85. https://doi.org/10.1111/emr.12413

Visch W, Lush H, Schwoerbel J, Hurd CL (2023) Nursery optimization for kelp aquaculture in the southern hemisphere: the interactive effects of temperature and light on growth and contaminants. Appl Phycol 4(1):44–53. https://doi.org/10.1080/26388081.2023.2174903

Visch W, Biancacci C, Sanderson JC, Schwoerbel J, Hurd L, Bellgrove A, Macleod C (2024) Aquaculture production of Australian Laminarian kelps: a manual and research recommendations for *Ecklonia radiata, Lessonia corrugata*, and *Macrocystis pyrifera*. Technical Rep. Univ Tasmania

Waller JD, Reardon KM, Caron SE, Jenner BP, Summers EL, Wilson CJ (2021) A comparison of the size at maturity of female American lobsters (*Homarus americanus*) over three decades and across coastal areas of the Gulf of Maine using ovarian staging. ICES J Mar Sci 78(4):1267–1277. https://doi.org/10.1093/icesjms/fsab034

Watanuki A, Aota T, Otsuka E, Kawai T, Iwahashi Y, Kuwahara H, Fujita D (2010) Restoration of kelp beds on an urchin barren: removal of sea urchins by citizen divers in southwestern Hokkaido. Bull Fish Res Agen 32:83–87

Wernberg T, Filbee-Dexter K (2019) Missing the marine forest for the trees. Mar Ecol Prog Ser 612:209–215. https://doi.org/10.3354/meps12867

Wernberg T, Kendrick GA, Toohey BD (2005) Modification of the physical environment by an *Ecklonia radiata* (Laminariales) canopy and implications for associated foliose algae. Aquat Ecol 39(4):419–430. https://doi.org/10.1007/s10452-005-9009-z

Wernberg T, Thomsen MS, Tuya F, Kendrick GA, Staehr PA, Toohey BD (2010) Decreasing resilience of kelp beds along a latitudinal temperature gradient: potential implications for a warmer future. Ecol Lett 13(6):685–694. https://doi.org/10.1111/j.1461-0248.2010.01466.x

Wernberg T, Smale DA, Tuya F, Thomsen MS, Langlois TJ, De Bettignies T, Bennett S, Rousseaux CS (2013) An extreme climatic event alters marine ecosystem structure in a global biodiversity hotspot. Nat Clim Chang 3(1):78–82. https://doi.org/10.1038/nclimate1627

Wernberg T, Bennett S, Babcock RC, De Bettignies T, Cure K, Depczynski M, Dufois F, Fromont J, Fulton CJ, Hovey RK, Harvey ES, Holmes TH, Kendrick GA, Radford B, Santana-garcon J, Saunders BJ, Smale DA, Thomsen MS, Tuckett CA, Tuya F, Vanderklift MA, Wilson S (2016) Climate-driven regime shift of a temperate marine ecosystem. Science 353(6295):169–172. https://doi.org/10.1126/science.aad8745

Wernberg T, Krumhansl K, Filbee-Dexter K, Pedersen MF (2019) Status and trends for the world's kelp forests. World seas: an environmental evaluation volume III: ecological issues and environmental impacts: 57–78. https://doi.org/10.1016/B978-0-12-805052-1.00003-6

Wernberg T, Thomsen MS, Baum JK, Bishop MJ, Bruno JF, MelindaA C, Filbee-Dexter K, Gagnon K, He Q, Murdiyarso D, Rogers K, Silliman BR, Smale DA, Starko S, Vanderklift MA (2024) Impacts of climate change on marine foundation species. Ann Rev Mar Sci 16:18.1–18.36. https://doi.org/10.1163/9789004204225_008

Westermeier R, Murúa P, Patiño DJ, Muñoz L, Müller DG (2016) Holdfast fragmentation of *Macrocystis pyrifera* (integrifolia morph) and *Lessonia berteroana* in Atacama (Chile): a novel approach for kelp bed restoration. J Appl Phycol 28(5):2969–2977. https://doi.org/10.1007/s10811-016-0827-2

Whitton Z, Waterford L (2023) Nature finance focus: tracking global trends in nature investment. https://pollinationgroup.com/wp-content/uploads/2025/06/Pollination_NatureFinanceFocus_2025.pdf

Wilding CM, Earp HS (2024) Attaching seaweed spores to used scallop shells could restore UK'S coastal kelp forests. The Conversation

Wilding CM, Earp HS, Cooper CN, Lubelski A, Smale DA (2023) British Kelp Forest Restoration: Feasibility Report

Wood G, Marzinelli EM, Coleman MA, Campbell AH, Santini NS, Kajlich L, Verdura J, Wodak J, Steinberg PD, Vergés A (2019) Restoring subtidal marine macrophytes in the Anthropocene: trajectories and future-proofing. Mar Freshw Res 70(7):936–951. https://doi.org/10.1071/MF18226

Wood G, Marzinelli EM, Vergés A, Campbell AH, Steinberg PD, Coleman MA (2020) Using genomics to design and evaluate the performance of underwater forest restoration. J Appl Ecol 57(10):1988–1998. https://doi.org/10.1111/1365-2664.13707

Wood G, Marzinelli EM, Campbell AH, Steinberg PD, Vergés A, Coleman MA (2021) Genomic vulnerability of a dominant seaweed points to future-proofing pathways for Australia's underwater forests. Glob Chang Biol 27(10):2200–2212. https://doi.org/10.1111/gcb.15534

Wood GV, Filbee-Dexter K, Coleman MA, Valckenaere J, Aguirre JD, Bentley PM, Carnell P, Dawkins PD, Dykman LN, Earp HS, Ennis LB, Francis P, Franco JN, Hayford H, Lamb JB, Ling SD, Layton C, Lis E, Masters B, Miller N, Moore PJ, Neufeld C, Pocklington JB, Smale D, Stahl F, Starko S, Steel SC, Verbeek J, Vergés A, Wilding CM, Wernberg T (2024a) Upscaling marine forest restoration: challenges, solutions and recommendations from the Green gravel action group. Front Mar Sci 11:1–15. https://doi.org/10.3389/fmars.2024.1364263

Wood GV, Griffin KJ, van der Mheen M, Breed MF, Edgeloe JM, Grimaldi C, Minne AJP, Popovic I, Filbee-Dexter K, van Oppen MJH, Wernberg T, Coleman MA (2024b) Reef adapt: a tool to inform climate-smart marine restoration and management decisions. Commun Biol 7:1368. https://doi.org/10.1038/s42003-024-06970-4

Young CN, Schopmeyer SA, Lirman D (2012) A review of reef restoration and coral propagation using the threatened genus *Acropora* in the Caribbean and western Atlantic. Bull Mar Sci 88(4):1075–1098. https://doi.org/10.5343/bms.2011.1143

Dr. Hannah S. Earp is a postdoctoral researcher at the Norwegian Institute of Marine Research. Her research investigates the impact of climate change and anthropogenic pressures on the structure and resilience of kelp forests and their associated biodiversity using a combination of field and laboratory approaches. She is also actively involved in kelp restoration research. To date she has worked on kelp forests in the United Kingdom, Norway, Chile and Peru. Her background includes research on opportunistic macroalgal blooms, ecological engineering and marine citizen science.

Dr. Georgina V. Wood is an Early Career Australian Research Council Fellow at Flinders University and Adjunct Research Fellow at the University of Western Australia. Originally from Sydney, she has a predilection for temperate marine forests and has conducted research spanning Australia's Great Southern Reef and as far afield as Norway. George works at the frontier of ecology and evolutionary biology, has led several marine restoration projects in collaboration with industry and is passionate about developing and applying novel techniques to transform the way we understand and manage marine ecosystems under climate change. George is now pioneering marine restoration decision-making tools, starting with the Reef Adapt online webtool for climate-smart marine restoration provenancing. She is currently an editor of Restoration Ecology and a coordinator of the Green Gravel Action Group for kelp forest restoration.

Dr. Karen Filbee-Dexter is an Australian Research Council Future Fellow based at the University of Western Australia's School of Biological Sciences and a researcher at the Norwegian Institute of Marine Research. Her research focusses on the drivers and patterns of long-term change in coastal ecosystems, particularly kelp forests. She has expertise in shifts to alternative stable states (e.g. turf reefs, urchin barrens), carbon cycling, restoration approaches and the expansion of kelp forests into the Arctic. She also tackles interdisciplinary questions such as links between social and ecological systems during regime shifts, and ethical questions associated with emerging use of genetic restoration techniques.

Prof. Thomas Wernberg is an Australian Research Council Laureate Fellow and a professor of Marine Botany who studies seaweed forests and climate change. He heads the Wernberg Lab (wernberglab.org), at the University of Western Australia, and he is a Principal Scientist at the Norwegian Institute of Marine Science. Thomas is an editor-in-chief for the journal Aquatic Botany, he is a Clarivate highly cited scientist and ranked among the top 200 most influential climate scientists in the world by Reuters.

Navigating Legal Waters: Seaweed Product Regulations in the United States and Canada

26

Spencer Serin

Abstract

The production and marketing of novel seaweed products in the United States and Canada for human, animal, and plant applications has increased in the past decade. Companies are looking to utilize domestic supply—e.g., seaweed farms—as a sustainable source of high-value raw material for bioproduct manufacture. Adherence to historical regulatory compliance frameworks can be difficult as new product types continue to be identified. There is little guidance on seaweed-specific regulatory requirements in North America beyond certain niche applications (dried whole plants, fertilizer, and animal feed). Consumers expect that seaweed products manufactured in the United States and Canada should at a minimum meet the regulations defined by their application—human, animal, or plant—and provide nutritional or health benefits (or both) consistent with product claims. This chapter discusses the relevant sections of existing guidelines for producers and regulators to be familiar with when determining new product innovation strategies or evaluating product compliance.

Keywords

North America seaweed regulations · Seaweed food ingredients · Biostimulant · Feed additives · Quality assurance/control

The domestic production and marketing of seaweed products in the United States and Canada for human, animal, and plant applications has greatly increased in the past decade as more companies emerge looking to utilize this sustainable feedstock with staggering chemical complexity. Plaguing the new developers of these products is the uncertainty surrounding the definition and regulatory compliance strategies to best support new innovations while mitigating the risk of shifting policy framework. Most products in these countries result from importation of seaweed derivatives from other locations, primarily Asia, sourced via cultivation or wild harvest. As such, there are few "seaweed-specific" guidelines available as opposed to larger comparable industries like seafood, fruits/vegetables, and other natural ingredients. Efforts have been made to define where seaweed could fit within federal frameworks, such as US Food (Janasie 2023), but these continue to be speculative; at the time of this writing, there is very little guidance on seaweed-specific regulatory requirements beyond certain niche applications (dried whole plants, fertilizer, and animal feed). Seaweed products manufactured in the United States and Canada should at a minimum meet the regulations defined by their application—human, animal, or plant—and provide nutritional or health benefits (or both) consistent with claim requirements. This chapter discusses the relevant sections of these guidelines for producers and regulators to be familiar with when determining new product innovation strategies or evaluating product compliance.

Regulatory guidelines for the United States and Canada products are contained within different documents depending on the intended application. US rules are broadly published in the Code of Federal Regulations (CFR), with Title 21 containing the rules for Food and Drugs that are managed by the Food and Drug Administration (FDA). This code provides guidance for mandatory and/or acceptable practices in the production, administration, and marketing of products intended for human or animal use. This includes acceptable labeling and product definitions that can inform business activities (product testing, quality management practices, and advertisement). Information is provided for operators of processing facilities describing mandatory practices that must be followed to permit the sale of products. The FDA

S. Serin (✉)
Marine Biologics, San Fransisco, CA, United States
e-mail: spencer@marinebiologics.com

A. T. Critchley et al. (eds.), *Advances in Temperate Phyconomy: Algal Harvest and Cultivation in Globally Distributed Temperate Waters*, Developments in Applied Phycology 15, https://doi.org/10.1007/978-3-032-02955-3_26

oversees the broad range of product types—namely those intended as food and drugs for humans or animals—and is specifically focused on process regulations that ensure these products are safe, properly labeled, and produced under sanitary conditions. Subchapters within 21 CFR refer to more specific product applications such as Food for Human Consumption (B), Cosmetics (G), and Animal Drugs, Feeds, and Related Products (E).

In Canada, Health Canada is responsible for setting standards and providing advice and information on the safety and nutritional value of food, drugs, and plant products. The Canadian Food Inspection Agency (CFIA) enforces the food safety and nutritional quality standards established by Health Canada. The applicable regulations for these products are provided in the Safe Food for Canadians Regulations (SFCR), and other associated regulations listed in Table 26.1. Like the US CFR Title 21, the Canadian regulation documents describe mandatory and/or acceptable practices in the production, administration, and marketing of products. Where possible, examples are made throughout this chapter to the relevant CFR or Canadian regulation sections that describe current policies and how they may be interpreted for seaweed. Drug and cosmetic applications are omitted.

Table 26.1 Select regulations in the US and Canada applicable to seaweed products

User	Product example	US reference	Canada reference
Human	Food ingredient	21 CFR Subchapter B: Food for Human Consumption	Safe Food for Canadians Regulations: SOR/2018-108
Animal	Feed ingredient	21 CFR Subchapter E: Animal Drugs, Feeds, and Related Products	Feeds Regulations: SOR/2024-132
	Veterinary drug/modifier		
Plant	Fertilizer	State-level	Fertilizers Regulations: C.R.C., c. 666

26.1 Human Food

26.1.1 Relevant Definitions

Food—(1) articles used for food or drink for man or other animals, (2) chewing gum, and (3) articles used for components of any such article (Federal Food, Drug, and Cosmetic Act).

Food additive—substances the intended use of which results or may reasonably be expected to result, directly or indirectly, either in their becoming a component of food or otherwise affecting the characteristics of food (21 CFR § 170.3).

Processed Food—any food other than a raw agricultural commodity and includes any raw agricultural commodity that has been subject to processing, such as canning, cooking, freezing, dehydration, or milling (Federal Food, Drug, and Cosmetic Act).

Brown Algae—Seaweeds of the species: Analipus japonicus, Eisenia bicyclis, Hizikia fusiforme, Kjellmaniella gyrata, Laminaria angustata, Laminaria claustonia, Laminaria digitata, Laminaria japonica, Laminaria longicruris, Laminaria longissima, Laminaria ochotensis, Laminaria saccharina, Macrocystis pyrifera, Petalonia fascia, Scytosiphon lomentaria, and Undaria pinnatifida. They are harvested principally in coastal waters of the northern Atlantic and Pacific oceans. The material is dried and ground or chopped for use in food (21 CFR § 184.1120).

Red Algae—Seaweeds of the species: Gloiopeltis furcata, Porphyra crispata, P. deutata, Porphyra perforata, P. suborbiculata, P. tenera and Rhodymenia palmata. Porphyra and Rhodymenia are harvested principally along the coasts of Japan, Korea, China, Taiwan, and the East and West coasts of the United States. Gloiopeltis is harvested principally in southern Pacific coastal waters. The material is dried and ground or chopped for use in food (21 CFR § 184.1121).

Spice—any aromatic vegetable substance in the whole, broken, or ground form, [...] whose significant function in food is seasoning rather than nutritional; that is true to name; and from which no portion of any volatile oil or other flavoring principle has been removed (21 CFR § 101.22).

Dietary Supplement—means a product (other than tobacco) intended to supplement the diet that bears or contains one or more of the following dietary ingredients: (A) a vitamin; (B) a mineral; (C) an herb or other botanical; (D) an amino acid; (E) a dietary substance for use by man to supplement the diet by increasing the total dietary intake; or (F) a concentrate, metabolite, constituent, extract, or combination of any described above (Federal Food, Drug, and Cosmetic Act).

26.1.2 Product Types

Seaweed use in food has spanned millennia as it contains a wide range of bioactive compounds, flavor, and nutritional benefits (McHugh 2003). Globally consumed brown seaweed is primarily types of kelp—the largest species being Kombu (*Saccharina japonica*) and Wakame (*Undaria pinnatifida*)—which are most often found as whole dried or flaked products.

The culinary approaches are based around rehydration which will cook, impart flavor to stocks (e.g., dashi), and/or convert the seaweed to a form more suitable for consumption. The other major brown seaweed application is in additive manufacturing where it is extracted to create products such as sodium/calcium alginate and alginic acid. Globally, red seaweed mariculture exceeds brown in terms of total biomass and is dominated by the species *Eucheuma*, *Gracilaria*, and Nori (*Porphyra/Pyropia* spp.). Red species are primarily used to produce the hydrocolloids carrageenans (*Eucheuma/Kappaphycus*) and agar (*Gracilaria*) but may be consumed in dried forms such as Nori sheets. Seaweed flakes and powders can be found as food ingredients or consumer products often labeled as a "spice" and marketed with language closely resembling a "dietary supplement." Brown seaweed is specifically described as a dietary supplement intended as an iodine source (21 CFR § 172.365).

26.1.3 Regulations

The United States and Canada have a reciprocal approach to food safety, i.e., they have reached a recognition arrangement regarding the operation and oversight of the food safety system of the United States, as implemented by FDA, and the food safety system of Canada, as implemented by CFIA and Health Canada. Therefore, references are primarily made to CFR with respect to product definitions, but these are expected to be equally valid in Canada. A noticeable exception is made with respect to dietary supplements and natural health products which fall under country specific regulations.

Selected dried species of brown and red algae meet the Generally Recognized as Safe (GRAS) standard under the category "direct food substances" and are noted above. The seaweed names listed in these categories may be antiquated but to date this is the complete list of dried seaweed recognized as GRAS in the CFR. Brown seaweed intended for this use must meet requirements as established in Food Chemicals Codex summarized in Table 26.2 (National Research Council 1981). Red algae have similar chemical standards with the exception for iodine and moisture levels. Notably omitted from the brown and red algae definitions are species intended for hydrocolloid extraction (alginates, carrageenans, agar) such as Laminaria *hyperborea, Eucheuma/Kappaphycus,* and *Gracilaria*. The brown and red algae definitions apply to dried compounds exclusively for use as flavor enhancers or adjuvants—i.e., spice—per the CFR § 101.22(a)(2).

Extracts or common seaweed derivatives have specific definitions that meet GRAS standards per CFR that describe a specific use or manufacturing requirement. For some additives, this includes a maximum level of use in specific types of food (as served). For example, sodium alginate can only present at a certain level in condiments and relishes (1%) which differs to its maximum inclusion rate in hard candy (10%). More information is available by reviewing the pertinent CFR section for a given seaweed-based food additive (Table 26.3).

Table 26.2 Specified chemical requirements for brown and red algae according to CFR

Analysis	*Brown algae*	Red algae
Arsenic (as As, inorganic)	3 ppm	3 ppm
Ash (Total)	45%	45%
Heavy Metals (as Pb)	0.004%	0.004%
Iodine	0.1–0.5%	0.05%
Lead	10 ppm	10 ppm
Loss on drying	13%	20%

Kelp as a direct food additive for dietary and nutritional iodine is defined under 21 CFR § 172.365. Specific seaweed extracts, such as fucoidan manufactured from *Undaria pinnatifida* (GRN No. 565) and *Fucus vesiculosus* (GRN No. 661) for use in baked goods, soups, snack foods, imitation dairy products, and seasoning/flavors at levels up to 30 mg per serving, have received GRAS approval from FDA. It is important to note that novel ingredients refined from seaweed may require a GRAS or New Dietary Ingredient (NDI) notification to the FDA, as seen above for fucoidan. If an ingredient is produced using a new production method, it may also trigger a need for a GRAS or NDI notification to the FDA.

A raw agricultural commodity is any food in any food in "its raw or natural state" (Federal Food, Drug, and Cosmetic Act). According to the FDA, this definition applies for harvested seaweed (Janasie 2023). Distribution of fresh seaweed is complicated by its highly perishable nature. Seaweed may still be considered a raw agricultural commodity if dried whole; however, as soon as the seaweed is cut, dried, packaged, or otherwise handled in a way for sale to consumers, it will likely be considered a processed food and manufacturers would be subject to compliance under the Food Safety Modernization Act (USA) or Safe Food for Canadians Regulations (Canada). This rule requires facilities to have comprehensive food safety policies in place such as Current Good Manufacturing Practices (CGMPs) which includes hazard analysis and risk-based preventive controls. This document only focuses on seaweed-specific hazards which can support the development of CGMPs.

26.1.4 Hazard Analysis

An operator must perform a detailed hazard analysis of all process steps to ensure preventive controls that are in place to control food safety are adequate. Hazards are separated

Table 26.3 Seaweed-based food additives generally recognized as safe (GRAS)

Category	Additive	21 CFR Section
Alginic acid + derivatives	Alginic acid	184.1011
	Sodium alginate	184.1724
	Ammonium alginate	184.1133
	Calcium alginate	184.1187
	Potassium alginate	184.1610
	Propylene Glycol Alginate	184.1610
Carrageenan + red seaweed derivates	Carrageenan	172.858
	Salts of carrageenan	172.620
	Carrageenan with polysorbate 80	172.626
	Furcelleran[a]	172.623
	Chondrus extract	182.7255
	Agar-agar	184.1115

[a]Prepared by aqueous extraction of *Furcellaria fastigiata*

into three classes: chemical, biological, and physical. Below, examples of these hazards in seaweed manufacturing are provided. Detailed overviews of the food safety hazards associated with seaweeds are summarized elsewhere (Banach et al. 2020; FAO and WHO 2022).

26.1.4.1 Chemical Hazards

The chemical hazards most associated with seaweed are toxic metals and iodine. The heavy metals—As, Cd, Hg, Pb—can be sequestered during growing and their presence has attracted the greatest attention from policy makers. Arsenic, in oxidation states 3+ and 5+, is incredibly common in marine environments and exists in two major forms, inorganic and organic. The inorganic forms are generally more hazardous than organic, and higher oxidation states further increase the toxicity (Kuivenhoven and Mason 2023). Arsenic can be found at high levels (>40 ppm) in dried seaweed, but only a small percentage of that is inorganic (Yanli et al. 2024). Arsenic speciation testing, which determines inorganic and organic forms using separation methods such as High-Performance Liquid Chromatography (HPLC), is substantially more expensive than total arsenic testing using higher throughput methods like Inductively Coupled Plasma (ICP) in tandem with Mass Spectrometry (MS) or Optical Emission Spectrometry (OES).

Although legislation exists for heavy metals in seaweed used in/as animal feed (discussed below), no such standards have been developed yet for seaweed used as food. Nevertheless, any ingredient specification prepared for seaweed should describe acceptable levels of heavy metals in any raw ingredient or finished processed forms. French recommendations for maximum tolerable levels (MTLs) of inorganic As, Cd, Pb, and Hg were published for edible seaweeds by the French Agency for Food, Environment and Occupational Health and Safety (ANSES) in 2020 (ANSES 2020). This provides some scientifically vetted MTLs that may be adopted by producers to document compliance of finished products.

Acute iodine toxicity, typically caused by overconsumption of supplements containing iodine, represents a considerable issue for food and feed manufacturers using seaweed, particularly kelp. The recommended dietary allowance varies: 150 μg/day for adults, 220–250 μg/day for pregnant women, and 250–290 μg/day for breastfeeding women. In general, 1 mg/day is considered safe for most individuals. For kelp intended as a dietary supplement, daily ingestion of the additive must be limited to a maximum amount of iodine of 225 ug (21 CFR § 172.365). The iodine level of most dried kelp is 1–4 mg/g DW, with exception of *Alaria* spp. which consistently contains less (Roleda and Skjermo 2018). At this value, kelp intended to supplement iodine could only be added to food so that each serving contains 225 mg or less of the dried ingredient. This may be suitable amounts to add seasoning characteristics but does not permit consumption of much seaweed if intended as a source of macronutrients such as fiber or protein. There is tremendous focus in the food chemistry community to identify processing methods that reduce iodine levels in kelp. Success has been observed for warm water treatment, boiling, and fermentation shown to remove up to 90% of iodine (Krook et al. 2024). This is accompanied by a drastic reduction in other minerals. Documented adherence of iodine limits is, and will continue to be, a major component of a company's quality control protocol.

Food allergens, present as proteins in common food types, are classified as chemical hazards and seaweed, due to its marine environment, has potential for contamination with two major sources: fish and shellfish. In a document prepared by the state of Connecticut for seaweed producers, the authors state that seaweed cultivated on longlines could be exposed to fouling organisms, including crustacean shellfish (Concepcion et al. 2020). Still, there have been few documented allergic reactions to seaweed and the Connecticut advisory group only suggested adding additional information to the label as a preventative method. This can be accomplished with a "may contain [allergen]" statement. According to the FDA: "such statements are not required by law and can be used to address unavoidable "cross-contact," only if manufacturers have incorporated good manufacturing processes in their facility and have taken every precaution to avoid cross-contact that can occur when multiple foods with different allergen profiles are produced in the same facility using shared equipment or on the same production line, as the result of ineffective cleaning, or from the generation of dust or aerosols containing an allergen" (FDA 2025). Therefore, standard GMPs that are designed to handle facility or employee allergen cross-contamination must still be maintained.

26.1.4.2 Biological Hazards

For seaweed, there is concern of contamination during growing and harvest with pathogens generally associated with fishery products: *Salmonella* spp., *Bacillus* spp., *E. coli*, *Listeria* spp., *Staphylococcus aureus*, and *Vibrio* spp. (Løvdal et al. 2021). These can present a hazard within seaweed sold as a raw agricultural product. These microbes can be controlled by processing—such as cooking, drying, or blanching—which are verifiable control steps that reduce the amount of many biological contaminants to acceptable levels. Producers should maintain specifications for raw ingredients, in-process, and finished products including biological standards as these can be required by purchasers or local health authorities. Microbiological testing of in-process or finished products presents an opportunity to verify preventive controls in a food safety plan. Producers should consult relevant guidance documents provided by the FDA (National Advisory Committee 2022) or Health Canada (2010) when determining the appropriate testing strategy.

26.1.4.3 Physical Hazards

Common physical hazards include foreign material collected during growth/harvest, metal introduced during cutting/milling, and other usual contaminants (glass, plastic, wood). These are commonly handled with GMPs that limit, detect, or remove contaminants during processing. A Connecticut guidance document recommends preventative maintenance of all cutting equipment and inspection of glass containers as sufficient control steps to limit physical hazards (Concepcion et al. 2020). Other potential sources of physical hazards will be operation specific and identified during routine risk analysis of each process step.

26.2 Animal Health

26.2.1 Product Types

The College for Veterinary Medicine (CVM), an agency of the FDA, monitors the safety of all types of animal food for all types of animals—i.e., livestock and companion pet. As of 2018, it is a requirement that animal food facilities in the United States have a Food Safety Plan in place that includes an analysis of hazards to determine which ones need control and risk-based preventive controls (21 CFR Subchapter E Part 507). Therefore, it is a general requirement that manufacturers of animal feed maintain CGMPs. These newer requirements, which fall under the FSMA in the United States, closely mirror those required for food such as preventive controls and hazard analysis, described previously. Within Canada, a similar CGMP requirement has been adopted for animal feed production under the Feeds Regulations, 2024.

Seaweed-derived food additives (sodium alginate, carrageenan) are commonplace as binders or texturizers for animal feed producers. The manufacturing specifications and use limits of these ingredients within animal feed are defined in 21 CFR Subchapter E Part 582. Other common applications of seaweed with animal feed are as single or mixed species seaweed blends prepared by drying and milling. They are often marketed as mineral supplements but are increasingly being considered as a natural source of functional health chemicals that can substitute antibiotic usage in certain animals, increase performance, and/or improve general animal health (Morias et al. 2020). In Canada, a range of seaweed-based animal feed ingredients are defined in the Canadian Feed Ingredient Table (CFIT). In the United States, label guidance comes from an industry organization: Association of American Feed Control Officials (AAFCO) who provide a similar set of definitions for seaweed derivatives. Relevant seaweed definitions are provided in Table 26.4.

26.2.2 Regulations

Both CFIT and AAFCO have clearly stated labelling requirements. They require that the seaweed family or families (for example, Sargassaceae) be clearly indicated. Further, the product shall be labelled with guarantees for maximum percentage of salt (NaCl) and the minimum percentage of potassium (K). A minimum or maximum guarantee for iodine must also be stated. Additional CFIT label requirements include maximum percent crude fiber and maximum percent sulfate. Seaweed extracts (dried or stabilized), a category unique to the CFIT, are not defined by AAFCO. Any seaweed extracts or derivatives must have percent moisture clearly stated for the ingredient.

Readers are reminded that only selected species of dried seaweed are GRAS for human food applications according to 21 CFR sections 184.1120 (brown) and 184.1121 (red). The list of "non-toxic macroscopic marine algae" described in CFIT and AAFCO is considerably more expansive than the CFR and includes several distinct orders of red, brown, and green algae suitable for animal feed. There are subtle differences with the CFIT and AAFCO lists, with the former including five additional seaweed orders. Conspicuous omissions from the Red Algae category are the emerging methane-reducing seaweed genus *Asparagopsis* (order: Bonnemaisoniales), reviewed in detail as a ruminant feed ingredient (Wasson et al. 2022; Camer-Pesci et al. 2023). This seaweed, when appropriately dried or processed, has been shown to reduce enteric methane emissions >80% when

Table 26.4 Seaweed-based single feed ingredients for animals in the United States and Canada

Name	Definition	Reference
Canadian feed ingredients table		
Seaweed extract liquid	Stabilized product resulting from the extraction of the stable, solid material obtained from the solubilization of non-toxic macroscopic marine algae.	1-503-090
Pelleted whole seaweed	Product resulting from drying, grinding, and pelleting non-toxic macroscopic marine algae	1-503-091
Dehydrated seaweed extract	Product resulting from drying the extract of the stable, solid material obtained from the solubilization of non-toxic macroscopic marine algae	1-503-092
Seaweed meal whole dehydrated	Product resulting from drying and grinding non-toxic macroscopic marine algae	1-503-093
Non-toxic macroscopic marine algae	Red Algae: *Bangiaceae, Gelidiaceae, Gigartinaceae, Gracilariaceae, Palmariaceae, Solieriaceae; Brown Algae: Alariaceae, Fucaceae, Laminariaceae, Lessoniaceae, Sargassaceae; Green Algae: Monostromataceae, Ulvaceae*.	
AAFCO (2025)		
Dried kelp	Dried seaweed of the families Laminariacae and Fucaceae. If the product is prepared by artificial drying, it may be called "Dehydrated Kelp."	60.19
Dried seaweed meal	Product resulting from drying and grinding non-toxic macroscopic marine algae (marine plants). If the product is prepared by artificial drying, it must be labeled as: Dehydrated Seaweed Meal	60.76
Non-toxic macroscopic marine algae	Red Algae: *Gelidiaceae, Endocladiaceae, Gigartinaceae, Gracilariaceae, Phyllophoraceae, Soliereiaceae, Hypneaceae, Palmariaceae, Bangiaceae; Brown Algae: Chordaceae, Laminariaceae, Lessoniaceae, Alariaceae, Fucaceae, Sargassaceae, Durvillaeceae; Green Algae: Monostromataceae, Ulvaceae*.	

fed at surprisingly low concentrations (<1% DM). Still, this is not a native US seaweed and further environmental and health studies must be completed before it can be sold under the dried seaweed meal definition.

The methane-reduction mode-of-action of *Asparagopsis* spp. as a feed additive is an illustrative example for another reason: it is an example of a gut modifier (Canada) or zootechnical animal food substance (the United States)—a product that affects the gastrointestinal (GI) tract of an animal with the intended purpose of affecting the byproducts of the digestion. When appropriately grown and processed, this seaweed creates a specific response in its animal host that shifts microbial communities and reduces methanogenesis. In this way, *Asparagopsis* lives at the intersection of veterinary drug (biological effect) and animal feed (nutritional benefit)—with the methane-reduction claims leaning more closely towards drug. Animal drug products require applicants to provide considerably more safety evidence and registration to either Health Canada or CVM (the United States) depending on where the product is distributed. Zootechnical animal food substances are still not a recognized category of feed ingredient in the United States. In 2023, a bill was introduced to congress—Innovative Feed Enhancement and Economic Development (FEED) Act of 2023—to preclude these products from the definition of substances that are used to treat or prevent diseases in animals, hormones, or active ingredients of animal drugs. Under this act, labels for zootechnical animal food substances must include a disclaimer that the substance may not be used to treat or prevent diseases in animals.

According to the CFIA, animal feed in Canada has three permissible types of claims: general, nutritional, or specialty. General claims describe the presence/absence of an ingredient or use of specific manufacturing method (organic, pesticide-free). Nutritional claims describe if the feed provides any important nutrient. Specialty claims are much more diverse and include pH adjustment, anticaking/antioxidant properties, and preservation. With the passing of Feeds Regulations 2024, single-feed ingredients can now be registered under the gut modifier category pending suitable supporting information can clearly demonstrate production/performance effects and/or gut modifying effects (CFIA 2024a). This evidence, collected via in-house research or from scientific literature, is required to demonstrate the product's efficacy for each effect when fed as directed per the target livestock species and may include the production class if the claim necessitates that specification. A minimum of one to three studies are required for registration of these types of products, with at least one in vivo study provided. Included in the label must be a minimum guarantee for each of the active ingredients in the gut modifier product. An example of active ingredient in *Asparagopsis* spp. may be bromoform ($CHBr_3$), which has been linked to anti-methanogenesis (Romero et al. 2023), and this chemical could presumably be claimed on the label with an active dose being established through science-based precedence. As in the United States, "gut modifier" products that directly target the reduction of pathogens or aid in the treatment or mitigation of disease in the animal are considered veterinary drugs in Canada.

26.2.3 Hazard Analysis

As previously discussed, seaweed is a potential source of heavy metals and, depending on the species, iodine. A feed formulator must ensure that the amount of these potential

Table 26.5 Maximum tolerable levels of certain metals in complete (as-fed) animal feeds

Source	Arsenic (mg/kg)	Cadmium (mg/kg)	Mercury (mg/kg)	Lead (mg/kg)	Iodine (mg/kg)
AAFCO	30	0.5	0.2	10	5–400
CFIA(2024b)	8	0.2 (horses), 0.4	–	8	–
NRC (2005)	30	10	0.2	10	5–400
FDA (2011)	12.5	10	0.3	10	–
Cali. Code § 2680	50	–	2	30	–

hazards in the complete feed is aligned with a suitable reference. A recent CVM report describes the tolerances associated with specific minerals in animal feed (Deemy 2019). A selection of heavy metals MTLs from federal or industry sources is provided for reference (Table 26.5). Seaweed is rarely expected to be a major component of an animal's diet (<10%, as fed) and therefore makes up only a small portion of the complete feed—it is important to factor ingredient dilution into a quality control protocol. Nevertheless, a comprehensive mineral testing strategy and descriptive labeling should be used to minimize any potential for hazard exposure to animals. Iodine recommendations are provided as well, and these can be animal species or production class specific. For example, AAFCO and National Research Council (NRC) both set the maximum daily inclusion of iodine much lower for horses (5 mg/kg complete feed) as compared with swine (400 mg/kg complete feed). Cattle, for comparison, have a maximum of 50 mg/kg iodine in complete feed. This further underscores the importance of declarative labelling of major minerals in seaweed animal feed products.

26.3 Plant Health

26.3.1 Relevant Definitions

Fertilizer—any organic or inorganic material of natural or synthetic origin which is added to soil to provide nutrients, including nitrogen, phosphorus, and potassium, necessary to sustain plant growth (USDA 2012).

Plant Biostimulant—a substance or micro-organism that, when applied to seeds, plants, or the rhizosphere, stimulates natural processes to enhance or benefit nutrient uptake, nutrient efficiency, tolerance to abiotic stress, or crop quality and yield (USDA 2019a, b).

Pesticide—any substance or mixture of substances intended for preventing, destroying, repelling, or mitigating any pest (USDA 2019a, b).

Supplement—products other than fertilizers that improve the physical condition of the soil or aid plant growth or crop yield (CFIA 2023).

Plant Regulator—any substance or mixture of substances intended, through physiological action, for accelerating or retarding the rate of growth or rate of maturation, or for otherwise altering the behavior of plants or the produce thereof, but shall not include substances to the extent that they are intended as plant nutrients, trace elements, nutritional chemicals, plant inoculants, and soil amendments (Federal Insecticide, Fungicide, and Rodenticide Act).

26.3.2 Product Types

Seaweed products for plants are generally classified under three categories: fertilizers, biostimulants, or pesticides. Broad definitions of these, according to the United States Department of Agriculture (USDA), are provided above. The classification of seaweed products for plant health as fertilizers seems suitable: initially, seaweed use in crops was explored due to its relatively high macro- (K) and micronutrient (Cl, Fe, Zn, Cu) levels. The ash-rich fraction present after combustion can be used as a plant nutrient source. It was this property that saw a drastic increase in the refining of Kelp during World War 1 in California when potassium supplies were cut off to the United States (Neushul 1989).

More recently, seaweed extracts have gained popularity as plant biostimulants. Extracts of brown seaweeds are widely used in crops largely for their plant growth-promoting effects and for their enhancement of crop tolerance to abiotic stresses such as salinity, extreme temperatures, nutrient deficiency, and drought (Battacharyya et al. 2015). Research in this topic reveals that the organic components of seaweed, e.g., carbohydrates and polyphenols, display remarkable bioactivity in plant–soil systems and are responsible for these plant health benefits. A detailed evaluation of mode-of-action and efficacy falls beyond the scope of this work but is referenced in detail elsewhere (Ali et al. 2021). It is important to understand that while the term biostimulant may be evoked with respect to a product's application, it remains an unregulated term in the United States and Canada—as opposed to fertilizer which has had regulated markets for years.

The use of seaweed extracts in pesticide products is comparatively limited to the other categories. Laminarin, an oligosaccharide produced by brown seaweeds as an energy storage molecule, stimulates the natural defense reactions of a wide variety of agricultural crops, including vegetables and fruits, against plant pathogens such as bacteria and fungi. For example, laminarin derived from *Laminaria digitata* elicited defense responses in grapevine cells against pathogens *B.*

cinerea and *P. viticola* (Aziz et al. 2003). This property has been utilized in the development of new pesticide products.

26.3.3 Regulations

26.3.3.1 Fertilizers

In the United States, a seaweed product intended to be used with crops must adhere to fertilizer regulatory programs established at the state level. In most cases, these agencies are strictly concerned with contaminants—heavy metals and other minerals—in fertilizer products at unsuitable levels. Three states have published requirements for specific metals in fertilizers: California, Oregon, and Washington. These describe mandatory testing requirements such as metal type and, in some cases, analysis method used. The results for these metal tests are published online by the Association of American Plant Food Control Officials (AAPFCO).

In comparison, all fertilizers imported into or sold in Canada must meet the standards outlined in the Fertilizers Regulations in compliance with the Fertilizer Act—which often requires some aspect of product registration. Plant products may either be registered as a fertilizer or supplement, with both requiring an applicant to provide CFIA with detailed manufacture practice, quality management system, and results of analysis—among other information. There are set safety limits established for trace metals and indicator organisms (*Salmonella* spp. and fecal coliforms) (CFIA 2024c). This closely aligns with the GMP requirements required for human food and animal feed including hazard analysis and documented preventive controls.

26.3.3.2 Biostimulants

Scientific discoveries continue to reveal interesting interactions between seaweed chemicals and plants (or perhaps the plant-associated microbiomes), but it is important to focus on the domestic regulations that control the sale and marketing of these types of products. Some products being sold as plant biostimulants in the United States may meet the Federal Insecticide, Fungicide, and Rodenticide Act (FIFRA) definition of plant regulators. The term "pesticide" includes plant regulator products which alter the rate of plant growth or maturation and therefore fall under the guidance of the Environmental Protection Agency (EPA). Product registration with the EPA is a daunting task.

According to 40 C.F.R. § 152.6, a product of any of the following types, intended only to aid the growth of desirable plants, is not a "plant regulator" and therefore is not a pesticide:

1. A plant nutrient product consisting of one or more macronutrients or micronutrient trace elements necessary to normal growth of plants and in a form readily usable by plants.
2. A plant inoculant product consisting of microorganisms to be applied to the plant or soil for the purpose of enhancing the availability or uptake of plant nutrients through the root system.
3. A soil amendment product containing a substance or substances intended for the purpose of improving soil characteristics favorable for plant growth.

There is a potential issue with certain seaweed products that claim the presence of plant hormones such as auxins, gibberellins, and cytokinins (Ali et al. 2021). These rather innocuous label claims carry added weight in the US market. Exogenous plant hormones are plant regulators and therefore require registration under FIFRA. It is perhaps more beneficial to identify and guarantee other components of a seaweed product before phytohormones. A recent publication of the European Biostimulants Industry Council (EBIC) suggested "that seaweed extracts would be unlikely to induce any phytohormone responsive phenotypes at field application because the minimum concentrations required for phenotypic changes in the plant are reported to be significantly higher than concentrations reported to be found in some commercial seaweed extracts" (EBIC 2023).

Per guidelines in the 2018 Farm Bill, the USDA and EPA published a report on regulatory recommendations pertinent to plant biostimulant products describing potential hurdles to their definition under existing guidelines (USDA 2019a, b). A draft guidance was prepared by the EPA in 2020 to further clarify the definition of biostimulant including differentiation from current guidance on plant regulator claims (EPA 2020). Biostimulant is not a regulated term in the United States, despite language of the type appearing in a bill introduced to the Senate that proposes to exclude all plant biostimulants from regulation under FIFRA (S. 802).

Per 40 CFR 152.6, seaweed products classified as either plant nutrient products, plant inoculants, or soil amendments can be reasonably excluded from the definition of pesticide. There are several examples of product claims generally considered "non-pesticidal" (i.e., non-plant regulator claim) that are suggested by the EPA (2022) and some are provided below:

- Enhances/aids/supports/helps/improves abiotic stress tolerance.
- Improves soil/seed nutrient conditions for root growth.
- Optimizes soil/foliar/seed nutrient conditions for plant growth.
- Improves/supports mycorrhizal/rhizobial association/symbiosis with plant roots/seeds.
- Optimizes nutrient use efficiency.
- Increases/improves/optimizes soil conditions for increased plant vigor.

It is important to remember that a seaweed product, liquid or solid, that guarantees a specific plant nutrient or effect must be produced in a manner that ensures consistency with label requirements and delivers a certain aspect of confidence to the end user. As AAPFCO suggests, "if you claim it, you must guarantee it."

26.3.3.3 Biopesticides

As mentioned, the only widespread example of seaweed as a pesticide is with laminarin extracted from certain species of brown macroalgae. In 2010, the CFR were update with an exemption from the requirement of a tolerance for residues of laminarin on food commodities when applied preharvest (40 CFR § 180.1295) thus paving the way for these products in agriculture. With respect to safety, the EPA considers the toxicological and ecological databases complete for characterizing hazard and assessing risk relative to the current pesticidal uses of laminarin—there have been no reported incidents associated with laminarin (EPA 2022). The Fungicide Resistance Action Committee (FRAC) lists fungicides, mainly for use in plant protection, according to their mode of action and resistance risk. Laminarin is classified under the "host plant defense induction" mode-of-action and "P04 - Polysaccharide elicitors" category (FRAC 2024). In 2015, a petition was submitted to include laminarin on the National List of permitted organic substances (Goëmar 2015). There is some contention with this description as the described extraction procedure doesn't align with the current National Organic Policy on aquatic plant extracts, which is how a laminarin product would likely be defined. Nevertheless, laminarin presents a recognized pesticidal activity and may be sourced from seaweed.

26.4 Conclusion

It is uncertain what product types will prove the most marketable for the North American seaweed industry. Certainly, this resource will play a role in future innovations within food, animal, and plant health. A concerted effort will need to be made by government, industry, and academic stakeholders to achieve sizable impact. Still, there remains uncertainty around the regulatory landscape facing new product entries which can dull investment into fundamental research and development. Within this document, a throughline is that manufacturing considerations play a major role in emerging regulatory frameworks. Revised animal feed production regulations in both the United States and Canada now require GMPs, hazard analysis, and risk-based preventive controls—the same manufacturing standard that human food is required to meet. The classification of seaweed within the current ingredient framework—plant, seafood, other—remains unclear. Nevertheless, producers can be assured that a firm understanding of associated seaweed hazards, especially chemical and microbial contaminants, and carefully documented methods on their control will be necessary. How these resulting products are labelled and marketed may present some challenges but as a fundamental criterion, they should all be safe. Seaweed products for more specialized applications will require further manufacturing consistency validated through quality system. It will be through the discovery of biological mode-of-actions for chemicals found in seaweed that support new and enhanced use in food and agriculture. Modern science increasingly permits rapid quantitation of trace metabolites or microbes in a microbiome using so-called omics techniques (Dai and Shen 2022)—which will lead to deeper understanding of seaweed potential. The onus will remain on the seaweed processor to prove these new products are safe and the product developer to ensure efficacy for desired use through detailed research and repeated trials. Only then can you truthfully market the benefits of seaweed products. To borrow from the AAPFCO: if you claim it, you must guarantee it.

References

AAFCO (2025) Online database of ingredients. https://www.aafco.org/resources/regulatory/online-database-of-ingredients/

Ali O, Ramsubhag A, Jayaraman J (2021) Biostimulant properties of seaweed extracts in plants: implications towards sustainable crop production. Plants 10(3):531

ANSES (2020) Maximum cadmium levels for seaweed intended for human consumption. Request No 2017-SA-0070

Aziz A, Poinssot B, Adrian M et al. (2003) Laminarin elicits defense response in grapevine and induces protection against *Botrytis cinerea* and *Plasmopara viticola*. Mol Plant Microbe Interact 16(12):1118–1128

Banach JL, Hoek-van den Hil EF, van der Fels-Klerx HJ (2020) Food safety hazards in the European seaweed chain. Compr Rev Food Sci Food Saf 19(2):332–364

Battacharyya D, Babgohari MZ, Rathor P et al. (2015) Seaweed extracts as biostimulants in horticulture. Sci Hortic 196(30):39–48

Camer-Pesci B, Laird DW, van Keulen M et al. (2023) Opportunities of *Asparagopsis* sp. cultivation to reduce methanogenesis in ruminants: a critical review. Algal Res 76:103308

Health Canada (2010) Microbial guidelines for ready-to-eat foods—A guide for the conveyance industry and environmental health officers (EHO). https://publications.gc.ca/collections/collection_2014/sc-hc/H164-167-2013-eng.pdf

CFIA (2023) Fertilizer or supplement registration: overview. https://inspection.canada.ca/en/plant-health/fertilizers/overview-0

CFIA (2024a) RG-1 regulatory guidance: chapter 3—guidance on data requirements for feed approval and registration. 3.31 registration requirements for gut modifier products. https://inspection.canada.ca/en/animal-health/livestock-feeds/regulatory-guidance/rg-1/chapter-3#c31

CFIA (2024b) RG-8 regulatory compliance: contaminants in feed. https://inspection.canada.ca/en/animal-health/livestock-feeds/regulatory-guidance/rg-8

CFIA (2024c) Guide to submitting applications for registration under the Fertilizers Act. https://inspection.canada.ca/en/plant-health/fertilizers/overview/guide

Concepcion A, Derosia-Banick K, Balcom N (2020) Seaweed production and processing in Connecticut: a guide to understanding and controlling potential food safety hazards. NOAA. https://repository.library.noaa.gov/view/noaa/36892

Dai X, Shen L (2022) Advances and trends in omics technology development. Front Med 9:911861

Deemy M (2019) CVM CY15–17 Report on heavy metals in animal food. https://www.fda.gov/media/132046/download

EBIC (2023) Recent insights into the mode-of-action of seaweed-based plant biostimulants. https://biostimulants.eu/wp-content/uploads/2024/11/20230116-EN-Seaweed-WhitePaper-v11-final.pdf

EPA (2020) Draft guidance for plant regulator products and claims, including plant biostimulants. https://www.epa.gov/sites/default/files/2020-11/documents/pbs-guidance-updated-draft-guidance-document-2020-11-13_0.pdf

EPA (2022) Laminarin PC Code 123200 interim registration review. Decision Case Number 6309

FAO and WHO (2022) Report of the expert meeting on food safety for seaweed—Current status and future perspectives. Food Safety and Quality Series No. 13. Rome. https://doi.org/10.4060/cc0846en

FDA (2011) Target animal safety review memorandum. https://www.fda.gov/media/81895/download

FDA (2025) Food allergies. https://www.fda.gov/food/nutrition-food-labeling-and-critical-foods/food-allergies

FRAC (2024) FRAC code list 2024: fungal control agents sorted by cross-resistance pattern and mode of action. FRAC code list. https://nyfzx.cn/pdf/use%20etc/FRAC%20Code%20List%202024.pdf

Janasie CM (2023) Federal food safety framework: where does seaweed fit in? J Food Pol 18(2):74–116

Krook JL, Duinker A, Larssen WE et al. (2024) Approaches for reducing the iodine content of the brown seaweed *Saccharina latissima*: effects on sensory properties. J Appl Phycol 36:783–796

Kuivenhoven M, Mason K (2023) Arsenic toxicity. In: StatPearls [Internet]. Treasure Island (FL). https://www.ncbi.nlm.nih.gov/books/NBK541125/

Laboratoires Goëmar SA (2015) Petition to include laminarin on the National List. https://www.ams.usda.gov/sites/default/files/media/Laminarin%20Petition.pdf

Løvdal T, Lunestad BT, Myrmel M et al. (2021) Microbiological food safety of seaweeds. Foods 10(11):2719

McHugh DJ (2003) Seaweeds used as human food. In: A guide to the seaweed industry. FAO, Rome. https://marineagronomy.org/sites/default/files/McHugh.%202003.%20Guide%20to%20Seaweed%20Industry.pdf

Morias T, Inacio A, Ministro M et al. (2020) Seaweed potential in the animal feed: a review. J Mar Sci Eng 8(8):559

National Advisory Committee on Microbiological Criteria for Foods (2022) Microbiological testing by industry of ready-to-eat foods under FDA's jurisdiction for pathogens (or appropriate indicator organisms): verification of preventive controls. J Food Prot 85(11):1646–1666

National Research Council (1981) Food chemicals codex, 3rd edn. The National Academies Press, Washington, DC, p 15

National Research Council (2005) Mineral tolerance of animals, 2nd edn. The National Academies Press, Washington DC. https://natureslogic.com/wp-content/uploads/2018/01/NRC-2005.pdf

Neushul P (1989) Seaweed for war: California's World War I kelp industry. Technol Cult 30(3):561–583

Roleda MY, Skjermo J (2018) Iodine content in bulk biomass of wild-harvested and cultivated edible seaweeds: inherent variations determine species-specific daily allowable consumption. Food Chem 254:333–339

Romero P, Belanche A, Jiménez E et al. (2023) Rumen microbial degradation of bromoform from red seaweed (*Asparagopsis taxiformis*) and the impact on rumen fermentation and methanogenic Archaea. J Anim Sci Biotechnol 14:133

USDA (2012) National Agricultural Library Thesaurus Concept Space – Fertilizers. https://lod.nal.usda.gov/nalt/5568

USDA (2019a) Report to the President of the United States and United States Congress on plant biostimulants. https://agriculture.house.gov/uploadedfiles/usda_report_on_plant_biostimulants_12.20.2019.pdf

USDA (2019b) National Agricultural Library Thesaurus Concept Space—Pesticides. https://lod.nal.usda.gov/nalt/en/page/287

Wasson DE, Yarish C, Hristov AN (2022) Enteric methane mitigation through *Asparagopsis taxiformis* supplementation and potential algal alternatives. Front Anim Sci 3:999338. https://doi.org/10.3389/fanim.2022.999338

Yanli Y, Alba M-R, Guangming Z et al. (2024) Survey of arsenic content in edible seaweeds and their health risk assessment. Food Chem Toxicol 187:114603

Spencer Serin is a seaweed chemical and regulatory compliance expert. He completed his PhD in chemistry in 2016 and has developed seaweed products and market research for 7 years covering a range of applications - food, agriculture (animal feed and plant biostimulants), and chemical extracts. In 2021, he co-founded Spoitz Enterprises which has provided scientific consulting services to a range of industry and government clients across North America. His firm was acquired by Marine Biologics in 2024 and he took the role of Chief Scientific Officer. His primary work area is seaweed chemical characterization, extraction, and regulatory compliance.

ASC-MSC Seaweed Standard: A Review

27

Miki Takada and Tim. K. Davies

Abstract

Seaweeds has been cultivated for centuries (Pereira L. Edible seaweeds of the world. CRC Press, Boca Raton, 2016), and while the impact of seaweed farming has mostly been reported as being beneficial or benign (Sultana et al. Aquacult Fish 8:463–480, 2023), it may also be associated with negative ecological and social impacts (Campbell et al. Front Mar Sci 6:107, 2019). To address these impacts, the ASC and MSC have created a Seaweed Standard that encompasses environmental and social best practice in the seaweed industry. This chapter identifies the four key features of certification programmes (Auld. Constructing private governance: the rise and evolution of forest, coffee, and fisheries certification. Yale University Press, New Haven and London, 2014) within the ASC-MSC Seaweed Programme. It also provides information on how this standard came about, outlines the certification process, and reviews the conditions that have been placed on seaweed production units in the programme.

Keywords

Seaweeds · Algae · Aquaculture · Sustainability · Certification · Standards

27.1 Introduction

According to Auld (2014), certification programmes typically have four features in common: (1) a governance structure and policies for managing operations; (2) a standard that regulates environmental and/or social impacts of the process in question, as well as a chain of custody (CoC) process; (3) a monitoring and inspection scheme, usually involving a third-party process of verification; and (4) labels and/or logos to be applied to products for end consumers. This chapter will explore the background of how the ASC-MSC Seaweed Standard came about, as well as how these four features are included in the Seaweed Programme.

27.2 Background

27.2.1 Justification for the Standard

While seaweed cultivation and farming are generally considered environmentally beneficial or benign (Sultana et al. 2023), these activities may result in negative ecological and social impacts (Campbell et al. 2019). Some of these negative impacts include disease transmission (Bhuyan 2023), impacts on population genetics (Bhuyan 2023), changes to the physicochemical environment (Campbell et al. 2019), light and noise pollution, and marine debris (Spillias et al. 2022).

The Aquaculture Stewardship Council (ASC) is an independent, not-for-profit, third-party certification and labelling programme for responsible farmed seafood founded in 2009 by WWF and IDH, the Dutch Sustainable Trade Initiative (WWF n.d.). The Marine Stewardship Council (MSC) is a similar scheme but for wild-caught seafood, founded in 1997 by WWF and Unilever, becoming independent in 1999 (Marine Stewardship Council n.d.-a). Building on each other's expertise in standard setting and seafood certification, the ASC and MSC jointly developed the environmentally sustainable and socially responsible seaweed standard. This is the first joint standard developed by these two organisations (Marine Stewardship Council 2017a), and the standard became effective on 1 March 2018 (Aquaculture Stewardship Council and Marine Stewardship Council 2018a).

M. Takada (✉) · T. K. Davies
Marine Stewardship Council (MSC), London, UK
e-mail: miki.takada@msc.org

A. T. Critchley et al. (eds.), *Advances in Temperate Phyconomy: Algal Harvest and Cultivation in Globally Distributed Temperate Waters*, Developments in Applied Phycology 15, https://doi.org/10.1007/978-3-032-02955-3_27

The ASC and MSC use a market mechanism to drive change in the aquatic environment and global seafood market, incentivising best practice for sustainable fishing in the aquaculture and wild capture sectors. Working with scientists and marine experts, the seaweed standard was developed to address environmental and social issues (Aquaculture Stewardship Council and Marine Stewardship Council 2018b).

27.2.2 Standard Development Process

In 2013, recognising the increase in seaweed production and the demand for MSC certified seaweed products, the Technical Advisory Board (TAB) of the MSC decided to move forward with developing a seaweed standard (Marine Stewardship Council 2014). A Memorandum of Understanding (MOU) was signed between the MSC and ASC in 2015 to 'further… their collaboration and commitment to find additional efficiencies and added value for partners handling both ASC and MSC certified products (Aquaculture Stewardship Council 2015)'. The ASC-MSC Seaweed Standard was developed following a participatory process, in accordance with the ISEAL Standard Setting Social and Environmental Standards (the Standard Setting Code) and FAO Guidelines, and following the MSC-ASC Joint Seaweed Standard Setting Procedure version 1.0 (Aquaculture Stewardship Council and Marine Stewardship Council 2016). Stakeholder workshops were conducted in Asian nations where seaweed consumption is common, namely, Indonesia, Japan, and China. Two 60-day online consultations were held to seek views on drafts of the standard and associated assessment process (Aquaculture Stewardship Council and Marine Stewardship Council 2018a), the first from 1 March 2016 to 30 April 2016 (Marine Stewardship Council 2016), and the second from 1 March 2017 to 30 April 2017. In this second consultation, stakeholders were asked to provide feedback on a draft of the seaweed standard, as well as requirements the assessors would need to follow when assessing against the standards (Marine Stewardship Council 2017b). These consultations were supported by public webinars, meetings, and workshops, and local outreach to the seaweed industry and affected stakeholders (Aquaculture Stewardship Council 2016). Technical input was provided by experts and academics, and a workshop with Conformity Assessment Bodies (CABs) was held in the United Kingdom (Aquaculture Stewardship Council and Marine Stewardship Council 2018a).

27.2.3 Governance Structure

The governance structure for the ASC-MSC Seaweed Standard was revised in 2025, coinciding with the start of the ASC-MSC Seaweed Standard review. The updated governance structure, subject to review, is outlined in Fig. 27.1 . Final approval of the Seaweed Standard is provided by the ASC Board of Trustees, with the MSC Board of Trustees endorsing the ASC Board's approval. The Seaweed Revision Technical Working Group, made up of stakeholders, technical experts, MSC Technical Advisory Board (TAB) representatives, and an ASC Technical Advisory Group (TAG) sponsor, provide input during the standard review (Aquaculture Stewardship Council and Marine Stewardship Council 2025).

A Joint Notice of Complaint Form can be found on the seaweed standard website. Stakeholders with complaints about the standard may complete and submit this form, and the complaints will be investigated by seaweed staff at the ASC and MSC (Aquaculture Stewardship Council and Marine Stewardship Council n.d.).

27.2.4 Independent Third-Party Certification by Accredited Conformity Assessment Body (CABs)

The ASC-MSC Seaweed (Algae) Programme is a third-party certification programme, with neither the ASC nor MSC directly involved in assessing the seaweed production units, but rather setting the standard against which applicants are assessed. CABs, along with environmental and social impact experts, carry out the assessments. Neither the MSC nor ASC is financially compensated for certifications or assessments of seaweed production units (Marine Stewardship Council n.d.-b). Instead, funding comes from royalties paid for ecolabels placed on consumer-facing products and license holders, who pay annual fees (Aquaculture Stewardship Council n.d.-a).

Assurance Services International (ASI) was selected as the assurance body for CABs for the seaweed standard. This means that for a CAB to certify production units against the seaweed standard, they must first be accredited by ASI. As of February 2025, there are four accredited CABs approved for certification (Assurance Services International n.d.).

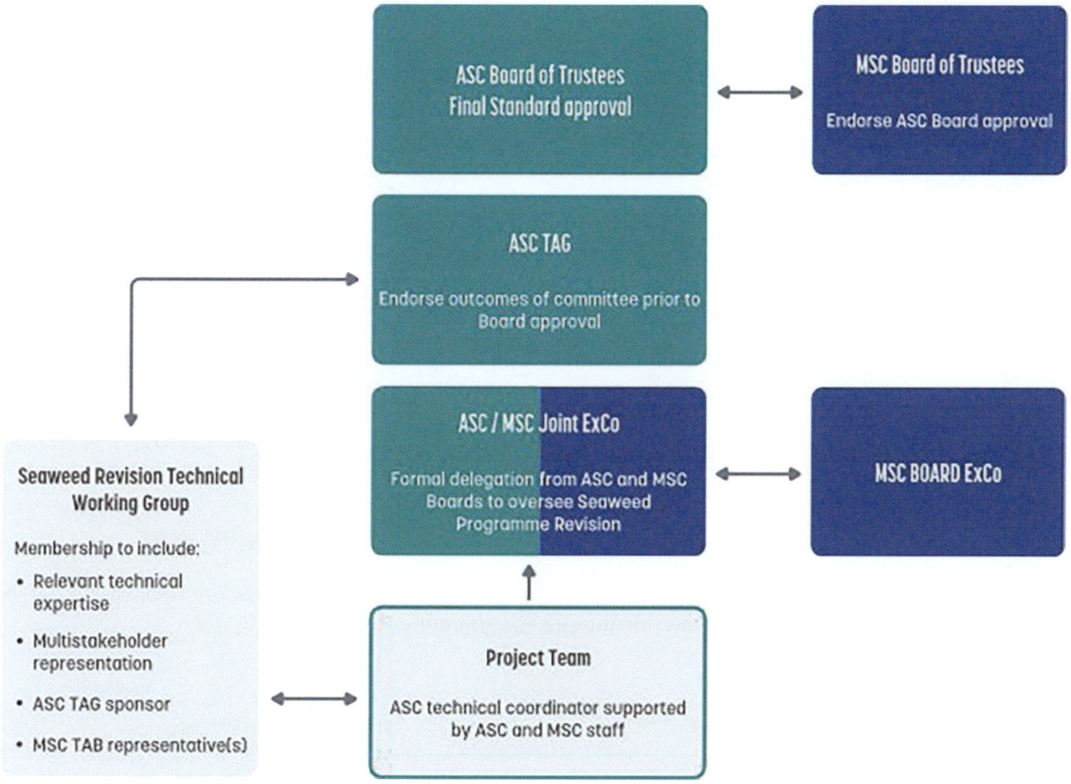

Fig. 27.1 The ASC-MSC Seaweed Standard governance structure (From Aquaculture Stewardship Council and Marine Stewardship Council 2025)

27.2.5 Engaged Seaweed Production Units

Since the inception of the seaweed standard on 22 November 2017, 12 seaweed production units have engaged[1] with the programme. Figure 27.2 shows the amount (in mt) of certified and in-assessment products by country. Table 27.1 shows the amount of certified microalgal and macroalgal products (in mt) produced, by country, in 2024.

27.3 The ASC-MSC Seaweed Certification Requirements

27.3.1 The Substantive Requirements: The ASC-MSC Seaweed Standard

There are five seaweed production categories in the standard—Table 27.2 lists these categories and how they are determined.

Each seaweed production unit is classified into one of these five categories. Using a 'skip-logic' approach, categorisation determines which performance indicators (PIs, described below) are assessed for the production unit. Table 27.3 outlines which PIs are assessed, depending on specific attributes of the production unit.

The standard is made up of an assessment tree, with 31 PIs and 67 scoring issues (SIs) within the five principles that comprise the standard. Figure 27.3 shows the assessment tree for Principle 1. Table 27.4 lists the PIs that make up the seaweed standard.

27.3.1.1 Principle 1

The purpose of Principle 1 is to ensure that seaweed farming and harvesting are conducted in a way that preserves the productivity of the wild seaweed populations. Principle 1 is made up of three PIs (Table 27.4)—PI 1.1 evaluates stock status, PI 1.2 examines the harvest strategy, and PI 1.3 assesses the genetic impact on wild stock. These PIs are included in the standard to ensure that the status of the wild stock and the genetic impact of the production unit on it are known, and that there is a harvest strategy in place that helps maintain the availability of the seaweed resource. This information helps to determine whether the stock is maintained at a level that promotes sustainable development.

The negative impact that both wild seed sourcing and seaweed farming can have on the population dynamics and genetics of wild seaweed populations is well known (Valero

[1]The Marine Stewardship Council considers production units that are certified, in assessment, and suspended to be 'engaged (Marine Stewardship Council 2024)'.

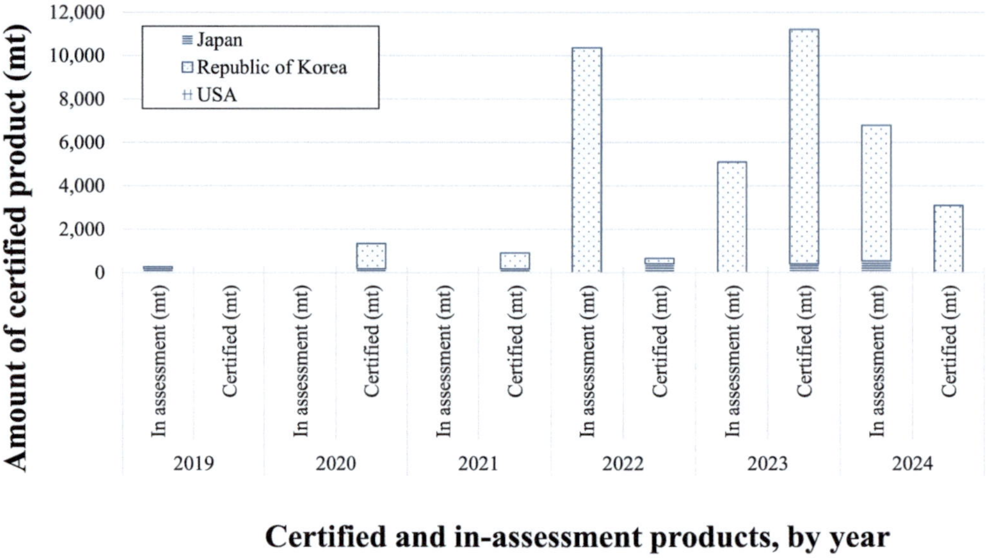

Fig. 27.2 Amount of certified and in-assessment product by year (The United States does not appear on this graph because assessment reports do not provide information on the amount produced by the production unit)

Table 27.1 Amount of certified macroalgal and microalgal product, by species per country in 2024[a]

	Target species			
	Macroalgae		Microalgae	
	Species	Amount (mt)	Species	Amount (mt)
Japan	Kombu	107.1	*Euglena* (Euglenophyta)	145.1
	Wakame	174.8	*Chlorella* (Chlorophyta)	127.0
Republic of Korea	Arame	480.0		
	Kombu	136.1		
	Laver	4582.5		
	Wakame	1046.6		
USA			*Schizochytrium* spp. (Fungi)	

[a]The United States does not appear on this graph because assessment reports do not provide information on the amount produced by the production unit

Table 27.2 Seaweed production categories, after Table 2 in the standard (Aquaculture Stewardship Council and Marine Stewardship Council 2018a)

Category	Production location and type	Linkages to wild stock
A	Harvest of natural populations	Wild stocks harvested
Bi	Cultivation at sea (including production units which require some stages cultivated in land-based hatcheries followed by grow-out at sea)	Seed supplied from wild stocks
Bii		Supply of seed from wild stocks NOT required or negligible
Ci	Cultivation entirely in land-based systems	Seed supplied from wild stocks
Cii		Supply of seed from wild stocks NOT required or negligible

et al. 2017; Campbell et al. 2019). The aim of these PIs is for the production units in assessment to understand the impact that they have on wild seaweed populations.

27.3.1.2 Principle 2

Principle 2 assesses the production unit's impact on the environment. Nine PIs make up Principle 2, with seven PIs evaluating the direct impact of the farming or harvesting practices. These are PI 2.1, which assesses the production unit's impact on habitats; PI 2.2, which examines the impact on ecosystem structure and function; PI 2.3 on Endangered, Threatened, and Protected (ETP) species; PI 2.4 on other species (species other than ETP species); PI 2.6, which examines the production unit's pest and disease management, PI 2.8, which relates to translocation; and PI 2.9, which examines the introduction of alien species. Two other PIs examine the production unit's sustainability practices—PI 2.5 evaluates waste management and pollution control by the production unit being assessed; and PI 2.7 examines energy efficiency.

Seaweed farming can create eutrophication and nutrient imbalance in an ecosystem (Tucker and Hargreaves 2008); as such, these are examined in two separate PIs in the standard. In PI 2.2 (ecosystem), nutrient depletion, nutrient balance (especially of carbon, nitrogen, phosphorus, and their ratios), and nutrient ratio divergence as a result of farming

Table 27.3 Attributes that determine which PIs are assessed, after Table 3 of the standard (Aquaculture Stewardship Council and Marine Stewardship Council 2018a)

PIs	Criteria	Yes/No	Action
PI 1.1—Stock status	Does the activity depend on wild stocks of seaweed or on seed supplied from them (categories A, Bi, Ci)?	Yes	Score these PIs
PI 1.2—Harvest strategy		No	Do not score these PIs
PI 1.3—Genetic impact on wild stock	Is translocation occurring or the activity requires stages cultivated in hatcheries?	Yes	See next criteria
		No	Do not score this PI
	Is it a land-based system (category C)	Yes	See next criteria
		No	Score this PI
	Is there contact with, extraction from, or impact on the marine environment which cannot be considered negligible?	Yes	Score this PI
		No	Do not score this PI
PI 2.1—Habitat	Is it a land-based system (category C)?	Yes	See next criteria
		No	Score these PIs
PI 2.2—Ecosystem structure and function	Is there contact with, extraction from, or impact on the marine environment which cannot be considered negligible?	Yes	Score these PIs
		No	Do not score these PIs
PI 2.7—Energy efficiency	Is it a microfamily business?	Yes	Do not score this PI
		No	Score this PI
PI 2.8—Translocations	Is translocation occurring?	Yes	See next criteria
		No	Do not score this PI
	Is it a land-based system (category C)?	Yes	See next criteria
		No	Score this PI
	Is there contact with, extraction from, or impact on the marine environment which cannot be considered negligible?	Yes	Score this PI
		No	Do not score this PI
PI 2.9—Introduction of alien species	Is the UoA targeting an alien species?	Yes	Score this PI
		No	Do not score this PI
PI 3.1—Legal and/or customary framework	Is the production of such low intensity, scale, and level of development as to be considered to not yet need a national legal framework?	Yes	Do not score this PI
		No	Score this PI
PI 5.4—Visibility, positioning, and orientation of farms or water-based structures	Is it a land-based system (category C)?	Yes	See next criteria
		No	Score these PIs
PI 5.5—Identification and recovery of substantial gear	Does the activity require the use of substantial gear or structures in the wild aquatic environment?	Yes	Score these PIs
		No	Do not score these PIs

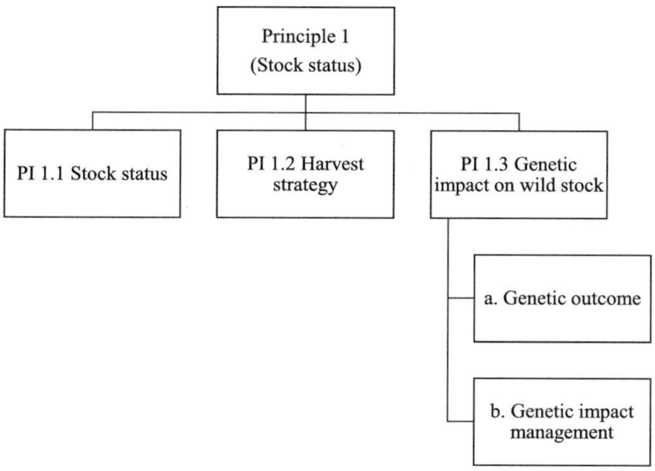

Fig. 27.3 Principle 1 assessment tree structure, after Fig. 1 of the ASC-MSC Seaweed (Algae) Standard v1.01. The top row (Principle 1 (Stock status)) is one of the five principles in the standard. The second row shows the three Performance Indicators (PIs) in Principle 1. The third and fourth rows are the scoring issues for Principle 1.3. (Aquaculture Stewardship Council and Marine Stewardship Council 2018a)

activity are evaluated, while waste and nutrient discharge are examined in PI 2.5 (waste management and pollution control).

The use of inorganic debris, such as floats made of styrofoam and plastic bottles, and 'tie-ties' and plastic straws can also lead to pollution (Zemke-White and Smith 2005). Therefore, production units should have a system in place for waste disposal and reduction.

And finally, alien and invasive species introduction is another negative impact that seaweed production can have (Bhuyan 2023). Therefore, production units should have a strategy in place that would prevent the furtherance of impacts on the ecosystem by alien species.

27.3.1.3 Principle 3

The focus of Principle 3 is effective management, whereby each production unit in assessment should have a framework in place that can deliver sustainable harvesting and farming outcomes as outlined in Principles 1 and 2. Three PIs make up Principle 3: PI 3.1 examines the legal and/or customary

Table 27.4 List of Performance Indicators, from Table 1 in Get certified! Your guide to the ASC-MSC Seaweed Standard audit process (Aquaculture Stewardship Council and Marine Stewardship Council 2017)

Principle 1: Sustainable wild populations	Population 2: Environmental impacts	Population 3: Effective management	Population 4: Social responsibility	Population 5: Community relations and interaction
PI 1.1 Stock status	PI 2.1 Habitat	PI 3.1 Legal and/or customary framework	PI 4.1 Child labour	PI 5.1 Community impacts
PI 1.2 Harvest strategy	PI 2.2 Ecosystem structure and function	PI 3.2 Decision-making processes	PI 4.2 Forced, bonded, or compulsory labour	PI 5.2 Conflict resolution
PI 1.3 Genetic impact on wild stock	PI 2.3 ETP species	PI 3.3 Compliance and enforcement	PI 4.3 Discrimination	PI 5.3 Rights of indigenous groups
	PI 2.4 Other species		PI 4.4 Health, safety, and insurance	PI 5.4 Visibility, positioning and orientation of farms or water-based structures
	PI 2.5 Waste management and pollution control		PI 4.5 Fair and decent wages	PI 5.5 Identification and recovery of substantial gear
	PI 2.6 Pest(s) and disease(s) and management		PI 4.6 Freedom of association and collective bargaining	PI 5.6 Noise, light, and odour
	PI 2.7 Energy efficiency		PI 4.7 Disciplinary practices	PI 5.7 Decommissioning of abandoned production units
	PI 2.8 Translocations		PI 4.8 Working hours	
	PI 2.9 Introduction of alien species		PI 4.9 Environmental and social training	

framework, PI 3.2 assesses decision-making processes, and PI 3.3 reviews compliance and enforcement. At a minimum, seaweed production units must abide by all relevant national and international laws and regulations.

27.3.1.4 Principle 4

Nine PIs make up Principle 4. PI 4.1 focuses on child labour; PI 4.2 on forced, bonded or compulsory labour; PI 4.3 on discrimination; PI 4.4 on health, safety, and insurance; PI 4.5 on fair and decent wages; PI 4.6 on freedom of association and collective bargaining; PI 4.7 on disciplinary practices; PI 4.8 on working hours; and PI 4.9 on environmental and social training.

With child labour previously being found in seaweed farming (UNHCR 2009), and the ever-increasing role of migrants in the seaweed industry in Asia (Asri et al. 2021), it is important to ensure that production units in assessment conduct operations in a socially responsible manner in a way that benefits both workers and the communities in which they operate.

27.3.1.5 Principle 5

Principle 5 evaluates whether production units minimise the negative impact they have on the surrounding communities. There are seven PIs that make up Principle 5: PI 5.1 evaluates community impacts; PI 5.2 conflict resolution; PI 5.3 rights of indigenous people; PI 5.4 visibility, positioning, and orientation of production units or water-based structures; PI 5.5 identification and recovery of substantial gear; PI 5.6 noise, light, and odour; and PI 5.7 decommissioning of abandoned production units.

The intent of this principle is for production units to interact with the local community in a transparent and responsive manner, while maintaining farm and harvesting sites that are clean and orderly.

27.3.2 The Procedural Requirements: The ASC-MSC Seaweed Certification and Accreditation Requirements (CAR)

27.3.2.1 Scope Criteria

The ASC-MSC Seaweed Standard applies to wild harvest and farmed seaweed operations of any scale and in any location around the globe. Only harvesting or farming activities that target seaweed species are eligible for certification. Production units that use carcinogenic, mutagenic, or teratogenic pesticides, or other chemicals that persist as toxins, either in the environment in which the operation operates or on the farmed seaweed product, are not eligible for certification (Aquaculture Stewardship Council and Marine Stewardship Council 2018b).

27.3.2.2 Notice of Initial Audit

The initial audit of a seaweed production unit is formally announced at least 30 days prior to the site visit via an announcement document that is posted on the ASC-MSC website. Within 5 days of the announcement being published, potential stakeholders are notified by the CAB of the upcoming site visit and invited to participate.

27.3.2.3 Site Visit—Stakeholder Consultation and Information Collection

Management locations and production sites are visited during a site visit to ensure that the impacts and management of the production unit can be reviewed and verified. The site visit is also where stakeholder interviews are conducted to ensure that CABs are aware of concerns or additional information the stakeholders have.

27.3.2.4 Setting Conditions

Once the assessment team has reviewed and analysed all the evidence collected at the site visit, the production unit is assessed against the Performance Indicators (PIs) and scoring issues of the standard. Any PIs that do not meet the 'minimum level' are assigned a critical condition that must be closed out in 3 months, before the assessment can continue. If a PI does not meet the 'target level', a condition is assigned that must be closed out within 1 year. Along with the condition, the client is required to prepare a client action plan (CAP), which includes:

- A root cause analysis of the condition.
- The corrective action(s) to be taken and an explanation of how these actions will resolve the condition.
- Who will complete these actions.
- The timeframe within which the condition will be closed.
- How progress to meet/close out the condition will be shown to the CAB.
- Any funding needs to close out the condition.

27.3.2.5 Publishing Reports

Transparency and stakeholder involvement are fundamental in ASC-MSC seaweed assessments. The CAB must publish a series of reports, starting with the Client Draft Report (shared with the client, which includes all written stakeholder submissions, the CAB's responses to these submissions, raised conditions, and references), the Public Comment Draft Report (includes the client action plan and the draft determination on whether the CAB recommends that the applicant be certified), and Final Report (includes any stakeholder and quality assurance comments received during the consultation period, any changes to scoring, rationales, conditions, or determinations, and amended client action plans if they exist). While the Client Draft Report remains unpublished, both the Public Comment Draft Report and Final Report are published and are open to stakeholder comments or submissions. Stakeholders are given 30 days to provide comments on the Public Comment Draft Report. Once the Final Report is published, a 10-day objection period starts, where stakeholders who have been involved in the assessment process to date and who are dissatisfied with the outcome of the assessment can lodge an objection. If no objections are received, a Public Certification Report is published on the ASC-MSC website. Once the PCR is published, the CAB publishes a certificate, valid for 3 years. During this three-year certification period, annual surveillance audits are conducted to determine whether conditions can be closed out and whether the production unit continues to be in compliance with the standard.

27.3.2.6 Chain of Custody

As with the ASC, the MSC Chain of Custody (CoC) Standard is used by the seaweed standard. A traceability system is required to be in place for every distributor, processor, and retailer that handles ASC-MSC certified seaweed. This ensures that certified seaweed remains traceable along the supply chain and segregated from non-certified product (Aquaculture Stewardship Council n.d.-b).

27.3.2.7 Ecolabel Use

Once the seaweed production unit is certified and the entire supply chain of the seaweed production unit in question is certified, an ASC and/or MSC ecolabel can be placed on menus or product packages (Aquaculture Stewardship Council and Marine Stewardship Council 2017). Whether the ASC or MSC (or both) ecolabel is used depends on to what production category the seaweed production unit belongs (production categories are described in Table 27.2).

Initially, only wild harvest (Category A) production units carried the MSC ecolabel, while farmed (Categories Bii and Cii) units carried the ASC ecolabel, and enhanced (Bi or Ci) or mixed source carried both the MSC and ASC ecolabels (Aquaculture Stewardship Council and Marine Stewardship Council 2017). However, a derogation[2] (Marine Stewardship Council 2023) was published in 2023 that explained that products in Category Bii shall carry both the ASC and MSC ecolabels (Table 27.5).

[2]The MSC defines a derogation as a temporary normative measure that allows for an MSC requirement to be applied differently or disregarded (Marine Stewardship Council 2022).

Table 27.5 Production categories and ecolabel use, from Derogation 9 (Aquaculture Stewardship Council and Marine Stewardship Council 2023)

Production category	Production identification category	Ecolabel use
Category A (wild harvest)	MSC	CERTIFIED SUSTAINABLE SEAFOOD MSC www.msc.org TM
Category B (all cultivation at sea, regardless of whether seeds are supplied from the wild)	ASC + MSC	CERTIFIED SUSTAINABLE SEAFOOD MSC www.msc.org TM AND FARMED RESPONSIBLY asc CERTIFIED ASC-AQUA.ORG TM
Category Ci (land-based enhanced production units, where seeds are supplied from the wild)	ASC + MSC	CERTIFIED SUSTAINABLE SEAFOOD MSC www.msc.org TM AND FARMED RESPONSIBLY asc CERTIFIED ASC-AQUA.ORG TM
Category Cii (land-based aquaculture systems that do not require or rely on supply form wild stock)	ASC	FARMED RESPONSIBLY asc CERTIFIED ASC-AQUA.ORG TM

27.4 Conditions

27.4.1 Seaweed Production Units in the ASC-MSC Seaweed Programme

Of the 12 seaweed production units that have been engaged in the programme, three have been categorised as Bi (cultivation at sea (including production units which require some stages cultivated in land-based hatcheries followed by grow-out at sea), with seed supplied from wild stock), seven have been categorised as Bii (cultivation at sea (including production units which require some stages cultivated in land-based hatcheries followed by grow-out at sea), with supply of seed from wild stocks not required or negligible), and two have been categorised as Cii (cultivation entirely in land-based systems, with supply of seed from wild stocks not required or negligible). Both seaweed production units that are categorised as Cii produce microalgal and heterotrophic species (*Euglena*, *Chlorella*, and *Schizochytrium* species) (Table 27.6).

Table 27.6 Production units by country, production category, and species

Country	Production category	Species	Number of production units
Japan	Bii	Kombu	1
		Wakame	1
	Cii	Chlorella	1
		Euglena	1
Republic of Korea	Bi	Laver	3
	Bii	Arame	1
		Hijiki	1
		Kombu	3
		Laver	3
		Wakame	2
USA	Cii	*Schizochytrium* spp.	1

27.4.2 Review of Conditions

Although all the production units that have entered the programme are unique, it is insightful to reflect upon which PIs have had conditions placed against them. The PI with the highest number of conditions placed against it was PI 4.4 (Health, Safety, and Insurance), followed by PI 2.5 (Waste Management and Pollution Control) and PI 5.2 (Conflict Resolution). Most of the production units that had these conditions were in category Bii (Cultivation at sea with supply of seed from wild stocks not required/negligible)—this is to be expected as most production units fall in this category. Looking at whether these conditions were critical conditions or non-critical conditions, they appear to be almost equally split, with some production units having these conditions assigned as a critical condition, while others were assigned as non-critical.

It is also useful to examine which PIs have never been assigned a condition, either critical or non-critical. No Principle 1 PIs have been assigned a condition, which may be expected as there are no category A production units (those that harvest natural populations) and very few Bi (cultivation at sea with seed supplied from wild stock) and Ci (cultivation in land-based systems with seed supplied from wild stocks) production units. There have also never been conditions assigned PI 2.4 (Other Species [Species Other than ETP Species]), PI 2.9 (Introduction of Alien Species), PI 3.1 (Legal and/or Customary Framework), PI 4.6 (Freedom of Association and Collective Bargaining), PI 5.3 (Rights of Indigenous People), PI 5.4 (Visibility, Positioning, and Orientation of Production Units or Water-Based Structures), or PI 5.7 (Decommissioning of Abandoned Production Units). This may mean either that all the production units that have entered the programme have been in compliance with these requirements, or that these requirements are not fit for purpose—for instance, perhaps the bar is too low.

Figure 27.4 outlines the number of conditions (refer to Sect. 27.2.4 for condition setting) that were set against seaweed production units in the programme by seaweed production category. No seaweed production units have had conditions set against any PIs in Principle 1 (sustainable wild populations). Principle 2 (environmental impacts) has nine PIs, and seaweed production units have had conditions set against seven of those (all but PIs 2.4 and 2.9). Three PIs make up Principle 3 (effective management), and two of the three PIs have had conditions placed against them. Principle 4 (social responsibility) consists of nine PIs, and all but PI 4.6 have had conditions placed against them. The last principle, Principle 5 (community relations and interaction), has seven PIs, and four of the seven PIs have had conditions placed against them.

Figure 27.5 outlines the number of conditions, set against seaweed production units by condition type (either condition or critical condition). Within Principle 2, PI 2.5 (performance indicator: waste management and pollution control) has an especially high number of conditions set against it—drilling down to the Scoring Issue (SI) level, and SI 2.5.c has a greater number of conditions set against it than the other SIs in PI 2.5. SI 2.5.c pertains to 'chemicals and hydrocarbon spills', and at the minimum level, seaweed production units are expected to have '…some measures in place that can help to prevent spills of chemicals and hydrocarbons originating from the UoA'. At the target level, '…there is a spill prevention and response plan in place for chemicals and hydrocarbons originating from the UoA'.

Within Principle 4, PI 4.4 (performance indicator: health, safety, and insurance) has the highest number of conditions, with scoring issue a and c having the greatest number of conditions (four each). At the minimum level for SI a, the employer should have 'a safe and healthy working and living environment (where accommodation is provided) for workers'. There is nothing at the target level. At the minimum level for scoring issue c, 'personnel are trained effectively on health and safety topics related to their role, responsibilities and activities'. At the target level, 'formal and regular training courses are undertaken. Risk assessments are documented and/or certified personnel are appointed to apply

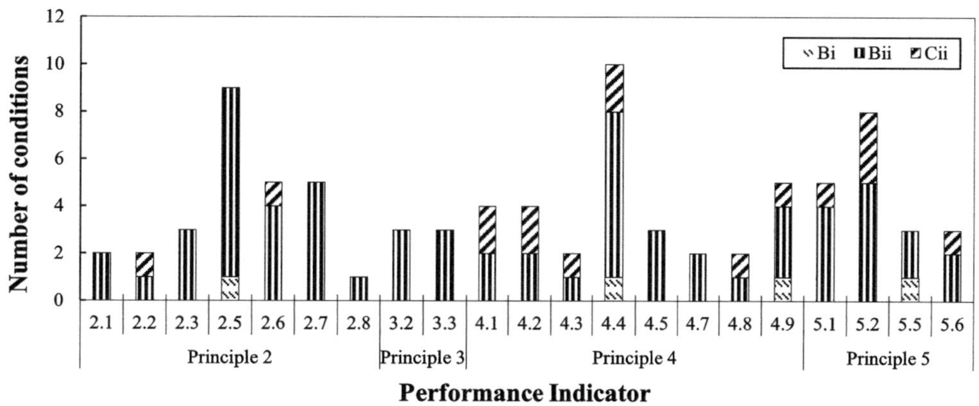

Fig. 27.4 Number of conditions (at the PI level) set against seaweed production units by seaweed production category

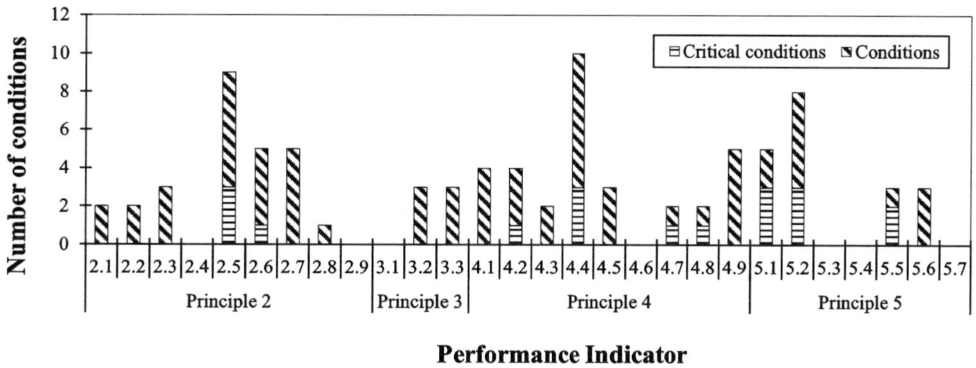

Fig. 27.5 Number of conditions (critical conditions and conditions) set against seaweed production units

health and safety risk assessment, which may include an overseeing management committee, first aiders, and/or fire marshals. Special risks associated with changing workplace or worker condition, such as expectant mothers, will have due consideration'.

In Principle 5, PI 5.2 has the greatest number of conditions, with each scoring issue assigned a condition. At the minimum level for scoring issue a, 'the management system incorporates or is subject by law to a mechanism for the resolution of legal disputes arising within the system', whereas at the target level, 'the management system incorporates or is subject by law to a transparent mechanism for the resolution of legal disputes, which is considered to be effective in dealing with most issues and that is appropriate to the context of the UoA'. At the minimum level for scoring issue b, 'organisations and individuals involved in the management process have been identified. Functions, roles and responsibilities are generally understood', while at the target level, 'functions, roles, and responsibilities are explicitly defined and well understood for key areas of responsibility and interaction/s'. At the minimum level for scoring issue c, 'the management system includes consultation processes that obtain relevant information from the main affected parties, including local communities and knowledge, to inform the management system', while at the target level, 'the management system includes consultation processes that regularly seek and accept relevant information, including local communities and knowledge. The management system demonstrates transparency and consideration of the information obtained'. At the minimum level for scoring issue d, 'the consultation process provides opportunity for all interested and affected parties to be involved. There is nothing at the target level'.

27.5 Conclusions

Most certification schemes have four features in common—(1) a governance structure; (2) a chain of custody process and a standard that regulates the social or environmental impacts of a process; (3) a monitoring and inspection scheme; and (4) ecolabels. This chapter outlines how these four features make up the ASC-MSC Seaweed Programme.

The first version of the ASC-MSC Seaweed Standard was published in 2017 and became effective in 2018. Since then, 12 seaweed production units have entered the programme, mainly in the Republic of Korea and Japan. Most of these production units produce macroalgae (laver, kelp, nori, wakame, hijiki), while some produce microalgal and heterotrophic species (*Chlorella*, *Euglena*, *Schizochytrium* spp.).

The ASC-MSC Seaweed Standard is a relatively new standard. Other seaweed certification schemes existed before the ASC-MSC Seaweed Standard was published in 2017, such as Friend of the Sea's FOS—Seaweeds and algal products requirements (Friends of the Sea 2020); others have been created since, such as Global Seafood Alliance's Seaweed Farm Standard (Critchley et al. 2023). With the world's population increasing daily, and as nations work to achieve the United Nations Sustainable Development Goals (SDGs) by the 2030 deadline, it will be interesting to see whether the demand for seaweed certification and certified seaweed products grows in future.

Author Contributions Conceptualisation: Miki Takada; Formal analysis and investigation: Miki Takada; writing—original draft preparation: Miki Takada; Writing—review and editing: Miki Takada; Supervision: Tim K. Davies.

Funding Not applicable

Declarations

Competing Interests Both authors are currently employed by the Marine Stewardship Council, the subject of this chapter.

References

Aquaculture Stewardship Council (ASC) (2015) MS and ASC sign MOU and commit to explore further opportunities to add value to partners. https://asc-aqua.org/news/msc-and-asc-sign-mou-and-commit-to-explore-further-opportunities-to-add-value-to-partners/. Accessed 26 Dec 2024

Aquaculture Stewardship Council (ASC) (2016) MSC and ASC collaborate to set global standard for seaweed. https://asc-aqua.org/news/msc-and-asc-collaborate-to-set-global-standard-for-seaweed/. Accessed 26 Dec 2024

Aquaculture Stewardship Council (ASC) (n.d.-a) Funding and finances. https://asc-aqua.org/about-asc/funding-finances/. Accessed 10 Feb 2025

Aquaculture Stewardship Council (ASC) (n.d.-b) The ASC-MSC Seaweed Standard. https://asc-aqua.org/producers/asc-standards/seaweed-standard/. Accessed 19 Feb 2025

Aquaculture Stewardship Council and Marine Stewardship Council (2016) MSC-ASC joint seaweed standard setting procedure version 1.0. https://asc-aqua.org/wp-content/uploads/2019/03/MSC-ASC-Joint-Seaweed-Standard-Setting-Procedure.pdf. Accessed 26 Dec 2024

Aquaculture Stewardship Council and Marine Stewardship Council (2017). Get certified! Your guide to the ASC-MSC Seaweed Standard audit process. https://asc-aqua.org/wp-content/uploads/2019/09/BC2040_Seaweed_GetCertified_Guide_2017_Interactive_Final.pdf. Accessed 26 Dec 2024

Aquaculture Stewardship Council and Marine Stewardship Council (2018a) ASC-MSC Seaweed (Algae) Standard v1.01. https://asc-aqua.org/wp-content/uploads/2017/11/ASC-MSC-Seaweed-Algae-Standard-v1.01.pdf. Accessed 2 Jan 2025

Aquaculture Stewardship Council and Marine Stewardship Council (2018b) ASC-MSC Seaweed (Algae) Certification and Accreditation Requirements v1.01. https://asc-aqua.org/wp-content/uploads/2017/11/ASC-MSC-Seaweed-Algae-Certification-and-Accreditation-Requirements-v1.01.pdf. Accessed 2 Jan 2025

Aquaculture Stewardship Council and Marine Stewardship Council (n.d.) The ASC-MSC Seaweed Standard. https://asc-aqua.org/wp-content/uploads/2020/07/MSC-ASC-PRO-002-Complaints--v1.0.docx. Accessed 2 Jan 2025

Asri A, Le Masson V, Montalescot V, Lim PE, Nor AM, Hussin H, Shaxson L (2021) The role of migrants in the Malaysian seaweed value chain. Mar Policy 134:104812
Assurance Services International (n.d.) Find a CAB. https://www.asi-assurance.org/s/find-a-cab. Accessed 2 Jan 2025
Auld G (2014) Constructing private governance: the rise and evolution of forest, coffee, and fisheries certification. Yale University Press, New Haven and London
Bhuyan MS (2023) Ecological risks associated with seaweed cultivation and identifying risk minimization approaches. Algal Res 69:102967
Campbell I, Macleod A, Sahlmann C, Neves L, Funderud J, Øverland M, Hughes AD, Stanley M (2019) The environmental risks associated with the development of seaweed farming in Europe—prioritizing key knowledge gaps. Front Mar Sci 6:107
Critchley A, Ask E, Chopin T, Duan D, Cottier-Cook E, Hayashi L, Kim J, Kraan S, Lim PE, Montalescot V, Msuya F, Neish I, Rebours C, Roleda M, Roberson L, Sanderson J, Sato Y, Yarish C (2023) Global Seafood Alliance - seaweed farm standard best aquaculture practices certification standard. https://www.researchgate.net/publication/373072306_Global_Seafood_Alliance_Seaweed_Farm_Standard_Best_Aquaculture_Practices_Certification_Standard
Friends of the Sea (2020) Friends of the Sea Standard: FOS—Seaweeds and algae products requirements. https://friendofthesea.org/wp-content/uploads/FoS-Seaweed-and-algae-v3-27102020.pdf. Accessed 2 Jan 2025
Marine Stewardship Council (2022) MSC Fisheries Certification Process v2.3. https://www.msc.org/docs/default-source/default-document-library/for-business/program-documents/fisheries-program-documents/msc-fisheries-certification-process-v2-3.pdf?sfvrsn=96c69f6a_39. Accessed 11 Feb 2025
Marine Stewardship Council (2023) Derogation 9—Use of both ASC and MSC labels on products coming from seaweed production category Bii. https://mscstandards.my.site.com/interpret/s/interpretation/a01N200000E4CXDIA3/int0002. Accessed 11 Feb 2025
Marine Stewardship Council (MSC) (2014) Industry update: MSC to develop sustainable seaweed standard. https://www.msc.org/media-centre/press-releases/press-release/industry-update-msc-to-develop-sustainable-seaweed-standard. Accessed 26 Dec 2024
Marine Stewardship Council (MSC) (2016) MSC and ASC collaborate to set global standard for seaweed. https://www.msc.org/media-centre/press-releases/press-release/msc-and-asc-collaborate-to-set-global-standard-for-seaweed. Accessed 26 Dec 2024
Marine Stewardship Council (MSC) (2017a) ASC and MSC release joint seaweed standard. https://www.msc.org/media-centre/press-releases/press-release/asc-and-msc-release-joint-seaweed-standard. Accessed 26 Dec 2024
Marine Stewardship Council (MSC) (2017b) MSC seeks stakeholder input on sustainable fishing and traceability standards. https://www.msc.org/media-centre/press-releases/press-release/msc-seeks-stakeholder-input-on-sustainable-fishing-and-traceability-standards. Accessed 26 Dec 2024
Marine Stewardship Council (MSC) (2024) Celebrating leadership in sustainable fishing. msc-annual-report-2023-2024.pdf. Accessed 10 Feb 2025
Marine Stewardship Council (MSC) (n.d.-a) Our history. https://www.msc.org/about-the-msc/our-history. Accessed 2 Jan 2025
Marine Stewardship Council (MSC) (n.d.-b) How we meet best practice. https://www.msc.org/about-the-msc/how-we-meet-best-practice. Accessed 26 Dec 2024
Pereira L (2016) Edible seaweeds of the world. CRC Press, Boca Raton
Spillias S, Cottrell RS, Kelly R, O'Brien KR, Adams J, Bellgrove A, Kelly B, Kilpatrick C, Layton C, Macleod C, Roberts S, Stringer D, McDonald-Madden E (2022) Expert perceptions of seaweed farming for sustainable development. J Clean Prod 368:133052
Sultana F, Wahab MA, Nahiduzzaman M, Mohiuddin M, Iqbal MZ, Shakil A, Mamun A, Khan MSR, Wong L, Asaduzzaman M (2023) Seaweed farming for food and nutritional security, climate change mitigation and adaptation, and women empowerment: a review. Aquacult Fish 8:463–480
Tucker CS, Hargreaves JA (2008) Environmental best management practices for aquaculture. Blackwell, Ames
UNHCR (2009) 2008 Findings on the worst forms of child labor—Tanzania. https://www.refworld.org/reference/annualreport/usdol/2009/en/70020. Accessed 11 Feb 2025
Valero M, Guillemin M-L, Destombe C, Jacquemin B, Gachon CMM, Badis Y, Buschmann AH, Camus C, Faugeron S (2017) Perspectives on domestication research for sustainable seaweed aquaculture. Perspect Phycol 4(1):33–46
World Wildlife Federation (WWF) (n.d.) Farmed seaweed. https://www.worldwildlife.org/industries/farmed-seafood#:~:text=In%202009%2C%20WWF%20co%2Dfounded,audit%20and%20certify%20compliant%20farms. Accessed 2 Jan 2025
Zemke-White WL, Smith JE (2005) Environmental impacts of seaweed farming in the tropics. http://www.marineagronomy.org/sites/default/files/Environmental%20Impacts%20of%20Seaweed%20Farming%20in%20the%20Tropics.pdf. Accessed 19 Feb 2025

Miki Takada obtained her Bachelor of Arts in Politics and Film from the University of California, Santa Cruz. She received her Master of Pacific and International Affairs (MPIA, now Master of International Affairs, Career Track: International Environmental Policy) from the University of California, San Diego, School of Global Policy and Strategy, and an MS in Fisheries from Humboldt State University (now California State Polytechnic University, Humboldt). She currently works at the Marine Stewardship Council (MSC) as a Fisheries Standard Manager.

28 Algae at the Final Frontier: A Review of the Potential Applications of Macroalgae in Space

Haley K. Murphy, Jessica Carneiro, Frances Edmonson, Trenton P. Causey, Steven G. Elsaid, Rafael Loureiro, and Andrew G. Palmer

Abstract

In 1947, fruit flies were launched into Earth's orbit, giving birth to the field of space biology. Seeds and animals quickly followed, and 1982 saw the first plant, *Arabidopsis thaliana,* grown in space. In 2015, NASA astronauts consumed space-grown crops for the first time. Since, experiments have taken place on the International Space Station and on CubeSats which aim to reveal the roles of gravity within fundamental biological processes here on Earth. Additionally, these experiments explore how best to leverage the benefits of many of these organisms as we embark on one of humanity's next great steps—visiting and settling on the Moon and Mars. Plans are underway to build long-term bases on each, and Bioregenerative Life Support Systems (BLSS) will be crucial to the sustainability and success of these settlements. Built using biological and ecological approaches, BLSS will facilitate effective resource management, improve mission security, reduce costs, and limit the environmental impact of these settlements both on-site and on our own planet. Considerable thought has been spent on the roles various photosynthetic microorganisms and plants can play in BLSS. Here, we explore how macroalgae could contribute to BLSS and how biology will fuel the next great leap.

H. K. Murphy · F. Edmonson · T. P. Causey · S. G. Elsaid · A. G. Palmer
Florida Institute of Technology, Melbourne, FL, USA
e-mail: murphyh2019@my.fit.edu

J. Carneiro
Federal University of the State of Rio de Janeiro, Rio de Janeiro, Brazil

R. Loureiro (✉)
Winston-Salem State University, Winston-Salem, NC, USA
e-mail: loureirorr@wssu.edu

28.1 Introduction

Not since the "Space Race" has humanity witnessed the rapid expansion beyond Earth that we see today. The next decade will see the return of crewed missions to the Moon and preparation for Martian missions in the next 30 years. However, we have never undertaken such missions alone. Indeed, we have always brought other life forms with us beyond Earth's atmosphere. The present day is no different; one only needs to examine the International Space Station (ISS), which is teeming with life beyond astronauts, including plants, mice, spiders, yeast, microalgae, and a variety of bacteria (Hill 2016; Ronca et al. 2019; Johnson et al. 2021; Mapstone et al. 2022; Ball et al. 2023; Justiniano et al. 2024). Aboard the ISS, experiments with these organisms demonstrate the changes made to biology by microgravity and ultraviolet radiation (i.e., the Low Earth Orbit (LEO) environment), elucidate the role of gravity in a variety of biological phenomena on our own planet, and offer astronauts a reminder of home. While these organisms were brought intentionally, others are "stowaways"—microorganisms associated with the human body and experimental organisms which have escaped sterilization efforts (Justiniano et al. 2024). Given that these stowaways will come with us anyway, the true question is not how to eliminate them, but how can they be contributing participants in the spaceflight exploration endeavor?

28.1.1 Leveraging the Power of Photosynthesis

When considering organisms to incorporate into spaceflight missions, photoautotrophs have consistently been ranked the most critical for their potential contributions to food security, life support, and psychological well-being, and as possible "molecular factories" engineered to produce pharmaceuticals on-site (Yan et al. 2016; Fahrion et al. 2021; Revellame

A. T. Critchley et al. (eds.), *Advances in Temperate Phyconomy: Algal Harvest and Cultivation in Globally Distributed Temperate Waters*, Developments in Applied Phycology 15, https://doi.org/10.1007/978-3-032-02955-3_28

et al. 2021; Verseux et al. 2021; Sukhinov et al. 2023). While numerous studies have considered the potential benefits of terrestrial plants and photosynthetic microbes, little to no literature has explored the incorporation of macroalgae (Escobar and Escobar 2017; Mapstone et al. 2022; Deuri et al. 2024). Here, we will consider some of the critical roles for photosynthetic organisms which are already being explored, as well as the requirements and challenges considered when selecting organisms for spaceflight missions. Building on the body of work for these other photoautotrophs and the predicted future of space exploration and settlement, we will consider how and when macroalgae could join us beyond Earth. Our goal is to inspire creative discussion by exploring the potential benefits of these organisms, which have largely been ignored by the current space revolution.

28.1.2 Photosynthetic Organisms Supporting the Mission

The National Aeronautics and Space Administration (NASA) (USA) began using the Vegetable Production System, called VEGGIE, aboard the ISS in 2014. The goals for this system were to conduct plant studies and eventually provide astronauts with access to fresh food (Fig. 28.1). This is of particular interest for long-distance spaceflight missions, as there are several nutrients which are not shelf stable, most notably the B vitamins, (Cooper et al. 2017; Tang et al. 2021; Sweet et al. 2025), and are therefore incompatible with a prepackaged diet from Earth. In addition, there are well documented benefits to morale from the presence of plants in these extremely isolated environments (Mauerer et al. 2016; Li et al. 2022a, b). Ultimately, however, the main benefit of these studies is the development of sustainable food production systems, which improve food security and the independence of astronauts beyond Earth. These benefits derive not only from plants, but also from edible cyanobacteria and microalgae, which can support nutrient fixation. Efforts to establish systems of independent, space-based food production will support the health and safety of astronauts on long-term missions to deep space and establishment of bases on the Moon and Mars (Fig. 28.2).

The economic benefits of on-site food production should not be ignored in the development of long-term extraterrestrial missions. Currently, shipments deliver necessities every 3 months to the ISS in Low Earth Orbit (LEO at an altitude of less than or equal to 2000 kilometers (*Commercial Space Frequently Asked Questions,* NASA). Here, resupply demands cost an estimated 1.6 to 1.7 billion dollars annually based on NASA projections from 2023 to 2030, which include cargo and crew rotations (*FY 2025 Budget Request,* NASA). Lunar and Martian resupply missions will be even more expensive, less frequent, and more prone to technical challenges and unforeseen complications. With current technology, resupply missions to the Moon would only take about 3 days, and as a result, emergency resupply missions are reasonable to factor into settlement planning (Burks 2021). However, Mars is approximately 9 months away (Burks 2021), making "emergency" shipments unrealistic. As a result, it is necessary that Mars settlement plans require the development of infallible supply chains and sustainable systems on-site, as missed resupply missions may well be catastrophic (Do et al. 2019).

As a result, developing systems with increased productivity, resilience, and sustainability is mission critical. While systems have been envisioned for BLSS at these sites, there are sufficient gaps in terms of how plants can support these missions for us to consider solutions beyond terrestrial plants. It is here, at the drawing board, where we are offered a chance to envision alternatives, and discussion of macroalgae might be fruitful.

Cargo and infrastructure will provide food, clean water and air, clothing, shelter, and other necessities to visitors of lunar and Martian bases, both in transit and at their destinations. To extend these resources and the lifetime of these settlements, regenerative life support systems incorporating photosynthetic organisms have been explored for over three decades. These systems leverage the inherent relationship present between producers and consumers to introduce edible biomass to the biome and convert inedible biomass into usable materials while assisting in life support systems through the natural carbon cycle. Theoretically, these systems should lower the costs and risks associated with failed resupply missions. With systems like these in place, each resupply mission will build upon the previous rather than replacing it, improving astronaut well-being and confidence in the program. Such efforts are not isolated in their benefits to off-world settlers, as sustainable systems and the development of related technologies may promote sustainability on Earth in a variety of emerging areas, including advanced urban farming (Atri et al. 2024).

28.1.3 In Situ Resource Utilization (ISRU)

Rather than accepting the closed-loop approach originally conceived for BLSS, sustainability researchers and mission planners have begun to consider the resources on-site which may prove useful; this collection of approaches is generally classified as in situ resource utilization (ISRU) (Fig. 28.3). Both the Moon and Mars contain many elements which may support these missions, including iron, potassium, calcium, and phosphorous concentrated in the regolith (unconsolidated layer of surface material above bedrock) (Taylor et al. 2000; Ehlmann and Edwards 2014; Morrison et al. 2018; Li

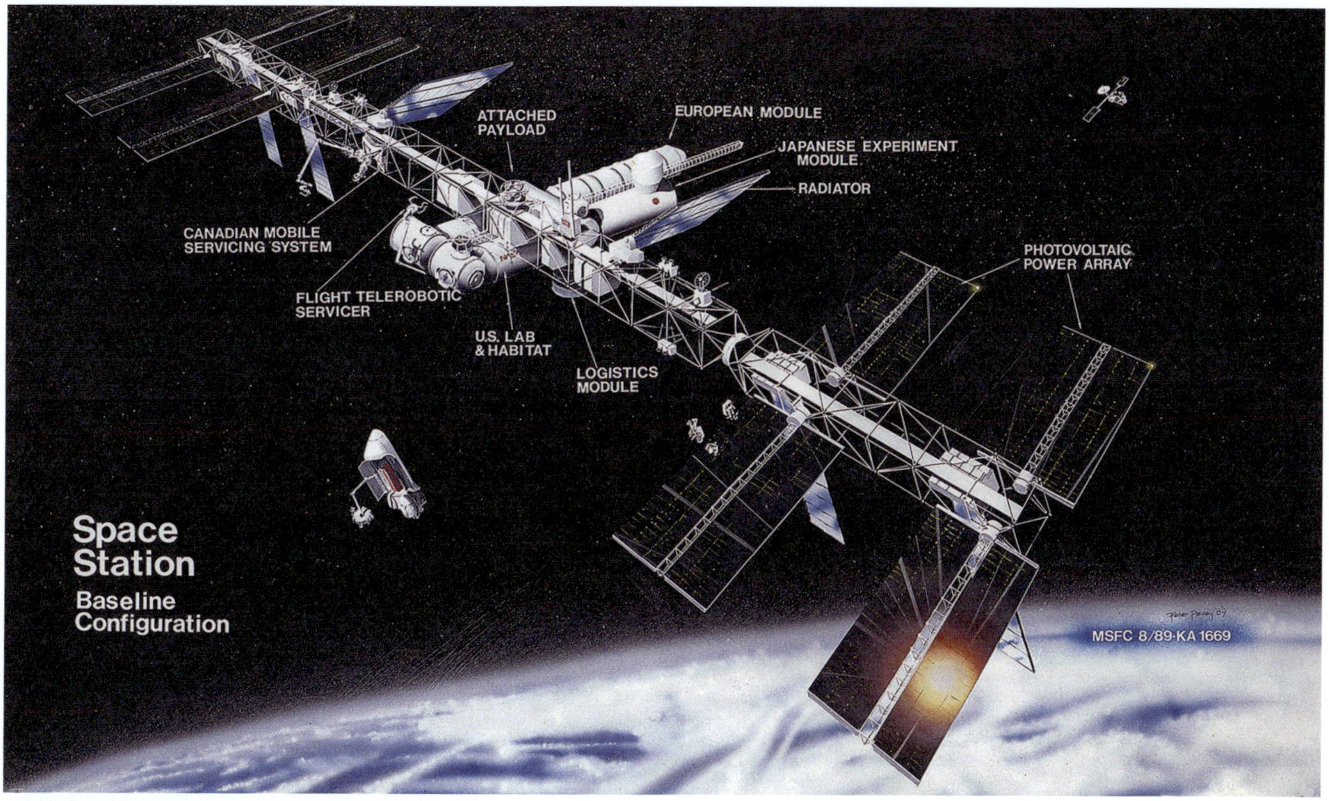

Fig. 28.1 Art depicting the configuration of the International Space Station created five years after its concept was announced in 1984 (NASA Image and Video Library 1989)

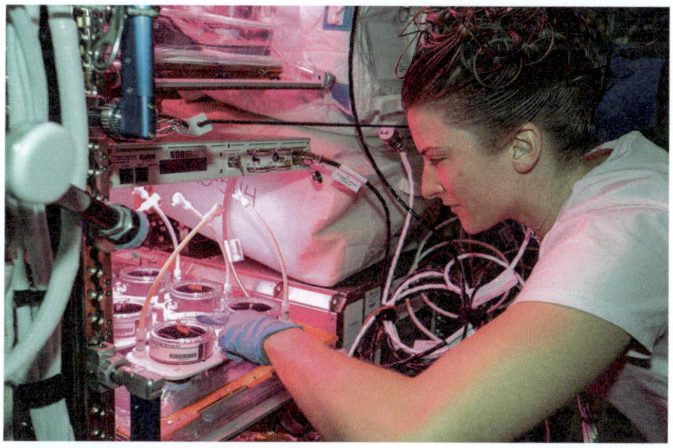

Fig. 28.2 Astronaut Kayla Barron working in the Veggie Unit aboard the ISS, 2022. Pictured is the passive orbital nutrient system (PONDS) (NASA Image and Video Library 2022)

et al. 2022a, b). Regolith is a candidate for structural use in building astronaut habitats, launch pads, and other infrastructure (Kading and Straub 2015; Isachenkov et al. 2021), and mineral mining is not only feasible but also financially favorable (Sommariva et al. 2020).

Crucial nutrients for crop production can be mined from the regolith (Duri et al. 2022), although there are critical remediating steps which need to be taken beforehand, like the removal of perchlorates on Mars or amendments, such as the addition of nitrogen at either location (Trainer et al. 2019; Misra et al. 2021; Rzymski et al. 2024). Regolith-based lunar and Martian agriculture has been explored using Earth-made simulants (Eichler et al. 2021; Toklu et al. 2023; Fackrell et al. 2024), which can make regolith a valuable "input" in isolated settlements. Finally, the atmosphere of Mars, while exceptionally thin relative to Earth's (approximately 2%), is roughly 96% carbon dioxide, providing a potential carbon source for increasing biomass in the system via photosynthesis (American Meteorological Society 2024; Williams 2024b). With an atmospheric density of approximately 0.00001%, the exosphere of the Moon is highly unlikely to provide any usable resource.

Bioregenerative Life Support Systems (BLSS) Having established that many of the resources necessary for life persist at these sites, what sorts of systems can they be incorporated into? Here, we consider the nature of bioregenerative life support systems. These recycling systems, inspired by Earth's own ecological processes, have great potential to incorporate in -situ resources. Sustainable space practices incorporate microbes (Santomartino et al. 2023), plants (Mortimer and Gilliham 2022), and even animals (Liu et al. 2021) to produce food, water, oxygen, and other necessities

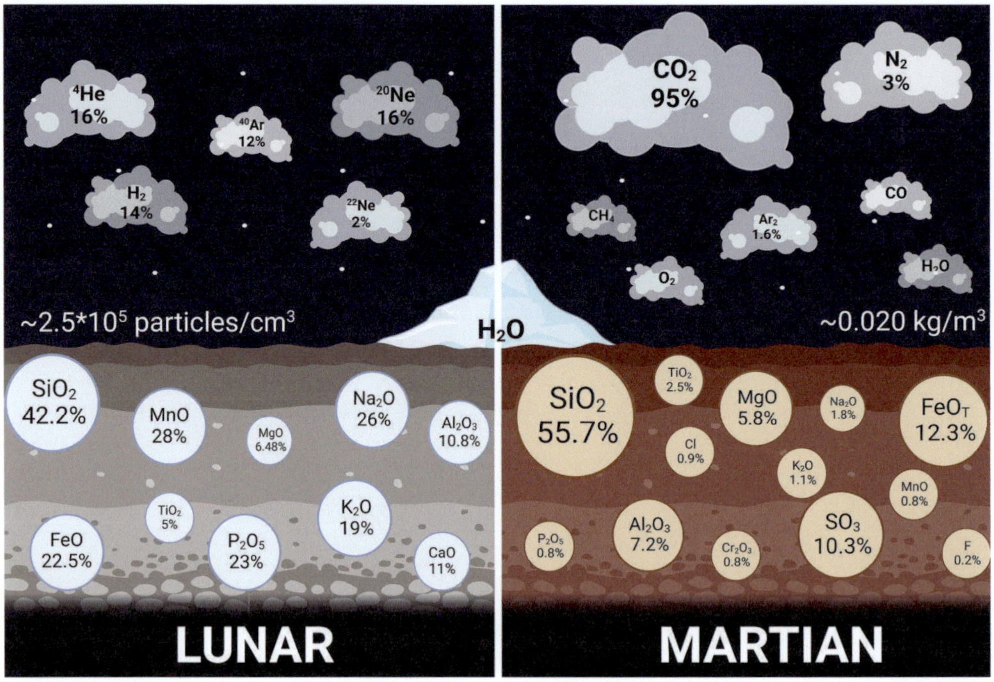

Fig. 28.3 In situ resources available on the Moon and Mars. Atmospheric resources are minimal on both (Williams 2024a, b) in comparison to Earth, where the atmospheric density of pure, dry air is 1.293 kg m^{-3} (American Meteorology Society 2024). Like Earth's crust, both the lunar and Martian surface minerals are heterogenous and vary in composition depending on location. Lunar regolith XRD data is sourced from mid-latitude at the northeastern Oceanus Procellarum by China's Chang'E-5 mission in 2020 (Taylor et al. 2000; Morrison et al. 2018), and Martian regolith XRD data is taken by the Chemistry and Mineralogy (CheMin) instrument aboard the Curiosity rover in the Murray formation of the Gale Crater (Ehlmann and Edwards 2014). The Gale Crater is a fluviolacustrine site and was theorized to contain a body of water in the past. Collection of these resources on the Moon and Mars will supplement materials shipped with and to astronauts. (Created in BioRender, Murphy HK (2024))

from shipped and in situ resources. High-yield bioreactors could produce fertilizers, nutritional supplements, pharmaceuticals, and specific molecular cocktails produced by waste processing and biomining using mechanisms which already occur on Earth (Shaw et al. 2022). Meanwhile, "greenhouses" lit by LED lights and utilizing these processed materials could provide additional life support, fresh food, and a relaxing reminder of home to astronauts (Jeong and Park 2021). Advances in genetic engineering will ensure our ability to widen the capabilities of these plants and microbes to fit changing nutritional and health needs (Berliner et al. 2021). Astronauts' mental and physical health, as well as mission success, will improve with their agency and confidence that lunar and Martian systems can support them.

Improving the sustainability of BLSS can be inspired by Earth's ecosystems. For example, biodiversity ensures multiple species are available to perform niche specific tasks, improving reliability, which is critical in the space environment (Hong et al. 2022). Simply put, overlapping functions in organisms within the BLSS encourage system resilience. The question at hand is how to meet the range of a settlement's needs, simultaneously limit material and energy costs, and ensure resilience of the system (Frossard et al. 2024). Incorporating macroalgae might well improve BLSS adaptability and longevity.

Projects like the European Space Agency's MELiSSA (Fig. 28.4), NASA's CELSS (Controlled Environment Life Support System), and the United Nations' Biosphere 2 have leveraged microbial metabolism to fuel a closed ecosystem, an interest which began in the latter half of the twentieth century and continues today (Prince and Knott 1989; Marino and Odum 1999; Lasseur et al. 2010). To understand how algae can supplement space habitation, one must examine the roles photosynthetic organisms play in the design of these manufactured ecosystems. In MELiSSA, cyanobacteria and higher plants provide food, oxygen, and potable water to astronauts. Macroalgae are quite capable of performing the same functions and can be cultivated in bioreactors similar to the design of MELiSSA. However, we must consider the prerequisites which spaceflight and long-term space habitation demand of BLSS and ISRU organisms.

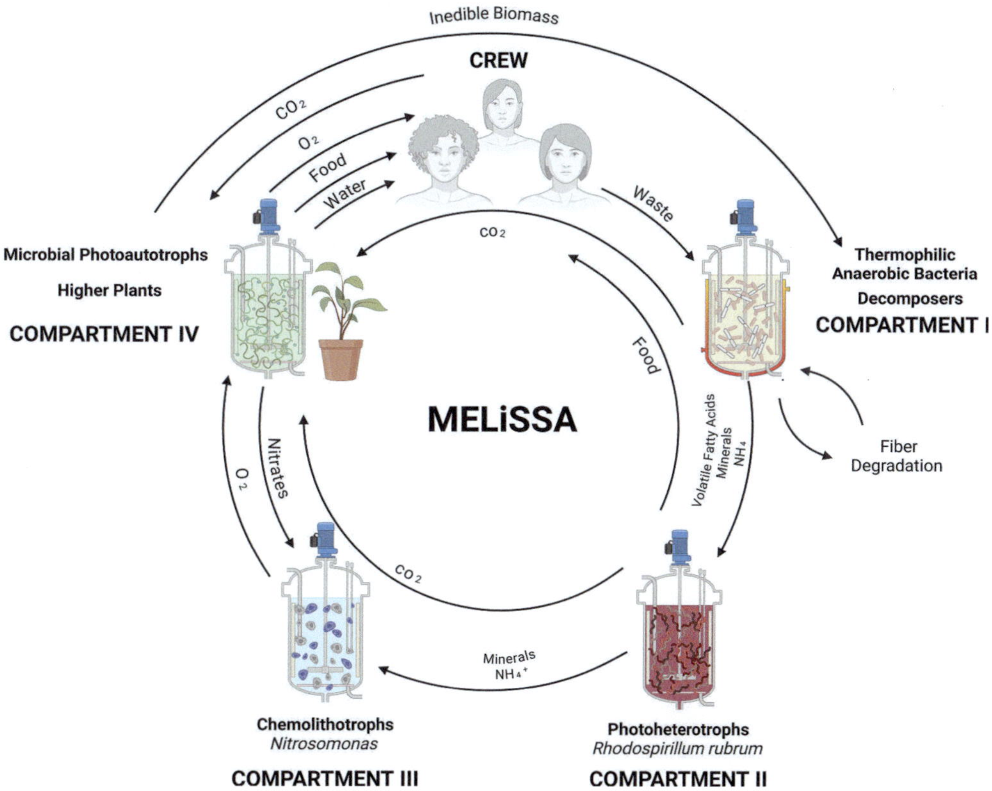

Fig. 28.4 MELiSSA (Micro-ecological Life Support System Alternative), which is inspired by lacustrine ecosystems. Efficiency as a closed system is currently being tested using rats in Earth-based facilities. Testing of the cyanobacteria *Limnospira platensis* (formerly *Arthrospira platensis*) growth occurs in the ESA's Columbus lab on the ISS (*MELiSSA Loop Diagram*, Lasseur et al. 2010; Paradiso et al. 2017; Created in BioRender, Murphy HK (2024))

28.2 Factors in Space

Along with their potential BLSS applications, another consideration in long-term spaceflight is the viability of the organism under long-duration exposure to spaceflight stressors. Even with tolerance for these stimuli, the functionality of organisms may be altered. The gravity we take for granted here on Earth supports many of the processes within microbial, plant, and animal life. Currently, ground-based simulations of microgravity, ultraviolet (UV) radiation, and galactic cosmic rays (GCR) are improving accessibility of this research and providing insight into the severity to which these stressors will affect survivability in space.

28.2.1 Microgravity

Entering LOE places travelers into a state of weightlessness, as they are constantly in free-fall toward Earth. Microgravity is the term for this, as a small amount of gravity is still experienced. Earth's gravity, or "g," is commonly referenced for comparison to environments in space and is an average of 9.81 m s^{-2}. The ISS experiences 0.89 g, although this is not felt by passengers. Lunar and Martian missions experience "fractional gravity" at 0.165 g and 0.380 g, respectively. The directionality of plant growth (Kiss 2025), flow of blood in mammals (Marshall-Goebel et al. 2019), and biofilm formation and sedimentation of bacteria (Sathasivam et al. 2021) are affected if the gravitational vector is limited compared to Earth. Studies exploring these effects have begun on the ISS, where microgravity is present and centrifugal force is applied to mimic the fractional gravity experienced on the Moon and Mars.

28.2.2 Radiation

Energized particles have limited access to Earth due to its magnetosphere. However, this radiation threatens astronauts and their fellow biological passengers by ionizing atoms throughout the body and forming reactive oxygen species (ROS) by water hydrolysis (Arena et al. 2014). Direct damage to essential molecules by UV and galactic cosmic rays in deep space can prevent DNA replication and other functions necessary for survival and reproduction (Dobney et al. 2023). Even the UV experienced in LEO is inhibiting most organisms, including some extremophiles, demonstrating the challenges raised in confronting radiation (Fig. 28.5; Olsson-Francis et al. 2009; Cockell et al. 2011).

28.2.3 Algae's Role in Spaceflight: Previously Tested Species

There is no shortage of spaceflight experiments using cyanobacteria and microalgae (Niederweiser et al. 2018; Ahmad

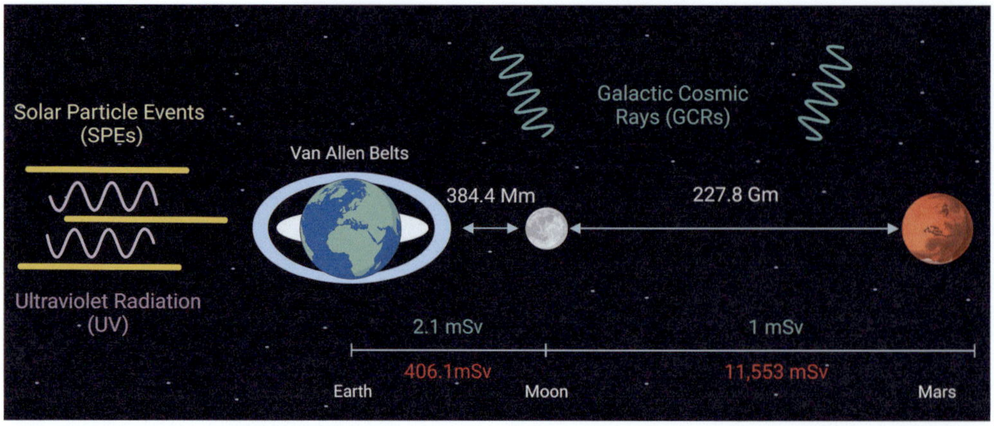

Fig. 28.5 Radiation levels in spaceflight pose risks to the crew and biological samples aboard, especially after long-term accumulation. Ionizing radiation comes from SPEs, GCRs, and the Van Allen Belts, while non-ionizing radiation is in the ultraviolet wavelengths. During solar minimum years, the Van Allen belt provides the largest source of radiation in travel beyond low Earth orbit (Märki 2020; Created in BioRender, Murphy HK and Causey TP (2024))

and Yin 2024). Considering the lack of experimentation performed on macroalgae thus far, the versatility of photosynthetic microorganisms offers a standard of comparison when judging the suitability of macroalgae for spaceflight. Although little genetic relation is found between the two groups, some of the qualities of macroalgae are shared with cyanobacteria and microalgae, including nutrient richness and plant growth promoting properties. With discretion, studies featuring cyanobacteria and microalgae can inspire new ideas for macroalgae in sustainable space systems.

Many macroalgae harbor a diverse microbiome, forming holobionts in natural habitats. Cyanobacteria and microalgae exchange nutrients with the host macroalga (Egan et al. 2017). Developing bioreactors which support this symbiosis could provide an opportunity to expand biodiversity in lunar and Martian habitats, where reliability is of the utmost importance, as well as diversify the functions performed by a single bioreactor within the system.

Examples of cyanobacterial applications in space include the highly nutritious *Limnospira platensis* (spirulina) featured in the MELiSSA system due to its photosynthetic efficiency and protein content around 60–70% (Fais et al. 2022). Many species of diazotrophic bacteria from the genera *Anabaena* and *Nostoc* have been considered as well to support the nitrogen cycle in space habitation (Tomita-Yokotani et al. 2021; Ahmad and Yin 2024). Cyanobacteria have demonstrated hardiness in simulated Martian conditions, including survival in Martian regolith simulants (Adams and Carr 1981; Bothe 2019) and some ultraviolet radiation in LEO (Cockell et al. 2011). Additionally, multiple species have been identified which efficiently remove urea from waste (Yang et al. 2008; Fais et al. 2022).

Fig. 28.6 Photobioreactor containing *Chlorella vulgaris,* which arrived on the ISS in 2019 for air regeneration and edible biomass production (European Space Agency 2019)

Over 50 spaceflight experiments studying microalgae have occurred, demonstrating the interest which surrounds their potential contributions to BLSS (Niederweiser et al. 2018). Microalgae *Chlamydomonas reinhardtii* and *Chlorella vulgaris* (Fig. 28.6) (Chlorophyta) have flown aboard the ISS frequently, where they are studied for oxygen production and carbon dioxide removal, and *C. vulgaris* has been examined for its high potential as a space-grown nutrient supplement (Niederweiser et al. 2018; Revellame et al. 2021).

Discussing the many ways microalgae and cyanobacteria have been studied for space applications highlights opportunities for macroalgae to potentially benefit BLSS. Algae, both micro- and macro-, present species with a quick growth time and accumulation of many nutrients required for human health.

28.3 Applications for Macroalgae in Space

Numerous properties of macroalgae shown in Fig. 28.7 are exploited here on Earth. To apply these functions to spaceflight, we will examine these common uses with emphasis on how they may improve extraterrestrial lifestyles. Edible genera like *Porphyra/Pyropia* (Rhodophyta) and *Saccharina* (Phaeophyceae), commonly called Nori and Kombu, respectively, are highly regarded for their nutrient content and flavor (Yamagishi et al. 2023), two key qualities to support astronaut health. Potential pollution resulting from space exploration should be considered when striving to settle the Moon and Mars responsibly, and biofuels and bioplastics can aid in this goal. Their lesser environmental impact makes them a favorable option when compared with petroleum-based products, which are not quickly degradable or recyclable (Kanagesan et al. 2022). Wastewater treatment and generation of oxygen will be necessary components within space-based life support, and chemical or mechanical methods could potentially be supplemented by efficient photosynthetic macroalgae (Cole et al. 2016). Bioremediation of toxic metals may be useful when reducing contamination of in -situ resources and has been performed using macroalgae (Santoro et al. 2022). Plant growth, whether done hydroponically or using the lunar and Martian regolith, will require nutrient supplementation. This could be performed by macroalgae, which have been used for biofertilization by farmers in the Middle East for thousands of years (Zafar et al. 2022). Finally, the many bioactive compounds macroalgae produce promote the health of humans and plants. These include phytohormones (Sunarpi et al. 2021), antioxidants (Tziveleka et al. 2021), vitamins (Biris-Dorhoi et al. 2020), and pharmaceuticals (Kaur et al. 2024).

28.3.1 Nutritional and Supplemental Value

Macroalgae and some cyanobacteria are consumed across the world as "functional foods" for their simultaneous health benefits and flavor (Aumeerun et al. 2019). Concentrated within macroalgal biomass is a robust offering of minerals, vitamins, soluble dietary fibers, essential amino acids, omega-3 fatty acids, and antioxidants (Rajapakse and Kim 2011; Murai et al. 2019). In Japan, each person consumes about 4.7 g of macroalgae on average per day, demonstrating the popularity of this food (Fig. 28.8; Zava and Zava 2011; Murai et al. 2019).

Macroalgae have been considered in the development of a healthy astronaut diet. Miyeoguk, a Korean seaweed soup containing the species *Undaria pinnatifida* (Phaeophyceae), has been examined as a freeze-dried meal to ship to astronauts. All materials sent to the ISS must be sterilized to prevent contamination with harmful microbes, so bringing Miyeoguk as part of the astronaut diet would rely on it to be both sterilized as well as safe after sterilization. To determine if this is feasible, freeze-dried Miyeoguk was irradiated with 10 kGy. In this case, Miyeoguk was not only sterilized but also retained its organoleptic qualities. Further, 30 kGy of radiation did not cause the seaweed soup to become carcinogenic and was safe for consumption (Song et al. 2012). Many macroalgae, like Miyeoguk, are considered a delicacy and would be of high interest to the astronaut diet. Developing

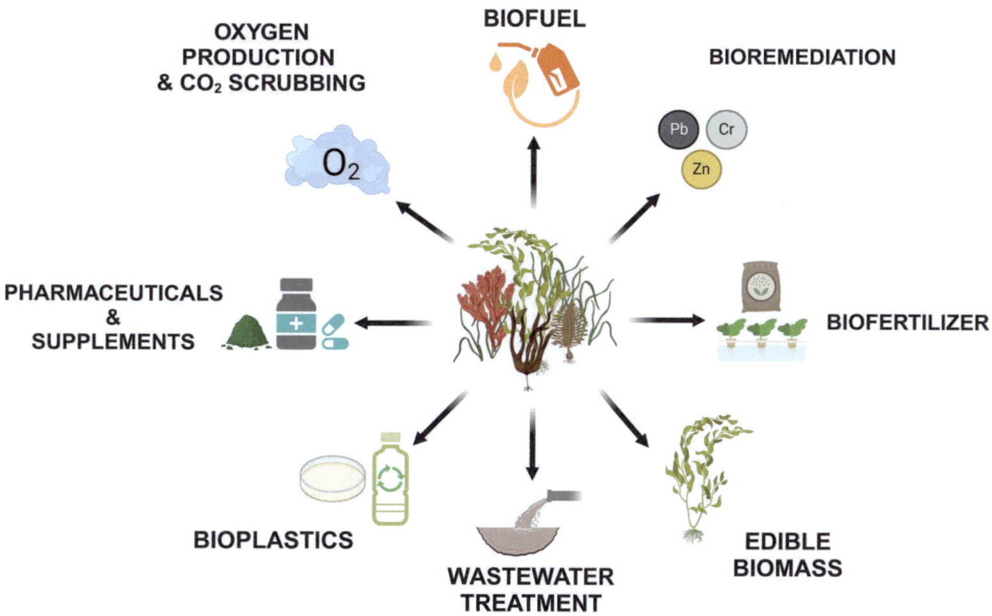

Fig. 28.7 Some of the applications for macroalgae as seen here on Earth (Hermann et al. 2015; Kanagesan et al. 2022; Priyadarsini et al. 2022; Santoro et al. 2022; Zafar et al. 2022; Created in BioRender, Causey TP (2025))

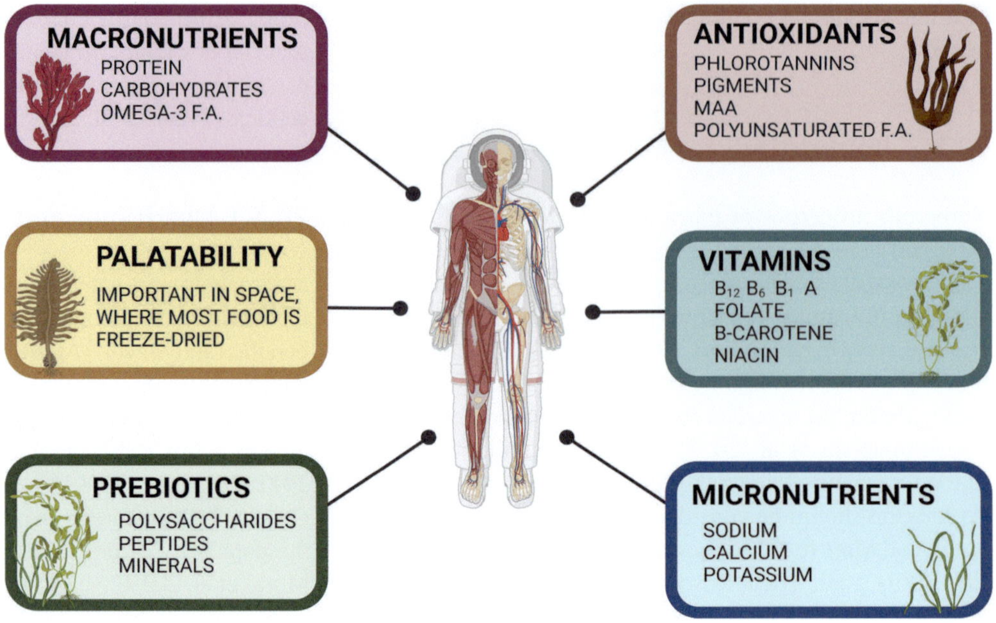

Fig. 28.8 Summary of the nutritional and health benefits of macroalgae as discussed here. Application of these benefits to spaceflight will enable construction of a well-rounded diet with the inclusion of fresh food for astronauts (Rajapakse and Kim 2011; Murai et al. 2019: Li et al. 2022a, 2022b; Machado et al. 2020; Kumar et al. 2021; Watanabe et al. 1999; Machado et al. 2020; Mouritsen et al. 2019; Cian et al. 2015; Xie et al. 2024; Created in BioRender, Murphy HK (2025))

procedures to safely bring edible macroalgae to space, like that created for Miyeoguk, will allow astronauts to benefit not only from the high nutrition seaweed contains but also from improvements to the quality of freshness and flavor of their food. Here, we will discuss some of the nutrients provided by macroalgae, with special attention paid to those which are particularly beneficial for long-term missions to space.

28.3.1.1 Macronutrients

Carbohydrates Carbohydrates are a necessary part of diet to promote short-term energy. Analysis of macroalgal dry weight has revealed around 81% of their biomass is carbohydrates, and about 50% of these are polysaccharides (including crude fiber) (Li et al. 2022a, 2022b). Polysaccharides produced by macroalgae have demonstrated numerous additional benefits for human health, including anticoagulant, anti-inflammatory, antioxidant, anticarcinogenic, and antiviral properties (Tanna and Mishra 2019).

Fiber Production of fiber in macroalgae is comparable to vegetables like cabbage, wheat bran, and sugar beet pulp (Burtin 2003). High levels of fiber derived from the macroalgal cell wall comprise about 65% of dry biomass on average and decrease the bioavailability of protein in red and green algae (MacArtain et al. 2007). Methods to degrade this fiber enzymatically are of interest to improve the absorption of macroalgal proteins in digestion (Patarra et al. 2011).

Proteins Among macroalgae, protein content is lower in brown algae when compared to red or green algae (Fleurence 1999). Measuring the amino acid content of the red algae *Pyropia dioica, Porphyra umbilicus,* and *Gracialaria vermiculophylla,* it was found these species contain all the essential amino acids required for human health (Machado et al. 2020).

Fats Macroalgae are typically observed to have a low lipid content. However, some species among brown and red algae display a high total lipid content of unsaturated and saturated fats, respectively (Sánchez-Machado et al. 2004). Both brown and red algae are additionally a good source of omega-3 and -6 fatty acids, both of which aid in cardiovascular health, mental health, and reduction of inflammation (van Ginneken et al. 2011).

28.3.1.2 Micronutrients

Vitamins As previously noted, the vitamin content of foods on long-term space missions lacks stability and will require supplementation by fresh foods. Many species of macroalgae are rich in water-soluble vitamins A, β-carotene, folate, niacin, and riboflavin, often in higher concentrations than many land vegetables (Kumar et al. 2021). Brown algae are generally rich in B_1 (thiamine), B_2 (riboflavin), B_6, and nicotinic acid (Kumar et al. 2021). The genera *Porphyra/Pyropia* (Rhodophyta) and *Ulva* (Chlorophyta), popular in sushi recipes, provide a substantial amount of vitamin B_{12} (cobalamin), a nutrient which is rare in terrestrial plants and is essential for cognitive functioning (Watanabe et al. 1999; Weir and Scott 1999; Huang et al. 2024). Additionally, *Porphyra/Pyropia* and *Ulva* do not generally contain the dangerous levels of iodine observed among many macroalgae (Watanabe et al. 1999).

Minerals Macroalgae accumulate many macro- and microminerals. Green macroalgae are generally rich in magnesium and iron, whereas brown and red algae demonstrate higher levels of sodium, potassium, and zinc (Circuncisão et al. 2018). Macroalgae display bioaccumulation and biosorption, which are of a much higher efficiency than the same processes in terrestrial plants and can contain 10 to 100 times the mineral content of land plants (Circuncisão et al. 2018).

28.3.1.3 Antioxidants

Due to ROS formed by the combination of galactic cosmic rays and ultraviolet radiation in space, astronauts face a serious threat to health. At the upper limit of NASA's acceptable level of radiation, the risk of developing cancer is increased by about 3% (Townsend and Fry 2002). Major groups of antioxidant compounds in macroalgae include carotenoids, phenolic compounds, phycobilin pigments, polyphenols, sulphanated polysaccharides, and vitamins (Cornish and Garbary 2010). Some macroalgae produce phlorotannins, which are a class of polyphenols with antioxidant properties (Simón et al. 2023). Extracts containing phlorotannins have displayed inhibition of cancer progression, and although the mechanism has not been fully elucidated, antioxidant activity is involved (Simón et al. 2023). Additionally, loss of bone density is a problem for astronauts, who have been shown to experience a greater than 10% loss of areal bone mineral density following a 6-month mission (Sibonga 2013). Antioxidant enzymes like those which are present in macroalgae have demonstrated inhibition of bone loss in mice (Surget et al. 2017), indicating they may be useful in decreasing astronauts' risk of developing osteoporosis.

28.3.2 Palatability and Organoleptic Traits

Although nutrition and preservation of astronaut health are very important factors, there are other considerations to be made when constructing a space diet. Most of the food which will be brought to the Moon and Mars will be freeze dried, like that of the ISS. Packaged meals prepared in this way can become bland and monotonous over time. Such foods can be supplemented with fresh produce grown in off-world agriculture to provide astronauts with a wider variety of flavors and textures. In response to this, organoleptic testing of potential space crops is underway to determine the taste and texture of the food astronauts may grow (Spencer et al. 2021). Macroalgae *Porphyra/Pyropia* spp. and *Gracilaria vermiculophylla* produce high concentrations of three amino acids, which are associated with umami flavor, including glutamate (Machado et al. 2020). However, volatile compounds found within seaweed have been found to influence their flavor and the satisfaction of the consumer over glutamate (Mouritsen et al. 2019), making their identification a necessary test in determining organoleptic favorability.

28.3.2.1 Prebiotics

Prebiotics are non-digestible compounds which support the growth of a healthy gut microbiome (O'Sullivan et al. 2010). Dysbiosis of the intestinal microbiome is linked to many gastrointestinal diseases like irritable bowel syndrome and colon cancer (Olvera-Rosales et al. 2021). Varieties of red algae, including *Sarcodiotheca gaudichaudii* and *Chondrus crispus* (Rhodophyta), produce polysaccharides, peptides, and minerals, which together conduct prebiotic activities to aid in supporting a healthy gut microbiome (Cian et al. 2015). Oligosaccharides from the species *Eucheuma denticulatum* (Rhodophyta) have been shown to support the growth of five different probiotics relevant to human digestion (Xie et al. 2024).

28.3.2.2 Wastewater Treatment and Recycling

Water is one of the most crucial resources for long-duration space missions, where a nearly closed system will be expected for sustainability of the BLSS. Current systems like the ISS Water Recovery System (WRS) recycle over 90% of the water on board, although further efficiency is needed for prolonged off-world operations. Algae have been identified as potential biological water purifiers capable of removing contaminants while concurrently producing oxygen and edible biomass. Wastewater results directly from astronauts as well as the "gray water" they produce from washing reusable items, forming a similar composition to the municipal wastewater, which is processed by bacteria here on Earth. The freshwater macroalgae *Spirogyra* spp. (Chlorophyta) has demonstrated the removal of up to 90.6% of nitrogen and 99.1% of phosphorus present in municipal wastewater (Ge et al. 2018). Such nutrient recovery could prove invaluable in space habitats, where waste streams from human metabolism, food processing, and hygiene must be recycled efficiently to extend the lifetime of the settlement.

Additionally, algae may be effective in removing contaminants resulting from water interacting with lunar or Martian regolith. Henriques et al. (2017) demonstrated that the green macroalga *Ulva lactuca* (sea lettuce) can effectively remove heavy metals such as cadmium, lead, and mercury from water. Even in multi-metallic solutions, *U. lactuca* exhibited high removal efficiencies, confirming its potential for use in regenerative water filtration subsystems. Macroalgae have also been employed to remove organic pollutants. Ultrafiltration techniques have been demonstrated which separate oligosaccharides and peptides from red macroalgae hydrolysates and are of significant interest to off-world applications, as controlled membrane filtration integrated into BLSS can optimize biomass utilization while maintaining water quality. Incorporating macroalgae to renew the potable water supply could optimize the process by simultaneously maintaining purity and reducing reliance on physiochemical, consumable filtration systems (Fig. 28.9).

28.3.3 Air Regeneration

Micro- and macroalgae have emerged as prime candidates for BLSS oxygen production and air revitalization due to their exceptional photosynthetic efficiency, ability to fix carbon dioxide, and capacity to integrate seamlessly into closed-loop ecological systems. Decades of research have indicated the feasibility of algae-based air revitalization, with terrestrial and spaceflight experiments that emphasize their potential to sustain long-term human space missions (Niederweiser et al. 2018).

28.3.3.1 Photosynthetic Efficiency and Oxygen Yield

Algae, particularly microalgae such as *Chlorella* and *Limnospira,* possess some of the highest known photosynthetic efficiency among oxygenic organisms. Their ability to rapidly convert carbon dioxide into oxygen surpasses that of many higher plants, making them an attractive option for closed-loop life support (Bartsch 1961). In waste treatment systems, algae have proven a reliable source of oxygen, highlighting this potential contribution to controlled ecological life support. A meta-analysis by Niederweiser et al. (2018) reviewed over 50 spaceflight experiments involving microalgae, which confirm their ability to thrive in microgravity and produce sufficient oxygen to support human respiration in closed habitats. The study emphasized that algae could theoretically achieve nearly complete atmospheric recycling, significantly reducing dependence on limited stored oxygen supplies for long-duration space travel. One key advantage of algal systems in BLSS is their tolerance and adaptability to environmental conditions. Microalgae *Chlorella* and *Dunaliella* demonstrate high productivity even under variable light, carbon dioxide, and temperature conditions. However, reduced photorespiration displayed in macroalgae could limit biomass accumulation relative to C3 or C4 terrestrial plants. High concentrations of antioxidants in many macroalgae, although beneficial to an astronaut diet, may also limit photooxidation relative to terrestrial plants.

28.3.3.2 Methods of Closed-Loop Carbon and Oxygen Cycling

The ability to efficiently cycle carbon and oxygen within a closed habitat is a fundamental requirement to sustain long-term missions in space. Macroalgae, traditionally considered less suitable for space applications due to their significant biomass requirements, have been investigated for their role in carbon sequestration, and research has suggested their superior carbon fixation rates could be leveraged in space environments if cultivation techniques are adapted to confined systems. Recent studies have underscored the importance of the microalgae's oxygen-generation efficiency in space applications, showing that certain strains can achieve oxygen production rates that surpass terrestrial plants by an order of magnitude (Niederweiser et al. 2018). Macroalgae-based carbon sequestration has been shown to sufficiently regulate the atmosphere with macroalgae cultivation as a global carbon dioxide scrubbing strategy (Wu et al. 2023). This research has direct implications for BLSS, where similar mechanisms could be employed on a controlled scale to regulate atmospheric conditions in space habitats.

28.3.3.3 Algal Bioreactors for Integrated Air Regeneration

Current spaceflight systems use physiochemical methods such as solid amine scrubbers and lithium hydroxide canisters, which limit the longevity of the system and rely heavily on resupply missions to sustain. In contrast, algae-based bioreactors have been proposed as a regenerative alternative. A notable advantage of algal carbon dioxide scrubbers is their ability to function autonomously with minimal maintenance. Ward et al. (1963) discussed the design of automated algal growth chambers that could be incorporated into closed-loop life support systems, reducing the need for frequent crew intervention while continuously purifying air. Concurrent cultures of cyanobacteria, microalgae, and macroalgae can form a consortium within bioreactors to raise efficiency.

Fig. 28.9 Contaminants which could potentially be removed by macroalgal growth in space-based BLSS, including astronaut waste and regolith leachates (Henriques et al. 2017; Ge et al. 2018; Created in BioRender, Murphy HK 2025)

28.3.4 ISRU and Regolith-Based Agriculture

While much of our discussion regarding incorporation of macroalgae into closed life support has focused on atmosphere and water recycling, there is a strong case to be made for further incorporation of macroalgae into regolith-based agriculture (RBA). RBA is intended to supplement hydroponic-based food production systems on the Moon and Mars by taking advantage of the most plentiful in situ resources available on both to develop a plant growth substrate. Lunar and Martian regolith vary in particle size but are generally considered to be extremely fine and categorized as dust. This fine material is prone to compaction, especially upon watering, which limits both root penetration and aeration. As a result, plants grown in simulated lunar and Martian regolith typically have a shallow root architecture and are affected long term by a decrease in biomass and yield. The absence of significant organic carbon or nitrogen deposits in these substrates means additional supplements are required. Removal of toxic components, such as perchlorates, is currently under development. With a combination of the amendments currently being developed, lunar and Martian regolith is expected to become a soil-like substrate and benefit the settlement as an in situ "input" of mineral resources (Fig. 28.10).

28.3.4.1 Biofertilization

Many macroalgae form symbiotic relationships with diazotrophic bacteria and provide a habitat for cyanobacteria or other nitrogen fixing bacteria (Tilstra et al. 2017). Capitalizing on these symbioses could potentially help address the lack of in situ nitrogen on the Moon and Mars. An example of this is frequently observed in rice paddies, where the freshwater environment allows macroalgal growth. In these paddies, the macroalga *Chara vulgaris* (Charophyta) houses many nitrogen fixing cyanobacteria from the genera *Anabaena, Nostoc,* and *Calothrix* (Adams et al. 2012), which provide nitrogen not only to the macroalgae but to the rice crop as well. These additional nutrients will be necessary supplementation for plants grown under the many abiotic stressors in spaceflight, space habitation, and growth in lunar and Martian regolith.

Symbiosis is not the only way algae have demonstrated nitrogen fixation. A recent discovery of the nitroplast, a new organelle with the capacity to fix nitrogen, was found in the marine microalga *Braarudosphaera bigelowii* (Haptophyta) and relatives (Coale et al. 2024). This case demonstrates a relatively recent (~100 Mya) evolution of an organelle from endosymbiosis. This discovery establishes *B. bigelowii* as an important algal model species for studying the nitroplast and could provide an additional space-based source of nitrogen or an opportunity to engineer macroalgae which independently fix nitrogen.

Addition of macroalgal biohumus to lunar and Martian regolith may also improve its nutrient content. Biohumus is typically produced from inedible biomass discarded after crop production, which is then processed by decomposers, typically fungi and worms, sequentially, to improve nutrient availability of soil (Muhamediyeva and Nurumova 2023). This has not yet been executed using macroalgae as organic matter, but it holds potential to produce nutrient-rich soil-like substrates when practiced on lunar and Martian regolith.

28.3.4.2 Biospacers: Organic Matter for Textural Amendments

The transformation of lunar and Martian regolith to a viable plant growth substrate does not end with fertilizer. The dusty texture of these substrates raises challenges in plant growth, including soil compaction and water delivery. For healthy plant growth, amendment of these physical characteristics of regolith is also required. "Spacers" used here on Earth, such as perlite, zeolite, and vermiculite, confer higher water retention and aeration to compact or well-drained soils. Alternatively, the application of organic matter improves many physical characteristics of soil, such as lowering the bulk density, improving aeration, and improving water infiltration (Bruce et al. 1990). Therefore, a sustainable alternative in a closed ecosystem is to implement the biomass produced from hydroponics systems, bioreactors, and previous crop growth to alter the texture of lunar and Martian regolith. Here, the organic matter can act as a "biospacer"—a renewable method of soil physical amendment and a valuable path of least resistance for roots, air, and water to travel. Macroalgal compost implemented as a biospacer has improved nutrient availability, soil aggregate stability, increased porosity, and increased water holding capacity (Ma et al. 2022).

Organic matter additionally provides a location for bacterial colonization. A healthy soil microbiome provides numerous benefits to plants by improving nutrient uptake, resistance to abiotic stressors and disease, and promoting plant growth (Yadav et al. 2021; Ali et al. 2023; Handy 2024). In space, where abiotic stressors will be constant and food security is imperative, these benefits could greatly ease the burden of required maintenance on astronauts to farm. Macroalgae rise as a primary option to harbor a healthy soil microbiome due to their many polysaccharides with prebiotic potential (Patel et al. 2021). Primarily, inedible biomass could serve as a textural amendment to improve lunar and Martian regolith physical characteristics as well as promote beneficial bacteria.

28.3.4.3 Seed Priming

Seed priming generally involves applying a treatment to seeds before planting to improve germination and seedling

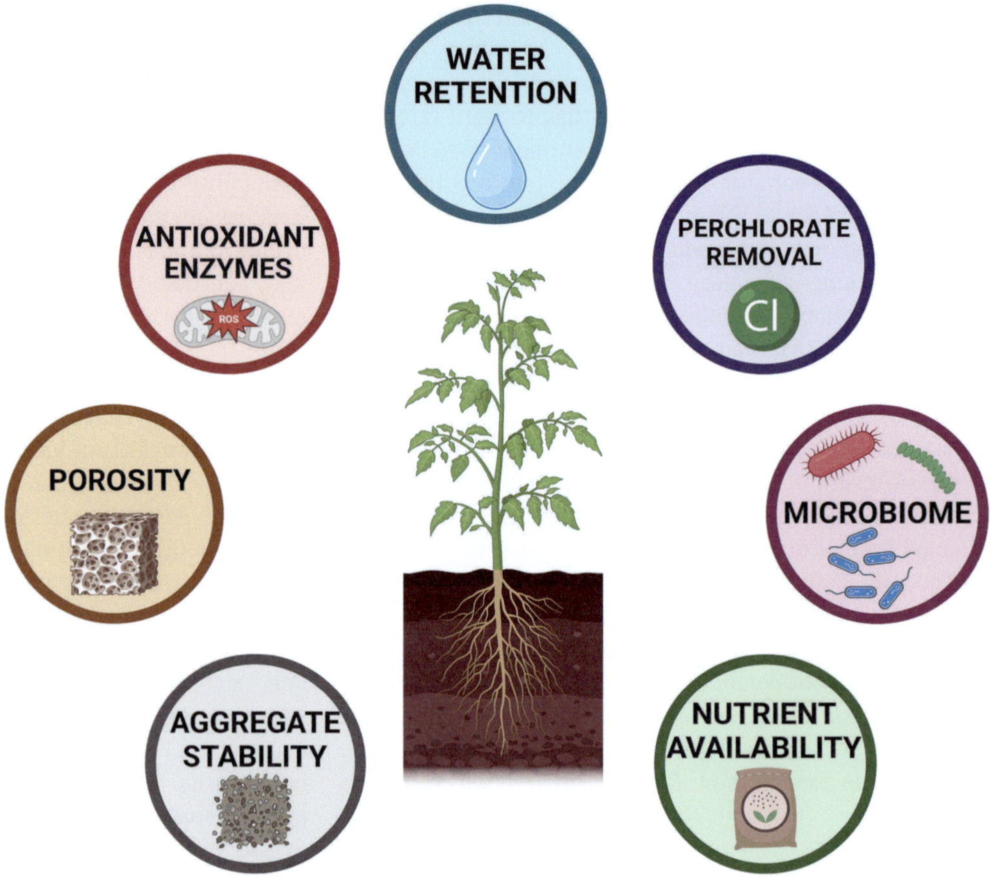

Fig. 28.10 Amendments to lunar and Martian regolith which can be performed by macroalgae, which may promote plant growth (Created in BioRender, Murphy HK (2025))

health. Methods of seed priming using macroalgae have shown the presence of bioactive compounds promoting seed germination. Oliveira et al. 2024 demonstrated this effect in chickpeas, peas, and lettuce using *Ulva lactuca,* which produces phytohormones that stimulate the metabolism and emergence of seedlings. In this study, optimal concentrations of the macroalgal extract were identified and subsequently promoted germination and growth in Martian Global simulant. More benefits conferred to plants by seed priming include a higher resistance to abiotic stressors, higher germination and seedling emergence, and quicker recovery from damage (El Boukhari et al. 2021; Hamouda et al. 2022).

Additionally, the spaceflight-relevant function of antioxidant production was observed to increase in tomatoes grown from seeds soaked in macroalgal extracts (El Boukhari et al. 2021). In wheat and mung bean, multiple species of brown algae were found to promote germination, growth, and kernel size of *Triticum aestivum* (common wheat) (Kumar and Sahoo 2011; Kalaivanan and Venkatesalu 2012). Extracts from green algae like *U. lactuca* are found to improve germination, yield, and quality of the plant upon seed soak (Vijayakumar et al. 2019), while red algae improved germination, storage stability, and fruit yield in *Capsicum frutescens* (wild chili pepper) (Dutta et al. 2019).

Another species of brown alga, *Ascophyllum nodosum,* improved growth rate, chlorophyll production, shoot, and root biomass in soybean seedlings (Oliveira et al. 2024). An additional benefit is seen in salt-stressed plants, where both seed priming and foliar application of algal extracts displayed a decrease in soil conductivity while increasing nutrient availability (Hamed et al. 2018). These results are relevant to Martian regolith-based agriculture, as one of the primary challenges in this media is the high concentrations of salts (Ramírez et al. 2019). Further studies should be performed to assess how the variation between extract concentration, species of macroalgae, and seedling species interacts (Fig. 28.11).

28.3.4.4 Foliar Spray

Spraying foliage with macroalgal extracts is another common method of distribution to plants. These extracts benefit plant growth by supplying several phytohormones. Among *Sargassum* spp., indole acetic acid (IAA) is the most common growth hormone produced (Sunarpi et al. 2021). *Sargassum ilicifolium* has been observed to produce 0.591 mg mL^{-1} of IAA and traces of naphthalene acetic acid (NAA) and gibberellic acid (GA). After foliar spray with extracts from brown algae, improvements were seen in

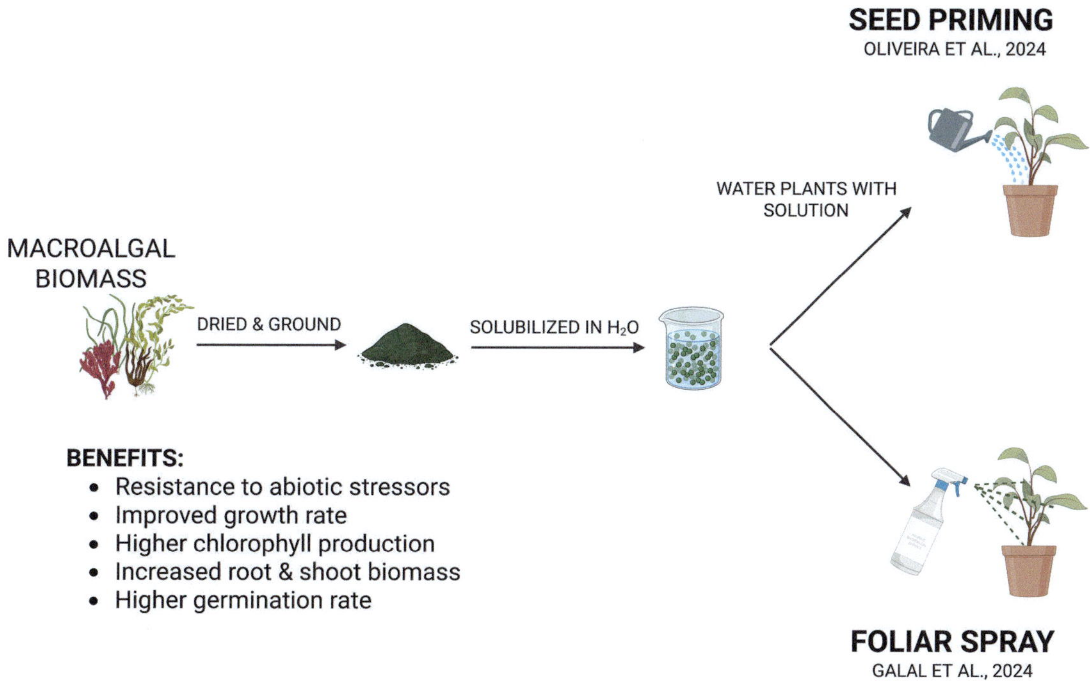

Fig. 28.11 Seed priming and foliar spray application of macroalgal extracts confers many improvements to plant growth, including in Martian regolith simulant (Azzam et al. 2024; Oliveira et al. 2024)

growth, biochemical yield, crop yield, mineral uptake, free amino acids, soluble protein, chlorophyll, fruit yield and quality, antioxidant enzymes, and raised nitrogen use efficiency. (Ali et al. 2016; Kocira et al. 2018; Chen et al. 2021; Ghafouri et al. 2024). Foliar application of brown algae to kiwifruit increased fruit yield and quality, along with antioxidant activity (Ghafouri et al. 2024). Another brown alga, *Ascophyllum nodosum,* was engineered to produce a biostimulant which improved the nitrogen usage efficiency of barley by 27% (Goñi et al. 2021). *Rhodophyta* and *Chlorophyta* extracts showed similar benefits as a foliar spray and were observed to confer higher yield and nutrient contents, as well as micronutrient and protein concentrations (Shah et al. 2013; Layek et al. 2018).

28.3.4.5 Hydroponics

Integrating macroalgae into space-based hydroponic agriculture could present a promising improvement to sustainability by promoting nutrient recovery from wastewater and producing plant growth promoting compounds (Lemos 2024). Nutrient-rich wastewater can be routed to macroalgae as excess nutrients are managed by species like *Ulva lactuca* (Chlorophyta) and *Gracilaria birdiae* (Rhodophyta) (Chopin et al. 2001; Castelar et al. 2014). Here, the edible biomass of the system is increased by macroalgal and plant growth, resources are recovered, and waste is processed (Oñez et al. 2025). Many of the compounds produced by macroalgae are useful for promoting plant growth, including macro- and micronutrients, vitamins, phytohormones, and polysaccharides (Amano and Noda 1994; Stirk et al. 2003; Khan et al. 2009; Shuuluka et al. 2013; Mamede et al. 2023).

Implementing hydroponics systems in space can reduce water usage where resources are limited, as well as allow for customization of nutrient formulations for each type of plant (Sharma et al. 2023; Thapa et al. 2024). Seaweed extracts from *Gracilaria edulis* and *Sargassum wightii* have been shown to enhance plant nutritional quality, abiotic stress tolerance, and crop quality characteristics regardless of the nutrient content of the seaweed, making them an attractive option for hydroponics (Vinoth et al. 2012). A study evaluating the macroalga *Sargassum polycystum* as a biostimulant in hydroponic lettuce systems demonstrated promising results after partially replacing commercial hydroponic nutrient solutions with seaweed extract. When replacing half of the commercial solution with seaweed extract, lettuce yield significantly increased as indicated by greater plant height (10.53%), leaf width (14.29%), plant weight (22.41%), and root length (44.83%) (Oñez et al. 2025). However, substitutions exceeding this amount were detrimental to growth, identifying limitations to the system. The authors recommended a 50% replacement of commercial solution with algal extract for profitable hydroponic lettuce production (Oñez et al. 2025).

Direct application of alginates, which are polysaccharides produced in the cell wall of brown algae, has shown significant benefits to plant growth, which may benefit space-based hydroponic systems. Depolymerized alginates at concentrations of 0.5–1 mg mL^{-1} improved the growth of rice and peanut plants, which were cultivated hydroponically (Hien et al. 2000). Integration of algae-derived biostimulants and bioremediation into hydroponics shows promise for optimization of sustainability and productivity. Reduction of dependence on synthetic fertilizers for crop production will naturally occur in a system like BLS, where most materials are meant to be reused. Fortunately, macroalgae have proven to be a useful tool in the process of recycling nutrients (Fig. 28.12).

28.3.4.6 Aquaponics

Aquaculture is broadly defined as the production of aquatic organisms and can be integrated into closed ecological systems to recycle waste, produce oxygen, and manage carbon dioxide (Castelar et al. 2014; Przybyla 2021). Aquaponics combines the principles of aquaculture with hydroponics to exchange nutrients between fish and plants to produce both simultaneously (Stoyanova et al. 2024). In aquaponic systems, fish waste is converted by bacteria into ideal nutrients for plant growth, creating an efficient closed cycle (Manan et al. 2025; Indriani and Ulaan 2024). The variety and versatility of macroalgae make them viable candidates for these integrated multitrophic aquaculture systems (IMTA) (Castelar et al. 2014). Algal growth and its resulting biomass yield benefit from these systems, as demonstrated by *Porphyra/Pyropia* spp., from which a single serving produces a higher protein content than one serving of fish, eggs, or soybeans when produced in an IMTA (Machado et al. 2020). This finding suggests that incorporating macroalgae into space-based aquaponic systems can help maintain water quality while contributing to food production, two essential elements for extending the longevity of planetary settlements. Additionally, fish are rich in amino acids, omega-3 fatty acids, and other essential nutrients for human health, making them ideal for an astronaut's diet (Przybyla 2021).

28.3.4.7 Biofuels

Fuels on the Moon and Mars will be indispensable, as they are required for return to Earth. Biogas and bioethanol are the two most efficient biofuels produced from macroalgae. Production of biogas requires destructive pretreatments, often by grinding, digesting, desiccation, or mechanical separation (Chen et al. 2015; Ghadiryanfar et al. 2016). Anaerobic digestion of macroalgal biomass produces methane with high efficiency due to the low lignocellulose content of macroalgae and conversion of all organic matter to biogas in the process (Chen et al. 2015). Methane yield is highest in the species *G. vermiculophylla* at 481 L CH_4 kg^{-1} (Chen et al. 2015). Biohydrogen, which can potentially be used as rocket fuel if converted into liquid form, is produced by several species of *Laminaria* (Borines et al. 2013). *Sargassum* has a particularly high rate of conversion to bioethanol at 89% efficiency (Borines et al. 2013). Making and using biofuels offers the opportunity to produce fuel on-site on the Moon and Mars, which is not available with petroleum products, improving the agency and independence of extraterrestrial settlements (Fig. 28.13).

28.3.4.8 Bioplastics

Resupply missions, which carry many of the materials which are a challenge to synthesize off-Earth. By increasing the number of materials which can be produced by the settlement, resupply missions are reduced or can supplement the settlement further. Bioplastics are developed as a sustainable alternative to petroleum-based plastics on Earth. Polysaccharides derived from macroalgae have been

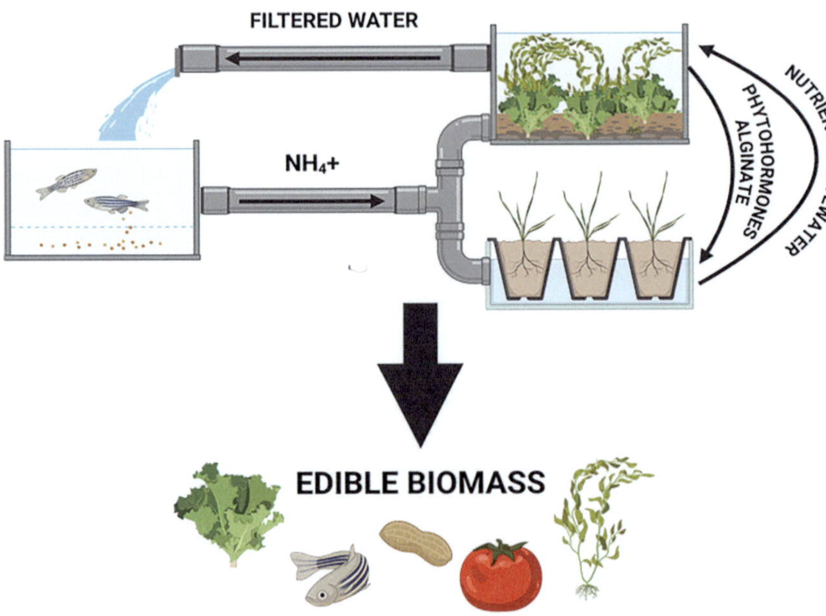

Fig. 28.12 Aquaponics system with integrated macroalgae. Fish produce waste, an excellent fertilizer, which is delivered to the macroalgae and plants. Macroalgae and plants exchange various nutrients and bioactive compounds. The system works concurrently to produce edible biomass for astronauts (Hien et al. 2000; Created in BioRender, Murphy HK (2025))

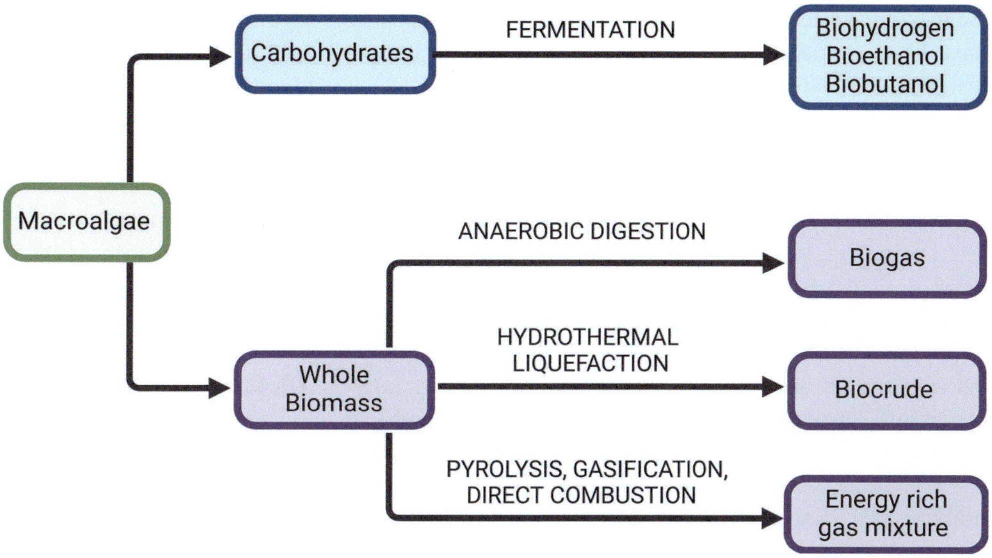

Fig. 28.13 Pathways of biofuel production from macroalgae (Kumar et al. 2016; Shi et al. 2011; Created in BioRender, Murphy HK (2025))

proposed as a source for bioplastic production, as they are highly abundant on Earth (Elkaliny et al. 2024). Although this process is not yet efficient (Elkaliny et al. 2024), further exploration of the topic may enable the production of compostable plastics for lunar and Martian missions. Even if bioplastics are produced on Earth and shipped to the settlement, the capacity for the materials to be biodegraded permits the continuous recycling of materials, a favorable approach for a closed system when compared to single-use plastics. Application of this degraded material to crops can promote their growth (Zhang et al. 2024), providing a valuable resource for crop growth. Production of bioplastics here on Earth for shipment to the Moon and Mars would bypass the challenges of their synthesis off-Earth while still maintaining the reuse of resources within lunar and Martian bases and preventing unnecessary waste from occupying the settlement or extraterrestrial geology. Limiting the human impact on the Moon and Mars in this way will preserve the landscape for future exploration.

28.3.4.9 Pharmaceuticals and Biotechnology

The spaceflight environment, which includes stressors like radiation, mental stress, and living in a confined space, has been found to lower immune responsiveness in astronauts (Cogoli 1993). With lowered immunity, injuries and illnesses will be more closely monitored for the astronauts' safety. Many macroalgae produce compounds which can be stored in lunar and Martian settlements for the purpose of healthcare. The brown algae *Sargassum* spp. produce several antibacterial substances, which include phenols, terpenes, acerogenins, indoles, fatty acids, and volatile halogenated hydrocarbons. A combination of these compounds has been shown to inhibit the growth of the antibiotic-resistant bacteria *Staphylococcus aureus* and *Klebsiella pneumoniae* (El Shafay et al. 2016). Sulfonated polysaccharides produced by macroalgae, including fucoidan, carrageenan, and sesquiterpene hydroquinone, aid in preventing host cell colonization of herpes virus, human immunodeficiency virus, hepatitis B, and dengue fever (Biris-Dorhoi et al. 2020).

Alginate, an oligosaccharide produced by brown algae, can be used as a wound dressing or drug delivery mechanism for nutritional supplements, probiotics, or other drugs by encapsulation (Puscaselu et al. 2020). Red and green algae have demonstrated the production of cancer-fighting compounds, which inhibit tumor growth by antioxidant activity (Ismail and Alotaibi 2020). Anti-inflammatory compounds are produced in macroalgae, including phlorotannins, carotenoids, sterols, alkaloids, and polyunsaturated fats (Rocha et al. 2022).

28.3.5 Choosing Macroalgal Species for BLSS and ISRU

Many of the desirable traits of organisms in space-based BLSS and ISRU mentioned above occur in species of macroalgae. Versatility in the form of multiple BLSS and ISRU functions enables widespread use of the organism, gleaning more benefits from maintenance of a single crop. Many macroalgae are edible and nutritious, offering a much-needed supply of fresh food and stable nutrients. As producers, macroalgae can potentially provide oxygen, remove carbon dioxide, and produce food and fibers (Jain et al. 2024; Clauwaert et al. 2017) as well as phytohormones and biofertilizers (Nabti et al. 2016), and produce pharmaceuticals, nutritional supplements, and antioxidants (Pathan et al. 2023; Cornish and Garbary 2010). All of these potential benefits to astronauts on long-term missions to the Moon and Mars are considered when identifying ideal species for spaceflight. Separately from the potential benefits we've discussed here, the feasibility and protocols for bringing macroalgae to space must be examined, including viability in transport and the resources they will consume.

28.4 Challenges of Bringing Macroalgae to Space

28.4.1 Long-term Storage: Survivability in Spaceflight and Reliability in Storage

As we consider how macroalgae may be successfully introduced into off-world life support, it is important to consider how they will be stored in spaceflight. Organisms which persist in environments that routinely experience fluctuating conditions, such as seasonal temperature changes, variations in light availability, or exposures to wind gusts or waves, must develop adaptive strategies to avoid these abiotic stressors. Typically, this is a perennial change in morphology which improves their persistence (Edwards 2000). Dormant stages may be exploited as a method of ensuring safe dispersal of the next generation in spaceflight (Slusarczyk 1995; Edwards 2000). In spores, seeds, and other presentations of dormancy, growth, metabolism, and movement are reduced to nearly imperceptible levels to conserve energy and protect from environmental harm, with normal activity resuming upon the return of favorable conditions (Withers and Cooper 2008).

Dormancy is not limited to the responses which occur before predictable environmental conditions, such as seasonal changes; many organisms can be "tricked" into reaching this state after the introduction of an abiotic stressor (Yamamoto 1976), and this is classified as "consequential dormancy" (Withers and Cooper 2008). Endospores and cysts can be formed in bacteria following stimulation by irradiation, desiccation, extreme temperatures, or nutrient deprivation (Jones and Lennon 2010; Hori et al. 2003). The remarkably efficient responses to oxidative stress in cyanobacteria enable some species of *Nostoc* to survive over 100 years in desiccated conditions (Cameron Roy 1962), while many cyanobacteria surpass at least 5 years. Microalgae also display production of spore-like cells called "cysts," which have been observed to remain stable for up to 100 years (Ribeiro et al. 2011).

Macroalgae display mechanisms with similar functioning to propagule production in the form of asexual reproduction, or cloning, but these forms are sensitive to desiccation and are only viable for a few days at a time when dehydrated (Santelices 1990; Bollen et al. 2017). Using cryopreservation, macroalgal spores survive 2–8 weeks, although they require resources like freezer space and energy consumption by the freezer during long-term spaceflight (Day 2018). Hydrated spores germinate almost immediately upon attachment to a substrate, preventing the storage of a liquid culture (Jung et al. 2018). However, a relatively novel method encapsulating oospores of the green alga *Chara* in low density agar retains spore viability after 6 months of cold storage (Rybak 2021). Therefore, specialized preparation of algae could be useful on lunar missions and even allow time for astronauts to settle in before implementing macroalgae, but would require attention on Martian journeys.

28.4.2 Resource Cost of Macroalgal Cultivation in Space

Macroalgae will require tanks to be grown in a lunar or Martian environment. Considering this, significant amounts of water will need to be transported and dedicated to algal growth. Shipping water to the Moon and Mars will be expensive, and it will be stored mostly for astronaut hydration. Both planetary bodies offer a source of water in situ, but this will require equipment and time to collect, analyze, and filter. The presence of toxic perchlorates and metals on Mars will require additional remediation. However, many of the beneficial elements which are present in the regolith and useful for plant growth or astronaut health may be solubilized, including calcium, iron, magnesium, manganese, potassium, and phosphorus (Morrison et al. 2018). This mixture may be sufficient for macroalgal growth alone, limiting the need to transport additional resources to create nutrient media. Additional resources required for growth will be the same as most photosynthetic species: light, nitrogen, carbon dioxide, and harvest/maintenance (Fig. 28.14).

These requirements are like the microalgae and cyanobacteria, which have captivated the space biology community, creating potential to co-culture the organisms and exploit the improved growth resulting from their symbiotic relationship.

28.5 Conclusions

The vision of Bioregenerative Life Support Systems on the Moon and Mars aims to construct efficient, closed-loop ecosystems which will support long-term settlements in space. As missions move beyond short duration lunar stays to long-term bases and interplanetary travel, the role of algae in BLSS becomes indispensable. Integrating biological components into space habitats offers enhanced resilience and sustainability, improved atmospheric control, and resource autonomy. The macroalgae discussed in this review are summarized in Table 28.1.

Fig. 28.14 Nick Hague, NASA astronaut, with microalgae culture bags aboard the ISS. microalgae produces resources from these bioreactors. Here, *Haematococcus pluvialis* is being studied for the production of supplementary antioxidant astaxanthin for astronaut consumption in microgravity (NASA Image and Video Library 2019)

Table 28.1 Examples of biospacers and their applications

Function	Spacer	Reference
Water Retention	*Ulva* spp., *Palmaria palmata Porphyra* spp., *Ascophyllum nodosum*, *Undaria pinnatifida*, *Enteromorpha* spp., *Gracilaria vermiculophylla*	Watanabe et al. (1999), Monteiro et al. (2022), Čmiková et al. (2024), Hoa et al. (2023), Machado et al. (2020)
Prebiotic activity	*Sarcodiotheca gaudichaudii*, *Chondrus crispus*, *Eucheuma denticulatum*	Cian et al. (2015), Xie et al. (2024)
Pharmaceutical properties	*A. nodosum*, *Fucus vesiculosus*, *Palmaria palmata*	Hejna et al. (2024); Hoa et al. (2023)
Antioxidant production	*Ulva* spp., *Gracilaria* spp., *Enteromorpha* spp.	Simón et al. (2023), Mawi et al. (2020), Yu et al. (2021)
Oxygen production	*Ulva* spp., *F. vesiculosus*, *Gracilaria* spp.	Hirayama et al. (2024), Torres-Romero et al. (2024)
Wastewater treatment	*Ulva* spp., *Sargassum* spp., *Gracilaria* spp., *Enteromorpha* spp.	Simón et al. (2023), Arumugam et al. (2018), Martínez-Ruiz et al. (2022)
Biofuel production	*Gracilaria* spp., *Sargassum* spp., *Ulva* spp., *Ectocarpus* spp.	Kumar et al. (2013), Orozco-González et al. (2022)
Fertilizers	*A. nodosum*, *Laminaria* spp., *G. vermiculophylla*, *Sargassum* spp., *Laminaria* spp.	Shukla et al. (2019), Katakula et al. (2020), Borines et al. (2013)
Hydroponics and Aquaculture	*Ulva fasciata*, *Gracilaria birdiae*, *Gracilaria edulis*, *Sargassum wightii*, *S. polycystum*, *Enteromorpha* spp., *Porphyra* spp.	Simón et al. (2023), Mawi et al. (2020); Yu et al. (2021), Chopin et al. (2001), Castelar et al. (2014), Vinoth et al. (2012), Oñez et al. (2025)
Extract supplementation	*U. lactuca*, *A. nodosum*	Oliveira et al. (2024)

28.5.1 Recommendations for Future Studies and Broader Impacts

One of the challenges for macroalgal growth presented here is the requirement for care during long-duration travel and consistent harvesting requirements. Prior to implementation of extraterrestrial macroalgal crops, autonomous systems would likely need to be developed to manage many of these requirements. Automated systems for many microorganisms and plants are already under development, so this is not a unique challenge to the incorporation of macroalgae into space-based systems.

In considering species which should be incorporated into BLSS, we recommend focusing on developing methods of preparation with high viability after long-term storage. An example of this, cryopreservation, has demonstrated that macroalgae can withstand a period analogous to advanced storage periods for lunar missions.

Among the species discussed here, we identify *Porphyra/Pyropia* spp. as being of the greatest interest for incorporation into BLSS for its versatile potential for use, which includes high nutrient, and, specifically protein content; production of the essential amino acids and vitamin B_{12}; umami flavor; and lack of dangerous iodine bioaccumulation. Providing astronauts with fresh food will ensure that nutritional needs are met, which will be one of the most important aspects of long-term spaceflight missions. The environment encountered by astronauts in space contains a unique combination of stressors which will negatively affect the health of astronauts, especially on upcoming missions to the Moon and Mars, which will be some of the longest in history.

Although live macroalgae are unlikely to be on the first trip to the Moon and Mars, the many benefits they can offer to a late stage, well established settlement are not to be ignored. In missions of the near future, production of macroalgae on Earth to supplement astronaut diets should be considered for their nutrient densities and antioxidant properties to help combat the stressors astronauts encounter in spaceflight.

References

Adams DG, Carr NG (1981) Heterocyst differentiation and cell division in the cyanobacterium *Anabaena cylindrical* effect of high light intensity. J Cell Sci 49(9):341–352. https://doi.org/10.1242/jcs.49.1.341

Adams DG, Duggan PS, Jackson O (2012) Cyanobacterial symbioses. Ecology of Cyanobacteria II. https://doi.org/10.1007/0-306-46855-7_1

Ahmad F, Yin DC (2024) Cyanobacteria for the global space biology program: challenges and opportunities. In: Mehmood MA, Malik S, Musharraf SG, Boopathy R (eds) Cyanobacteria biotechnology. Springer, Cham. https://doi.org/10.1007/978-3-031-70698-1_12

Ali N, Farrell A, Ramsubhag A, Jayaraman J (2016) The effect of *Ascophyllum nodosum* extract on the growth, yield, and fruit quality of tomato grown under tropical conditions. J Appl Phycol 28:1353–1362. https://doi.org/10.1007/s10811-015-0608-3

Ali S, Tyagi A, Mir RA, Rather IA, Anwar Y, Mahmoudi H (2023) Plant beneficial microbiome a boon for improving multiple stress tolerance in plants. Front Plant Sci 14:1266182. https://doi.org/10.3389/fpls.2023.1266182

Amano H, Noda H (1994) Effects of plant growth regulators, organic acids, and sugars on growth and chemical constituents in the tissue culture of sea lettuce *Ulva pertusa*. Fish Sci Res 60(4):449–454. https://www.jstage.jst.go.jp/article/fishsci1994/60/4/60_4_449/_pdf

American Meteorological Society (2024) Air, glossary of meteorology. https://glossary.ametsoc.org/wiki/Air

Arena C, de Micco V, Macaeva E, Quintens R (2014) Space radiation effects on plant and mammalian cells. Acta Astron 104(1):419–431. https://doi.org/10.1016/J.ACTAASTRO.2014.05.005

Arumugam N, Chelliapan S, Kamyab H, Thirugnana S, Othman N, Nasri NS (2018) Treatment of wastewater using seaweed: a review. Int J Environ Res Public Health 15(12):2851. https://doi.org/10.3390/ijerph15122851

Azzam E, El-Howeity M, Galal H, Nofal A (2024) Effect of Foliar Application of Macroalgae Aqueous Extracts on the Nutrient Status and Fruit Quality of "Valencia" Orange. Egyptian J of Botany 64(3):801–813. https://doi.org/10.21608/ejbo.2024.256269.2617

Atri D, Umansky P, Sreenivasan KR (2024) Sustainability as a core principle of space and planetary exploration. Space Policy 70. https://doi.org/10.1016/J.SPACEPOL.2024.101636

Aumeerun S, Soulange-Govinden J, Driver MF, Rao AR, Ravishankar GA, Neetoo H (2019) Macroalgae and microalgae: novel sources of functional food and feed. In: Handbook of Algal Technologies and Phytochemicals: Volume I: Food, Health and Nutraceutical Applications, vol 1, pp 207–219. ISBN: 9780429054242

Ball N, Anderson L, Downing S, Gresser A, Hindupur A, Hogan J, Kagawa H, Kostakis A, Paddock M, Ray H, Roque O, Settles M, Sharif S, Sims K, Soundararajan M, Villanueva A, Vu S, Donovan F (2023) BioNutrients: microbial production of on-demand nutrients on the international Space Station. Oral presentation at American Society for Gravitational and Space Research. https://doi.org/10.1111/j.1750-3841.2009.01265.x

Bartsch AF (1961) Algae as a source of oxygen in waste treatment. J Water Pollut Control Fed 33(3):239–249. https://www.jstor.org/stable/25034371

Berliner AJ, Hilzinger JM, Abel AJ, McNulty MJ, Makrygiorgos G, Averesch NJH, Sen Gupta S, Benvenuti A, Caddell DF, Cestellos-Blanco S, Doloman A, Friedline S, Ho D, Gu W, Hill A, Kusuma P, Lipsky I, Mirkovic M, Luis Meraz J, et al. (2021) Towards a biomanufactory on Mars. Front Astron Space Sci 8. https://doi.org/10.3389/FSPAS.2021.711550/PDF

Biris-Dorhoi ES, Michiu D, Pop CR, Rotar AM, Tofana M, Pop OL, Socaci SA, Farcas AC (2020) Macroalgae—a sustainable source of chemical compounds with biological activities. Nutrients 12(10):3085. https://doi.org/10.3390/NU12103085

Bollen M, Battershill CN, Pilditch CA, Bischof K (2017) Desiccation tolerance of different life stages of the invasive marine kelp *Undaria pinnatifida*: potential for overland transport as invasion vector. J Exp Mar Biol Ecol 496:1–8. https://doi.org/10.1016/J.JEMBE.2017.07.005

Borines MG, de Leon RL, Cuello JL (2013) Bioethanol production from the macroalgae *Sargassum* spp. Bioresour Technol 138:22–29. https://doi.org/10.1016/J.BIORTECH.2013.03.108

Bothe H (2019) The cyanobacterium Chroococcidiopsis and its potential for life on Mars. J Astrobiol 2:398–411. http://journalofastrobiology.com/CyanobacteriumChroococcidiopsisMarsFinalBothe.pdf

Bruce RR, Langdale GW, Dillard AL (1990) Tillage and crop rotation effect on characteristics of a sandy surface soil. SSSA 54(6):1744–1747. https://doi.org/10.2136/SSSAJ1990.03615995005400060039X

Burks D (2021) NASA's Space Launch System. https://www.boeing.com/content/dam/boeing/boeingdotcom/space/space_launch_system/source/space-launch-system-flip-book-040821.pdf

Burtin P (2003) Nutritional value of seaweeds. Electron J Environ Agric Food Chem 2(4):498–503

Cameron Roy E (1962) Species of *Nostoc* Vaucher occurring in the Sonoran Desert in Arizona. Trans Am Micros Soc 8(4):379–384

Castelar B, Reis RP, dos Santos Calheiros AC (2014) *Ulva lactuca* and *U. flexuosa* (Chlorophyta, Ulvophyceae) cultivation in Brazilian tropical waters: recruitment, growth, and ulvan yield. J Appl Phcol 26(5). https://doi.org/10.1007/S10811-014-0329-Z/TABLES/6

Chen H, Zhou D, Luo G, Zhang S, Chen J (2015) Macroalgae for biofuels production: Progress and perspectives. Renew Sust Energ Rev 47:427–437. https://doi.org/10.1016/J.RSER.2015.03.086

Chen D, Zhou W, Yang J, Ao J, Huang Y, Shen D, Jiang Y, Huang Z, Shen H (2021) Effects of seaweed extracts on the growth, physiological activity, cane yield and sucrose content of sugarcane in China. Front Plant Sci 12. https://doi.org/10.3389/FPLS.2021.659130/BIBTEX

Chopin T, Buschmann AH, Halling C, Troell M, Kautsky N, Neori A, Kraemer GP, Zertuche-González JA, Yarish C, Neefus C (2001) Integrating seaweeds into marine aquaculture systems: a key toward sustainability. J Phycol 37(6):975–986. https://doi.org/10.1046/J.1529-8817.2001.01137.X

Cian RE, Drago SR, de Medina FS, Martínez-Augustin O (2015) Proteins and carbohydrates from red seaweeds: evidence for beneficial effects on gut function and microbiota. Mar Drugs 13(8):5358–5383. https://doi.org/10.3390/MD13085358

Circuncisão AR, Catarino MD, Cardoso SM, Silva AMS (2018) Minerals from macroalgae origin: health benefits and risks for consumers. Mar Drugs 16(11):400. https://doi.org/10.3390/MD16110400

Clauwaert P, Muys M, Alloul A, de Paepe J, Luther A, Sun X, Ilgrande C, Christiaens MER, Hu X, Zhang D, Lindeboom REF, Sas B, Rabaey K, Boon N, Ronsse F, Geelen D, Vlaeminck SE (2017) Nitrogen cycling in Bioregenerative Life Support Systems: challenges for waste refinery and food production processes. Prog Aerosp Sci 91:87–98. https://doi.org/10.1016/J.PAEROSCI.2017.04.002

Čmiková N, Kowalczewski P, Kmiecik D, Tomczak A, Drozdzynska A, Slachicinski M, Krolak J, Kacaniova M (2024) Characterization of selected microalgae species as potential sources of nutrients and antioxidants. Foods 13(13):2160. https://doi.org/10.3390/foods13132160

Coale TH, Loconte V, Turk-Kubo KA, Vanslembrouck B, Mak WKE, Cheung S, Ekman A, Chen JH, Hagino K, Takano Y, Nishimura T,

Adachi M, le Gros M, Larabell C, Zehr JP (2024) Nitrogen-fixing organelle in a marine alga. Science 384(6692):217–222. https://doi.org/10.1126/SCIENCE.ADK1075/SUPPL_FILE/SCIENCE.ADK1075_MOVIES_S1_TO_S3.ZIP

Cockell CS, Rettberg P, Rabbow E, Olsson-Francis K (2011) Exposure of phototrophs to 548 days in low earth orbit: microbial selection pressures in outer space and on early earth. ISME J 5(10):1671–1682. https://doi.org/10.1038/ismej.2011.46

Cogoli A (1993) Space flight and the immune system. Vaccine 11(5):496–503. https://doi.org/10.1016/0264-410X(93)90217-L

Cole AJ, Neveux N, Whelan A, Morton J, Vis M, de Nys R, Paul NA (2016) Adding value to the treatment of municipal wastewater through the intensive production of freshwater macroalgae. Algal Res 20:100–109. https://doi.org/10.1016/J.ALGAL.2016.09.026

Commercial Space Frequently Asked Questions. (2024) National Aeronautics and Space Administration. https://www.nasa.gov/humans-in-space/leo-economy-frequently-asked-questions/

Cooper M, Perchonok M, Douglas GL (2017) Initial assessment of the nutritional quality of the space food system over three years of ambient storage. npj Microgravity 3. https://doi.org/10.1038/s41526-017-0022-z

Cornish ML, Garbary DJ (2010) Antioxidants from macroalgae: potential applications in human health and nutrition. Algae 25(4):155–171. https://doi.org/10.4490/ALGAE.2010.25.4.155

Day JG (2018) Cryopreservation of macroalgae. In: Protocols for macroalgae research, pp 79–94. https://doi.org/10.1201/B21460-4

Deuri B, Singha AD, Pathak J (2024) Role of algae in space nutrition: a biotechnological approach to life support system. AgriGate 4(9)

Do S, Shishko R, Antonelli D, Cichan T, Collom R, Conrad P, Coverstone V, Davis RM, Edwards CM, Fuller M, Goodliff KE, Hoffman SJ, Saikia S, Sheppard P, Shull S, Stone D, Whetsel C (2019) Logistics is a key enabler of sustainable human missions to mars. Executive Summary https://assets.pubpub.org/1al9pba3/51579295653594.pdf

Dobney W, Mols L, Mistry D, Tabury K, Baselet B, Baatout S (2023) Evaluation of deep space exploration risks and mitigations against radiation and microgravity. Front Nucl Med 3. https://doi.org/10.3389/FNUME.2023.1225034/PDF

Duri LG, Caporale AG, Rouphael Y, Vingiani S, Palladino M, de Pascale S, Adamo P (2022) The potential for lunar and Martian regolith simulants to sustain plant growth: a multidisciplinary overview. Front Astron Space Sci 8. https://doi.org/10.3389/FSPAS.2021.747821/BIBTEX

Dutta SK, Layek J, Akoijam RS, Boopathi T, Vanlalhmangaiha SS, Singh SB, Lungmuana PN (2019) Seaweed extract as natural priming agent for augmenting seed quality traits and yield in *Capsicum frutescens* L. J Appl Phycol 31(6):3803–3813. https://doi.org/10.1007/S10811-019-01871-0

Edwards MS (2000) The role of alternate life-history stages of a marine macroalga: a seed Bank analogue? Ecology 81(9):2404–2415. https://esajournals.onlinelibrary.wiley.com/doi/abs/10.1890/0012-9658%282000%29081%5B2404%3ATROALH%5D2.0.CO%3B2

Egan S, Kumar V, Nappi J, Gardiner M (2017) Microbial diversity and symbiotic interactions with macroalgae. Algal and Cyanobacteria Symbioses, pp 493–546. https://doi.org/10.1142/9781786340580_0016

Ehlmann BL, Edwards CS (2014) Mineralogy of the Martian surface. Annu Rev Earth Planet Sci 42:291–315. https://doi.org/10.1146/annurev-earth-060313-055024

Eichler A, Hadland N, Pickett D, Masaitis D, Handy D, Perez A, Batcheldor D, Wheeler B, Palmer A (2021) Challenging the agricultural viability of martian regolith simulants. Icarus 354. https://doi.org/10.1016/j.icarus.2020.114022

El Boukhari M, Barakate M, Choumani N, Bouhia Y, Lyamlouli K (2021) *Ulva lactuca* extract and fractions as seed priming agents mitigate salinity stress in tomato seedlings. Plants 10(6):1104. https://doi.org/10.3390/plants10061104

El Shafay SM, Ali SS, El-Sheekh MM (2016) Antimicrobial activity of some seaweeds species from Red sea, against multidrug resistant bacteria. Egypt J Aquat Res 42(1):65–74. https://doi.org/10.1016/J.EJAR.2015.11.006

Elkaliny NE, Alzamel NM, Moussa SH, Elodamy NI, Madkor EA, Ibrahim EM, Elshobary ME, Ismail GA (2024) Macroalgae bioplastics: a sustainable shift to mitigate the ecological impact of petroleum-based plastics. Polymers 16(9):1246. https://doi.org/10.3390/POLYM16091246

Escobar CM, Escobar AC (2017) Duckweed: a tiny aquatic plant with enormous potential for bioregenerative life support systems. In: 47th International Conference on Environmental Systems. http://hdl.handle.net/2346/73062

European Space Agency (2019) PhotoBioreactor, Universistät Stuttgärt/ESA. https://www.esa.int/ESA_Multimedia/Images/2019/06/PhotoBioreactor

Fackrell LE, Humphrey S, Loureiro R, Palmer AG, Long-Fox J (2024) Overview and recommendations for research on plants and microbes in regolith-based agriculture. Sustain Agric 2(1):1–23. https://doi.org/10.1038/s44264-024-00013-5

Fahrion J, Mastroleo F, Dussap CG, Leys N (2021) Use of Photobioreactors in regenerative life support systems for human space exploration. Front Microbiol 12:699525. https://doi.org/10.3389/FMICB.2021.699525/BIBTEX

Fais G, Manca A, Bolognesi F, Borselli M, Concas A, Busutti M, Broggi G, Sanna P, Castillo-Aleman YM, Rivero-Jiménez RA, Bencomo-Hernandez AA, Ventura-Carmenate Y, Altea M, Pantaleo A, Gabrielli G, Biglioli F, Cao G, Giannaccare G (2022) Wide range applications of *Spirulina*: from earth to space missions. Mar Drugs 20(5):299. https://doi.org/10.3390/MD20050299

Fleurence J (1999) Seaweed proteins: biochemical, nutritional aspects and potential uses. Trends Food Sci Technol 10(1):25–28. https://doi.org/10.1016/S0924-2244(99)00015-1

Frossard E, Crain G, de Azcárate G, Bordóns I, Hirschvogel C, Oberson A, Paille C, Pellegri G, Udert KM (2024) Recycling nutrients from organic waste for growing higher plants in the Micro Ecological Life Support System Alternative (MELiSSA) loop during long-term space missions. LSSR 40:176–185. https://doi.org/10.1016/J.LSSR.2023.08.005

FY 2025 Budget Request (2024). https://www.nasa.gov/fy-2025-budget-request/

Ge S, Madill M, Champagne P (2018) Use of freshwater macroalgae *Spirogyra* sp. for the treatment of municipal wastewaters and biomass production for biofuel applications. Biomass Bioenergy 111:213–223. https://doi.org/10.1016/J.BIOMBIOE.2017.03.014

Ghadiryanfar M, Rosentrater KA, Keyhani A, Omid M (2016) A review of macroalgae production, with potential applications in biofuels and bioenergy. Renew Sustain Energ 54:473–481. https://doi.org/10.1016/J.RSER.2015.10.022

Ghafouri M, Razavi F, Arghavani M, Gheshlaghi EA (2024) Enhancing mineral uptake and antioxidant enzymes activity of kiwifruit via foliar application of brown macroalga extract. JHPR **7**(1):15–30. https://doi.org/10.22077/JHPR.2024.6926.1343

Goñi O, Łangowski Ł, Feeney E, Quille P, O'Connell S (2021) Reducing nitrogen input in barley crops while maintaining yields using an engineered biostimulant derived from *Ascophyllum nodosum* to enhance nitrogen use efficiency. Front Plant Sci 12:664682. https://doi.org/10.3389/FPLS.2021.664682/BIBTEX

Hamed SM, Abd El-Rhman AA, Abdel-Raouf N, Ibraheem IBM (2018) Role of marine macroalgae in plant protection and improvement for sustainable agriculture technology. BJBAS 7(1):104–110. https://doi.org/10.1016/J.BJBAS.2017.08.002

Hamouda MM, Saad-Allah KM, Gad D (2022) Potential of seaweed extract on growth, physiological, cytological and biochemical parameters of wheat (*Triticum aestivum* L.) seedlings. J Soil Sci Plant Nutr

22(2):1818–1831. https://doi.org/10.1007/S42729-022-00774-3/TABLES/6

Handy D (2024) Investigating microbe-plant symbioses in space. https://repository.fit.edu/etd

Hejna M, Dell'Anno M, Liu Y, Rossi L, Aksmann A, Pogorzelski G, Jóźwik A (2024) Assessment of the antibacterial and antioxidant activities of seaweed-derived extracts. Sci Rep 14. https://doi.org/10.1038/s41598-024-71961-8

Henriques B, Rocha LS, Lopes CB, Figueira P, Duarte AC, Vale C, Pardal MA, Pereira E (2017) A macroalgae-based biotechnology for water remediation: simultaneous removal of Cd, Pb and Hg by living *Ulva lactuca*. J Env Manag 191:275–289. https://doi.org/10.1016/J.JENVMAN.2017.01.035

Hermann C, FitzGerald J, O'Shea R, Xia A, O'Kiely P, Murphy JD (2015) Ensiling of seaweed for a seaweed biofuel industry. Bioresour Technol 196:301–313. https://doi.org/10.1016/j.biortech.2015.07.098

Hien NQ, Nagasawa N, Tham LX, Yoshii F, Dang VH, Mitomo H, Makuuchi K, Kume T (2000) Growth-promotion of plants with depolymerized alginates by irradiation. Radiat Phys Chem 59(1):97–101. https://doi.org/10.1016/S0969-806X(99)00522-8

Hill DE (2016) Jumping spiders in outer space (Araneae:Salticidae). Peckhamia 146(1):1–7. https://peckhamia.com/peckhamia/PECKHAMIA_146.1.pdf

Hirayama S, Urata K, Kusumoto M, Ikegami Y (2024) Producing supersaturated dissolved oxygen seawater by photosynthesis of *Ulva* sp. and its applications. Aquac Res. https://doi.org/10.1155/2024/9411516

Hoa NX, Thuy TT, Huyen VT, Vinh PT, Duc PT, Ai DTT (2023) Inhibitory effects of *Ludwigia octovalvis* (Jacq.) raven extracts on the growth of *Microcystis aeruginosa*. Viet J Agri Sci 6(4):1969–1977. https://doi.org/10.31817/VJAS.2023.6.4.07

Hong P, Schmid B, de Laender F, Eisenhauer N, Zhang X, Chen H, Craven D, de Boeck HJ, Hautier Y, Petchey OL, Reich PB, Steudel B, Striebel M, Thakur MP, Wang S (2022) Biodiversity promotes ecosystem functioning despite environmental change. Ecol Lett 25(2):555–569. https://doi.org/10.1111/ELE.13936

Hori K, Okamoto J, Tanji Y, Unno H (2003) Formation, sedimentation and germination properties of *Anabaena* akinetes. Biochem Eng J 14(1):67–73. https://doi.org/10.1016/S1369-703X(02)00136-5

Huang QN, Watanabe F, Koseki K, He RE, Lee HL, Chiu THT (2024) Effect of roasted purple laver (nori) on vitamin B12 nutritional status of vegetarians: a dose-response trial. Eur J Nutr 63(8):3269–3279. https://doi.org/10.1007/S00394-024-03505-9/FIGURES/4

Indriani Y, Ulaan M (2024) Analysis of the effects aquaponic cultivation systems on the growth of water spinach (*Ipomoea aquatica*) and Pak Choy (*Brassica rapa* L.). J Biol Trop 24(4):596–603. https://doi.org/10.29303/JBT.V24I4.7730

Isachenkov M, Chugunov S, Akhatov I, Shishkovsky I (2021) Regolith-based additive manufacturing for sustainable development of lunar infrastructure—an overview. Acta Astron 180:650–678. https://doi.org/10.1016/J.ACTAASTRO.2021.01.005

Ismail MM, Alotaibi BS (2020) Therapeutic uses of red macroalgae. Molecules 25(19):4411. https://doi.org/10.3390/molecules25194411

Jain P, Ramanan H, R. Hari Priya, B. Vishalakshi, Soundararajan Mohan Kumar, S. Kavitha Rao, Sarvepalli Lokesh, Ravi (2024) Functional food development using Ulva fasciata and its acute and chronic health benefits on *Drosophila melanogaster*. Nutr 49(2). https://doi.org/10.1186/s41110-024-00289-0

Jeong JE, Park SA (2021) Physiological and psychological effects of visual stimulation with green plant types. Int J Environ Res Public Health 18(24). https://doi.org/10.3390/IJERPH182412932

Johnson CM, Boles HO, Spencer LSE, Poulet L, Romeyn M, Bunchek JM, Fritsche R, Massa GD, O'Rourke A, Wheeler RM (2021) Supplemental food production with plants: a review of NASA research. Front Astron Space Sci 8. https://doi.org/10.3389/FSPAS.2021.734343/BIBTEX

Jones S, Lennon J (2010) Dormancy contributes to the maintenance of microbial diversity. Proc Natl Acad Sci 107(13):5881. https://pmc.ncbi.nlm.nih.gov/articles/PMC2851880/

Jung SM, Lee JH, Lee HJ, Jeon JY, Park TH, Yoon JH, Shin HW (2018) The growth of alginate-encapsulated macroalgal spores. Aquaculture 491:333–337. https://doi.org/10.1016/j.aquaculture.2018.03.033

Justiniano YAV, Ledford ME, Castro-Wallace SL, Nguyen HN, Li W, Irwin T, Sysoeva TA (2024) More than a decade of international Space Station microbial sampling in the environmental control and life support systems. In: 53rd International Conference on Environmental Systems

Kading B, Straub J (2015) Utilizing in-situ resources and 3D printing structures for a manned Mars mission. Acta Astron 107:317–326. https://doi.org/10.1016/j.actaastro.2014.11.036

Kalaivanan C, Venkatesalu V (2012) Utilization of seaweed *Sargassum myriocystum* extracts as a stimulant of seedlings of *Vigna mungo* (L.) Hepper. SJAR 10(2):466–470. https://doi.org/10.5424/SJAR/2012102-507-10

Kanagesan K, Abdulla R, Derman E, Sabullah MK, Govindan N, Gansau JA (2022) A sustainable approach to green algal bioplastics production from brown seaweeds of Sabah, Malaysia. J King Saud Univ Sci 34(7). https://doi.org/10.1016/J.JKSUS.2022.102268

Katakula AAN, Gawanab W, Itanna F, Mupambwa HA (2020) The potential fertilizer value of Namibian beach-cast seaweed (*Laminaria pallida* and *Gracilariopsis funicularis*) biochar as a nutrient source in organic agriculture. Sci Afr 10. https://doi.org/10.1016/J.SCIAF.2020.E00592

Kaur M, Shitanaka T, Surendra KC, Khanal SK (2024) Macroalgae-derived bioactive compounds for functional food and pharmaceutical applications—a critical review. Crit Rev Food Sci Nutr 65:4172–4194. https://doi.org/10.1080/10408398.2024.2384643

Khan W, Rayirath UP, Subramanian S, Jithesh MN, Rayorath P, Hodges DM, Critchley AT, Craigie JS, Norrie J, Prithiviraj B (2009) Seaweed extracts as biostimulants of plant growth and development. J Plant Growth Regul 28:386–399. https://doi.org/10.1007/s00344-009-9103-x

Kiss JZ (2025) Statoliths function in gravity perception in plants: yes, no, yes! Planta 261(2):45. https://doi.org/10.1007/s00425-025-04631-2

Kocira A, Świeca M, Kocira S, Złotek U, Jakubczyk A (2018) Enhancement of yield, nutritional and nutraceutical properties of two common bean cultivars following the application of seaweed extract (*Ecklonia maxima*). Saudi J Biol Sci 25(3):563–571. https://doi.org/10.1016/J.SJBS.2016.01.039

Kumar G, Sahoo D (2011) Effect of seaweed liquid extract on growth and yield of *Triticum aestivum* var. Pusa Gold. J Appl Phycol 23(2):251–255. https://doi.org/10.1007/S10811-011-9660-9/TABLES/1

Kumar S, Gupta R, Kumar G, Sahoo D, Kuhad RC (2013) Bioethanol production from *Gracilaria verrucosa*, a red alga, in a biorefinery approach. Bioresour Technol 135:150–156. https://doi.org/10.1016/J.BIORTECH.2012.10.120

Kumar K, Ghosh S, Angelidaki I, Holdt SL, Karakashev DB, Morales MA, Das D (2016) Recent developments on biofuels production from microalgae and macroalgae. Renew Sustain Energ 65:235–249. https://doi.org/10.1016/J.RSER.2016.06.055

Kumar Y, Tarafdar A, Badgujar PC (2021) Seaweed as a source of natural antioxidants: therapeutic activity and food applications. J Food Qual 1. https://doi.org/10.1155/2021/5753391

Lasseur C, Brunet J, de Weever H, Dixon M, Dussap G, Godia F, Leys N, Mergeay M, Van Der Straeten D (2010) Melissa: the European project of a closed life support system. Gravit Space Biol 23(2):3–12

Layek J, Das A, Idapuganti RG, Sarkar D, Ghosh A, Zodape ST, Lal R, Yadav GS, Panwar AS, Ngachan S, Meena RS (2018) Seaweed extract as organic bio-stimulant improves productivity and quality of rice in eastern Himalayas. J Appl Phycol 30(1):547–558. https://doi.org/10.1007/S10811-017-1225-0/FIGURES/3

Lemos MFL (2024) Beyond earth: harnessing marine resources for sustainable space colonization. Mar Drugs 22(11):481. https://doi.org/10.3390/md22110481

Li C, Hu H, Yang MF, Pei ZY, Zhou Q, Ren X, Liu B, Liu D, Zeng X, Zhang G, Zhang H, Liu J, Wang Q, Deng X, Xiao C, Yao Y, Xue D, Zuo W, Su Y et al (2022a) Characteristics of the lunar samples returned by the Chang'E-5 mission. Natl Sci Rev 9(2). https://doi.org/10.1093/NSR/NWAB188

Li Z, Zhang W, Wang L, Liu H, Liu H (2022b) Regulating effects of the biophilic environment with strawberry plants on psychophysiological health and cognitive performance in small spaces. Built Environ 212:108801. https://doi.org/10.1016/j.buildenv.2022.108801

Liu H, Yao Z, Fu Y, Feng J (2021) Review of research into bioregenerative life support system(s) which can support humans living in space. LSSR 31:113–120. https://doi.org/10.1016/J.LSSR.2021.09.003

Ma C, Song W, Yang J, Ren C, Du H, Tang T, Qin S, Liu Z, Cui H (2022) The role and mechanism of commercial macroalgae for soil conditioner and nutrient uptake catalyzer. J Plant Growth Regul 97(3):455–476. https://doi.org/10.1007/S10725-022-00819-8

MacArtain P, Gill CIR, Brooks M, Campbell R, Rowland IR (2007) Nutritional value of edible seaweeds. Nutr Rev 65(12):535–543. https://doi.org/10.1111/J.1753-4887.2007.TB00278.X

Machado M, Machado S, Pimentel FB, Freitas V, Alves RC, Oliveira MBPP (2020) Amino acid profile and protein quality assessment of macroalgae produced in an integrated multi-trophic aquaculture system. Foods 9(10):1382. https://doi.org/10.3390/FOODS9101382

Mamede M, Cotas J, Bahcevandziev K, Pereira L (2023) Seaweed polysaccharides in agriculture: a next step towards sustainability. Appl Sci 13(11):6594. https://doi.org/10.3390/APP13116594

Manan NA, Mohamed SJ, Kamaruzzan AS, Razman MMA, Kasan NA, Ikhwanuddin M (2025) Aquaponics: a sustainable technology for aquaculture and agriculture food security. Planet Sustain 3(1). https://doi.org/10.46754/ps.2025.01.004

Mapstone LJ, Leite MN, Purton S, Crawford IA, Dartnell L (2022) Cyanobacteria and microalgae in supporting human habitation on Mars. Biotechnol Adv 59. https://doi.org/10.1016/J.BIOTECHADV.2022.107946

Marino BDV, Odum HT (1999) Biosphere 2: introduction and research progress. Ecol Eng 13:3–14

Märki A (2020) Radiation analysis for moon and mars missions. IJASS 8(3):16–26. https://doi.org/10.11648/j.ijass.20200803.11

Marshall-Goebel K, Laurie SS, Alferova IV, Arbeille P, Auñón-Chancellor SM, Ebert DJ, Lee SMC, MacIas BR, Martin DS, Pattarini JM, Ploutz-Snyder R, Ribeiro LC, Tarver WJ, Dulchavsky SA, Hargens AR, Stenger MB (2019) Assessment of jugular venous blood flow stasis and thrombosis during spaceflight. JAMA Netw Open 2(11). https://doi.org/10.1001/jamanetworkopen.2019.15011

Martínez-Ruiz M, Martínez-Gonzalez CA, Kim D, Santiesteban-Romero B, Reyes-Pardo H, Villaseñor-Zepeda KR, Meléndez-Sánchez ER, Ramirez-Gamboa D, Diaz-Zamorano AL, Sosa-Hernandez JE, Coronado-Apodaca KG, Gamez-Mendez AM, Iqbal HMN, Parra-Saldivar R (2022) Microalgae bioactive compounds to topical application products—a review. Molecules 27(11):3512. https://doi.org/10.3390/molecules27113512

Mauerer M, Schubert D, Zabel P, Bamsey M, Kohlberg E, Mengedoht D (2016) Initial survey on fresh fruit and vegetable preferences of Neumayer Station crew members: input to crop selection and psychological benefits of space-based plant production systems. Open Agric 1(1):179–188. https://doi.org/10.1515/opag-2016-0023

Mawi S, Krishnan S, Din MFM, Arumugam N, Chelliapan S (2020) Bioremediation potential of macroalgae *Gracilaria edulis* and *Gracilaria changii* co-cultured with shrimp wastewater in an outdoor water recirculation system. Environ Technol Innov 17. https://doi.org/10.1016/J.ETI.2019.100571

MELiSSA loop diagram (2015) European Space Agency. https://www.esa.int/ESA_Multimedia/Images/2015/06/MELiSSA_loop_diagram

Misra G, Smith W, Garner M, Loureiro R (2021) Potential biological remediation strategies for removing perchlorate from Martian Regolith. New Space 9(4):217–227. https://doi.org/10.1089/SPACE.2020.0055

Monteiro JP, Costa E, Melo T, Domingues P, Fort A, Domingues MR, Sulpice R (2022) Lipidome in-depth characterization highlights the nutritional value and species-specific idiosyncrasies of different *Ulva* species. Algal Res 64. https://doi.org/10.1016/J.ALGAL.2022.102694

Morrison SM, Downs RT, Blake DF, Vaniman DT, Ming DW, Hazen RM, Treiman AH, Achilles CN, Yen AS, Morris RV, Rampe EB, Bristow TF, Chipera SJ, Sarrazin PC, Gellert R, Fendrich KV, Morookian JM, Farmer JD, DesMarais DJ, Craig PI (2018) Crystal chemistry of martian minerals from Bradbury landing through Naukluft Plateau, Gale crater, Mars. Am Min 103(6):857–871. https://doi.org/10.2138/AM-2018-6124/ASSET/GRAPHIC/J_AM-2018-6124_FIG_012.JPG

Mortimer JC, Gilliham M (2022) SpaceHort: redesigning plants to support space exploration and on-earth sustainability. COBIOT 73:246–252. https://doi.org/10.1016/J.COPBIO.2021.08.018

Mouritsen OG, Duelund L, Petersen MA, Hartmann AL, Frøst MB (2019) Umami taste, free amino acid composition, and volatile compounds of brown seaweeds. J Appl Phycol 31(2):1213–1232. https://doi.org/10.1007/S10811-018-1632-X/FIGURES/8

Muhamediyeva DK, Nurumova AY (2023) Enhancing soil fertility through the application of biohumus. E3S Web Conf 411. https://doi.org/10.1051/E3SCONF/202341102044

Murai U, Yamagishi K, Sata M, Kokubo Y, Saito I, Yatsuya H, Ishihara J, Inoue M, Sawada N, Iso H, Tsugane S (2019) Seaweed intake and risk of cardiovascular disease: the Japan Public Health Center–based Prospective (JPHC) study. Am J Clin Nutr 110(6):1449–1455. https://doi.org/10.1093/AJCN/NQZ231

Nabti E, Jha B, Hartmann A (2016) Impact of seaweeds on agricultural crop production as biofertilizer. Int J Environ Sci Technol 14(5):1119–1134. https://doi.org/10.1007/S13762-016-1202-1

NASA Image and Video Library (2022) Astronaut Kayla Barron checks out plants growing inside the Veggie facility National Aeronautics and Space Administration. https://www.nasa.gov/image-article/astronaut-kayla-barron-checks-out-plants-growing-inside-veggie-facility/

NASA Image and Video Library, MicroAlgae (2019) National Aeronautics and Space Administration. https://images.nasa.gov/details/iss059e059230

NASA Image and Video Library, Space Station (1989) National Aeronautics and Space Administration. https://images.nasa.gov/details/0102173

Niederweiser T, Kociolek P, Klaus D (2018) A review of algal research in space. Acta Astron 146:359–367. https://doi.org/10.1016/J.ACTAASTRO.2018.03.026

O'Sullivan L, Murphy B, McLoughlin P, Duggan P, Lawlor PG, Hughes H, Gardiner GE (2010) Prebiotics from marine macroalgae for human and animal health applications. Mar Drugs 8(7):2038–2064. https://doi.org/10.3390/MD8072038

Oliveira JC, Pereira RC, Johnson TS, Loureiro R (2024) Seed priming with *Ulva lactuca* L. in cultivars grown in Martian and lunar regolith analogues. Gravit Space Res:77–93. https://doi.org/10.2478/gsr-2024-0006

Olsson-Francis K, de la Torre R, Towner MC, Cockell CS (2009) Survival of akinetes (resting-state cells of cyanobacteria) in low earth orbit and simulated extraterrestrial conditions. J Int Soc Stud Orig Life 39(6):565–579. https://doi.org/10.1007/S11084-009-9167-4

Olvera-Rosales LB, Cruz-Guerrero AE, Ramírez-Moreno E, Quintero-Lira A, Contreras-López E, Jaimez-Ordaz J, Castañeda-Ovando A, Añorve-Morga J, Calderón-Ramos ZG, Arias-Rico J, González-

Olivares LG (2021) Impact of the gut microbiota balance on the health–disease relationship: the importance of consuming probiotics and prebiotics. Foods 10(6):1261. https://doi.org/10.3390/FOODS10061261

Oñez LJP, Catubis KML, Cabillo RA, Pascual PRL (2025) Improved morphological and yield responses of green lettuce (*Lactuca sativa* L.) grown with seaweed extract as a hydroponic nutrient solution. J Plant Nutr 48:1–12. https://doi.org/10.1080/01904167.2024.2411402

Orozco-González JG, Amador-Castro F, Gordillo-Sierra AR, García-Cayluela T, Alper HS, Carrillo-Nieves D (2022) Opportunities surrounding the use of *Sargassum* biomass as precursor of biogas, bioethanol, and biodiesel production. Front Mar Sci 8. https://doi.org/10.3389/fmars.2021.791054

Paradiso R, de Micco V, Aronne G, de Pascale S (2017) The European Space Agency MELiSSA project: Food Production and Preparation - The state of the art. https://www.iris.unina.it/handle/11588/670383

Patarra RF, Paiva L, Neto AI, Lima E, Baptista J (2011) Nutritional value of selected macroalgae. J Appl Phycol 23(2):205–208. https://doi.org/10.1007/S10811-010-9556-0/TABLES/2

Patel AK, Singhania RR, Awasthi MK, Varjani S, Bhatia SK, Tsai ML, Hsieh SL, Chen CW, Dong C (2021) Emerging prospects of macro- and microalgae as prebiotic. Microb Cell Factories 20(1):1–16. https://doi.org/10.1186/S12934-021-01601-7

Pathan A, Mahajan MM, Jain PG, Zambad SP, Bhandari GS, Darak AD, Ahire ED, Rajora AK, Surana KR (2023) Nutraceutical properties of seaweed vitamins. In: Vitamins as nutraceuticals: recent advances and applications, pp 167–184. https://doi.org/10.1002/9781394175543.CH7

Prince RP, Knott WM (1989) CELSS breadboard project at the Kennedy Space Center. In: Lunar base agriculture: soils for plant growth, pp 155–163. https://doi.org/10.2134/1989.LUNARBASEAGRICULTURE.C12

Priyadarsini M, Raichel Nivetha X, Mathimani T, Anto S, Hareesh Krishnan H, Glivin G, Premalatha M, Mariappan V, Sekhar J (2022) Omega-3-fatty acids from algae for health benefits. Mater Today Proc 66:1514–1518. https://doi.org/10.1016/J.MATPR.2022.07.177

Przybyla C (2021) Space aquaculture: prospects for raising aquatic vertebrates in a bioregenerative life-support system on a lunar base. Front Astron Space Sci 8. https://doi.org/10.3389/fspas.2021.699097

Puscaselu RG, Lobiuc A, Dimian M, Covasa M (2020) Alginate: from food industry to biomedical applications and management of metabolic disorders. Polymers 12(10):2417. https://doi.org/10.3390/polym12102417

Shukla PS, Mantin EG, Adil M, Bajpai S, Critchley AT, Prithiviraj B (2019) Ascophyllum nodosum-Based Biostimulants: Sustainable Applications in Agriculture for the Stimulation of Plant Growth Stress Tolerance and Disease Management. Front Plant Sci 10:655. https://doi.org/10.3389/fpls.2019.00655

Rajapakse N, Kim SK (2011) Nutritional and digestive health benefits of seaweed. Adv Food Nutr Res 64:17–28. https://doi.org/10.1016/B978-0-12-387669-0.00002-8

Ramírez DA, Kreuze J, Amoros W, Valdivia-Silva JE, Ranck J, Garcia S, Salas E, Yactayo W (2019) Extreme salinity as a challenge to grow potatoes under Mars-like soil conditions: targeting promising genotypes. Int J Astrobiol 18(1):18–24. https://doi.org/10.1017/S1473550417000453

Revellame ED, Aguda R, Chistoserdov A, Fortela DL, Hernandez RA, Zappi ME (2021) Microalgae cultivation for space exploration: assessing the potential for a new generation of waste to human life-support system for long duration space travel and planetary human habitation. Algal Res 55. https://doi.org/10.1016/J.ALGAL.2021.102258

Ribeiro S, Berge T, Lundholm N, Andersen TJ, Abrantes F, Ellegaard M (2011) Phytoplankton growth after a century of dormancy illuminates past resilience to catastrophic darkness. Nat Commun 2(1):1–7. https://doi.org/10.1038/ncomms1314

Rocha DHA, Pinto DC, Silva AMS (2022) Macroalgae specialized metabolites: evidence for their anti-inflammatory health benefits. Mar Drugs 20(12):789. https://doi.org/10.3390/MD20120789

Ronca AE, Moyer EL, Talyansky Y, Lowe M, Padmanabhan S, Choi S, Gong C, Cadena SM, Stodieck L, Globus RK (2019) Behavior of mice aboard the international Space Station. Sci Rep 9(1):1–14. https://doi.org/10.1038/s41598-019-40789-y

Rybak AS (2021) Microencapsulation with the usage of sodium alginate: a promising method for preserving stonewort (Characeae, Charophyta) oospores to support laboratory and field experiments. Algal Res 54. https://doi.org/10.1016/J.ALGAL.2021.102236

Rzymski P, Losiak A, Heinz J, Szukalska M, Florek E, Poniedziałek B, Kaczmarek Ł, Schulze-Makuch D (2024) Perchlorates on Mars: occurrence and implications for putative life on the Red Planet. Icarus 421. https://doi.org/10.1016/J.ICARUS.2024.116246

Sánchez-Machado DI, López-Cervantes J, López-Hernández J, Paseiro-Losada P (2004) Fatty acids, total lipid, protein and ash contents of processed edible seaweeds. Food Chem 85(3):439–444. https://doi.org/10.1016/J.FOODCHEM.2003.08.001

Santelices B (1990) Patterns of reproduction, dispersal, and recruitment in seaweeds. Oceanogr Mar Biol Annu Rev 28:177–276

Santomartino R, Averesch NJH, Bhuiyan M, Cockell CS, Colangelo J, Gumulya Y, Lehner B, Lopez-Ayala I, McMahon S, Mohanty A, Santa Maria SR, Urbaniak C, Volger R, Yang J, Zea L (2023) Toward sustainable space exploration: a roadmap for harnessing the power of microorganisms. Nat Commun 14(1):1–11. https://doi.org/10.1038/s41467-023-37070-2

Santoro C, González-Fernandez C, Znad H, Awual MR, Martini S (2022) The utilization of algae and seaweed biomass for bioremediation of heavy metal-contaminated wastewater. Molecules 27(4):1275. https://doi.org/10.3390/MOLECULES27041275

Sathasivam M, Hosamani R, Swamy BK, Kumaran S (2021) Plant responses to real and simulated microgravity. Life Sci Space Res 28:74–86. https://doi.org/10.1016/J.LSSR.2020.10.001

Shah MT, Zodape ST, Chaudhary DR, Eswaran K, Chikara J (2013) Seaweed sap as an alternative liquid fertilizer for yield and quality improvement of wheat. J Plant Nutr 36(2):192–200. https://doi.org/10.1080/01904167.2012.737886

Sharma S, Lishka B, Shahi A, Shubham, Kaushal S (2023) Hydroponics: the potential to enhance sustainable food production in non-arable areas. Curr Appl Sci Technol 42(39):13–23. https://doi.org/10.9734/cjast/2023/v42i394253

Shaw M, Humbert M, Brooks G, Rhamdhani A, Duffy A, Pownceby M (2022) Mineral processing and metal extraction on the lunar surface - challenges and opportunities. Miner Process Extr Metall Rev 43(7):865–891. https://doi.org/10.1080/08827508.2021.1969390

Shi X, Jung KW, Kim DH, Ahn YT, Shin HS (2011) Direct fermentation of *Laminaria japonica* for biohydrogen production by anaerobic mixed cultures. Int J Hydrog Energy 36(10):5857–5864. https://doi.org/10.1016/J.IJHYDENE.2011.01.125

Shuuluka D, Bolton JJ, Anderson RJ (2013) Protein content, amino acid composition and nitrogen-to-protein conversion factors of *Ulva rigida* and *Ulva capensis* from natural populations and Ulva lactuca from an aquaculture system, in South Africa. J Appl Phycol 25(2):677–685. https://doi.org/10.1007/S10811-012-9902-5

Sibonga JD (2013) Spaceflight-induced bone loss: is there an Osteoporosis Risk? Curr Osteoporos Rep 11(2):92–98. https://doi.org/10.1007/S11914-013-0136-5/FIGURES/1

Simón L, Arazo-Rusindo M, Quest AFG, Mariotti-Celis MS (2023) Phlorotannins: novel orally administrated bioactive compounds that induce mitochondrial dysfunction and oxidative stress in cancer. Antioxidants 12(9):1734. https://doi.org/10.3390/ANTIOX12091734

Slusarczyk M (1995) Predator-induced diapause in daphnia. Ecology 76(3):1008–1013. https://doi.org/10.2307/1939364

Sommariva A, Gori L, Chizzolini B, Pianorsi M (2020) The economics of moon mining. Acta Astron 170:712–718. https://doi.org/10.1016/J.ACTAASTRO.2020.01.042

Song BS, Park JG, Kim JH, Choi J, Ahn DH, Hao C, Lee JW (2012) Development of freeze-dried miyeokguk, Korean seaweed soup, as space food sterilized by irradiation. Radiat Phys Chem 81(8):1111–1114. https://doi.org/10.1016/J.RADPHYSCHEM.2011.10.025

Spencer LE, Sirmons TA, Romeyn MW, Wheeler RM (2021) Production, nutritional and organoleptic analysis of Solanaceous crops for space. In: 50th International Conference on Environmental Systems. https://ttu-ir.tdl.org/server/api/core/bitstreams/0dfc91a7-c5b2-482a-96dd-0b205ef7204a/content

Stirk WA, Novák O, Strnad M, van Staden J (2003) Cytokinins in macroalgae. J Plant Growth Regul 41:13–24. https://link.springer.com/article/10.1023/A:1027376507197

Stoyanova S, Sirakov I, Velichkova K (2024) Sustainable production: integrating medicinal plants with fish farming in aquaponics—a mini review. Sustainability 16(15):6337. https://doi.org/10.3390/SU16156337

Sukhinov DV, Gotovtsev PM, Sergeeva YE (2023) Phototrophic microorganisms in bioregenerative life support systems for long-term crewed expeditions: prospects and challenges. Acta Astron 211:518–538. https://doi.org/10.1016/J.ACTAASTRO.2023.07.006

Sunarpi H, Nikmatullah A, Ambana Y, Ilhami BTK, Abidin AS, Ardiana N, Kirana IAP, Kurniawan NSH, Rinaldi R, Jihadi A, Prasedya ES (2021) Phytohormone content in brown macroalgae Sargassum from Lombok coast, Indonesia. IOP Conf Ser 712(1). https://doi.org/10.1088/1755-1315/712/1/012042

Surget G, Roberto VP, le Lann K, Mira S, Guérard F, Laizé V, Poupart N, Cancela ML, Stiger-Pouvreau V (2017) Marine green macroalgae: a source of natural compounds with mineralogenic and antioxidant activities. J Appl Phycol 29(1):575–584. https://doi.org/10.1007/S10811-016-0968-3/FIGURES/4

Sweet P, Ball N, Vu S, Anderson L, Downing S, Gresser A, Hindupur A, Hogan J, Kagawa H, Kostakis A, Paddock M, Ray H, Roque O, Sharif S, Sims K, Soundararajan M, Villanueva A, Donovan F, Settles AM (2025) Bionutrients-1: utilizing genomics and Transcriptomics to assess the reliability of microorganisms for in situ nutrient production on Long duration missions. In review.

Tang H, Rising HH, Majji M, Brown RD (2021) Long-term space nutrition: a scoping review. Nutrients 14(1):194. https://doi.org/10.3390/NU14010194

Tanna B, Mishra A (2019) Nutraceutical potential of seaweed polysaccharides: structure, bioactivity, safety, and toxicity. Compr Rev Food Sci Food Saf 18(3):817–831. https://doi.org/10.1111/1541-4337.12441

Taylor LA, Morris RV, Pieters C, Patchen A, Taylor DH, Keller LP, Wentworth S, McKay DS (2000) Chemical characterization of lunar mare soils. Lunar and Planetary Science XXXI 31st Annual Lunar and Planetary Science Conference March 13-17 2000 Houston Texas abstract no. 1865

Thapa U, Hansda NN, Kundu S, Giri A (2024) Advancements in hydroponic systems: a comprehensive review. Arch Curr Res Int. https://doi.org/10.9734/acri/2024/v24i11973

Tilstra A, Bednarz VN, Cardini U, van Hoytema N, Al-Rshaidat MMD, Wild C (2017) Seasonality affects dinitrogen fixation associated with two common macroalgae from a coral reef in the northern Red Sea. Mar Ecol Prog Ser 575:69–80. https://doi.org/10.3354/meps12206

Toklu YC, Çalış Açıkbaş N, Açıkbaş G, Çerçevik AE, Akpinar P (2023) Production of a set of lunar regolith simulants based on Apollo and Chinese samples. Adv Space Res 72(2):565–576. https://doi.org/10.1016/J.ASR.2023.03.035

Tomita-Yokotani K, Kimura S, Ong M, Tokita M, Katoh H, Abe T, Hashimoto H, Sonoike K, Ohmori M (2021) Investigation of Nostoc sp. HK-01, cell survival over three years during the Tanpopo Mission. Astrobiology 21(12):1505–1514. https://doi.org/10.1089/AST.2021.0152/ASSET/IMAGES/AST.2021.0152_FIGURE8.JPG

Torres-Romero I, Zhang H, Wijker RS, Clark AJ, McLeod RE, Jaggi M, Stoll HM (2024) Hydrogen isotope fractionation is controlled by CO2 in coccolithophore lipids. Proc Natl Acad Sci USA 121(26):e2318570121. https://doi.org/10.1073/PNAS.2318570121/SUPPL_FILE/PNAS.2318570121.SD01.XLSX

Townsend LW, Fry RJM (2002) Radiation protection guidance for activities in low-earth orbit. Adv Space Res 30(4):957–963. https://doi.org/10.1016/S0273-1177(02)00160-6

Trainer MG, Wong MH, McConnochie TH, Franz HB, Atreya SK, Conrad PG, Lefèvre F, Mahaffy PR, Malespin CA, Manning HLK, Martín-Torres J, Martínez GM, McKay CP, Navarro-González R, Vicente-Retortillo Á, Webster CR, Zorzano MP (2019) Seasonal variations in atmospheric composition as measured in gale crater. Mars J Geophys Res 124(11):3000–3024. https://doi.org/10.1029/2019JE006175

Tziveleka LA, Tammam MA, Tzakou O, Roussis V, Ioannou E (2021) Metabolites with antioxidant activity from marine macroalgae. Antioxidants 10(9):1431. https://doi.org/10.3390/ANTIOX10091431

van Ginneken VJT, Helsper JPFG, de Visser W, van Keulen H, Brandenburg WA (2011) Polyunsaturated fatty acids in various macroalgal species from North Atlantic and tropical seas. Lipids Health Dis 10. https://doi.org/10.1186/1476-511X-10-104

Verseux C, Heinicke C, Ramalho TP, Determann J, Duckhorn M, Smagin M, Avila M (2021) A low-pressure, N2/CO2 atmosphere is suitable for cyanobacterium-based life-support systems on Mars. Front Microb 12. https://doi.org/10.3389/FMICB.2021.611798/BIBTEX

Vijayakumar S, Durgadevi S, Arulmozhi P, Rajalakshmi S, Gopalakrishnan T, Parameswari N (2019) Effect of seaweed liquid fertilizer on yield and quality of *Capsicum annum* L. Acta Ecol Sin 39(5):406–410. https://doi.org/10.1016/J.CHNAES.2018.10.001

Vinoth S, Gurusaravanan P, Jayabalan N (2012) Erratum to: effect of seaweed extracts and plant growth regulators on high-frequency in vitro mass propagation of *Lycopersicon esculentum* L (tomato) through double cotyledonary nodal explant. J Appl Phycol 24(5):1339–1340. https://doi.org/10.1007/S10811-011-9748-2/FIGURES/2

Ward CH, Wilks SS, Craft HL, Wilks CH (1963) Use of algae and other plants in the development of life support systems. ABT 25(7):512–521

Watanabe F, Takenaka S, Katsura H, Masumder SA, Abe K, Tamura Y, Nakano Y (1999) Dried green and purple lavers (Nori) contain substantial amounts of biologically active vitamin B12 but less of dietary iodine relative to other edible seaweeds. J Agr Food Chem 47(6):2341–2343. https://doi.org/10.1021/JF981065C/ASSET/IMAGES/LARGE/JF981065CF00001.JPEG

Weir DG, Scott JM (1999) Brain function in the elderly: role of vitamin B 12 and folate. Br Med Bull 55(3):669–682. https://doi.org/10.1258/0007142991902547

Williams DR (2024a) Moon fact sheet. Goddard Space Flight Center, National Aeronautics and Space Administration. https://nssdc.gsfc.nasa.gov/planetary/factsheet/moonfact.html

Williams DR (2024b) Mars fact sheet. Goddard Space Flight Center, National Aeronautics and Space Administration. https://nssdc.gsfc.nasa.gov/planetary/factsheet/marsfact.html

Withers PC, Cooper CE (2008) Dormancy. Encyclopedia of ecology, Five-Volume Set, pp 952–957. https://doi.org/10.1016/B978-008045405-4.00503-6

Wu J, Keller DP, Oschlies A (2023) Carbon dioxide removal via macroalgae open-ocean mariculture and sinking: an Earth system modeling study. ESD 14:185–221. https://doi.org/10.5194/esd-14-185-2023

Xie C, Lee ZJ, Ye S, Barrow CJ, Dunshea FR, Suleria HAR (2024) A review on seaweeds and seaweed-derived polysaccharides: nutrition, chemistry, bioactivities, and applications. Food Rev Int 40(5):1312–1347. https://doi.org/10.1080/87559129.2023.2212055

Yadav AN, Kour D, Kaur T, Devi R, Yadav A, Dikilitas M, Abdel-Azeem AM, Ahluwalia AS, Saxena AK (2021) Biodiversity, and biotechnological contribution of beneficial soil microbiomes for nutrient cycling, plant growth improvement and nutrient uptake. Biocatal Agric Biotechnol 33. https://doi.org/10.1016/J.BCAB.2021.102009

Yamagishi K, Sun W, Iso H (2023) Seaweed in traditional diets and relationship with cardiovascular disease incidence and mortality. Ancient and traditional foods, plants, herbs and spices used in cardiovascular health and disease, pp 65–74. ISBN: 9781003220329

Yamamoto Y (1976) Effect of some physical and chemical factors on the germination of Akinetes of *Anabaena cylindrica*. J Gen Appl Microbiol 22:311–323

Yan N, Fan C, Chen Y, Hu Z (2016) The potential for microalgae as bioreactors to produce pharmaceuticals. Int J Mol Sci 17(6):962. https://doi.org/10.3390/IJMS17060962

Yang C, Liu H, Li M, Yu C, Yu G (2008) Treating urine by *Spirulina platensis*. Acta Astron 63(7–10):1049–1054. https://doi.org/10.1016/J.ACTAASTRO.2008.03.008

Yu J, Li Q, Wu J, Yang X, Yang S, Zhu W, Liu Y, Tang W, Nie S, Hassouna A, White WL, Zhao Y, Lu J (2021) Fucoidan extracted from Sporophyll of *Undaria pinnatifida* grown in Weihai, China—chemical composition and comparison of antioxidant activity of different molecular weight fractions. Front Nutr 8. https://doi.org/10.3389/FNUT.2021.636930/BIBTEX

Zafar A, Ali I, Rahayu F (2022) Marine seaweeds (biofertilizer) significance in sustainable agricultural activities: a review. IOP Conf Ser 974(1):012080. https://doi.org/10.1088/1755-1315/974/1/012080

Zava TT, Zava DT (2011) Assessment of Japanese iodine intake based on seaweed consumption in Japan: a literature-based analysis. J Thyroid Res 4(1):1–7. https://doi.org/10.1186/1756-6614-4-14/TABLES/2

Zhang S, Ju P, Yin X, Liu H, Zhang B, Li H, Meng X, Ai S (2024) Conversion of macroalgae into environmentally friendly bioplastics by noncovalent bond assembly. ACS Sustain Chem Eng 12(47):17044–17416. https://doi.org/10.1021/ACSSUSCHEMENG.4C07767/ASSET/IMAGES/LARGE/SC4C07767_0006.JPEG

Haley K. Murphy is a graduate student pursuing her Master of Science in Ecology at the Florida Institute of Technology. She is conducting research on spaceflight applications for cyanobacteria at the Palmer Lab. Haley received a B.Sc. in Astrobiology in Spring of 2024.

Jessica Carneiro has a master's degree in Marine Biotechnology, focusing on algae as biostimulants for plants growing in regolith. She is a Ph.D. student at UNIRIO, Brazil. Her research involves the interaction between microorganisms identified on the ISS and plants germinating in Martian and lunar regolith simulants in collaboration with FIT and WSSU.

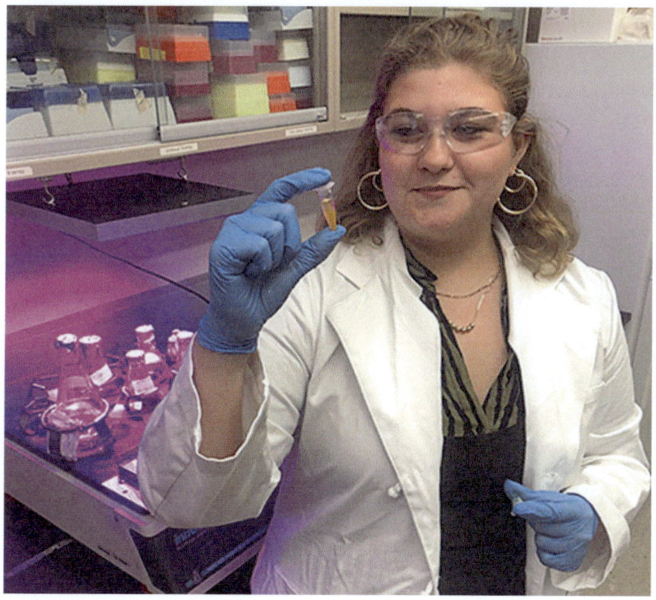

Frances Edmonson (Frannie) is a Biological Sciences PhD student at the Florida Institute of Technology working in the Palmer Lab of Astrobiology and Chemical Ecology. Her research includes microbial remediation of Martian and lunar regolith simulants and space agriculture.

Trenton P. Causey is an undergraduate student in Astrobiology at the Florida Institute of Technology who will be pursuing a M.S in Ecology. He researches space biology with a focus on lignocellulosic waste recycling and Martian and lunar regolith amendment strategies.

Steven G. Elsaid is a graduate student pursuing his Ph.D in Biological Sciences at the Florida Institute of Technology. His research uses bacteria isolated from the ISS to promote plant growth at the Palmer Lab at FIT.

Dr. Rafael Loureiro is an Associate Professor of Botany and Plant Physiology at Winston-Salem State University (WSSU), where he specializes in plant stress physiology and astrobotany research. He earned his Ph.D. in Botany from Brazil's National School of Tropical Botany and has extensive experience in controlled environment agriculture. His research focuses on plant-microbe interactions, stress responses, and in situ resource utilization to support sustainable food production in space.

Dr. Andrew G. Palmer is an Associate Professor of Biological Sciences at the Florida Institute of Technology (FIT) where he has been appointed for over a decade. He received his B.S. in Biochemistry from Florida State University, his Ph.D. in Chemistry from Emory University, and was an NIH postdoctoral fellow at the University of Wisconsin-Madison. Among the research projects in his lab is the development of sustainable space agriculture systems

Correction to: Advances in Temperate Phyconomy: Algal Harvest and Cultivation in Globally Distributed Temperate Waters

Alan T. Critchley, Anicia Q. Hurtado, Iain C. Neish
and Leonel Pereira

Correction to:
Chapters 9 and 20 in: A. T. Critchley et al. (eds.), *Advances in Temperate Phyconomy: Algal Harvest and Cultivation in Globally Distributed Temperate Waters*, Developments in Applied Phycology 15, https://doi.org/10.1007/978-3-032-02955-3

The original version of Chapter 20 was inadvertently published with the incorrect figure legend for Figure 20.1, which has been updated as below.

In addition, the MSC mission statement (in the 2nd paragraph of Section 9.8) of Chapter 9 was missing and has now been updated as mentioned below.

Fig. 20.1:
Projected kelp species richness in BC. Figures created using ArcGIS using data modeled by Assis et al. (2024). Data was exported using R (version 4.4.2; R Core Team 2024) after being manually subset to species in Laminariales. Species richness is presented as a percentage out of the 44 species modeled to keep the scale consistent between figures. (**a**) indicates baseline (years 2010–2020) species richness, (**b**) indicates species richness by the end of the century (2090–2100) under projected SSP1-1.9 and (**c**) indicates species richness by the end of the century (2090–2100) under projected SSP3–7

Section 9.8
The Maine Seaweed Council strives to protect the ecosystems of Maine's marine algae and adhere to sustainable cultivation and harvest practices, promote the use of Maine seaweeds, educate the public, regulators, and elected officials, and provide a collaborative forum for its members.

The updated versions of these chapters can be found at
https://doi.org/10.1007/978-3-032-02955-3_9
https://doi.org/10.1007/978-3-032-02955-3_20

A. T. Critchley et al. (eds.), *Advances in Temperate Phyconomy: Algal Harvest and Cultivation in Globally Distributed Temperate Waters*, Developments in Applied Phycology 15, https://doi.org/10.1007/978-3-032-02955-3_29

MIX
Papier aus verantwortungsvollen Quellen
Paper from responsible sources
FSC® C105338

If you have any concerns about our products,
you can contact us on
ProductSafety@springernature.com

In case Publisher is established outside the EU,
the EU authorized representative is:
Springer Nature Customer Service Center GmbH
Europaplatz 3, 69115 Heidelberg, Germany

Printed by Libri Plureos GmbH
in Hamburg, Germany